Communications in Computer and Information Science 837

Commenced Publication in 2007
Founding and Former Series Editors:
Phoebe Chen, Alfredo Cuzzocrea, Xiaoyong Du, Orhun Kara, Ting Liu,
Dominik Ślęzak, and Xiaokang Yang

Editorial Board

More information about this series at http://www.springer.com/series/7899

Ivan Zelinka · Roman Senkerik
Ganapati Panda
Padma Suresh Lekshmi Kanthan (Eds.)

Soft Computing Systems

Second International Conference, ICSCS 2018
Kollam, India, April 19–20, 2018
Revised Selected Papers

Springer

Editors
Ivan Zelinka
Department of Computer Science
Faculty of Electrical Engineering and
    Computer Science VŠB-TUO
Ostrava-Poruba
Czech Republic

Roman Senkerik
Faculty of Applied Informatics
Tomas Bata University in Zlín
Zlín
Czech Republic

Ganapati Panda
School of Electrical Sciences
Indian Institute of Technology Bhubaneswar
Bhubaneswar, Odisha
India

Padma Suresh Lekshmi Kanthan
Baselios Mathews II College of Engineering
Kerala
India

ISSN 1865-0929   ISSN 1865-0937   (electronic)
Communications in Computer and Information Science
ISBN 978-981-13-1935-8   ISBN 978-981-13-1936-5   (eBook)
https://doi.org/10.1007/978-981-13-1936-5

Library of Congress Control Number: 2018953175

This Springer imprint is published by the registered company Springer Nature Singapore Pte Ltd.
The registered company address is: 152 Beach Road, #21-01/04 Gateway East, Singapore 189721, Singapore

Preface

This CCIS volume contains the papers presented at the Second International Conference on Soft Computing Systems 'ICSCS 2018' held during April 19–20, 2018, at Baselios Mathews II College of Engineering, Sasthamcotta, India. ICSCS 2018 is a prestigious international conference that aims at bringing together researchers from academia and industry to report and review the latest progress in cutting-edge research on soft computing systems, to explore new applicational areas, to design new nature-inspired algorithms for solving hard problems, and finally to create awareness about these domains to a wider audience of practitioners.

ICSCS 2018 received 439 paper submissions from 10 countries across the globe. After a rigorous double-blind peer-review process, 87 full-length articles were accepted for oral presentation at the conference. This corresponds to an acceptance rate of 19.8% and is intended to maintain the high standards of the conference proceedings. The papers included in this CCIS volume cover a wide range of topics in soft computing systems, imaging science, machine learning, neural networks, data mining, communication protocols, security and privacy, artificial intelligence, and hybrid techniques and their real-world applications to problems occurring in diverse domains of science and engineering.

The conference featured two distinguished keynote speakers: Prof. Ganapati Panda, Indian Institute of Technology Bhubaneswar, and Prof. Dr. Swagatam Das, Electronics and Communication Sciences Unit, Indian Statistical Institute, Kolkata.

We take this opportunity to thank the authors of the submitted papers for their hard work, adherence to the deadlines, and patience with the review process. The quality of a refereed volume depends mainly on the expertise and dedication of the reviewers. We are indebted to the Program Committee/Technical Committee members, who produced excellent reviews within short time frames.

We would also like to thank our sponsors for providing logistical support and financial assistance. First, we are indebted to Baselios Mathews II College of Engineering Management and Administration for supporting our cause and encouraging us to organize the conference at the college. In particular, we would like to express our heartfelt thanks for their financial support and infrastructural assistance. Our sincere thanks to H. G Zachariah Mar Anthonios, Manager; Rev. Fr. Thomas Varghese, Administrator; Dr. F. V. Albin, Director; Prof. Oommen Samuel, Dean (academic); Rev. Fr. Dr. Koshy Vaidyan, Dean (student affairs); and Rev. Fr. Abraham Varghese, Project Manager. We thank Dr. Mirtha Nelly Aldave, West Hartford, Connecticut, USA and Prof. Dr. Mihir Narayan Mohanty, Siksha 'O' Anusandhan (Deemed to be University), Bhubaneswar, Odisha, for providing valuable guidelines and inspiration to overcome various difficulties in the process of organizing this conference.

We would also like to thank the participants of this conference. Finally, we would like to thank all the volunteers for meeting the deadlines and arranging every detail to make sure that the conference could run smoothly. We hope the readers of these proceedings find the papers inspiring and enjoyable.

April 2018

Ivan Zelinka
Roman Senkerik
Ganapati Panda
Padma Suresh Lekshmi Kanthan

Organization

Chief Patron

H. H. Baselios Marthoma Paulose II	BMCE, India
H. G. Zachariah Mar Anthonios	BMCE, India
Thomas Varghese	BMCE, India

Patron

F. V. Albin	BMCE, India
Oommen Samuel	BMCE, India
Koshy Vaidyan	BMCE, India
Abraham Varghese	BMCE, India

General Chairs

Roman Šenkeřík	Tomas Bata University, Czech Republic
Ivan Zelinka	Technical University of Ostrava, Czech Republic
Ganapati Panda	IIT, Bhubaneswar, India
Padma Suresh Lekshmi Kanthan	BMCE, India

Program Chairs

Swagatam Das	Indian Statistical Institute, India
B. K. Panigrahi	IIT, Delhi, India

Organizing Chairs

S. S. Dash	SRM University, India
Syed Abdul Rahman	Universiti Putra Malaysia, Malaysia

Special Session Chairs

P. N. Suganthan	Nanyang Technological University, Singapore
Akhtar Kalam	Victoria University, Australia
Pradip K. Das	IIT, Guwahati, India

Conference Coordinators

D. H. Manjiah	Mangalore University, India
Vivekananda Mukherjee	Indian School of Mines, India
M. P. Somasundaram	Anna University, India

Organizing Secretary

Krishna Veni   BMCE, India
Rusli Abdullah   Universiti Putra Malaysia, Malaysia

Technical Program Committee

K. Shanti Swarup   IIT Madras, India
R. Rama   IIT Madras, India
N. P. Padhy   IIT Roorkee, India
R. K. Behera   IIT Patna, India
A. K. Pradhan   IIT Kharagpur, India
K. S. Easwarakumar   Anna University, India
Thanga Raj Chelliah   IIT Roorkee, India
Shiva Shankar B. Nair   IIT Guwahati, India
Arun Tangirala   IIT Chennai, India
Bharat Bikkajji   IIT Chennai, India
Goshaidas Ray   IIT Kharagpur, India
Jayant Pal   IIT Bhubaneswar, India
Khaparde S. A.   IIT Mumbai, India
Laxmidhar Behera   IIT Kanpur, India
Manish Kumar   Banaras Hindu University, India
Ahmad Farid bin Abidin   Universiti Teknologi MARA, Malaysia
M. Nasir Taib   Universiti Teknologi MARA, Malaysia
Wahidah Mansor   Universiti Teknologi MARA, Malaysia
P. D. Chandana Perera   University of Ruhuna Hapugala, Sri Lanka
Ajith Abraham   MIR Labs, USA
Damian Flynn   University College Dublin, Ireland
Radha Raj   University of Luxembourg, Luxembourg
Akhtar Kalam   Victoria University, Australia
Rozita Jallani   Universiti Teknologi MARA, Malaysia
Yiu-Wing Leung   Hong Kong Baptist University, Hong Kong
Rishad A. Shafik   University of Southampton, UK
Sumeet Dua   Louisiana Tech University, USA
Yew-Soon Ong   Nanyang Technological University, Singapore
Syed Abdul Rahman   Universiti Putra Malaysia, Malaysia
Tan Kay Chen   National University of Singapore, Singapore
Tariq Rahim Soomro   Al Ain University of Science & Technology, UAE
Ashutosh Kumar Singh   Curtin University, Malaysia
Liaqat Hayat   Yanbu Industrial College, KSA
Raj Jain   Washington University, USA
Kannan Govindan   University of Southern Denmark, Denmark
K. Baskaran   Shinas College of Technology, Sultanate of Oman
Sathish Kannan   Cambridge University, UK
Arijit Bhattacharya   Dublin City University, Ireland
Raghu Korrapati   Walden University, USA

Abdel-Badeeh M. Salem	Ain Shams University, Egypt
Imre J. Rudas	Óbuda University, Hungary
Ramana G. Reddy	The University of Alabama, USA
Gopalan Mukundan	Chrysler Group LLC, USA
Wahyu Kuntjoro	Universiti Teknologi MARA, Malaysia
Gerasimos Rigatos	University Campus, Croatia
Balan Sundarakani	University of Wollongong, Dubai
Farag Ahmed Mohammad Azzedin	KFUPM, Saudi Arabia
A. M. Harsha S. Abeykoon	University of Moratuwa, Sri Lanka
Kashem Muttaqi	University of Wollongong, Australia
Ahmed Faheem Zobaa	Bournemouth University, UK
Alfredo Vaccaro	University of Sannio, Italy
David Yu	University of Wisconsin–Milwaukee, USA
Dmitri Vinnikov	Tallinn University of Technology, Estonia
Gorazd Štumberger	University of Maribor, Slovenia
Hussain Shareef	Universiti Kebangsaan Malaysia, Malaysia
Joseph Olorunfemi Ojo	Texas Tech University, USA
Ilhami Colak	Gazi University, Turkey
Ramazan Bayindir	Gazi University, Turkey
Junita Mohamad-Saleh	US, Malaysia
Dan M. Ionel	University of Kentucky, USA
Murad Al-Shibli	EMET, Abu Dhabi
Nesimi Ertugrul	University of Adelaide, Australia
Omar Abdel-Baqi	University of Wisconsin–Milwaukee, USA
Adel Nasiri	University of Wisconsin–Milwaukee, USA
Richard Blanchard	Leeds Beckett University, UK
Shashi Paul	De Montfort University, UK
A. A. Jimo	Tshwane University of Technology, South Africa
Zhao Xu	HKPU, Hong Kong
Mohammad Lutfi Othman	University Putra Malaysia, Malaysia
Ille C. Gebeshuber	UKM-Malaysia & TU Wein, Austria
Tarek M. Sobh	University of Bridgeport, USA
Amirnaser Yazdani	Ryerson University, Canada
Asim Kaygusuz	Inonu University, Turkey
Fathi S. H.	Amirkabir University of Technology, Iran
Gobbi Ramasamy P.	Multimedia University Cyberjaya Campus, Malaysia
Josiah Munda	Tshwane University of Technology, South Africa
Loganathan N.	Nizwa College of Technology, Sultanate of Oman
Ramesh Bansal	University of Pretoria, South Africa
Varatharaju V. M.	Ibra college of Technology, Sultanate of Oman
Xavier Fernando	Ryerson University, Canada
Priya Chandran	NIT Calicut, India
R. Sreeram Kumar	NIT Calicut, India
Vadivel A.	NIT Trichy, India

S. Selvakumar	NIT Trichy, India
Anup Kumar	Panda National Institute of Technology, India
Kumaresan N.	NIT Trichy, India
Mathew A. T.	NIT Calicut, India
Chithra Prasad	TKM College of Engineering, India
Rajasree M. S.	IIITMK, Technopark Campus, India
S. Arun	TKM Institute of Technology, India
Benz Raj	Annamalai University, India
A. Marsalin Beno	St. Xavier's Catholic College of Engineering, India
S. Kannan	Kalasalingam University, India
D. H. Manjiah	Mangalore University, India
S. Siva Balan	Noorul Islam Univeristy, India
V. Kalaivani	National Engineering College, India
S. T. Jaya Christa	Mepco Schlenk Engineering College, India
B. V. Manikandan	Mepco Schlenk Engineering College, India
R. S. Shaji	Noorul Islam University, India
I. Jacob Raglend	Noorul Islam University, India
P. Jeno Paul	St. Thomas College of Engineering, India
R. S. Rajesh	Manonmaniyam Sundaranar University, India
K. L. Shunmuganathan	R.M.K Engineering College, India
P. Somsundram	Anna University, India
S. Deva Raj	Kalasalingam University, India
M. Madeeswaran	Mahendra Engineering College, India
K. A. Mohamed Junaid	R.M.K Engineering College, India
A. Suresh	SMK Fomra Institute of Technology, India
Gnana Dhas	Pondicherry Engineering College, India
V. Kavitha	University College of Engineering, India
B. Sankara Gomathy	National Engineering College, India
S. Velusami	Annamalai University, India
K. A. Janardhanan	Noorul Islam University, India
D. P. Kothari	J.B. Group of Educational Institution, India
E. G. Rajan	Pentagram Research Centre Pvt. Ltd., India
J. Sheeba Rani	Indian Institute of space Science and Technology, India
I. A. Chidambaram	Annamalai University, India
G. Wiselin Jiji	Dr. Sivanthi Aditanar Engineering College, India
S. Ashok	NIT Calicut, India
T. Easwaran	Alagappa University, India
V. Vaidehi	Madras Institute of Technology, India
Vivekananda Mukherjee	Indian School of Mines, India
M. P. Somasundaram	Anna University, India
N. K. Mohanty	SVCE, India
K. Vijayakumar	SRM University,India
C. Bharathi Raja	SRM University, India
Suresh Chandra Satapathy	ANITS, India
Manimegalai Rajkumar	Park Institute of Technology, India

M. R. Rashmi	Amirtha University, India
P. Jegatheswari	Ponjesly Engineering College, India
Christopher Columbus	PSN College of Engineering, India
N. Nirmal Singh	VV College of Engineering, India
P. Muthu Kumar	Care School of Engineering, India
Velayutham Ramakrishnan	Einstein Engineering College, India
Ruban Deva Prakash	Sree Narayana Gurukulam college of Engineering, India
N. Krishna Raj	Sri Sasta Institute of Engineering and Technology, India
R. Kanthavel	Velammal Engineering College, India
S. S. Kumar	Noorul Islam University, India
Suja Mani Malar	PET Engineering College, India
M. Willjuice Iruthaya Rajan	National Engineering College, India
T. Vijayakumar	Sri Eshwar College of Engineering, India
Rusli bin Abdullah	Universiti Putra Malaysia, Malaysia
Nattachote Rugthaicharoenc	Rajamangala University of Technology, Thailand
Faris Salman Majeed Al-Naimy	Technical College of Engineering, Sultanate of Oman
K. Nithiyananthan	BITS Pilani, Dubai
G. Saravana Elango	NIT Trichy, India
Sishaj P. Simon	NIT Trichy, India
S. Vasantha Ratna	Coimbatore Institute of Technology, India
S. Baskar	Thiagarajar college of Engineering, India
K. K. Thyagarajan	RMD Engineering College, India
S. Joseph Jawahar	Arunachala College of Engineering for Women, India
Seldev Christopher	St. Xaviers Catholic College of Engineering, India
P. Prathiban	National Institute of Technology, India
R. Saravanan	Vellore Institute of Technology, India
A. Abudhair	National Engineering College, India
N. S. Sakthivel Murugan	Park College of Engineering and Technology, India
S. Edward Rajan	Mepco Schlenk Engineering College, India
S. V. Muruga Prasad	KVM College of Engineering, India
T. Sree Rengaraja	Anna University, India
S. S. Vinsly	Lourdes Mount College of Engineering and Technology, India
N. Senthil Kumar	Mepco Schlenk Engineering College, India
S. V. Nagaraj	R.M.K Engineering College, India
K. SelvaKumar	Annamalai University, India
S. Padma Thilagam	Annamalai University, India
Arun Shankar	PSG College of Technology, India
Karuppanan P.	Motilal Nehru National Institute of Technology, India
Udhayakumar K.	Anna University, India
Uma Maheswari B.	Anna University, India

Contents

Image Processing

VLSI

Cloud Computing

Network Communication

Power Electronics

Green Energy

Soft Computing

Genic Disorder Identification and Protein Analysis Using Soft Computing Methods

J. Briso Becky Bell[1(✉)] and S. Maria Celestin Vigila[2]

[1] Computer Science Department, DMI Engineering College,
Aralvoimozhi 629606, India
brisobell@gmail.com
[2] Information Technology Department, Noorul Islam Centre for Higher
Education, Kumaracoil 629180, India
celesleon@yahoo.com

Abstract. The field of Omics [1] has produced a large amount of research data, which is desirable for processing and estimating the discriminant classes and disordered sequences, usually the gene and protein play an vital role in controlling the biological process of the human body, with the use of genic data one can easily able to find the mutated gene causing disease and by the use of protein data the intrinsic disorder protein causing defective parts activity can be traced out. This paper brings out the soft computational machine learning research efforts in the genomic [2] and proteomic [3] data, thus providing easier machine intelligence disease classifier [4] with discriminant feature selection. Then the disease features are effective in selecting the optimal disorder enzyme causing protein [5], so that the relevant biological process activities [6] affected due to the various protein enzyme causing effects can be effectively comprehended.

Keywords: Genetic algorithm · Support vector machine · K nearest neighbor
Fuzzy C mean · Gene ontology

1 Introduction

Genomics and proteomics have led to various researchers in estimating the discriminant disease classes. As, gene and Protein play a vital role in controlling the biological process of the human body. So with the use of various soft computing [1] approaches and pattern recognition principles, one can easily implement the information learning system for processing the continuous disease data sets in classifying the Inter-related diseases.

Machine learning [1] uses various statistical soft computing approaches for learning or training the sample data, and then creates a mathematical model to classify the test data sample to relevant class. In supervised learning the sample data are available with a label class, during training stage of learning labeled class is used along with sample data. E.g. Artificial Neural Network (ANN), Support Vector Machine (SVM), Genetic Algorithm (GA), etc. In unsupervised learning no labeled data are provided so algorithms can be used to predict previously unknown patterns. Pattern recognition is a

© Springer Nature Singapore Pte Ltd. 2018
I. Zelinka et al. (Eds.): ICSCS 2018, CCIS 837, pp. 3–13, 2018.
https://doi.org/10.1007/978-981-13-1936-5_1

clustering approach, which attempts to assign each input sample data to one of a given set of classes. E.g. K Nearest Neighbor (KNN), Fuzzy C Mean (FCM), etc.

This paper could lead a way in identification of disease causing mutated genes. It also helps in classification of disease samples from normal samples in disease datasets. It enables to find Inter-related diseased gene by referring the ontology of genes. In order to identify the disorder protein sequence in relevance to genes is searched. By analyzing the disease stage syndrome internal disease affected body parts can be easily identified.

The rest of the paper is organized as follows: Sect. 2 provides a summary of the general selection and classification processes. Section 3 explains some of the methods used for gene selection, enhancement and sample classification. Section 4 deals with the induced principle and in the Sect. 5 sets out the inference and comments.

2 Literature Review

DNA MAMS Deoxyribo Nucleic Acid Micro Array Mass Spectrometry [2] is an significant technology for gene expression analyzing. Usually the microarray data are represented as images, which have to be converted into gene expression matrix in which columns represent various samples, rows represent genes. Enormously used by the Physicians for identification of many diseases when compared with clinical or morphological data. Genic disease sample datasets are available access from NCBI National Center for Biotechnology Information Databases.

Gene selection [2] uses certain statistical approaches, so one can easily select the set of highly expressive genes. By taking the $n \times m$ Microarray data matrix with a set of gene vectors of the form shown in (1) The resultant is a $n \times d$ microarray data matrix with a set of meaningful gene vectors, where $d < m$.

$$G = \{g_1, g_2, \ldots, g_m\} \tag{1}$$

By using machine learning algorithms and statistical soft computing classification [4] approach, one can easily classify a set of disease sample classes. In the set of Microarray samples of the form shown in (2), The resultant is a sampler classifying $h:$ $S \to C$ which maps a sample 'S' to its classification label 'C'. The class label can be either a majority class or a minority class for a binary class dataset samples.

$$D = \{(S_1, C_1), (S_2, C_2), \ldots, (S_n, C_n)\} \tag{2}$$

Gene Ontology (GO) [5] is a collection of organized vocabularies describing the biology of a gene product in any organism. These vocabularies can be basically categorized into three types. The first, Molecular Function (MF) represents the elemental activity/task, the examples of these function are carbohydrate binding and ATPase activity. The next is Biological Process (BP) which describes the biologically related occurring mechanisms; the examples of such are mitosis or purine metabolism. The last one is Cellular Component (CC), it denotes the location or complexes or structure of a cell component, the examples for this is nucleus, telomere, and RNA polymerase II.

The importance of protein sequencing [6] is that, they provide pictures of molecular level disease process, so it is needed most as prerequisite for structure based drug design. The protein are sequenced or constructed by transcriptions of genes, as gene is the basic functional unit (microscopic) the protein is the next level cellular constituent (macro molecular) in the atomic human body. Shortest path analysis of Protein-Protein interaction networks, the functional protein association network has always been used to study the mechanism of diseases.

3 Methods

As a soft computing approach, it is proposed to use some of the linear classifiers as SVM, Naïve Bayes and KNN [7] algorithms. And for computing the feature selection Pearson Correlation Coefficient (PCC) and Feature Assessment by Information Retrieval (FAIR) can be applied. Thus for finding the optimality of protein sequence the GA can be used. Also gene enrichment can be provided by GO Analysis [6] in finding the CC, BP and MF of associated genes.

3.1 Gene Selection Methods

Gene selection can be computed using some linear statistical algorithms PCC and FAIR. PCC is a statistical test, which is used to measure the quality and strength of the relationship of two variables. The range of correlation coefficients R_{xy} can vary -1 to 1. The coefficient value closer to 1 indicates the strength of the relation; absolute values indicate a stronger relationship. The direction of the relationship is symbolized by sign of the coefficient value. If the variables increase together or either the variables decrease together, it takes positive value, and if one variable increases as the other decreases then, it takes negative value. The correlation is found using 3, where x, y are the two variables and the μ, σ are their mean and variances.

$$ Rxy = \frac{1}{N-1} \sum (\frac{X - \mu X}{\sigma X})(\frac{Y - \mu Y}{\sigma Y}) \tag{3} $$

FAIR is a single feature classifier in which the decision boundary is set at the Midpoint between the two class means. This possibly is not the apt choice for the decision boundary. But by sliding the decision boundary, one can increase the number of true positives at the expense of classifying more false positives. Here it is accomplished by examining P-R curves built by starting from each direction and taking the maximum of the two areas. For the P-R curve, we take a parallel tabled value of the precision and recall given by (4) and (5) for the majority class. Then build the P-R curve, by taking the maximum area from these values. Where, tp (True Positives), fp (False Positives) and fn (False Negatives)

$$ Precision = \frac{tp}{(tp + fp)} \tag{4} $$

$$Recall = \frac{tp}{(tp + fn)} \tag{5}$$

3.2 Sample Classification Technique

The classification of genic samples can be carried out using some integrated statistical algorithms such as KNN and SVM. KNN [4] is an instance classifier; working on by relating the unknown to the known instances according to similarity measure or some distance, So that unknown instances can be easily identified. If both the instances are set far apart in the instance space measured by the distance function, it is less likely to be in the same class rather than two closely located instances. For continuous variables the distance measures used are given by (6) and (7). They are Euclidean distance, and Mahalanobis distance. Where x and y are the unknown sample and class label respectively.

$$d(x, y) = \sqrt{\sum_{i=1}^{n} (xi - yi)^2} \tag{6}$$

$$d(x, y) = \sqrt{(x - y)^T S^{-1}(x - y)} \tag{7}$$

SVM [7] is a statistical learning technique; it is used to classify data points by assigning it to one of the two, half disjoint spaces. With the use of kernel functions (specific to the datasets), it can easily classify Non-linear relationships between data. It uses a Non-linear mapping for transforming the original training data into higher dimension data. Within the new dimension it searches for the optimal linear separating Hyper-plane given by (8). In this method it correctly classifies as many possible samples by maximizing the margin for correctly classified samples. Where, W^T is weighing factor.

$$g(x|w, w_0) = w^T x + w_0 \tag{8}$$

3.3 Gene Enhancement Method

This method is an unsupervised method used to enhance the existing genes by clustering similar genes. It has been used successfully to feature analysis, clustering, and classifier designs in fields medical imaging. A data can be represented in various feature spaces. Here the FCM algorithm [8] classifies the data by grouping similar data points in the feature space into clusters. In clustering, one iteratively minimize an objective function that symbolizes the distance from any given data point to a cluster center weighted by that data point's membership rank. Therefore the result of such a clustering is regarded as prime solution with a determined degree of the accuracy.

3.4 Protein Classification Method

In order to classify the genes in terms of protein variations an optimization cum classification principle is used here. In GA [4] first, generate random population of N chromosomes. Then, for each chromosome x in the random population, assess the fitness function f(x). Thus creating a new population of chromosomes by iterating the GA operations until the new population completes selection process of two parent chromosomes from the current population based on their fitness value (if better fitness i.e. best selection) With a crossover probability crossover the parents to form a new child. If crossover operation was not done, child is an exact copy of parents. In mutation a probable new child is mutated at any locus place of existing parent in a current population for evaluation. The new population is used for the future iterations. If the end condition is satisfied, end the process as the optimal solution is reached in current population.

4 Data Sampling and Induced Principle

The prime applications of this paper is to develop informative software, which help medical professionals in diagnostics of syndrome [1] and new chromosomal aberration diseases, It can also be used in identification of certain disease abstractions and analyses the bodily parts affected due to the disease gene effects [5]. The data used are mainly gene data which are available as sample wise in datasets. The population encloses various microarray and mass spectrometry methods in observational data collection of sample data on various genes. These data can be identified as frames of diseases by various types. The similitude structure of datasets and the disease datasets having observed sample sizes are specified in Tables 1 and 2 respectively.

Table 1. Structure of a gene dataset.

S	G_1	G_2	...	G_{m-1}	G_m	Class
S1	96.42	21.43	...	71.59	40.71	0
S2	38.42	29.19	...	37.06	31.15	1
S3	98.6	43.12	...	54.7	12.4	0
...	...	...	...	...	...	...
Sn-1	54.25	67.52	...	16.46	37.68	1
Sn	21.72	38.05	...	12.42	26.41	1

Here, the datasets have sample size S_n samples and the gene features ranges from G_1 to G_m genes and each sample is subjected to a class, as each sample may belong to any one class of the two in functional aspect of belonging truth. This sample dataset is a binary class datasets which only contain any two class values denoted by either 1 or 0. Here, the leukemia binary class disease dataset has 7129 genes and 72 samples, where there are 47 Acute Lymphoblast type samples and 25 Acute Myeloid type samples, likewise colon cancer binary class disease datasets has 62 samples and 2000 genes in which normal samples are 22 and tumor samples were 40 The major challenges facing samples are mainly data In-Sufficiency problem, high dimensionality problem and class imbalance problem.

Table 2. Gene expression dataset details.

Dataset type	No. of samples	No. of genes	Label of class	No. of samples in class
Colon cancer	62	2000	Normal	22
			Tumor	40
Leukaemia	72	7129	Acute Lymphoblast	47
			Acute Myeloid	25

4.1 Data in-Sufficiency Problem

As disease data are highly Un-shareable in nature and diseased patient samples [9] are very less in available due to lacking gene treatment facility. Only few samples are collected for the minority classed samples. So by various random over sampling and under sampling this problem can be rectified to limits.

4.2 High Dimensionality Problem

As these data are skewed [10] towards genes that is, the abundant number of genes per sample. As there are number of gene observations ranges above 2000 genes, it is hard for data collection and pre-processing. In these many genes only few genes are vital for bringing out the disease syndrome, so it has to be carefully analysed and selected by statistical measures.

In an experiment of numerous genes, only a few genes show high relation with the targeted disease. Research works have concluded that numerous genes vary much between different diseases; only a few genes are more than enough for diagnosing the disease accurately. Thus with dimensionality reduction, computation is much reduced with prediction accuracy increased.

4.3 Class Imbalance Problem

The class imbalance problem [11] is a complex challenge faced by machine learning algorithms, a classifier affected by this problem for a specific dataset shows strong accuracy but performance is degraded much on the minority class. As the ratio difference in availability of majority class samples to the minority class samples is so high, there is a systematic bias in classifier in biasing towards the majority class samples.

In an imbalanced dataset, Re-sampling methods [12] strategically add minority samples and remove majority samples bringing the distribution of the dataset near to optimal distribution. In New algorithms approach [13] handles the class imbalance problems a way different than the standard machine learning algorithms; these

introduced boosting, bagging methods and One-Class, Cost-Sensitive learners algorithms which maximize statistics rather than accuracy. Gene selection includes
approaches which [14] sub select small gene set from the original large gene set thus
reducing the class Imbalance problem of the dataset.

The conceptual architecture of the proposed system is shown in Fig. 1. In this
system the gene and protein datasets are input to the system and the gene selection
methods [15] are acted on those dataset so that an constrained set of selected genes
were extracted out of all genes and by learning those selected genes on various classifiers [16–18] the samples are classified and evaluated for truth and accuracy of
classification.

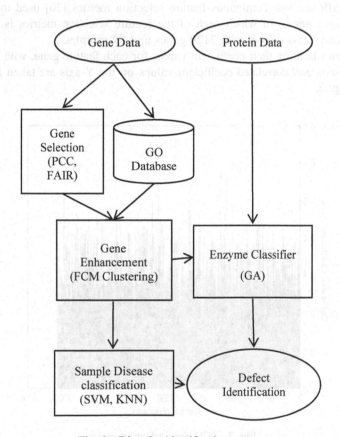

Fig. 1. Disorder identification system

The GO represent the hierarchical orderly arrangement of GO terms, which represent the various relation of one gene with other gene. As each gene is compost of
various GO terms and each GO terms has a natural lineage relation with each other.
The GO databases or human genes are available online, so by which one can find the
relation of two or more genes.

5 Inferences

In this system the binary class datasets [19] are given as inputs and the feature selection metrics are acted on those dataset so that an constrained set of selected features were extracted out of all features and samples were classified by learning those selected genes on classifiers. This system has been developed and implemented in MATLAB tool and it is worked under Windows 7 OS with Core i3 Processor environment. In this experiment some of the results had only presented here.

5.1 Gene Ranking

PCC and FAIR are two continuous feature selection metrics [20] used to select the most expressive genes. In which each of the feature selection metrics is trained on leukemia binary class dataset with 7129 genes and 72 samples.

The metrics holding their coefficient values for each feature gene, with number of genes on X-axis and correlated coefficient values on the Y-axis are taken for PCC is shown in Fig. 2.

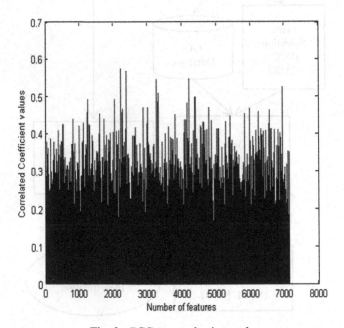

Fig. 2. PCC gene selection ranks

In FAIR metrics the threshold area values are taken for each of the gene. Here by taking the number of genes on X-axis the relative threshold area values are plotted on the Y-axis and the high threshold is taken as high ranked genes feature is depicted in Fig. 3. The most expressive gene features in the classifier can induce higher accuracy scores, so we took top 10 features for various gene selection metrics in Leukemia data.

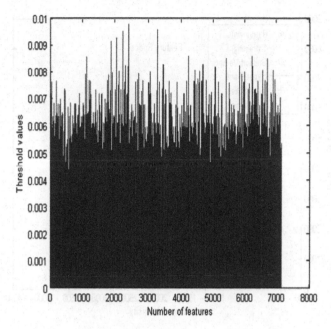

Fig. 3. FAIR gene selection ranks

5.2   Sampling Classification

The gene scores of the top 10 genes are separated in 50:50 ratio based on number of samples and the class of the samples. Here, the first half is taken as training set [21] and the next half is taken as test set, the test & train samples for leukemia data is given in Table 3.

Table 3. Test and train samples for leukemia data.

Data/sample	Total sample	Class1 samples	Class2 samples
Total data	72	47	25
Train data	37	24	13
Test data	35	23	12

The classifiers used for classification task is SVM technique. While training and testing the Leukemia data's top 10 genes with higher ranking score is classified using SVM as classifier model. In Fig. 4, genes of the same sample set are plotted in X-axis and Y-axis respectively, during training support vectors are generated and an optimized hyper-plane is created and classification of test samples are classified based on the predicted support vectors on either side of Hyper-plane as positives and negatives.

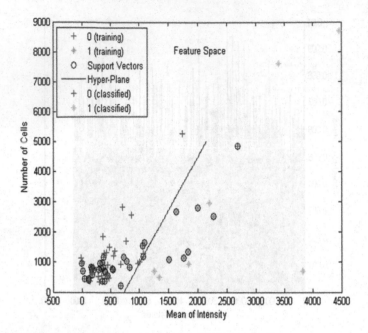

Fig. 4. SVM Sample Classification

6 Conclusion

In this paper, the various soft computational analytical algorithms are examined in relation to the proposed schemes. The gene data sampling have been analyzed and human gene GO databases are identified for gene enrichment. Thus far this soft computational based pattern recognition scheme can boost the existing information in machine learning.

Acknowledgments. I give my sincere thanks to my research guide and my fellow research students, for their high motivation towards this work. And I thank the NICHE research center for continuing the research work and finally I thank God for providing through a good parental support.

References

1. Vanitha, D., Devaraj, D., Venkatesulu, M.: Gene expression data classification using support vector machine and mutual information-based gene selection. Proc. Comp. Sci. **47**, 13–21 (2015). Elsevier
2. Chanchal, K., Matthias, M.: Bioinformatics analysis of mass spectrometry-based proteomics data sets. FEBS Let. **583**, 1703–1712 (2009). Elsevier
3. Maji, P., Paul, S.: Scalable Pattern Recognition Algorithms: Applications in Computational Biology and Bioinformatics. Springer, Cham (2014). https://doi.org/10.1007/978-3-319-05630-2. 22

4. Gunavathi, C., Premalath, K.: Performance analysis of genetic algorithm with KNN and SVM for feature selection in tumour classification. Int. J. Comput. Control. Quant. Inf. Eng. **8**, 1397–1404 (2014)
5. Kristain, O., Marko, L., Sampsa, H.: Fast gene ontology based clustering for microarray experiments. Bio-Data Min. **01**, 1–8 (2008). Bio-Med Central Ltd
6. Jianzhen, J., Yongjin, K.: Discovering disease-genes by topological features in human protein-protein interaction network. Bioinform. Sys. Biol. **22**, 2800–2805 (2007)
7. Mundra, P.A., Rajapakse, J.C.: SVM-RFE with MRMR filter for gene selection. IEEE Trans. Nanobiosci. **9**, 31–37 (2010)
8. Ganeshkumar, P., Victoire, T.A.A., Renukadevi, P.: Design of fuzzy expert system for microarray data classification using a novel genetic swarm algorithm. Expert Syst. Appl. **39**, 1811–1821 (2012)
9. Wasikowski, M., Chen, X.: Combating the small class imbalance problem using feature selection. IEEE Trans. Knowl. Data Eng. **22**, 1388–1400 (2010)
10. Chawla, N., Japkowicz, N., Kotcz, A.: Editorial: special issue on learning from imbalanced datasets. ACM SIGKDD Explor. Newsl. **6**, 1–6 (2004)
11. Zheng, Z., Wu, X., Srihari, R.: Feature selection for text categorization on imbalanced data. ACM SIGKDD Explor. Newsl. **6**, 80–89 (2004)
12. Kotsiantis, S., Kanellopoulos, D., Pintelas, P.: Handling imbalanced datasets: a review. GESTS Int. Trans. Comput. Sci. Eng. **30**, 15–23 (2006)
13. Visa, S., Ralescu, A.: The effect of imbalanced data class distribution on fuzzy classifiers - experimental study. In: FUZZIEEE 2005, Reno, Nevada, USA, vol. 5, pp. 749–754. IEEE, Nevada (2005)
14. Weiss, G., Provost, F.: Learning when training data are costly: the effect of class distribution on tree induction. J. Artif. Intell. Res. **19**, 315–354 (2003)
15. Han, J., Kamber, M.: Data Mining Concepts and Techniques. Morgan Kaufmann, San Francisco (2006)
16. Ben-Dor, A., Bruhn, L., Friedman, N., Nachman, I., Schummer, M., Yakhini, Z.: Tissue classification with gene expression profiles. J. Comput. Biol. **7**, 559–584 (2000)
17. Mitchell, T.: Machine Learning. McGraw Hill, New York (1997)
18. Fawcett, T.: An introduction to ROC analysis. Pattern Recogn. Lett. **27**, 861–874 (2006)
19. National Center for Biotechnology Information. http://www.ncbi.nlm.nih.gov
20. Bell, J.B.B., Kumar, P.G.: Using continuous feature selection metrics to suppress the class imbalance problem. Int. J. Sci. Eng. Res. **3**, 27–35 (2012)
21. Alpaydin, E.: An Introduction to Machine Learning. The MIT Press, Massachusetts (2004)

A Weight Based Approach for Emotion Recognition from Speech: An Analysis Using South Indian Languages

S. S. Poorna$^{(\boxtimes)}$, K. Anuraj, and G. J. Nair

Department of Electronics and Communication Engineering,
Amrita Vishwa Vidyapeetham, Amritapuri, Kollam, India
poorna.surendran@gmail.com

Abstract. A weight based emotion recognition system is presented to classify emotions using audio signals recorded in three south Indian languages. An audio database with containing five emotional states namely anger, surprise, disgust, happiness, and sadness is created. For subjective validation, the database is subjected to human listening test. Relevant features for recognizing emotions from speech are extracted after suitably pre-processing the samples. The classification methods, K-Nearest Neighbor, Support Vector Machine and Neural Networks are used for detection of respective emotions. For classification purpose the features are weighted so as to maximize the inter cluster separation in feature space. An inter performance comparison of the above classification methods using normal, weighted features as well as feature combinations are analyzed.

Keywords: Emotion from speech · Short time energy · Formant frequencies
LPC · Weights · Centroid · KNN · SVM · Neural network · Accuracy
Precision · Recall

1 Introduction

One of the major challenges in human computer interfaces (HCI) is recognizing the emotions to respond back in a suitable way. Introducing emotions to the HCI makes their behavior closer to human. Speech is an important mode for identifying the affective state of a person. It also contains the information about the speaker's identity and his health conditions. The features extracted from the vocal utterances conveyed by the human can be used to identify the affective states of the person. Speech, for emotion recognition finds applications in many fields like call centers, medicine, defense, lie detection, e-learning, gaming etc. A review of different speech databases, features and classifiers used in speech emotion recognition given in the published works of El Ayadi et al. [1].

According to Linguistic Survey [2] of India, there are 122 official languages. Some of the recently developed speech emotion recognition systems (ERS) are in Indian languages (Rao and Koolagudi [3]; Kamble et al. [4]; Firoz Shah et al. [5]; Rajisha et al. [6]; Renjith [7]). Studies on emotion recognition for different Indian languages

I. Zelinka et al. (Eds.): ICSCS 2018, CCIS 837, pp. 14–24, 2018.
https://doi.org/10.1007/978-981-13-1936-5_2

and language combinations is cited in the published articles of Swain et al. [8] and Kandali et al. [9]. A text independent database in two languages spoken by natives of Odisha: Cuttacki and Sambalpuri, was analysed by Swain et al. [8]. Speech recordings in emotions - anger, happiness, disgust, fear, sadness, surprise and neutral were classified with HMM and SVM. Based on the accuracy measure, they made a comparative study with different features using the above classifiers. The study showed that MFCC feature along with SVM classifier gave better results (82.17%), compared to other individual features as well as feature combinations. Kandali et al. [9] analyzed six emotions. They used a data base of 140 acted emotional utterances of each speaker, in five native languages of Assam viz, Assamese, Bodo, Dimasa, Karbi and Mishing. The features from speech, MFCC, wavelet packet cepstral coefficients and teager energy of MFCC & WPCC were classified using GMM. Analysis using each feature gave recognition success of 87.95% for MFCC, 90.05% for tMFCC, 94% for WPCC2 and 100% for tWPCC2.

In the present work five basic emotions namely happy, anger, sad, surprise and disgust are considered. The multi lingual and multi modal database 'Amrita emo' [10, 11] is used. This paper aims at designing a feature based, speaker independent multi language emotion classifier, trained on three South Indian languages, Malayalam, Tamil and Telugu. The subsequent sections briefly explains the methodologies adopted for extending the database, pre-processing techniques involved, and feature extraction & classification methods employed for the study. Performance of the system will be tested using supervised learning methods namely KNN, SVM and NN using individual features and feature combinations for samples within languages. Further in order to improve the classification accuracy, the features will be weighted for analysis.

2 Methodology

2.1 Data Acquisition and Database

The database 'Amrita emo' containing emotional speech samples in anger, happy and sad is further updated to include additional emotions viz. surprise and disgust. An omni- directional microphone at sampling rate 44 kHz, is used for acquiring the audio samples. The samples are digitized with 16 bit resolution. The emotional speech recordings considered for analysis were recorded for approximately 1.5 s duration. The content of the audio included the sentence translations of 'Oh my God' in respective languages. Emotional speech samples were gathered from healthy male subjects in the age group 20–30 who were fluent in speaking their native languages. 7 Malayalam speakers, 5 Telugu speakers and 4 Tamil speakers volunteered for this study. The subjects were emoted by showing sensitive video clips before collecting the actual samples. Trial recordings were taken before the actual sample acquisition. The multi lingual emotional speech database included 350 Malayalam, 200 Tamil and 250 Telugu emotional short speech segments. A total of 800 speech samples are used for subsequent analysis.

2.2 Preprocessing

Since the analysis involves real time applications, the speech samples are recorded in presence of noise. The samples are low pass filtered using an IIR filter of cut off frequency 4 kHz for noise removal. Further a high pass filtering for pre-emphasizing the frequencies above 1 kHz [10] is carried out. Further the speech samples are segmented and hamming windowed with duration 20 ms, an overlapping of 10 ms between frames is also provided.

2.3 Extraction of Features

The features for classification are extracted from the pre-processed speech samples. Human brain carryout emotion recognition mainly with the help of visual as well as auditory cues. Since we are dealing with speech, the response of vocal tract, medium through which it traverses and the auditory system are significant in recognizing emotions. In this work, we are considering the features associated with the vocal tract response alone for emotion recognition. The features considered for the work are energy contour, formant frequencies and liner predictive coefficients. Energy of the short segment of speech could be related to the loudness variation of speech. The energy corresponding to each frame is calculated as the sum of squared absolute value of the signal [12]. The spectral peaks in speech are the regions of the vocal tract frequency responses matching the input frequencies. The resonant frequencies of the vocal tract are extracted and observed as formants. In the case of text dependent speech samples, the locations of these resonant frequencies remains the same, but the magnitude of these varies with emotions [3]. The distinct peaks in the windowed Fourier transform of speech provide the information regarding the location of these resonances. For this work, peaks are extracted from the smoothened spectra of the speech signal to obtain the formants. The first five formants were considered in this work for classifying the emotions. The human vocal tract can be approximated to an all pole filter excited by either periodic pulse train or random noise inputs. Extraction of these filter coefficients is based on linear prediction. The behavior of this filter varies according to the emotional variation of speech. This will be reflected in the filter coefficients or Linear Prediction Coefficients (LPC). The LPC calculations for this work was based on Levinsons' Recursion. The amplitude of LPC's decreases, as the number of coefficients increases. Hence the first ten LPC's are considered for recognizing emotions.

2.4 Secondary Statistical Parameters and Feature Vector

Mean value is removed from the extracted features for normalization. In order to reduce the dimensionality of the feature vector, the secondary statistical parameters are computed for these features based on their relevance. For energy contour, mean and variance formed the feature vector rather than using the entire set of values. Hence a total of 17 features - two energy based parameters, first five formats and first ten linear prediction coefficients formed the feature vectors for classification.

3 Classification

This section describes about the standard classifiers used, a new method of weighting for separating the features and the effect these weighted features on individual languages. Three classification methods K-Nearest Neighbor (KNN), Support Vector Machine (SVM) and Artificial Neural Network Classifier (ANN) are used for analysis. These classifiers were chosen since they perform classification of the underlying data in different perspectives. KNN is a clustering algorithm that captures the correlation of the features to classify a new instant, SVM [13] on the other hand tries to partition the classes without overlap and ANN uses a biological model which can separate a mixture of dissimilar data. In the ensuing sessions of this paper, the features without weighting will be specified as normal features and those weighted as wfeatures. KNN classifies the new emotion in the test set based on the majority vote of its nearest neighbors. Five nearest neighbors were considered for this work. SVM separates the emotions by fitting hyper planes [10], with the help of support vectors. One against all method of multi class SVM classification, with Radial Basis Functions (RBF) kernel is used. RBF fits smooth separating planes and closed boundaries and hence improves the classification accuracy. A Neural Network Classifier uses a set of input layers hidden layers and output layers to classify the emotions, with the help of an activation function. The classifier uses a biological model which mimics human nervous system. A multi-layer feed forward neural network classifier is used in this work

3.1 Weights to the Feature Vectors-Wfeatures

To aid better visualization and to check whether the emotions clusters in feature space, all possible three dimensional plots of the feature vectors were taken. Scattered feature points, past 3 sigma radius from the cluster centroids were removed manually to avoid overlapping. K-Means clustering is applied these labelled training features find the cluster centroids. For data clusters which are still overlapping, a weight factor was multiplied with the overlapping features, so that the clusters splits up. The weight calculation is as given in Eq. 1,

$$w_{mj} = \frac{[C_m(i) - C_m(j)]^r}{f} \tag{1}$$

where w is the weight applied to the feature vectors of emotion j, $C_m(i)$ and $C_m(j)$ are the centroids of emotions i and j with feature m respectively, r corresponds to the power of the distance measure and $1/f$ is a factor that depends on the spread of the cluster in the hyperspace. This factor is selected in such a way, so as to maximize the inter-cluster distance and minimize the intra cluster spread.

For weighting the features in respective languages during training, Eq. 1 is used and the misclassification rate was analyzed using an SVM classifier for different values of r and $1/f$. The factor $1/f$ is fractional values so that it will not disturb the distribution of data in hyperspace. Figure 1 shows the plot of misclassification vs. $1/f$ for values of $r = 1, 2$ and 3. It can be seen form the above figure that for higher powers of centroid distance, the misclassification increases. The minimum values of misclassification was obtained for $r = 1$. Since misclassification increased with increase in r, higher powers were ignored. From Fig. 1(a) the values of $1/f$ which gave minimum values of misclassification are 0.4 for Telugu, 0.6 for Tamil and 0.8 for Malayalam. Wfeatures in this work are obtained by weighting the features with the above mentioned r and $1/f$ for respective languages.

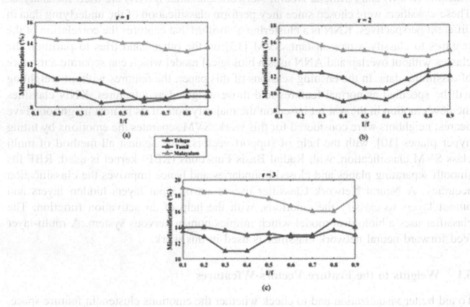

Fig. 1. Plot of misclassification vs. $1/f$ for languages Telugu, Tamil and Malayalam (a) for $r = 1$ (b) $r = 2$ and (c) $r = 3$.

A sample 3 dimensional plot of five emotions in Telugu, with features energy mean, energy variance and first LP coefficient is shown in Fig. 2. Figure 2(a) gives the feature space with the emotions anger, happy, disgust, sad and surprise, before applying weights. Figure 2(b) gives the same space after applying weights (wfeatures). We can see that applying weights makes the respective emotions cluster around their centroids, with minimum overlap between the neighbors. In this paper, both the normal features as well as wfeatures were analyzed for emotion recognition.

3.2 Training and Testing

The feature vector is 17 dimensional with 350 Malayalam, 250 Telugu and 200 Tamil text dependent data points. The performance comparison of the emotion recognition system is done for normal and wfeatures. The evaluation is done in individual languages for different cases of feature combinations as well as for all features. Table 1 shows the different models assumed for analysis by combining different features. In all the cases considered, 60% of the random population is used for training and the remaining 40% for testing.

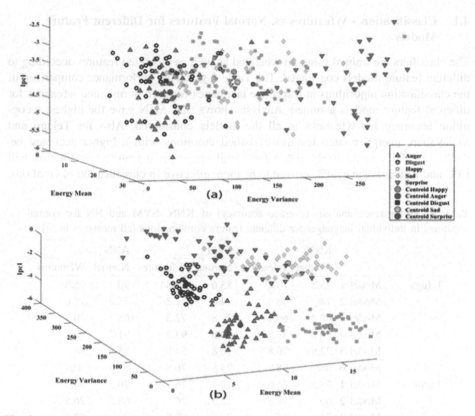

Fig. 2. A sample 3D plot of energy mean-energy variance - LPC1 of five Telugu emotions (a) before and (b) after applying weights

Table 1. Different cases of feature combinations used for comparing the performance of normal and wfeatures

Model	Features
Model 1	Energy related features
Model 2	LPC
Model 3	Formants
Model 4	Energy based features + LPC
Model 5	Energy based features + formants
Model 6	LPC + formants

4 Simulation Experiments and Results

Based on the test results, the performance of these classification methods are evaluated. The analysis was done for two cases using normal and weighted features (i) for different feature models and (ii) using all extracted features, in all the three languages. The statistical measures chosen to compare the classification method are accuracy, precision and recall.

4.1   Classification - Wfeatures vs. Normal Features for Different Feature Models

The classifiers are trained using un-weighted as well as weighted features according to different feature models considered. The Table 2 gives the performance comparison of the classification algorithms in respective languages for both normal and wfeatutes for different feature models assumed. Analysis shows that ANN gave the highest recognition accuracy for wfeatures in all the models considered. Also for Telugu and Malayalam, energy related features classified emotions with a higher accuracy i.e. 96.8% and 97% respectively, while for Tamil language, a feature combination will LPC and formants gave 94%, proved to be more effective in classification of emotions.

Table 2. Performance analysis (average accuracy) of KNN, SVM and NN for normal vs. wfeatures in individual languages for different feature combinations (all measures in %)

		KNN		SVM		ANN	
		Normal	Wfeatures	Normal	Wfeatures	Normal	Wfeatures
Telugu	Model 1	75.2	76.8	83.6	84.84	90	96.8
	Model 2	74	76	74.6	77.2	75.2	75.6
	Model 3	68.4	69.8	70.8	72.3	68	70
	Model 4	76.8	77.2	90.8	93.2	91.2	96
	Model 5	72.8	76.8	76.8	84.8	85.6	86
	Model 6	68	68.8	74.8	76.8	68.8	70.4
Tamil	Model 1	68.5	75.5	72	76	70.5	80.5
	Model 2	69	78.5	72.5	76	68.5	76.5
	Model 3	74	86	72.5	90.5	73	82.5
	Model 4	68	76	71.5	81.5	68	84.5
	Model 5	67	75.5	72.5	84.5	71	82.5
	Model 6	74	86	72	90.5	71	94
Malayalam	Model 1	88.3	90	80.33	84.67	74.33	97
	Model 2	62	69.3	68	70	69.67	70
	Model 3	63	69	65	67.5	70.33	71.67
	Model 4	64	67	63	67.5	73	85.33
	Model 5	80	87	82	90.1	76	91.67
	Model 6	63	67	66.6	72.3	74	76

4.2   Classification - Wfeatures vs. Normal Features for All Features

The weight based approach versus the normal features were compared with all the extracted features: Energy based, LPC and Formants for the languages Telugu, Tamil and Malayalam. Tables 3, 4 and 5 shows the performance measures evaluated for different classifiers. The one against all method of classification was adopted for evaluation of emotions in respective languages. The experimental results shows that the weighted system gave superior classification results compared to the one with normal

Table 3. Performance of emotion recognition system using Telugu speech for normal and wfeatures (in percentage)

		Accuracy		Precision		Recall	
		Normal	Wfeatures	Normal	Wfeatures	Normal	Wfeatures
KNN	Anger	82	81	71.87	85.57	71.87	86.86
	Happy	66	89	51.69	64.52	53.13	86.25
	Sad	54	75	46.61	60.40	45	60
	Surprise	54	75	80	80	100	100
	Disgust	77	85	64.47	76.51	65	79.38
	Average	**66.6**	**81**	**62.93**	**73.4**	**67**	**82.49**
SVM	Anger	91	90	69.46	90.7	63.75	76.87
	Happy	83	80	54.04	74.2	78.6	85.6
	Sad	85	100	77.78	100	90	100
	Surprise	81	100	89.41	100	90	100
	Disgust	81	100	51.3	100	90	100
	Average	**84.2**	**94**	**68.39**	**92.98**	**82.47**	**92.49**
ANN	Anger	82	93	95.6	92.9	81.3	98.8
	Happy	82	86	86.7	100	81.6	97.5
	Sad	78	99	100	100	72.5	98.8
	Surprise	80	100	80	100	100	96
	Disgust	78	96	100	95.2	72.5	100
	Average	**80**	**94.8**	**92.46**	**97.62**	**81.58**	**98.22**

Table 4. Performance of emotion recognition system using Tamil speech for normal and wfeatures (in percentage)

		Accuracy		Precision		Recall	
		Normal	Wfeatures	Normal	Wfeatures	Normal	Wfeatures
KNN	Anger	80	70	66.67	55.02	59.38	55.47
	Happy	77.5	80	59.72	67.15	55.47	61.72
	Sad	70	76.25	48.04	63.45	48.44	64.07
	Surprise	63.75	97.50	48.04	75.3	48.44	68.75
	Disgust	72.5	85	45.84	92.11	47.66	62.5
SVM	**Average**	**72.75**	**81.75**	**53.66**	**70.61**	**51.88**	**62.50**
	Anger	63.5	87.5	58.76	80.77	63.28	92.19
	Happy	75	97.5	57.84	94.45	56.25	98.44
	Sad	67.5	92.5	57.98	87.27	60.94	90.62
	Surprise	63.75	97.50	57.37	96.09	60.94	96.09
	Disgust	48.75	98.75	48	99.23	46.88	96.87
	Average	**63.7**	**94.75**	**55.99**	**91.56**	**57.66**	**94.84**
ANN	Anger	75	92.5	81.4	100	89.1	90.61
	Happy	78.8	97.5	84.1	100	90.6	96.9
	Sad	73.8	92.5	80.3	98.3	89.1	92.2
	Surprise	82.5	95.5	82.1	96.3	100	100
	Disgust	75	97.5	80.6	97	90.6	100
	Average	**77.02**	**95.1**	**81.7**	**98.32**	**91.88**	**95.94**

features. Above mentioned tables shows that, while wfeatures were considered ANN gave the highest accuracy i.e. 94.8% for Telugu, 95.1% for Tamil 91.98% for Malayalam. In the case of weighted features, the recognition accuracy in the respective languages shows a definite pattern viz. accuracy (ANN) > accuracy (SVM) > accuracy (KNN). Precision and recall rates of individual emotions also showed an increase with wfeatures compared to the normal ones.

In all the languages considered, the average accuracy measure shows an increase of more than 9% for wfeatures irrespective of the classification methods compared to normal case. The highest improvement in average recognition accuracy was obtained using ANN. Similarly the average precision and recall rates also showed an improvement for wfeatures in all the languages considered.

Table 5. Performance of emotion recognition system using Malayalam speech for normal and wfeatures (in percentage)

		Accuracy		Precision		Recall	
		Normal	Wfeatures	Normal	Wfeatures	Normal	Wfeatures
KNN	Anger	83.33	87.5	75.8	80.77	84.9	92.19
	Happy	73.33	82.5	78.57	72.62	91.67	67.19
	Sad	73.33	100	59.33	100	60.05	100
	Surprise	78	80.83	50.81	84.19	50.67	64.59
	Disgust	78.33	83.33	62.94	74.52	58.34	67.71
	Average	**77.26**	**86.83**	**65.49**	**82.42**	**69.13**	**78.34**
SVM	Anger	85	92	92.5	98	69	95
	Happy	75	83	82	83	90	100
	Sad	90	96	83	93	95	100
	Surprise	86	89.5	90	90	82	85
	Disgust	70	92.5	48.04	97	82	96
	Average	**81.2**	**90.6**	**79.11**	**92.2**	**83.6**	**95.2**
ANN	Anger	84	100	86	100	95	100
	Happy	83.3	99	83.3	99	99	100
	Sad	86.7	94.2	87	98.9	97.9	93.8
	Surprise	80	86.7	82.7	87	94.8	97.9
	Disgust	77.5	80	81.1	81	93.8	97.9
	Average	**82.3**	**91.98**	**84.05**	**93.18**	**96.1**	**97.92**

5   Conclusion and Future Work

The paper gives a comprehensive review of emotion recognition from speech using a novel method of weighting the features. Normally, the performance of speech emotion recognition systems depends on various factors such as the emotional eliction of the subjects from whom the recordings are being taken, the presence of noise while recording, choice of feature extraction methods, the classifiers used etc. A method of minimal misclassification by scaling the features using suitable weights was adopted here. SVM classifier was used to fix the weights for respective languages. This classifier was chosen since it gives unaltered results unlike KNN where the number of nearest neighbors and ANN where the number of layers were variables. The method was tested with emotions recorded in three south Indian Languages. Our analysis shows that the weight factors are language dependent for the features considered in this work.

A significant increase in recognition accuracy is obtained while evaluating with wfeatures. This is because applying weight can address the issues of overlap between emotions. For example, in some cases where the subject is surprised, there is a chance that he will also be happy. Also disgust speech can have a content of anger or sad. Weights to the features have the advantage of increasing the separation of the emotion clusters in the feature space. Although in some cases of objective evaluation using wfeatures, the performance measures for individual emotions were slightly lower compared to normal features, the average evaluation of performance measures of the system showed increase in all the measures.

Different models of feature combinations were analysed using normal features as well as using wfeatures. In all the cases considered, the recognition rate using wfeatures was higher compared to normal features. In all the models considered, ANN gave best recognition accuracy compared to other classification methods adopted. The analysis shows that for the languages Telugu and Malayalam, energy related features were more accurate in recognizing emotions (96.8% and 97% respectively), while in Tamil language, a feature combination will LPC and formants gave more accuracy (94%). In the speech emotion recognition system with the features considered, in general the wfeatures gave an increased recognition in terms of average recognition accuracy, precision and recall. ANN classifier gave the highest average classification accuracy i.e. 94.8% for Telugu, 95.1% for Tamil 91.98% for Malayalam, compared the other classifiers used. It is usually seen that ANN classifier gives better recognition rate than other classifiers. This is due to the adaptive learning properties of NN Classifier, to learn from errors. The SVM Classifier gave better classification accuracy compared to KNN in most the cases since the RBF kernel was able to fit smooth separating planes for classification. In the case of analysis with individual languages in Tables 3, 4 and 5, there was an increase of more than 9% for wfeatures classified using ANN.

The system may be further extended by including samples of female speakers and text independent samples. A further improvement in accuracy could be obtained by using a hierarchical classification approach. This system can be adopted in automated call center applications, which uses South Indian languages as their medium of communication. This could be also used in real time emotion identification applications where people speak a blend of languages, rather than sticking on to one. Also Emotion recognition systems with more than 90% accuracy are required for companion robots which can respond to emotions, and uses speech and facial images for recognizing the emotions. The emotion recognition system for South Indian languages may be further customized by including other Indian languages as well as foreign languages.

References

1. El Ayadi, M., Kamel, M.S., Karray, F.: Survey on speech emotion recognition: features, classification schemes, and databases. Pattern Recogn. **44**(3), 572–587 (2011)
2. http://peopleslinguisticsurvey.org/
3. Rao, K.S., Koolagudi, S.G.: Robust Emotion Recognition Using Spectral and Prosodic Features. SpringerBriefs in Speech Technology. Springer, New York (2013). https://doi.org/10.1007/978-1-4614-6360-3
4. Kamble, V.V., Deshmukh, R.R., Karwankar, A.R., Ratnaparkhe, V.R., Annadate, S.A.: Emotion recognition for instantaneous Marathi spoken words. In: Satapathy, S.C., Biswal, B. N., Udgata, S.K., Mandal, J.K. (eds.) Proceedings of the 3rd International Conference on Frontiers of Intelligent Computing: Theory and Applications (FICTA) 2014. AISC, vol. 328, pp. 335–346. Springer, Cham (2015). https://doi.org/10.1007/978-3-319-12012-6_37
5. Firoz Shah, A., Raji Sukumar, A., Babu Anto, P.: Automatic emotion recognition from speech using artificial neural networks with gender-dependent databases. Published in World Congress on Nature and Biologically Inspired Computing NABIC (2009)
6. Rajisha, T.M., Sunija, A.P., Riyas, K.S.: Performance analysis of Malayalam language speech emotion recognition system using ANN/SVM. In: International Conference on Emerging Trends in Engineering, Science and Technology (2016). Elsevier Procedia Technology
7. Renjith, S., Manju, K.G.: Speech based emotion recognition in Tamil and Telugu using LPCC and hurst parameters—a comparitive study using KNN and ANN classifiers. In: International Conference on Circuit, Power and Computing Technologies (2017)
8. Swain, M., Sahoo, S., Routray, A., Kabisatpathy, P., Kundu, J.N.: Study of feature combination using HMM and SVM for multilingual Odiya speech emotion recognition. IJST **18**(3), 387–393 (2015)
9. Kandali, A.B., Routray, A., Basu, T.K.: Vocal emotion recognition in five native languages of Assam using new wavelet features. IJST **12**, 1–13 (2009)
10. Poorna, S.S., Jeevitha, C.Y., Nair, S.J., Santhosh, S., Nair, G.J.: Emotion recognition using multi-parameter speech feature classification. In: IEEE International Conference on Computers, Communications, and Systems, 2–3 November 2015, India (2015)
11. Poorna, S.S., et al.: Facial emotion recognition using DWT based similarity and difference features. In: IEEE 2nd International Conference on Inventive Computation Technologies (2017)
12. Jittiwarangkul, N., Jitapunkul, S., Luksaneeyanawin, S., Ahkuputra, V., Wutiwiwatchai, C.: Thai syllable segmentation for connected speech based on energy. In: IEEE Proceedings of Asia-Pacific Conference on Circuits and Systems (1998)
13. Bhaskar, J., Sruthi, K., Nedungadi, P.: Hybrid approach for emotion classification of audio conversation based on text and speech mining. Procedia Computer Science **46**, 635–643 (2015)

Analysis of Scheduling Algorithms in Hadoop

Juliet A. Murali[✉] and T. Brindha

Noorul Islam University, Thuckalay, Tamil Nadu, India
julietamurali@gmail.com, Brindha@niuniv.com

Abstract. Distributed system consists of networked computers that provide a
coherent system view to its users. Distributed computing is the use of distributed
system to solve complex computational problems. Cloud is a distributed environment, having large capacity data centers. It needs parallel processing and task
scheduling. MapReduce is programming model for processing this big data.
Hadoop is a Java based implementation of MapReduce framework. The task
scheduling in MapReduce framework is an optimization problem. This paper
describes about some advantages and disadvantages used in different Hadoop
MapReduce scheduling algorithms. It also gives the important of performance
metrics considered in different scheduling algorithms. This shows that each
scheduling algorithms have different performance objectives.

Keywords: Cloud computing · Distributed computing · Hadoop · Map reduce
Scheduling

1 Introduction

Big data represent a large volume of data that are stored in distributed data centres. It
also includes techniques and technologies that are used for extracting hidden values of
large data set as the requirements of the users. Cloud is a distributed environment that
contains data centres with large capacity [1, 2].

Cloud computing is a paradigm that makes available the on-demand accessing
configurable shared resources over the internet. The cloud deployment model comes in
six types: Private Clouds, Public Clouds, Hybrid Clouds, Community Clouds, Federated
Clouds and Multi-clouds and Inter-clouds. A public cloud is a publicly accessible cloud
environment with the help of cloud vendors. The Community Cloud infrastructure is
used by a specific organizations. Hybrid Cloud is a collection of cloud infrastructures
like private, community, or public. A federated cloud is the deployment and management
of multiple external and internal cloud computing services to match business needs.

This paper mainly deals with scheduling of cloud resources. The rest of the paper is
presented as follows. Section 2 gives a description about distributed system. Sections 3
and 4 deals with cloud computing models and different performance metrics associated
with scheduling respectively. Section 5 deals with some Hadoop scheduling algorithms
used in Job Trackers and its analysis with performance objectives.

© Springer Nature Singapore Pte Ltd. 2018
I. Zelinka et al. (Eds.): ICSCS 2018, CCIS 837, pp. 25–34, 2018.
https://doi.org/10.1007/978-981-13-1936-5_3

2 Distributed System

Distributed system is a one in which the sharable hardware and software resources are placed on networked computers. The resource communication is attained by the help of message passing and to achieve good performance. The main problems that are to be considered in distributed systems are the synchronization problem, security-related problem, authentication problem, job scheduling and so on. Among these problems, scheduling is considered here. In a distributed environment the resource scheduling means the decision making about the effective utilization and allocation of computing resources that may be hardware or software resources. Cloud is one of the distributed computing models [7] (Fig. 1).

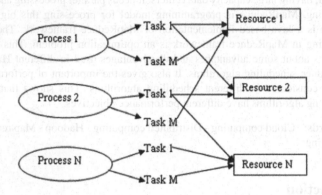

Fig. 1. Distributed system resource sharing.

3 Cloud Computing

Cloud is a distributed environment that contains data centres with large capacity [1, 2]. Cloud computing provides on-demand accessing computing power and resources over the internet. It is a parallel and distributed environment contains a large number of interconnected computers across the internet. This cloud model have five essential characteristics, three service models, and four deployment models. The main characteristic of cloud computing includes on-demand service, distribution, virtualization elasticity and resource pooling. Three popular service models are Application/Software as a Service (SaaS), Platform as a Service (PaaS), and Infrastructure as a Service (IaaS) [5]. Other prominent Service models incudes Data Analytics as a Service and High Performance Computing/Grid as a Service.

The large volume of data called big data, that are stored in distributed data centres. It also includes techniques and technologies that are used for extracting hidden values from large data set as the requirements of the users. Cloud is a distributed environment that contains data centres with large capacity. Parallel processing and task scheduling are required for accessing cloud computing services. Traditional data processing applications are not suitable for processing Big data. Map Reduce is one of the programming models that are used for processing Big data in data centres.

3.1 Map Reduce Framework

Map Reduce is the widely used big data processing platform proposed by google. From the name itself, the data processes is done in two phases - Map and Reduce [2]. Parallel processing and task scheduling are required for accessing cloud computing services. Traditional data processing applications are not suitable for processing Big data. MapReduce is one of the programming models that are used for processing Big data in data centres. The data centres can include more than one map reduce jobs that are running simultaneously. It processes the data by dividing the job into independent tasks [3, 4].

The execution of MapReduce system starts by dividing the input data into small pieces. The map task is allocated to slave nodes by the master nodes. Data locality is one of the main concern during the task allocation. The map function processes a key/value pair to generate intermediate key/value pair [6]. The input to the map function is a split file contains key/value pair. The map function produces the intermediate key/value pair and is stored locally. The shuffling of key/value pair is also done based on a general key by the slave nodes. After the map task, the master got information about the location of intermediate key/value pair generated by map function. This key/value pair is accessed by reduce function as the direction of the master node. Copies processes are introduced in between map and reduce tasks [16]. The reduce function merge all intermediate values. The reduce function make use of an intermediate key. Figure 2 shows the Map Reduce Process.

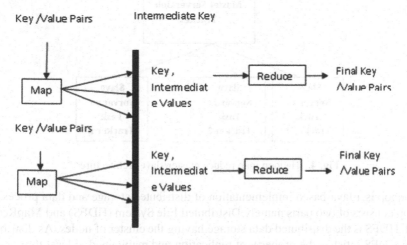

Fig. 2. Map reduce process

The reduced function merge all intermediate code, it can be available locally or remotely. The general grouping of key/Value pair is done in reduces function according to a general key, which in turn reduce the output size. The final output is generated after the completion of the map and reduces tasks. The Map-Reduce execution steps are shown in Fig. 3.

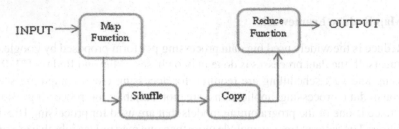

Fig. 3. Map-reduce execution steps

3.2   Hadoop Map Reduce Framework

Hadoop is the Java based open source implementation of MapReduce programming model. It is the distributed parallel programming model. The Map Reduce framework has a Master/Slave architecture. The master consists of Job Tracker and the slave contains TaskTracker. The JobTracker accepts jobs, divides it into task and distributes it to TaskTrackers. TaskTrackers execute the task assigned by JobTrackers as shown in Fig. 4. The JobTrackers are in need of task scheduling. The Map Reduce task scheduling is considered as an optimization problem. Mainly the optimization of scheduling is based on the completion time of tasks that are to be scheduled.

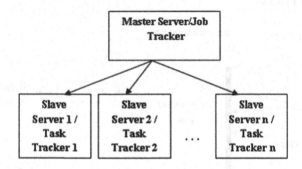

Fig. 4.   Hadoop map reduce master/slave architecture

Hadoop is a Java-based implementation of distributed storage and data processing. Hadoop consists of two parts namely Distributed File System (HDFS) and MapReduce model. HDFS is the distributed data storage having the cluster of nodes. As distributed system HDFS satisfies the property of replication and maintained at least three copies same date and are distributed over the nodes.

The MapReduce model act as big data processing model, across the distributed system. Hadoop has a master/slave architecture. The master node is named as Name Node and worker or slave as Data Nodes. The name node maintains metadata regarding the data scatted over slave nodes. These data are made use at the time of task scheduling when allocating the task to the slave node [15, 17].

4 Performance Metrics of Task Scheduling

Scheduling is an optimization problem. Performance metrics help to evaluate the effectiveness and performance of schedulers. Some of them are discussed as follows [5].

4.1 Execution Time

The execution time or CPU time of a given task is the time spent by system to execute the task. The mechanism used to measure execution time is implementation defined. It also specifies how much amount of time the task actively utilize the resources.

4.2 Response Time

Response time is the amount of time from a request was submitted until the first response is produced. It is also termed as the time difference between tasks becomes active and the time it completes.

4.3 Data Locality

Data locality is a measure of task data localization. It make sure that the data is accessed from the local drive during processing of task. The data locality ensures considerable amount of improvement in performance of schedule.

4.4 Resource Utilization

Resource utilization is the usage of sharable resources. Good resource utilization make sure that most of the time the resources were used by any processes. It means the idle time of resource is reduced.

4.5 Deadline

The completion of task execution within a specified time limit it is known as deadline. The scheduler tried to complete most of the task within a time limit for better performance.

4.6 Makespan

The amount of time required from start of first task to the end of final task in a schedule. The main aim of the scheduling algorithm is to minimize the make span in order to attain better performance.

4.7   Throughput

The number of tasks that are to be completed in given amount of time is known as throughput. The schedulers are tried to maximize the throughput.

5   Hadoop Task Scheduling Algorithms

Hadoop is used for implementing Map and Reduce activities. Hadoop scheduling algorithms are categorized as Inbuilt Scheduling Algorithms and User Defined Scheduling Algorithms [10, 11].

5.1   Inbuilt Scheduling Algorithms

5.1.1   FIFO Scheduler
Hadoop uses FIFO as the default scheduling algorithm. The jobs are prioritized in first come, first served basis. The main drawbacks are the short jobs are to be waited for long jobs, low performance for multiple jobs types; reduce data locality, not pre-emptive. The main advantages are simple and efficient, jobs are executed at the order they come [4].

5.1.2   Fair Scheduler
In fair scheduler all jobs get average equal share of resources. It makes a group of job pools. It provides fairness in sharing resources in the pool. It allows quick response to small jobs among large jobs. It is complicated configuration and may lead to unbalanced performance because the job weights are not considered [8].

5.1.3   Capacity Scheduler
It ensures the fair management of the recourses among a large number of users. They are similar to fair schedulers, uses queues instead of resource pools. It maximizes the resource utilization and throughput. This is the most complex among three schedulers [8].

5.2   User Defined Scheduling Algorithms

Many user defined scheduling algorithms are available. Some of them are as follows.

5.2.1   Delay Scheduling
In delay scheduling data is not available locally the task tracker will wait for a fixed amount of time [9, 12]. The next task in the queue is scheduled if the above constraint is satisfied. It is mainly used in the homogeneous environment.

5.2.2   Matchmaking Scheduler

The matchmaking scheduler contains a locality marker, which identifies the local map tasks. During scheduling slave nodes considers local tasks before any other non-local task. This scheduler ensures data locality of slave nodes [12].

5.2.3   Longest Approximate Time to End (LATE)

LATE schedulers identifies slow running tasks and create a speculative copy and complete the task in some other resources. The speculative tasks are very slow tasks, it may due to contention for resources, overloaded CPU, etc. It improves the execution time of scheduling and response time of a job [3, 12, 13].

5.2.4   Deadline Constraint Scheduler (DCS)

In DCS [3, 12, 14] the deadline is getting as part of the input. Jobs are scheduled if the specified deadline is met is known as schedulability test. It is a dynamic scheduling scheme and can be used in both Homogeneous and Heterogeneous environments. Here two data processing models namely job execution cost model and a constraint-based Hadoop scheduler are used.

5.2.5   Resource Aware Scheduler (RAS)

RAS includes two activities user free slot filtering and dynamic free slot advertisement [12]. It is a dynamic and it is for Homogeneous and Heterogeneous environments. RAS mainly concentrate on efficient utilization of resources like CPU, IO, disk, network and so on [18].

5.2.6   A Self-adaptive Map Reduce Scheduling Algorithm (SAMR)

SAMR [13] dynamically identifies slow tasks by examining historical information recorded on each node. It also considers the remaining time of task at the time of scheduling.

5.2.7   Multi-objective Earliest Finish Time (MOEFT) Scheduling

In multi-objective EFTS [10], map tasks scheduled first followed by reduce tasks. These tasks may be scheduled on different VMs, according to the availably of slots. The tasks are to be selected as from queue in order. This is an iterative process. The map task without a parent is placed on the top of the queue. The earliest finish time depends on the number of tasks in each job, scheduling scheme and decision model. The workload information got updated as a result of completion of the map and reduce task. The scheduling decision is made on the basis of completion time, constraints and cost with deadline constraint.

Table 1 give the comparison of different Hadoop schedulers with performance metrics. The scheduling strategy is categorized as static and dynamic. The allocation of resources to the task is done before the start of task execution is called static scheduling strategy. In dynamic scheduling strategy, the resource allocation is done during the execution time on task. It also describes about the environment in which the scheduling algorithms are active, it may homogeneous, heterogeneous or both.

Table 1. Comparison of hadoop schedulers.

Scheduler	Strategy	Environment	Performance metric relation
FIFO	Static	Homogeneous	Reduce data locality Poor response time for short jobs
Fair	Static	Homogeneous	Fast response time for small jobs
Capacity	Static	Homogeneous	Better resource utilization Throughput
Delay	Static	Homogeneous	Achieve data locality
Matchmaking	Static	Homogeneous	Reduce jobs misses deadline
LATE	Static	BOTH	Minimize job response time Improves execution time
Deadline constraint	Dynamic	BOTH	Meet deadline Improve system utilization
Resource aware	Dynamic	BOTH	Good resource utilization
SAMR	Dynamic	BOTH	Improve execution time Reduces the marksman
MOEFT	Dynamic	BOTH	Minimize execution time Reduces the marksman

Figure 5 shows the different performance matrices considered by different scheduling algorithms. Most of the scheduling algorithms aim to reduce response time of jobs, minimization of execution time, and efficiency of resource utilization and so on. The static scheduling algorithm mainly deals with the performance objectives like data locality, response time, execution time, etc., where as the dynamic scheduling algorithms concentrate on resource utilization, execution time, make span etc. Multi-objective algorithms are more efficient.

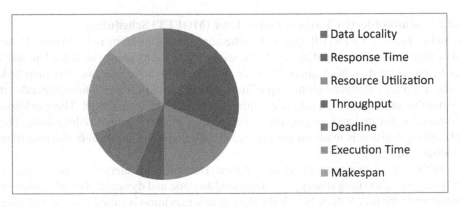

Fig. 5. Performance metric relationship of various scheduling algorithms.

Most of the scheduling algorithms have their own performance objectives; they are having some advantages and disadvantages so that each one got some importance. It is better to use scheduling algorithms have multi-objective. Select the objectives for the schedule based on the requirement of the problem or the environment in which the scheduling strategy is introduced. From the performance metric relationship of the various scheduling algorithms, it is found that most of the algorithms consider Response Time, Resource Utilization, Execution time etc. as the performance objectives.

6   Conclusion

Task scheduling is a combinatorial problem in cloud computing environment. Nowadays different scheduling algorithms are available for both cloud and Hadoop environment. They are having different performance objectives. They have their own advantages and disadvantages. The users and researchers use these algorithms directly or with some modifications according to their requirements. The analysis shows that most of the scheduling algorithms aim to reduce response time of jobs, minimization of execution time, and efficiency of resource utilization and so on. In the future include the comparative study of more multi-objective algorithm and its selection strategy.

References

1. Li, R., Haibo, H., Li, H.: MapReduce parallel programming model: a state-of-the-art survey. Int. J. Parallel Prog. **44**(4), 832–866 (2016)
2. Johannessen, R., Yazidi, A., Feng, B.: Hadoop MapReduce scheduling paradigms. In: IEEE 2nd International Conference on Cloud Computing and Big Data Analysis (ICCCBDA), pp. 175–179 (2017)
3. Mohamed, E., Hong, Z.: Hadoop MapReduce job scheduling algorithms survey. In: IEEE Conference Publications, pp. 237–242 (2016)
4. Makwe, A., Kanungo, P.: Scheduling in cloud computing environment using analytic hierarchy process model. In: IEEE International Conference on Computer, Communication and Control (2015)
5. Ali, S.A., Alam, M.: A relative study of task scheduling algorithms in cloud computing environment. In: 2nd International Conference on Contemporary Computing and Informatics (IC3I), pp. 105–111 (2016)
6. Kashyap, R., Louhan, P., Mishra, M.: Economy driven real-time scheduling for cloud. In: 10th International Conference on Intelligent Systems and Control (ISCO) (2016)
7. Al-Najjar, H.M., Hassan, S.S.N.A.S.: A survey of job scheduling algorithms in distributed environment. In: IEEE International Conference on Control System, Computing and Engineering, pp. 39–44, November 2016
8. Gautam, J.V., Prajapati, H.B., Dabhi, V.K., Chaudhary, S.: A survey on job scheduling algorithms in Big data processing. In: IEEE International Conference on Electrical, Computer and Communication Technologies (ICECCT) (2015)
9. Zaharia, M., Borthakur, D., Sarma, J.S., Elmeleegy, K., Shenker, S., Stoica, I.: Delay scheduling: a simple technique for achieving locality and fairness in cluster scheduling. In: Proceedings of the EuroSys, pp. 265–278 (2010)

10. Hashem, I.A.T., Anuar, N.B., Marjani, M.: Multi-objective scheduling of map reduce jobs in big data processing. Multimed. Tools Appl. **77**, 9979–9994 (2017)
11. Yazdanpanah, H., Shouraki, A., Abshirini, A.A.: A comprehensive view of MapReduce aware scheduling algorithms in cloud environments. Int. J. Comput. Appl. **127**(6), 10–15 (2015)
12. Usama, M., Liu, M., Chen, M.: Job schedulers for big data processing in Hadoop environment: testing real-life schedulers using benchmark programs. Digit. Commun. Netw. **3**, 260–273 (2017)
13. Zaharia, M., Konwinski, A., Joseph, A.D., Katz, R.H., Stoica, I.: Improving mapreduce performance in heterogeneous environments. In: Proceedings of the OSDI, vol. 8, no. 4, pp. 29–42 (2008)
14. Kc, K., Anyanwu, K.: Scheduling hadoop jobs to meet deadlines. In: IEEE Second International Conference on Cloud Computing Technology and Science, pp. 388–392 (2010)
15. Dhingra, S.: Scheduling algorithms in big data: a survey. Int. J. Eng. Comput. Sci. **5**(8), 17737–17743 (2016). ISSN 2319-7242
16. Singh, R.M., Paul, S., Kumar, A.: Task scheduling in cloud computing: review. (IJCSIT) Int. J. Comput. Sci. Inf. Technol. **5**, 7940–7944 (2014)
17. Senthilkumar, M., Ilango, P.: A survey on job scheduling in big data. Cybern. Inf. Technol. **16**(3), 35–51 (2016)
18. Soualhiaa, M., Khomhb, F., Tahar, S.: Task scheduling in big data platforms: a systematic literature review. J. Syst. Softw. **134**, 170–189 (2017). ISSN 0164-1212

Personalized Recommendation Techniques in Social Tagging Systems

Priyanka Radja[(✉)]

Delft University of Technology, Mekelweg 2, 2628 CD Delft, The Netherlands
radja.priyanka@gmail.com

Abstract. Prior to the advent of the social tagging systems, different traditional approaches like content based filtering and collaborative filtering were employed in recommender systems. The content based filtering approach recommended resources to users based on the resources the same target user liked in the past. The collaborative filtering technique recommended resources to users if other users with similar preferences had liked them. These approaches did not consider the reason why the user liked a resource i.e. the context. Hence, the recommendation provided to the user is not catered to his interests i.e. the recommendation is not personalized. Social tagging systems allow users to append a tag to the resources. The users are free to choose these tags. Therefore, these tags already have the context information as the users choose tags which help them remember the resource for future use. Hence, the users implicitly include the reason why they like the target resource as the tag for that resource. This paper highlights different approaches used in social tagging systems to provide personalized recommendation of resources for each user. Experimental evaluation of these approaches on data collected from different social tagging systems is analyzed. Moreover, some improvements to these approaches are suggested to improve the efficiency, accuracy and novelty of the personalized recommendations.

Keywords: Personalized recommendation · Tagging · Clustering
Integrated diffusion · Collaborative filtering · Content based filtering · FolkRank
Tag Expansion based Personalized Recommendation

1 Introduction

With the advent of the internet, the number of users who actively rely on the social web for multimedia data retrieval has increased drastically. The migration of the users from the use of traditional television sets to the websites on the internet for multimedia information generation and retrieval has caused the problem of information overload.

As more people start using the internet, the amount of information that is generated also increases. This huge amount of information must be handled properly and referenced for future use. The success of many social media, blogging and tagging websites is due to the engaging content these websites provide to the users. Unlike in the earlier times, when the television and radio were the only source of multimedia information, which restricted the users to a few fixed options in the form of channels, the internet

© Springer Nature Singapore Pte Ltd. 2018
I. Zelinka et al. (Eds.): ICSCS 2018, CCIS 837, pp. 35–45, 2018.
https://doi.org/10.1007/978-981-13-1936-5_4

makes it possible for innumerous options to be available at the users' end. If irrelevant information is provided to the users by the different social media, blogging and tagging websites, the users will eventually lack the interest to actively participate in these websites. Therefore, personalized recommendation is inevitable in the social web. Personalized recommendation of multimedia content aids the social websites to stay relevant and not fade into their non-existence. It also helps in promoting their services without boring the users with information that is not applicable to them. This paper reviews how personalized recommendation of information can be provided to each user in social tagging systems in particular.

Social tagging systems allow users to add tags to resources like web pages (Delicious1), pictures (Flickr2/Pinterest3) etc. The tags are labels that represent the content of the information briefly. Therefore, the tags, generated by the user for his resources, help him to find the resources in the future. They also allow other users of the system to view the resources, organized and categorized under a label denoted by the tag name. Therefore, social tagging systems facilitate users to select a previously used tag of another user to retrieve all the resources annotated by that tag.

This paper is a survey on the different personalized recommendation techniques adopted in social tagging systems. Personalized recommendation for resources in social tagging systems can be provided to the users by different methods like Contextual Collaborative Filtering recommendation [1], Hierarchical Clustering [2], Tag Expansion based Personalized Recommendation TE-PR [3], FolkRank [4] and Integrated Diffusion [5]. These topics will be addressed in detail in Sect. 3. There are a lot of existing approaches to recommend tags to users of a social tagging system. Although this is related to recommending resources to a user, the existing approaches like content based similarity [6], GRoMO [7], Latent Dirichlet allocation [8], Tensor Dimensionality Reduction [9] and Pairwise Interaction Tensor Factorization (PITF) [10] to recommend tags to users in social tagging systems are not reviewed in this paper.

The organization of the paper is as follows. The different traditional approaches adopted in information retrieval and recommendation prior to personalized recommendation in social tagging systems like collaborative filtering [11] and content based filtering [11] are discussed in Sect. 2. The recent advances in providing personalized recommendation to users in social tagging systems are discussed in Sect. 3 and the analysis of the experimental evaluation of these recent advances is given in Sect. 4. Finally, the future scope of this paper and conclusion is provided in Sect. 5.

2   Basic Methodologies in Recommender Systems

Prior to social tagging systems, recommendations in other systems were provided by different recommendation approaches like popularity-based, content-based, collaborative filtering, association-based, demographics-based and reputation-based approaches [12]. The two basic, traditional approaches – content-based filtering and collaborative filtering will be discussed in this section.

Content based filtering suggests resources to users based on the content of these resources. The degree of relevance of a resource to a user is determined by the features or the contents of the resource as given by [13]. The resources that the user has liked in the past are also considered while recommending resources to a user according to [14]. Therefore, the features of resources and the resources that the user has liked in the past filter out the available resources to recommend the most relevant one to the user in content based filtering technique. All recommendations made to the target user are independent of the other users of the system. The users of the system with preferences similar to that of the target user are not considered. This additional information which can improve the recommendation is not leveraged in this approach.

Collaborative filtering technique exploits the user-to-user relations to recommend resources to the target user. Users with similar preferences to the target user are identified and the resources that these similar users have liked will be recommended to the target user [14]. The reason why a user likes a resource is not taken into account in this approach. Therefore, two users who may like a resource for different reasons will be considered as users with similar preferences. This is a major drawback of this technique. In the most common neighborhood-based collaborative filtering, neighbors for a target user are identified by analyzing the preferences of the target user and all other users of the system. A similarity score such as cosine similarity, mean squared difference or Pearson correlation coefficient is chosen to determine the similarity between the target user and the other users of the system. All the users with a similarity greater than a threshold are chosen as the neighbors in weight thresholding, aggregate based best k neighbors or center-based best k neighbors [12]. Once the neighbors for a target user are determined, the preference score for an item is determined as the sum of a weighted average of the preference scores provided by all the neighbors of the target user and a weighted average deviation from the mean preference score for all neighbors. The similarity between the target user and the neighbors acts as the weight in the above calculation [15].

The traditional or basic methodologies employed in recommender systems for providing recommendations to their users were addressed in this chapter. The popular approaches employed in providing personalized recommendation for social tagging systems will be discussed in the following chapter.

3   Recent Advances

The existing approaches for personalized recommendation in social tagging systems are briefly discussed in this section.

According to Halpin et al. [1], a social tagging system is viewed as a tripartite model with 3 entities – users, tags and resources. These three entities constitute the user space, tag space and resource space respectively. The user space consists of all users of the social tagging system. The tag space consists of all the unique tags generated thus far that act as a label for categorizing the different resources. The resource space consists of the URIs of the different resources. A successful tag instance is when there are two links; one from user to tag and another from the same tag to a single or multiple resources.

In Fig. 1, the link from user 1 to tag 1 does not have a complementary link from tag 1 to any resource hence it is an unsuccessful tag instance and hence the user 1 will have no recommended resources to view. All other tag instances are successful in Fig. 1.

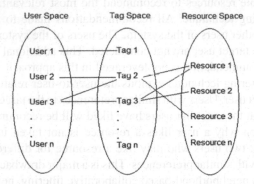

User Space   Tag Space   Resource Space

Fig. 1. Successful and unsuccessful tag instances

Note that a user may be associated with many tags as he may like many things like books, music and dance, each of which forms a separate tag. Also note that a resource may have different tags associated with it as a single resource can fall under different categories like how a Shih-Tzu is both a dog and an animal and hence both these tags are associated with it.

Collaborative filtering (CF) recommends a resource to a user if another user with similar preferences had liked that resource in the past [2]. However, CF does not consider the reason why the user likes the resource i.e. the traditional collaborative filtering does not consider the context of the tags implied by the users. Contextual collaborative filtering [16] leverages this information. Contextual collaborative filtering considers why a user liked or prefers a resource whereas traditional collaborative filtering is only based on the numerical ratings of a resource [16]. In contextual collaborative filtering, tags are considered to contain the context information or the reason why a user liked the resource that was tagged.

In Integrated diffusion, all the resources associated with a user are collected and this information is distributed to the user space and tag space. Now, the resources, similar to this information, that are located on the user and tag space are reflected back and recommendation scores are allotted. The resource with the highest score is recommended to the user [2]. In simple terms, the resources either viewed or uploaded by the target user previously are collected and their information i.e. the attributes and tags associated with them are cross-referenced with existing tags in the tag space and with the profiles of other users in the user space which contain the preferences and characteristics information of each user. Once there is a match, a score is allotted to each of the matched resources based on the relevance or frequency and the resource with the highest score will be recommended to the user.

Context-based Hierarchical clustering [3] can eliminate tag redundancy and tag ambiguity. The former refers to the case where many tags have the same meaning and the latter to the case where a single tag might have multiple meanings.

By clustering tags which are related to each other, redundancy is eliminated because redundant tags are aggregated into a single cluster. Ambiguity is also eliminated as a tag takes the meaning shared by all the other tags in a cluster. Therefore, the ambiguity between Mercury (element) and Mercury (planet) is eliminated as all elements form a cluster and all planets form a separate cluster. Therefore, Mercury in the cluster consisting of all planets can be easily identified as Mercury (planet).

In Fig. 2, the different user profiles and resources are connected by the cluster of tags in the middle which is represented by the circles. Note that the edges connecting the resources and the clusters will be constant regardless of the users. Resource 1 is a Saint Bernard dog located in Greece and Resource 2 is an online chess web application developed using html. These properties of the two resources are always constant and hence the edges connecting the resources with the clusters are also constant. Although user 1 and user 3 have the similarity in their liking for dogs and JavaScript, user 3 also likes Greece. Recommending the user 1, resources related to Greece because the similar user 3 has liked it will not yield a successful personalized recommendation. For this reason context-based approaches must be adopted as explained below.

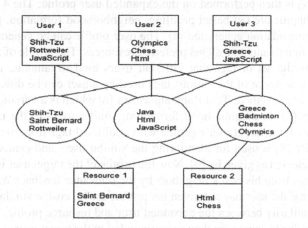

Fig. 2. Relationship between user profiles, clusters of tags and resources

In context based hierarchical clustering algorithm proposed by [3], a folksonomy is created which represent 4 tuples – users, resources, tags and annotations where annotations denote a triple consisting of a user, a resource and a tag value. Therefore the folksonomy can be viewed as a tripartite hypergraph where users, resources and tags form the nodes and annotations form the hyper edges [3]. The hierarchical clustering algorithm uses a cosine relation between the query with the target tag and the various resources associated with a tag to find a similarity coefficient. The user's profile is used to find the relevance of resources in the system to that of the user's query. The similarity and relevance coefficient are then combined to give the personalized rank score which determines the resource(s) to be recommended. In this approach, clustering phase is independent of the recommendation phase.

The traditional agglomerative clustering requires each tag to be placed in a singleton cluster initially. The tags are then combined to form a hierarchical tree based on the similarity coefficient. A division coefficient is also used to dissect the tree or a branch comprising of a cluster into smaller, individual clusters. The user's query is applied over the entire tree structure. However, the hierarchical clustering algorithm identifies the user's tag in the tree, navigates upwards to broaden the breadth of clusters to which the user's query applies also called the generalization level and then applies division coefficient [3]. Once the clustering phase is complete, the recommendation phase uses the personalized rank score to obtain the most relevant resource(s). Since the user's profile is the input to find the relevance coefficient, this approach generates recommendations that are different for each user i.e. generates personalized recommendation based on the user's profile.

Tag Expansion based Personalized Recommendation TE-PR [4] finds some similar users to the target user and expands the target user's profile by appending the tags of the similar users' found using the relevance feedback method as given in [5]. The tags associated with a resource are also expanded by the relevance feedback method. The recommendation is then performed on the expanded user profile. The 4 main steps of the TE-PR technique are tag-based profiling, neighborhood formation, profile expansion, and recommendation generation [4]. The user profile can be generated by the set of tags used by him to annotate all his preferred resources. The profile of a resource can be generated by the set of tags that different users used to annotate that particular resource. The importance of the various tags used by a user can be differentiated by a graph based ranking method, Text Rank algorithm [6] which is a variation of the Page Rank algorithm [7]. In neighborhood formation, similar users to the target user are identified to enrich the target user's profile with additional tags from the similar users. Cosine similarity [8] is used for identifying the similar users and center-based best k neighbors are selected as given by [9]. Now the profile of the target user is expanded by including the tags from his best k neighbors by the relevance feedback method [5]. For each resource that the user has not given his preference to, cosine similarity is used to calculate the similarity between the expanded user and resource profile. The resources with the top similarity scores are then recommended to the target user.

In Fig. 3, the profile of user 1 is expanded by appending the tags from the profiles of similar users 2, 7 and 9 found using the cosine similarity followed by center-based best k neighbors using relevance feedback method.

FolkRank [17], an algorithm adapted from the traditional PageRank [7], is specific to the folksonomies. The traditional PageRank cannot be applied to a folksonomy because the graph structure in folksonomies differs from the web structure in that a folksonomy has undirected triadic hyper edges and not directed binary edges like the web [17]. The hyper edges between the user space, tag space and resource space in a folksonomy are converted to undirected, weighted, tripartite graph and a version of PageRank that takes into account the weights of the edges is applied to the tripartite graph. This version of PageRank called Adapted PageRank provides a global ranking irrespective of the preferences of the user but the FolkRank provides ranking of resources based on a topic in the preference vector i.e. the FolkRank provides a topic-specific ranking for each preference vector [17].

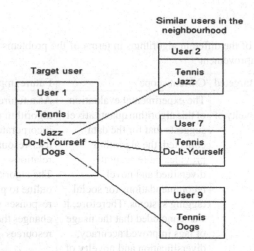

Fig. 3. Tag expansion of a target user's profile based on the similar neighbors.

4 Discussion

The experimental evaluation of the recent advances in personalized recommendation of resources for social tagging systems will be analyzed in this section. The most popular techniques like Hierarchical clustering, FolkRank, Tag Expansion based Personalized Recommendation TE- PR and Integrated Diffusion when applied to real data sets collected from different social tagging systems like Delicious, Flickr etc., are briefly discussed as follows.

For integrated diffusion, data sets were collected from Delicious, MovieLens and BibSonomy. The data sets were cleaned to remove outliers like any isolated nodes which represented users who did not tag any resource. This ensured that all resources were collected by at least two users and annotated by at least one tag. The area under the ROC curve, recall, diversification and novelty were calculated for integrated diffusion on these data sets from Delicious, MovieLens and BibSonomy. The calculations showed that using tags improved the accuracy, diversification and novelty of the recommendations [2].

For context based hierarchical clustering, data sets from Delicious and Last.FM were collected. The effectiveness of the algorithm was tested by calculating the difference between inverse of the ranks of the target resource [3] for a basic recommendation using only the tag and a personalized recommendation using the tag and the user profile. "Leave one out" approach was used while calculating the rank of the target resource for personalized recommendation where the target resource-tag pair was removed from the user's profile and the rest of the user's profile was used to generate the recommendation [3]. For comparison, the best k means clustering technique was also implemented and the experiment revealed that the recommendations by the hierarchical clustering showed more improvement than the k means based algorithm. This improvement was more significant in Last.FM than Delicious. One reason behind this may be the fact that Delicious has sparser data than Last.FM and sparser data will not provide precise

Table 1. Comparison of the different algorithms in terms of the problems targeted, the current scope and the future improvement

Algorithm & problem targeted	Current scope	Future improvement
Integrated diffusion - diversification and novelty of recommendation	The experimental evaluation of this algorithm quantitatively suggests that for the data collected, the algorithm provides more accurate, diversified and novel recommendations for social tagging systems. Therefore, it was concluded that the usage of tags improved accuracy, diversification and novelty of the recommendations	As a future improvement, the algorithm may be modified to incorporate weights between user-resource and resource-tag relations The algorithm must be made online to provide real-time responses when the user changes the tags or selects new resources
Context-based hierarchical clustering - tag redundancy and tag ambiguity	The algorithm, when applied to datasets crawled from Delicious and Last.Fm, proved that the context based clustering technique significantly improves the effectiveness of the personalized recommendation in sparser datasets than in dense datasets. This also solves tag redundancy and ambiguity to some extent	Natural language processing and semantic analysis can be further applied to eliminate tag ambiguity and redundancy. Moreover, users and resources may also be clustered in addition to tags and these clusters can be used to further improve the personalized recommendations
Tag Expansion based Personalized Recommendation TE-PR - relevance in general, tag completeness	The experiment on CiteULike proved that the tag completeness is important in tag based recommendation techniques. The results of the experiment proved that the expansion of the resource profiles have positive contributions to personalized recommendations but the expansion of user profiles is useless [4]	The experiment must be carried out for multiple social tagging systems to compare and contrast between the performances of the algorithm for each system. As graph based ranking is a critical step in TE-PR, the algorithm may be modified in the future to incorporate an alternative approach to graph based ranking to outperform the present version of the TE-PR
FolkRank - serendipitous browsing	Since the FolkRank provides the personalized recommendation based on the tags associated with a resource and not the content, it can be effectively applied to multimedia resources	As folksonomy based systems grow larger, organization of the internal structure of the systems must be improved using semantic web technologies without bothering untrained users [17]

measurements for the connection between tags when generating tag clusters. When context based hierarchical clustering was applied, Last.FM did not benefit significantly from the use of context for cluster selection due to the higher data density when compared to Delicious.

The Tag Expansion based Personalized Recommendation TE-PR was applied to the data crawled from the social bookmarking site CiteULike. Before the algorithm was applied, the data collected was pre-processed to remove outliers. Average Recall rate (recall@N), when n resources are recommended to every user and Mean Average Precision (MAP) were used as performance metrics. Traditional Collaborative Filtering (TCF) was used as benchmark to compare the improvement of TE-PR with. The effect of number of neighbors for TCF was examined and 25 neighbors achieved best performance for TCF technique [4]. The effect of weighted and unweighted edges was then studied for TE-PR and the results showed that weighted edge method outperforms the unweighted method [4]. The effect of neighbors for user profile expansion and resource profile expansion for TE-PR was also studied and this showed that the profile expansion does not have any contribution for users but improves the quality of the tag-based resource profiles [4]. Comparing the benchmark TCF and TE-PR methods, TE-PR outperformed the benchmark in both MAP and recall@N values.

FolkRank and his benchmark Adapted PageRank were both applied to data collected from Delicious. The data was preprocessed before applying these algorithms. The result showed that the Adapted PageRank algorithm recommended globally frequent tags and FolkRank provided more personal tags. FolkRank provided good results for topically related resources in the folksonomy [17]. The results also show that a small perturbation can alter the size and structure of the folksonomy drastically i.e. a single user can provide sufficient if not all the points for a topic that has not been collected yet [17]. FolkRank algorithm promotes serendipitous browsing by suggesting useful resources that the users did not even know existed [17] (Table 1).

5 Conclusion and Future Work

The recent advances for providing personalized recommendation in social tagging systems were studied. The results when these approaches were applied to different social tagging systems were analyzed to prove the efficiency of these algorithms in terms of accuracy and novelty when compared to appropriate benchmarks. As a future work, these algorithms may be applied to the same data collected from a single social tagging system so that the different approaches like integrated diffusion, Context-based Hierarchical Clustering, Tag Expansion based Personalized Recommendation TE-PR and FolkRank can be compared with each other. In order to facilitate such a comparison a single metric must be established. Since there is a lack of similarity in the implementation of these different approaches, creation of a single metric to compare the effectiveness of each of these approaches will be problematic. Note that the different approaches work better for different datasets and hence different social tagging systems. Context-based Hierarchical Clustering works better for sparser datasets and for denser datasets, TE-PR is more efficient because the data in the dense dataset related to the user profile expansion are useless and hence are neglected. Only the expansion of resource profiles results in positive contributions. These facts have to be considered while creating a single metric to analyze and compare the different approaches of providing personalized recommendation in social tagging systems.

References

1. Halpin, H., Robu, V., Shepard, H.: The dynamics and semantics of collaborative tagging. In: Proceedings of the 1st Semantic Authoring and Annotation Workshop (SAAW 2006), vol. 209, November 2006
2. Zhang, Z.K., Zhou, T., Zhang, Y.C.: Personalized recommendation via integrated diffusion on user–item–tag tripartite graphs. Phys. A: Stat. Mech. Appl. **389**(1), 179–186 (2010)
3. Shepitsen, A., Gemmell, J., Mobasher, B., Burke, R.: Personalized recommendation in social tagging systems using hierarchical clustering. In: Proceedings of the 2008 ACM Conference on Recommender systems, pp. 259–266. ACM, October 2008
4. Yang, C.S., Chen, L.C.: Enhancing personalized recommendation in social tagging systems by tag expansion. In: 2014 International Conference on Information Science, Electronics and Electrical Engineering (ISEEE), vol. 3, pp. 1695–1699. IEEE, April 2014
5. Rocchio, J.: Relevance feedback in information retrieval. In: Salton, G. (ed.) The SMART Retrieval System: Experiments in Automatic Document Processing, pp. 313–323. Prentice-Hall, Englewood Cliffs (1971)
6. Mihalcea, R., Tarau, P.: TextRank: bringing order into texts. In: Association for Computational Linguistics, July 2004
7. Brin, S., Page, L.: Reprint of: the anatomy of a large-scale hypertextual web search engine. Comput. Netw. **56**(18), 3825–3833 (2012)
8. Basu, C., Hirsh, H., Cohen, W.: Recommendation as classification: using social and content-based information in recommendation. In: AAAI/IAAI, pp. 714–720, July 1998
9. Herlocker, J.L., Konstan, J.A., Borchers, A., Riedl, J.: An algorithmic framework for performing collaborative filtering. In: Proceedings of the 22nd Annual International ACM SIGIR Conference on Research and Development in Information Retrieval, pp. 230–237. ACM, August 1999
10. Byde, A., Wan, H., Cayzer, S.: Personalized Tag Recommendations via Tagging and Content-Based Similarity Metrics (2007)
11. Guan, Z., Bu, J., Mei, Q., Chen, C., Wang, C.: Personalized tag recommendation using graph-based ranking on multi-type interrelated objects. In: Proceedings of the 32nd International ACM SIGIR Conference on Research and Development in Information Retrieval, pp. 540–547. ACM, July 2009
12. Wei, C., Shaw, M., Easley, R.: A survey of recommendation systems in electronic commerce. In: Rust, R.T., Kannan, P.K. (eds.) E-Service: New Directions in Theory and Practice. M. E. Sharpe (2001)
13. Alspector, J., Kolcz, A., Karunanithi, N.: Comparing feature-based and clique-based user models for movie selection. In: Proceedings of the Third ACM Conference on Digital Libraries, pp. 11–18. ACM, May 1998
14. Balabanović, M., Shoham, Y.: Fab: content-based, collaborative recommendation. Commun. ACM **40**(3), 66–72 (1997)
15. Shardanand, U., Maes, P.: Social information filtering: algorithms for automating "word of mouth". In: Proceedings of the SIGCHI Conference on Human factors in Computing Systems, pp. 210–217. ACM Press/Addison-Wesley, May 1995

16. Tso-Sutter, K.H., Marinho, L.B., Schmidt-Thieme, L.: Tag-aware recommender systems by fusion of collaborative filtering algorithms. In: Proceedings of the 2008 ACM Symposium on Applied Computing, pp. 1995–1999. ACM, March 2008
17. Hotho, A., Jäschke, R., Schmitz, C., Stumme, G.: Information retrieval in folksonomies: search and ranking. In: Sure, Y., Domingue, J. (eds.) ESWC 2006. LNCS, vol. 4011, pp. 411–426. Springer, Heidelberg (2006). https://doi.org/10.1007/11762256_31

Hybrid Crow Search-Ant Colony Optimization Algorithm for Capacitated Vehicle Routing Problem

K. M. Dhanya[1]($\boxtimes$), Selvadurai Kanmani[2], G. Hanitha[2], and S. Abirami[2]

[1] Department of Computer Science and Engineering,
Pondicherry Engineering College, Puducherry, India
dhanyakm@pec.edu
[2] Department of Information Technology,
Pondicherry Engineering College, Puducherry, India
{kanmani,hanitha14it114,abirami14it102}@pec.edu

Abstract. Capacitated Vehicle Routing Problem (CVRP) is a NP-Hard problem in which the optimal set of paths taken by the vehicles is determined under the capacity constraint. Ant Colony Optimization (ACO) is a metaheuristic method incorporating the ant's ability to find the shortest path from source to destination using the concept of pheromone trails. It has been used to solve CVRP. However, it exhibits stagnation property, due to which, the exploration probability of new route is reduced. Crow Search Algorithm (CSA) is a recently developed metaheuristic method inspired from crow's food hunting behavior. This paper provides a hybrid relay algorithm which involves ACO and CSA to solve CVRP. The hybridization is done to get a consistent solution with optimal execution time. The experimentation with Augerat instances shows betterment in the optimal solution at the earliest time.

Keywords: Crow search algorithm · Ant Colony Optimization
Metaheuristic · Capacitated Vehicle Routing Problem · Hybridization
Optimal route distance

1 Introduction

Capacitated Vehicle Routing Problem (CVRP) comes under Vehicle Routing Problem (VRP) variants in which the optimal set of paths taken by the vehicles is determined under capacity constraint [1]. CVRP is represented as a graph $G = (A, B)$ where $A = \{a_0, a_1, \ldots, a_y\}$ is the set of nodes and $B = \{(a_i, b_j) \mid a_i, b_j \in A, i < j\}$ is the set of edges. The node a_0 represents depot with x homogeneous delivery vehicles of capacity C, to serve the demands c_i of y customers, $i = 1, 2, \ldots, y$. The edge set, B defines the distance matrix between customers or between customer and depot. A CVRP solution comprises a set of paths of the vehicles. One of the objectives of CVRP is to minimize the total distance covered by the vehicles.

To solve CVRP, the distance matrix is generated using the Euclidean distance formula for two vertices.

© Springer Nature Singapore Pte Ltd. 2018
I. Zelinka et al. (Eds.): ICSCS 2018, CCIS 837, pp. 46–52, 2018.
https://doi.org/10.1007/978-981-13-1936-5_5

CVRP has been solved using many metaheuristic algorithms such as Simulated Annealing algorithm (SA), Variable Neighborhood Search algorithm (VNS), Particle Swarm Optimization algorithm (PSO), Ant Colony Optimization algorithm (ACO), Artificial Bee Colony algorithm (ABC) and Genetic Algorithm (GA) to obtain optimal solutions [1–6]. Among them, ACO which has good exploitation property exhibits stagnation behavior. CSA is a recently introduced algorithm, which produces better solutions with a reasonable span of time. The hybrid implementation of CSA and ACO is carried out with the intention of producing optimal solutions by making use of both exploitation and exploration.

2 Existing Work

Recently, an improved version of SA was applied on CVRP with loading constraints and a VNS algorithm on CVRP and both of the algorithms have obtained good solutions compared to other existing algorithms [2, 3]. PSO and its variants are also found to be more effective in solving CVRP [4]. One of the research works, in which ACO algorithm is used to solve CVRP has shown good performance for instances up to 50 nodes [5]. An improved version of ABC algorithm has also achieved good solutions for CVRP [6]. An optimized crossover operator was also used to solve CVRP and results produced by it were competitive to other algorithms [1].

Some hybridization metaheuristics have also been utilized for solving CVRP. The most recent ones are Artificial Immune System (AIS) hybridized with Artificial Bee Colony (ABC) algorithm and hybrid Genetic-Ant Colony Optimization algorithm [7, 8]. A hybridized version of Intelligent Water Drops (IWD) and Advanced Cuckoo Search Algorithm (ACS) was also applied on CVRP [9]. In this study, CSA is hybridized with ACO algorithm to give a newer dimension of solutions.

3 Proposed Work

This section first gives a brief overview of crow search algorithm and ant colony optimization followed by the proposed crow search-ant colony optimization algorithm.

3.1 Crow Search Algorithm

Crow search algorithm (CSA) is a population-based metaheuristic algorithm inspired from crow's good memory and awareness capabilities for finding food sources. The crows reserve the obtained food for future need [10–18]. The greedy crows chase each other also to obtain better food.

CSA has shown promising results in many applications. CSA consists of a flock of crows positioned in random locations. Crows possess memories initialized with same random positions. Let $x^{a,itr}$ be the location of crow a at time itr, then next location of crow a at itr + 1 time is selected depending on the awareness probability ($AP^{b,itr}$) of crow b in two ways as specified below.

1. If the awareness probability of Crow b is less, then Crow a will follow Crow b and
   its new position will be calculated using Eq. 1 given below.

$$x^{a,itr+1} = x^{a,itr} + r_a * fl^{a,itr} * (mem^{b,itr} - x^{a,itr})$$ (1)

where r_a is a number selected randomly from the range 0 and 1 and $fl^{a;itr}$ specifies
the flight length of crow a at time itr and $mem^{b,itr}$, is the memory of crow b at time
itr.

2. If Crow b possess high awareness probability, then Crow a cannot chase it. In that
   case, Crow a will move to different position randomly.

Once the next position is selected, its feasibility will be evaluated. Then, the crow's
memory is updated with new position, if its fitness is found to be better than crow's
memory.

In CSA, awareness probability (AP) controls intensification and diversification.
Diversification which helps CSA to explore globally in search space can be achieved
by increasing the value of awareness probability. Intensification comes into role when
CSA search locally by making use of smaller value of awareness probability.

3.2 Ant Colony Optimization

Ant Colony Optimization (ACO) is a population based metaheuristic algorithm whose
basic idea is to obtain optimal solution for an optimization problem based on behaviour
of ants [5]. When ants move at random, it deposits pheromone on its path. The shortest
path of ants is determined by more pheromone trails. This natural behaviour of ants to
find shortest path makes ACO suitable for vehicle routing problems. In ACO method,
pheromone trail updating is done according to the Eq. 2 specified below.

$$pt_{i,j} = (1 - pe) \, pt_{i,j} + \Delta pt_{i,j}$$ (2)

where $pt_{i,j}$ is the pheromone trail on an edge i, j and pe is the pheromone evaporation
rate. The pheromone trail deposited by an ant k when it travels from node i to node j,
$\Delta pt_{i,j}$ is given by Eq. 3 otherwise 0.

$$\Delta pt_{i,j}^k = 1/d_k$$ (3)

where d_k is the distance covered by ant k.

ACO algorithm converges fast into the optimal solution. Hence, it has solved
various NP-hard problems like routing problems achieving good optimal results.

The proposed work aims at hybridizing CSA and ACO and apply it on CVRP. The
hybridization will take into account the good features of both the algorithms to produce
optimal solutions.

3.3 Hybridizing CSA with ACO

CSA and ACO are hybridized on a relay-based technique. In relay technique, the
output of a metaheuristic algorithm is given as an input to another metaheuristic

algorithm [19]. Initially, ACO algorithm is implemented for CVRP and the best ant solution is generated. The best ant solution generated is stored as the initial memory of the crows in CSA algorithm. This memory is used for computation of the solution in CSA algorithm. The solution generated by the CSA algorithm is compared with the previous memory and if the new one is better, the memory is updated. Finally, the best crow solution is considered as the output of the proposed hybrid algorithm (Fig. 1).

Input CVRP Instance
Find solution using ACO:
  Initialize ACO parameters
  Repeat
        for each ant do
                Construct solution using pheromone trail
        end for
        Update the pheromone trail
  Until stopping criteria
  Obtain Best Ant solution
Find solution using CSA:
  Initialize position of crows with the Best Ant solution
  Initialize CSA parameters
  Assess position of the crows
  Initialize each crow's memory with its initial position
  Repeat
        for each crow a do
                Choose a random crow b to follow
                Select new position based on awareness probability of
                crow b
        end for
        Check the feasibility of new positions
        Assess new position of the crows
        If new position is better, then update memory of the crows.
  Until stopping criteria
  Output Best Crow solution

Fig. 1. Hybrid crow search-ant colony optimization algorithm

Hybridisation is done to eradicate the stagnation property of ACO by using the exploration property of CSA and to obtain the best results.

4 Implementation

The proposed method has been implemented in Java using Eclipse Kepler IDE on a machine with 4 GHz Intel Core i5 processor, 4 GB RAM and Windows 10 Operating System.

The parameter values used by the algorithms which are hybridized are given in Tables 1 and 2 below.

Table 1. Parameters of CSA

Sl. no.	Parameter	Value
1	Crows	25
2	Runs	25
3	Iterations	100
4	Flight length	[1.5, 2.5]
5	Awareness probability	[0, 1]

Table 2. Parameters of ACO

Sl. no.	Parameter	Value
1	Ants	100
2	Runs	1
3	Iterations	100
4	Pheromone trail	0.0
5	Relative influence of pheromone trail	1
6	Relative influence of heuristic information	5
7	Evaporation rate	0.1

To carry out the experiments, Augerat instances have been used as the input dataset. Dataset contains number of nodes, vehicles and capacity limit of each vehicle. The location coordinates of each node and the demand of each customer will also be indicated.

5 Experimental Results

The output indicates the optimal route distance and the run and iteration at which it was obtained. The execution time taken by the algorithm is also denoted. The results are tabulated and shown in Table 3.

The output shows that in most of the cases, the total distance and computing time seems to increase as the number of customers increases.

Table 3. Obtained output

Sl. no.	Dataset name	Best result (distance)	Run	Iteration	Computation time (in ms)
1	A-n32-k5	1100	5	62	828
2	A-n34-k5	950	4	8	844
3	A-n44-k6	1367	9	7	1317
4	A-n45-k6	1548	16	69	1339
5	A-n46-k7	1643	4	62	1409
6	A-n48-k7	1971	4	91	1531
7	A-n53-k7	1753	22	66	1620
8	A-n55-k9	1965	24	35	1661
9	A-n60-k9	2418	23	84	1963
10	A-n61-k9	1486	25	44	2086
11	A-n63-k10	2381	13	37	2272
12	A-n63-k9	2726	12	70	2181
13	A-n64-k9	2379	14	55	2226
14	A-n65-k9	2539	6	82	2280
15	A-n69-k9	2374	4	68	2374

6 Conclusion

A hybrid relay CSA-ACO method has been proposed in this work to solve CVRP. The hybridization is mainly carried out to eradicate the stagnation property of ACO by incorporating the exploration property of CSA. It has been tested on Augerat CVRP instances. Computational results have shown better results with an acceptable computational time. Mostly, the optimal route distance rises as the number of customers increases. The research can be extended by using the proposed algorithm to solve VRP under dynamic conditions.

References

1. Nazif, H., Lee, L.S.: Optimised crossover genetic algorithm for capacitated vehicle routing problem. J. Appl. Math. Model. **36**, 2110–2117 (2012)
2. Wei, L., Zhang, Z., Zhang, D., Leung, S.C.: A simulated annealing algorithm for the capacitated vehicle routing problem with two-dimensional loading constraints. Eur. J. Oper. Res. **265**(3), 843–859 (2018)
3. Amous, M., Toumi, S., Jarboui, B., Eddaly, M.: A variable neighborhood search algorithm for the capacitated vehicle routing problem. Electron. Notes Discret. Math. **58**, 231–238 (2017)
4. Kachitvichyanukul, V.: Particle swarm optimization and two solution representations for solving the capacitated vehicle routing problem. Comput. Ind. Eng. **56**(1), 380–387 (2009)
5. Mazzeo, S., Loiseau, I.: An ant colony algorithm for the capacitated vehicle routing. Electron. Notes Discret. Math. **18**, 181–186 (2004)
6. Szeto, W.Y., Wu, Y., Ho, S.C.: An artificial bee colony algorithm for the capacitated vehicle routing problem. Eur. J. Oper. Res. **215**(1), 126–135 (2011)

7. Zhang, D., Dong, R., Si, Y.W., Ye, F., Cai, Q.: A hybrid swarm algorithm based on ABC and AIS for 2L-HFCVRP. Appl. Soft Comput. **64**, 468–479 (2018)
8. Kuo, R.J., Zulvia, F.E: Hybrid genetic ant colony optimization algorithm for capacitated vehicle routing problem with fuzzy demand—a case study on garbage collection system. In: 4th International Conference on Industrial Engineering and Applications (ICIEA), pp. 244–248. IEEE (2017)
9. Teymourian, E., Kayvanfar, V., Komaki, G.M., Zandieh, M.: Enhanced intelligent water drops and cuckoo search algorithms for solving the capacitated vehicle routing problem. Inf. Sci. **334**, 354–378 (2016)
10. Askarzadeh, A.: A novel metaheuristic method for solving constrained engineering optimization problems: crow search algorithm. Comput. Struct. **169**, 1–12 (2016)
11. Abdelaziz, A.Y., Fathy, A.: A novel approach based on crow search algorithm for optimal selection of conductor size in radial distribution networks. Eng. Sci. Technol. Int. J. **20**(2), 391–402 (2017)
12. Hinojosa, S., Oliva, D., Cuevas, E., Pajares, G., Avalos, O., Gálvez, J.: Improving multi-criterion optimization with chaos: a novel Multi-Objective Chaotic Crow Search Algorithm. Neural Comput. Appl. 1–17 (2017)
13. Marichelvam, M.K., Manivannan, K., Geetha, M.: Solving single machine scheduling problems using an improved Crow Search Algorithm. Int. J. Eng. Technol. Sci. Res. **3**, 8–14 (2016)
14. Nobahari, H., Bighashdel, A.: MOCSA: a multi-objective crow search algorithm for multi-objective optimization. In: 2nd Conference on Swarm Intelligence and Evolutionary Computation (CSIEC), pp. 60–65. IEEE (2017)
15. Oliva, D., Hinojosa, S., Cuevas, E., Pajares, G., Avalos, O., Gálvez, J.: Cross entropy based thresholding for magnetic resonance brain images using Crow Search Algorithm. Expert Syst. Appl. **79**, 164–180 (2017)
16. Rajput, S., Parashar, M., Dubey, H.M., Pandit, M.: Optimization of benchmark functions and practical problems using Crow Search Algorithm. In: Fifth International Conference on Eco-friendly Computing and Communication Systems, pp. 73–78. IEEE (2016)
17. Sayed, G.I., Hassanien, A.E., Azar, A.T.: Feature selection via a novel chaotic crow search algorithm. Neural Comput. Appl. 1–18 (2017)
18. Turgut, M.S., Turgut, O.E.: Hybrid artificial cooperative search-crow search algorithm for optimization of a counter flow wet cooling tower. Int. J. Intell. Syst. Appl. Eng. **5**(3), 105–116 (2017)
19. Talbi, E.G.: Metaheuristics: From Design to Implementation, vol. 74. Wiley, London (2009)

Smart Transportation for Smart Cities

Rohan Rajendra Patil$^{(\boxtimes)}$ and Vikas N. Honmane

Computer Science and Engineering,
Walchand College of Engineering, Sangli, India
rprohanpatil2@gmail.com, vhonmane@gmail.com

Abstract. Bus transportation is an important mode of public transportation as it is preferred by many people every day. This mode of transportation plays a huge role in everyday life. But even when so much is dependent of bus transportation, currently there is no system which makes this journey easy and convenient. People face various problems while travelling by bus. Over-Crowded buses and their unpredictable timings make the bus journey very inconvenient. So to provide the bus passengers a convenient way to travel, this system can be used. This system provides crowd information and expected arrival time of the buses to the user's smart phone. The user can be anywhere and with the help of the mobile application, the user can find out the crowd in the buses, their arrival timings etc. which helps the user to take better decisions. Also other features like nearby bus stops is available in user's application. This will reduce the inconvenience and provide systematic way to travel.

Keywords: Smart transportation · Location based crowd calculation
E-ticket using passenger's biometric · Bus travel
Travelling smart phone application

1 Introduction

Many People need to travel every day. Millions of people choose buses as their mode of transportation. But even when buses play such an important role in public transportation, there is no system which makes the bus journey convenient. Even today many people dislike the bus journey due to the over-crowded buses and their unpredictable arrival timings. Pollution is one of the major issues of today. Looking at the pollution levels around the world, we need to fight pollution whenever and wherever possible. Vehicles cause lot of pollution and we can reduce the number of vehicles on the road by diverting people towards buses. Buses have the capacity to accommodate many people. The proposed system makes the bus journey easy and convenient for the passengers. This directly lead to less cars and less pollution.

This system provides estimated arrival time along with the crowd in the bus to the passengers who are using the mobile application. The passengers can be anywhere and they can see which bus to take, when will that bus arrive, how much crowd is there in that bus etc. The user can decide which bus to take and avoid the crowd and inconvenience caused during bus travel. When a passenger enters a bus, this system only needs his/her destination stop and mobile number or fingerprint. The person is

© Springer Nature Singapore Pte Ltd. 2018
I. Zelinka et al. (Eds.): ICSCS 2018, CCIS 837, pp. 53–61, 2018.
https://doi.org/10.1007/978-981-13-1936-5_6

identified and the ticket is sent to his/her phone via sms. This system saves tons of paper everyday.

2 Related Work

Few researchers have explored this sector and few systems have been developed. But all these systems are far from prefect.

In [1], a Wi-Fi based crowd detection system is explained. This system is deployed in Madrid. The major problem with this system is it detects crowds based on the number of people connected to the bus router. This is unreliable method as many people avoid connecting to the bus Wi-Fi. In [2], a system which is deployed in Pune and Ahmedabad is explained. This system provides the distance and time after which the bus will arrive to the user's bus stop.

In [3], the real time challenges in tracking the bus using GPS is explained. The problem increases when adjacent roads are present. The GPS may show the bus on the wrong road. In [4], explains an application called OneBusAway. This was the first app to bring the estimated arrival time on the user's smart phone. The results clearly showed that access to arrival time increased the satisfaction with bus journey. In [7], QR codes are used to identify the bus stops and search the buses according to the QR code scanned.

In [5, 6, 8–10], tracking algorithms are explained. All the papers explain how a bus or vehicle can be tracked and the estimated arrival time can be sent to the user's smart phone.

3 Methodology

In this system a central server plays the most important role. In short this server is responsible to track and collect all the information about all the buses. The server then manipulates this information and send it to the user's smart phone application.

The system consists of three main components:

3.1 Mobile Applications of Conductor and User

The conductor in each bus has a mobile application. This mobile application can replace the traditional working of the conductor. The mobile application can issue tickets to the customer and also scan the monthly passes. After issuing the tickets or scanning any pass, the ticket and monthly pass information (source and destination stops) is passed to the central server. The location of the bus is constantly updated to the central server. When a ticket is issued, a sms is sent to the passenger which contains the ticket details (Fig. 1).

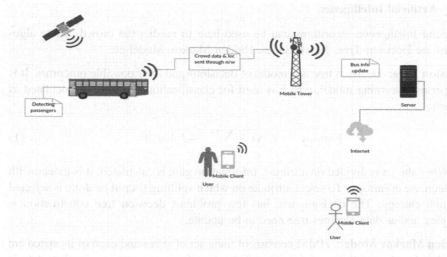

Fig. 1. System diagram

3.2 Central Server

The central server keeps track of the buses at all the time. Different buses are on different routes at a given time, all the buses have their bus id to distinguish them from others. The conductor's application constantly sends the ticket and pass information along with the location of the bus to the central server. The server collects this information and performs its task of crowd detection. Then this information is made available on the user's smart phone application through internet. The user can find out the crowd in the bus, also the server has information like all the bus stops and routes and timings of all the buses etc. so user can find out next bus, nearest bus stops etc. To inform the user about estimated arrival time, Google maps api is used and by entering source destination as the bus's current location and destination as the preferred stop, the system finds out the estimated arrival timing. With all this information, the app also provides crowd prediction based on previous crowd levels. For this Artificial Intelligence is used.

The central server calculates crowd by maintaining a counter. Initially, the counter is initialized to 0. Whenever new tickets are issued or pass is scanned, the counter is incremented based on the number of tickets and monthly passes. The server constantly tracks the bus and it also has the ticket information containing source stop and destination stop. So whenever the bus reaches a stop, it checks for tickets and passes whose destination stop is the current stop. These number of tickets and passes are stored in temporary variable and the temporary variable is deducted from the counter. Thus we get accurate crowd information everytime.

- Initially, Counter $C_i = 0$;
- Counter $C_i = C_i + $ Tickets issued + Monthly passes scanned.
- Temp $t_i = $ All the passengers whose destination stop is the current stop.
- Counter $C_i = C_i - t_i$.

3.3   Artificial Intelligence

Artificial Intelligence algorithms can be used here to predict the crowd. Some algorithms are Decision Tree, Naive Bayes, Hidden Markov Model etc.

Decision Tree: Decision tree is a model of decisions and their possible outcomes. It is a supervised learning model, which is used for classification. Entropy is calculated as follows.

$$\text{Entropy: } - E(S) = \sum_{i=1}^{c} -P_i \log_2 P_i \qquad (1)$$

When the set is divided on attribute, Information gain is calculated, it is gained with the decrease in entropy. To select attribute on which splitting should be done is selected by high entropy. The decision tree has few problems decision tree construction is complex and as data changes tree need to be update.

Hidden Markov Model: HMM consists of finite set of states and each of its stated are associated with probability distribution. Transition form one state to another state is given by transition probabilities. In a specific state, an output is generated according to associated probabilities.

Naïve Bayes: Naive Bayes algorithm is used for constructing classifiers: models that give labels to instances of problem, demonstrated as vectors of feature values, here the labels are taken from some specific set.

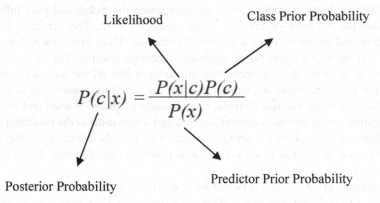

- P(c|x) is the predicted crowd levels in the bus according to the month.
- P(x|c) is the Likelihood of the crowd levels in that bus.
- P(c) is the past experience of the crowd levels in the bus.
- P(x) is the past experience of the crowd levels in the month.

4 Internal Data Processing

The internal data processing of the system is showed below in Fig. 2. When a passenger enters a bus, the conductor asks for destination stop of the passenger. The conductor can either enter the mobile number of the passenger or scan the finger of the passenger. The fingerprint details are sent to Aadhar database. The mobile number of the passenger is fetched from the aadhar database. The ticket is sent to the passenger's mobile phone through sms. If the fingerprint is used instead of entering the mobile number manually then lot of time is saved while issuing tickets. The conductor can issue ticket in three simple steps, the conductor has to enter source stop, destination stop and scan the finger of the passenger. The ticket will be automatically sent to the passenger via sms. The conductor can also scan monthly passes of the passengers. The source and destination stops of each passenger is known to the conductor's app. The application sends this data to the central server along with the location of the bus. This information is sent using internet. The server acquires this information from all the buses. The server then keeps track of the crowd in all the buses. Initially the crowd counter is initialized to zero. When a ticket is issued or pass is scanned, the counter is increased and when the bus reaches a bus stop, all the passengers whose destination stop is the current reached stop are reduced from the counter. Thus we can accurately find out the crowds in all the buses. The mobile application uses Google Map api. Thus by entering the current position of the bus as source stop and by entering passenger's stop as destination, estimate arrival time is found out. The crowd information and the arrival time is sent to the user's smart phone app. If the user thinks that the arriving bus is too crowded, then the user can see the list of all the buses which travel to the user's destination. The user can select any bus from the list which suits their timing and view the crowd in that bus and also the estimated arrival time.

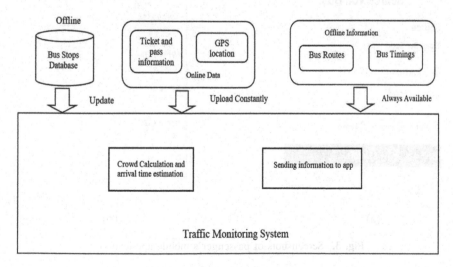

Fig. 2. Internal data processing

All the information about all the buses, their timetable, their routes, all the stops is stored in database. This information is stored on central server and is available offline. This information is very rarely updated. Online data is collected in buses. As tickets are issued their source and destination stops and source and destination stops of the monthly pass passengers is collected and sent to the server along with the location of the bus. This information is constantly updated to the central server. The updated information is collected by the server and is processed with offline information. The user can be anywhere and with the help of the mobile application the user can find out the appropriate bus, the crowd in that bus, location of the bus, timing of next bus etc. The app also provides many other features like nearby bus stops.

5   Results and Discussions

The user application needs only source, destination stops and timing of travel. The application appropriately finds out the buses and provides a list of available buses. The user can find out the appropriate bus from the list.

Figures 3a, b and 4c show the mobile application of user. The buses are fetched perfectly. The user can see the whole list of the buses. After choosing a bus, the user can see all the details of that bus like crowd in that bus, estimate arrival time of the bus, duration of the journey etc. The user can find out the suitable bus based on these details.

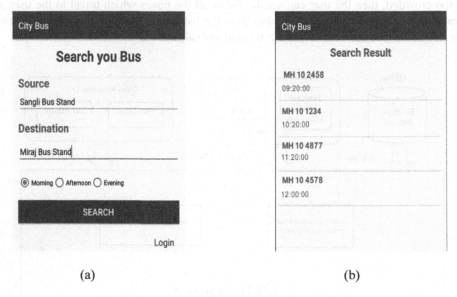

(a)   (b)

Fig. 3. Screenshots of passenger's mobile application

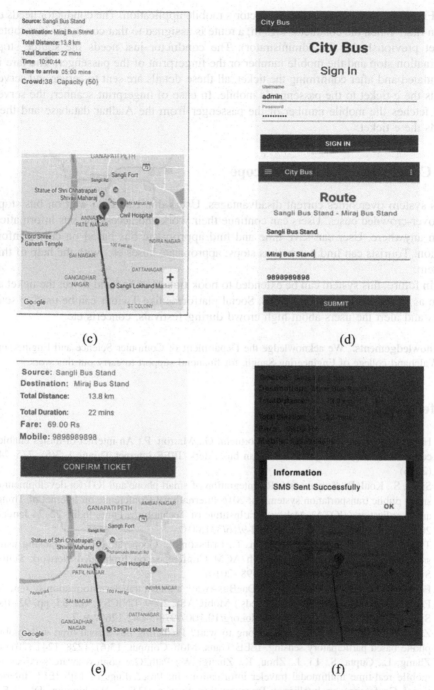

(c)

(d)

(e)

(f)

Fig. 4. Screenshots of passenger and conductor mobile applications.

Figure 4d, e and f show the conductor's mobile application. The conductor needs to login first. When the conductor logs in, a route is assigned to that conductor (the router is set previously by the administrator). The conductor just needs the source stop, destination stop and the mobile number or the fingerprint of the passenger. The fare is calculated and after confirming the ticket, all these details are sent to server and server sends the e-ticket to the passenger's mobile. In case of fingerprint scanner, the server first fetches the mobile number of the passenger from the Aadhar database and then sends the e-ticket.

6 Conclusion and Future Scope

This system overcomes current disadvantages. Users don't need to wait on bus stops for over-crowded buses. Users can continue their work and access the bus information from anywhere. User can save time and find appropriate bus based on crowd information. Tourists can find nearby bus stops, appropriate buses etc. with the help of this system.

In future, this system can be extended to book tickets online and detect the ticket as soon as the passenger enters the bus. Social platforms like Twitter can be used to scan twits and alert the users about high crowd during festivals, concerts etc.

Acknowledgements. We acknowledge the Department of Computer Science and Engineering of Walchand college of Engineering Sangli, for financial support to carry out this work.

References

1. Handte, M., Foell, S., Wagner, S., Kortuem, G., Marron, P.: An internet-of-things enabled connected navigation system for urban bus riders. IEEE Internet Things J. 3(5), 735–744 (2016)
2. Sutar, S., Koul, R., Suryavanshi, R.: Integration of smart phone and IOT for development of smart public transportation system. In: 2016 International Conference on Internet of Things and Applications (IOTA), Maharashtra Institute of Technology, Pune, India, 22–24 January 2016. IEEE (2016). 978-1-5090-0044-9/16/$31.00©
3. Thiagarajan, A., Biagioni, J., Gerlich, T., Eriksson, J.: Cooperative transit tracking using smart-phones. In: Proceedings of 8th ACM Conference on Embedded Network Sensor System, Zurich, Switzerland, pp. 85–98 (2010)
4. Ferris, B., Watkins, K., Borning, A.: OneBusAway: a transit traveler information system. In: Phan, T., Montanari, R., Zerfos, P. (eds.) MobiCASE 2009. LNICST, vol. 35, pp. 92–106. Springer, Heidelberg (2010). https://doi.org/10.1007/978-3-642-12607-9_7
5. Zhou, P., Zheng, Y., Li, M.: How long to wait? Predicting bus arrival time with mobile phone based participatory sensing. IEEE Trans. Mob. Comput. 13(6), 1228–1241 (2014)
6. Zhang, L., Gupta, S., Li, J., Zhou, K., Zhang, W.: Path2Go: context-aware services for mobile real-time multimodal traveler information. In: Proceedings of 14th IEEE International Conference on Intelligent Transportation System (ITSC), Washington, DC, USA, pp. 174–179 (2011)

7. Eken, S., Sayar, A.: A smart bus tracking system based on location-aware services and QR codes. In: 2014 IEEE International Symposium on Innovations in Intelligent Systems and Applications (INISTA) Proceedings. IEEE (2014). 978-1-4799-3020-3/14/$31.00©
8. Sujatha, K., Sruthi, K., Rao, N., Rao A.: Design and development of android mobile based bus tracking system. In: 2014 First International Conference on Networks and Soft Computing. IEEE (2014). 978-1-4799-3486-7/14/$31.00_c
9. Singla, L., Bhatia, P.: GPS based bus tracking system. In: IEEE International Conference on Computer, Communication and Control (IC4-2015) (2015)
10. Pholprasit, T., Pongnumkul, S., Saiprasert, C., Mangkorn-ngam, S., Jaritsup, L.: LiveBus-Track: high-frequency location update information system for shuttle/bus riders. In: 2013 13th International Symposium on Communications and Information Technologies (ISCIT). IEEE (2013). 978-1-4673-5580-3/13/$31.00©

A SEU Hardened Dual Dynamic Node Pulsed Hybrid Flip-Flop with an Embedded Logic Module

Rohan S. Adapur[1](✉) and S. Satheesh Kumar[2](✉)

[1] VLSI, VIT Vellore, Vellore, India
rohans.adapur2016@vitstudent.ac.in
[2] VIT Vellore, Vellore, India
satheeshkumar.s@vit.ac.in

Abstract. In this paper we study the operation and working of a Dual Dynamic node hybrid flip-flop (DDFF-ELM) with an embedded logic module. It is one of the most efficient D-Flip-flops in terms of power and delay as compared to other dynamic flip flops. A double exponential current pulse is passed to the sensitive nodes of the circuit to model a radiation particle strike in the circuit. The faulty output is then corrected using a radiation hardening by design technique. All the circuits are implemented using Cadence 90 nm technology and a comparison is made between the power and delay of already implemented D- flip-flops.

Keywords: DDFF-ELM · SEU (Single event upset)
RHBT technique (Radiation hardening by design) · Particle strike

1 Introduction

When a charged particle hits a circuit, the respective node of the circuit may get charged and discharged depending upon the particle. Thus there is a single event effect which leads in soft errors. Soft errors are the errors that can be rectified in due course of the circuit operation and corrected unlike hard errors which result in permanent damage to the circuit. Radiation effects occur mainly due to gamma rays, solar flares and cosmic rays.

In the magnetosphere especially in the zone of Von Allen Belts, high energy particle present in outer space come in contact with the air molecules in the atmosphere. These particles collide with air molecules to form high energy protons and neutrons. When these particles strike the sensitive nodes of circuits in satellites, faulty outputs are obtained. This effect is known as Single Event Upset (SEU) [3, 6]. Also due to the creation of charged particles due to ions in the nodes of a circuit we see a sudden spike in the output of the circuit. This effect is called Single Event Transient (SET) [4, 5]. Soft errors can be corrected by redesigning the circuits to make the circuit free from radiation effects. Some of the Radiation Hardening by design techniques adopted in the circuit are Redundant Latches [6], DICE cell [2, 3], Triple modular Redundancy [7], and Dual Modular Redundancy. We can also harden the circuit by adopting methods like temporal hardening [8]. But usually radiation hardening by design methods are used since they mitigate the cost constraints.

© Springer Nature Singapore Pte Ltd. 2018
I. Zelinka et al. (Eds.): ICSCS 2018, CCIS 837, pp. 62–68, 2018.
https://doi.org/10.1007/978-981-13-1936-5_7

In this paper we first study the functioning of a Dual dynamic node hybrid flip flop with an embedded logic module and study its advantages. Then by applying a double exponential pulse at the sensitive nodes of the flip-flop we model the particle strike and obtain the Single Event Upset (SEU) in the circuit. Then we apply Radiation Hardening by design (RHBT) technique and obtain the radiation hardened waveform in the circuit using Cadence 90 nm technology. Finally we compare the Area, Power and Delay values in the flip-flop with the already implemented flip-flop of Xi Che [2] and study them.

2 DDFF-ELM Flip-Flop

The name dual dynamic node hybrid flip-flop [1] comes from the fact that it has two main sensitive nodes X1 and X2 as shown in Fig. 1. It is hybrid because it is a combination of both static and dynamic flip-flops. It has an attached embedded logic module to perform the function of a D-Flip flop. As in any dynamic logic circuits the operation of the flip flop is mainly in two stages. (1) Pre-charge phase and evaluation phase.

The actual working of the flip-flop occurs when there is an overlap between CLK and CLKB. When the input is high during the pre-charge phase the node X1 is pulled low. Due to this the PM1 transistor is high and T2 also high. Because PM1 is high X2 node is pulled high and PM3 is not conducting. Since T2 is high FB is pulled low and therefore F is high. Similarly during the evaluation phase when clock is low PM0 is high. So X1 gets charged high. But PM1 is already high and blocks the 1 from X1. The outputs will be floating and hence retain their previous values. Hence the circuit works like an ideal flip-flop.

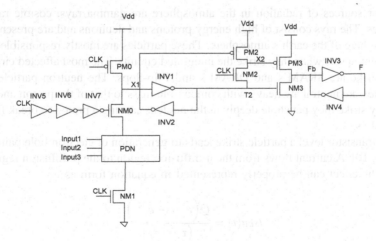

Fig. 1. DDFF-ELM flip-flop

When input is low and clock is high X1 gets charged in the pre-charge phase. T2 will be low. Since Clock is high node X2 will be pulled low due to NM2. Hence PM3 will be high. So FB will be high and hence F will be low. Similarly in the evaluation phase the outputs will be floating and the circuits will function as an ideal D-Flip-flop as shown in Fig. 2.

The main advantages of the DDFF-ELM flip-flop is it computes the data quickly and also it pull-up and pull-down networks. These circuits will function independently and help in reduction of power. By the addition of Inverters INV5, INV6, INV7 large bits of information can be continuously through the flip-flop. Thus this circuit works effectively.

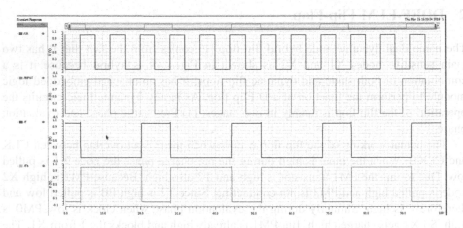

Fig. 2. DDFF-ELM waveform

3 Modelling of a Particle Strike

The major sources of radiation in the atmosphere are gamma rays, cosmic rays and solar flares. The rays consist of high energy protons and neutrons and are present in the magnetosphere of the earth's atmosphere. These particles are mostly responsible for the single event upsets when they strike the integrated circuits. The most affected circuits in particle strike are SRAM's and DRAM's and Flip-flops. The neutron particles even though they are neutral are heavy with comparable mass to that of the proton and hence when they strike they penetrate deeply in the nucleus causing a proton to break free and cause SEU's.

On a transistor level a particle strike leads to generation of electron-hole pairs in the device [9, 10]. A current flows from the n-diffusion region to the p-diffusion region due to this. The effect can be properly represented in equation form as

$$Iseu(t) = \frac{Q\left(e^{-\frac{t}{\tau_\alpha}} - e^{-\frac{t}{\tau_\beta}}\right)}{(\tau_\alpha - \tau_\beta)} \tag{1}$$

Where Q is the amount of charge deposited due to particle strike, τ_α is the deposition time constant and τ_β is the ion track establishment constant.

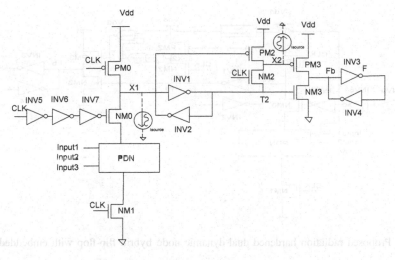

Fig. 3. Particle strike using a current pulse

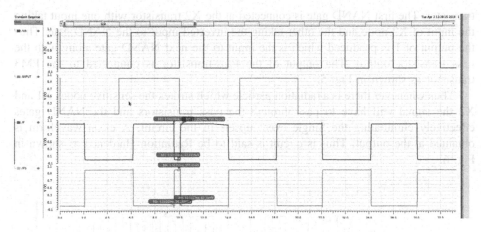

Fig. 4. Radiation affected waveform of DDFF-ELM flip-flop shown in pointer

In modelling of a particle strike in cadence 90 nm technology we take into consideration a current pulse of very high amplitude to cause a particle strike on the sensitive nodes of the flip-flop as shown in Fig. 3. This causes a bit flip which results in the single event upset as shown in the Fig. 4.

4 Proposed Radiation Hardened DDFF-ELM Flip-Flop (DDFF-ELM)

In the radiation hardened flip- flop we use pass-transistor along with two NAND gates in the circuit. The operation of the circuit is quite normal as that of normal DDFF-ELM flip flop. The pass transistor is connected with the input as the input of the pass

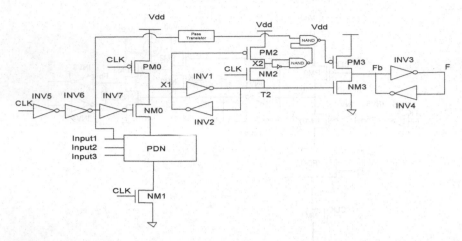

Fig. 5. Proposed radiation hardened dual dynamic node hybrid flip-flop with embedded logic module

transistor. The first NAND gate is connected at the X2 transistor with one input taking the input of X2 node and the other taking the inverted input to the NAND gate. Thus the output of 1 is produced which is the input to the next NAND gate along with the pass transistor output. The output of the Pass-transistor is connected to the PM3 transistor as shown in Fig. 5.

Thus whenever there is a radiation particle which strikes the sensitive nodes X1 and X2 the signal will have to pass through the pass transistors and the NAND gates effectively eliminating the Single event upsets in the circuit. A clean waveform is obtained at the output. Thus is circuit is said to be Radiation Hardened as shown in Fig. 6.

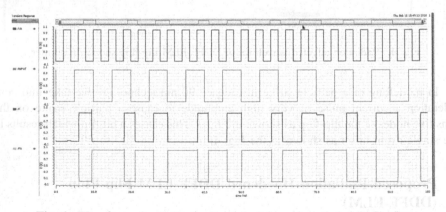

Fig. 6. Waveform of radiation hardened dual dynamic node hybrid flip-flop

The radiation hardening techniques used in this paper cost effective and they consume lesser area. The power consumption along with the area in the circuit is much less as compared to the already implemented circuit of Xi Che. This technique can be effectively employed to any dynamic flip-flops like Conditional Data mapping flip-flop (CDMFF), Hybrid latch flip-flop (HLFF), Cross charge control Flip-flops (XCFF).

5 Tabulation and Simulation

In this section we list out the values of power (average), area (no. of transistors) and delay. We list out the delays of the individual flip flops and also its radiation hardened circuit. We also make a comparison of a similar radiation hardened dynamic flip flop written by author Xi che [2].We implement the radiation hardened Master-Slave D-Flip-flop in Cadence 90 nm technology and compare the results with the proposed DDFF-ELM flip-flop. The Tabular column is listed below

Flip-flop	Area (no of transistors)	Power (uW)	Delay (ns)
DDFF ELM	22	2.56×10^{-6}	0.185
DDFF RH	32	364.7×10^{-3}	155.3×10^{-12}
Dynamic logic	16	648×10^{-9}	33.57×10^{-12}
Dynamic logic RH	40	76.42×10^{-3}	76.74×10^{-12}

6 Conclusion

In this paper we first obtained the waveform of the dynamic flip flop without radiation hardening. By applying radiation hardening techniques we established that the circuit can be mitigated of its Single event Upset (SEUs). Comparison was made between the proposed DDFF-ELM latch and the already implemented Master-Slave D- Flip flop. We established the fact that the proposed model has lesser area overhead and its power is also less in comparison with the already existing design.

References

1. Absel, K., Manuel, L., Kavitha, R.K.: Low-power dual dynamic node pulsed hybrid flip-flop featuring efficient embedded logic. IEEE Trans. Very Large Scale Integr. (VLSI) Syst. 21(9), 1693–1704 (2013)
2. Xuan, S.X., Li, N., Tong, J.: SEU hardened flip-flop based on dynamic logic. IEEE Trans. Nucl. Sci. 60(5), 3932–3936 (2013)
3. Jahinuzzaman, S.M., Islam, R.: TSPC-DICE: a single phase clock high performance SEU hardened flip-flop. In: 53rd IEEE International Midwest Symposium on Circuits and Systems (MWSCAS), pp. 73–76. IEEE (2010)
4. She, X., Li, N., Carlson, R.M., Erstad, D.O.: Single event transient suppressor for flip-flops. IEEE Trans. Nucl. Sci. 57(4), 2344–2348 (2010)

68   R. S. Adapur and S. Satheesh Kumar

5. She, X., Li, N., Erstad, D.O.: SET tolerant dynamic logic. IEEE Trans. Nucl. Sci. **59**(2), 434–438 (2012)
6. Fazeli, M., Patooghy, A., Miremadi, S.G., Ejlali, A.: Feedback redundancy: a power efficient SEU-tolerant latch design for deep sub-micron technologies. In: 37th Annual IEEE/IFIP International Conference on Dependable Systems and Networks, DSN 2007, pp. 276–285. IEEE (2007)
7. Nicolaidis, M., Perez, R., Alexandrescu, D.: Low-cost highly-robust hardened cells using blocking feedback transistors. In: 26th IEEE VLSI Test Symposium, VTS 2008, pp. 371–376. IEEE (2008)
8. Zhao, C., Zhao, Y., Dey, S.: Constraint-aware robustness insertion for optimal noise-tolerance enhancement in VLSI circuits. In: Proceedings of the 42nd annual Design Automation Conference, pp. 190–195. ACM (2005)
9. Barnaby, H.J., McLain, M.L., Esqueda, I.S., Chen, X.J.: Modeling ionizing radiation effects in solid state materials and CMOS devices. IEEE Trans. Circuits Syst. I Regul. Pap. **56**(8), 1870–1883 (2009)
10. Fulkerson, D.E., Nelson, D.K., Carlson, R.M., Vogt, E.E.: Modeling ion-induced pulses in radiation-hard SOI integrated circuits. IEEE Trans. Nucl. Sci. **54**(4), 1406–1415 (2007)

Soft Computing and Face Recognition: A Survey

J. Anil[1]([✉]), Padma Suresh Lekshmi Kanthan[2],
and S. H. Krishna Veni[2]

[1] Department of Electrical and Electronics Engineering,
Noorul Islam Centre for Higher Education,
Kumaracoil, Kanyakumari District, Thuckalay, Tamil Nadu, India
aniljayamohan@gmail.com
[2] Baselios Mathews II College of Engineering Sasthamcotta,
Kollam 690 520, Kerala, India
padmasuresh77@gmail.com, shkrishnaveni@gmail.com

Abstract. Soft computing has found profound application in the challenging areas like pattern recognition, classification, optimization etc. Face recognition is basically a pattern recognition problem. This paper reviews some of the efficient algorithms that uses soft computing techniques along with conventional methods like Principal Component Analysis, Radial Basis Functions etc. for face recognition. Conventional methods for pattern recognition are very efficient even without introducing soft computing techniques. But this is not the case with face recognition as it is a complex problem owing to the various challenges like illumination variation, ageing, expression changes, occlusion etc. Due to these factors face recognition becomes unpredictable and so soft computing techniques can be applied, which are very efficient in solving unpredictable and incomplete problems. Soft computing copies human mind. In other words, soft computing thinks like humans. The algorithms discussed in this paper employs various soft computing techniques like Neural networks, genetic algorithm and fuzzy logic.

Keywords: Soft computing · Face recognition · Principle component analysis
Radial basis functions · Pattern recognition problem · Neural networks
Genetic algorithm · Fuzzy logic

1 Introduction

Soft computing is one of the recent technique for solving most complex problems whose solutions are mostly uncertain. Also in most of the cases even the problem will not be defined completely. Fuzzy logic, Genetic algorithm and Neural networks are the components of Soft Computing. Soft computing techniques try to replicate human mind. Human mind is a very complex system. Human mind has the power to solve problems even when the problem is not completely defined. Also most complex problems can be solved easily by human mind. Soft computing techniques are trying to copy this ability of human mind. It is not based on a perfect solution which depends entirely on the statistical data available. It is applied when the solution of a problem is uncertain or

© Springer Nature Singapore Pte Ltd. 2018
I. Zelinka et al. (Eds.): ICSCS 2018, CCIS 837, pp. 69–79, 2018.
https://doi.org/10.1007/978-981-13-1936-5_8

unpredictable. The answer to the problem will be an approximation. Soft computing methods can be applied when conventional methods fail to find any solution [1].

Soft computing methods find wide applications in solving intractable problems, Pattern matching, Approximation, Classification, Optimization, In mobile Adhoc Network, Driver Drowsiness Detection etc.

Various techniques employ soft computing along with conventional face recognition methods to increase the efficiency and accuracy of face recognition system. Combination of Principal Component Analysis with Two-layer feed forward network, Combination of soft computing by adding stochasticity to conventional Radial Basis Function Neural Network, Concept of Surface curvatures with Cascade Architectures of Fuzzy Neural Networks, Progressive switching pattern and soft computing and Fuzzy clustering techniques are discussed in this paper.

The paper is structured as follows. Section 2 gives a brief description of the various steps involved in face recognition. Section 3 reviews the need for soft computing in face recognition. Section 4 gives a basic idea about the working of neural network. In Sect. 5 different algorithms for face recognition employing soft computing are discussed. This section gives a brief insight of the various steps involved in each method. This section also gives an idea of how to incorporate soft computing in face recognition so as to improve the efficiency of the system. Section 6 gives the result of the study in a tabular form with a brief description. The study is concluded in Sect. 7.

2 Face Recognition

Face recognition is one of the practical application of pattern recognition problem. It is a very strong biometric verification method. Now days a lot of mobile phones rely on face recognition as a security feature for unlocking. Apart from this face recognition find wide application in Human computer interaction, law enforcement, face tagging etc. Face recognition is a complex problem. This is due to the fact that the human face is flexible. It can change with a lot of factors. So normal pattern matching may give erroneous results.

The basic steps in face recognition are (i) preprocessing (ii) Face detection (iii) Face area cropping (iv) Feature Extraction (v) Classification (vi) Recognition [2].

Face feature extraction can be done using Appearance based methods or Geometric feature based methods. Both methods have their own advantages and disadvantages. Some algorithms use a hybrid method.

Once the features are extracted the extracted features can be used for classification.

3 Need for Soft Computing in Face Recognition

Soft Computing Techniques are applied when the problem is complex or solution to the problem is uncertain due to the complexity of the problem. Face recognition is a very complex and unpredictable problem under unconstrained conditions. There can be a lot of uncertainties in face recognition due to following challenges [3].

a. Pose change
b. Expression changes
c. Presence of Occlusions
d. Illumination changes
e. Ageing

Any of these challenges introduces uncertainty in the problem of face recognition. So in order to address any of these challenges application of soft computing is found to be useful. More over the application of soft computing has increased the accuracy level of face recognition under unconstrained conditions.

4 Neural Networks

As stated earlier neural network is a Soft Computing Technique. Neural network has the ability of self-learning and adaptability [4]. That is NN has the ability to make its own organization and there by solve most complex problems. It also deals with data which is imprecise or incomplete and it can derive meaningful information from this incomplete data. Neural networks are based on the Human Brain structure. Computers can do large scale of repeated calculations very well but when it comes to complex problems like pattern recognition the performance of computer is far behind human brain. Artificial neural network's role becomes very relevant in such conditions. In Human brain the information is stored as patterns. Human brain recognizes faces based on the patterns stored in the brain. Using ANN computers are trying to copy the same method for face recognition.

Basically the topologies of all the Artificial neural networks are same. There are 3 basic layers for ANN. They are the input layer, hidden layers and Output layer. The input layer accepts the information from outside. This information will be passed on to the hidden layers through the connectors for further processing. These layers consist of neurons and weights. The final result will be obtained from the output layer after processing.

Based on the connections ANN can be classified as Backpropagation Neural Network (BPNN) and Feedforward neural network. BPNN is used to train Multilayer Perceptron. Multilayer Perceptron is a network constituted by input layer, hidden layers and output layer. Input layer consist of a set of sensory units. BPNN is computationally heavy and time taken for training is more.

Feedforward neural network is a unidirectional network in which the information will be transferred only in one direction. There will not be any feedback loop. This network also consists of input layer, hidden layers and an output layer. Here the inputs will be applied to the input layer and the output produced by this layer will be transferred to the first level of the hidden layer. Similarly, the output of each layer is input to the next layer. In this network the final output solely depends on the current inputs and weights. There is no memory for this neural network.

The neural networks will be trained by using training images. The neural network studies the problem from these training images. Once the neural network is trained it can be used to Identify the test images.

5   Different Algorithms for Face Recognition Employing Soft Computing

5.1   Face Recognition Using Principal Component Analysis and Artificial Neural Network of Facial Images Datasets in Soft Computing

Satonkar, Pathak and Khanale presents a face recognition using Principal Component Analysis and Two-Layer Feed Forward Neural Network in their paper [5]. In the paper dimensionality reduction is done by using Principal Component Analysis (PCA) there by obtaining the feature vector. These input feature vectors are fed into a two-layer feed forward neural network. The neural network is trained first using the training images and then tested using the test images.

The Eigen faces are obtained from the input image by applying PCA. This causes considerable reduction in the dimension while retaining the important details in the image. The Eigen vector values are sorted from high value to low value. The highest Eigen value gives the principal component. Once the high valued Eigen vectors are chosen they are used to form the feature vector. This feature vector is given as input to the neural Network.

The artificial neural network (ANN) consists of information processing units called neurons. Connection links and weights are other components of the ANN. Connection links are used to transmit the input signals through the neural network. The connection links consists of weights and these weights are multiplied to the input signals. Thus the net input is obtained. Activations are applied to these net input signal to achieve the output signal. The result of neural network will depend on the weights. The value of the weights are assigned during the training session. The neural network will also be having a bias which is fixed as one in the proposed method. Here a two-layer feed forward neural network is used instead of a single layer one. The weights and bias are updated based on gradient descent and adaptive learning rate. The performance of the system is expressed in terms of mean squared error. Once the training is completed the neural network will be ready for recognition. The network was trained for 700 epochs. 71 images were selected from Face95 database and 5 local images were used for testing. For the 71 images from the Face95 database the performance was 0.087516 with 337 epochs and for the local images it was 0.029753 with 127 epochs. For the both the databases the neural network took only few seconds for execution. Also in both the cases the accuracy is reported as 100%.

5.2   Conventional Radial Basis Function Neural Network and Human Face Recognition Using Soft Computing Radial Basis Function

In this paper Pensuwon, Adams and Davey proposed a new method in which the conventional Radial Basis Function Neural Network is clubbed with Soft computing to achieve a more efficient face recognition system [6]. Here Soft computing acts as an intelligent system which strengthens the RBF neural network by making it to work like human mind. By using soft computing, it is possible to achieve a robust and low cost solution to a problem. Here the principle is to add uncertainty, approximation and partial truth and exploit the tolerance of imprecision.

In Conventional RBFN the input space is divided into subclasses and each class is assigned a value in the center. This value is called the prototype value. When an input vector is initiated a function is used to find the distance of the input vector with the prototype or center value of each class. This membership function value is calculated for all the sub classes. The membership function should attain maximum value in the center i.e. zero distance. After attaining the membership values of the input vector in each subclass the results are combined to find the membership degree.

In the proposed system soft computing is implemented by the introduction of stochasticity into the problem of calculating the output of RBF units. The stochastic value of n center values of RBF units introduced is given by the sigmoid function

$$y'_n = 0.5 * (1 + (\tanh(0.5 * y_n))) \qquad (1)$$

The decision of whether to keep the new center value or not depends on the comparison of the new value to a random value between 0 and 1. If the new value is larger, the new value is taken as the RBF center value. If it's not the case, then the original value of RBF units is kept as such.

Addition of stochasticity to the RBF units has resulted in better classification. The improved RBFN method has shown improvement in the recognition rate, reduced training time and testing time when compared with the traditional RBFN method.

5.3 Soft Computing Based Range Facial Recognition Using Eigen Faces

This is a 3D facial recognition system proposed by Lee, Han and Kim which takes into consideration the surface curvatures [7]. In order to reduce the dimensionality Eigen faces are considered. Eigen faces reduces data dimension without much loss in the original information contained in the image.

The 3D image is taken by using a laser scanner. The laser scanner image can have accurate depth information. This is owing to the fact that the laser scanner uses a filter and laser. The 3D image obtained by using laser scanner is least affected by lighting illuminations.

In order to increase the accuracy of face recognition the normalized face images are considered. For normalizing the facial image nose end is considered as reference. This is because of two reasons. In a 3 D image nose is the most protruded element in a face. Also nose is placed in the middle of the face while considering the frontal face image. Thus nose can play a vital role in normalization of the face. The nose point is extracted by using iterative selection method. Here normalization means to place the face in the standard special position. For this panning, tilting and rotation are done as per required for the particular image [8].

In the proposed algorithm the surface type of each point in the face is determined by applying Principal, Gaussian and Mean curvatures. These curvatures are calculated along with the sign to determine the surface type. The values can be positive, negative or zero. Then z(x, y) is found out which gives the surface depth information. Once z (x, y) is found first fundamental form and second fundamental forms are calculated using the formalism introduced by Peet and Sahota [9]. The first fundamental form gives the arc length of surface of the point under consideration. The second fundamental form gives the curvature of these curves at the point under consideration in the given direction. Then the minimum and maximum curvatures represented by k1 and k2 are calculated. These are called the Principal curvatures. They are invariant to the motion of the surface. The Gaussian curvature and Mean curvature are calculated using the values of principal curvatures k1 and k2. For the characterization of the facial image the Principal curvatures and Gaussian curvatures are the most suited values.

Next step is to find the Eigen face. Eigen faces are calculated in order to reduce the dimension. This reduces the complexity of the overall problem. Face Identification is a pattern recognition problem. Once the Eigen faces are calculated the pattern recognition is carried out in the Eigen faces instead of the original image. The usual method of identification is to use Euclidian distance to calculate the difference of the test image with a predefined face class. In the proposed method instead of using Euclidian Distance Cascade Architectures of Fuzzy Neural Networks (CAFNN) is used.

CAFNN was originally introduced by Pedrycz, Reformat and Han [10]. CAFNN is comprised of Logic Processors. These logic processors are cascaded and they consist of fuzzy neurons. Here memetic algorithm is used to optimize the input subset and connections. Thus a close fisted knowledge base is constructed. Even though the knowledge base is parsimonious it is an accurate one [11]. Memetic algorithm is used since it is a very effective algorithm. The output class of the problem is fuzzified as binary for classification. Here Winner-take-all method is used to find out to which class the test data belongs. For example, if there are 5 face images i.e. there are 5 classes, the number of output crisp dataset is 5. Suppose the test image belongs to class 3 then the Boolean output will be "0 0 1 0 0". The "1" in 3rd position represents that the test image belongs to class 3. This is decided based on the membership value. The test data belongs to the class were the membership value is maximum.

5.4 Recognition of Human Face from Side-View Using Progressive Switching Pattern and Soft-Computing Technique

Raja, Sinha and Dubey has proposed a method for recognition of human face from the side view pose. Most of the face recognition problems deals with the frontal face images [12]. Of the Face recognition methods developed so far only a small percentage of face recognition methods deals with side poses. So the work done by Raja et.al. is very relevant in face recognition. In the proposed method both frontal image and profile views are used but in different aspects. The frontal views are used for learning while the profile views are used for understanding.

As discussed for learning purpose the frontal images are considered. The feature vector is formed from the features extracted from the frontal image. For understanding the features are extracted from the side view. This is a very tedious process because it is very knowledge intensive. For this the Progressive switching pattern and Soft-Computing are employed. Built, Complexion, hair and texture are the categories of features that are extracted from the side view for understanding purpose.

In this work front face analysis and side face analysis is done. Front face analysis employs Statistical methods like Cross correlation and Auto correlation using 4 neighborhoods and 8 neighborhoods. Also the mean clusters are calculated using Fuzzy-C means clustering methods. The extracted features are stored as trained dataset. In the side face analysis, a progressive switching angle is introduced and its value is initialized to 0. Before extracting the features morphological operations such as thinning and thickening are done. As in the case of frontal face here also cross correlation and auto correlation are applied. Also to obtain the mean of the clusters Fuzzy-C means clustering method is used. Then the distance measure of the extracted features of the test image is stored. Forward-Backward dynamic programming of neural network is used to find the best fitting patterns. The process is validated using genetic algorithm. If the best fit testing fails, the progressive switching angle is increased by one step and the whole process is repeated starting from feature extraction. This is done until a best fit is found out. The steps usually used are of size 5. If the speed of processing is to be increased the step size can be incremented to 10 instead of 5. Once the best fit is found out Support Vector Machine is used for classification and characterization.

In the proposed method nineteen parameters are extracted from the frontal face image and they are stored in the corpus. Also the distances measured between these features are stored. From the 19 parameters few of the parameters are used in the understanding part. Some of the parameters are forehead width, eyes to nose distance, lips to chin distance, eyes to lips distance, number of wrinkles, Texture of face, normal behavior pattern etc. The results of this method were found to be remarkable.

5.5 Face Recognition Using Fuzzy Clustering Technique

Aradhana, Karibasappa, Reddy had discussed in their paper about how fuzzy clustering technique can be employed in face recognition [13]. In the proposed method a cognitive model of the human face is designed. Fuzzy clustering can be done in different methods. Some of them are Hierarchical clustering [14], Interactive clustering [15], Fuzzy-C means clustering [16, 17], Partitional clustering [18] etc. In the proposed method Fuzzy clustering is utilized for recognizing human face. For this a R-dimensional matrix is created using the face image database. For this initially the face image is divided into segments such as eyes, nose and mouth. The representative nodes for these segments, i.e. eyes, nose and mouth will be present in the database. There will be representative node for various types of eyes such as normal, short, long. Similarly, there will representative nodes for various types of nose and mouth also. For nose the categories may be normal, flat, small, long, moderately small etc. Mouth can be categorized as normal, long, short. Each type of eyes, nose or mouth will be grouped into a cluster and the representative nodes will be the cluster center. Thus the no. of comparisons may be reduced since the comparison is made with the representative nodes and not the whole database. While comparing the distance of descent of each feature i.e. eyes, nose and mouth of the test image with the representative nodes is calculated and stored in a matrix called the fuzzy scatter matrix. The matching is done based on the distance of descent. For matching the distance should be minimum. Fuzzy If Then rule is used for the final face recognition. The most critical factor in this method is the selection of the representative node because the accuracy of the system depends entirely on the representative node. Any change in the representative node will affect the whole process. In the proposed method a heuristic approach is made which cover all variations of the facial features. The proposed method used Yale database and a local database which consisted of 400 images with 40 classes. The facial features are extracted by using cross-correlation method. The results show that the fuzzy clustering method is better than PCA and PCA BPNN when the image set is big since in the fuzzy clustering technique only representative nodes are considered. When the acceptance ratio is compared the proposed methods is better than PCA and it is giving a performance as strong as PCA with BPNN. For larger image set when the execution times are considered the proposed method is better than PCA and PCA with BPNN. So overall performance of the proposed technique is better than PCA and PCA with BPNN.

6 Result

A study of the importance of Soft Computing in Face recognition is done in this paper. The paper also examines how soft computing techniques are incorporated with the conventional face recognition methods to achieve better result. Various methods and Techniques used in face recognition for employing the soft computing are discussed in the Table 1. The key points give the advantages of using soft computing techniques.

Table 1. Review of application of soft computing techniques in face recongnition

Sl. No.	Title	Data base used	Methods and Techniques	Key points
1	Face recognition using principal component analysis and artificial neural network of facial images datasets in soft computing	Face 95 database	PCA and Two-layer Feed Forward Neural Network is used. ANN is trained using the training images. The performance of the system is expressed in terms of mean squared error	Eigen faces reduce dimensionality. The processing time was very less. Also very high accuracy rate is reported
2	Radial basis function neural network	ORL face database	Input space divided into subclasses. A prototype vector is assigned to every class in the center of it. Membership of every input vector in the subclasses is calculated	It is a well-established model for classification. It can provide a fast, linear algorithm for complex nonlinear mappings
3	Human face recognition using soft computing RBF	BioID face database	Soft computing is implemented by the introduction of stochasticity into the problem of calculating the output of RBF units. The decision of whether to keep the new center value or not depends on the comparison of the new value to a random value between 0 and 1	Addition of stochasticity to the RBF units has resulted in better classification. The improved RBFN method has shown improvement in the recognition rate reduced training time and testing time

(continued)

Table 1. (*continued*)

Sl. No.	Title	Data base used	Methods and Techniques	Key points
4	Soft computing based range facial recognition using eigen faces	Images obtained using 3D laser scanner	A 3D laser scanner is used to acquire the face image. The surface curvature of regions of interest are found out. Instead of using original Image Eigen faces are used. CAFNN is used for identification	Since memetic algorithm is used the outputs are more reliable. Winner-take-all method is used
5	Recognition of human face from side-view progressive switching pattern and soft computing technique	Local images of ten subjects, both frontal and side pose	Frontal images are considered for learning and side views are considered for understanding. Cross correlation and auto correlation are employed. Mean clusters are calculated using Fuzzy-C means. Forward Backward Dynamic programming of NN is used to find the Best fitting patterns	Uses neural network method to find the Best fitting pattern and is validated using genetic algorithm. Progressive switching angle is also used which increases the accuracy of the system
6	Face recognition using fuzzy clustering technique	Yale database and local database	Fuzzy clustering is used for classification of face segments such as eye, nose and mouth. Matching is done based on the distance of descent. Fuzzy If Then rule is used for final face recognition	Results show that Fuzzy clustering is better than PCA and PCA BPNN. Fuzzy clustering is computationally lighter since only representative nodes are used

7 Conclusion

This paper examines the application of soft computing techniques in face recognition. Six methods which gives efficient output are reviewed in this paper. Importance is given to how soft computing is incorporated in each method. Studies have proved that employing soft computing methods have improved the efficiency of the conventional methods. Application of genetic algorithm and fuzzy logic are also reviewed apart to neural network. Some methods use combination of these techniques i.e. neural network, fuzzy logic and genetic algorithm. The importance of Eigen faces are also discussed. Using the soft computing techniques have helped in overcoming some of the challenges like illumination differences, change in poses etc.

Acknowledgment. The authors would like to thank the editor in chief and anonymous reviewers of this conference for their constructive feedback which has helped in improving the contents of this paper. The authors would also like to express their gratitude towards Noorul Islam Centre for Higher Education, Kumaracoil, Kanyakumari District, Thuckalay, Tamil Nadu for providing the facilities.

References

1. Kantharia, K., Prajapati, G.: Facial behaviour recognition using soft computing techniques: a survey. In: International Conference on Advanced Computing and Communication Technologies (2015)
2. Anil, J., Suresh, L.P.: Face recognition. In: International Conference on Circuit, Power, Computing Technologies (2016)
3. Sahu, A.K., Dewangan, D.: Soft computing approach to recognition of human face. Int. Res. J. Eng. Technol. **3**, 65–69 (2016)
4. Bhandiwad, V., Tekwani, B.: Face recognition and detection using neural networks. In: International Conference on Trends Electron Informatics, pp. 879–882 (2017)
5. Satonkar, S.S., Pathak, V.M., Khanale, P.B.: Face recognition using principal component analysis and artificial neural network of facial images datasets in soft computing. Int. J. Emerg. Trends Technol. Comput. Sci. **4**, 110–116 (2015)
6. Pensuwon, W., Adams, R.G., Davey, N.: Human face recognition using soft computing RBF. In: 2006 IEEE Region 10th Conference on TENCON 2006 (2006). https://doi.org/10.1109/tencon.2006.344206
7. Lee, Y.-H., Han, C.-W., Kim, T.-S.: Soft computing based range facial recognition using eigenface. In: Alexandrov, V.N., van Albada, G.D., Sloot, P.M.A., Dongarra, J. (eds.) ICCS 2006. LNCS, vol. 3994, pp. 862–869. Springer, Heidelberg (2006). https://doi.org/10.1007/11758549_115
8. Lee, Y.: 3D face recognition using longitudinal section and transection. In: Proceeding of DICTA (2003)
9. Peet, F.G., Sahota, T.S.: Surface curvature as a measure of image texture. IEEE Trans. Pattern Anal. Mach. Intell. **1**, 734–738 (1985)
10. Pedrycz, W., Reformat, M., Han, C.W.: Cascade architectures of fuzzy neural networks. Fuzzy Optim. Decis. Mak. **3**, 5–37 (2004)
11. Ciaramella, A., Tagliaferri, R., Pedrycz, W., Di Nola, A.: Fuzzy relational neural network. Int. J. Approx. Reason **41**, 146–163 (2006). https://doi.org/10.1016/j.ijar.2005.06.016
12. Raja, R., Sinha, T.S., Dubey, R.P.: Recognition of human-face from side-view using progressive switching pattern and soft-computing technique key words. Adv. Model Ser. B Signal Process. Pattern Recognit. **58**, 14–34 (2015)
13. Aradhana, D., Karibasappa, K., Reddy, A.C.: Feature extraction and recognition using soft computing tools. Int. J. Sci. Eng. Res. **6**, 1436–1443 (2015)
14. Ho, T.K., Hall, J.J., Srihavi, S.N.: Decision combination in multiple classifier system. IEEE Trans. Pattern Anal. Mach. Intell. **16**, 75–77 (1994)

15. Young, A.N., Ellis, H.D.: Handbook of Research on Face Processing. North, Holland, Amsterdam (1989)
16. Hathaway, R.J., Bezdek, J.C.: Fuzzy c-means clustering of incomplete data. IEEE Trans. Syst. Man Cybern. Part B **31**, 735–744 (2001)
17. Hung, N.-C., Yang, D.-L.: An efficient fuzzy C-means clustering algorithm. In: IEEE International Conference on Data Mining (2002)
18. Grover, N.: A study of various fuzzy clustering algorithms. Int. J. Eng. Res. **3**, 177–181 (2014)

Development of Autonomous Quadcopter for Farmland Surveillance

Ramaraj Kowsalya$^{(\boxtimes)}$ and Parthasarathy Eswaran

Electronics and Communication Engineering,
SRM Institute of Science and Technology, Chennai, India
kowsalya.ramaraj@gmail.com,
eswaran.p@ktr.srmuniv.ac.in

Abstract. There are various technologies available to surveillance the farmland through which the farmer can assess the conditions of the crops. Among all technology the autonomous quadcopter is an efficient, small size and cheaper tool to take farmland images. Autonomous quadcopter is advantageous because of its automatic navigation without human interaction. In this work, the paddy field considered with four predefined coordinates. The sensor circuit is deployed at the four coordinates to measure the water level and moisture condition in the field. Here, Way point GPS navigation algorithm is proposed and it allows the quadcopter to fly on its own with its destination. GPS navigation algorithm were implemented and simulated in Proteus. Camera is interfaced with quadcopter to capture the images and that images are analyzed through image processing and send the results to farmers mobile using Wi-Fi.

Keywords: Autonomous quadcopter · GPS navigation · PIC microcontroller Proteus simulation

1 Introduction

In general, farming plays a vital role in supplying foods and other fields such as clothing, medicine and it was the primary source in terms of economy before industrial resolution. The high production of crop is always at risk because of shortage of water, disease and irrigation. So it is necessary to observe the field regularly. Nowadays this can be achievable using available assistive technology to assist the farmer. As farmers are still using traditional methods for farming, that leads to low yielding in crops. So, automatic machineries with modern science and technologies are needed to increase the crop yield. The most commonly used technologies are VRT (Variable Rate Technology), GPS (Global Positioning System), Various Maps, UAV (Unmanned Aerial Vehicles), Guidance Software. Here, the emerging technology of quadcopter is used to surveillance the farmland. It is used to assist the farmer in terms of counting plants, predict crop yields, analyses the water level, analyses the crop that is affected by diseases and examine the soil moisture level through which, we can increase the production. Autonomous quadcopter is an unmanned aerial vehicle that can be used for examining the area where the human unable to surveillance.

© Springer Nature Singapore Pte Ltd. 2018
I. Zelinka et al. (Eds.): ICSCS 2018, CCIS 837, pp. 80–87, 2018.
https://doi.org/10.1007/978-981-13-1936-5_9

The quadcopter come under category of rotorcraft UAV and has distinguishable features such as vertical take-off, hovering at particular place over fixed wing UAV. Because of this features it is used in numerous application such as transportation, forest fire detection, defense surveillance, areas hit by natural calamities, delivery system, crop spraying in agriculture etc. [1]. In the beginning, the quadcopter can be navigated towards its path using RC (remote control). But developing quadcopter with autonomous navigation is complex task in the outdoor environment. Generally, camera based or GPS based navigation systems are used for autonomous quadcopter [2]. This paper proposes the method in which, water level in the paddy field and moisture conditions are measured by using simple sensor circuit and that can be monitored using quadcopter. The GPS navigation algorithm is proposed and that is preprogrammed in the PIC microcontroller through which the quadcopter navigate towards its path automatically and will reach the base station.

In early 2002, researchers made quadcopter to fly along with its trajectory using waypoint guidance algorithm. In this, vehicle made to fly along with current waypoint by deriving line following guidance and finding optimal waypoint changing point [3]. In 2009, the navigation algorithm was proposed by using RTK-GPS and encoders. Sometimes GPS fails to receive signal [4]. To overcome this problem DR navigation method was used and position errors were decreased using RTK-GPS units. Finally, they proved that RTK-GPS was efficient for position data accuracy compare to DGPS unit. In 2011, autonomous navigation system was proposed with GPS receiver, inertial sensor that is gyro, compass and encoders. The acquired information from GPS was fused using GPS/DR fusing algorithm [5]. Trajectory linearization algorithm was designed to navigate line path based on tracking the heading angle. In 2013, Rengarajan [6] proposed a method for quadrotor navigation GPS and Atmega328P on board microcontroller. The current and target locations are already loaded in the microcontroller.

This paper is organized as follows; Sect. 1 describes relative works done. Section 2 describes the proposed method, GPS navigation algorithm, flow diagram. Section 3 shows Experiments and Results. Finally, proposed work is concluded in Sect. 4.

2 Proposed Method

2.1 Proposed Design Methodology

In this proposed design methodology we are developing the air frame of quadcopter with X-shape structure connected with four Brush less (BLDC) Motor at the four corners of the quadcopter. This X-shape quadcopter is used in this work to surveillance the farmland as shown in Fig. 1. Here, the paddy field is considered with five nodes where the sensors are placed to measure the water level and moisture level. The nodes are represented with its GPS coordinates which is denoted as (x, y). The quadcopter will get initialized from its base station by receiving message from PC through wireless. Once the quadcopter start will move in the predefined path autonomously that is pre-programmed in the controller. Once it reaches to node 1 position will hover for some time to capture image. Then it will move to node 2 and hover to capture image

[8]. Like this it will cover the entire node and capture the image at the respective nodes and reaches to base station as shown in schematic diagram. Once reaches to base station the acquired images from quadcopter transferred to PC through wireless or using USB cable. Acquired images are analyzed using image processing techniques and the results are coordinated to farmers mobile through messages.

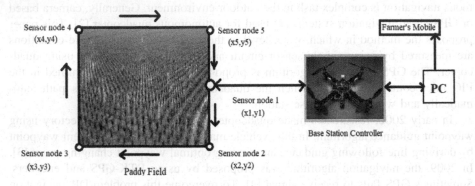

Fig. 1. Schematic diagram for farmland surveillance.

2.2   Design of Quadcopter

The Autonomous Quadcopter is designed for its surveillance application because of it lifting, hovering ability and high level of stability. We are developing the X-shape quadcopter and mounting is provided to fix four BLDC motor to its corner. The complete quadcopter design includes camera, controller and telemetry which is used observe the images or videos on laptop from quadcopter that is located far away [7]. The airframe is the body of quadcopter and it has four rotors at its end. Here, X-shape Airframe is made as shown in Fig. 2 [10].

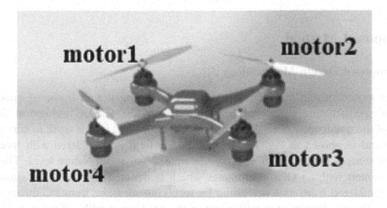

Fig. 2.   Quadcopter model.

2.3 Degree of Freedom of Quadcopter

The quadcopter contains four degrees of freedom that is yaw, pitch, roll and altitude. It will be controlled by adjusting the speed of the motor. The propellers are connected with motors, in which two motors adjacent to each other will rotate in opposite direction. The thrust produced by the quadcopter is twice that of total quadcopter weight. The quadcopter will takeoff only if the quadcopter produce the optimum thrust.

Table 1. Quadcopter movement condition

Degrees of freedom	Motor1	Motor2	Motor3	Motor4
Pitch	Positive	Positive	Negative	Negative
Roll	Positive	Negative	Negative	Positive
Yaw	Negative	Positive	Negative	Positive

Positive: Increasing the Motor Speed.
Negative: Decreasing the Motor Speed.

Table 1 shows various conditions for four motors. For pitch movement increase the speed of Motor 1 and 2, decrease the speed of Motor 3 and 4. For Roll movement increase the speed of Motor 1 and 4, decreasing the speed of Motor 2 and 3. For Yaw movement increase the speed of Motor 2 and 4, decrease the speed of Motor 1 and 3.

2.4 GPS Navigation Algorithm

The navigation process in desired area using GPS module can be attained by acquiring current GPS location with its coordinates i.e. Latitude, longitude and comparing it with all target GPS position by calculating distance from current GPS position to target GPS position. Generally, GPS module will display the acquired information in standard NMEA data format. It represented in sentence form and data in this is separated by commas. Example: GPGGA Format [11]

$GPGGA,hhmmss,llll.ll,b,yyyyy.yyy,b,X,XX,x.x,xxx.x,M,xx.x,M,xxxx,*xx
Always this format start with $ symbol.

- hhmmss = data taken at hh: mm: ss UTC
- llll:ll = latitude- position
- b = N or S
- Yyyyy. yyy = longitude position
- b = E or W
- X = GPS quality indicator (0 = invalid; 1 = GPSFix; 2 = DifGPSFix:)
- XX = number of satellites being used
- x:x = horizontal dilution of position
- xxx:x = Altitude above mean sea level
- M = units of altitude; meters
- xx:x = geoidal height
- M = units of geoidal height; meters
- Xxxx = Differential reference station ID number
- xx = check sum data.

The distance between current waypoint and target waypoint is calculated using Eq. 1.

Distance between two GPS position [9].

$$2\sin^{-1}\left(\left(\sqrt{\cos(lat_a)\cdot\cos(lat_b)\cdot\left(\sin\left(\frac{lon_a-lon_b}{2}\right)\right)}\wedge 2\right)+\left(\sin\left(\frac{lat_a-lat_b}{2}\right)\right)^2\right)$$

(1)

2.5 GPS Location Tracking Flow Diagram

See Fig. 3.

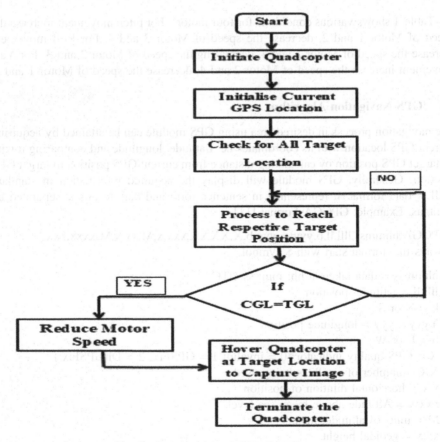

Fig. 3. GPS location tracking flow diagram

2.6 Navigation Algorithm

Step 1 Activate the quadcopter from base station to current waypoint Location using RC control. (quadcopter in stabilize mode)

Step 2 From current waypoint the quadcopter get in to Auto mode. (quadcopter start navigate autonomously).

Step 3 GPS module read current GPS position.

Step 4 Quadcopter will move to next predefined Waypoint through calculated distance between current waypoint and next Target Waypoint.

Step 5 Once it reaches the first target position will hover for 10 min to capture image.

Step 6 Now the first target position becomes as Current position and will reach to second Target position by comparing calculated distance between them.

Step 7 Similarly the quadcopter complete its Navigation and at last reach to its base Station.

3 Results and Discussions

In order to implement this algorithm, code was written in C language on PIC Microcontroller IDE (Integrated Development Environment). For experiment, the GPS locations of SRM Institute of Science and Technology near MBA block was selected. The GPS waypoint navigation was simulated in Proteus using circuit diagram as shown in Fig. 4.

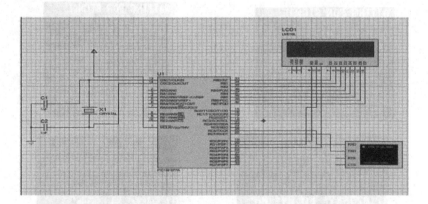

Fig. 4. GPS location tracking simulation circuit

In this, PIC16f778 microcontroller is used and virtual terminal is connected with UART port of PIC microcontroller. Once the GPS sensor reads the coordinates values it will send to the microcontroller. After performing the applied algorithm, microcontroller will give the waypoint location and displayed in the virtual terminal

Table 2. Waypoint coordinates

Waypoints	Latitude	Longitude
Home location	12.823865	80.044290
Target location 1	12.824367	80.044290
Target location 2	12.824367	80.044779
Target location 3	12.823887	80.044779
Target location 4	12.823887	80.044650
Home Location	12.823865	80.044290

Table 2 describes all base and GPS location latitude and longitude values. The distance between base station location to target 1 location is 55 m, between target 1 location to target 2 location is 53 m, between target 2 location to target 3 location is 53 m, between target 3 to target 4 location is 14 m and target 4 to again base station location is 39 m (Fig. 5).

(a)

(b)

(c)

(d)

(e)

(f)

Fig. 5. GPS coordinates simulation result. (a): Shows GPS Coordinates of Home Location and Current Locations. (b): Shows navigation process to reach target point 1. (c): Shows navigation process to reach target point 2. (d): Shows navigation process to reach target point 3. (e): Shows navigation process to reach target point 4. (f): shows navigation process to reach Home Location

4 Conclusion

The GPS based waypoint navigation algorithm has been implemented. The algorithm was tested with PIC microcontroller and simulation was carried out in Proteus. The current and target GPS coordinates are predefined in the quadcopter. The navigation was carried out by calculating the distance between current GPS location and active GPS location. If the calculated distance is matched with predefined distance, quadcopter will hover at target position and made to capture the image. Like this, the quadcopter will navigate through all waypoints and reaches the base station.

In the near future, the proposed algorithm will be implemented in hardware for real time application. The accuracy of this algorithm has been checked and improved using GPS sensor data and proposed techniques

References

1. Tripathi, V.K., Behera, L., Vema, N.: Design of sliding mode and backstepping controllers for a quadcopter. In: 2015 39th National Systems Conference (NSC), pp. 1–6, December 2015
2. Krajnk, T., Nitsche, M., Pedre, S., Peuil, L., Mejail, M.E.: A simple visual navigation system for an UAV. In: International Multi-Conference on Systems, Signals and Devices, pp. 1–6, December 2012
3. Whang, I.H., Hwang, T.W.: Horizontal waypoint guidance design using optimal control. IEEE Trans. Aerospace Electron. Syst. **38**(3), 1116–1120 (2012)
4. Woo, H.-J., Yoon, B.J., Cho, B.-G., Kim, J.H.: Research into navigation algorithm for unmanned ground vehicle using real time kinemtatic (RTK)-GPS. In: 2009 ICCAS-SICE, pp. 2425–2428, August 2009
5. Zhang, J.: Autonomous navigation for an unmanned mobile robot in urban areas. In: 2011 IEEE International Conference on Mechatronics and Automation, pp. 2243–2248, August 2011
6. Anitha, G., Rengarajan, M.: Algorithm development and testing of low cost waypoint navigation system. IRACST Eng. Sci. Technol.: Int. J. (ESTIJ) **3**, 411–415 (2013)
7. Kumar, P.V., Challa, A., Ashok, J., Narayanan, G.L.: GIS based fire rescue system for industries using quad copter x2014; a novel approach. In: 2015 International Conference on Microwave, Optical and Communication Engineering (ICMOCE), pp. 72–75, December 2015
8. Leong, B.T.M., Low, S.M., Ooi, M.P.-L.: Low-cost microcontroller-based hover control design of a quadcopter. In: 2012 International Symposium on Robotics and Intelligent Sensors, pp. 458–464 (2012)
9. Rajesh, S.M., Bhargava, S., Sivanathan, S.B.M.K.: Mission planning and waypoint navigation of a micro quad copter by selectable gps coordinates. Int. J. Adv. Res. Comput. Sci. Softw. Eng. (IJARCSSE) **4**, 143–152 (2014)
10. https://grabcad.com/library/drone-quadcopter-3
11. http://www.gpsinformation.org/dale/nmea.htm

Evolutionary Algorithms

Performance Evaluation of Crow Search Algorithm on Capacitated Vehicle Routing Problem

K. M. Dhanya[1(✉)] and S. Kanmani[2]

[1] Department of Computer Science and Engineering,
Pondicherry Engineering College, Puducherry, India
dhanyakm@pec.edu
[2] Department of Information Technology,
Pondicherry Engineering College, Puducherry, India
kanmani@pec.edu

Abstract. Crow Search Algorithm is a novel Metaheuristic method based on the intelligent behavior of crows. It has been used to solve some optimization problems like engineering design problems, feature extraction and classification problems but has not been applied to vehicle routing problem. In this paper, crow search algorithm has been utilized to solve capacitated vehicle routing problem and the performance of it on various size-capacitated vehicle routing problem instances is analyzed. The factorial design ANOVA is used to determine the performance of crow search algorithm under different parameter settings on capacitated vehicle routing problem instances.

Keywords: Crow Search Algorithm · Metaheuristic method
Optimization problem · Vehicle routing problem
Capacitated vehicle routing problem · Factorial design ANOVA

1 Introduction

Vehicle Routing Problem (VRP) is a challenging problem in logistics and transportation where the optimal paths for routing vehicles to different customers are to be determined [1, 2]. Capacitated Vehicle Routing Problem (CVRP) is a variant of VRP introduced by Dantzig and Ramser in 1959 [3]. The main aim of the optimization problem, CVRP is to minimize the total distance travelled by vehicles or total cost incurred on vehicles when they are routed to meet the requirements of customers under capacity constraints. To solve CVRP, various exact, heuristic and metaheuristic methods have been successfully used [4–10]. Exact methods such as integer linear programming and dynamic programming suitable for small-scale problems were applied on CVRP. Clarke and Wright Savings algorithm was one of the heuristic methods used to handle CVRP. Some of the metaheuristic methods utilized on CVRP were Artificial Bee Colony (ABC), Ant Colony Optimization (ACO), Genetic Algorithm (GA), Particle Swarm Optimization (PSO), Simulated Annealing (SA) and Variable Neighborhood Search Algorithm (VNS). Among the solution methods,

I. Zelinka et al. (Eds.): ICSCS 2018, CCIS 837, pp. 91–98, 2018.
https://doi.org/10.1007/978-981-13-1936-5_10

metaheuristic methods that are inspired from nature are found to be more efficient in producing optimal solutions to solve real world optimization problems [11, 12, 26]. Crow Search Algorithm (CSA) is a metaheuristic method that establishes good balance between exploitation and exploration [13–21]. It also has good convergence rate since it produced solutions on constrained engineering design problems within one second. These distinguishing properties of CSA provide an insight to make use of it for solving CVRP.

Crow Search Algorithm is a simple population based metaheuristic method with only two adjustable parameters, flight length and awareness probability. This study examines whether CSA performed similarly under same parameter settings on CVRP instances of varying sizes and also evaluates the parameter effects on performance of CSA in solving CVRP using factorial design ANOVA.

The rest of the paper is organized as follows: Sect. 2 presents Crow Search Algorithm on Capacitated Vehicle Routing Problem. Section 3 investigates the Parameter Effects of CSA on CVRP instances. Section 4 handles Results and Discussion of the study and Sect. 5 is the conclusion.

2   Crow Search Algorithm on Capacitated VRP

Capacitated Vehicle Routing Problem is an optimization problem whose one of the objectives is to minimize the total distance travelled by vehicles [6]. In this study, CSA is used to handle CVRP. CSA considers a flock of crows whose position and memory are initialized with randomly selected customers. Each crow is also assigned an awareness probability and a flight length. The crow solutions are generated initially by considering a vehicle from the depot. The vehicle can select a customer to be served in two manners based on the awareness probability of a randomly selected crow. In the first manner, the crow can choose a customer by following the selected crow whereas in the other manner, the crow will choose a customer randomly. The crow can move to the chosen customer only if it has not visited earlier. In case, the selected unvisited customer has a demand, which violates the capacity limit of the vehicle, the vehicle must return to the depot and another vehicle is to be send from the depot to serve the selected customer. The process is to be repeated until all the customers are served exactly once. Once the crow solution is generated, the memory of crow is to be updated with it, if the new solution has the lowest total distance covered by vehicles than the solution in the memory.

To evaluate the performance of CSA on varying size CVRP instances, three instances were randomly selected from Christofides and Eilon CVRP instances [23]. The selected instances, E-n23-k3, E-n51-k5 and E-n76-k8 were considered as small, medium and large instances respectively. The experiment was carried out on Intel Core i7 processor with 2.4 GHz and 4 GB RAM. The algorithms were implemented in C++ using Microsoft Visual Studio 2010 on 64-bit Windows 10 operating system. The CSA algorithms were executed 20 times for a maximum number of iterations, 225. The flock size of crows was initialized as 25 in each case. One of the crow solutions along with the total distance (Dist.) obtained by CSA under same parameter settings for each of the CVRP instances is shown in Table 1.

Table 1. Performance of CSA on CVRP instances

Instance	Crow solution	Dist.
E-n23-k3	0→7→4→5→11→0→10→12→0→18→9→13→15→3→16→2→1→19→20→14→17→22→21→8→6→0	902
E-n51-k5	0→17→19→24→25→18→32→36→16→30→27→0→38→9→42→40→13→12→5→34→21→0→41→22→ 46→14→37→31→47→6→29→2→0→43→26→23→7→8→50→11→49→44→39→10→0→48→45→4→ 33→15→28→20→35→1→3→0	1290
E-n76-k8	0→45→71→63→7→48→13→49→16→18→46→55→42→39→0→4→66→27→37→20→36→61→28→0→ 58→29→11→47→62→22→64→24→0→43→41→15→57→10→51→68→69→6→60→ 70→30→0→34→53→23→25→50→9→32→40→0→21→19→75→3→56→1→ 73→5→35→44→0→74→2→14→31→52→54→59→0→67→38→26→17→33→8→65→12→72→0	2086

3 Parameter Analysis of Crow Search Algorithm on CVRP

Parameters are the performance determining factors in an algorithm, which can have different values [27, 28]. CSA has only two parameters, awareness probability and flight length. The goal of the study was to find out whether different parameter settings affect the performance of CSA algorithm on CVRP instances. Hypothesis was formulated as the first step for testing the effects of parameters on performance of CSA [29]. Three set of Hypothesis for each CVRP instances were tested; the first two sets were used to determine the influence of each parameters and the third set to evaluate the interaction of the parameters on the performance of CSA. Each set comprised a Null Hypothesis (H_0) and an Alternative Hypothesis (H_1). The hypothesis sets considered in this study were as follows:

Flight Length (FL):

H_0: There is no significant difference on performance of CSA based on flight length.
H_1: There is significant difference on performance of CSA based on flight length.

Awareness Probability (AP):

H_0: There is no significant difference on performance of CSA based on awareness probability.
H_1: There is significant difference on performance of CSA based on awareness probability.

Interaction (FLxAP):

H_0: There is no significant interaction of awareness probability and flight length on performance of CSA.
H_1: There is significant interaction of awareness probability and flight length on performance of CSA.

Then, a significance level of 0.05 was set for this testing. Factorial Design Analysis of Variance was determined as the test static to analyze the effect of CSA parameters and their interactions on CVRP instances [22, 25, 27, 30]. Here, two factors, awareness probability and flight length determined the performance of the algorithm. Each factor considered different levels and the levels used by the parameters in this work were as follows: Awareness probability with five levels represented by numbers starting from 0.1 at an interval of 0.2 and Flight Length with five levels represented by numbers starting from 0.5 at an interval of 0.5.

94   K. M. Dhanya and S. Kanmani

In this study, four replications of CSA were considered by running the algorithm four times with different random seeds under same parameter settings and their mean performance was calculated. The mean of the performances of CSA obtained under varying parameter configurations on each of the three instances are shown in Table 2.

Table 2. Mean performance of CSA on CVRP instances

	Flight length	Awareness probability				
		0.1	0.3	0.5	0.7	0.9
E-n23-k3	0.5	925.5	930.75	951.25	954	936.25
	1.0	876	899.75	902.5	922	925
	1.5	904.5	922	921.5	928.75	948
	2.0	881	909.25	923.75	922.5	921
	2.5	904.25	921.25	912	910.5	927.75
E-n51-k5	0.5	1327.5	1303	1328	1303.75	1329.5
	1.0	1344.75	1330	1299.75	1330.25	1339.25
	1.5	1334.25	1324.25	1302.5	1334.5	1327.5
	2.0	1328.75	1309.5	1328	1313	1327.75
	2.5	1318	1315.5	1320	1322	1307.25
E-n76-k8	0.5	2108	2116.25	2111.5	2124	2129.75
	1.0	2138	2121.75	2095.5	2140	2154
	1.5	2130.5	2141	2092.5	2132	2129.5
	2.0	2126	2123	2101.5	2140.5	2141.75
	2.5	2087.25	2143.5	2094.5	2126.75	2129

Analysis of Variance (ANOVA) for factorial design of each of the CVRP instances was prepared and tabulated in Table 3. The first column of the table represents sources of variation (SV); that is the variation may be due to flight length, awareness probability and their interaction or due to chance (Error) and the second column provides their degree of freedom (df). For each instance, the sum of squares (SS) and the corresponding mean sum of squares (MSS) for sources of variation were computed. Then, F-value (F) of each factors and their interaction were calculated.

The F-static table values of $_{75}F^4$ and $_{75}F^{16}$ at 5% level of significance are 2.494 and 1.78 [24]. In case of E-n23-k3 instance, the computed value for the hypothesis concerning FL and AP were greater than the corresponding tabulated values. Hence, the first two null hypotheses were rejected and it could be interpreted that the performance of CSA was affected by FL and AP. The value computed for the interaction hypothesis was smaller than the value obtained from the F-statistic table. So, that null hypothesis was accepted which implied that the performance of CSA was not influenced by the interaction of FL and AP.

The hypothesis testing performed on E-n51-k5 instance produced values, which were smaller than that of F-static table values. Therefore, the null hypothesis was acceptable in each cases leading to the conclusion that there was no significant difference on the performance of CSA by parameters and their interaction.

Table 3. ANOVA result on CVRP instances

SV	df	E-n23-k3			E-n51-k5			E-n76-k8		
		SS	MSS	F	SS	MSS	F	SS	MSS	F
FL	4	14461.84	3615.46	5.081	1927.34	481.83	0.988	2722.26	680.57	0.852
AP	4	13562.74	3390.68	4.765	3313.64	828.41	1.699	18348.26	4587.06	5.745
FL x AP	16	5728.16	358.01	0.503	9208.16	575.51	1.181	10150.54	634.41	0.795
Error	75	53363.50	711.51		36562.50	487.5		59883.50	798.44	
Total	99	87116.24			51011.64			91104.56		

On E-n76-k8 instance, it was found that the values computed for hypothesis set concerning AP were greater than the table value and the values obtained for the other two hypothesis sets were smaller than the table value. Therefore, the null hypothesis of AP was rejected and that of other two cases were accepted. It could be concluded that the performance of CSA was influenced by the parameter, AP only.

4   Results and Discussion

CSA is a swarm based intelligent method with only two parameters and can be implemented easily. It is so simple compared to other metaheuristic methods like Ant Colony Optimization, Genetic Algorithm and Particle Swarm Optimization, which possess more number of parameters. As the number of parameters increases, parameter space of the algorithm increases in exponential manner, which makes the algorithm more complex. The parameter effects on the performance of CSA for varying size CVRP instances are illustrated in Figs. 1, 2 and 3.

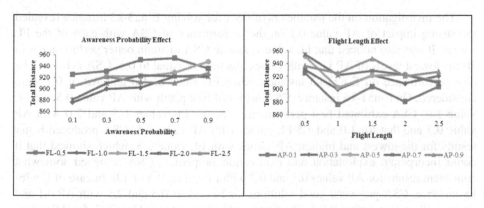

Fig. 1. Parameter effects on performance of CSA on E-n23-k3 instance

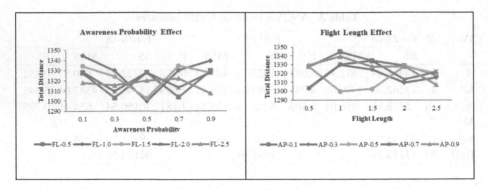

Fig. 2. Parameter effects on performance of CSA on E-n51-k5 instance

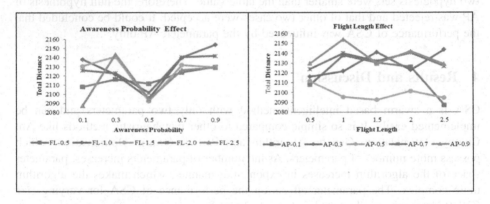

Fig. 3. Parameter effects on performance of CSA on E-n76-k8 instance

The investigation on the parameters of CSA for solving E-n23-k3 instance revealed the strong impact of AP value 0.1 on the performance of CSA with each of the FL values. It was also noticed that FL value 1.0 made CSA to attain better performances in all the lowest levels of AP i.e., with values less than or equal to 0.5. CSA achieved the best performance on E-n23-k3 instance, when FL value of 1.0 and AP of 0.1 were considered. On E-n51-k5 instance, CSA achieved best result with AP value 0.5 and FL value 1.0. CSA exhibited best performances for the FL values 0.5 and 2.0 with AP value 0.3 and that for 1.0 and 1.5 FL values with AP value 0.5. CSA produced better results for the lowest and highest AP values with FL value 2.5, which showed that it could incorporate exploitation and exploration properties. CSA achieved somewhat consistent results for AP values 0.3 and 0.7 with FL values 0.5 or 1.0. In case of E-n76-k8 instance, CSA produced good solutions for FL values 0.5 and 2.5 with AP 0.1 and in other FL values with AP 0.5. The best results were achieved by CSA for AP values 0.3 and 0.7 with FL 0.5 and that for AP values 0.5 and 0.9 with FL values 1.5 and 2.5 respectively. CSA achieved better performance with AP 0.1 and FL 2.5 on that instance.

5 Conclusion

CSA is a simple metaheuristic method with only two parameters and the influence of those parameters on the performance of it in solving various size-CVRP instances is investigated by using factorial design ANOVA. The performance analysis carried out on CVRP instances has shown that both the parameters of CSA have great significance on the performance of it on small CVRP instance whereas they do not produce any significant difference on its performance on medium size CVRP instance. In case of large CVRP instance, AP has more significance on the performance of CSA than FL. Hence, it can be concluded that parameter effects on the performance of CSA is problem dependent. As future work, the performance of CSA on other variants of VRP can be measured by computing time or performance deviations.

References

1. Lin, C., Choy, K.L., Ho, G.T.S., Chung, S.H., Lam, H.Y.: Survey of green vehicle routing problem: past and future trends. Expert Syst. Appl. **41**, 1118–1138 (2014)
2. Mazzeo, S., Loiseau, I.: An ant colony algorithm for the capacitated vehicle routing. Electron. Notes Discret. Math. **18**, 181–186 (2004)
3. Hosseinabadi, A.A.R., Rostami, N.S.H., Kardgar, M., Mirkamali, S., Abraham, A.: A new efficient approach for solving the capacitated vehicle routing problem using the gravitational emulation local search algorithm. Appl. Math. Model. **49**, 663–679 (2017)
4. Szeto, W.Y., Wu, Y., Ho, S.C.: An artificial bee colony algorithm for the capacitated vehicle routing problem. Eur. J. Oper. Res. **215**(1), 126–135 (2011)
5. Ng, K.K.H., Lee, C.K.M., Zhang, S.Z., Wu, K., Ho, W.: A multiple colonies artificial bee colony algorithm for a capacitated vehicle routing problem and re-routing strategies under time-dependent traffic congestion. Comput. Ind. Eng. **109**, 151–168 (2017)
6. Nazif, H., Lee, L.S.: Optimised crossover genetic algorithm for capacitated vehicle routing problem. Appl. Math. Model. **36**(5), 2110–2117 (2012)
7. Kachitvichyanukul, V.: Particle swarm optimization and two solution representations for solving the capacitated vehicle routing problem. Comput. Ind. Eng. **56**(1), 380–387 (2009)
8. Wei, L., Zhang, Z., Zhang, D., Leung, S.C.: A simulated annealing algorithm for the capacitated vehicle routing problem with two-dimensional loading constraints. Eur. J. Oper. Res. **265**(3), 843–859 (2018)
9. Amous, M., Toumi, S., Jarboui, B., Eddaly, M.: A variable neighborhood search algorithm for the capacitated vehicle routing problem. Electron. Notes Discret. Math. **58**, 231–238 (2017)
10. Kir, S., Yazgan, H.R., Tüncel, E.: A novel heuristic algorithm for capacitated vehicle routing problem. J. Ind. Eng. Int. **13**(3), 323 (2017)
11. Kar, A.K.: Bio inspired computing – a review of algorithms and scope of applications. Expert Syst. Appl. **59**, 20–32 (2016)
12. Boussaïd, I., Lepagnot, J., Siarry, P.: A survey on optimization metaheuristics. Inf. Sci. **237**, 82–117 (2013)
13. Askarzadeh, A.: A novel metaheuristic method for solving constrained engineering optimization problems: crow search algorithm. Comput. Struct. **169**, 1–12 (2016)

14. Abdelaziz, A.Y., Fathy, A.: A novel approach based on crow search algorithm for optimal selection of conductor size in radial distribution networks. Eng. Sci. Technol. Int. J. **20**(2), 391–402 (2017)
15. Hinojosa, S., Oliva, D., Cuevas, E., Pajares, G., Avalos, O., Gálvez, J.: Improving multi-criterion optimization with chaos: a novel Multi-Objective Chaotic Crow Search Algorithm. Neural Comput. Appl. 1–17 (2017)
16. Marichelvam, M.K., Manivannan, K., Geetha, M.: Solving single machine scheduling problems using an improved crow search algorithm. Int. J. Eng. Technol. Sci. Res. **3**, 8–14 (2016)
17. Nobahari, H., Bighashdel, A.: MOCSA: a multi-objective crow search algorithm for multi-objective optimization. In: 2017 2nd Conference on Swarm Intelligence and Evolutionary Computation (CSIEC), pp. 60–65. IEEE (2017)
18. Oliva, D., Hinojosa, S., Cuevas, E., Pajares, G., Avalos, O., Gálvez, J.: Cross entropy based thresholding for magnetic resonance brain images using Crow Search Algorithm. Expert Syst. Appl. **79**, 164–180 (2017)
19. Rajput, S., Parashar, M., Dubey, H.M., Pandit, M.: Optimization of benchmark functions and practical problems using Crow Search Algorithm. In: 2016 Fifth International Conference on Eco-Friendly Computing and Communication Systems, pp. 73–78. IEEE (2016)
20. Sayed, G.I., Hassanien, A.E., Azar, A.T.: Feature selection via a novel chaotic crow search algorithm. Neural Comput. Appl. 1–18 (2017)
21. Turgut, M.S., Turgut, O.E.: Hybrid artificial cooperative search-crow search algorithm for optimization of a counter flow wet cooling tower. Int. J. Intell. Syst. Appl. Eng. **5**(3), 105–116 (2017)
22. Pabico, J.P., Albacea, E.A.: The interactive effects of operators and parameters to GA performance under different problem sizes. arXiv preprint arXiv:1508.00097 (2015)
23. [Dataset] Neo Networking and Emerging Optimization Research Group (2013). Capacitated VRP Instances. http://neo.lcc.uma.es/vrp/vrp-instances/capacitated-vrp-instances/
24. F Distribution Table. http://www.itl.nist.gov/div898/handbook/eda/section3/eda3673.htm
25. Two way ANOVA. https://www3.nd.edu/~rwilliam/stats1/x61.pdf
26. Talbi, E.G.: Metaheuristics: From Design to Implementation, vol. 74. Wiley, Hoboken (2009)
27. Bohrweg, N.: Sequential parameter tuning of algorithms for the vehicle routing problem (2013)
28. Silberholz, J., Golden, B.: Comparison of metaheuristics. In: Gendreau, M., Potvin, J.Y. (eds.) Handbook of Metaheuristics. International Series in Operations Research & Management Science, vol. 146, pp. 625–640. Springer, Boston (2010). https://doi.org/10.1007/978-1-4419-1665-5_21
29. Kothari, C.R.: Research Methodology: Methods and Techniques. New Age International, New Delhi (2004)
30. Héliodore, F., Nakib, A., Ismail, B., Ouchraa, S., Schmitt, L.: Performance evaluation of metaheuristics. In: Metaheuristics for Intelligent Electrical Networks, pp. 43–58 (2017)

Ultrasonic Signal Modelling and Parameter Estimation: A Comparative Study Using Optimization Algorithms

K. Anuraj$^{(\boxtimes)}$, S. S. Poorna, and C. Saikumar

Department of Electronics and Communication Engineering, Amrita Vishwa Vidyapeetham,
Amritapuri, Kollam, India
anuraj19@gmail.com, poorna.surendran@gmail.com

Abstract. The parameter estimation from ultrasonic reverberations is used in applications such as non-destructive evaluation, characterization and defect detection of materials. The parameters of back scattered Gaussian ultrasonic echo altered by noise: Received time, Amplitude, Phase, bandwidth and centre-frequency should be estimated. Due to the assumption of the nature of noise as additive white Gaussian, the estimation can be approximated to a least square method. Hence different least square cure-fitting optimization algorithms can be used for estimating the parameters. Optimization techniques: Levenberg-Marquardt(LM), Trust-region-reflective, Quasi-Newton, Active Set and Sequential Quadratic Programming are used to estimate the parameters of noisy echo. Wavelet denoising with Principal Component Analysis is also applied to check if it can make some improvement in estimation. The goodness of fit for noisy and denoised estimated signals are compared in terms of Mean Square Error (MSE). The results of the study shows that LM algorithm gives the minimum MSE for estimating echo parameters from both noisy and denoised signal, with minimum number of iterations.

Keywords: Gaussian echo · Estimation · Wavelet denoising
Maximum likelihood estimation · Least square · Optimization · Wavelet
Denoising · Principal component analysis · MSE

1 Introduction

The information from back scattered ultrasonic echo carries very crucial details pertaining to the characterization and nature of the materials used as the reflectors. These methods usually assumes a model for the echo signals and based on the nature of the back scattered echo from the reflector material, various parameters of the echo will be estimated and hence the characterization of the material. This method is also used in flaw or defect identification and to quantitatively evaluate the structural integrity of the material [1]. These techniques are also applied to medical applications in tissue characterization like accessing bone quality and risk of fracture [2]. Kyung and Young Jhang [3] provide a brief review of various ultrasonic techniques for nondestructive evaluation of damages in materials. The echo is nonlinear, characterized using a Gaussian model,

© Springer Nature Singapore Pte Ltd. 2018
I. Zelinka et al. (Eds.): ICSCS 2018, CCIS 837, pp. 99–107, 2018.
https://doi.org/10.1007/978-981-13-1936-5_11

with parameters: Received time, Amplitude, Phase, bandwidth and centre frequency. An initial guess for these parameters are assumed and white noise is added to the same to characterize the reflected signal. The noisy backscattered signal will undergo suitable pre-processing techniques and further subjected to various iterative estimation algorithms. Similar study of signal modelling and parameter estimation using Gaussian echo, using Levenberg-Marquardt (LM) curve fitting algorithm was carried out by Laddada et al. [4]. In their work, an improved mean squared error was obtained for LM algorithm. This work was compared with the results obtained using the optimization algorithms viz. Gauss Newton and simplex algorithms by Demirli et al. [5]. Another derivative free method of estimation- simplex, which provides a better global minimum, also applies to this problem of least square estimation [4, 5, 8]. Noninvasive estimation can also be done effectively without initial guess using Genetic algorithm [6].

Even though the optimization on LM and Gauss Newton had been tried on ultrasonic parameter estimation problems, in this paper we aim to implement the least square estimation using the optimization methods: Trust-region-reflective, Quasi-Newton, Active Set and Sequential Quadratic Programming. A detailed review on nonlinear least square optimization methods are described in the article by Dennis [7]. The paper is organized as follows: Sect. 2 gives the theoretical description of signal modeling and maximum likelihood estimation. A brief description of wavelet denoising and optimization methods are given in Sects. 3 and 4. Section 5 gives the comparison of results and discussion, followed by conclusion.

2   Signal Modelling and Maximum Likelihood Based Parameter Estimation

Noninvasive testing of material defects makes use of ultrasonic reverberations from the analyzing material. The parameters of back scattered echoes can covey some valuable information regarding the material features. In order to capture these hidden material features, echo parameters are to be estimated by Maximum Likelihood Estimation (MLE) which is less complicated and capable of handling huge data sets. Since this MLE estimation requires a predefined model, we assume the ultrasonic reverberations as Gaussian shaped and is represented as,

$$r(t) = x(\theta, t) + w(t) \tag{1}$$

where r(t) is the received echo, w(t) is assumed to be white Gaussian noise since the estimation and optimization is done by using the simplest AWGN channel model. The parameters of transmitted ultrasonic pulse are represented in vector form as given in Eq. 2.

$$\theta = \left[A, \beta, \tau_d, \varphi_d, f_0, t \right] \tag{2}$$

where A the amplitude, β bandwidth, τ_d received time of echo after reflection, φ_d phase of the echo and time of flight t are the parameters to be estimated. The ultrasonic reverberations, which has a Gaussian form is modeled as

$$x(\theta, t) = Ae^{\beta(t-\tau_d)^2}\cos\left(2\pi f_0\left(t-\tau_d\right) + \varphi_d\right) \tag{3}$$

Considering AWGN to be independent and identically distributed, the received observations follow a normal distribution with joint probability density function given in Eq. 4.

$$f(r, \theta) = \frac{1}{(2\pi)^{\frac{N}{2}}|\text{cov}(\theta)|^{\frac{1}{2}}}e^{\left(-\frac{1}{2}(r-\mu(\theta))^T\text{cov}^{-1}(\theta)(r-\mu(\theta))\right)} \tag{4}$$

For white Gaussian noise with zero mean, the vector representation of mean $\mu(\theta)$ is $x(\theta)$ and hence the covariance is

$$\text{cov}(\theta) = E\left[(r - x(\theta))(r - x(\theta))^T\right] \tag{5}$$

MLE estimation is performed by maximizing the log likelihood function as given in Eq. 6.

$$\theta_{\text{estimated}} = \arg\ \max_{\theta} f(r, \theta) \tag{6}$$

Logarithm of likelihood function, $f(r, \theta)$ attains its maximum at the same point as it has a monotonically increasing nature.

$$\ln f(r, \theta) = -\frac{N}{2}\ln(2\pi) - \frac{1}{2}\ln(\text{cov}(\theta) - \frac{1}{2}\text{cov}^{-1}(\theta)D(\theta) \tag{7}$$

where

$$D(\theta) = (r - x(\theta)(r - x(\theta))^T = \|r - x(\theta)\|_2^2 \tag{8}$$

is the objective function. Hence MLE reduces to a least square minimization problem and Eq. (6) can be written as

$$\theta_{\text{estimated}} = \arg\ \min_{\theta} D(\theta) \tag{9}$$

Here the parameter estimation problem simplifies to a nonlinear function ($D(\theta)$) minimization problem and can solve by applying various optimization methods.

3   Wavelet Denoising

Wavelet denoising technique uses a multi resolution analysis technique, in which set of approximation and wavelet filters followed by down samplers are used in the analysis path. The filters in this work used Symlet-29 wavelet and the denoising was done at level 4. Denoising method usually uses a threshold function in the wavelet coefficients, retain the signal of interest at a sparse level and remove the noise components. Choosing the components, with maximum variance using PCA, still lowers the variability of noise, thereby providing denoising effect [11].

4   Brief Description of Optimization Methods

Optimization algorithms are applied to least square curve fitting problems, since an optimum solution is not possible to these problems on single iteration. Usually these algorithms will not converge to a global minima unless the initial values are wisely chosen and adequate number of iterations are performed. Many algorithms are available for iteratively solving this linear and non-linear least square curve fit and obtain the global minimum solution. Thrust region reflective, Levenberg- Marquardt, Active set, Quasi Newton and Sequential quadratic programming are some of these. The following sections briefly explains the different optimization algorithms:

4.1   Trust-Region Reflective (TRR) Least Squares Algorithm

This algorithm minimizes the non-linear objective function, assuming it to be bounded. Here for finding the optimized value of the objective function $D(\theta)$ we approximate it to a quadratic function $r(y)$, which actually represents $D(\theta)$ in its neighbourhood, so called as the thrust region sub problem [9]. Now minimizing the area $r(y)$ in iterative steps such that if $f(x + y) < f(x)$ then x is updated to $x + y$ else x remain unchanged and the thrust region dimension is adjusted. Thrust region problem is mathematically given in Eq. (10).

$$\min\left\{\frac{1}{2}y^T Hy + y^T g\right\} \quad | \quad \|\text{Diag}(y)\| \le \in \tag{10}$$

where H is the Hessian, g the gradient, Diag the diagonal scaling matrix, $\in$ positive scalar and $\|.\|$ the L2 norm.

4.2   Levenberg- Marquardt (LM) Algorithm

LM algorithm [4] starts from a set of initial assumption of parameters and redefining them iteratively by successive approximation,

$$\theta_{i+1} \approx \theta_i + \in \tag{11}$$

where parameter vector in i-th iteration and shift vector are represented by θ_i and $\in$ respectively. To estimate the shift vector, the algorithm tries to linearize the vector by a first order taylor series approximation during each iteration as given in equations below

$$D(\theta + \in) = \left\|r - x(\theta_i + \in)\right\|_2^2 \tag{12}$$

$$x(t, \theta_{i+1}) = x(t, \theta_i) + J_i \in \tag{13}$$

Where J_i is the Jacobian matrix. By substitution of Eq. (13) in Eq. (12) and approximate the derivative with respect to the shift vector as zero yields Eq. (14)

$$(r - x(t, \theta_i)) = J_i^T J_i \in \tag{14}$$

The above equation is linear in nature and can be solved to get $\in$. A damped version is introduced by Levenberg. Marquardt modified the relation by giving maximum scaling along a direction with minimum gradient there by eliminating slow convergence along minimum gradient. With the above assumptions the shift vector $\in$ can be derived as given in Eq. (15),

$$\in = (J_i^T J_i + \lambda \mathrm{diag}(J_i^T J_i))^{-1} J_i e_i \tag{15}$$

Where e_i is error vector and λ represents the damping factor. Hence the Levenberg-Marquardt algorithm is represented in Eq. (16)

$$\theta_{i+1} = \theta_i - (J_i^T J_i + \lambda \mathrm{diag}(J_i^T J_i))^{-1} J_i e_i \tag{16}$$

4.3 Active Set (aS) Method

This method finds the most feasible solution for the constrained objective function, with the help of inequality constraints. A feasible region near the possible solutions of x is identified such that $c_i(x) \geq 0$, where c_i's are the constraint functions. A matrix, called active set S_k can be from these constraint functions. A global solution to this problem can be achieved with the help of Lagrange Multipliers of S_k, by solving Karush-Kuhn-Tucker (KKT) [10] equations.

4.4 Quasi Newton (QN) Method

This method is a sub class of variable metric methods and is based on Newton's method. To find the extrima of an objective function where the gradient is zero. Quasi Newton method is used in cases where the Hermissian matrix computation is expensive. Rob Haelterman [12] gives a secant based method for least squared optimization using Quasi Newton Method.

4.5 Sequential Quadratic Programming(SQP)

SQP can be applied to optimization problems in cases where the objective function is twice differentiable with respect to the constraints. This method reduces Newton's method, when the gradient are removed. This method uses a quadratic approximation for the objective function [13].

5   Results and Discussion

The performance of the above explained algorithms are analyzed on an echo signal, with Gaussian nature, generated according Eq. 3. given in Fig. 1 (a). The parameters are chosen for the echo as amplitude $A = 1$, bandwidth $\beta = 25(MHz)^2$, received time of echo after reflection $\tau_d = 1\mu s$, center frequency $f_0 = 5\,MHz$, and phase of the echo $\varphi_d = 1$ rad and hence the vector is $X = [1\ 25\ 1\ 5\ 1]$. The Noise altered echo with white noise of SNR 5 dB is as given in Fig. 1(b). The parameters of this noisy signal are to be estimated, using the above optimization methods and the performance of these methods are compared, assuming the initial condition for the parameters as $X_0 = [1\ 15\ 0.7\ 3\ 0.8]$. The same optimization methods were also analyzed after denoising the noise corrupted Gaussian echo. Similar to the work specified in [4], we also used the multivariate wavelet denoising using principal component analysis (PCA). Different mother wavelets were tried out and Symlet-29 wavelet function was found to be appropriate. Soft threshold function was applied at level 4 decomposition for denoising. The denoised echo is shown in Fig. 1(c). Estimated signal using the optimization techniques: Active set, Sequential quadratic programming, Quasi Newton and Trust region reflective are indicated in Fig. 3(a to d) respectively. Mean square error (MSE) based comparisons for the estimated echo, was done with and without denoising, using different optimization algorithms.

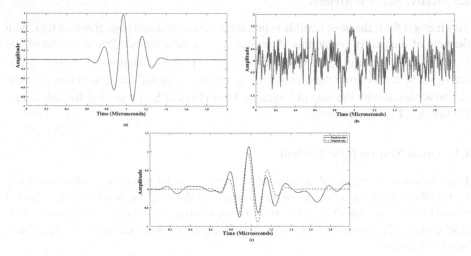

Fig. 1. (a) Gaussian echo (b) AWGN with SNR 5 dB (c) denoised echo

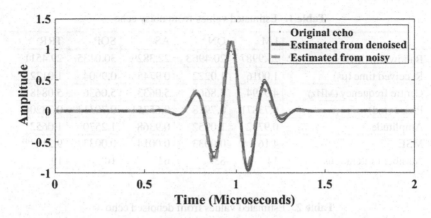

Fig. 2. Estimated signal using Levenberg-Marquardt (LM) Algorithm

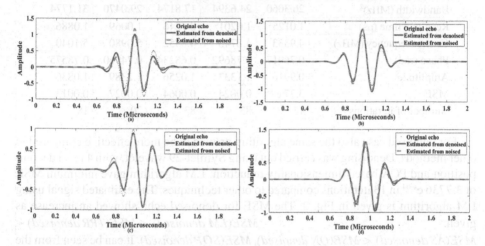

Fig. 3. Estimated signal using optimization algorithms (a) Active Set (b) Sequential Quadratic Programming (c) Quasi Newton (d) Trust Region Reflective

Tables 1 and 2 gives the results of Least square curve fit for noisy and denoised Gaussian echo, using different algorithms: Levenberg-Marquardt, Trust-region-reflective, Quasi-Newton, Active Set and Sequential Quadratic Programming. The goodness of fit was evaluated in terms of Mean Square Error (MSE) between the estimated noisy or denoised signal and the original signal. From Table 1, it can be inferred that among the optimization techniques used for noisy echo, Levenberg- Marquardt Algorithm gave the lowest MSE of $4.16\,e^{-04}$ in 14 iterations. Comparing this result with the other techniques, estimated noisy echo, MSE increases as: $MSE(LM) < MSE(TRR) < MSE(QN) < MSE(SQP) < MSE(AS)$.

Table 1. Estimated values from noisy echo

	LM	QN	AS	SQP	TRR
Bandwidth (MHz)^2	23.9987	20.4963	22.3829	30.0135	29.4511
Received time (µs)	1.0016	1.0222	0.9745	0.9904	1.0172
Centre frequency (MHz)	4.9194	4.8669	5.0633	5.0636	5.0848
Phase (rad)	0.73871	0.7918	0.65351	0.70314	0.78206
Amplitude	0.9732	1.0952	0.9368	1.2570	1.0652
MSE	4.16 e^{-04}	0.0033	0.0014	0.0033	9.21 e^{-04}
Number of iterations	14	83	62	10	19

Table 2. Estimated values from denoised echo

	LM	QN	AS	SQP	TRR
Bandwidth (MHz)^2	26.3060	24.6394	17.8174	29.0470	31.1774
Received time (µs)	1.0125	1.0201	0.9846	1.0069	1.0886
Centre frequency (MHz)	4.9333	4.8740	4.6352	5.0980	5.1040
Phase (rad)	0.76917	0.78692	0.68234	0.74960	0.78575
Amplitude	0.9916	1.1337	1.0256	1.2389	1.0836
MSE	3.77 e^{-04}	0.0024	0.0064	0.0027	0.0013
Number of iterations	13	83	83	20	18

For denoised case also the same algorithm proved to be more effective compared to other methods. Denoising was carried out using Symlet-29 wavelet, with 4 level decomposition and PCA based dimensionality reduction. LM algorithm gave minimum MSE of 3.7736 e^{-04} in 13 iterations compared to other techniques. The estimated signal using LM algorithm is given in Fig. 2. The MSE for denoised echo showed an increase as given: $MSE(LM\ denoised) < MSE(TRR\ denoised) < MSE(AS\ denoised) < MSE(QN\ denoised), MSE(SQP\ denoised)$. It can be seen from the MSE values that denoising method will reduce the MSE for most of the algorithms considered. Even though SQP uses less number of iterations for estimating the noisy signal parameters, the MSE is high when compared to LM. Further in LM method, the relative error in amplitude decreases from 2.68% to 0.84%, while denoising.

6   Conclusion

The paper gives a comparison of different optimization techniques for parameter estimation. The parameters, received time, amplitude, phase, bandwidth and centre frequency are estimated for a Gaussian echo. Different optimization algorithms: Levenberg-Marquardt, Trust-region-reflective, Quasi-Newton, Active Set and Sequential Quadratic Programming are applied for estimation. The above optimization methods were compared using Mean square error for goodness of fit of the estimated signal. The Levenberg Marquardt algorithm gave the best estimate in less number of iterations with minimum MSE for both noisy and denoised echo. MSE reduced when a wavelet based denoising was applied before estimation. The initial assumptions of the parameters always comes with practice and prior knowledge. These techniques will take more number of iterations to converge and hence need additional computational time unless the initial assumptions regarding the parameters are accurate. This work would be extended to the analysis using other types of optimization techniques as well as by using different signal models. Further we can investigate the estimation and optimization of noisy echoes parameters with various SNRs.

References

1. Lu, Y., Saniie, J.: Model-based parameter estimation for defect characterization in ultrasonic NDE applications. In: IEEE International Ultrasonics Symposium (IUS), Taipei, pp. 1–4 (2015). https://doi.org/10.1109/ultsym.2015.0342
2. Marutyan, K.R., Anderson, C.C., Wear, K.A., Holland, M.R., Miller, J.G., Bretthorst, G.L.: Parameter estimation in ultrasonic measurements on trabecular bone. In: AIP Conference Proceedings, vol. 954, no. 1, pp. 329–336. AIP (2007)
3. Jhang, K.-Y.: Nonlinear ultrasonic techniques for nondestructive assessment of micro damage in material: a review. Int. J. Precis. Eng. Manuf. 10(1), 123–135 (2009)
4. Laddada, S., Lemlikchi, S., Djelouah, H., Si-Chaib, M.O.: Ultrasonic parameter estimation using the maximum likelihood estimation. In: 4th International Conference on Electrical Engineering (ICEE), Boumerdes, pp. 1–4 (2015). https://doi.org/10.1109/intee.2015.7416791
5. Demirli, R., Saniie, J.: Model-based estimation of ultrasonic echoes. part I: analysis and algorithms. IEEE Trans. Ultrason. Ferroelectr. Freq. Control 48(3), 787–802 (2001). https://doi.org/10.1109/58.920713G
6. Liu, Z., Bai, X., Pan, Q., Li, Y., Xu, C.: Ultrasonic echoes estimation method using genetic algorithm. In: IEEE International Conference on Mechatronics and Automation, Beijing, pp. 613–617 (2011). https://doi.org/10.1109/icma.2011.5985731
7. Dennis Jr., J.E.: Nonlinear least-squares. In: Jacobs, D. (ed.) State of the Art in Numerical Analysis. Academic Press, London (1977)
8. Nelder, J.A., Mead, R.: A simplex method for function minimization. Comput. J. 7, 308–313 (1967)
9. Le, T.M., Fatahi, B., Khabbaz, H., Sun, W.: Numerical optimization applying trust-region reflective least squares algorithm with constraints to optimize the non-linear creep parameters of soft soil. Appl. Math. Model. 41, 236–256 (2017)
10. Feng, G., Lin, Z., Yu, B.: Existence of an interior pathway to a Karush-Kuhn-Tucker point of a nonconvex programming problem. Nonlinear Anal. Theory Method. Appl. 32(6), 761–768 (1998)
11. Aminghafari, M., Cheze, N., Poggi, J.-M.: Multivariate denoising using wavelets and principal component analysis. Comput. Stat. Data Anal. 50(9), 2381–2398 (2006)
12. Haelterman, R., Degroote, J., Van Heule, D., Vierendeels, J.: The quasi-Newton least squares method: a new and fast secant method analyzed for linear systems. SIAM J. Numer. Anal. 47(3), 2347–2368 (2009). https://doi.org/10.1137/070710469
13. Nocedal, J., Wright, S.: Numerical Optimization. Springer, Heidelberg (2006). https://doi.org/10.1007/978-0-387-40065-5

Image Processing

A Histogram Based Watermarking for Videos and Images with High Security

P. Afeefa[⊠] and Ihsana Muhammed

Computer Science Engineering, RCET, Thrissur, Kerala, India
afeefa4449@gmail.com, ihsanamuhd@royalcet.ac.in

Abstract. The world of digital Images and videos have more importance in past two decades. Hence it is very important to provide high security for multimedia content. We can use digital watermarking for this purpose. Histogram based watermarking can be used to provide highly secure and robust watermarking for images. The same technic is applicable for videos also. Videos. The shape of histogram of image is manipulated to for the efficient insertion of digital co watermarks to the images. The movement of pixels from one region to other will remove the side effect of Gaussian filtering. The proposed system is highly robust as it makes changes to the histogram of the image. The use of secret key improves the security of the proposed system. Also the imperceptiblity of watermarking makes the scheme perfect. Here the three step scenario for watermarking makes a good protection system for digital contents. In video watermarking the video frames are separated before inserting the watermarks. In the case of images the histogram of the image is computed in various gray levels. The proposed system is powerful tool against attacks.

Keywords: Histogram · Digital watermarking · Gaussian filtering

1 Introduction

Nowadays Digital world of photography is an important field that grows rapidly. Communication using multimedia content is very usual thin gin this era. Hence we have to provide security and integrity for the multimedia content. When sharing the images through internet we should provide high security i.e. Integrity, Authentication, Confidentiality to the contents presents in the images there are variety of technics to provide security of multimedia content.

Usually the digital Images and videos are need protection methods which are different in nature. Here we are providing a good watermarking scheme which can be used for both videos and audios. Hence the computation complexity and overheads to provide integrity for multimedia content can be minimised.

Usually a watermarking scheme have two main phases watermark embedding phase and watermark Detection phase. The watermark embedding phase include insertion of watermark to the image. Detection of watermark is the process of extraction of watermark from the transmitted data.

By using encryption and digital watermarking we can provide the confidentiality, authenticity and to the images. Rank based watermarking, quantization based

© Springer Nature Singapore Pte Ltd. 2018
I. Zelinka et al. (Eds.): ICSCS 2018, CCIS 837, pp. 111–115, 2018.
https://doi.org/10.1007/978-981-13-1936-5_12

watermarking, watermarking using moment, etc. are various methods of image watermarking. In this paper we are concentrated on histogram based watermarking. The watermark is embedded in to the image by changing the shape of the histogram of image.

Watermarking is used to hiding the information such as hide secret information in digital media like photographs, digital music, or digital video. Nowadays which has seen a lot of research images are need to be stored for future reference. For medical image security, when the image is interest.

The proposed system can be used in application like medical image protection, image authentication purpose, and other image protection systems. The proposed system is very power full tool in image protection schemes.

A digital Image watermark is robust if it can be extracted from the received image reliably. That is without any change as compared with the data inserted. And the watermark is said to be Imperceptible if the watermarked content is perceptually equal to the un watermarked content. The proposed system have this two properties well.

The video protection schemes can be used in authentication purpose to avoid illegal transformation and manipulation of multimedia content.

2 Related Works

Hyoung Joong Kim explains an image watermarking Based on Invariant Region [1]. It is a watermarking scheme based on the distinct regions of images. They are identifying the invariant regions based on the harris detector. It is a good watermarking method. In this watermarking the transformation for the image is chosen based on the shape of the invariant regions. Thus it cant be used for object with complex shapes. Ping Dong, Jovan G in [2] explains two digital watermarking. One method is based on mesh elements. That is image patch elements are considered as the mesh elements. This watermarking needs original host image to compare it with the received image. The other method is based on the normalisation. Image is normalised to attain a threshold value of moment. This methods are only used for private watermarking. They are not capable of handling public watermarking schemes.

3 Proposed System

The proposed system of watermarking is a very powerful tool against attacks like signal processing attacks and cropping and random blending attacks. Here the host image is considered as a signal. The main phase of the watermark embedding system is histogram construction and selection of pixels to embed the watermarks.

In this proposed system the watermark embedding and detection phases have some sub modules. There are mainly four phases in this system:

1. Image preprocessing
2. Histogram construction
3. Select pixels
4. Embedding watermark

Similarly the watermark detection phase also have the following phases like image preprocessing, histogram construction, identification of watermarked pixels and watermark extraction. The block diagram shows the watermarking phases (Fig. 1).

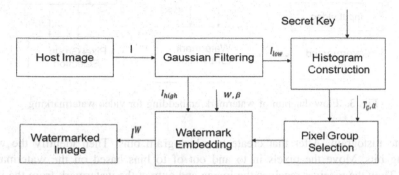

Fig. 1. Flow diagram of watermark embedding for image watermarking

The proposed algorithm performs as follows: First we select an image for transmission over the internet, then perform image preprocessing using Gaussian filter. Then compute histogram of the image. After that create two histogram bins. Then identify the watermarking bits. Move the pixels in to and out of to bins based on the watermarking pixels. Then the receiver receives the image and extract the watermark from the image using private key. Finally we calculate the integrity of the received image using PSNR calculation.

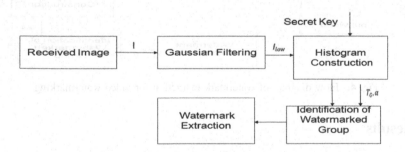

Fig. 2. Flow diagram of watermark detection for image watermarking

In the watermark decoding process the received image is preprocessed by using gaussian filter. After that construct histogram of the received image. Then identify the watermarked region. Then the watermark is extracted by using the rules of watermark embedding (Fig. 2).

In the case of video watermarking the video is preprocessed by extracting the audio and video frames separately. After that the same precess of image watermarking is applied to the video frames. That is image preprocessing using Gaussian filter. Then

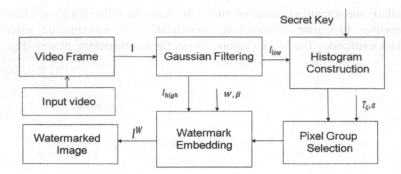

Fig. 3. Flow diagram of watermark embedding for video watermarking

compute histogram. After that create two histogram bins. Then identify the water-marking bits. Move the pixels in to and out of to bins based on the watermarking pixels. Then the receiver receives the image and extract the watermark from the image using private key (Fig. 3).

In the watermark decoding process the video is preprocessed by extracting the audio and video frames separately. After that the same precess of image watermarking is applied to the video frames. By using gaussian filter. After that construct histogram of the received image. Then identify the watermarked region. Then the watermark is extracted by using the rules of watermark embedding (Fig. 4).

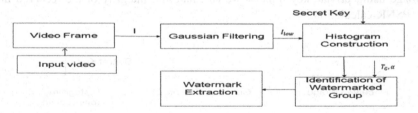

Fig. 4. Flow diagram of watermark extraction for video watermarking

4   Results

The proposed system of watermarking scheme performs well against various attacks. It is robust and imperceptible. The following screenshots shows the result of this proposed method. The input image and watermarked image is exactly same. The water-marked content can be extracted by using the secret key in the decoding phase. The manipulation by attackers can be identified by using the watermark. Figure 5(a) Gives the idea of input of watermarking images Fig. 5(b) gives the idea of output of watermarking images.

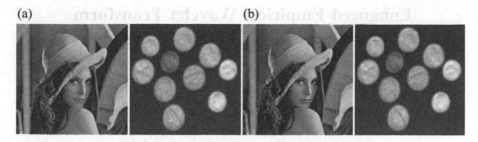

Fig. 5.   (a) un watermarked Image, (b) watermarked Image

In the case of video the watermarked content is perceptibly equivalent to the un watermarked content. The watermarked data can be extracted successfully. Figure 6 Gives the comparison of input and output watermarking of video frames.

Fig. 6.   Comparison of watermarked frames

5   Conclusion

Emerging multimedia applications needs sophisticated protection schemes. Thus the proposed system provides a secure digital watermarking which is capable to handle both video and image applications simultaneously. This will leads to improve security of multimedia content in real world applications. The integrity of digital images and videos will be consistent in communicative systems.

References

1. Xiang, S., Kim, H.J., Huang, J.: Invariant image watermarking based on statistical features in the low-frequency domain. IEEE Trans. Circuits Syst. Video Technol. **18**(6), 777–790 (2008)
2. Dong, P., Brankov, J.G., Galatsanos, N.P., Yang, Y., Davoine, F.: Digital watermarking robust to geometric distortions. IEEE Trans. Image Process. **14**(12), 2140–2150 (2005)
3. Zong, T., Xiang, Y., Guo, S., Rong, Y.: Rank-based image watermarking method with high embedding capacity and robustness digital object identifier, May 2016. https://doi.org/10.1109/ACCESS.2016.255672

Enhanced Empirical Wavelet Transform for Denoising of Fundus Images

C. Amala Nair and R. Lavanya$^{(\boxtimes)}$

Department of Electronics and Communication Engineering,
Amrita School of Engineering, Coimbatore,
Amrita Vishwa Vidyapeetham, Coimbatore, India
amalanairc@gmail.com, r_lavanya@cb.amrita.edu

Abstract. Glaucoma is an ophthalmic pathology caused by increased fluid pressure in the eye, which leads to vision impairment. The evaluation of the Optic Nerve Head (ONH) using fundus photographs is a common and cost effective means of diagnosing glaucoma. In addition to the existing clinical methods, automated method of diagnosis can be used to achieve better results. Recently, Empirical Wavelet Transform (EWT) has gained importance in image analysis. In this work, the effectiveness of EWT and its extension called Enhanced Empirical Wavelet Transform (EEWT) in denoising fundus images was analyzed. Around 30 images from High Resolution Fundus (HRF) image database were used for validation. It was observed that EEWT demonstrates good denoising performance when compared to EWT for different noise levels. The mean Peak Signal to Noise Ratio (PSNR) improvement achieved by EEWT was as high as 67% when compared to EWT.

Keywords: Denoising · Glaucoma · Empirical Wavelet Transform
Enhanced Empirical Wavelet Transform · Peak Signal to Noise Ratio

1 Introduction

Glaucoma is the second common source of vision loss usually seen in the age group of 40–80 years. It is characterised by very high eye pressure, which damages the Optic Nerve Head (ONH) causing peripheral vision loss and finally leading to blindness [1]. Approximately 64.3 million people in the world were suffering from glaucoma in 2013. By 2020 this number might rise to 76 million and by 2040 it might affect 111.8 million people [2]. Although glaucoma cannot be cured, timely treatment will help to hold back its progression. Therefore, diagnosis of this disease is important to avoid preventable vision loss [3].

The diagnosis necessitates regular eye tests, which is expensive and time - consuming. Conventional diagnosis techniques are based on manual observations and hence restricted by the expertise of ophthalmologists in the domain and prone to inter observer variability [1]. These limitations impose the need for automated methods which offer consistency, objective analysis and time efficiency. Among the various imaging modalities used for glaucoma detection, digital fundus photography is preferred for automated diagnosis since it is cost effective and captures a large retinal field.

© Springer Nature Singapore Pte Ltd. 2018
I. Zelinka et al. (Eds.): ICSCS 2018, CCIS 837, pp. 116–124, 2018.
https://doi.org/10.1007/978-981-13-1936-5_13

Clinical information suggests that the ONH examination is the most beneficial method for diagnosing glaucoma structurally. Proper segmentation of structures in and around ONH requires precise identification of the border between the retina and the rim which has a number of limitations [4]. The accuracy of the system developed relies on the accuracy of segmentation performed.

Among the various techniques used for image analysis, wavelet transform has shown to have an upper hand. The drawback of wavelet analysis is that it has fixed basis and hence is non-adaptive with respect to signal characteristics [5]. Huang et al. [6] propounded Empirical Mode Decomposition (EMD) which is adaptive in nature. It decomposes the non- stationary signal into modes known as Intrinsic Mode Functions which acts as the bases. EMD makes use of a process known as sifting for signal decomposition. However, there are a few shortcomings of EMD, such as lack of a strong theoretical background, no robust stopping criterion for sifting process, mode mixing and end effects [7].

To overcome the limitations of EMD, Empirical Wavelet Transform (EWT) was suggested [8] and it is shown to have an upper hand over other time-frequency analysis methods [9, 10]. By combining the time-frequency localisation properties of wavelets and adaptability of Empirical Mode Decomposition (EMD), EWT is found to be apt for analysing fundus images. However, EWT does not take spectrum shape into consideration while performing segmentation of the spectrum to decompose images into different modes. The drawback of EWT was identified and a new approach was proposed by Hu et al. [11] known as Enhanced Empirical Wavelet Transform (EEWT). It makes use of an envelope-based approach using the Order Statistics Filter (OSF) for segmentation of the Fourier spectrum. EEWT is found to have better performance for non-stationary signal analysis [10].

Fundus images are generally affected by additive, multiplicative noise and a mixture of these two [12]. EWT has shown to have an upper hand in Computer Aided Detection (CAD) systems for diagnosing glaucoma. In such systems, as a preprocessing step, techniques like Median and Gaussian filters are used for denoising the fundus images [13]. Rather than making use of an additional preprocessing step, the inherent denoising ability of EWT and EEWT can be utilized. The purpose of the work is to study the effects of EWT and EEWT in denoising of fudus images.

This paper is arranged as follows. Section 2 gives a summary of the two techniques used and the methodology adopted. Section 3 covers the results and discussions. Finally, the paper concludes in Sect. 4.

2 Methodology

This work focuses on analyzing EWT and its extension EEWT on fundus images. EWT and EEWT were applied on the fundus images to form sub images from low to high frequency. Next, the modes were thresholded to eliminate the effect of noise. Inverse transform was performed on these modes to reconstruct the signal. EWT, EEWT and the denosing method employed are explained in Sects. 2.1, 2.2 and 2.3.

2.1   Empirical Wavelet Transform

Gilles [8] proposed EWT in order to analyse signals such that adaptability of EMD and time frequency localisation of wavelets can be combined together. EWT decomposes the signal into modes using wavelet filter banks, whose supports are derived from the location of information in the signal. The main steps involved in EWT are segmentation of the spectrum followed by construction of EWT basis and their application on the segments formed. For N segments, a bank of filters will be defined; N−1 band pass filters and a low pass filter. The procedure involved in EWT is illustrated in Fig. 1.

Segmentation of the signal spectrum requires dividing it into N continuous segments given as $\lambda_n = [\omega_{n-1}\omega_n]$, where ω_n is the frequency at any point n. Centered on each ω_n, a transition phase T_n is defined with a width of $2\tau_n$. Excluding 0 and π, N−1 boundaries should be detected. They are obtained by finding all the local maxima of the spectrum and sorting them in descending order. The first N−1 maxima are then selected. Boundaries are found as the average between the positions of two consecutive maxima. Empirical scaling and wavelet functions are obtained similar to Littlewood-Paley wavelets and Meyers wavelets. For the signal spectrum $v(\omega)$, empirical scaling function and empirical wavelet function are shown as Eqs. (1) and (2) respectively.

The empirical scaling function is defined as:

$$\phi_n(\omega) = \begin{cases} 1 & if \ |\omega| \leq \omega_n - \tau_n \\ \cos\left[\frac{\pi}{2}v\left(\frac{1}{2\tau_n}(|\omega| - \omega_n + \tau_n)\right)\right] & if \ \omega_n - \tau_n \leq |\omega| \leq \omega_n \\ 0 & otherwise \end{cases} \tag{1}$$

The empirical wavelet function is defined as:

$$\psi_n(\omega) = \begin{cases} 1 & \\ \cos\left[\frac{\pi}{2}v\left(\frac{1}{2\tau_{n+1}}(|\omega| - \omega_{n+1} + \tau_{n+1})\right)\right] & if \ \omega_{n-1} - \tau_{n-1} \leq |\omega| \leq \omega_{n-1} - \tau_{n-1} \\ \sin\left[\frac{\pi}{2}v\left(\frac{1}{2\tau_n}(|\omega| - \omega_n + \tau_n)\right)\right] & if \ \omega_{n+1} - \tau_{n+1} \leq |\omega| \leq \omega_{n+1} + \tau_{n+1} \\ 0 & if \ \omega_n - \tau_n \leq |\omega| \leq \omega_n + \tau_n \\ & otherwise \end{cases}$$

$$\tag{2}$$

By taking the inner product between signal and empirical wavelet function, wavelet coefficients can be obtained. Similarly, the inner product of signal with empirical scaling function gives the scaling coefficient.

This concept was extended to images in [14] by Gilles et al. The empirical counterpart of Tensor wavelets, Curvelets, Littlewood-Paley wavelets and Symlets were built. In 2 dimensional Littlewood-Paley wavelet transform, images are filtered using wavelets with annuli supports. Hence, Polar FFT is preferred in this case. Pseudo polar FFT is a method which helps to do this with less computational complexity since FFT is computed on a square grid rather than polar grid [15].

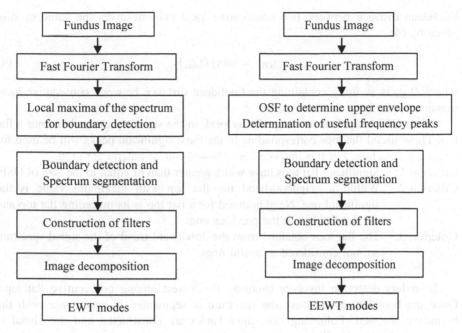

Fig. 1. Flowchart of EWT   Fig. 2. Flowchart of EEWT

2.2 Enhanced Empirical Wavelet Transform

EWT is limited by the fact that it can be used to analyze signals with well separated frequencies. For signals that are noisy, or non-stationary in nature, EWT may not perform well, as the boundary detection may result in errors. The local maxima, which might be significant may not be considered while those which are part of the noise might be considered. Segmentation performed with such boundaries will be incorrect. This drawback is a result of spectrum shape not being considered by EWT. On the other hand, EEWT considers the spectrum shape for segmentation of the signal.

In EEWT, OSF is first applied on the input signal to obtain the upper envelope from which the major peaks are found. The procedure involved in EEWT analysis is illustrated in Fig. 2.

FFT of the signal is taken in order to obtain the spectrum. OSF is performed on the spectrum using Max filter for upper envelope detection. A sliding window of size s_{OSF} centered at a point is used to determine the upper envelope (U) at that point as the maximum value of elements in that region. The upper envelope is given by (3)

$$U(n) = \max_{k \in A_n}(D(k)) \tag{3}$$

where D denotes the sequence of data and A_n denotes the sliding window. At any point n, the value of U is the maximum value of D over the region A_n. The size of sliding window is found by considering all the local maxima. The minimum value of the

Euclidean distance between two consecutive local maxima gives the value of s_{OSF} given by (4)

$$s_{OSF} = \min\{D_{max}\} \tag{4}$$

where D_{max} is an array containing the Euclidean distance between consecutive local maxima of the data.

When OSF is applied on a signal, any peak in the signal spectrum becomes a flat top. These useful flat tops corresponding to the most significant peaks will be used for boundary detection. For this purpose, the following three criteria are used.

Criterion 1:   Significant flat tops have width greater than or equal to the size of OSF.

Criterion 2:   Within a neighbourhood the flat top with maximum value is the significant one. Neighborhood for a flat top is its preceding flat top and the flat top before the previous one.

Criterion 3:   The flat tops obtained from the downward trend of the signal spectrum are not considered as useful ones.

Boundary detection involves choosing the lowest among consecutive flat tops. Once the boundary is detected, the spectrum is segmented in accordance with the boundary obtained. Following this, filter banks are constructed and the signal is decomposed using the filters.

2.3   Denoising Using EWT/EEWT

Soft thresholding is applied on EWT modes to remove noise. The threshold value was found using Eq. (5).

$$\tau = \sigma\sqrt{2\log(M)} \tag{5}$$

where τ is the universal thresholding, M represents the total number of pixels in the image and σ gives the noise level estimate. It is given by Eq. (6).

$$\sigma = \frac{median\{w\}}{0.6745} \tag{6}$$

where w is the wavelet coefficient.

EEWT can distinguish noise and meaningful components effectively boundaries detected are optimal. Thus denoising was performed by removing the mode containing the highest frequency.

3   Results and Discussions

A total of 30 retinal fundus images were acquired from High Resolution Fundus (HRF) image database [16]. MATLAB R2017a on Windows platform was used for the implementation of this work. For validation of the denoising performance, Speckle and

Gaussian noise were added to the images, followed by decomposition of the images into EWT and EEWT modes. Figure 3 shows the green channel of fundus image decomposed into modes using EWT. Figure 4 shows the decomposition of images using EEWT.

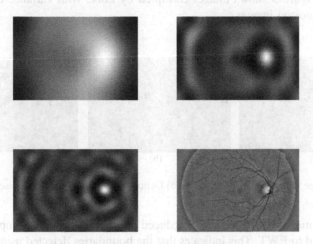

Fig. 3. Decomposition using EWT

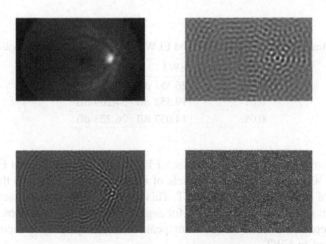

Fig. 4. Decomposition using EEWT

Denoising was performed using EWT and EEWT methods. By comparing the sub-band images of EWT and EEWT, it is clear that noise is more distinguishable in EEWT than EWT. Further, in EEWT, the low frequency content was retained without much noise in the first intrinsic mode function. High frequency components clearly show the influence of noise. On the other hand, in EWT, it is difficult to distinguish between noise and meaningful content since noise gets mixed in all modes. This shows that spectral boundaries detected for EEWT are optimal as compared to EWT.

These cases were considered with different values for noise variance. Images were corrupted by Speckle and Gaussian noise with variance of 0.001, 0.01 and 0.05 respectively. Figure 5 shows images corrupted by noise with variance of 0.05, EWT and EEWT denoised images.

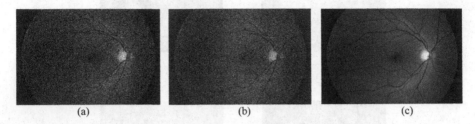

(a)   (b)   (c)

Fig. 5. (a) Image with noise variance 0.05, (b) Denoised using EWT, (c) Denoised using EEWT

Denoised images of EEWT shows reduced noise levels and thus improved quality when compared to EWT. This indicates that the boundaries detected using EWT could not effectively separate noise from image content. To quantitatively analyze the denoising performances of the two methods, Peak Signal to Noise Ratio (PSNR) was calculated.

Table 1. Mean PSNR values of EWT and EEWT denoising for different noise variance

Noise variance	EWT	EEWT
0.001	26.351 dB	39.256 dB
0.01	19.552 dB	34.209 dB
0.05	14.057 dB	26.723 dB

Table 1 shows the mean PSNR values for EWT based denoising and EEWT based denoising on 30 images for different levels of noise variance. It is seen that the mean performance of EEWT is better than EWT. This can be attributed to the fact that EEWT takes spectrum shape into consideration for segmentation of the spectrum. As a result, EEWT can separate out the insignificant peaks caused by noisy components better when compared to EWT.

It is further observed that EWT is inefficient especially when greater amount of noise is present in the image. On the other hand, EEWT shows good performance even when noise levels are high. Hence EEWT is suitable in analysing glaucoma using fundus images which are susceptible to noise during acquisition process. The inherent denoising capability of the technique alleviates the need for denoising as a pre-processing step.

4   Conclusion

Though much work has been based recently on EWT for image analysis, it can be seen that EEWT is a better alternative for non-structural approach. EEWT has better inherent noise removing capability resulting in an overall improved performance compared to EWT. Thus EEWT, with its combined noise robustness, adaptability and time-frequency localisation is a promising technique for computer aided glaucoma diagnosis.

References

1. Haleem, M.S., Han, L., Hemert, J.V., Li, B.: Automatic extraction of retinal features from colour retinal images for glaucoma diagnosis: a review. Comput. Med. Imaging Graph. **37**, 581–596 (2013)
2. Tham, Y.C., Li, X., Wong, T.Y., Quigley, H.A., Aung, T., Cheng, C.Y.: Global prevalence of glaucoma and projections of glaucoma burden through 2040: a systematic review and meta analysis. Ophthalmology **121**, 2081–2090 (2014)
3. Zhang, Z., et al.: A Survey on computer aided diagnosis for ocular diseases. BMC Med. Inform. Decis. Mak. **14**, 80 (2014)
4. Almazroa, A., Burman, R., Raahemifar, K., Lakshminarayanan, V.: Optic disc and optic cup segmentation methodologies for glaucoma image detection: a survey. J. Ophthalmol. **2015**, 581–596 (2013)
5. Lei, Y., Lin, J., He, Z., Zuo, M.J.: A review on empirical mode decomposition in fault diagnosis of rotating machinery. Mech. Syst. Signal Process. **35**, 108–126 (2013)
6. Huang, N.E., et al.: The empirical mode decomposition and the hilbert spectrum for non-linear and non-stationary time series analysis. Proc. Roy. Soc. Lond. **454**, 903–995 (1998)
7. Xuan, B., Xie, Q., Peng, S.: EMD sifting based on bandwidth. IEEE Signal Process. Lett. **14**, 537–540 (2007)
8. Gilles, J.: Empirical wavelet transform. IEEE Trans. Biomed. Eng. **61**, 3999–4010 (2013)
9. Jambholkar, T., Gurve, D., Sharma, P.B.: Application of empirical wavelet transform (EWT) on images to explore brain tumor. In: 3rd International Conference on Signal Processing, Computing and Control, pp. 200–204. IEEE (2015)
10. Maheshwari, S., Pachori, R.B., Acharaya, U.R.: Automated diagnosis of glaucoma using empirical wavelet transform and correntropy features extracted from fundus images. IEEE J. Biomed. Health Inform. **21**, 803–813 (2017)
11. Hu, Y., Li, F., Li, H., Liu, C.: An enhanced empirical wavelet transform for noisy and non-stationary signal processing. Digit. Signal Process. **60**, 220–229 (2017)
12. Hani, A.F.M., Soomro, T.A., Fayee, I., Kamel, N., Yahya, N.: Identification of noise in the fundus images. In: 3rd IEEE International Conference on Control System, Computing and Engineering, pp. 191–196. IEEE CSS Chapter, Malaysia (2013)
13. Dharani, V., Lavanya, R.: Improved microaneurysm detection in fundus images for diagnosis of diabetic retinopathy. In: Thampi, S.M., Krishnan, S., Corchado Rodriguez, J. M., Das, S., Wozniak, M., Al-Jumeily, D. (eds.) SIRS 2017. AISC, vol. 678, pp. 185–198. Springer, Cham (2018). https://doi.org/10.1007/978-3-319-67934-1_17

14. Gilles, J., Tran, G., Osher, S.: 2D empirical transforms. Wavelets, ridgelets and curvelets revisited. SIAM J. Imaging Sci. **7**, 157–186 (2014)
15. Averbuch, A., Coifman, R.R., Donoho, D.L., Elad, M., Israeli, M.: Fast and accurate polar fourier transform. Appl. Comp. Harmon. Anal. **21**, 145–167 (2006)
16. Budai, A., Bock, R., Maier, A., Hornegger, J., Michelson, G.: robust vessel segmentation in fundus images. Int. J. Biomed. Imaging **2013**, 11 (2013)

Kernelised Clustering Algorithms Fused with Firefly and Fuzzy Firefly Algorithms for Image Segmentation

Anurag Pant[✉], Sai Srujan Chinta, and Balakrushna Tripathy

School of Computing Science and Engineering,
VIT Vellore, Vellore 632014, Tamil Nadu, India
{anurag.pant2014, chintasai.srujan2014,
tripathybk}@vit.ac.in

Abstract. The aim of our research is to combine the conventional clustering algorithms based on rough sets and fuzzy sets with metaheuristics like firefly algorithm and fuzzy firefly algorithm. Image segmentation is carried out using the resultant hybrid clustering algorithms. The performance of the proposed algorithms is compared with numerous contemporary clustering algorithms and their firefly fused counter-parts. We further bolster the performance of our proposed algorithm my using Gaussian kernel in place of the traditional Euclidean distance measure. We test the performance of our algorithms using two performance indices, namely DB (Davis Bouldin) indexand Dunn index. Our experimental results highlight the advantages of using metaheuristics and kernels over the existing clustering algorithms.

Keywords: Fuzzy firefly · Gaussian kernel · Data clustering · RFCM
Dunn index

1 Introduction

When we segment an image, we partition it into several meaningful homogeneous regions without overlap. A popular method to achieve image segmentation is by employing data clustering algorithms. In our previous research, [1, 8, 9, 11] we have fused Firefly algorithm [13] with clustering algorithms such as Fuzzy C-Means [7] and Intuitionistic Fuzzy C-Means [8]. We have shown that doing so improves the clustering quality and the segmentation quality. Data clustering algorithms tend to depend on the random initialization of centroid values at the beginning of the algorithm. This severely undermines the consistency of output as well as the performance of the algorithm. Metaheuristics such as Firefly algorithm can be used to compute near-optimal centroids. These values can then be passed to the clustering algorithms thereby circumventing the problems associated with random initialization of centroids. In our previous work [2], we made use of the Fuzzy Firefly algorithm [5] in place of the Firefly algorithm, to prevent the solution from getting stuck at the local optima by influencing the movement of each firefly by a set of fireflies which glow with an intensity above a threshold value. The main drawback of using Euclidean distance is that there is heavy

I. Zelinka et al. (Eds.): ICSCS 2018, CCIS 837, pp. 125–132, 2018.
https://doi.org/10.1007/978-981-13-1936-5_14

dependency on the initial centres and only linearly separable clusters can be identified. We eliminate this drawback by making use of Gaussian Kernel instead of Euclidean distance.

2 Fuzzy C-Means Algorithm (FCM)

The algorithm for Fuzzy C-Means is derived from the idea of fuzzy sets [14]. Within the algorithm, we initialize cluster centers using randomized values. Then we calculate the distance d_{ik} of each cluster center i to each pixel k. The Membership Matrix is calculated as given:

$$\mu_{ik} = \frac{1}{\sum_{j=1}^{c} \left(\frac{d_{ik}}{d_{jk}}\right)^{\frac{2}{m-1}}} \tag{1}$$

c denotes the number of clusters while m denotes the fuzzifier (which we take to be 2). The cluster centers are computed as follows:

$$v_i = \frac{\sum_{j=1}^{N} (\mu_{ij})^m x_j}{\sum_{j=1}^{N} (\mu_{ij})^m} \tag{2}$$

IFCM is derived from the intuitionistic fuzzy set model [3]. In this algorithm, we compute the hesitation degree and add it to the membership matrix. This helps to improve the clustering process. In our research, we calculate the non-membership values using Yager's intuitionistic fuzzy complement [6]:

$$f(x) = (1 - x^\alpha)^{\frac{1}{\alpha}} \tag{3}$$

In our paper, we take α to be 2. The hesitation degree of data point x in cluster center A is given as:

$$\pi_A(x) = 1 - \mu_A(x) - (1 - \mu_A(x)^\alpha)^{\frac{1}{\alpha}} \tag{4}$$

μ' represents the modified membership matrix which is given as follows:

$$\mu'_A(x) = \mu_A(x) + \pi_A(x) \tag{5}$$

The concept of Rough sets was introduced in 1982 [10]. In this paper, we use Rough Fuzzy C-Means. This algorithm combines both fuzzy sets and rough sets. In the context of image segmentation, the pixels are adjudged to be part of the lower or upper approximations of rough sets depending on whether $\mu_{ik} - \mu_{jk} < \varepsilon$ or not for some predefined value of ε. If this condition is satisfied, then $x_k \in \overline{B}U_i$ and $x_k \in \overline{B}U_j$ and x_k cannot belong to any lower approximation. If this condition is not satisfied, then $x_k \in \underline{B}U_i$. The cluster update formula employed in RFCM is given in (6).

$$
v_i = \begin{cases} w_{low} \dfrac{\sum_{x_k \in \underline{B}U_i} x_k}{|\underline{B}U_i|} + w_{up} \dfrac{\sum_{x_k \in \overline{B}U_i \setminus \underline{B}U_i} \mu_{ik}^m x_k}{\sum_{x_k \in \overline{B}U_i \setminus \underline{B}U_i} \mu_{ik}^m}, & \text{if } \underline{B}U \neq \phi \wedge \overline{B}U_i \setminus \underline{B}U_i \neq \phi; \\[3ex] \dfrac{\sum_{x_k \in \overline{B}U_i \setminus \underline{B}U_i} \mu_{ik}^m x_k}{\sum_{x_k \in \overline{B}U_i \setminus \underline{B}U_i} \mu_{ik}^m}, & \text{if } \overline{B}U_i \setminus \underline{B}U_i \neq \phi; \\[3ex] \dfrac{\sum_{x_k \in \underline{B}U_i} x_k}{|\underline{B}U_i|}, & ELSE. \end{cases} \tag{6}
$$

3 Firefly Algorithm

Firefly Algorithm was proposed in 2009 [12]. It is a bio-inspired meta-heuristic which mimics the behaviour of fireflies. Biologically, fireflies are attracted to luminous objects. In this algorithm, each firefly has a brightness of its own. All the fireflies are attracted to the brightest firefly, which has a random movement. The attraction between two fireflies is inversely proportional to the distance between the two fireflies. The brightness of each firefly is calculated using the objective function which is problem-specific. The attractiveness function β is determined by the following formula:

$$
\beta(r) = \beta_0 e^{-\gamma r^2} \tag{7}
$$

Here, β_0 denotes the default value of attractiveness, γ is the light absorption coefficient and $r_{i,j}$ is the Euclidean distance between the two fireflies i and j:

$$
x_i = x_i + \beta_0 e^{-\gamma r_{i,j}^2}(x_i - x_j) + \alpha\left(rand - \frac{1}{2}\right) \tag{8}
$$

Here, α is the randomization parameter and 'rand' denotes a function for generating random numbers in the interval [0, 1].

4 Fuzzy Firefly Algorithm

The fuzzy firefly algorithm was introduced in 2014 with the goal of increasing the search area of each firefly and decreasing the number of iterations [4]. When iterating, k-brighter fireflies are chosen to influence the less brighter fireflies. Here, k is a user-defined parameter which depends on how complex the problem is and the population of the swarm. The attractiveness $\psi(h)$ of the firefly h (one of the k brighter fireflies) is given by:

$$
\psi(h) = \frac{1}{\left(\frac{f(p_h) - f(p_g)}{\beta}\right)} \tag{9}
$$

Here, $f(p_h)$ is the fitness of the firefly h, while $f(p_g)$ denotes the fitness of the local optimum firefly. β is defined as:

$$\beta = \frac{f(p_g)}{l} \qquad (10)$$

l is a user-set parameter. Firefly i moves towards firefly h, one of the k-brighter fireflies using the equation given below:

$$X_i = x_i + \left(\beta_0 e^{-\gamma r_{j,i}^2}(x_j - x_i) + \sum_{h=1}^{k} \psi(h)\beta_0 e^{-\gamma r_{h,i}^2}(x_h - x_i) \right) \times \alpha \left(rand - \frac{1}{2} \right) \qquad (11)$$

5 Distance Measures

Usually the distance between the pixels of the image and the pixels of the cluster centers is calculated using Euclidean distance or Manhattan distance. However, the Euclidean distance relies heavily on the initial cluster centers and only allows us to identify linearly separable clusters. Fusing Firefly and Fuzzy Firefly with the clustering algorithms allows us to eliminate this drawback. To further deal with the issue of linear separability, we make use of kernels in this paper. We also make use of functions for non-linear mapping since they allow us to carry out clustering in feature space and allow us to transform the problem into a linear problem by changing the separation problem from image space to kernel space.

$$K(x, y) = \exp(-\frac{\|x - y\|^2}{\sigma^2}) \qquad (12)$$

The equation for Gaussian Kernel is given in (12). σ denotes the standard deviation of x.

6 Proposed Algorithms

Throughout this paper, grayscale values of the pixels are considered to be the data points and the cluster centers. Each firefly is representative of a set of cluster centers (grayscale values) that are initialized to random values. The intensity of each firefly is calculated using the objective function which is specific to each clustering algorithm. Once converged, we pass the best firefly values to the clustering algorithm. Furthermore, as explained in the previous section, we replace the Euclidean distance measure with Gaussian kernel to further bolster the performance of our algorithms. The experimental results show that the clustering algorithms perform better when combined with the metaheuristics.

7 Results and Discussions

An index name 'Error' has been used by us to evaluate the proximity of the final centroid values provided by the metaheuristics to the terminal centroid values output when the clustering algorithms are executed without metaheuristics. Furthermore, DB index and Dunn index have been used to evaluate how well the image has been clustered and segmented. DB index is inversely proportional to the accuracy of segmentation whereas Dunn index is directly proportional to the accuracy of segmentation.

7.1 Brain MRI Segmentation

It can be observed in Fig. 1 that the brain and the tumor are segmented into different clusters by the use of clustering algorithms. It can be observed in Table 1 that RFCM performs much better clustering as compared to FCM or IFCM. The clustering performed by IFCM is slightly better than FCM. The number of iterations taken to converge is the least for RFCM, and then for IFCM and then FCM. The kernelized versions of the clustering algorithms follow a similar pattern but their clustering is by far superior to their original versions. The fuzzy firefly implementation of the clustering algorithms manages to improve the clustering quality of the images as can be observed from looking at the DB and Dunn indices. The fuzzy firefly further manages to outperform the firefly metaheuristic by reducing the error as well as the number of iterations by a larger margin.

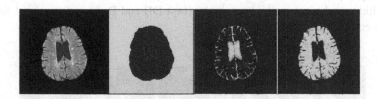

Fig. 1. The Brain MRI image on the left is segmented into 3 different clusters on the right

Table 1. DB and Dunn index values for Brain MRI image

Centroids	3				4			
Algorithm	Iterations	DB	Dunn	Error	Iterations	DB	Dunn	Error
FCM	9	7.2851	0.1801		32	7.652	0.1151	
IFCM	13	7.1724	0.1846		23	7.4327	0.1181	
RFCM	10	3.1889	0.3294		17	2.596	0.1648	
GKFCM	12	0.0008	25.7878		37	0.0006	12.3972	
GKIFCM	14	0.0503	25.7498		34	0.053	12.4158	
GKRFCM	16	0.0379	30.8501		27	0.0304	10.0311	
FCMFA	9	7.2857	0.1801	76.4305	27	7.6517	0.115	86.4999
FCMFFA	7	7.2825	0.1801	7.1194	9	7.6519	0.115	30.8863

(continued)

Table 1. (*continued*)

Centroids	3				4			
Algorithm	Iterations	DB	Dunn	Error	Iterations	DB	Dunn	Error
IFCMFA	11	7.1735	0.1846	54.3553	19	7.4326	0.1181	44.2845
IFCMFFA	9	7.1734	0.1846	11.9443	15	7.4123	0.1178	30.8579
RFCMFA	10	3.1889	0.3294	46.7503	12	2.6093	0.1486	53.9519
RFCMFFA	9	3.1889	0.3294	15.1053	10	2.5889	0.1714	27.4305
GKFCMFA	9	0.0008	25.7879	48.4196	23	0.0006	12.3725	64.7224
GKFCMFFA	8	0.0008	25.7887	23.474	20	0.0006	12.3747	16.7579
GKIFCMFA	8	0.0503	25.7498	33.686	27	0.053	12.4151	59.6881
GKIFCMFFA	8	0.0503	25.7498	13.7125	14	0.053	12.4152	16.1798
GKRFCMFA	8	0.0379	30.8462	35.2052	23	0.0304	10.0311	78.0239
GKRFCMFFA	7	0.0379	30.8501	13.1132	11	0.0304	10.0311	19.0157

7.2   Lena

It can be observed in Fig. 2 that the clustering algorithms have successfully segmented the image into different clusters. The results in Table 2 follow the same pattern as was observed in the brain MRI image, with the kernelized versions of the clustering algorithms giving better results as compared to the original clustering algorithms. The fuzzy firefly implementation again manages to outperform the firefly implementation of the algorithms and gives us the best results (best DB and Dunn indices, least iterations and least error value).

Fig. 2. The Lena image on the left is segmented into 4 different clusters on the right

Table 2. DB and Dunn index values for Lena image

Centroids	3				4			
Algorithm	Iterations	DB	Dunn	Error	Iterations	DB	Dunn	Error
FCM	36	11.3992	0.1426		19	7.0308	0.2055	
IFCM	25	11.0121	0.1493		25	6.8666	0.2105	
RFCM	15	3.3744	0.5219		16	2.2168	0.6006	
GKFCM	46	0.0028	13.0463		13	0.0011	26.0287	
GKIFCM	28	0.1207	13.2926		19	0.0475	25.5933	
GKRFCM	27	0.0398	38.0985		28	0.0158	78.6379	

(*continued*)

Table 2. (*continued*)

Centroids	3				4			
Algorithm	Iterations	DB	Dunn	Error	Iterations	DB	Dunn	Error
FCMFA	25	11.3991	0.1426	21.7257	22	7.0302	0.2059	24.7372
FCMFFA	7	11.3931	0.1429	5.5837	9	7.0301	0.2057	18.9851
IFCMFA	24	11.0129	0.1492	22.989	19	6.87	0.2106	37.6213
IFCMFFA	14	11.0129	0.1492	8.7429	20	6.8666	0.2105	25.8002
RFCMFA	14	3.3545	0.5107	26.7541	14	2.2168	0.6006	19.709
RFCMFFA	7	3.3744	0.5219	13.715	6	2.1595	0.6142	13.9215
GKFCMFA	19	0.0028	13.0964	26.2751	12	0.0011	26.0244	42.2044
GKFCMFFA	18	0.0028	13.0974	7.4856	10	0.0011	26.0255	18.0598
GKIFCMFA	19	0.1196	13.5077	51.2936	18	0.0475	25.5936	37.6778
GKIFCMFFA	13	0.1196	13.5077	6.5014	18	0.0473	26.1401	30.6011
GKRFCMFA	16	0.0393	40.0028	29.0793	18	0.0158	78.6379	61.2994
GKRFCMFFA	14	0.0393	40.0028	14.6707	12	0.0158	78.6375	30.5762

8 Conclusion

The kernelized versions of the clustering algorithms perform much superior segmentation of the image as compared to the traditional clustering algorithms. The Fuzzy Firefly implementation manages to improve the clustering process by improving the DB and Dunn indices. It also manages to outperform the Firefly metaheuristic by further reducing the number of iterations and by returning cluster center values that are much closer to the actual cluster centers than the ones returned by the Firefly implementation. By increasing the coverage of the solution space, Fuzzy Firefly metaheuristic manages to eliminate the problem of getting stuck at the local optima. In the future, we plan to extend our research by utilizing the Fuzzy Firefly to find the optimal fuzzification parameter and by making use of the hybrid clustering algorithms in Big Data clustering.

References

1. Jain, A., Chinta, S., Tripathy, B.K.: stabilizing rough sets based clustering algorithms using firefly algorithm over image datasets. In: Satapathy, S.C., Joshi, A. (eds.) ICTIS 2017. SIST, vol. 84, pp. 325–332. Springer, Cham (2018). https://doi.org/10.1007/978-3-319-63645-0_36
2. Anurag, P., Chinta, S.S., Tripathy, B.K.: Comparative analysis of hybridized C-Means and fuzzy firefly algorithms with application to image segmentation. In: Presented in 2nd International Conference on Data Engineering and Communication Technology (ICDECT) 2017 (2018)
3. Atanassov, K.T.: Intuitionistic fuzzy sets. Fuzzy sets Syst. **20**(1), 87–96 (1986)
4. Chaira, T.: A novel intuitionistic fuzzy C means clustering algorithm and its application to medical images. Appl. Soft Comput. **11**(2), 1711–1717 (2011)

5. Hassanzadeh, T., Kanan, H.R.: Fuzzy FA: a modified firefly algorithm. Appl. Artif. Intell. **28** (1), 47–65 (2014)
6. Yager, R.R.: On the measures of fuzziness and negation part II lattices. Inf. Control **44**, 236–260 (1980)
7. Ruspini, E.H.: A new approach to clustering. Inf. Control **15**(1), 22–32 (1969)
8. Tripathy, B.K., Namdev, A.: Scalable rough C-Means clustering using firefly algorithm. Int. J. Comput. Sci. Bus. Inf. **16**(2), 1–14 (2016)
9. Chinta, S.S., Jain, A., Tripathy, B.K.: Image segmentation using hybridized firefly algorithm and intuitionistic fuzzy C-Means. In: Somani, A.K., Srivastava, S., Mundra, A., Rawat, S. (eds.) Proceedings of First International Conference on Smart System, Innovations and Computing. SIST, vol. 79, pp. 651–659. Springer, Singapore (2018). https://doi.org/10.1007/978-981-10-5828-8_62
10. Pawlak, Z.: Rough sets. Int. J. Parallel Prog. **11**(5), 341–356 (1982)
11. Chinta, S., Tripathy, B.K., Rajulu, K.G.: Kernelized intuitionistic fuzzy C-Means algorithms fused with firefly algorithm for image segmentation. In: 2017 International Conference on Microelectronic Devices, Circuits and Systems (ICMDCS), Vellore, pp. 1–6 (2017)
12. Yang, X.: Firefly algorithm, stochastic test functions and design optimization. Proc. IJBIC **2**, 78–84 (2010)
13. Yang, X.-S.: Firefly algorithms for multimodal optimization. In: Watanabe, O., Zeugmann, T. (eds.) SAGA 2009. LNCS, vol. 5792, pp. 169–178. Springer, Heidelberg (2009). https://doi.org/10.1007/978-3-642-04944-6_14
14. Zadeh, L.A.: Fuzzy sets. Inf. Control **8**(3), 338–353 (1965)

Performance Analysis of Wavelet Transform Based Copy Move Forgery Detection

C. V. Melvi[✉], C. Sathish Kumar, A. J. Saji, and Jobin Varghese

Department of Electronics and Communication Engineering,
Rajiv Gandhi Institute of Technology, Kottayam, India
cvmelvil@gmail.com, kumarcsathish@gmail.com,
aj.saji@gmail.com, jobinvarghese@gmail.com

Abstract. In the modern world, digital images are the sources of information. These sources can be manipulated by image processing and editing software. Image authenticity becomes a socially relevant issue in image forensics. Copy move is a main digital image forgery attack where the region of image is copied and pasted in the same image at different locations for hiding the information. This paper presents an analysis of accuracy in detecting copy move forgery based on different types of wavelet transform. For each wavelet transform, analysis is done at different levels of decomposition. The result indicates that both stationary wavelet transform (SWT) and lifting wavelet transform (LWT) work more effectively as compared to discrete wavelet transform (DWT).

Keywords: Digital image forgery · Wavelet transform

1 Introduction

The well developed computer technology made digital images a part of the human life. Availability of image editing software has also increased in the digital world which facilitates the digital image forgery. This leads to serious problems in various fields like medical imaging, journalism etc. As a consequence, digital images cannot be considered as evidence in court and this necessitates the need for an image forensic tool to discriminate forged form from original images. In the past few years, researchers are focusing on the solution in detection of such forgeries made in the image [1–4].

Copy-move is one of the most commonly used forgery technique in which the portion of the image is copied and pasted into another region of the same image. This forgery includes geometric transforms and post-processing operations such as JPEG compression, rotation, scale and noise addition. This type of tampering is very difficult to detect due to the local similarity. To check integrity of the digital images, active and passive techniques are used. In active technique, the authenticity of the image is detected by embedding the watermark and the image will be authentic if the extracted watermark is similar to original one. In passive technique, forgery is detected by analyzing the contents of the image and does not have any prior knowledge about the image. In the literature, different techniques of copy move forgery detection using passive method have been discussed. They are classified as block–based and keypoint–

© Springer Nature Singapore Pte Ltd. 2018
I. Zelinka et al. (Eds.): ICSCS 2018, CCIS 837, pp. 133–140, 2018.
https://doi.org/10.1007/978-981-13-1936-5_15

based methods. In block-based method the image is divided into blocks and these blocks are used for further processing. In keypoint-based method, keypoint features are used for forgery detection. The block–based methods can be classified based on different features such as intensity, frequency, dimension, moment and texture. In intensity based method the entropy of luminance channel is used as feature vector and correlation coefficient finds the matching blocks as in [1].

A frequency based method to detect copy move forgery based on DCT coefficients is suggested in [2]. A block based approach based on texture was proposed in [3]. The five texture descriptors such as statistical, Tamura, Haralick, edge histogram, and Gabor Descriptors are extracted from each block. Then these blocks are sorted and computed distance between spatial coordinates of blocks to find matching ones. Dimensionality reduction based method using principal component analysis (PCA) is introduced and is robust to additive noise or lossy compression as reported in [4]. For detecting copy-move regions moment based method using Zernike moments is discussed in [5].

2 Overview of the Proposed System

Analysis has three phases where in each phase the wavelet coefficients are extracted by applying DWT, SWT and LWT respectively. Four subbands of the forged image such as low-low (LL), low-high (LH), high-low (HL) and high-high (HH) are obtained after passing through low-pass and high-pass filters. LL subband gives fine approximation of the image. LH, HL, and HH represent the coarse level approximation of the original image. At each level, the LL subimage is decomposed into four subimages at next level.

Size of the image is reduced at every level in the case of DWT and LWT transform [6]. In the case of SWT, dimension is not reduced since down sampling is not performed [7]. The coefficients of corresponding transform are generated upto five levels using different wavelets such as haar, daubechies4 (db4), daubechies6 (db6) and symlets8 (sym8). Then non-overlapping blocks of the LL subimage with fixed size 8×8 are generated. These blocks of coefficients are stored in a coefficient matrix and sorted lexicographically. Euclidean distance between each block is calculated to get matching blocks of forged image. The blocks with less distance are assumed as forged which is shown in Fig. 1.

LWT gives faster implementation of the wavelet transform. As compared to DWT, it requires less number of computations. LWT includes three operations namely split, lifting, and scaling as in [8]. In split phase, the signal is divided into even indexed sample and odd indexed sample using lazy wavelet transform. Even pixel coefficients are predicted with primal lifting coefficients and detail coefficients are obtained by adding these coefficients to odd pixel coefficients. Approximation coefficients get by updating detail coefficients with lifting coefficients and added into even pixel coefficients.

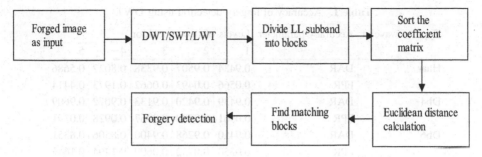

Fig. 1. Block diagram of the proposed system

3 Experimental Results and Analysis

3.1 Performance Evaluation

The performance of the system is evaluated at pixel level which measures how accurately the forged regions are detected as in [9]. Detection accuracy rate (DAR) and false positive rate (FPR) are considered for performance evaluation as in Eqs. (1) and (2) respectively. DAR indicates the algorithm which correctly detects pixels of copy-move locations in the forged image. FPR reflects the percentage of pixels which are not in forged region. For ideal case, DAR is close to 1 and FPR is close to 0.

$$\text{DAR} = \frac{|\psi_S \cap \overline{\psi_S}| + |\psi_T \cap \overline{\psi_T}|}{|\psi_S| + |\psi_T|} \tag{1}$$

$$\text{FPR} = \frac{|\overline{\psi_S} \cap \psi_S| + |\overline{\psi_T} \cap \psi_T|}{|\psi_S| + |\psi_T|} \tag{2}$$

ψ_S and ψ_T represent pixels of original region and forged regions in original image respectively. $\overline{\psi_s}$ and $\overline{\psi_T}$ represent pixels of original and forged regions in detected image respectively.

Performance Evaluation Using DWT. Performance of copy move forgery detection using DWT is shown in Table 1. DWT of forged image is applied upto five levels and accuracy rates DAR and FPR are measured for each wavelet. Average DAR and FPR rate for each wavelet based on DWT is shown in Figs. 2 and 3 respectively. Haar wavelet shows minimum DAR and db4 shows maximum DAR with minimum FPR. This result indicates that the accuracy rate depends on the type of wavelet used.

Table 1. Accuracy of forgery detection using DWT

Type of wavelet	Accuracy parameter	Levels of decomposition				
		1	2	3	4	5
Haar	DAR	0.9434	0.9507	0.9338	0.8027	0.5686
	FPR	0.0566	0.0493	0.0662	0.1973	0.4414
Db4	DAR	0.9439	0.9420	0.9133	0.9072	0.9609
	FPR	0.0561	0.0580	0.0867	0.0928	0.0391
Db6	DAR	0.9450	0.9268	0.9400	0.8506	0.8351
	FPR	0.0550	0.0732	0.0600	0.1494	0.1649
Sym8	DAR	0.9436	0.9322	0.9425	0.8800	0.8403
	FPR	0.0678	0.0575	0.1200	0.1500	0.1597

Performance Evaluation Using SWT. Performance of SWT is shown in Table 2. SWT is applied on the forged image to get the LL subband at each level. This LL band is used for further processing. The performance is evaluated at five levels using different wavelets. For every wavelet at each level, system achieves DAR value of about 90% and minimum FPR as shown in Figs. 2 and 3. In SWT, lines of both DAR and FPR remain as almost constant for every wavelet. System based on SWT is more effective than DWT.

Performance Evaluation Using LWT. Performance evaluation of the system is shown in Table 3. LWT is applied for the forged image using different wavelets with five levels of decomposition. The system shows high accuracy rate about 95% upto first four levels of decomposition as compared to DWT and SWT. The graphical representation of LWT shows that wavelet sym8 has high DAR with minimum FPR as shown in Figs. 2 and 3.

Table 2. Accuracy of detection using SWT

Type of wavelet	Accuracy parameter	Levels of decomposition				
		1	2	3	4	5
Haar	DAR	0.9316	0.9422	0.9434	0.9250	0.9299
	FPR	0.0684	0.0578	0.0566	0.0750	0.0701
Db4	DAR	0.9435	0.9479	0.9454	0.9356	0.9371
	FPR	0.0565	0.0521	0.0546	0.0644	0.0629
Db6	DAR	0.9457	0.9427	0.9456	0.9370	0.9314
	FPR	0.0543	0.0523	0.0544	0.0630	0.0686
Sym8	DAR	0.9471	0.9446	0.9445	0.9373	0.9407
	FPR	0.0529	0.0554	0.0555	0.0627	0.0593

Table 3. Accuracy of detection using LWT

Type of wavelet	Accuracy parameter	Levels of decomposition				
		1	2	3	4	5
Haar	DAR	0.9507	0.9377	0.9473	0.9326	0.8789
	FPR	0.0493	0.0623	0.0527	0.0674	0.1211
Db4	DAR	0.9478	0.9504	0.9460	0.9410	0.8759
	FPR	0.0522	0.0496	0.0540	0.0590	0.1445
Db6	DAR	0.9519	0.9512	0.9490	0.9500	0.9023
	FPR	0.0481	0.0488	0.0510	0.0500	0.0977
Sym8	DAR	0.9519	0.9515	0.9504	0.9453	0.9258
	FPR	0.0481	0.0485	0.0496	0.0547	0.0742

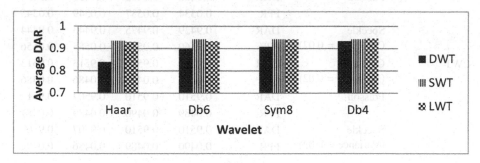

Fig. 2. Graphical representation of average DAR for DWT, SWT and LWT

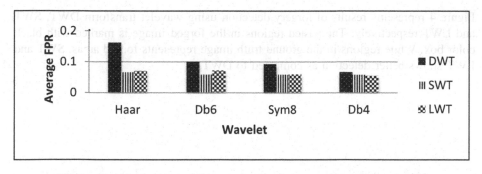

Fig. 3. Graphical representation of average FPR for DWT, SWT and LWT

Performance Evaluation with Presence of Noise. Performance analysis of DWT, SWT, and LWT with presence of noise is shown in Table 4. Three types of noises such as Gaussian, Poisson and Speckle are introduced into the forged image. Gaussian noise and speckle noise with variance 0.02 is added to the image. The results indicate that the system has robustness against noise since the average value of DAR is above 90% in all levels of decomposition for every wavelet. So the system can detect forgery occurred in the image even in the presence of noise.

Table 4. Accuracy of detection with presence of noise

Type of wavelet transform	Type of noise	Accuracy parameter	Type of wavelet			
			Haar	Db4	Sym8	Db6
DWT	Gaussian (Variance = 0.02)	DAR	0.9418	0.9334	0.9354	0.9300
		FPR	0.0581	0.0666	0.0645	0.0700
	Poisson	DAR	0.9433	0.9321	0.9395	0.9380
		FPR	0.0567	0.0678	0.0608	0.0675
	Speckle (Variance = 0.02)	DAR	0.9353	0.9269	0.9392	0.9324
		FPR	0.0646	0.0730	0.0608	0.0675
SWT	Gaussian (Variance = 0.02)	DAR	0.9465	0.9468	0.9442	0.9451
		FPR	0.0534	0.0531	0.0558	0.0549
	Poisson	DAR	0.9466	0.9462	0.9450	0.9454
		FPR	0.0534	0.0537	0.0549	0.0545
	Speckle (Variance = 0.02)	DAR	0.9439	0.9475	0.9443	0.9444
		FPR	0.0560	0.0525	0.0555	0.0556
LWT	Gaussian (Variance = 0.02)	DAR	0.9507	0.9510	0.9513	0.9513
		FPR	0.0492	0.0490	0.0486	0.0486
	Poisson	DAR	0.9510	0.9510	0.9504	0.9512
		FPR	0.0489	0.0490	0.0489	0.0487
	Speckle (Variance = 0.02)	DAR	0.9510	0.9510	0.9507	0.9507
		FPR	0.0490	0.0489	0.0486	0.0492

3.2   Results of Forgery Detection

Figure 4 represents results of forgery detection using wavelet transform DWT, SWT and LWT respectively. The pasted regions in the forged image is marked with black color box. White regions in the ground truth image represents forged areas. SWT and LWT shows better detection as compared to DWT.

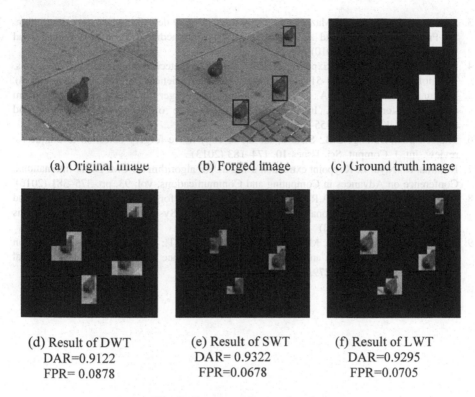

(a) Original image   (b) Forged Image   (c) Ground truth image

(d) Result of DWT
DAR=0.9122
FPR= 0.0878

(e) Result of SWT
DAR= 0.9322
FPR=0.0678

(f) Result of LWT
DAR=0.9295
FPR=0.0705

Fig. 4. Results of forgery detection

4 Conclusion

The detection of digital image forgery is a challenging research topic in image forensics. Copy move forgery detection based on wavelet transform have been performed and analysis is done with wavelet transforms DWT, SWT and LWT. Investigations have been performed with different wavelets namely haar, db4, db6 and sym8 for five levels of decomposition. It is observed that stationary and lifting wavelet transforms perform more efficiently than DWT in the forged image. Noises like Gaussian, speckle and poisson added in the image to show robustness of the system against noise. The implemented method can detect copy move forgery efficiently with less computational time.

References

1. Solorio, B., Nandi, A.K.: Exposing duplicated regions affected by reflection, rotation and scaling. In: Proceedings of International Conference on Acoustics Speech and Signal Processing, pp. 1880–1883 (2011)
2. Gupta, A., Saxena, N., Vasistha, S.K: Detecting copy move forgery using DCT. Int. J. Sci. Res. Publ. **3**, 1–4 (2013)

3. Ardizzone, E., Bruno, A., Mazzola, G.: Copy-move forgery detection via texture description. In: Proceedings of the 2nd ACM Workshop on Multimedia in Forensics, Security and Intelligence, pp. 59–64 (2010)
4. Popescu, A.C., Farid, H.: Exposing digital forgeries by detecting duplicated image regions. Technology report TR2004-515, Department Computer Science, Dartmouth College (2004)
5. Mohamadian, Z., Pouyan, A.: Detection of duplication forgery in digital images in uniform and non-uniform regions. In: International Conference on Computer Modelling and Simulation (UKSim), pp. 455–460 (2013)
6. Thajeel, S.A., Sulong, G.B.: State of the art of copy-move forgery detection techniques: a review. Int. J. Comput. Sci. Issues **10**, 174–183 (2013)
7. Reshma, R., Niya, J.: Keypoint extraction using SURF algorithm for CMFD. In: International Conference on Advances in Computing and Communications, vol. 93, pp. 375–381 (2016)
8. Hashmi, M.F., Hambarde, A.R., Keskar, A.G.: Copy move forgery detection using DWT and SIFT features. In: International Conference on Intelligent Systems Design and Applications (ISDA), pp. 188–193 (2013)
9. Mahmooda, T., Irtazab, A., Mehmood, Z., Mahmood, M.T.: Copy move forgery detection through stationary wavelets and local binary pattern variance for forensic analysis in digital images. Forensic Sci. Int. **279**, 8–21 (2017)

High Resolution 3D Image in Marine Exploration Using Neural Networks - A Survey

R. Dorothy$^{(\boxtimes)}$ and T. Sasilatha

Department of EEE, AMET Deemed to be University, Chennai, India
rdorothyjaj.dgl@gmail.com, deaneeem@ametuniv.ac.in

Abstract. Stereovision is a system used to remake 3D perspective of a protest from two or extra 2d visual observation by using either neighborhood-based generally or features based extraction strategies. This paper proposes a local stereo matching algorithmic rule for correct disparity estimation by using the salient features and novel back propagation maximum neural network. The 3D image is obtained by using different types of algorithms. Once the 3D picture is gotten the improvements in sub-base recognizable proof brings 3D reflection seismic, constantly utilized in a natural compound investigation, to the shallow study advertise by down-scaling the regular strategies to acknowledge decimeter determination imaging of the most noteworthy several meters of the sub-surface in three measurements. Shallow high determination sub-base profiling depends for the most part on single-channel 2d techniques. In qualification of the 2d strategies that produce singular vertical cross areas of the sub-surface, the 3D strategy joins information gathered over the review space into a data volume. The information will then be seen in any introduction autonomous of the obtaining course, depicting structures and questions in three measurements with expanded quality and determination.

Keywords: Adaptive expectation maximization algorithm
Back propagation maximum neural network · Local stereo matching
Hybrid neural network · Multiple fitting algorithms

1 Introduction

A short description of the papers surveyed for the emergence of the planned model is given. The subsequent are the papers gave an updated plan about the present work. Numerous methodologies being performed to design 3D image in marine and different papers associated with the development of the image in numerous fields are being presented. In this paper, we have a tendency to propose neural network novel model for similarity measure that is robust to disparity mapping and Stereo Correspondence.

1.1 2D Hybrid Bilateral Filter

Image restoration refers to the technique that expects to recuperate a best quality unique picture from an adulterated adjustment of that picture given a particular model for debasement technique. The hybrid bilateral filter (HBF) is utilized for sharpness change

© Springer Nature Singapore Pte Ltd. 2018
I. Zelinka et al. (Eds.): ICSCS 2018, CCIS 837, pp. 141–146, 2018.
https://doi.org/10.1007/978-981-13-1936-5_16

and clamor expulsion. The HBF is utilized to hone a photo by expanding the borders while not producing overshoot or undershoot.

The ABF doesn't include edge recognition either their introduction of the picture or extraction of edges. In the ABF, the sting of a slant is produced by revising the reference diagram by means of a variable channel with counterbalance and expansiveness. HBF repaired pictures are swindler than those repaired by the respective channel. The 2d hybrid bilateral needs a 4-double loop, hence it's not brisk unless modest channels are utilized.

1.2   Neural Network Mode

The neural system ought to be prepared with the preparation methodology before processing the coordinating degree for each part (Fig. 1). To beat them over previously mentioned drawback, the neural system is utilized. A neural system could be a system of neurons. The system has an information relate degreed a yield. It might be prepared to pass on the best possible yield for a chose input. The nerve cell is in-charge of clear activities, yet the entire framework will make parallel checks on account of the result of its wide parallel structure. The input sources zone unit summed up and roused into an assumed trade in this manner known as exchange works. The output of the exchange operator is considered as a result of the output of the nerve cell. This output may again associate with the contributions of various neurons leading to a large network of neurons. The network that provides the matching degree price on the brink of one permanently match to try and also the price on the brink of zero for a foul match try.

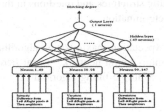

Fig. 1. Block diagram of neural network

1.2.1   Feature Extraction – Reliable Multiple Fitting Algorithms

The most important component for a feature point is that it can differ from its neighboring image points. On the off chance that, it wouldn't be possible to match it with a corresponding point in another image. Therefore, the features are differentiated by the neighboring image points obtained after a small displacement. Reliable multiple fitting algorithms are used to calculate median variance and standard deviation.

2 Literature Survey

2.1 Deep Learning-Based Recognition of Underwater Target [1]

Underwater target recognition remains a difficult task because of the advanced and changeable surroundings. There is an enormous range of strategies to handle this drawback. However, most of them fail to hierarchically extract deep features. in this paper, a unique deep learning framework for underwater target classification is planned. First, rather than extracting features hoping on professional data, sparse auto encoder (AE) is used to find out invariant options from the spectral information of underwater targets. Second, stacked auto encoder (SAE) is employed to induce high-level options as a deep learning technique. At last, the joint of SAE and softmax is projected to classify the underwater targets. Experiment results with the received signal data from three totally different targets on the ocean indicated that the projected approach will get the very best classification accuracy compared with support vector machine (SVM) and probabilistic neural network (PNN).

2.2 Applications for Advanced 3D Imaging, Demonstrating, and Printing Systems for the Organic Sciences [2]

This paper represents several zoological activities, together with scientific expeditions, or in academic settings, necessitates troubled removal of specimens from their natural setting. The anthropogenetic impact theory present by Cour champ clearly indicates the unsustainability of current practices, with dramatic changes necessary for the welfare and property of our ecosystems. For each public, and tutorial, education and analysis, there's a transparent link between increasing rarity and access limitation sure enough specimens, resulting in the decrease in each public awareness and capability for education and analysis. From this research's perspective, an associate idealized scenario is wherever totally digital specimens will be created to represent 3D geometry, visual textures, mechanical properties and specimen practicality, granting precise replicas to be generated from this digital specimen once needed. Combining this with the utilization of video game increased reality, and mixed reality will satisfy the academic and analysis desires however additionally the property of our ecosystems.

2.3 Peer-Reviewed Technical Communication Prudent Image Preprocessing of Digital Holograms of Marine Plankton [3]

This paper exhibits a gathering of pictures preprocessing approaches unit produced for the strategy being film remade from advanced multi-dimensional images. Initial, a limit based equation of picture division is arranged and connected to separate the areas of life form from the principal computerized pictures. To support the execution of picture division, relate adequate channel is received to decrease the foundation motion from the picture and furthermore the picture dark level is changed in accordance with strengthens the picture refinement. Second, we tend to build up a special and practical edge location technique deliberately for the double pictures. Third, we tend to propose and utilize a simple affix code-based equation to take out the single-pixel branches on the frame limit, which can encourage limit following work steadily. At that point, relate equation is enhanced and connected to follow the limits of the life form districts. This equation is streamlined bolstered the association between two successive chain-codes such it's fast on execution. At last, break purposes of the shape limit zone unit quickly recognized upheld chain-codes and furthermore the limit is drawn compactly by a plane figure contained these focuses. When pictures territory unit pre-prepared by these methodologies, some excess information of the frame is lessened which will quicken the running rates of extra picture process and help distinguishing proof and characterization of a living being at the species level. We tend to break down the exactness and intensity of our calculations. The outcomes demonstrate that our equation of picture division includes a brilliant execution in precision. Our edge discovery strategy furthermore beats the customarily utilized edge recognition systems as far as confinement execution and furthermore the timeframe.

2.4   Towards Real-Time Underwater 3D Reconstruction with PlenopticCameras [4]

In Achieving continuous observation is basic to building up a completely self-sufficient framework that can detect, explore, and connect with its condition. Recognition errands like online 3D reproduction and mapping are strongly contemplated for earthly apply autonomy applications. Notwithstanding, attributes of the submerged space like lightweight weakening and lightweight scrambling abuse the steadiness limitation, that is the partner hidden suspicion in courses produced for arriving based generally applications. Furthermore, the confused idea of daylight proliferation submerged points of confinement or maybe keeps the subsea utilization of period profundity sensors utilized in dynamic earthbound mapping techniques. There are late advances inside the improvement of plenoptic (likewise called lightweight field) cameras that utilization a variety of little focal points catching every force and beam bearing to adjust shading and profundity estimating from a solitary uninvolved detecting component. This paper exhibits a conclusion to-end framework to tackle these cameras to give constant 3D reproductions submerged. Results are given for data assembled amid a water tank and along these lines, the anticipated method is substantial quantitatively through examination with a ground truth 3D show accumulated noticeable all around to exhibit that the arranged approach will create rectify 3D models of articles submerged continuously.

2.5   Recognition of Harms in the Submerged Metal Plate Utilizing Acoustic Backward Dissipating and Image Preparing Strategies [5]

In this paper Non-dangerous testing and basic well-being, observing are fundamental for security and unwavering quality of aqua active vitality related fields. This paper demonstrates the issue of harm location of submerged limited length plates utilizing acoustic backward dissipating and picture preparing strategies. Time arrangement signals and the 2D pictures got from these signs have been concentrated to enhance discovery precision. Ideal parameters are chosen for closed-end edge echoes disposal and linearization is utilized to lessen a computational intricacy. A hearty and straightforward strategy was proposed to identified and limit a conceivable harm in 2D pictures in light of picture handling and investigation. The trial comes about demonstrate that the discovery rate for break harm achieves 100% and restriction exactness achieves 96% by and large.

2.6 Minimal Effort 3D Submerged Surface Entertainment System by a Picture Preparing [6]

In this paper, a non-meddling strategy to reproduce submerged surfaces is portrayed in this work. As fundamental favorable circumstances, it just requires working some minimal effort material and subroutines for the outstanding Matlab programming. The strategy depends on a point design coordinating system and on the handling of a few pictures of anticipated lines at first glance to be reproduced. The technique has been approved by reproducing the shape and measurements of a seashell and a reversed spoon, and it is appeared to be legitimate for examining 3D surfaces with an exactness blunder relative to the anticipated line thickness and the quantity of them. After the approval test, the mistake did not surpass 1 mm, which gives a worldwide normal blunder of 2.4% in relative terms. The created programming likewise provides for the client the likelihood of getting quantitative information from the 3D surface, for example, the most extreme and least estimations of the remade surface, and the volume of various locales of the surface.

2.7 Low Cost a Stereo Framework for Imaging and 3D Recreation of Submerged Organisms [7]

This paper introduces a self-sufficient minimal effort gadget for submerged stereo imaging and 3D remaking of marine life forms (benthic, fishes, full scale, and uber zooplankton) and seabed with a high precision. The framework is intended for arrangements installed self-governing, settled and towed stages. The inward equipment comprises two Raspberry Pi scaled-down PCs and two Raspberry camera modules. The 3D imaging procurement framework is completely programmable in obtaining planning and catchsettings. At the point when tried on sets of pictures containing objects of known size, the framework restored exactness of metric estimations of the request of 2%. The framework is proposed as a model, and a joint effort with organizations has been built up keeping in mind the end goal to understand a total business item.

2.8 Submerged 3D Capture Using a Low-Cost Commercial Depth Camera [8]

This paper displays a submerged 3D catch for utilizing a business profundity camera. Submerged catch frameworks utilize standard cameras. Consistent is valid for a profundity camera being utilized submerged. We tend to portray an action system that redresses the profundity maps of refraction impacts Our approach offers energizing prospects for such applications. To the least complex of our data, our is that the first approach that with progress exhibits submerged 3D catch exploitation gleam value profundity cameras like Intel Real Sense. They portray a whole framework, and in addition securing lodging for the profundity camera that is proper for hand-held use by a jumper. Their primary commitment is Associate in nursing simple to-utilize action procedure that we tend to judge on show data furthermore as 3D reproductions amid a research center stockpiling tank.

3   Conclusion

In this paper, we have given a thorough study of the developing advancement on the 3D picture. This strategy demonstrates the utilization of an inventive engineering of Hopfield in view of the neural network – Hybrid-Maximum Network. The system presented here has been utilized as a part of the stereo coordinating procedure.

In our proposed mode back propagation system is utilized to accomplish proficient dissimilarity mapping with the reduced no occluded region and buried objects.

References

1. Zhang, X., Yu, Y., Niu, L.: Deep learning-based recognition of underwater target. IEEE Trans. **20**(1), 14–22 (2017)
2. Digumarti, S.T., Taneja, A., Thomas, A.: Disney Research Zurich, ETH Zurich Disney Research Zurich Walt Disney World "Applications for advanced 3D imaging, modelling, and printing techniques for the biological sciences". IEEE Trans. **20**(1), 14–22 (2017)
3. Liu, Z., Watson, J., Allen, A.: Peer-reviewed technical communication efficient image preprocessing of digital holograms of marine plankton. IEEE Trans. **30**(1), 22–44 (2017)
4. Skinner, K.A., Johnson-Roberson, M., et al.: Towards real-time underwater 3D reconstruction with plenoptic cameras. IEEE Multimed. **19**(2), 4–10 (2016)
5. Campos, R., Garcia, R., Alliez, P., Yvinec, M.: A surface reconstruction method for in-detail underwater 3D optical mapping. Int. J. Robot. Res. **34**(1), 64–89 (2015)
6. Garcia, P.R., Neumann, L.: Low cost 3D underwater surface reconstruction technique by image processing. Springer (2014)
7. Porathe, T., Prison, J., Man, Y.: Low cost stereo system for imaging and 3D reconstruction of underwater organisms. In: Human Factors in Ship Design & Operation, London, U.K., p. 93 (2014)
8. Bianco, G., Gallo, A., Bruno, F., Muzzupappa, M.: Underwater 3D capture using a low-cost commercial depth camera. Sens. (Basel) **13**(8), 11007–11031 (2013)
9. Pinto, T., Kohler, C., Albertazzi, A.: Regular mesh measurement of large free form surfaces using stereo vision and fringe projection. Opt. Lasers Eng. **50**(7), 910–916 (2012)
10. Hansen, R.E.: Synthetic aperture sonar technology review. Mar. Technol. Soc. J. **47**(5), 117–127 (2013)

Ship Intrusion Detection System - A Review of the State of the Art

K. R. Anupriya$^{(\boxtimes)}$ and T. Sasilatha

Department of EEE, AMET Deemed To be University, Chennai, India
anupriyakumaradhas09@gmail.com, deaneeem@ametuniv.ac.in

Abstract. Surveillance is a serious problem for border control, protection of sea surface areas, port protection and other security of commercial facilities. It is specifically challenging to secure the border areas, battlefields from human and nonhuman intruders and to protect sea surface areas from trespassing of unlicensed marine vessels. In this paper, a review is made on various ship intrusion detection systems. The review analyzes the whole active ship intrusion detection system. Through the extensive survey, the whole pose of the active ship intrusion detection system is analyzed. Since the security issues are at an increased level, the study and survey about ship intrusion detection system have paid a lot of attention.

Keywords: Border control · Intrusion detection system · Marine vessels
Wireless sensor network

1 Introduction

Intrusion detection is a major problem in border areas. It is very hard to detect the intervention in large areas because it is difficult to human to check out those areas often. Nowadays our society facing major problems like Terrorism, Insecurity and other crimes. In our society people are having a panic for being attacked by bandits, robbers, pirates, and crooks. Surveillance is the primary issue in today's world and 24 h human security is just not practical. To overcome above mentioned security problems, it is necessary to introduce a brilliant security system.

CCTV cameras also have an important role in maritime surveillance system used Ship Intrusion Security System based on CCTV camera can be used to produce video recordings for security purposes. Most commonly used surveillance techniques are Ship Intrusion Security System based on RADAR and Ship Intrusion Security System based on the Satellite image. In this Ship Intrusion Security System based on the Satellite imaging, to perform the monitoring task the system architecture based on an object-oriented methodology [2]. This Ship Intrusion Security System has a completely automated shoreline intrusion security detection device, completely or partially automated intrusion security detection device in seashore areas and a partially automated intrusion security detection device in border areas. The high security in the maritime harbour and border areas importance cannot be undervalued. In the world, 80 percent of world business trade operations are done with the help of sea transportation. Surveillance in

© Springer Nature Singapore Pte Ltd. 2018
I. Zelinka et al. (Eds.): ICSCS 2018, CCIS 837, pp. 147–154, 2018.
https://doi.org/10.1007/978-981-13-1936-5_17

seashore area is the major problem encountered by the whole world. Nowadays, video surveillance is necessary in the protection of port areas and border areas. In order to overcome the security issues, harbour areas request a recent advanced supervision camera technology.

Wireless Sensor Network (WSN) has been emerging in the last decade as a powerful tool for connecting the physical and digital world. (WSNs) are developed for terrestrial ship intrusion detection recently. These wireless sensor networks deploy sensors in the border area to monitor the intervention and to detect intrusions [16–18].

2   Evolution of Ship Intrusion Detection Security System

2.1   Ship Intrusion Security System Based on CCTV Camera

CCTV (Closed-circuit television) camera plays a important role in maritime security. Ship Intrusion Security System based on CCTV camera can be used to produce video recordings for security purpose. A Basic Closed-circuit television (CCTV) camera system architecture consists of a camera, which is straightly connected to a LCD display (Liquid Crystal Display) using a coaxial cable. The camera captured the information in the form of video, each video consist of several frames. This captured video and images can be displayed using the LCD display, which is used to detect trespassing unauthorized marine vehicles. Even if CCTV (Closed-circuit television) camera-based surveillance system is very easy and simple solution, but 24 h continuous checking of the video recording is not possible because of the human error (Fig. 1).

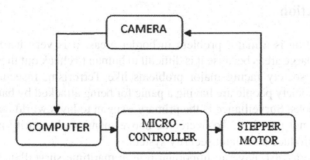

Fig. 1.   CCTV camera based surveillance system block diagram

Motion Detection and Tracking based Camera Surveillance System
In this method [2], the surveillance system has used the camera with artificial intelligence. This security system has the camera which has 360° rotation in order to monitor the movements of the intruders, which is called object tracking. The security system has a microcontroller and a computer along with the high-resolution camera which operates together with the system. To detect and track the intruders this security system uses some image processing techniques as well as some basics of microcontrollers. The integration of both tracking and motion-based methods detect intrusion smartly and provide better performance.

2.2   Ship Intrusion Security System Based on Radar

Ship Intrusion Security System based on Radar method is used to detect the trespassing of marine vessels. In this RADAR based detection system, seashore environment background is shown as dark and targets are shown as bright in the SAR images, which makes this method easy to detect trespassing of marine vessels. But when the wind is ferocious, large ocean waves will be stirred, due to this strong backscattering echo can be raised. This situation causes more difficulties. The overall accuracy of security system turns out to be poor, due to the worst weather conditions.

LSMDRK based PolSAR Ship Detection

In this method [3], the local scattering mechanism difference based on regression kernel (LSMDRK) is developed as a discriminative feature for ship detection. In this method, local scattering mechanism difference based on regression kernel (LSMDRK) is developed for ship intrusion detection. In this, the intrusion detection can be done by using a RADARSAT-2 data set. This method provides better detection on weak targets compared to some classical intrusion detection methods.

SAR Ship Detection based on Haar-Like Features

In worst weather conditions, the ship detection is at seashore environment is more complex due to the absence of night visibility, and wide areas of concern. The surveillance of an exclusive economic zone (EEZ) areas are a essential part of the world. Synthetic Aperture Radar (SAR) images can be effectively used to monitor an exclusive economic zone (EEZ) areas. In order to protect the border areas, scientific investigations on present and future methods for intrusion detection security systems are needed to be evaluated constantly. The multiple sources of SAR data can be used to create the data set, which is used for ship detection. Synthetic aperture radar (SAR) [4] images provides a required coverage of area at a poor resolution. A SAR based ship intrusion detection method is used standard constant false alarm rate (FAR) prescreening, which is $1.47 \times 10 - 8$ across a large swath Sentinel-1 with cascade classifier ship discriminator and processed with RADARSAT-2 newly created SAR data set. Ships detection is done by using adaptive boosting training on the classifier based Haar-like features with an accuracy of 89.38%.

CopSAR based Maritime Surveillance

In this intrusion detection method [5], a synthetic aperture radar (SAR) images based security technique used for maritime surveillance, in order to detect bright targets over a dark background, to reduce the amount of processed and stored data, to increase the range swath, with no geometric resolution loss. Accordingly, this method can be used for maritime surveillance. This method developed a new synthetic aperture radar acquisition mode, which is a simple processing technique. This new synthetic aperture radar acquisition mode used coprime array beamforming concept, in this two pulses which having Nyquist pulse repetition frequencies (divided to coprime integer number) transmitted separately, these sequence processed with standard synthetic aperture radar processing. After the synthetic aperture radar processing, the aliased images are combined in order to eliminate the aliasing. CopSAR based Maritime Surveillance provide the better performance in the ship intrusion detection.

DNN (Deep Neural Network) and ELM (Extreme Learning Machine) based ship detection on Spaceborne Optical Image
In maritime surveillance, spaceborne images [6] based ship detection is very attractive. Because of their higher resolution and more visualized contents, optical images based ship detections are more suitable compared to other remote sensing images. However, marine vehicle intrusion detection system based on spaceborne images has two shortcomings are available. (1) Spaceborne Optical Image-based ship detection results are affected by fog, clouds and sea surfs, when compared to infrared and SAR images. (2) due to their higher resolution, the ship detection is more difficult. In order to solve these problems, Deep Neural Network and Extreme Learning Machine algorithms can be used to detect a ship in seashore environment. In the Deep neural network algorithm, the extracted wavelet coefficients from compressed JPEG2000 image are combined with DNN and Extreme learning machine. Deep Neural Network can be used for high-level classification and representation of features and ELM can be used for decision-making and feature pooling. Deep Neural Network (DNN) and Extreme Learning Machine (ELM) based ship detection system has less detection time and achieves high detection accuracy.

Undersampled SAR based maritime surveillance
In surface monitoring scenarios, synthetic aperture radar (SAR) based intrusion detection systems need low-pulse repetition frequency (PRF) (which is smaller than the Doppler bandwidth) for wide swath image, depending upon the minimum antenna area constraint, which cause azimuth ambiguities. Undersampled SAR [7] based maritime surveillance system used to detect the intruding marine vehicles over the border areas. In this method azimuth ambiguity signals are adopted a range sub spectra concept, to misregister the azimuth ambiguity signals. In addition, undersampled SAR based maritime surveillance system uses both principal component analysis (PCA) and k-means clustering algorithms. By adjusting the ambiguities in the corresponding undersampled SAR image, it can be mitigated. This security system is only suitable for undersampled SAR images which having bright targets with dark backgrounds. Undersampled SAR based maritime surveillance system provides better performance compared to other traditional surveillance systems.

Maritime ship intrusion detection on high-resolution remote sensing images using RIGHT algorithm
In this ship detection method [8] RIGHT (Robust Invariant Generalized Hough Transform) algorithm can be used for the detection purpose. The ship-detection method is based on High-resolution remote sensing images. The RIGHT (Robust Invariant Generalized Hough Transform) algorithm is an extraction algorithm. In order to increase the adaptability of the RIGHT (Robust Invariant Generalized Hough Transform) algorithm, some iterative training methods are used for learning robust shape model automatically. This robust shape model can take target's shape variability, which is available in the training dataset. According to their importance, each targets used in this model equipped

with corresponding individual weights, which will reduce the false positive rate. In this RIGHT (Robust Invariant Generalized Hough Transform) based ship detection framework the effectiveness can be improved through the iteration process.

SVM based Ship Intrusion Detection Security System
Surveillance is a serious problem in border control, protection of sea surface areas, port protection and other security of commercial facilities. It is specifically challenging to secure the border areas, battlefields from human and nonhuman intruders and to protect ocean surface areas and active port areas from trespassing of unlicensed marine vehicles. Support vector machine (SVM) algorithm [9] is combined with image processing techniques, to detect trespassing of unauthorized marine vessels, to provide better detection. So, this SVM based Ship Intrusion Detection Security System used as a real-time surveillance system in seashore environments.

2.3 Ship Intrusion Security System Based on Satellite Imaging

In this Ship Intrusion Security System based on Satellite imaging, to perform the monitoring task the system architecture based on an object-oriented methodology [20]. This Ship Intrusion Security System has a completely automated shoreline intrusion security detection device, completely or partially automated intrusion security detection device in seashore areas and a partially automated intrusion security detection device in border areas. At the time of intrusion detection sometimes satellite images are not clear due to clouds. Due to this problem, this method cannot produce the better result. Apart from this, the Satellite imaging based Ship Intrusion Security System is very expensive.

2.4 Ship Intrusion Security System Using Terrestrial Sensor

Terrestrial sensor based Ship Intrusion Security System is widely discussed [14–16]. Wireless Sensor Network (WSN) has been emerging in the last decade as a powerful tool for connecting the physical and digital world. In order to improve the security level in the border areas, sensors can be deployed in the border area to monitor the intervention and to detect intrusions. Still, these wireless sensor networks may work well on the earth surface area, it is challenging to deploy these sensors on the sea surface for ship intrusion detection. When terrestrial sensors are deployed on the sea surface area, they move around randomly, because the sensors get tossed by ocean waves. When the sensors tossed by the ocean waves, the sensing operation will affect. Due to this above-mentioned problem, the intrusion detection task becomes difficult and this will reduce performance of the system (Fig. 2).

Fig. 2. Wireless sensor network deployment

Wireless Sensor-Based Ship Intrusion Detection
Wireless sensor network-based intrusion detection system [10] armed with three-axis accelerometer sensors. These sensors can be deployed on the sea shore areas to detect intrusion of unlicensed marine vehicles. In order to detect the trespassing of unauthorized marine vessels, the Wireless sensor network-based intrusion detection system is combined with signal processing techniques by distinguishing the ocean waves and ship-generated waves. To improve detection reliability, this ship intrusion detection system introduces spatial and temporal correlations of the intrusions. The real data obtained from this experiments are evaluated and from these evolution results, the intrusion detection system provides better detection ratio and detection latency.

Intruder ship tracking in the wireless environment
Intrusion detection is a challenging task for all Harbours or Naval Administration to restrict and monitor the movement of defence or commercial ships are challenging task for all port areas and naval administration. Most commonly used surveillance techniques RADAR based Ship Intrusion detection Security System and Satellite imaging based Ship Intrusion detection Security System. In this RADAR based detection system, seashore environment background is shown as dark and targets are shown as bright in the SAR images, which makes this method easy to detect trespassing of marine vessels. But when the wind is ferocious, large ocean waves will be stirred, due to this strong backscattering echo can be raised. This situation causes more difficulties. The overall accuracy of security system turns out to be poor, due to the worst weather conditions. At the time of intrusion detection sometimes satellite images are not clear due to clouds. Due to this problem, this method cannot produce the better result. Apart from this, the Satellite imaging based Ship Intrusion Security System is very expensive. This wireless based intrusion detection security system [11] introduces a reliable intrusion detection algorithm, Which classifies different kinds of objects approaching the experimental setup and that objects present out of phase with the ocean waves. The intrusion detection algorithm depends upon the superimposition of temporal and spatial correlation values of sensor nodes that are deployed in the sea surface up to a certain distance. This intrusion detection system detects intruder ship more efficiently.

Maritime Surveillance System Using LABVIEW
The main aim is to detect the this maritime surveillance system is used to detect the unlicensed marine vehicles, which cross the border areas in sea surface using axis sensors and ultrasonic sensor [12]. These sensors deployed on the grid, which is separated by the distance of 40 km. If the intruder ship crosses the border, the sensors sense the objects and measure the intruder distance and angle. This framework can be graphically displayed in the LabVIEW (Laboratory Virtual Instrumentation Engineers Workbench) in the form of graphical representation. If the intrusion is detected in the border area an alert message sends to the consent authorities using GSM (Global System for Mobile communication).

FPGA based Ship Intrusion Detection
This method points out the advantages of Wireless Sensor Networks (WSN) in ocean-ography, which introduce Reconfigurable SoC (RSoC) architecture [20] to detect ship intruders. The tri-axis digital accelerometer sensor is interfaced with FPGA-based Wire-less sensor node. To detect trespassing of ships, the ship-generated waves are distin-guished from the ocean waves, by using signal processing techniques. This framework is a three level detection system, Which can detect intrusion of unlicensed marine vehi-cles in the border areas. This framework uses Xilinx ISE simulator for simulation.

3 Conclusion

In this paper a survey of various intrusion detection security system based on CCTVs (Closed Circuit Television), RADAR (Radio Detection and Ranging), Satellite Imaging are discussed. In order to protect the border areas, harbor areas and secured industrial spaces from the intrusion of unauthorized marine vehicles, various researchers proposed various ship Intrusion detection security systems. Some of the Ship Intrusion Detection system has most advantages over intruder detection and some may have some chal-lenges. This review will help the researchers to know about the various ship intrusion detection techniques with its strength and challenges.

References

1. Nair, A., Saraf, R., Patil, A., Puliyadi, V., Dugad, S.: Electronic poll counter of crowd using image processing. Int. J. Innov. Res. Comput. Commun. Eng. **4**(3), 4249–4258 (2016)
2. Choudhari, A., Gholap, V., Kadam, P., Kamble, D.: Camera surveillance system using motion detection and tracking. Int. J. Innov. Res. Adv. Eng. (IJIRAE) **1**(4) (2014)
3. He, J., Wang, Y., Liu, H., Wang, N.: PolSAR ship detection using local scattering mechanism difference based on regression kernel. IEEE Geosci. Remote Sens. Lett. **14**(10), 1725–1729 (2017)
4. Schwegmann, C.P., Kleynhans, W., Salmon, B.P.: Synthetic aperture radar ship detection using haar-like features. IEEE Geosci. Remote Sens. Lett. **14**(2), 154–158 (2017)
5. Di Martino, G., Iodice, A.: Coprime synthetic aperture radar (CopSAR): a new acquisition mode for maritime surveillance. IEEE Trans. Geosci. Remote Sens. **53**(6), 3110–3123 (2015)

6. Tang, J., Deng, C., Huang, G.-B., Zhao, B.: Compressed-domain ship detection on spaceborne optical geoscience and remote sensing **53**(3) (2015)
7. Wang, Y., Zhang, Z., Li, N., Hong, F., Fan, H., Wang, X.: Maritime surveillance with undersampled SAR. IEEE Geosci. Remote Sens. Lett. **14**(8), 1423–1427 (2017)
8. Xu, J., Sun, X., Zhang, D., Fu, K.: Automatic detection of inshore ships in high-resolution remote sensing images using robust invariant generalized hough transform. IEEE Geosci. Remote Sens. Lett. **11**(12), 2070–2074 (2014)
9. Dugad, S., Puliyadi, V., Palod, H., Johnson, N., Rajput, S., Johnny, S.: Ship intrusion detection security system using image processing & SVM. In: International Conference on Nascent Technologies in the Engineering Field (ICNTE-2017). IEEE (2017)
10. Luo, H., Wu, K., Guo, Z., Gu, L., Ni, L.M.: Ship detection with wireless sensor networks. IEEE Trans. Parallel Distrib. Syst. **23**(7), 1336–1343 (2012)
11. Rao, M., Kamila, N.K.: Tracking intruder ship in wireless environment. Hum. Centric Comput. Inf. Sci. **7**, 14 (2017). https://doi.org/10.1186/s13673-017-0095-4
12. Madhumathi, R.M., Jagadeesan, A.: Int. J. Innov. Res. Electr. Electron. Instrum. Control Eng. **2**(10) (2014)
13. Latha, P., Bhagyaveni, M.A., Lionel, S.: A reconfigurable soc architecture for ship intrusion detection. J. Theor. Appl. Inf. Technol. **60**(1) (2014)
14. Gu, L., et al.: Lightweight detection and classification for wireless sensor networks in realistic environments. In: Proceedings of Third International Conference on Embedded Networked Sensor Systems (SenSys 2005), pp. 205–217 (2005)
15. Arora, et al.: A line in the sand: a wireless sensor network for target detection, classification, and tracking. Comput. Netw. **46**(5), 605–634 (2004)
16. Duarte, M., Hu, Y.H.: Vehicle classification in distributed sensor networks. J. Parallel Distrib. Comput. **64**(7), 826–838 (2004)
17. Latha, P., Bhagyaveni, M.A.: Reconfigurable FPGA based architecture for surveillance systems in WSN. In: Proceedings of IEEE International Conference on Wireless Communication and Sensor Computing (ICWCSC), pp. 1–6 (2010)
18. Kumbhare, A., Nayak, R., Phapale, A., Deshmukh, R., Dugad, S.: Indoor surveillance system in dynamic environment. Int. J. Res. Sci. Innov. **2**(10), 103–105 (2015)
19. Bergeron, A., Baddour, N.: Design and development of a low-cost multisensor inertial data acquisition system for saiing. IEEE Trans. Instrum. Meas. **63**(2), 441–449 (2014)
20. Jacob, T., Krishna, A., Suresh, L.P., Muthukumar, P.: A choice of FPGA design for three phase sinusoidal pulse width modulation. In: International Conference on Emerging Technological Trends (ICETT), pp. 1–6 (2016)

Novel Work of Diagnosis of Liver Cancer Using Tree Classifier on Liver Cancer Dataset (BUPA Liver Disorder)

Manish Tiwari[1](✉), Prasun Chakrabarti[2](✉), and Tulika Chakrabarti[3](✉)

[1] Department of Computer Science and Engineering, Mewar University, Chittorgarh 312901, Rajasthan, India
immanishtiwari@gmail.com

[2] Department of Computer Science and Engineering, ITM Universe Vadodara, Paldi 391510, Gujarat, India
dean.research@itmuniverse.ac.in

[3] Department of Chemistry, Sir Padampat Singhania University, Udaipur 313601, Rajasthan, India
tulika.chakrabarti@spsu.ac.in

Abstract. The classification plays a vital role towards diagnosis of liver cancer because still diagnosis of liver cancer is tedious job in early stages and late stage it is incurable. In this paper, BUPA liver disorder has been used and the Tree classifier used result is analyzed into WEKA Tool. LMT, J48, Random Forest, REP tree, Extra tree, Simple cart algorithms are have been utilized to investigate towards performance (accuracy, precision and recall) and error evaluation (Mean absolute error, Root mean squared error, Relative absolute error, Root relative squared error) performed.

Keywords: Accuracy · Precision · Recall · Error evaluation · LMT
J48 · Reptree · Extra tree · Simple cart algorithm

1 Introduction

Liver cancer related risk factors include hereditary, hepatitis B, hepatitis C virus. Tumors are of two types - benign and malignant. A benign tumor is not cancerous, it can be removed and it will not come back after removal. Malignant stage is critical and is also known as hepatocellular carcinoma or malignant hepatoma. Many works has been done based on the liver cancer patient data [1]. Various classification algorithms such as Naïve Bayes, Decision Tree, Multilayer Perceptron, k-NN, Random Forest etc. [2] have been utilized to investigate Accuracy, precision, recall sensitivity, specificity.

2 Literature Survey

The paper [3] was based on the classification algorithms e.g. Naïve Bayes classifier, C4.5, Back Propagation, Neural Network algorithm and Support Vector Machine on liver patient datasets(UCLA liver disorder and AP dataset using performance

© Springer Nature Singapore Pte Ltd. 2018
I. Zelinka et al. (Eds.): ICSCS 2018, CCIS 837, pp. 155–160, 2018.
https://doi.org/10.1007/978-981-13-1936-5_18

parameters such as Accuracy, Precision, Sensitivity and specificity). The work indicated best results for accuracy (71.59%), precision (69.74%), specificity (82%) in Back Propagation. NBC classifier presented sensitivity (77.95%) higher than other classifiers. Using common attributes (SGOT, SGPT, ALP) in AP Liver dataset KNN gives good accuracy compared to other algorithms.

In the work [4], the performance of the ANN and SVM were compared on different cancer datasets describing accuracy, sensitivity, specificity and area under curve (AUC). BUPA liver disorder training set (70%) and testing set (30%) were selected, after analysis SVM gave (Accuracy-63.11%, Sensitivity-36.67%, specificity-100.0% AUC-68.34%) and Artificial Neural Networks gave (Accuracy-57.28%, Sensitivity-75.00%, specificity-32.56% AUC-53.78%).

In the paper [5], only 78% of liver cancer patients related with cirrhosis dataset of two types HCC and non tumor livers was used and the data was divided for the training and testing purposes. The missing values were removed using K-nearest neighbor method. In this method author used the optimized fuzzy neural network using the principal component analysis and compared it with the GA search results. It showed that if lesser amount of genes were used then FNN-PCA could give 95.8% accuracy.

The study [6] was used to classify the liver and non-liver disease dataset. Medical data containing 15 attributes from Chennai medical hospital was applied for preprocessing. C4.5 and Naïve Bayes classifier were applied for analysis. C4.5 gave better accuracy than Naïve Bayes algorithm.

The research work [7] entails the algorithms C4.5 and Random Tree chosen for analysis. Accuracy of the both algorithms was excellent for diagnosis of liver disease disorder.

In the paper [8] authors carried out investigations on ILPD dataset analysis using IBM SPSS Modeler. Dataset was partitioned into training and testing in the ratio of 60% and 30% respectively and 10% for the validation. The data was preprocessed by cleaning method then was applied for mining based classifications such as Boosted C5.0 and CHAID algorithms for extraction of rules. The Boosted C5.0 algorithm elaborated training accuracy, testing accuracy and validation in 92.33%, 93.75% and 91.55% respectively. The CHAID algorithm gave 76.14% training accuracy, 65.00% testing accuracy and 69.01% validation.

The work [9] indicated performance analysis using software. The WEKA tool gave lowest results for Naïve Bayes algorithms. Using Knime tool decision tree algorithm gave best results (95.37% accuracy).

In the paper [10] various classifications methods such as decision tree, MLP and Bayesnet were used. In fact the individual classification does not produce the perfect accuracy and robust model. So it was combined with C4.5, CART and RF to produce 75.34% accuracy compared to individual models. Information gain feature selection was applied to improve performance of the model. However if three feature selections were chosen, the model became robust and provided 76.03% accuracy.

In the study [11] NCD prediction model played a role to yield optimum accuracy consisted k-means for clustering technique, feature selection using SVM and k-NN classifiers. Combining k-means, SVM and KNN it gave enhanced accuracy (Accuracy-97.97, AUC-0.998). The authors also worked on many dataset such as Pima Indian

Dataset, Breast Cancer Diagnosis Dataset, Breast Cancer Biopsy Dataset, Colon Cancer, ECG and Liver Disorder.

In the work [12] author used six techniques on ILPD (Indian Liver Patient) dataset have been discussed. It covered 72% liver patients and 28% non-liver patients. Algorithms were performed under sampling and over sampling for balancing dataset. If genetic programming was used under sampling (50%) then it produced 84.75% accuracy. If oversampling (200%) was performed then Random forest gave better accuracy (89.10%). In this paper all the algorithms were used after ten cross validations.

The paper [13] entails liver cancer diagnosis using information retrieval technique. C4.5, Naïve Bayes, Decision tree, Support Vector Machine, Back Propagation neural network and classification and regression tree and compared speed, accuracy, performance and cost. Among all algorithms C4.5 gave best results.

In the work [14] hybrid model construction was used to perform the relative analysis in three phases for enhancing the prediction accuracy. Firstly classification algorithms were applied on original datasets collected from UCI repository. In the second phase, a significant attributes subset was selected from dataset for feature selection then classification algorithm was applied on it. In third phase, results of classification algorithms were compared with feature selection and without feature selection. Without feature selection the SVM gave good accuracy but after applying feature selection the Random forest gave best accuracy (71.87%) among other algorithms.

3 Methodology

Based on the BUPA Liver Disorder Dataset, the data preprocessing has been performed. Next supervised filter has been applied using WEKA 3.8.1 Tool. The various Tree classifiers viz. LMT algorithm, J48, Random Forest, Reptree, Extra Tree, Simple Cart Algorithm have been used for simulation purpose. In the next phase, the performance and related error evaluation has been carried out. Finally the inference has been drawn based on the optimum results.

4 Result and Discussion

BUPA liver disorder is used as the dataset for analysis. It shows that the algorithm which classifies dataset more correctly accuracy is high and the error is very less. So it is inferred that the algorithm have more accuracy can give more correct the information about the liver cancer it may help in the early detection for liver cancer analysis as mention in following Table 1.

Table 1. Tree classifier's algorithm comparison based on error and performance

S. No.	Algorithm	Evaluation	Type of error and performance	Results
1	LMT algorithm	Error	Mean absolute error	0.0661
			Root mean squared error	0.2005
			Relative absolute error	13. 5633%
			Root relative squared error	40.6112%
		Performance	Accuracy	95.07%
			Precision	94.4%
			Recall	93.8%
2	J48 algorithm	Error	Mean absolute error	0.05
			Root mean squared error	0.1934
			Relative absolute error	10.2534%
			Root relative squared error	39.1719%
		Performance	Accuracy	95.9%
			Precision	95.8%
			Recall	94.5%
3	Random forest algorithm	Error	Mean absolute error	0.0556
			Root mean squared error	0.1393
			Relative absolute error	11.41%
			Root relative squared error	28.21%
		Performance	Accuracy	98.26%
			Precision	97.9%
			Recall	97.9%
4	Reptree algorithm	Error	Mean absolute error	0.0968
			Root mean squared error	0.2384
			Relative absolute error	19.87%
			Root relative squared error	48.30%
		Performance	Accuracy	93.62%
			Precision	94.2%
			Recall	90.3%
5	ExtraTree algorithm	Error	Mean absolute error	0.0406
			Root mean squared error	0.2014
			Relative absolute error	8.326%
			Root relative squared error	40.8094%
		Performance	Accuracy	95.94%
			Precision	94.6%
			Recall	95.9%
6	Simple Cart algorithm	Error	Mean absolute error	0.0683
			Root mean squared error	0.2204
			Relative absolute error	14.065%
			Root relative squared error	44.6471%
		Performance	Accuracy	94.78%
			Precision	96.4%
			Recall	91.0%

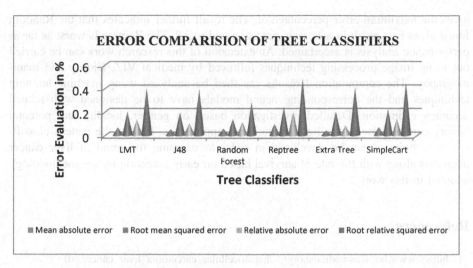

Fig. 1. Error evaluations of tree classifier

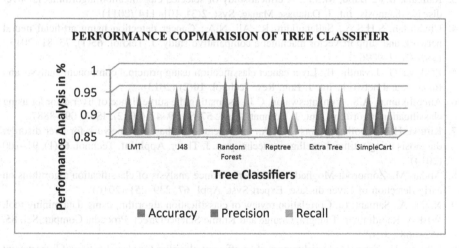

Fig. 2. Performance analysis of tree classifier

Above graph showing the comparison of Error evaluation and Performance analysis (Figs. 1 and 2).

5 Conclusion and Future Perspective

The classification algorithms are helpful for the early detection of the liver cancer. In present paper a thorough analysis of error rate of six tree classifiers (LMT, J48, Random Forest, Reptree, Extra Tree, Simple Cart) has been pointed out. Random forest tree algorithm gives minimum error rate than all other algorithms where as Reptree gives the maximum error percentage of. The result further indicates that the Random forest gives best result in accuracy, precision and recall. The Reptree is worst as far as performance analysis is ascertained. An extension of this research work can be carried out using image processing techniques followed by medical VIZ. Biopsy and mammography. The computational results can then be analyzed using machine learning techniques and the corresponding neural models have to be designed with related accuracy estimation. Detailed investigation based on gender, locality and parental history can be done in the light of statistical approaches. Finally the pattern classification techniques can be developed in order to examine the trend of liver cancer diagnosis along with the rate of survival based on early detection by the methodology adopted in this work.

References

1. https://www.hopkinsmedicine.org/.../hepatocellular_carcinoma_liver_cancer.pdf
2. Lesmana, C.R.A.: Alcoholic liver disease and alcohol in non-alcoholic liver disease: does it matter? J. Metab. Synd. **3**, 147 (2014). https://doi.org/10.4172/2167-0943.1000147
3. Ramana, B.V., Babu, M.S.P.: A critical study of selected classification algorithms for liver disease diagnosis. Int. J. Database Manag. Syst. **2**(3), 101–114 (2011)
4. Ubaidillah, S.H.S.A., Sallehuddin, R., Ali, N.A.: Cancer detection using artificial neural network and support vector machine: a comparative study. J. Teknol. **65**(1), 73–81 (2013). ISSN 0127-9696
5. Ilakkiya, G., Jayanthi, B.: Liver cancer classification using principal component analysis and fuzzy neural network. Int. J. Eng. Res. Technol. **10**(2) (2013)
6. Aneeshkumar, A.S., Venkateswaran, C.J.: Estimating the surveillance of liver disorder using classification algorithms. Int. J. Comput. Appl. **57**(6), 39–42 (2012). ISSN 095-8887
7. Kiruba, H.R., Tholkappiaarasu, G.: An intelligent agent based framework for liver disorder diagnosis using artificial intelligence techniques. J. Theor. Appl. Inf. Technol. **69**(1), 91–100 (2014)
8. Abdar, M., Zomorodi-Moghadam, M.: Performance analysis of classification algorithms on early detection of Liver disease. Expert Syst. Appl. **67**, 239–251 (2016)
9. Naika, A., Samant, L.: Correlation review of classification algorithm using data mining tool: WEKA, Rapidminer, Tanagra, Orange and Knime ScienceDirect. Procedia Comput. Sci. **85**, 662–668 (2016)
10. Pakhale, H., Xaxa, D.: Development of an efficient classifier for classification of liver patient with feature selection. Int. J. Comput. Sci. Inf. Technol. **7**(3), 1541–1544 (2016)
11. Sutanto, D.H., Ghani, M.K.A.: Improving classification performance of K-nearest neighbour by hybrid clustering and feature selection for non-communicable disease prediction. J. Eng. Appl Sci. **10**(16), 6817–6825 (2015)
12. Pahareeya, J., Vohra, R.: Liver patient classification using intelligence techniques. Int. J. Adv. Res. Comput. Sci. Softw. Eng. **4**(2), 295–299 (2014)
13. Sindhuja, D., Priyadarsini, R.J.: A survey on classification techniques in data mining for analyzing liver disease disorder. IJCSMC **5**(5), 483–488 (2016)
14. Gulia, A., Vohra, R.: Liver patient classification using intelligent techniques. Int. J. Comput. Sci. Inf. Technol. **5**(4), 5110–5115 (2014)

Performance Analysis and Error Evaluation Towards the Liver Cancer Diagnosis Using Lazy Classifiers for ILPD

Manish Tiwari[1(✉)], Prasun Chakrabarti[2(✉)],
and Tulika Chakrabarti[3(✉)]

[1] Department of Computer Science and Engineering,
Mewar University, Chittorgarh 312901, Rajasthan, India
immanishtiwari@gmail.com
[2] Department of Computer Science and Engineering,
ITM Universe Vadodara, Paldi 391510, Gujarat, India
dean.research@itmuniverse.ac.in
[3] Department of Chemistry, Sir Padampat Singhania University,
Udaipur 313601, Rajasthan, India
tulika.chakrabarti@spsu.ac.in

Abstract. This paper, entails the various Lazy classifiers such as IBKLG, LocalKnn algorithm, RseslibKnn algorithm used for diagnosis of the liver cancer. The results have been noted in terms of both performance and errors. The performance analyzed based on the accuracy, precision and recall and error evaluation are based on the Mean absolute error, Root mean squared error, Relative absolute error and Root relative squared error. The LocalKnn is best in terms of accuracy and recall while IBKLG indicates best precision.

Keywords: IBKLG · LocalKnn · RseslibKnn · Accuracy · Precision
Recall root mean squared error · Relative absolute error · Root relative squared

1 Introduction

Liver is the largest organ after the skin in our body. It perform many functions cleansing blood toxins, converting food into nutrients to control hormone level. The diagnosis of liver diseases at early stage can improve survival rate of patient life. Techniques are used to find pattern from the large dataset are called the data mining techniques. it have several function such as classification, association rules and clustering etc. classification is supervised learning technique used for dataset in dissimilar group of classes or in different levels. Classification method performs two steps one is dataset are used to trained to built model and in second it used for classification [1].

I. Zelinka et al. (Eds.): ICSCS 2018, CCIS 837, pp. 161–168, 2018.
https://doi.org/10.1007/978-981-13-1936-5_19

2 Literature Survey

In the paper [2] Indian liver patient dataset and UCLA dataset were used. Analysis was done by ANOVA and MANOVA to recognize difference among the groups. Authors took common attributes e.g. ALKPHOS, SGPT and SGOT for both datasets. Analysis of Variance (ANOVA) was done using multivariate tables. Author investigated 99% and 90% significant levels and found the good results.

The study [3] deals with two distinct feature combinations viz SGOT, SGPT, and Alkaline Phosphates of two datasets (ILPD and BUPA liver disorder). Error rate, sensitivity, prevalence and specificity were exponentially observed. The attributes like total bilirubin, direct bilirubin, albumin, gender, age and total proteins facilitate in liver cancer diagnosis.

The paper [4] indicated neural network to train adaptive activation function for extracting rules. OptaiNET, an Artificial Immune Algorithm (AIS) was used to set rules for liver disorders. Based on input attribute adaptive activation was trained to use neural network extract rules efficiently in hidden layer. ANN to performs the data coding, to classifies coding data and finally extracts rules. It correctly diagnosed 192 samples (out of 200) belonging to class 0 covering 96% and 135 samples (out of 145) belonging to class 1 covering 93%. Entire samples correctly diagnosed 94.8%.

The study [5] pointed out univariate analysis and feature selection for predicator attributes. Predictive data mining is a significant tool for researchers of medical sciences. ILPD dataset was chosen for men and women. The classification algorithms were trained to test and to perform some results for accuracy and error analysis. For men and women the SVM gave high accuracy 99.76% and 97.7% respectively.

In the survey [6] classification algorithm decision tree induction (J48 algorithm) employing dataset from the Pt. B.D. Sharma Postgraduate Institute of Medical Science, Rohtak was used. The dataset contained 150 instances (100 instances for training purpose and 50 instances for the test data), 8 attributes and 2 classes for the model using 10 fold cross validation in WEKA tool and J48 algorithms classified correctly 100% instances. The result was expressed in four categories e.g. cost/benefit of J48 for class YES = 44, cost/benefit of J48 for class NO = 56, classification accuracy for YES = 56%, classification accuracy for NO = 44%. Many other algorithms on this dataset were applied and J48 algorithms showed best results.

The publication [7] described classification using data mining approaches on ILPD. Naïve bayes, Random Forest and SVM. The algorithms were implemented using R tool and for improving the accuracy the hybrid neuro SVM that is the combination of the SVM and feedforward Neural Network (ANN) was used. Root mean square error (RMSE) and mean absolute percentage error were pointed out. This model gave 98.83% accuracy.

In the publication on [1] various decision tree algorithms were used based on the data mining concept such as AD Tree, Decision Tree, J48, Random Forest, Random Tree on the liver cancer dataset. They were used for the training purpose and pre-processing was applied for missing or noisy data. Classification algorithms were performed with feature selection and without using feature selection. Its performances were measured in terms of Accuracy, Precision, and Recall. The accuracy (71.35%) of the decision stump was very good compared to other algorithms and J48 and random forest gave 70.66% and 70.15% accuracy respectively.

The publication on [8] indicated PSO java to execute dataset and to categorize training attributes in order to retrieve pbest and gbest. The pbest was then compared with lbest to set the best solution for attribute selection. The PSO gave gammagt 4.60, alkphos 4.49, SGPT 3.91, SGOT 3.07, drinks 1.36. The selected dataset was applied to WEKA tool to perform the classification. Then it applied the Kstar algorithm. PSO-Kstar algorithm is the best data mining technique giving accuracy up to 100%.

The paper [9] described different clustering algorithms for predication on BUPA liver disorder and ILPD dataset for performance analysis. The simple BIZ model was selected effectively. Different attribute selections were done for accuracy, such as 5, 6, 7, 8 and 9. The logistic Regression and SVM (PSO) gave best results for the BUPA liver disorder as well as ILPD dataset, with accuracy 89.14% and 89.66% respectively.

3   Methodology

In this process the Indian liver patient dataset have been taken after the preprocessing is performed in this method the missing values problem are solved after the supervised filter are used in that resample method are used then Lazy classifier such as IBKLG, LocalKnn, RseslibKnn algorithms are used in WEKA tool for classification. 10 folds cross validations are used then performance and error evaluation is performed (Fig. 1).

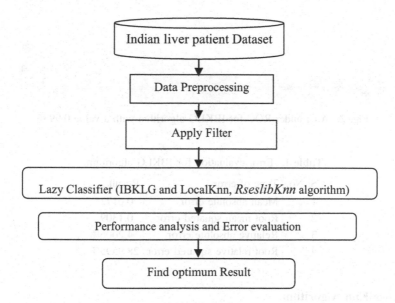

Fig. 1. Classification process

4 Result and Discussion

Lazy classifiers are used for analysis of the liver cancer disease. In this process any algorithm that gave better accuracy, precision and classified more correct instances is the good algorithm in term of early diagnosis of the liver cancer.

4.1 IBKLG Algorithm

IBKLG classifier is a part of lazy classifier. K-nearest neighbors classifier can select appropriate value of K based on cross-validation. It also performs distance weighting. It selects number of neighbor is one, The standard deviation set to 1.0, do not check capabilities to false, meanSquared value to false. It is based on nearest neighbor search algorithm using linearNNSearch algorithm. 10 folds cross validations are used for testing. It correctly classifies 573 instances (covering 98.28%) and incorrectly classifies 10 instances (covering 1.72%) out of 583 instances (Fig. 2, Tables 1 and 2).

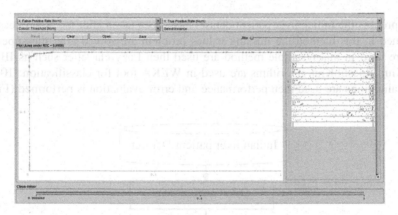

Fig. 2. Area under ROC for IBKLG algorithm with a value 0.9986

Table 1. Error evaluation for IBKLG algorithm.

Sr.No.	Type of error	Result
1	Mean absolute error	0.0172
2	Root mean squared error	0.1309
3	Relative absolute error	4.2006%
4	Root relative squared error	28.9595%

4.2 LocalKnn Algorithm

LocalKnn algorithm is based on K nearest neighbor classifier with local metric induction. It improves accuracy in relation to standard k-nn, particularly in case of data with nominal attributes. It works with reasonably 2000 + training instances. 100 batch size is selected. Do not check capabilities to set to false. Learning Optimal K values to

Table 2. Confusion matrix for IBKLG algorithm.

Performance vector:			
Confusion Matrix:			Accuracy: 98.28% (for class 1 malignant)
	M(T)	B(T)	Precision: 99.3% (for class 1 malignant)
M(P)	409	7	
B(P)	3	164	Recall: 98.3% (for class 1 malignant)
Class 1 is selected for the result because it mention positive in liver disorder			

true and number of neighbors used to vote for the decision to one, size of the local uses induce local metric to 100. The metric vicinity size for density based is 200. The voting for the decision by nearest neighbors is set to inverse square distance. It uses distance based weighting method. 10 fold cross validations are applied. It correctly classifies 576 instances (covering 98.80%) and incorrectly classifies 7 instances (covering 1.20%). Time taken to build model is 68.19 s (Fig. 3, Tables 3 and 4).

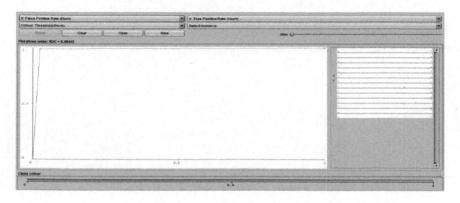

Fig. 3. Area under ROC for LocalKnn algorithm with a value 0.9844

Table 3. Error evaluation for LocalKnn Algorithm

Sr.No.	Type of error	Result
1	Mean absolute error	0.012
2	Root mean squared error	0.1096
3	Relative absolute error	2.9346%
4	Root relative squared error	24.2362%

Table 4. Confusion matrix for Local Knn Algorithm

Performance vector:		
Confusion Matrix:		Accuracy: 98.80% (for class 1 malignant)
	M(T)	B(T)	Precision 99.0% (for class 1 malignant)
M(P)	413	3	Recall 99.3% (for class 1 malignant)
B(P)	4	163	
Class 1 is selected for the result because it mention positive in liver disorder		

4.3 RseslibKnn Algorithm

RseslibKnn is a part of lazy classifier. It sets some properties defines such as batch size, learning optimal k value, do not check capabilities, cross validation, kernel setting, density based metric and so on. Time taken to building model is 1.3 s. 10 folds cross validations. It correctly classifies 571 instances (covering 97.94%) and incorrectly classifies 12 instances (covering 2.06%) out of 583 instances (Fig. 4, Tables 5 and 6).

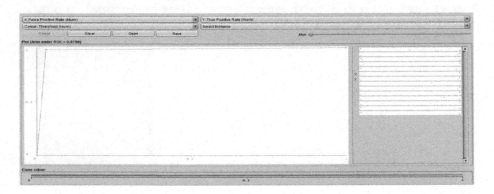

Fig. 4. Area under ROC for RseslibKnn algorithm with a value 0.9766

Table 5. Error evaluation for RseslibKnn algorithm

Sr.No.	Type of error	Result
1	Mean absolute error	0.0206
2	Root mean squared error	0.1435
3	Relative absolute error	5.0307%
4	Root relative squared error	31.7327%

Table 6. Confusion matrix for RseslibKnn algorithm

Performance vector:			
Confusion Matrix:			Accuracy: 97.94% (for class 1 malignant)
	M(T)	B(T)	Precision: 98.8% (for class 1 malignant)
M(P)	409	7	
B(P)	5	162	Recall: 98.3% (for class 1 malignant)

Class 1 is selected for the result because it mention positive in liver disorder

4.4 Comparison of Error Evaluation and Performance Analysis of Three Lazy Classifiers (RselibKnn, IBKLG, LocalKnn) for ILPD Dataset

See Figs. 5 and 6.

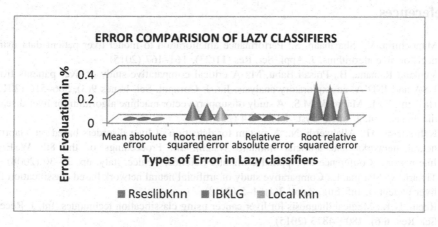

Fig. 5. Error evaluation of Lazy classifier

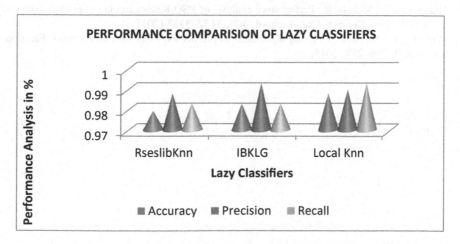

Fig. 6. Performance analysis of Lazy classifier

5 Conclusion and Future Perspective

A close assessment of error estimation of three Lazy classifiers (RseslibKnn, IBKLG, LocalKnn) has been performed whereby the minimum error value is achieved through LocalKnn. The LocalKnn is best in terms of accuracy and recall while IBKLG indicates best precision. It is evident that if any classification algorithm classifies instances accurately, then diagnosis of the liver cancer can be done easily and accurately in early stages.

Further research work or classifiers can be applied on different types of cancers such as Breast cancer, Prostate Cancer, Lung cancer etc. Appling these algorithms may generate better results. As an extension of this Biopsy and mammography images can be used for analysis using machine learning methods. Research can also be applied for analysis of survival rate of the patient.

References

1. Manochitra, V., Shajahaan, S.: Performance amelioration to model liver patient data using decision tree algorithms. J. Appl. Sci. Res. **11**(23), 161–167 (2015)
2. Venkata Ramana, B., Prasad Babu, M.: A critical comparative study of liver patients from USA and INDIA: an exploratory analysis. Int. J. Comput. Sci. Issues **9**(3), 506–516 (2012)
3. Hashem, E.M., Mabrouk, M.S.: A study of support vector machine algorithm for liver disease diagnosis. Am. J. Intell. Syst. **4**(1), 9–14 (2014)
4. Kahramanli, H., Allahverdi, N.: A system for detection of liver disorders based on adaptive neural networks and artificial immune system. In: Proceedings of the 8th WSEAS International Conference on Applied Computer Science, Venice, Italy, pp. 25–30 (2008)
5. Tiwari, A., Sharma, L.: Comparative study of artificial neural network based classification for liver patient. J. Inf. Eng. Appl. **3**(4), 1–5 (2013)
6. Reetu, N.K.: Medical diagnosis for liver cancer using classification techniques. Int. J. Recent Sci. Res. **6**(6), 4809–4813 (2015)
7. Nagaraj, K., Sridhar, A.: NeuroSVM: A Graphical User Interface for Identification of Liver Patients. Int. J. Comput. Sci. Inf. Technol. **5**(6), 8280–8284 (2014)
8. Thangaraju, P., Mehala, R.: Performance analysis of PSO-KStar classifier over liver diseases. Int. J. Adv. Res. Comput. Eng. Technol. **4**(7), 3132–3137 (2015)
9. Mazaheri, P., Norouzi, A.: Using algorithms to predict liver disease classification. Electron. Inf. Plan. **3**, 256–259 (2015)

Exploring Structure Oriented Feature Tag Weighting Algorithm for Web Documents Identification

Karunendra Verma[1](✉), Prateek Srivastava[1], and Prasun Chakrabarti[2]

[1] Department of CSE, Sir Padampat Singhania University, Udaipur, India
k.verma2006@gmail.com, prateek.srivastava@spsu.ac.in
[2] Department of Computer Science and Engineering, ITM Universe Vadodara,
Paldi 391510, Gujarat, India
dean.research@itmuniverse.ac.in

Abstract. There are various ways of web page classification but they take higher time to compute with lesser accuracy. Hence, there is a need to invent an efficient algorithm in order to reduce time and increase web page classification result. It is generally find that a few tags like title can contain the principle substance of text, and these patterns may have an impact on the adequacy of text classification. Although, the most widely recognized text weighting calculations, called term frequency inverse documents frequency (TF-IDF) doesn't consider the structure of website pages. To take care of this issue, another feature tags weighting calculation is put in advanced. It thinks about the web page structure data like title, Meta tags, head etc. also content the useful information. In this proposed study first web site pages data are pre-processed and find text weight using TFIDF, after that using feature tag weighting calculation, frequent and important tags will find; then on the basis of text weight and tags weight, web document will classify.

Keywords: Web page classification · Feature tags · TFIDF · Text weight

1 Introduction

Presently day by day Internet has turned out to be extremely well known and intuitive for exchanging data. The web is tremendous, differing and dynamic thus increase the versatility, interactive media information and fleeting issues. The development of the web has its result in a gigantic measure of data that is currently unreservedly offered for client's entrance. A few various types of information must be taken care of and composed in a way that they can be gotten to by the clients viably and productively.

The web is an accumulation of interrelated documents on at least one web servers. Web mining is the utilization of information mining methods to extricate learning from web information including web archives, hyperlinks between pages, usages logs of sites and so on.

Web mining is comprehensively partitioned into following classes: web content mining, web usage mining and web structure mining.

Text classification is a procedure of partitioning text into one or multiple classes. As the advancement of the web technology, objective is to achieve accurate web text

© Springer Nature Singapore Pte Ltd. 2018
I. Zelinka et al. (Eds.): ICSCS 2018, CCIS 837, pp. 169–180, 2018.
https://doi.org/10.1007/978-981-13-1936-5_20

from web documents. Web documents have clear identifier (i.e., HTML tags) to convey its structure information. Generally, in the content extricating process, HTML page structures are expelled and separate plain text from each website pages. In many cases, there is a lot helpful data regarding the content organization based on HTML tags. Numerous investigators demonstrate to the structural data, particularly HTML tags, similar to table design, hyperlink, be able to utilized to enhance viability of web content classification.

2 Review of Literature

Kovacevic et al. [9] proposed hierarchical representation that incorporates browser screen which facilitates with each HTML article in a page. Utilizing image data one can characterize heuristics for the acknowledgment of basic page zones, for example, footer, header, right and left menu, and focus of a web page. In the underlying examinations the creator demonstrates that utilizing heuristics characterized objects are perceived legitimately in 73% of cases. At long last, demonstrate that a Naive Bayes classifier, given that the proposed representation.

Zou et al. [16] have discovered that because of the proximity of the raucous information here is a requirement for characterization of the web page for true applications. A strategy which will appropriately guarantee the arrangement be the support vector machine since it has the ability of speculation. Creator's recommended strategy gives a way which will expand the precision of arrangement by joining the support vector machine idea among the K - nearest neighbor procedures.

Tomar et al. [4] present the idea of an order device for pages called Web Characterize, which utilizes changed customized naive Bayesian calculation with a multinomial form to arrange pages hooked on different classifications. In this exploration test result alongside the grouping exactness investigation with expanding vocabulary measure, was likewise appeared.

Ryan et al. [10] examined the region of classification arrangement has an accentuation on recovering the highlights, for example, content from the particular archives. Since the principle point of work is considered whether visual properties of HTML site page can altogether enhance the arrangement of pulverous sorts. Evidently, it appears that it would put a noteworthy test and will be likewise helpful to recover those visual attributes which getting the design highlight of types. The majority of site pages delivered from different business sites and physically sorted into types. The three unique qualities are thought about one next to the other (a). With the literary attributes (b). With the HTML qualities (c). Visual qualities. Creator's work can demonstrate that by utilizing HTML qualities and URL attributes helps in expanding the precision of characterization when contrasted with printed alone. In this way, it additionally appears that by including the visual attributes, it builds the pulverous grouping.

Kang et al. [7] exhibit an investigation on mining web information from the various accessible information on WWW. As the pages are not completely organized so it ends up noticeably hard deciding from the useful block techniques which give the valuable information extraction from the futile information, for example, promotions which is more vital. In this proposed strategy creator present a website page arrangement in type of pieces by building a tree arrangement demonstrate that show the HTML include and a vector display that speaks to an element of blocks. Hence, by building the single classifier it ends up noticeably hard to characterize a piece precisely. To defeat this issue in proposed strategy creator utilizes the various classifiers one for each preparation informational collection and characterization technique prevails by consolidating every one of them.

Mun et al. [11] found that the size of web page increases a lot as the number of offered services as well as link increases and then due to their accessing speed decreases. The author uses the link graph arrangements for troubleshoot this problem. By introducing this link graph system author enables to reduce the load of server to a greater extent.

Rathod [13] indicates frameworks of three unique methods of web pages mining, in particular web structure mining, web usage mining and web content mining. The advancement and utilization of Web mining strategies with regards to web content usage and structure information will prompt substantial enhancements in numerous web applications as of web crawlers and web specialists to web examination and personalization.

Gowri et al. [3] portrayed a short overview about the current approach in web administrations synthesis. The principle looks into regions in web administrations are identified with revelation, security, and creation. Among every one of these regions, web administrations organization ends up being a testing one in light of the fact that inside the administration arranged figuring area, Web benefit synthesis is a successful acknowledgment to fulfill the hurriedly changing prerequisites of business. In this manner, the Internet benefit creation has unfurled itself extensively in the exploration side. Be that as it may, the present endeavors to order Web benefit structure are not fitting to the targets. This article proposes a novel categorization matrix for Web service work, which recognizes the unique situation and innovation measurements. The setting measurement is gone for examining the QoS effect on the exertion of Web benefits creation, although the innovation measurement concentrates on the system impact on the exertion. At last, this paper gives a proposal to enhance the nature of administration determination which takes part in the arrangement procedure with C skyline approach utilizing operators.

Sarac et al. [14] worked on the firefly algorithm (FA) inspired by the flashing behavior of fireflies, which belongs to the category of Meta heuristic algorithm. It flashes primary intention to attract other fireflies through a signaling system.

Jain et al. [2] proposed another strategy "Intelligent Search Method (ISM)". In this technique creator proposes to index the web pages via an intelligent search approach. This new strategy incorporated with any of the page positioning calculations to deliver better and significant indexed lists.

Keller et al. [8] introduce a GRABEX strategy for removing navigational block pieces in light of the connection designs. The technique was connected to mine breadcrumb routes. Dissecting to which additional navigational chunk type the GRABEX strategy can be connected is additionally intriguing for prospect work. A creator trusts that paginations or past/next routes can be mined too if appropriate graph creation strategies are actualized. The GRABEX strategy can likewise be reached out to extract non-navigational page components if diagrams are not produced from hyperlinks but rather from different structures e.g. text or linked images.

Jose et al. [6] demonstrate the Rough set hypothesis applications in different areas like company, prescription, trade, media transmission and numerous different fields. The consequences of this approach can be utilized for target promoting on the grounds that sponsors can post their notices on content pages particularly pages in bring down estimate. This likewise distinguishes the most favored substance by a client since clients invested more energy in potential pages.

Ye et al. [15] enhanced and proposed a kind new technique of semantic relevancy algorithm based for semantic importance calculation in light of the Wikipedia hyperlink arrange, incorporating the semantic data in the paging system and the class organize sensibly to complete semantic relationship figuring.

Sadegh et al. [1] explored social tags as a novel confirmation to categorize objects on the web. A new linkage structure between objects and tags is investigated for categorization. Tags moreover work as bridges to attach the heterogeneous domains of objects.

He et al. [5] work in view of the way that the web is an accumulation of different web records. The grouping of a web record is implied for three things for the most part: indexing, search and retrieval. There is a distinction between web grouping and content characterization. This distinction is because of the structure of the web reports. These distinctions could be at least one of the accompanying: meta information, the title of the record and different connections accessible in the archive and so on. In this paper, creators have picked both of the accompanying strategies, for example, Information gain and $\chi 2$ - test for feature selection for classification. After feature selection, this paper uses Support Vector Machine (SVM) classifier for categorization. The strategy affirms, evaluates and broadens past study by presenting another structure-based technique for depiction and order of web archives. Contrasted with conventional web archive order strategies, consolidating the complete text among structure Information gain almost 6% exactness change on account of comparative classifications and 3.7% correctness enhancement in the case of different categories.

Qian et al. [12] worked on novel weighted Hamming distance based on Page Rank algorithm for anomaly intrusion detection. Using the Hamming distance with the Page Rank weight to estimate deformity degree of unusual system calls and focus on optimizing the algorithm complexity.

Chen et al. [19] characterized five best level type classifications and grew new techniques to mine 31 features from web database, which examined both features and contents. Their assessment results comes that extra features can help a classifier enhances its knowledge of the categorization.

Abramson et al. [20] exhibited a technique to facilitate uses data from URLs for website page database since a few URLs may include some text to shows the class. This approach can partially take care of the issue; however it is as yet not a general approach for all web pages genre.

3 Research Gap

The literature review entails that; the classification done on the dataset of web structure is optimized by structure based web document analysis. However, beside these described techniques there are various other ways also to perform web structure based classification. Certainly, web structure based classification gives better result in association with feature selection results because it finds various features in the record of dataset. But while using this technique with simple web structure based classification, there is a scope of improvement in the following two concerns.

- Web structure based classification itself takes longer time to compute.
- The result of simple web structure based classification is not that much optimized.

And the reasons behind these two concerns are the accuracy of the classification method. It is proposed to improvise these concerns to get better efficiency in this work.

4 Methodology and Used Algorithms

To overcome above limitations, work has divided into following steps (Fig. 1):

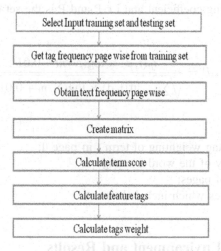

Select Input training set and testing set

Get tag frequency page wise from training set

Obtain text frequency page wise

Create matrix

Calculate term score

Calculate feature tags

Calculate tags weight

Fig. 1. Work flow

4.1 Term Frequency Inverse Document Frequency (TFIDF) [21]

$$TermScore_{ij} = \frac{n_{ij}}{\sum_k n_{kj}} \times \log \frac{|D|}{|\{d : t_i \in d\}|} \tag{1}$$

Where

n_{ij}, no. of presences of term t_i in page d_j;
n_{kj}, sum of presences of all term in page d_j;
D, total number of pages;
d, number of pages which incorporated term t_i.

4.2 Feature Tag Weighting Algorithm

Tag Frequency [18]

$$tf_t(t,d) = \sum_{i \in P} \left[tf_t(t_i, d) \times \frac{a_i}{\sqrt{\sum_{j=0}^{k} a_i^2}} \right] \qquad (2)$$

Where

t_{ft} (t, d), tag frequency of term t in page d;
t_{ft} (t_i, d), tag frequency of the term t in tag i;
a_i is the tag weighting coefficient and i ϵ P and P is the set of tags.

Tag weight [18]

$$W_t(t,d) = \frac{tf_t(t,d) \times \log(N/nt + 0.01)}{\sqrt{\sum_{t \in d} [\text{tf}(t,d) \times \log(N/n_t + 0.01)]^2}} \qquad (3)$$

Where

W_t (t, d) is feature tag weighting of term t in page d;
t_{ft} (t, d) is frequency of the word t in page d;
N is total number of pages;
n_t is number of pages which included term t.

5 Programming Environment and Results

To simulate above work we used Net Beans IDE 8.2 and JDK 1.8. Investigations utilize the Bank Search dataset [17], which is particularly intended to help an extensive variety of web pages processing tests. The database comprises of 2202 web archives arranged into ten uniformly sized classes like A: Commercial banks Banking and finance, B: Building society Banking and finance, C: Insurance agencies Banking and finance, D: Java Programming languages, E: C++/C Programming languages, F: VB Programming languages, G: Astronomy Science, H: Biology Science, I: Soccer Sport, J: Motor etc. and each contains 200 web archives (Figs. 2, 3, 4, 5, 6, 7 and 8).

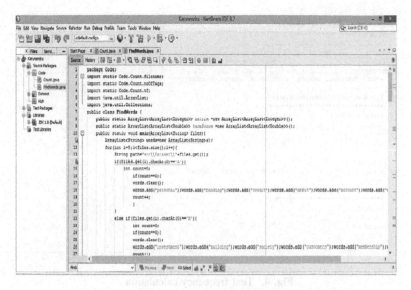

Fig. 2. Programming environment

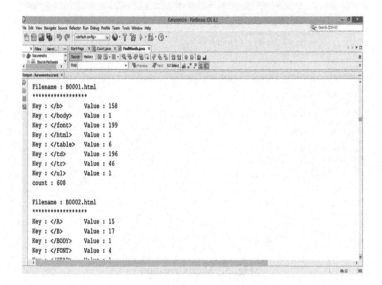

Fig. 3. Tag frequency calculation

The results include characterization of two classes from the dataset. The initial 1500 records are utilized as training set and the rest 500 records are utilized as testing set. a few classes are very similar, while a few classes are very distinct (e.g. class A: Commercial Banks and J: Motor Sport). Categorizing related classes is obviously a new troublesome machine-learning task.

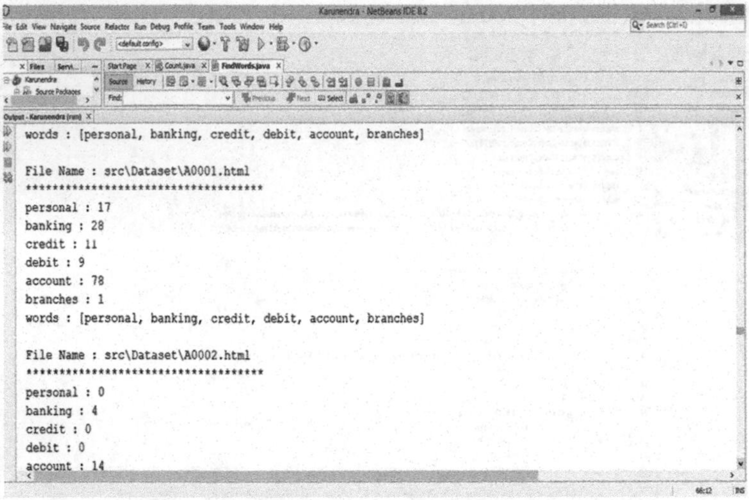

Fig. 4. Text frequency calculation

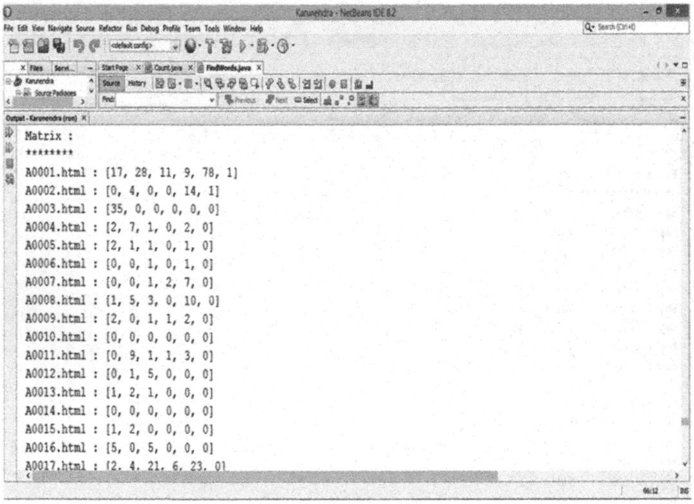

Fig. 5. Matrix creation

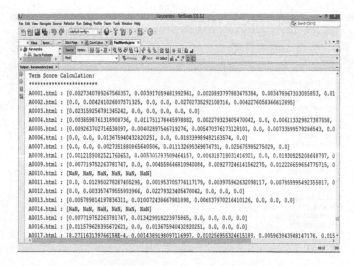

Fig. 6. Term score calculation

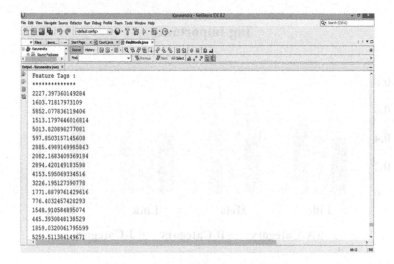

Fig. 7. Feature tags calculation

In "Fig. 9," four major tag types in a web pages (title, meta, link, and image) importance were compared from rest of all tags. Then this number was normalized against the web pages with respect to tag weighting function.

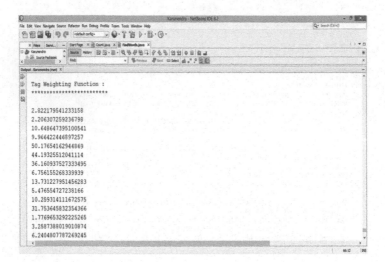

Fig. 8. Tag weight calculation

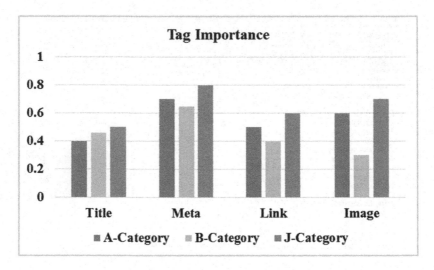

Fig. 9. Tag importance

6 Conclusion and Future Work

This article has exhibited a structure-based strategy for fabricating high precise web page categorization. It has shown in "Fig. 9", the handiness of thinking about structure data, which incorporates Links, META tags, TITLE and alternative texts of images. The process is assessed utilizing the Bank Search database, and the investigations show the reward of structure-based characterization for both similar categories and different categories. Pure text classification has not considered the difference in the web content, which has HTML tags to express further structure data of the content. In the event that we simply utilize TF-IDF which suits to the text classification, web text might be overlooked. The feature tag weighting calculation considers the impact of HTML labels on the web content classification. It performs superior to TF-IDF in the impact of tagged web content classification. As indicated by our test result, include tag weighting calculation gets the good precisions values.

In the meantime, our test advises us that while characterizing the web text, we can also consider HTML tags with a specific end goal to enhance the impact of web content classification. Be that as it may, there are as yet some limitations in this paper like only some HTML tags are considered for results we may include some more for more accuracy of the results. Furthermore we can also apply the appropriate classifier for web pages categorization.

Acknowledgements. I would like to thank all the people those who helped me to give the knowledge about these research papers. I am thankful to Dr. Prateek Srivastava & Dr. Prasun Chakrabarti to encourage and guided in this topic which helped me to speed up the work for structure based web page classification for fast search. Finally, I like to acknowledge all the websites and IEEE papers which I have gone through and referred to create this research paper.

References

1. Sadegh, A.H., Hossein, R., Behroo, N.: Web page classification using social tag. In: IEEE International Conference on Computational Science and Engineering, vol. 4, no. 1, pp. 588–593 (2009)
2. Jain, A., Sharma, R., Dixit, G., Tomar, V.: Page ranking algorithms in web mining, limitations of existing methods and a new method for indexing web pages. In: International Conference on Communication Systems and Network Technologies, vol. 3, no. 1, pp. 640–645. IEEE (2013)
3. Gowri, R., Lavanya, R.: A novel classification of web service composition and optimization approach using skyline algorithm integrated with agents. In: IEEE Computational Intelligence and Computing Research (ICCIC), pp. 26–28 (2013)
4. Tomar, G.S., Verma, S., Jha, A.: Web page classification using modified naïve bayesian approach. In: IEEE TENCON 2006, Hong Kong, pp. 14–17 (2006)
5. Kejing, H., Henyang, C.: Structure-based classification of web documents using support vector machine. In: Proceedings of CCIS 2016, pp. 215–219. IEEE (2016)
6. Jose, J., Lal, P.S.: A rough set approach to identify content and navigational pages at a website, pp. 5–9. IEEE (2008)
7. Kang, J., Choi, J.: Block classification of a web page by using a combination of multiple classifiers. In: IEEE Networked Computing and Advanced Information Management, vol. 2, no. 1, pp. 290–295 (2008)
8. Keller, M., Hartenstein, H.: GRABEX: a graph-based method for web site block classification and its application on mining breadcrumb trails. In: WIC/ACM International Conferences on Web Intelligence (WI) and Intelligent Agent Technology (IAT), pp. 290–297. IEEE (2013)
9. Kovacevic, M., Diligenti, M., Gori, M., Milutinovic, V.: Recognition of common areas in a web page using visual information: a possible application in a page classification. In: IEEE Data Mining, pp. 250–257 (2002)

10. Ryan, L., Michal, C., Lei, Y.: Using visual features for fine-grained genre classification of web pages. In: Proceedings of the 41st Annual IEEE Hawaii International Conference on System Sciences, vol. 1, no. 10, pp. 7–10 (2008)
11. Mun, Y., Lee, M., Cho, D.: Classification of web link information and implementation of dynamic web page using Link Map System. In: IEEE Granular Computing, pp. 26–28 (2008)
12. Qian, Q., Li, J., Cai, J., Zhang, R., Xin, M.: An anomaly intrusion detection method based on PageRank algorithm. In: International Conference on Green Computing and Communications and IEEE Internet of Things and IEEE Cyber, Physical and Social Computing, pp. 2226–2230. IEEE (2013)
13. Dushyant, R.: A review on web mining. Int. J. Eng. Res. Technol. (IJERT) (2012)
14. Sarac, E., Ozel, S.A.: Web page classification using firefly optimization. In: IEEE International Symposium on Innovations in Intelligent Systems and Applications (INISTA) (2013)
15. Ye, F., Zhang, F., Luo, X., Xu, L.: Research on measuring semantic correlation based on the Wikipedia hyperlink network, pp. 309–314. IEEE (2013)
16. Zou, J.Q., Chen, G.L., Guo, W.Z.: Chinese web page classification using no se-tolerant up port vector machines. In: Natural Language Processing and Knowledge Engineering, IEEE NLP-KE, pp. 785–790 (2005)
17. Sinka, M.P., Corne, D.W.: BankSearch dataset (2005). http://www.pedal.reading.ac.uk/bansearchdataset/
18. Lu, Y., Peng, Y.: Feature weighting improvement of web text categorization based on particle swarm optimization algorithm. J. Comput. 10(1), 260–269 (2006)
19. Chen, G., Choi, B.: Web page genre classification. In: Proceedings of the ACM Symposium on Applied Computing, pp. 2353–2357 (2008)
20. Abramson, M., Aha, D.M.: What's in a URL? Genre classification from URL. In: Workshops at the 26th Advancement of Artificial Intelligence (AAAI) Conference on Artificial Intelligence, pp. 1–8 (2012)
21. Zhu, J., Xie, Q., Yu, S.I., Wong, W.H.: Exploiting link structure for web page genre identification. Data Min. Knowl. Discov. 1–26 (2015)

MQMS - An Improved Priority Scheduling Model for Body Area Network Enabled M-Health Data Transfer

V. K. Minimol[1](✉) and R. S. Shaji[1,2]

[1] Noorul Islam University, Kanyakumari, India
minimoldeepak@gmail.com
[2] Department of Computer Science and Engineering, St. Xavier's Catholic
College of Engineering, Nagercoil, India

Abstract. Mobile health is a new area of technology that gives the health care
system a new face and place in the world. With the support of Body Area
Network the m-health application has to make a lot of changes in the area of
health support. There are so many research works has been conducted to make
the application efficient. As in the case of any network traffic the m-health
application also suffers problems. The paper put forward a new idea of
scheduling the vital signals from the body with the help of queuing theory. It
uses some analytical modeling, by considering the signal packets from sensors
are following poisons distribution and the packets are arriving randomly. From
the queuing theory uses some equations to find the average waiting time,
maximum number of packets waiting for the service, efficiency of the system
etc. Here the major issues while incorporating BAN with m-health is the number
of nodes and distance from the patients to the receiving station, Number of
servers in the receiving station, Priority of the signals etc.

Keywords: Body Area Networks · M-health · Queuing theory
Poisons distribution

1 Introduction

M-health the novel application of technology and new trend in the health care system,
incorporated Body Area Network (BAN) as a supporting infrastructure to make the
entire system so efficient and easy to handle. The assistance of BAN in m-health make
the diagnosis clearer and give opportunity to change the design of m-health from mere
mobile phone conversation from the patients to doctor to capturing signals from various
body parts and sending it to a receiving stations. Here the paper introduces an analytical
approach to find the efficiency of the queuing system by reducing delay time and
finding the average number of signals in the queue. The architecture has a two-layered
structure one is the internal BAN and other is the external network between the BAN
and receiving station. The existing studies focused on variety of priority scheduling
models in wireless sensor networks. The proposed study applies queuing models for
considering the priority of vital signals.

© Springer Nature Singapore Pte Ltd. 2018
I. Zelinka et al. (Eds.): ICSCS 2018, CCIS 837, pp. 181–192, 2018.
https://doi.org/10.1007/978-981-13-1936-5_21

The major objectives of this paper are scheduling the signals from BAN according to their priority, finding the delay in processing of signals and determine the efficiency of queuing service.

2 Related Work

The review has conducted on focusing on the popularity of m-health application, problems related with it and various scheduling methods adopted for efficient transfer of signals from BAN to receiving station. The BAN is key factor in the architecture of m-health. The vital signals from different sensors flows to the outside network. Whenever the number of signals arriving increases and they are not processed there will be severe traffic problem in the network. The vital signals from the sensors not reached in the receiving centre properly then the diagnosing of health problems could not be accurate. The recent surveys on m-health, web reports reveal the wide acceptance of this application as well as the view of people and society about the health care system. In Refs. [1–3] the latest applications of m-health were described. Now day M-health is a part of IOT so the acceptance of the application is increasing more and more. At the same time the anxiety of both the patients and physicians were also mentioned in the paper. The papers not mentioned any solution for the authenticity or security of data transfer. Reference [4] deals with the technological growth in e-Health services. The recent developments in the area of technology has vital role to make the m-health application popular. But the optimization of secure transmission cannot have achieved also by this technology. Research works are going on in the field of BAN as well as in the area of m-health. Most of the research works are in the area of BAN rather than m-health. The studies by Yankovic et al. [5] proposed semantic authorization model for pervasive healthcare by employing ontology technologies. It is a novel decision propagation model to enable fast evaluation and updating of concept-level access decision. The European countries mostly depend on m-health services in the area of health services. But the developing countries still have precincts in accepting and implementing the applications. The economy and lack of knowledge in technology is one obstruction in this area. In the case of BAN, to improve the quality of service an analytical mode has been designed by Worthington et al. [6] in his work he treated the signals in to different classes and analyze the various metrics like delay, through put, and packet loss rate etc. The paper analyze the queue traffic problem in terms of Markov chain transition probabilities for low latency queuing model. The paper could not give an efficient solution for the considered metrics. The life time and total energy consumed by WBAN were studied by Kumar et al. [7]. The paper proposed an energy efficient and cost-effective network design for WBAN the paper considered the metrics such as low latency, high throughput, guaranteeing multiple services etc. The routing of signals from BAN discussed by Liang et al. [8] in their studies by introducing the concept of collecting tree protocol and analyze metrics like reliability delay and energy conception. The application of queuing principles in health care were discussed by Sayed and Perrig [9] The paper introduced the basics of queuing principles and various queuing models. It explains the various situations in hospitals and prove the improvements in the efficiency by the application of queuing theory. The paper not considering the Body Area Network and their signals. In Ref. [10] Patwari et al. worked on authenticating loss data in body sensor health care monitoring. A network coding mechanism is used to retrieve the loosed packets during transmission. The paper not considering outdoor communication in BAN. The paper considered the packet losses in BAN are bursts, the assumption of point loss is not optimal. The theory of queuing is mathematically complex but the application of queuing theory to the analysis of performance is remarkably straight forward. The study on various scheduling techniques not considered the scheduling of signals based on priority. The paper proposes a new scheduling scheme considering the priority of signals from different sensors. Chang et al. [13] have explained a system architecture for a mobile health monitoring platform based on a wireless body area network (WBAN). They detail the WBAN features from either hardware or software point of view. The system architecture of this platform was three-tier system. Each tier was detailed. They had designed a flowchart of a use of the WBANs to illustrate the functioning of such platforms. They show the use of this platform in a wide area to detect and to track disease movement in the case of epidemic situation. Indeed, tracking epidemic disease was a very challenging issue. The success of such process could help medical administration to stop diseases quicker than usual. In this study, WBANs deployed over volunteers who agree to carry a light wireless sensor network. Sensors over the body will monitor some health parameters (temperature, pressure, etc.) and will run some light classification algorithms to help disease diagnosis. Alameen et al. [14] have stated a wireless and mobile communication technologies it had promoted the development of Mobile-health (m-health) systems to find new ways to acquire, process, transport, and secure the medical data. M-health systems provide the scalability needed to cope with the increasing number of elderly and chronic disease patients requiring constant monitoring. However, the design and operation of such systems with Body Area Sensor Networks (BASNs) was challenging in two-fold. They integrate wireless network components, and application layer characteristics to provide sustainable, energy efficient and high-quality services for m-health systems. In particular, they use an Energy-Cost-Distortion solution, which exploits the benefits of in-network processing and medical data adaptation to optimize the transmission energy consumption and the cost of using network services. Moreover, they present a distributed cross-layer solution, which was suitable for heterogeneous wireless m-health systems with variable network size. Zhang et al. [18] have proposed a way to improve sensing coverage and connectivity in unattended Wireless Sensor Networks. However, accessing the medium in such dynamic topologies raises multiple problems on mobile sensors. Synchronization issues between fixed and mobile nodes may prevent the latter from successfully sending data to their peers. Mobile nodes could also suffer from long medium access delays when traveling through congested areas.

3 Proposed Architecture

The proposed architecture of m-health uses two layered architecture. One is the intranet which is the inter connection of sensors within the body. The second one is the networking of this BAN and external nodes up to the receiving station. As in the case of any network BAN also suffers the problem of security, routing, authenticity, privacy etc. The paper deals with the problem of scheduling of signals from BAN. There is more than one sensor within the body; they capture signals from the different part of the body. The signals are collected to a sink node. From this the signals transmitted out to the external network and it is received by the mobile device in the patient's hand. From the device the signals captured by the intermediate node, from the nearest node the signal captured by the receiving server in the receiving station. The signals from diffcrent sensors may be of different types, and according to the priority the processing of signals can be arrange from the diagnosing centre with priority scheduling. This study proposes a multiple queue multiple server scheduling models to consider the priority of signals. The priority of signals is calculated by comparing with the prede-fined value of vital signals (Fig. 1).

The DFD of the proposed architecture is depicted below.

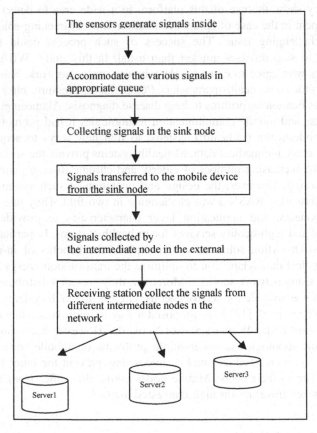

Fig. 1. Architecture of M-health with BAN

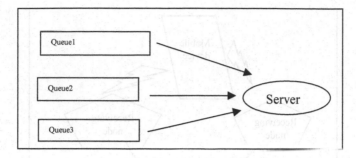

Fig. 2. Queuing of signals from different sensors.

3.1 Queuing Scheduling in BAN

Inside the BAN here we use multiple queue single servers scheduling. The signals from different sensors forms separate queue and collected by the receiving station. While this the arrival time per hour and service rate can be calculated. From this the utilization factor R can be calculated as $R = N/\mu$. It should be <1 for the steady state of the system. Here the arrival of signals as well as service rate follows poisons distribution. If the utilization factor >1, there is a need of additional servers. The queuing model can be shown as in Fig. 3 (Fig. 2).

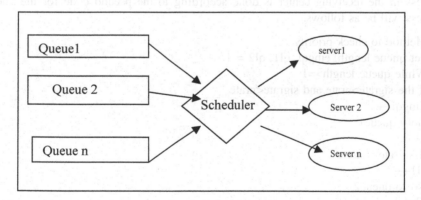

Fig. 3. Queue of signals outside BAN

3.2 Queuing Model in the Network Outside BAN

The above discussed simulation process is for the scheduling inside the patient's body. In the external network there may be more than one collecting node. The node has to determine the signal's priority and send them in to the receiving station. From the receiving center then accept the signal and channelize them in the required doctor's server. Here the scheduling model changes in to multiple queues multiple server models (Fig. 4).

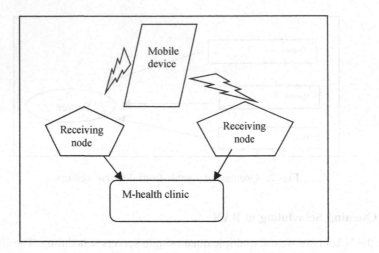

Fig. 4. Architecture of network outside the BAN

The queue model can be shown in Fig. 3.

The nodes collect signals from the mobile devices and arrange them in the queue according to priority. The signals having high deviation from the normal rate are arranged in the high priority queue and having normal rate are in the medium priority queue and the emergency messages are in the low priority queue. The scheduling process in the receiving center is done according to the pseudo code for the above process will be as follows.

Method to check priority
Set queue length, emgql, ql1, ql2 = 1
While queue length>=1
If the sigrate>nrate and sigrate<nrate
Emgql++
Goto queue1
Else
If sigrate = nrate
ql1++
Go to queue2
Else
ql2++
Goto queue3

Method to delivering required server
In server 1
Emgql = n
While emgql>=1
Send the signal in server1

In server 2
q11 = n
While q11>=1
Send the signal in queue2
In server 3
q12 = n
While q12>=1
Send the signal in queue3

The proposed model can be implemented through the multiple queue multiple server. In this model, the incoming signals are arranged in different queues based on priority as high, medium, and low. So it forms different queues and the signals from these queues are sent to different servers for processing and this is the second stage in the architecture. The general form of multiple queues multiple servers is **(M/M/C):GD/∞/∞)**.

The parameters are

(a) Arrival rate follows poisson distribution.
(b) Service rate follows poisson distribution.
(c) Number of servers is C.
(d) Service discipline is general discipline.
(e) Maximum number of signals permitted in the system is infinite.
(f) Size of calling source is infinite.

$$\emptyset = \frac{N}{C\mu} \tag{1}$$

For steady state the efficiency should be <1

$$Wq = \frac{Lq}{N} \tag{2}$$

$$Lq = \frac{\emptyset}{c - \emptyset} \tag{3}$$

where $\emptyset/c \approx 1$

$$Ws = Wq + \frac{1}{\mu} \tag{4}$$

From the intermediate servers the signals transferred to the server in the receiving station. It is the third stage and it follows single queue multiple server queue model.

3.3   Analytical Implementation of Queuing Model

For the implementation of priority scheduling the abnormal readings are arranged in the Tables 1 and 2 The arrival time and service rates are calculated in minutes. We take the readings for 30 min from 20 people. The abnormal readings are arranged in the high priority queue, and the rest of the signals are arranged in medium and low priority queue. The calculations are made with queuing theory principle of multiple queue multiple method. The waiting time of signals in the queue, the number of signals waiting in the queue are calculated. In this calculation the waiting time and queue length are reduced to the maximum. The calculations are represented in the following section.

Table 1. Signal details

Arrival rate in min.	6	7	6	5	9	10	4
Service rate	3	5	7	10	15	12	6

Table 2. Signal details

Arrival rate in min.	4	1	1	4	1	4	2	2	3	3	2	2	3	3	1	4	3	3	6	5
Service rate	3	4	3	3	3	1	2	2	3	2	2	4	3	4	2	2	3	3	3	3

From the table data
The mean arrival rate $\mu = 6.7 = 7$
The mean service rate $N = 8.28 = 8.3$
The inter arrival $= 1/7$ and inter service rate $= 1/8.3$
The efficiency $\emptyset = (1/N)/(1/\mu) = 7/8.3 = \mathbf{0.843}$
So the efficiency meet the steady state condition i.e., < 1.

The average number of customers in the queue Lq

$$Lq = \emptyset^2/1 - \emptyset = N^2/\mu(\mu - N) = \mathbf{4.54}$$

Around 5 signals are in the queue.
Average waiting time in the queue

$$Wq = N/\mu(\mu - N)$$
$$= \frac{1}{8.3} / \frac{1}{7}\left(\frac{1}{7} - \frac{1}{8.3}\right)$$
$$= 49/1.3 = \mathbf{37.69 \ min.}$$

So each signal has to wait around 38 min in the queue to get the service.
The same set of signals can be implemented through priority queue, While the signals are arriving, arrange them in different queues like high priority, medium priority

and low priority queue. Then the queue will be like multiple queue multiple server manner. The calculation can be done using multiple queue multiple server equations as follows and for each level of priority there is a server.

Mean arrival rate = 8.3 implies $\mu = 1/8.3$
Mean service rate = 6.66 $\approx$ 7 implies N = 1/7
Number of servers = 3

Applying the values in multiple queues multiple server model equations the following results obtained.

$$\emptyset = N/c\mu = 7/3 * 8.3 = \mathbf{0.2811}$$

$$p0 = \sum_{c=0}^{2} \frac{1}{c!}\left(\frac{N}{\mu}\right)^c + \frac{1}{c!(c\mu - N)/c\mu} * (N/\mu)^3$$

$$= 1/0!(7/8.3)^0 + 1 * (7/8.3)^1 + 1/2 * 7^2/8.3^2 + 1/3! * \frac{3 * 1/7}{\left(\frac{3}{7} - \frac{1}{8.3}\right)} * \left(\frac{7}{8.3}\right)^3$$

$$= 0.4769$$

$$Lq = \frac{(N\mu)(N/\mu)^c}{(c-1)!(c\mu - N)^2} * p0$$

$$= \left((1/7 * 8.3) * (7/8.3)^3 / \left(2 * (3/7 - 1/8.3)^2\right)\right)0.4769$$

$$= \mathbf{0.1304785}$$

$$Wq = Lq/N$$
$$= 0.1304785/1/8.3$$
$$= \mathbf{0.108297155}$$

Table 2 represents the second set of data in priority queue and calculate the waiting time and queue length in multiple queue multiple server manner and single queue single server manner.

Inter arrival rate = 1/2.9 = 0.3448
Service rate = 1/2.7 = 0.37037

For single server

$\emptyset$ = 0.93096
lq = 12.5705
Wq = 35.32

For multi-server

$\emptyset$ = 0.31032
lq = 0.1987
Wq = 0.5764

4 Parameter Evaluated

The proposed model considered the various metrics like arrival rate of the signals in the queue, waiting time in the queue, service rate and overall efficiency of the system. It is obvious from the result that the queue length, waiting time can be reducing to the maximum such that around 1 signal is in the queue for processing. That is, by applying the queuing equations to calculate the parameters, the result is too much better and it has proved by equations. In the normal case the priority is not considered and it follows the single queue single server manner. In the proposed model the queue length, waiting time and efficiency are calculated, by considering the priority of signals, in multiple queue multiple server manner. By analyzing the result, it is clear that average waiting time of signals in the queue, and number of signals in the queue can be reduced, such that no signals are waiting in the queue and the average waiting time is very less, i.e. 0.182. The graph 1 shows the comparison between single queue single server and priority-based scheduling. The Table 3 shows the comparison between various metrics evaluated. The results show that the priority-based scheduling using multiple queues multiple servers is more efficient than the normal scheduling.

Table 3. Comparing results of SQSS and MQMS

Metrics evaluated	Single queue single server	Multiple queue multiple server
Efficiency	Case 1 = 0.843 Case 2 = 0.93096	Case 1 = 0.2811 Case 2 = 0.31032
Average queue length	Case 1 = 4.54 Case 2 = 12.5705	Case 1 = 0.1304785 Case 2 = 0.1987
Average waiting time	Case 1 = 37.69 min. Case 2 = 35.32	Case 1 = 0.108297155 Case 2 = 0.5764

5 Result and Discussion

The paper tries to explain the scheduling scheme for signals coming out from the BAN with an analytical model using queuing theory. Here the scheduling uses multiple queue multiple server. In each case the average waiting time in the queue, average queue length and the utility factor all lies as per the steady state condition with the collected values. The values are calculated for single queue single server (SQSS) manner also for comparison so the study compare the proposed MQMS model with the FCFS (single queue single server) model. The results are shown below in the Table 3 and in Figs. 5 and 6.

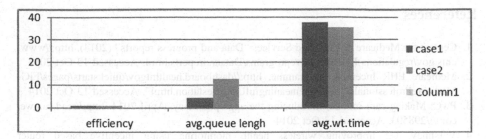

Fig. 5. Results in MQMS

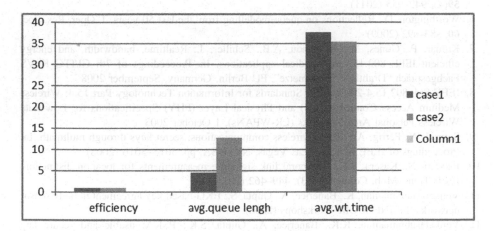

Fig. 6. Results in SQSS

6 Conclusion

The paper deals with the scheduling of signal from body sensors in m-health application. There is no research work in this area based on queuing theory. Here the network is considered as a two-layered architecture, one is the internal BAN architecture and the other is the external network. The signals are collected to the receiving station from there the scheduling begins. According to the recorded readings by applying the proposed algorithm the parameters like waiting time, number of packets in the queue are analyzed. The results are tabulated in Table 1. The study proved the efficiency of MQMS scheduling algorithm. In the present health care system, the mobile health can do much more in the public health care system. So, the future work in this study is focused on designing a frame work for pubic health care system with M-health by considering the prioritized scheduling of vital signals and emergency messages.

References

1. Centers for Medicare & Medicaid Services "Data and progress reports" (2014). http://www. cms.gov/regulations)EHRIncentiveprograms/Dataandreports.html. Accessed 13 Oct 2014
2. Medicare EHR Incentive programme. http://dashboard.healthitgov/quickstarts/pages/FIG-MedicareProfessionals-"stageonemeaningfulUseAttestation.html". Accessed 13 Oct 2014
3. PwC: Making care mobile, introducing the apps pharmacy, April 2014. http://read.ca.pwc.com/i/298503. Accessed 13 Oct 2014
4. Varshney, U.: Improving wireless health monitoring using incentive based router corporation. IEEE Comput. 41(5), 56–62 (2008)
5. Yankovic, N., Green, L.V.: Identifying good nursing levels: a queuing approach. Oper. Res. 59(4), 942–955 (2011)
6. Worthington, D.: Reflections on queue modelling from the last 50 years. J. Oper. Res. Soc. 60, s83–s92 (2009)
7. Kumar, P., Günes, M., Almamou, A.B., Schiller, J.: Realtime, bandwidth, and energy efficient IEEE 802.15.4 for medical applications. In: Proceedings of 7th GI/ITG KuVS Fachgespräch "Drahtlose Sensornetze", FU Berlin, Germany, September 2008
8. IEEE Std. 802.15.4-2003: IEEE Standards for Information Technology Part 15.4: Wireless Medium Access Control (MAC) and Physical Layer (PHY) Specifications for Low-Rate Wireless Personal Area Networks (LR-WPANs), 1 October 2003
9. Sayeed, A., Perrig, A.: Secure wireless communications: secret keys through multipath. In: Proceedings of IEEE ICASSP, Las Vegas, NV, USA, pp. 3013–3016 (2008)
10. Patwari, N., Kasera, S.K.: Temporal link signature measurements for location distinction. IEEE Trans. Mob. Comput. 10(3), 449–462 (2011)
11. Venkatasubramanian, K., Banerjee, A., Gupta, S.: EKG-based key agreement in body sensor networks. In: INFOCOM Workshops (2008)
12. Venkatasubramanian, K.K., Banerjee, A., Gupta, S.K.: PSKA: usable and secure key agreement scheme for body area networks. TITB 14(1), 60–68 (2010)
13. Cheng, W., Wu, D., Cheng, X., Chen, D.: Routing for information leakage reduction in multi-channel multi-hop ad-hoc social networks. In: Wang, X., Zheng, R., Jing, T., Xing, K. (eds.) WASA 2012. LNCS, vol. 7405, pp. 31–42. Springer, Heidelberg (2012). https://doi.org/10.1007/978-3-642-31869-6_3
14. Al Ameen, M., Liu, J., Kwak, K.: Security and privacy issues in wireless sensor networks for healthcare applications. J. Med. Syst. 36, 93–101 (2012)
15. Hu, C., Liao, X., Cheng, X.: Verifiable multi-secret sharing based on LRSR sequences. Theor. Comput. Sci. 445, 52–62 (2012)
16. Young, J.P.: The basic models. In: A Queuing Theory Approach to the Control of Hospital Inpatient Census, pp. 74–97. John Hopkins University, Baltimore (1962)
17. Alonso, L., Ferrús, R., Agustí, R.: WLAN throughput improvement via distributed queuing MAC. IEEE Commun. Lett. 9(4), 310–312 (2005)
18. Zhang, X., Campbell, G.: Performance analysis of distributed queuing random access protocol - DQRAP. DQRAP Research Group, August 1993
19. Nafees, A.: Queuing theory and its application: analysis of the sales checkout operation in ICA supermarket, June 2007
20. Galant, D.C.: Queuing Theory Models for Computer Networks. NASA Technical Memorandum 101056
21. Xie, Z., Huang, G., He, J., Zhang, Y.: A clique-based WBAN scheduling for mobile wireless body area network algorithm. Procedia Comput. Sci. 31, 1092–1101 (2014)
22. Ross, T.J.: Fuzzy Logic for Engineering Applications. Wiley, London (1998)

Data Compression Using Content Addressable Memories

Ashwin Santhosh[(⊠)] and Harish Kittur Malikarjun[(⊠)]

Department of Micro and Nano Electronics,
School of Electronics Engineering, VIT Vellore, Vellore, Tamil Nadu, India
ashwin.santhosh2016@vitstudent.ac.in,
kittur@vit.ac.in

Abstract. This paper presents a 16 bit CAM of 16 words for data compression. The input data is searched for match among the 16 stored data. The match produces compressed data output. The average compression time is also calculated for each search data and is compared with the existing CRAM compression design.

Keywords: Content-Addressable memory · NAND cell · XNOR CAM

1 Introduction

The access time for an item stored in memory is reduced if the memory is accessed by its content rather than by its address. Such type of memories are called Content Addressable Memories. The search data is compared with an array of stored data and its match location address is obtained within single clock cycle. High search speed makes content addressable memories a better choice for lookup operations. Speed of operation of the system is improved because of the parallel search operation. It is also possible to cascade CAM cells [1] so that the size of the look up table can be increased. They have a wide range of applications like ATM switches, network routers, data compression, IP filter, memory mapping etc. Even though it has these many applications it is best suitable for searching operations [5].

2 CAM Basics

The Fig. 1 shows an array of content addressable memory which contains four words each with four bits arranged in horizontal manner. Corresponding to each word there exist a matchline and search data is also given to each cell. Then all matchlines are precharged high, making them all temporarily in the match condition. Each cell compares its stored data against the data on its SL. Matchlines where all bits match discharge to ground and rest of the matchlines where there is a mismatch discharge to ground. Then the address of the match location is produced by the encoder [1].

© Springer Nature Singapore Pte Ltd. 2018
I. Zelinka et al. (Eds.): ICSCS 2018, CCIS 837, pp. 193–199, 2018.
https://doi.org/10.1007/978-981-13-1936-5_22

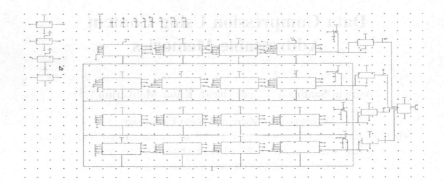

Fig. 1. CAM with 4 words having 4 bits each.

3 XNOR NAND Type Cell

The NAND cell does the comparison between the stored data and the search data using 3 comparison transistors shown in Fig. 2. The operation of the NAND cell is explained below with example. Consider SL = 1 and stored value = 1, the first transistor say MX is ON passing logic 1 to node D. Hence the middle transistor say MY is turned ON. When SL = 0 and stored value = 0 also MY is turned ON. Logic 1 is passed by transistor MZ in this case. Thus for both the match cases the node D is at logic 1 turning ON MY. The case when SL is not equal to stored value leads to the miss condition turning OFF MY. Thus the node D is XNOR function of SL and stored value [1]. When n cells are connected in series the match line ML1 to MLn are connected forming a word [1–4]. A match condition occurs for the entire word only if every cells in a word in the chain remain matched (Fig. 3).

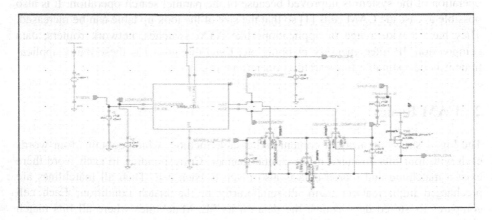

Fig. 2. NAND type CAM cell

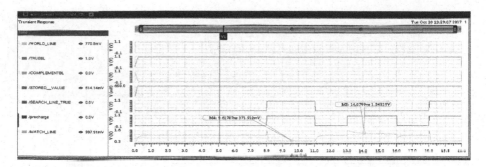

Fig. 3. Output waveforms showing the functionality of a single XNOR NAND type cell

4 Data Compression

Data compression eliminates the unwanted data that is present in an information, producing a shorter but same information. Content Addressable Memories [1] are the best choice for Data Compression because the movement of packets through LAN or WAN needs address translation. Majority of the compression algorithm time is taken for maintaining and searching these data structures. The algorithm throughput can be increased if they are replaced by some hardware search engine. In a Data Compression application, Content Addressable Memory lookup is performed after each word of the actual data is given. A CAM will generate a result in a one clock cycle regardless of the size of the table or length of search list. This makes Content Addressable Memory the best choice for Data Compression applications that make use of complex tables in their algorithm. Data compression helps to reduce the processing time of the system and also helps in resource optimization [5]. In the proposed design, the data bits are stored in the CAM. Then the search data is compared with the bits stored in each of the rows of CAM cells. For the match case, the lowest match location address (token) replaces the search data and hence the number of bits in the final output is reduced [2]. In the case of missing match, the next search data is searched in the array [2]. The existing CRAM design [3] consist of RAM, encoder/decoder in addition to Content Addressable Memories. To a single silicon array the compression and decompression engines (CAM and RAM) are added [3, 4]. They are fast but simultaneous compression and decompression cannot be achieved on separate data.

5 The 16 * 16 CAM Array

It consist of cascaded connection of 16 CAM cells in 16 rows to form the 16 match lines. At each of the matchline output the precharge transistor [4] puts the initial voltage of the match lines to VDD. For match case all the nmos transistors M1 to M16 in the corresponding row are on thus creating a path to ground. For mismatch case one of the series transistors M1 to M16 is off putting the corresponding match line in high state. All the matchline outputs are inverted and are given as input to the 16 * 4 encoder. At the encoder output, the compressed 4 bits are obtained (Fig. 4 and Table 1).

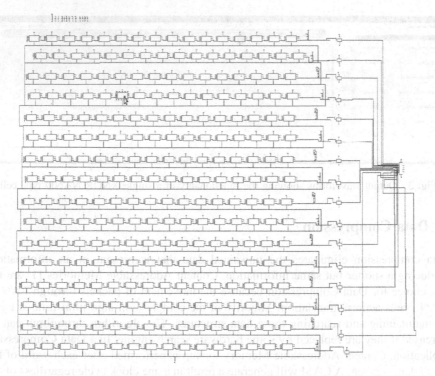

Fig. 4. The 16 * 16 CAM array

Table 1. The 16 bit stored words in each of the 16 rows.

Match line	Token	Stored data
1	0000	0000000000000000000000000000000
2	0001	0001000100010001000100010001
3	0010	0010001000100010001000100010
4	0011	0011001100110011001100110011
5	0100	0100010001000100010001000100
6	0101	0101010101010101010101010101
7	0110	0110011001100110011001100110
8	0111	1000100010001000100010001000
9	1000	1001100110011001100110011001
10	1001	1010101010101010101010101010
11	1010	1011101110111011101110111011
12	1011	1100110011001100110011001100
13	1100	1101110111011101110111011101
14	1101	1110111011101110111011101110
15	1110	1111111111111111111111111111
16	1111	0111011101110111011101110111

6   Simulation Results

The 16 * 16 CAM array is designed in 90 nm CMOS technology and the following simulation results are obtained in cadence virtuoso for four different search data (Figs. 5, 6, 7, 8 and Tables 2, 3).

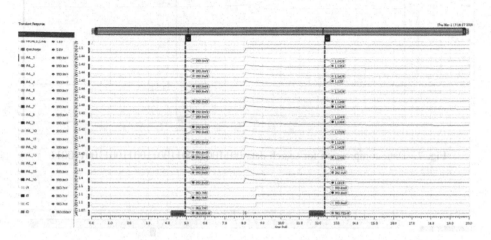

Fig. 5. Data compression for search data 1111111111111111111111111111111111

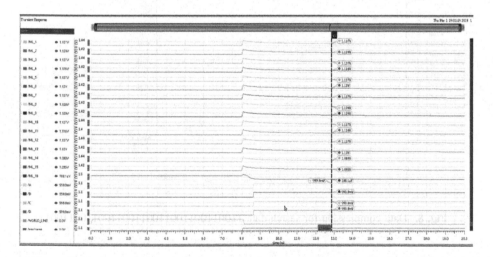

Fig. 6. Data compression for search data 011101110111011101110111011101110111

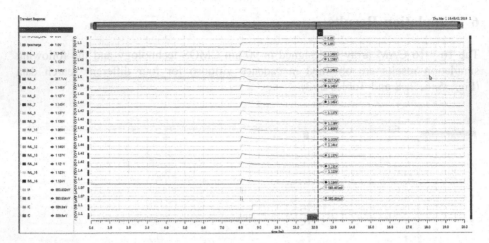

Fig. 7. Data compression for search data 0011001100110011001100110011

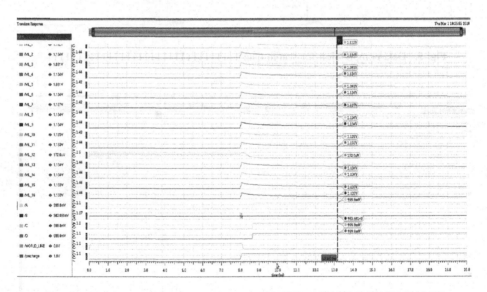

Fig. 8. Data compression for search data 1100110011001100110011001100

Table 2. Four compressions obtained for four different search input data.

Search input data	Compressed output
111111111111111111111111111111111	1110
0011001100110011001100110011	0011
1100110011001100110011001100	1011
0111011101110111011101110111	1111

Table 3. Comparison of compression time.

Design name	Compression time
CRAM compression design	100 MB/s
Proposed design	16 bits/0.5 ns

7 Conclusion

The 16 bit CAM of 16 words is designed for data compression. Data compression is successfully demonstrated. The compression time for the precharged match line is 0.5 ns. All the four compressions are obtained under 1 ns. Faster compression time is achieved as compared to CRAM design for data compression.

References

1. Pagiamtzis, K., Sheikholesami, A.: Content addressable memory (CAM) circuits and architectures: a tutorial and survey. IEEE J. Solid State Circ. **41**, 712–727 (2006)
2. Content Addressable Memory (CAM) Applications for ispXPLD devices
3. Craft, D.J.: A fast hardware data compression algorithm and some algorithmic extensions
4. McAuley, A.J., Francis, P.: Fast routing table lookup using CAMs. In: Proceedings of IEEE Infocom, vol. 3, pp. 1282–1391 (1993)
5. Peng, M., Azgomi, S.: Content-Addressable memory (CAM) and its network applications. Altera International Ltd

Heart Block Recognition Using Image Processing and Back Propagation Neural Networks

P. Asha(✉), B. Sravani, and P. SatyaPriya

Department of Computer Science and Engineering,
Sathyabama Institute of Science and Technology, Chennai, India
ashapandian225@gmail.com,
sravanichowdary909@gmail.com,
satyapriya.pinnamaneni30@gmail.com

Abstract. Nowadays in the health economy scenario, there has been a significant number of medicinal services data consisting of concealed details. The treatment for ischemic heart disease is a tough nut to crack as it calls for a widened comprehension and skill. One of the widely used conventional approaches to figure out heart ailment is through a doctor's assessment. Rather the disease can also be found out through a range of health checkups that includes Heart Magnetic Resonance Imaging (MRI) test, to arrive at an enhanced and useful output, original techniques are recommended, and the suggested methods are DTCWT and neural network (NN). The aforementioned method paves the way for an enhanced yield given identifying heart disease. A standout amongst the most regularly utilized image preparing system is Discrete Wavelet Transform (DTCWT) which best fits for changing images from the spatial domain into the frequency domain. The NN prepares the image with the help of extorted facet DTCWT feature. In this paper, the literature survey and the proposed methodologies, advantages and disadvantages of the proposed methodologies has been discussed. This is implemented in matlab.

Keywords: Backpropogation · DTCWT · Neural network · K-means

1 Introduction

Medical history contains a lot number of tests to diagnose a particular disease. With the help of the data mining tools, healthcare sector is can use it as an intelligent diagnosis tool for predicting diseases. One of the most widely used approach to figure out the heart problems is by doctors check up, other than this there are many approaches to predict are magnetic resonance imaging (MRI) test, stress test, electrocardiogram (ECG) etc.

In general image processing is one of the most vastly growing technologies. It is a method which is used to perform operations of an image, in order to get an enhanced image and also used to get useful information from it. Generally, an image processing, the input is an image and output can be image or characteristics associated with the image. Main features of image processing are to improve clarity and to remove noise

© Springer Nature Singapore Pte Ltd. 2018
I. Zelinka et al. (Eds.): ICSCS 2018, CCIS 837, pp. 200–210, 2018.
https://doi.org/10.1007/978-981-13-1936-5_23

and to prepare the image for discipline. Matlab is used to automate the image processing with scripts. It generally imports and visualizes image data. In the present times, heart disease is treated as one of the major causes of death in the world. It may be caused due to various factors, which cannot be easily predictable by medical practitioners who need higher knowledge of predictions for recognition.

2 Existing System

Singh et al. [1] proposed a method to control the heart attacks for improving the healthcare sector. In this paper, they said that the correct and precise prediction of heart we will get when we use an electrocardiogram (ECG) data and clinical data. The data must be sent to the nonlinear disease prediction model to detect arrhythmias tachycardia, bradycardia, infraction, a trial and many more. In this, they used ECG data and clinical data to train the artificial neural network to correctly check the heart and predict the abnormalities.

As the traditional approach to predict the heart disease is not good enough, a machine learning approach was used for prediction of heart disease. They said, that by developing a medical diagnosis for these diseases will give the most accurate result than the traditional method. In this, they proposed the methodology was by using the artificial neural network back proportion algorithm. In this, they took Cleveland dataset. The 13 clinical data that is obtained from the dataset are used by neural networks and the neural network was trained by using the back propagation algorithm to predict the heart disease with 95% accuracy [2].

Krishnaiah et al. [3] Presented a prediction of heart disease by using multilayer perceptron neural network for multilayer perception heart disease. The neural network accepts the 13 clinical features as the inputs. These inputs are trained using the back propagation technique. They used inputs from Cleveland heart disease database to feed as the input to the neural network. In this, the prediction system gives the better result with the correct technique for the getting the image by using self-organizing map and executed it.

The next approach attempted at predicting the heart disease by using risk factors of the patients. Risk factors such as family, hypertension, age, high cholesterol, etc. of the patients were collected. They used two major techniques that are genetic algorithm and neural network. This hybrid system uses an optimized genetic algorithm for initialization of neural network weights. This result shows that we can anticipate the disease by using the risk factors that are collected. The system is implemented in Matlab with the accuracy of 89% [4].

Jabber et al. [5] suggested a technique for early diagnosis of heart disease. In this, they have taken a data set of the sick individuals and classified them whether they are healthy or sick individuals here mainly they used naïve Bayes, genetic search methods. After taking datasets they are applying the chi-square/gain ratio for the feature selection measure. We will get the data in descending order. Next, apply genetic search on the dataset to optimize the dataset. The next, the step is to take dominate features and remove the least found attributes. Next, we should give the heart disease data to naive

Bayes [14] for testing data set fed to naive Bayes for classification. In this, they are using various feature selection methods to get the highest accuracy.

Princy et al. [6] proposed detection of heart disease such as global hypokinesia in this; they developed a computational method to predict global hypokinesia based on global hypokinesia through MRI images. These image prediction methods are done by using machine learning. They collected 30 patients of each category which are having set of 900–1200 Dicom images. They used cell profiler for feature extraction, in this CP-CHRAM is cast-off for the model building using MRI images by using this model they got 80–85% accuracy with 25 image set and 97% accuracy by merging imaging using amide. This is developed by using python and available in open source. Mane et al. [7] aimed at the development of associative classification techniques on heart dataset for easy and early diagnosis of heart diseases. The main focus of this paper is to propose and generate Classification Association Rules (CARS) efficiently and observe which one gives the more accurate predicted value of early prediction of heart disease. Association Algorithms like Apriori and FP-growth [15] are used to finds association rules and classification algorithm are used to predict the small set of relationships between attribute in a database to construct accurate classification. Mainly it is used to find high prediction accuracy. It also works on weka environment. The results of K-nearest with appropriate gives better results than others. This expression shows large no of rules support in the better diseases of heart disease and even the specialist in their diagnosis.

Pahwa et al. [8] focused on hidden Naive Baye's (HNB) can be applied to heart disease prediction. Results are shown that HNB can record 100% in terms of accuracy. In this input is heart disease data set. Initially, data set is loaded, applying preprocessing filter discretization, the partition of data into training and test set is done. Heart disease data set is trained by HNB and measures accuracy of HNB. They called heart analog dataset in ARF from the UCI repository and then adopted preprocessing techniques. And then Numeric attributes were discretized by discretization filter unsupervised login discretization and IQR to measure the variability by dividing the dataset into quartiles. Applied HNB classifier for diagnosis of heart disease and tested performance for analog data and also applied discretization and IQR filters to improve the efficiency of HNB.

The main approach of Sultana et al. [9] is to select features before classification in order to improve the performance of models and performing feature selection SVM_FREE and gain ratio algorithms are applied to the dataset. First stage is to preprocess of raw data. Data transformation is applied to raw data to make it a binary classification problem and at the second stage processed data input and applies feature selection approach to input to extract a subset of relevant features and at stage 3 classification algorithms are applied to features for prediction and dataset used is Cleveland heart disease database. And finally, the result of SVM-FREE and gain ratio is considered to the calculated final weight of each attribute and here attributes of value (weight) above threshold values are selected as the final set of attributes. The main result of this classifies the data in two classes either in positive or in a negative result for heart disease a hybrid approach of feature selection to optimize classification problem, results of SVM-free and gain ratio are used to get a subset of features and remove redundant features.

The aim was to find the rules to analyze the rate of risk of patients, according to parameters mentioned on their health. In this initially, they considered database of the heart disease which contains the medical data of patients. They preprocess the data to make it more efficient. After collecting data they have used KEEL tool. In primary preprocessing phase, they have used an MV algorithm to fill the value which is present in the dataset. They had used statistical method for finding the performance of the classifier [10]. The objective was to improve the efficiency and speed of the process and can be applied in information retrieval, image processing, and pattern matching [11, 12]. In this mainly work is integrated with feature selection using PCA (Principal Component Analysis) and feature selection using information gain ratio for selecting the relevant attributes. They removed data noise initially and next is to feature extraction, where the PCA is made to extract the critical feature which is the most relevant features. The attributes considered are of age, sex, slope etc. It is an overview of feature selection and feature extraction. Various methods like outlier detection, feature extraction using PCA is performed.

A better system was introduced for prediction of the heart disease by using a multilayer perceptron neural network. In this work, they considered 13 clinical features as input and it is trained using backpropagation algorithm, as they considered multi-layer perception architecture of the neural network. They are two steps, primary step is 13 clinical attributes is taken as input and then they trained them by back propagation learning algorithm. Cleveland heart disease database has been used to feed the input to the neural network. In this, they used accurate techniques for classification and retrieval of an image by self-organizing map (SOM). Image texture is actually performed in two phases [13].

3 Comparative Study on Existing Methodologies

See Table 1.

Table 1. Comparison of existing and proposed system

Method	Advantages	Disadvantages
Classification, Naive Bayes classifier, genetic search	1. Used to accurately predict target class for each case 2. This method eliminates useless data and reduces the diagnostic tests to be taken	Only applied for limited data set
Artificial neural network, backpropagation, multilayer perception, machine learning	1. Machine learning is used for prediction is more accurate 2. It is designed by using multilayer perceptron	It can be used by using hybrid model and with other classification algorithms to obtain a more accurate diagnosis

(continued)

Table 1. (*continued*)

Method	Advantages	Disadvantages
Multilayer perception, Machine learning, backpropagation	1. In this the system considers the 13 clinical features as input 2. It used self-organizing ma for classification of image	Not describes much about the SOM feature that is used
Data mining, risk factors	1. Prediction based on risk factors 2. Reduces the regular checkups hospital	Still, need a system predict in early stages
Artificial neural network, ECG, heart rate	Precise prediction due to the usage of ECG and clinical data	Only preliminary work is done and helpful in identifying risks
Bigdata, Clustering, Classification, improved k-means, map reduce	They used big data approach and Hadoop map-reduce platform	Need more accuracy in the prediction
Classification, Association, heart disease	Usage classification and association rules to give a small set of relationship between the attributes	Chance of missing relation between some attributes is more
Hidden navie Bayes, classification, heart disease	They used hidden navie Bayes for better classification of the data	Cannot assure 100% accuracy and efficiency
Feature selection, classifier, NavieBayes	All the features before classification are selected for more efficiency	There is a chance of missing some features for further classification
C4.5, decision tree, CVD, CAD	They had used statistical method for analyzing the classifier performance	As information is manually collected from patient
Dimensionality reduction, feature selection, PCA	Mainly feature selection and feature extraction is used	Cannot assure for 100% accuracy
Multilayer perceptron, backpropagation, machine learning	As it a stepwise procedure of collecting analyzing data all the features are taken and evaluated	Complicated as all the steps are taken to be considered

4 Proposed System

The implementation has been carried out in MATLAB. To perform this, we consider the image of heart as the input (Fig. 1). Initially, images of the heart are pre-processed (Fig. 2) to resize the heart and remove the noise. By removing the noise that is present in the image we can find the affected area more accurately. They are made to convert into RGB to gray scale image and many more according to your need.

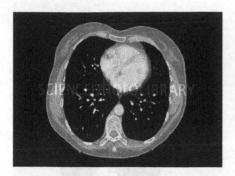

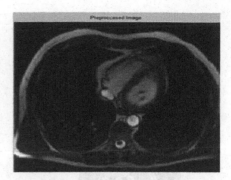

Fig. 1. Input image

Fig. 2. Preprocessed image

After this, feature extraction techniques are applied to the heart images to get features that will be useful in classifying and recognition of images. We will also get the other features of the heart such as entropy values, range etc. Filtering technique is used for enhancing and modifying the image of the heart. We adopted discrete wavelet transform (DTCWT) and dominant rotated local binary pattern (DRLBP) to find out the high intensity and low-intensity regions. Low regions are chosen and they are performed to predict block part accurately. If we take high intensity values noise level will be high therefore we can't find the exact blockage. Next, K-means clustering is used to group the images into the clusters and take those cluster parts of the heart where we can find the blockage exactly. In this we also use gray level co-occurrence matrix. It is based on orientation and distance between image pixels. Meaningful statistics are extracted from the matrix as a texture representation. The features which we will get by this are energy, correlation, homogeneity. With the help of these values we can compare the input images with the data set images more precisely.

Initially we keep the images of the heart in the database. Then again we will apply the feature extraction to those images to remove the noise. Then will train the images by the neural network. Then will apply back propagation technique to the images. Then we will compare the input images and the images in the database. Then will know whether it is a normal heart or the abnormal heart. If it is an abnormal heart then it will show what type of heart disease occurred and which area it is present in our heart (Fig. 3).

Suppose if the input image is not matched with the images in the database than the input image is directly added to the database so, that if any new heart diseases are found then they are directly added. Because of this new heart diseases can be identified and can be resolved (Fig. 4).

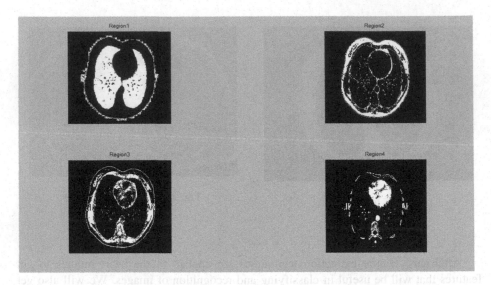

Fig. 3. Clustered image showing defected region

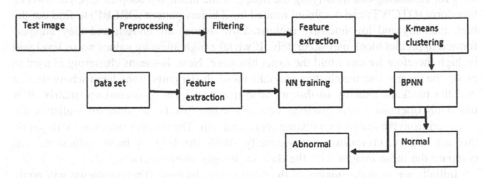

Fig. 4. Architectue diagram

5   Methodology Used

5.1   Filtering Method

In this we used Dominant rotated local binary pattern (DRLPB) and dual tree complex wavelet transform (DTCWT).

DRLPB: It is generally used for a different object texture and edge counter feature extraction process. In this, we have to differentiate the dominant part of the input image. It may be either by considering the pixel values or maybe by considering the feature values. By this, we differentiate the high and low-intensity signals and forward out the feature extraction.

DTCWT: Generally if is complex values extension of standard discrete wavelet transform (DWT). It is a 2D wavelet transform. Initially, usage of complex wavelets in image processing was implemented in 1995 by Lina and Gagnon. Generally by using DTCWT enhancement of image along with the spatial coordinates of the image, which improves the perception of an image. When we improve the resolution of an image high-frequency components are required to preserve other than enlarging the image. DTCWT has good directional selectivity and shifts invariant property. In this original input image is decomposed and produces six complex valued high-frequency sub-band images and low-frequency sub band images. This sub band image undergoes bi-cubic interpolation interpolated input image using covariance based interpolation goes through verse DTCWT to get the super-resolved image of an original image. The resultant output image is the super-resolution of the original image.

5.2 K-Means Clustering

This helps to partition of the data. This is simple and early procedure to class by a given data set through a certain number of clusters. This was first used by James queen in 1971. This is iterative algorithm, k-means is also known as a Lloyd's algorithm. Initially mean are randomly generated within the clusters. Centroid of each cluster becomes the new mean; this step is continued until convergence has been reached. This cannot assure for the optimum solution. If the input variables are more then we opt for the back propagation.

5.3 Back Propagation

Back propagation in a neural network is used in artificial neural network's which is used for training neural networks, it is a multilayer perception and is a sensible approach for dividing the contribution of each weight. In this they initialize network with random weight and present training inputs to network and calculate output. Comparison of network output (error function) is done. Feed-forwarding of the input pattern and calculating BP of associated error and then adjusting of the weights are done. BPA trains the output for any input, but not same as the one used in training.

Using all these methods, we are able to identify the abnormal heart (Fig. 5) and normal heart (Fig. 6).

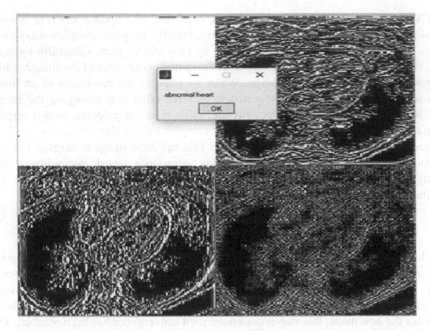

Fig. 5. Abnormal heart

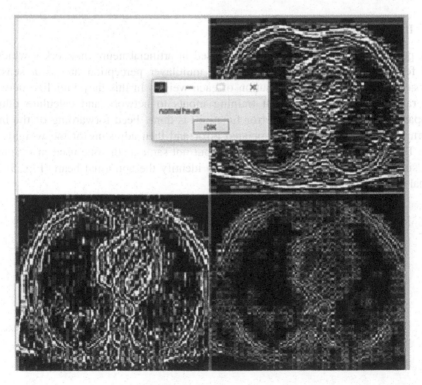

Fig. 6. Normal heart

6 Conclusion

This proposed heart disease prediction is used for diagnosing at the early stage. We have used neural networks and clustering. We have taken images as the input and compared those with the images in the database. We have trained the images and removed the noise from the images so that we can get the accurate output. We used median filter for the filtering process. Finally we will get whether it is a normal heart or the heart with the disease with an accuracy of 90%.

References

1. Singh, J., Kamra, A., Singh, H.: Prediction of heart diseases using associative classification. In: 2016 5th International Conference on Wireless Networks and Embedded Systems (WECON), pp. 1–7 (2016)
2. Jabbar, M.A., Deekshatulu, B.L., Chandra, P.: Computational intelligence technique for early diagnosis of heart disease. In: IEEE International Conference in Engineering and Technology (ICETECH), 2015, pp. 1–6, March 2015
3. Krishnaiah, V., Srinivas, M., Narsimha, G., Chandra, N.S.: Diagnosis of heart disease patients using fuzzy classification technique. In: International Conference on Computing and Communication Technologies, pp. 1–7 (2014)
4. Ravish, D.K., Shanthi, K.J., Shenoy, N.R., Nisargh, S.: Heart function monitoring, prediction and prevention of Heart Attacks: Using Artificial Neural Networks. In: 2014 International Conference on Contemporary Computing and Informatics (IC3I), Mysore, pp. 1–6. IEEE (2014)
5. Jabbar, M.A., Samreen, S.: Heart disease prediction system based on hidden naïve bayes classifier. In: 2016 International Conference on Circuits, Controls, Communications and Computing (I4C), Bangalore, pp. 1–5 (2016)
6. Princy, R.T., Thomas, J.: Human heart disease prediction system using data mining techniques. In: 2016 International Conference on Circuit, Power and Computing Technologies (ICCPCT) (2016)
7. Mane, T.U.: Smart heart disease prediction system using Improved K-Means and ID3 on big data. In: 2017 International Conference on Data Management, Analytics and Innovation (ICDMAI) Zeal Education Society, Pune, India, 24–26 February 2017 (2017)
8. Pahwa, K., Kumar, R.: Prediction of heart disease using hybrid technique for selecting features. In: 2017 4th IEEE Uttar Pradesh Section International Conference on Electrical, Computer and Electronics (UPCON), GLA University, Mathura, 26–28 October 2017 (2017)
9. Sultana, M., Haider, A., Uddin, M.S.: Analysis of data mining techniques for heart disease prediction. IEEE (2016)
10. Saxena, K., Sharma, R.: Efficient heart disease prediction system using decision tree. In: International Conference on Computing, Communication and Automation (ICCCA 2015) (2015)
11. Asha, P., Jebarajan, T.: Improved parallel pattern growth data mining algorithm. Int. Rev. Comput. Softw. 9(1), 80–87 (2014)
12. Sonawane, J.S., Patil, D.R.: Prediction of heart disease using multilayer perceptron neural network. In: ICICES 2014, S.A. Engineering College, Chennai, Tamil Nadu, India (2014)

13. Vineetha, V., Asha, P.: A symptom based cancer diagnosis to assist patients using naive bayesian classification. Res. J. Pharm. Biol. Chem. Sci. **7**(3), 444–451 (2016)
14. Pandian, A., Thaveethu, J.: SOTARM: size of transaction based association rule mining algorithm. Turk. J. Electr. Eng. Comput. Sci. **25**(1), 278–291 (2017)
15. Asha, P., Srinivasan, S.: Analyzing the associations between infected genes using data mining techniques. Int. J. Data Mining Bioinform. **15**(3), 250–271 (2016)

Design and Development of Laplacian Pyramid Combined with Bilateral Filtering Based Image Denoising

P. Karthikeyan[✉], S. Vasuki, K. Karthik, and M. Sakthivel

Department of Electronics and Communication Engineering,
Velammal College of Engineering and Technology, Madurai, India
kpkarthi2001@gmail.com

Abstract. This paper mainly deals with image denoising algorithm by applying bilateral filter in Laplacian subbands using raspberry pi. Bilateral filter is used for reducing various noises especially Additive White Gaussian Noise which occur more in standard test images. The important feature of this nonlinear bilateral filter is the preservation of the edges, while reducing the noise in the images. The main idea is to replace the pixel's intensity value by a weighted average of adjacent pixel intensity value. Euclidean distance and radiometric differences of pixels, are used for weight calculation. This calculation mainly preserves sharp edges by looping through each pixel and adjusting weights to the nearest pixel values. Our project aims to reduce cost and power consumption using Raspberry pi. It consists of series of microcomputer packed onto a single circuit board. These low power computers are mass produced at very low prices. The performance of Raspberry pi is equivalent to a personal computer. In order to perform noise reduction in Raspberry pi, python2 and opencv package is installed in real-time Raspbian Linux operating system and algorithm is executed using python2 programming language. The denoising method using Laplacian subbands provides better denoised images compared to Gaussian filter and Bilateral filter that applied in standard test images.

Keywords: Image denoising · Laplacian subbands · Gaussian filter
Bilateral filter · Raspberry pi

1 Introduction

A digital image is a representation of image using a finite set of digital values called pixels or picture elements. Digital image processing is a domain of signal processing which takes the image as the input, such as a photograph or an video frame; the output may be either an image or a set of characteristics or parameters related to the image.

The image enhancement algorithm aims at reducing noise and also preserves the edges. The images are affected by various noises like Impulse noise, Poisson noise, Additive White Gaussian Noise etc. For example, each pixel in an image with Additive White Gaussian Noise is distributed by Gaussian random variable with zero mean and variance σ^2. Poisson noise occurs in photon counting in optical devices, where it

© Springer Nature Singapore Pte Ltd. 2018
I. Zelinka et al. (Eds.): ICSCS 2018, CCIS 837, pp. 211–221, 2018.
https://doi.org/10.1007/978-981-13-1936-5_24

resembles with the particle nature of light. The quantization and amplifier noise occurs due to the conversion of electrons to pixel intensity [1].

The quality of overall image depends upon the combination of at least five factors: contrast, blur, noise, artifacts and distortion. While retrieving an image from some sources, it is mainly affected by additive noise, having Gaussian distribution. Image denoising plays a vital role in transmission, formation and display systems of image. Thus, numerous methods have been developed for image denoising. In [2], Bilateral filter smoothes the image by enhancing the edges, using a nonlinear combination of neighboring image intensity values. The intensity values used for combination may be gray or colors based on their photometric similarity and geometric closeness. The combination uses near values than distant values in both domain and range. The slope of the edges are increased to sharpen an image as in the case of adaptive bilateral filter [3].

Image denoising can also be performed using the combination of bilateral filter and wavelet thresholding called as multiresolution bilateral filtering [4]. In [5], Gaussian noise is removed using bilateral filter in natural images. It is implemented using Raspberry pi. In [6], guided image filtering has better behavior near edges to remove and enhance the appearance of an image according to the distance measure between adjacent pixels. It is the fastest edge preserving filter. The Weighted Least Squares (WLS) filter reduces halos by global intensity shifting.

In [7], Gradient Histogram Preservation (GHP) algorithm is proposed to denoise an image. In addition to image denoising, it enhances the texture structures. In [8], a generalized masking algorithm using individual treatment of the model components was proposed. It reduces halo effect using edge preserving filter. In [9], it performs two operation, contour detection and image segmentation.

Wavelet transforms have become a very powerful tool in the area of image denoising. The wavelet coefficients of the image are Wiener filtered to denoise an image degraded by an AWGN [10]. In [11], bilateral filtering in the Laplacian subbands is used to denoise the noisy image which is implemented using Raspberry pi. The pi has a single-core processor that runs at 700 MHz. It has a coprocessor for performing floating point calculations. It has 512 MB of RAM. In pi, the image gets onto a Micro Secure Digital (SD) card instead of a standard-sized SD card.

2 Materials

2.1 Gaussian Filter

Gaussian filters smoothes an image by calculating weighted averages in a filter box. This filter doesn't produce any overshoot to a step function input while minimizing the rise and fall time.

$$GF[I]_a = G_\sigma(\|a - b\|)I(b) \tag{1}$$

Where

$$G_\sigma(d) = \frac{1}{2\pi\sigma^2} e^{-\frac{d^2}{2\sigma^2}} \tag{2}$$

The spatial distance is defined by $G_\sigma(\|a - b\|)$ and σ is the standard deviation of the Gaussian distribution.

2.2 Bilateral Filter

Bilateral filter overcomes various problems of Gaussian filter. A bilateral filter is a non-linear filter, which reduces noise without affecting the image edges. Here, the intensity value of each pixel is replaced by a weighted average of intensity values from adjacent pixels. Mathematically, the output of a bilateral filter at a location a is given as follows,

$$BF(I)_a = \frac{1}{W} \sum_{q \in s} G_{\sigma_s}(\|a - b\|)G_{\sigma_r}(|I(a) - I(b)|)I(b) \tag{3}$$

Where $G_{\sigma_s}(\|a - b\|)$ is a geometric closeness function and it is given by

$$G_{\sigma_s}(\|a - b\|) = e^{-\frac{a - b^2}{2\sigma_s^2}} \tag{4}$$

$G_{\sigma_r}(|I(a) - I(b)|)$ is a gray level similarity function and it is defined as,

$$G_{\sigma_r}(|I(a) - I(b)|) = e^{-\frac{|I(a) - I(b)|^2}{2\sigma_r^2}} \tag{5}$$

W is a normalization constant given by,

$$W = \sum_{q \in s} G_{\sigma_s}(\|a - b\|)G_{\sigma_r}(|I(a) - I(b)|) \tag{6}$$

$\|a - b\|$ is the Euclidean distance between a and b, and S is a spatial neighbourhood of a.

The parameters σ_s and σ_r describes the bilateral filter behavior. Space: σ_s represent spatial extent of the kernel, size of the considered neighborhood. Range: σ_r minimum amplitude of an edge. The space parameter proportional to image size. The range parameter proportional to edge amplitude.

2.3 Laplacian Pyramid

The original image is filtered using Gaussian filter to produce a low band image. It is done by appropriate kernel bandwidth. The low band image is then removed from the original image to produce a high band image. This process is repeated to produce a set of subband images called Laplacian pyramid [11]. For an image I, the process is described as,

$$G_0 = I, \tag{7}$$

$$G_{k+1} = \downarrow 2, \left(Gaussian\left(G_k\right)\right) \ for \ k = 0, \dots n - 1$$

The Laplacian pyramid is given by,

$$L_{k+1} = G_k - \uparrow 2\left(G_{k+1}\right) \ for \ k = 0, \dots n - 1$$

$$L_{n+1} = G_n \tag{8}$$

2.4 Raspberry Pi

The Raspberry Pi is an ultra cheap minicomputer having length of 9 cm and width of 5.5 cm. The Raspberry Pi has CPU, RAM and GPU in one component called System on Chip (SoC). It uses an ARM1176JZF-S 700 MHz CPU which is single core and also has a co-processor for floating point calculations. The working memory in Raspberry pi is 512 MB SDRAM. The power consumption is low which consumes only 5 to 7 watts of electricity [12] (Fig. 1).

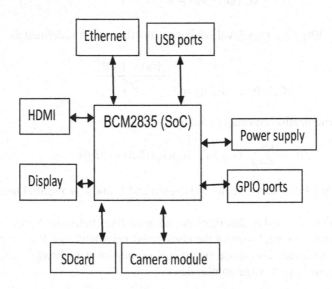

Fig. 1. Raspberry pi hardware architecture

2.5 Memory

The Raspberry Pi processor has two levels of cache memory. Level 1 is smaller in size and has 32 KB of cache memory. Level 2 is large in size and has 128 KB of cache memory. Cache memory has stored the recently used programs. The instructions are executed in the Arithmetic Logic Unit in CPU. CPU produces accurate integer calculation results (Table 1).

Table 1. Raspberry Pi specification

Chip	Broadcom BCM2835 SoC
Core architecture	ARM11
CPU	700 MHz Low Power ARM1176JZFS Applications Processor
Memory	512 MB SDRAM
Operating System	Boots from SD card, running a version of the Linux operating system
Dimensions	85.6 x 53.98 x 17 mm
Power	Micro USB socket 5 V, 1.2A

2.6 Requirements

The Raspberry pi is a tiny, versatile device and also integrates inside of other devices too. It should have both hardware and software requirements.

- The Raspberry pi need an SD card and power supply for connecting keyboard and mouse. It should have a display for running operating system like Linux, Microsoft Windows. The display
- The software requires a cross compiler that converts source code files into Raspberry Pi-compatible executable files which is in the SD card. The word length is 32 bit. A word is defined as a complete information that the CPU can execute. These instructions are executed by Arithmetic Logic Unit (ALU) which is an important part of the CPU.

2.7 Networking

The model A and A+ can be connected to a system by an external user supplied USB Ethernet or an wifi adapter due to lack of an 8P8C ("RJ45") USB port in Ethernet. In case of model B and B+ the USB port is having build-in Ethernet adapter.

2.8 Power Supply

The raspberry pi is powered by the supply voltage of 5v DC power supply. The power supply used by model B is 700 mA to 1200 mA. When no peripherals are attached model A uses 500 mA power supply. The raspberry pi mainly uses below 1A. The processor becomes unstable if the supply voltage is below 4.75v. The GPIO pins requires power of 50 mA and keyboard, mice requires below 100 mA or over 1000 mA.

2.9 Essential Peripherals

- High Definition Multimedia Interface (HDMI) by the use of a single cable it supports high-quality digital video and audio [13].
- Composite video do not support HDMI which mainly provides the Raspberry Pi to be connected to a Television. It does not produce output in high quality as HDMI and is an analogue standard.
- By using an converter, it is also possible to connect a computer monitor with a Digital Visual Interface (DVI) connection to the HDMI. The two of these standards are digital.

2.10 Keyboard and Mouse

The Raspberry Pi uses the USB Keyboard and Mouse by plugging into it also if paired wireless keyboard and mouse can also be used. Midi sized keyboards can also be integrated into the track pads. It requires less space than the standard keyboards (Fig. 2).

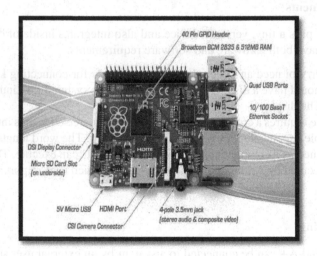

Fig. 2. Raspberry pi B + module

3   Methods

3.1 Denoising Algorithm

- Let the image be considered as I. The image I is affected as AWGN having variance σ that mainly occurs in natural images.
- The noisy image is filtered using Gaussian filter by downsampling to obtain a low band image $G_{k+1} =\downarrow 2G_K$.
- The Laplacian subbands are obtained from $L_{K+1} = G_K - \uparrow 2(G_{K+1})$.
- The process gets a set of subband signals called Laplacian pyramid.

- Applying bilateral filter in the Laplacian pyramid for denoising.
- Evaluate the Peak Signal to Noise Ratio and Image Quality Index values in Raspberry pi using Python2 programming.
- Compare the Peak Signal to Noise Ratio for Gaussian filter, bilateral filter in natural images and the combination of Laplacian pyramid and bilateral filter.

The performance of noise reduction at the slope and slowly varying areas which become flat in the high band. The Laplacian sub-band bilateral filtering can be a considerable choice for denoising the noisy images mainly for fast implementation of real-time systems. The proposed method improves Peak Signal to Noise Ratio of the image compared to the bilateral filtering and subband decomposition methods.

The subband bilateral filtering algorithm is compared with the original bilateral filtering $\sigma_{d} = 1.8$ and $\sigma_{r} = 2\sigma$. The images are corrupted by using 100 different noisy sequences of Gaussian distribution with variance of 20 to 50. The Peak Signal to Noise

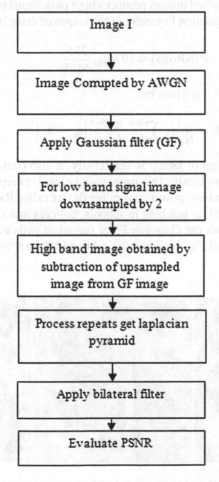

Fig. 3. Work flow of proposed denoising algorithm

Ratio is calculated and compared with the original bilateral filtering by python2 programming language using Raspberry pi [11] (Fig. 3).

4   Results and Discussion

Experiments were performed in images having Additive White Gaussian noise of zero mean and variance values σ_n varies from 20 to 50. The images are denoised by applying bilateral filter in the Laplacian subbands. The peak signal to noise ratio and image quality index values of the proposed method are calculated and compared with Gaussian and Bilateral filter. The proposed method produces high peak signal to noise ratio and Image Quality Index value compared to Gaussian and bilateral filter applied in standard test images.

The peak signal to noise ratio are calculated from the mean squared error value between the original image and compressed image.

The high quality denoised images produces high peak signal to noise ratio. The peak signal to noise ratio calculation in terms of mean squared error is given by

$$PSNR(dB) = 10 \log_{10} \frac{255^2}{MSE} \tag{11}$$

where, mean squared error is given by,

$$MSE = \frac{1}{NXN} \sum_{i=0}^{N-1} \sum_{j=0}^{N-1} \left| x_{i,j} - \hat{x}_{i,j} \right|^2 \tag{12}$$

The personal computer or laptop is very costly. It also consumes large amount of power, as well as are very costly. This paper mainly involves replacement of computer with an ultra-low cost and low power consuming device called Raspberry pi. It is a hand held device and it is used for teaching in schools, colleges and also for presentation in seminars. Figure 4 shows the Gaussian Filter output in python2. Figure 5 shows the bilateral filter output and Fig. 6 shows the subband bilateral filter output in python2 (Fig. 7, Table 2).

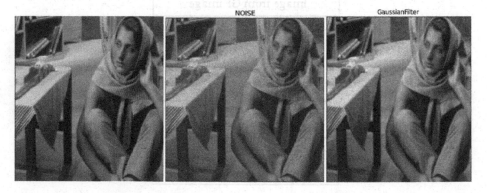

Fig. 4. Gaussian filter output of Barbara image in python2

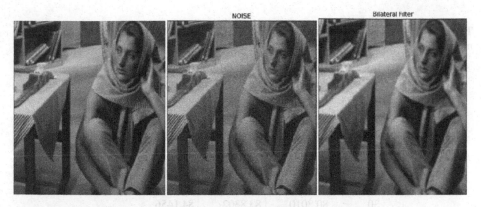

Fig. 5. Bilateral filter output of Barbara image in python2

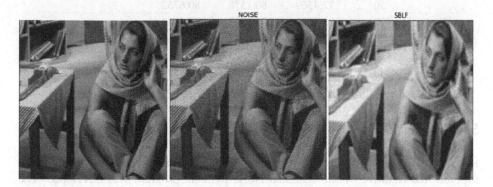

Fig. 6. Laplacian pyramid + BLF output of Barbara image in python2

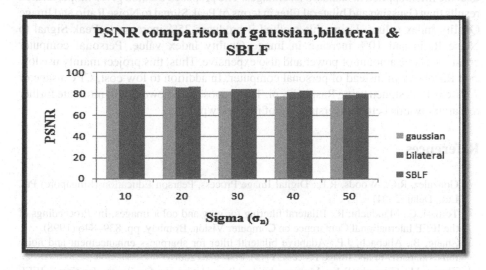

Fig. 7. Peak Signal to Noise Ratio comparison of Gaussian, bilateral & SBLF

Table 2. Peak Signal to Noise Ratio comparison between various denoising methods and proposed method

Test images	σ_n	Gaussian	Bilateral	Laplacian pyramid + bilateral
Barbara	10	89.5272	89.1661	89.6782
	20	86.9389	86.5595	87.1245
	30	82.1330	84.5762	85.0178
	40	77.5564	81.9093	83.1002
	50	73.6793	79.0986	80.1671
Lena	10	90.4026	90.9518	90.9743
	20	86.1894	86.5475	86.7864
	30	80.3010	83.8302	84.1456
	40	77.5714	83.3651	83.9543
	50	72.4569	80.1578	80.6752
Cameraman	10	82.3680	86.8563	87.2134
	20	77.1461	86.0642	86.9761
	30	75.8910	80.8823	81.7815
	40	75.3506	78.7431	79.4537
	50	70.4389	74.0023	75.0238

5 Conclusion

Raspbian wheezy OS was installed successfully on SD card in Raspberry Pi Kit. To suppress the Additive White Gaussian Noise python2 program was developed. The combination of Laplacian pyramid and bilateral filter method produce improved results than the filters applied in standard test images. The proposed method gives outperformed results than Gaussian and bilateral filter in terms of Peak Signal to Noise Ratio and Image Quality Index values. Using this method it produces 25% increase in Peak Signal to Noise Ratio and 10% increase in Image Quality Index value. Personal computer consumes large amount of power and also expensive. Thus, this project mainly use low cost Raspberry pi instead of personal computer. In addition to low cost, CPU usage of 7.3% can be designed using Raspberry pi. The proposed framework will motivate further research towards better understanding of raspberry pi.

References

1. Gonzalez, R.C., Woods, R.E.: Digital Image Process. Pearson Education (Singapore) Pte. Ltd., Delhi (2004)
2. Tomasi, C., Manduchi, R.: Bilateral filtering for gray and color images. In: Proceedings of the IEEE International Conference on Computer Vision, Bombay, pp. 839–846 (1998)
3. Zhang, B., Allebach, J.P.: Adaptive bilateral filter for sharpness enhancement and noise removal. IEEE Trans. Image Process. 17(5), 664–678 (2008)
4. Zhang, M., Gunturk, B.K.: Multiresolution bilateral filtering for image denoising. IEEE Trans. Image Process. 17(12), 2324–2333 (2008)

5. Karthikeyan, P., Vasuki, S., Boomadevi, R.: Effective noise removal in graylevel image using joint bilateral filter. Int. J. Appl. Eng. Res. 9(21), 4831–4836 (2014). ISSN 0973-4562
6. He, K., Sun, J., Tang, X.: Guided image filtering. IEEE Trans. Pattern Anal. Mach. Intell. 35(6), 1397–1409 (2013)
7. Zuo, W., Zhang, L., Song, C., Zhang, D.: Texture enhanced image denoising via gradient histogram preservation. In: IEEE Conference on Computer Vision and Pattern Recognition (CVPR), pp. 1203–1210 (2013)
8. Deng, G.: A generalized unsharp masking algorithm. IEEE Trans. Image Process. 20(5), 1249–1261 (2011)
9. Arbelaez, P., Maire, M., Fowlkes, C., Malik, J.: Contour detection and hierarchical image segmentation. IEEE Trans. Pattern Anal. Mach. Intell. 33(5), 898–916 (2011)
10. Jacob, N., Martin, A.: Image Denoising in the Wavelet Domain Using Wiener Filtering, December 2004
11. Jin, B., You, S.J., Cho, N.I.: Bilateral image denoising in the Laplacian subbands. EURASIP J. Image Video Process. 2015(1), 1–12 (2015)
12. https://www.slideshare.net/anija03/Raspberry_Pi
13. Quick Start Guide, the Raspberry Pi – Single Board Computer. https://www.farnell.com/datasheets/1524403.pdf

Deep Learning

Diabetes Detection Using Deep Neural Network

Saumendra Kumar Mohapatra, Susmita Nanda,
and Mihir Narayan Mohanty[⊠]

Biomedical and Speech Processing Lab, Department of Electronics and
Communication Engineering, ITER, Siksha 'O' Anusandhan
(Deemed to be University), Bhubaneswar, India
saumendramohapatra@soa.ac.in,
n.susmita@gmail.com, mihir.n.mohanty@gmail.com

Abstract. Diabetes is rapidly emerging worldwide issue with huge social, health and financial significances. Most of the people in the world suffer from Diabetes respective of new born child to old aged people including male and female. A diabetes patient has high blood sugar and it depend on the production of insulin in the body. Patients suffering from diabetes are treated with special diet and regular exercise. If diabetes is not controlled by the patient there is a chance of higher risk so for this a better treatment is required for this silent killer disease. Here in this paper authors have purposed Deep Neural Network (DNN) for the automatic identification of the disease. The experiment has done with the Pima Indian data set. The classification result has been presented in the result section.

Keywords: Diabetes · Deep neural network · Activation function
Classification · Accuracy

1 Introduction

The medical science has become more advanced these days due to the research oriented works. Diabetes is one of the common disease and it causes because of the insufficiency of insulin in human body. It is one kind of silent killer because it draws supplementary diseases in the human body. Diabetes is a prolonged disease which is directly proportional to the glucose present in the blood. It causes due to the insufficiency of insulin production. It has been classified into two sorts like type1 and type 2. The Pima are one of the foremost populations with relevant to Diabetes. The foremost population samples which are regarding are separate by two types i.e. type-2 positive and negative instances [1]. One of the most efficient ways for a patient to survive is to normalize the concentration of blood sugar level and it can be controlled by proper dieting and exercising along with specialized treatment. People suffering from diabetes needs to do better treatment from the early stage otherwise it can make many problem in the body. It is a kind of disease which arises many problems to the doctors. Doctors will take the risk for the patient to cure the disease by authenticating many records about the patient and desired disease. To enable the machines to educate, there is one efficient way known as machine learning. This method will help to solve the problem in easiest way

© Springer Nature Singapore Pte Ltd. 2018
I. Zelinka et al. (Eds.): ICSCS 2018, CCIS 837, pp. 225–231, 2018.
https://doi.org/10.1007/978-981-13-1936-5_25

to understanding and specification of the disease and also by helping to decrease the cost of unnecessary costly medical tests [2, 3].

Neural Network is one of the broadly used Machine learning techniques that work on the concept of the Biological Neural Network. It is a mixture of both processed information and interconnected bunch of neurons that enters into it. There is different neural network technique like DNN, RBFN, and SVM etc. for the classification purpose. Among all DNN is one of technique which can give better result as compare to others. Deep Learning methods are broadly connected to different areas of science and technology. Traditional data processing methods have some limitations of handling big volume of data. Also, Big Data analytics needs new and refined algorithms in view of machine and deep learning techniques to process data continuously with high precision and effectiveness [4–6]. Deep Learning uses both supervised and unsupervised techniques to learn multi-level data features for the classification and pattern recognition task. Big data gives incredible chances for broad of areas including web based business, industrial control and smart medical technology. It creates many challenging issue in Data Mining and information processing because of its features of big volume, huge assortment, huge speed and veracity [7]. Depp Neural network (DNN) needs a heavy amount of data for its training and it can be applied on classification, speech enhancement, image processing, signal processing, etc. [8–10].

Here in our paper we have used DNN for the classification of Diabetes. This technique has been implemented on the disease records of Pima Indian dataset of the University of California which are collected earlier [11]. Once the experiment is successfully completed in R Studio then the result is shown in the result section.

The remaining part of the paper is ordered in the distinct positions. In Sect. 2, the proposed DNN is shown. In Sect. 3, the outcome of the experiment is shown and the conclusion and further works are contained by Sect. 4.

2 Proposed Methodology

Here in this paper we have used DNN for the classification of the Pima Indian Diabetes data set. In Fig. 1 proposed work has been presented.

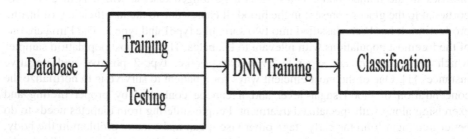

Fig. 1. Purposed method

2.1 Deep Neural Network

Deep learning is one type of structure based learning technique and consists of input layers, hidden layers and output layers. In the recent years most of the researchers are using a new advanced technique of neural network called deep neural network (DNN) which can be applied in data mining, image processing, pattern recognition, speech enhancement, biomedical informatics etc. Deep learning is a kind of artificial neural network system with various data representation layers that learn representation by expanding the level of deliberation from one layer to another layer [12, 13]. The structure of the DNN is represented in Fig. 2. In our purposed model we are taking the training and testing data set from the original data set which applied on the DNN model for the calculation of validation result.

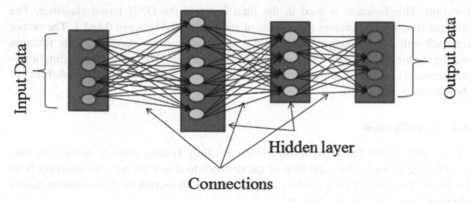

Fig. 2. Purposed method

2.2 Database

Here in this work the data has been taken from UCI repository. The data set contains total 768 female patient records among them 268 patients are suffering from Diabetes. Here in this record total eight features are taken for the experiment. In Table 1 the total description of the data set has been presented.

Table 1. Complete description of the database

Feature variable	Minimum value	Maximum value	Mean	Median
Number of times pregnant	0	17	3.84	3.0
Plasma glucose concentration	0	199	120.89	117.00
Diastolic blood pressure	0	122	69.10	72.00
Triceps skin fold thickness	0	99	20.53	23.00
2 h serum insulin	0	546	79.79	30.5
Body mass index	0	67.1	31.00	32.00
Diabetes pedigree function	.078	2.42	0.47	0.37
Age	21	81	33.24	29.00

The data set has been divided into training and testing set with 70% and 30% ratio. Total 534 and 234 data samples are taken for training and testing purpose respectably. Each data set contains class variable 0 for negative and 1 for positive.

2.3 Training of the DNN

The neural network is trained with eight numbers of hidden layers. In this purposed model RcLu (Rectified Linear Unit) activation function is used for the training of the network. ReLU activations are the simplest non-linear activation function and it can be used for the training of the large networks. It allows each unit in the network to express more data in the training of the network. As we know that our data set contains total 8 numbers of features so here the network is trained with eight input layer [14, 15]. In the output of the network it has two layer negative and positive with Softmax activation function. This function is used in the final layer of the DNN based classifier. The softmax function compresses the outputs of each unit to be between 0 and 1. The output for each unit can be compressed within 0 and 1 by softmax function. This function highlights the major value and destroys the smaller one. The yield of this function is proportional to an absolute probability distribution; it discloses to you the probability of the classes is valid [16].

2.4 Classification

In this experiment the classification is done using Testing dataset. Total 234 data samples are taken for the validation of the network to check the patients suffering from Diabetes. The classification result is shown in the result section by the confusion matrix obtained from the Testing data set.

3 Result Discussion

The Diabetes data has been obtained from the UCI repository and then we have checked for the missing value. The data set contains 768 patient's records without any missing value. In Fig. 3 the data set structure is presented. The data set is divided into Training and Testing set with the probability of 70% and 305 respectably. The Network is trained with the 534 numbers of patients sample. The training performance is presented in Fig. 4.

In this experiment the training accuracy is around 100%. The total number of epochs is 300, batch size is 32 and the validation split is 0.2. The loss in this experiment is 0.037. After successfully training of the network the next step is the validation step. The validation has been done with the previously stored 234 testing data set. In Table 2 confusion matrix that we have get is presented. Here we are getting the average

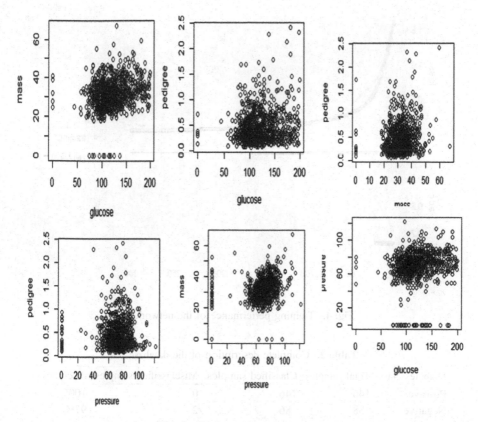

Fig. 3. Diabetes dataset structure

classification accuracy 98.5%. For calculation of the classification accuracy we can take the ratio of the accurately classified cases to the total number of cases. The accuracy formula is presented as follows:

$$Accuarcy = \frac{TP + TN}{N} \times 100 \qquad (1)$$

Here TP = truly positive, TN = truly negative and N is the total number of cases.

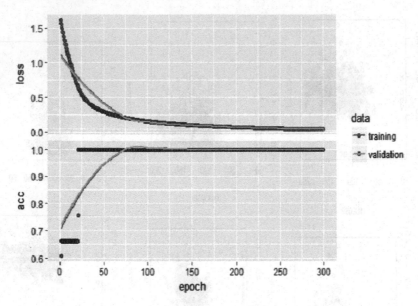

Fig. 4. Training perfermance of the network.

Table 2. Complete description of the database

Diabetes class	Total samples	Classified samples	Misclassified samples	Accuracy
Positive	146	146	0	100%
Negative	88	86	2	97%
Average accuracy = 98.5%				

4 Conclusion

Diabetes is one of the long term disease and it needs proper treatment at very early stage. If a machine can correctly classify the disease with the information then it can help doctors for better treatment. Deep learning has established to be a proficient technique for the analysis of the large volume of data. Here in this paper we have applied DNN based classification for the women affected by Diabetes. Pima Indian Diabetes data set has been taken for experiment. In future different models might be utilized for even exact outcome and further more improvement strategy may be connected for comparison.

References

1. Zhang, Q., Chen, L.T.Z., Li, P.: A survey on deep learning for big data. Inf. Fusion **42**, 146–157 (2018)
2. Goldberg, D.E., Holland, J.H.: Genetic algorithms and machine learning. Mach. Learn. **3**, 95–99 (1988)

3. Jan, B., et al.: Deep learning in big data analytics: a comparative study. Comput. Electr. Eng. (2017). https://doi.org/10.1016/j.compeleceng.2017.12.009
4. Craven, M.W., Shavlik, J.W.: Using neural networks for data mining. Future Gener. Comput. Syst. **13**, 211–229 (1997)
5. Jia, F., Lei, Y., Lin, J., Zhou, X., Lu, N.: Deep neural networks: a promising tool for fault characteristic mining and intelligent diagnosis of rotating machinery with massive data. Mech. Syst. Sig. Process. **72**, 303–315 (2016)
6. Al Rahhal, M.M., Bazi, Y., AlHichri, H., Alajlan, N., Melgani, F., Yager, R.R.: Deep learning approach for active classification of electrocardiogram signals. Inf. Sci. **345**, 340–354 (2016)
7. Yan, C., Xie, H., Yang, D., Yin, J., Zhang, Y., Dai, Q.: Supervised hash coding with deep neural network for environment perception of intelligent vehicles. IEEE Trans. Intell. Transp. Syst. **19**, 284–295 (2018)
8. Ram, R., Mohanty, M.N.: Enhancement of speech using deep neural network with discrete cosine transform. J. Intell. Fuzzy Syst. (2018, Accepted). Recent Adv. Mach. Learn. Soft Comput.
9. Ram, R., Mohanty, M.N.: Performance analysis of deep neural networks on use of speech enhancement task. Int. J. Inf. Technol. Proj. Manag. (IJPOM) (2018, Accepted)
10. Ram, R., Mohanty, M.N.: The use of deep learning in speech enhancement. Ann. Comput. Sci. Inf. Syst. **14**, 109–113 (2017). Proceedings of the First International Conference on Information Technology and Knowledge Management
11. Bache, K., Lichman, M.: UCI machine learning repository (2013)
12. Setlak, G., Bodyanskiy, Y., Vynokurova, O., Pliss, I.: Deep evolving GMDH-SVM-neural network and its learning for data mining tasks. In: Federated Conference on Computer Science and Information Systems (FedCSIS), pp. 141–145. IEEE, September 2016
13. Acharya, U.R., Fujita, H., Oh, S.L., Hagiwara, Y., Tan, J.H., Adam, M.: Application of deep convolutional neural network for automated detection of myocardial infarction using ECG signals. Inf. Sci. **415**, 190–198 (2017)
14. Clevert, D.A., Unterthiner, T., Hochreiter, S.: Fast and accurate deep network learning by exponential linear units (ELUs). arXiv preprint arXiv:1511.07289 (2015)
15. Nair, V., Hinton, G.E.: Rectified linear units improve restricted Boltzmann machines. In: Proceedings of the 27th International Conference on Machine Learning (ICML 2010), pp. 807–814 (2010)
16. Dunne, R.A., Campbell, N.A.: On the pairing of the softmax activation and cross-entropy penalty functions and the derivation of the softmax activation function. In: Proceedings of 8th Australian Conference on Neural Networks, Melbourne, vol. 181, p. 185 (1997)

Multi-label Classification of Big NCDC Weather Data Using Deep Learning Model

Doreswamy[1], Ibrahim Gad[1,2(✉)], and B. R. Manjunatha[3]

[1] Department of Computer Science, Mangalore University,
Konaje, Karnataka, India
doreswamyh@yahoo.com, gad_12006@yahoo.com
[2] Faculty of Science, Tanta University, Tanta, Egypt
[3] Department of Marine Geology, Mangalore University,
Konaje, Karnataka, India
omsrbmanju@yahoo.com

Abstract. Nowadays, analysis of weather data is remarkably considered the heart of the decision making in many different aspects such as Agricultural crops, scheduling of flights in airports, and marine navigation. One of the greatest challenges in big data analysis, as well as weather data, is to build a high accuracy multi-label classification model. The current state of the multi-label classification methods has many problems that need to be investigated by researchers in this area. Deep learning models have the potential to handles such problems in the big data area to extract the relation between data by reducing the dimension nullity. Specifically, the volume of weather data is massive and weather change is so dynamic because weather is influenced by many factors. In this work, we investigated a new deep learning model for multi-label weather classification for NCDC dataset in 16 years. This new model considers the combination of Long Short-Term Memory (LSTM) and fully connected models. The proposed deep learning model achieves excellent results in multi-label weather classification when it is compared with the other classical techniques as the results of the proposed model achieves 99.9% multi-label weather classification task.

Keywords: Weather classification · NCDC dataset · Deep learning
Long Short-Term Memory (LSTM) · Big data

1 Introduction

An observations of weather dataset plays vital role in habitat, events and life activities for humans. Thus, the process of weather forecasting plays a significant role in planning activities of human's life. Furthermore, the prior knowledge of weather status is more useful for making a good decision for a particular event such as extract crops or scheduling flights. Daily weather can be classified as beautiful, warm, neutral, very cold or cold. For the temperate climate, the low values of temperature correlate with cold and winter due to the winter's snow, high values of temperature related to summer and moderate values in spring, summer, and warm. These correlations presented based on human senses and subjective interpretations where snow is unlikely to fall in

© Springer Nature Singapore Pte Ltd. 2018
I. Zelinka et al. (Eds.): ICSCS 2018, CCIS 837, pp. 232–241, 2018.
https://doi.org/10.1007/978-981-13-1936-5_26

summer [1]. In general, the characteristics of weather dataset are a continuous, non-linear, multi-dimensional, and dynamic in nature [2]. The models are used in weather forecasting become more complex due to many factors collected for weather data and to get high accuracy in the predicted values.

In the last few decades, the importance of deep learning and big data is continuously increasing everyday consequently the number of published papers in these fields are growing significantly. Big data is very large data sets that can be analyzed by computational models to extract hidden patterns, trends, and information. There are many challenges in big data such as capturing, storage, updating, searching, sharing, transferring, querying, visualization, and analysis data. However, big data is very important for weather analysis because it provides the opportunity to improve the accuracy of weather forecasting by using different tools of big data. Massive amount weather data is collected daily basis from different stations around the world so the combination between deep learning and big data is one of the main advantages to analyze a huge amount of weather data [3].

Deep learning models have achieved high precision results in different areas such as weather, processing of weather image [4], semantic segmentation [5], speech recognition, remote sensing [6], and computer vision [1]. Thus, deep learning model is essential in getting the relevant information from the data, so is considered an appropriate approach for big data analytic. The deep learning model performs an efficient in supervised or unsupervised problems [7, 8]. So, deep learning models gives an opportunity to extract hidden relations between the data features by capture complex nonlinear representations between weather data.

The remainder of the paper is organized as follows: Sect. 2 shows different previous works related to our work. Section 3 presents the basic idea of the proposed work. Section 4 explains NCDC data sources and preparation methodology, parameters, evaluation, and finally provides the results of experiments. In the final Sect. 5, summarizes and suggests directions for future work.

2 Related Work

Weather classification problem is considered a big challenging for researchers because of the collected data of weather is huge and unpredictable [6]. In order to predict the class of weather or label depends on many variables like temperature, dew point, mean station pressure, visibility and wind speed. Many researches have been done in weather forecasting and classification [9–11]. Several machine learning methods including artificial neural networks [12, 13], support vector machines [14, 15], and more recently, deep neural networks [16–20] have been used to predict weather conditions.

Sankaralingam et al. [14], implemented a Support Vector Machine (SVM) model used for classifying weather dataset and also for predicting weather conditions where the accuracy of this model is 84.21%. Wu et al. [21], proposed the support vector machines (SVMs), artificial neural network, and classification decision tree are different machine learning models that are employed in predicting the soil texture classes. The results of the study showed that the support vector machines have the highest accuracy 94.3% to identify the classes of soil texture. Liu et al. [16] proposed deep

Convolutional Neural Network (CNN) classification model as alternative methodology in order to classify climate extreme events, where the proposed model achieved 89% accuracy in detecting Weather events.

3 Model

This section gives an explanation of the proposed model for a weather classification which is done by employing the Long Short-Term Memory (LSTM) which is implemented in deep learning library namely Theano.

A general deep learning model consists of a finite number of layers and neurons. The number of hidden layers is called the depth of network, while the width of the model is the maximum number of neurons in any layer. The three most common layers of the model are (1) single input layer which receives input data, (2) a fixed number of hidden layers in which data transformed in a nonlinear form through it, and (3) lastly the output layer that computes in the final result of the model [22, 23].

In Recurrent Neural Network (RNN) model the input data are dependent on each other such that data of the previous iterations will make improvement for prediction accuracy of the next iterations [7]. The main task of the memory in RNN model is to collect information from the calculations that have been done in previous computations. The most common problem in deep learning is vanishing gradient so it is also a problem in RNN that looks back just a few steps. So as to address the vanishing gradient issue, there is a ReLU function that can be used instead of tanh and sigmoid as activation function in the final layer [24].

Hochreiter and Schmidhuber [25] were proposed the LSTM model. In general, unlike RNN, LSTM has a gated cell that helps the system to save more information in compared with RNN. So, information in the gated cell can be stored, written, or read from it. The job of opening and closing gates is to remove or store information. The four main elements of memory cell are: (1) input gate, (2) a neuron with a self-recurrent connection, (3) a forget gate that allows the cell to remember or forget its previous state, and (4) an output gate which allows the state of the memory cell to have an effect on other neurons or prevent it.

Figure 1 shows the proposed model of weather classification. First, the input layer receives the weather variables (TEMP, VISIB, ..., MIN). Then the hidden layers are consists of LSTM that are employed to extract hidden features of the data. Finally, the result of the LSTM is fed to final fully connected layers that have Softmax activation function, that serves as classifier and outputs a probability distribution over class labels so as to predict the class label for the current input data, and the error is calculated between the predicted label and the true label.

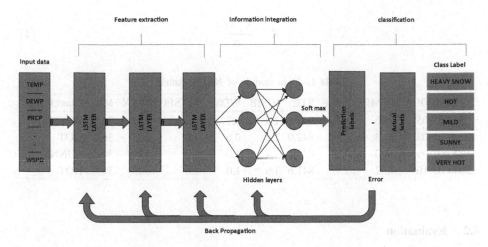

Fig. 1. Architecture of the proposed model.

4 Results

In this section, experiments for classifying NCDC dataset were demonstrated by applying deep learning that improved the accuracy of classification of weather data. Experimental results of the proposed model is discussed in the following subsections.

4.1 Dataset

NCDC has weather data collected from different stations placed all over the world and stored in one repository which is accessible publicly [26, 27]. Each station has many sensors which are used for collecting weather variables [6]. Although more than ten various weather elements were collected, only the following variables were considered temperature, dew point, station pressure, visibility, precipitation, maximum sustained wind speed, and wind speed. The first step of our work started by downloading and cleaning up data of the selected weather stations in India from the repository of NCDC.

Recent data gathered in one-day interval from weather stations of India starting from 2000 to 2016 consists of the following attributes: temperature, dew point, station pressure, visibility, precipitation, maximum sustained wind speed, and wind speed. The raw data was used as input data after pre-processed and cleaned it from missing data. Next, a table of size 9 columns is created where by first eight attributes of data corresponding to daily data of the selected variables (temperature, ..., wind speed) and the last attribute corresponds to the label of that day at the ninth column for each entry. Table 1 shows a sample consists of a few numbers of rows for the data collected of India stations. Because, different variables have different values and scales, it is necessary to normalize before using the values of variables before training the deep learning model. All variables values are normalized in the range of [−1, 1] by applying standard scaler technique that removes the mean and scales of the data to unit variance as shown in Eq. 1.

$$Z = \frac{X - \mu}{\sigma} \tag{1}$$

Table 1. Data sample of NCDC dataset.

YEARMODA	TEMP	DEWP	STP	VISIB	WDSP	MXSPD	MAX	MIN	Class label
2000-01-05	36.6	25.2	844.1	1.2	0.7	1.0	53.6	20.7	SUNNY
2000-01-08	43.3	33.8	842.0	1.2	0.6	1.0	50.9	36.5	HOT
2000-01-09	38.0	34.7	839.6	1.4	1.0	1.9	53.4	30.2	SUNNY
2000-01-10	40.2	32.7	841.6	1.9	1.0	1.0	50.0	30.0	HOT

4.2 Evaluation

Evaluation metrics such as accuracy, precision, recall, and F_1 score are employed to evaluate each deep learning model [29]. In the binary classification problem, there exists only two output classes that needs to distinguish: positive and negative. Thus, there are four possible outcomes: True Positive (TP), False Positive (FP), True Negative (TN), and False Negative (FN).

The *accuracy* is defined by the following formula:

$$\text{Accuracy} = \frac{TP + TN}{TP + TN + FP + FN} \tag{2}$$

While *precision* is formulated as:

$$\text{precision} = \frac{TP}{TP + FP} \tag{3}$$

and *recall* is formulated as:

$$\text{recall} = \frac{TP}{TP + FN} \tag{4}$$

F_1 score can be computed by using both of precision and recall, where the best value of F_1 score is 1 and the worst score is 0. The formula for the F_1 score is described as:

$$F_1 = 2 \cdot \frac{precision \cdot recall}{precision + recall} \tag{5}$$

4.3 Experiments

Deep neural network's experimental work requires large number of tuning of different parameters to get the best accuracy. Several parameters affect the effectiveness of a deep learning models such as learning rate, number of hidden layers, number of neurons in each layer and number of epochs. In our experiment we selected the following combination of parameters number of epochs = 1500, batch size = 365, and number of neurons in each layer = 64 are used for the proposed model which produced the best prediction of labels and minimize error rates. Moreover, We use the following configuration to compile the model: categorical cross-entropy is used for loss function, 'Adam' for optimizer, and accuracy, MSE for metrics to track the F1 score during training.

Experiments were done by using the NCDC variables dataset first described in the previous section. The data includes time series of 12 variables related to the dew point, pressure, temperature, and precipitation, measured on a daily basis from 2000 to 2016 in 79 different locations of India. The dimension of original data (439391, 9) is divided into training and testing data (307573-131818 split) and saved in separate variables see Table 2.

Table 2. The total number of each class label for train and test data.

Class label	All	Train	Test
VERY HOT	375694	281770	112807
HEAVY SNOW	42563	31898	12673
HOT	20088	15089	6027
SUNNY	1033	776	308
MILD	13	10	3
Total	439391	307573	131818

In deep learning model implementation depends on Theano [24], which is a famous library in Python for deep learning. The basic model ran for 1500 epochs and achieved an F_1 score of 1, and for average precision score, micro-averaged over all classes is 1. In order to evaluate model, we considered the size of window 7 in addition, a dropout regularization is used to solve overfitting problem of the proposed model. The last step of the process is a Softmax layer that is used to classify the final results.

It is clear that using an effective LSTM model that produces a significant increase in classification accuracy. Table 3 shows the full results of applying LSTM neural network. We conclude that after 600 epochs the performance of the proposed work achieves more than 99% accuracy which is considerably higher value as shown in Fig. 2. Table 4 details the F_1 score, precision, and recall for each class, where the proposed model achieved a 99.9% F_1 score on the test dataset. The proposed model is highly predicting clear labels for the training dataset that has the most training examples of these labels. The model performs well on most of the uncommon labels which achieves 83% F_1 for nearly all of the rare labels for example in our case 'MILD' class. The limitation of the proposed model is the failure to identify uncommon labels.

The last task of the model is classifications, for which we measure performance with Area Under the ROC Curve (AUC). Table 3 presents the average performance of the proposed model: We can see that LSTM reduces overall MSE by 0.000174 and improves overall AUC by 99%. To show that the LSTM model have the ability to learn from features, we used the backward layer that reduces the error between the predicted labels and the actual labels. As it is clearly indicated in Fig. 3b, there is a higher prediction for class labels, where almost all predicted and true labels are matched excluding 'MILD' class. Figure 3a depicts AUC curve for the model, which compares AUC of applying the proposed model for each class.

As shown in Table 4 and Fig. 3a the ROC curve is approaching the top left corner of the diagram. This shows the accuracy of prediction increases gradually. The proposed approach achieves 99.9% classification accuracy comparison to 84.21% [14], 94.3% [21], 89.4% [16] in the weather classification task. According to the results, it proves that the proposed model is an effective model for predicting the label of weather day.

Table 3. Model classification performance.

	Model evaluation			
	MSE	RMSE	MAE	Correlation
Test	0.000174	0.013209	0.000068	0.9999
	Accuracy	Precision	Recall	F_1
	0.99996	1.0	1.0	1.0

Table 4. Classification report.

	Precision	Recall	F_1-score	Support
HEAVY_SNOW	1.00	1.00	1.00	12673
HOT	1.00	1.00	1.00	6027
MILD	1.00	0.67	0.80	3
SUNNY	1.00	1.00	1.00	308
VERY_HOT	1.00	1.00	1.00	112807

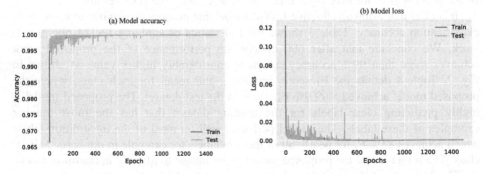

Fig. 2. The accuracy and loss values of the proposed model.

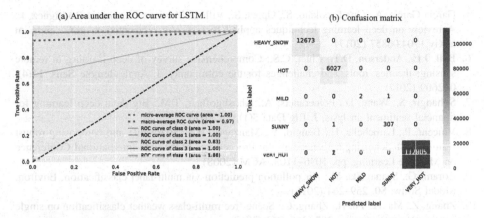

Fig. 3. Non normalized confusion matrix and ROC curve for the proposed model.

5 Conclusion

Deep learning models have the ability to handle the problems of learning in the area of big data. It transformed inputs using more than one layer, where hidden layers are applied to extract hidden information from features of input data. Therefore, consecutive process in deep learning prepares the way to find the hidden relationships in weather data. Thus, these properties are sufficient to choose deep learning to be the most recommendable models for weather forecasting. This paper presented experimental results that demonstrate the possibility of deep learning for predicting classes or labels for weather data. In addition to it provided a clear interpretation how deep learning models are appropriate for weather classification problem. In this work, deep LSTM model is designed for weather classification and good results are achieved on multi-label classification of NCDC data, and overcomes data mining approach in weather classification problem. Although the proposed model is performed well on majority class labels, it should also achieve similar with minority class labels. Furthermore, it is possible to combine the best models into an ensemble model so as to predict minority class labels.

References

1. Volokitin, A., Timofte, R., Van Gool, L.: Deep features or not: temperature and time prediction in outdoor scenes, pp. 63–71 (2016)
2. Adeyemo, A.: Soft computing techniques for weather and climate change studies. Afr. J. Comput. ICT 6(2), 77–90 (2013)
3. Liu, J.N., Hu, Y., He, Y., Chan, P.W., Lai, L.: Deep neural network modeling for big data weather forecasting. In: Pedrycz, W., Chen, S.M. (eds.) Information Granularity, Big Data, and Computational Intelligence, pp. 389–408. Springer, Heidelberg (2015). https://doi.org/10.1007/978-3-319-08254-7_19
4. Lin, D., Lu, C., Huang, H., Jia, J.: RSCM: region selection and concurrency model for multi-class weather classification. IEEE Trans. Image Process. 26, 4154–4167 (2017)

5. Garcia-Garcia, A., Orts-Escolano, S., Oprea, S., Villena-Martinez, V., Garcia-Rodriguez, J.: A review on deep learning techniques applied to semantic segmentation. arXiv preprint arXiv:1704.06857 (2017)
6. Ball, J.E., Anderson, D.T., Chan, C.S.: Comprehensive survey of deep learning in remote sensing: theories, tools, and challenges for the community. J. Appl. Remote Sens. **11**(4), 042609 (2017)
7. Sohangir, S., Wang, D., Pomeranets, A., Khoshgoftaar, T.M.: Big data: deep learning for financial sentiment analysis. J. Big Data **5**(1), 3 (2018)
8. Vincent, P., Larochelle, H., Bengio, Y., Manzagol, P.A.: Extracting and composing robust features with denoising autoencoders. In: Proceedings of the 25th International Conference on Machine Learning, pp. 1096–1103. ACM (2008)
9. Corani, G., Scanagatta, M.: Air pollution prediction via multi-label classification. Environ. Model Softw. **80**, 259–264 (2016)
10. Zhang, Z., Ma, H., Fu, H., Zhang, C.: Scene-free multi-class weather classification on single images. Neurocomputing **207**, 365–373 (2016)
11. Xu, J., Liu, J., Yin, J., Sun, C.: A multi-label feature extraction algorithm via maximizing feature variance and feature-label dependence simultaneously. Knowl.-Based Syst. **98**, 172–184 (2016)
12. Choi, S., Kim, Y.J., Briceno, S., Mavris, D.: Prediction of weather-induced airline delays based on machine learning algorithms. In: 2016 IEEE/AIAA 35th Digital Avionics Systems Conference (DASC), pp. 1–6. IEEE (2016)
13. Sawaitul, S.D., Wagh, K., Chatur, P.: Classification and prediction of future weather by using back propagation algorithm-an approach. Int. J. Emerg. Technol. Adv. Eng. **2**(1), 110–113 (2012)
14. Sankaralingam, B.P., Sarangapani, U., Thangavelu, R.: An efficient agro-meteorological model for evaluating and forecasting weather conditions using support vector machine. In: Satapathy, S.C., Das, S. (eds.) Proceedings of First International Conference on Information and Communication Technology for Intelligent Systems: Volume 2. SIST, vol. 51, pp. 65–75. Springer, Cham (2016). https://doi.org/10.1007/978-3-319-30927-9_7
15. Rani, R.U., Rao, T.: An enhanced support vector regression model for weather forecasting. IOSR J. Comput. Eng. (IOSR-JCE) **12**, 21–24 (2013)
16. Liu, Y., et al.: Application of deep convolutional neural networks for detecting extreme weather in climate datasets. arXiv preprint arXiv:1605.01156 (2016)
17. Di, C., Yang, X., Wang, X.: A four-stage hybrid model for hydrological time series forecasting. PLoS ONE **9**(8), e104663 (2014)
18. Qin, S., Liu, F., Wang, J., Song, Y.: Interval forecasts of a novelty hybrid model for wind speeds. Energy Rep. **1**, 8–16 (2015)
19. Grover, A., Kapoor, A., Horvitz, E.: A deep hybrid model for weather forecasting. In: Proceedings of the 21th ACM SIGKDD International Conference on Knowledge Discovery and Data Mining, pp. 379–386. ACM (2015)
20. Jin, L., Chen, M., Jiang, Y., Xia, H.: Multi-traffic scene perception based on supervised learning. IEEE Access **6**, 4287–4296 (2018)
21. Wu, W., Li, A.D., He, X.H., Ma, R., Liu, H.B., Lv, J.K.: A comparison of support vector machines, artificial neural network and classification tree for identifying soil texture classes in southwest china. Comput. Electron. Agric. **144**, 86–93 (2018)
22. LeCun, Y., Bengio, Y., Hinton, G.: Deep learning. Nature **521**(7553), 436 (2015)
23. Goodfellow, I., Bengio, Y., Courville, A., Bengio, Y.: Deep Learning, vol. 1. MIT Press, Cambridge (2016)
24. Lipton, Z.C., Berkowitz, J., Elkan, C.: A critical review of recurrent neural networks for sequence learning. arXiv preprint arXiv:1506.00019 (2015)

25. Hochreiter, S., Schmidhuber, J.: Long short-term memory. Neural Comput. **9**(8), 1735–1780 (1997)
26. NCDC: National climatic data center. NOAA's National Centers for Environmental Information (NCEI) (2016)
27. Lawrimore, J.H., et al.: An overview of the global historical climatology network monthly mean temperature data set, version 3. J. Geophys. Res.: Atmos. **116**(D19) (2011)
28. Doreswamy, Gad, I., Manjunatha, B.: Performance evaluation of predictive models for missing data imputation in weather data. In: 2017 International Conference on Advances in Computing, Communications and Informatics (ICACCI), pp. 1327–1334. IEEE, September 2017
29. Xie, Y., Zhu, C., Zhou, W., Li, Z., Liu, X., Tu, M.: Evaluation of machine learning methods for formation lithology identification: a comparison of tuning processes and model performances. J. Pet. Sci. Eng. **160**, 182–193 (2018)
30. Theano Development Team: Theano: a python framework for fast computation of mathematical expressions. arXiv e-prints abs/1605.02688, May 2016

Object Recognition Through Smartphone Using Deep Learning Techniques

Kiran Kamble[✉], Hrishikesh Kulkarni, Jaydeep Patil,
and Saurabh Sukhatankar

Department of Computer Science and Engineering,
Walchand College of Engineering, Sangli, Sangli, India
kirankamble5065@gmail.com,
kulkarnihrishi97@gmail.com,
jaydeeppatil3232@gmail.com,
sukhatankarsaurabh1997@gmail.com

Abstract. Object recognition technology has matured to a point at which exciting applications have become possible. Indeed, industry has created a variety of computer vision products and services from the traditional area of machine inspection to more recent applications such as object detection, video surveillance, or face recognition. This paper is about achieving the goal of object recognition through advanced techniques like deep learning on handy devices like smartphones and tablets. Deep learning algorithms (Convolutional Neural Networks (CNN)) are used for the primary aim of object recognition. Images are clicked through the camera of the smartphone during experimentation and are fed to the CNN network. The top four results predicted by the network are depicted on the smartphone screen in the audio and the visual form i.e. predicted object name and the probability of predicted object being the one actually clicked in the decreasing order of probabilities. The accuracy obtained in object recognition is about 93% through the application.

Keywords: Modern object recognition technique · Deep learning application
CNN · Mobile application

1 Introduction

Object detection and recognition is the first step in computer vision based research fields. Most of the applications are real time applications, thus object recognition system should be robust, fast and efficient. Object recognition system usually separates out the background and finds the dominating objects present in particular visual media like images or videos. Such systems can be implemented and deployed on wearable devices to facilitate the visually challenged people to ease their life through artificial vision. Obtaining 3-dimensional objects from 2D images and videos is a challenging task. To perform this at the start formulation of the problem is essentially: with input of pre knowledge of how certain objects may seem, plus an image of a scene probably having those objects, to detect which objects are existing in the scene and where. Credit is accomplished by identical features of an image and model of an object. The two most

© Springer Nature Singapore Pte Ltd. 2018
I. Zelinka et al. (Eds.): ICSCS 2018, CCIS 837, pp. 242–249, 2018.
https://doi.org/10.1007/978-981-13-1936-5_27

important concerns that a method must address are the definition of a feature, and how the identical is found. Obviously these desires are generally difficult to accomplish, for example difficult to recognize objects in images occupied in complete darkness. Even for traditional machine learning models, designing a feature extraction algorithm was essential which generally involved a lot of thick mathematics (complex design), wasn't very efficient, and didn't execute too sound at all (accuracy level just wasn't suitable for real-world applications). This was followed by designing a whole classification model to classify the input given the mined features [1]. With the rise of autonomous vehicles, smart video surveillance, facial detection and various people counting applications, fast and accurate object detection systems started rising in demand. These systems involve not only recognizing and classifying every object in an image, but localizing each one by drawing the appropriate bounding box around it. This makes object detection a significantly harder task than its traditional computer vision predecessor, image classification [2]. Deep learning models have become fairly laid-back to implement, particularly with high-level open source libraries such as Keras, Pytorch, and Tensor Flow [3]. The smartphones provide the hosting platform for the objective of object identification and recognition. Considering the increase in the use of smartphones, they serve as the handy and perfect platform to demonstrate object recognition through deep learning. The paper is ordered as tracks: Sect. 2 discusses previous work with respect to object recognition. In Sect. 3 the proposed method is presented with necessary details. Section 4 gives experimental results. Future scope is provided in Sect. 5.

2 Previous Work

Since the object recognition has gained much influence as one of the significant applications of deep learning, many systems, applications and software have been embedded with this feature. Here are some of the systems exhibiting object recognition [4].

A. YOLO: Provides the users with real time recognition application. The application uses a sole neural network to the whole image. The network splits the image into sub regions and predicts bounding boxes and likelihoods for each section. These bounding boxes are weighted by the predicted likelihoods. Though better than many of the applications, the application hasn't made its way in many of the android smartphones.

B. Glass: Is an android application available on the Google Play Store. It recognizes the dominant object from the image clicked or uploaded. The application suffers drawbacks as far as server speed and real time computation of prediction is concerned.

C. Object Recognition-Free Computer Vision: This is yet another android application on Google Play Store which serves the users with the feature of object recognition from the clicked image. The application though helps to recognize the objects along with the probabilities denoting the accuracy, there are always limited number of categories in which the object could be classified in and all the categories (about 10) and the probability of object being belonging to that is displayed to the user.

D. Machine Learning detection: The android application on Google Play Store serves users with real time object recognition feature. It provides the results with the recognized object being bounded along with the predicted object name and accuracy. The application size is about 90 mb and lists single probability when the object is focused upon by the camera.

E. Click2Know (C2K): is a very handy, an optimal size Android application which facilitates the users with multiple object recognition from the clicked image. It helps users to get the results computed real timely (about 2 s from clicking and strong internet connection) and lists out four results based on predicted objects and corresponding accuracies in decreasing order of significance.

3   Proposed System

The main aim of proposed system is to recognize objects through smart phone for visually challenged people using the deep learning on Android platform. A flowchart in the Fig. 1 shows the actual methodology. Initially user (visually challenged) has to create an account simply by speaking create an account. Using account user can capture the photo of object to be recognized. Within fraction of seconds produced result will be spoken as well as displayed on the screen.

The first one is the firebase which is used as real time database so that real time result can be computed. The image which is captured by camera of user is uploaded to real time database in the format of byte image. Database schema is represented in the Fig. 2.

The schema of users is comprises of unique UID (user id) which is generated according to firebase account of the individual user along with the image which is uploaded at that instant of time multiple requests from different user at same instance of time.

Image which is in the database (firebase) is taken and the results are pushed back into the database. The image which is in the byte format needs to be converted into regular .PNG format so that further proceeding can be done on the image.

Conversion of the image to 299 × 299 pixel is performed. Before feeding image to Convocational Neural Network Inception V3 model it is necessary to convert linear structure like Array which is vectored form of corresponding image in .rbg format.

Feeding these values directly into a network may lead to numerical overflows. It also turns out that some choices for activation and objective functions are not compatible with all kinds of input. The wrong combination results in a network doing a poor job at learning which is done by pre-processing technique Dimensionality Reduction [4]. Dimensionality Reduction helps in transforming vectored image data into a compressed space with less dimensions, that can be useful to control the amount of loss and it uses as input to CNN. After performing dimensional reduction it is necessity to adequate image to the format the model requires. The image is fed to CNN model for prediction (Fig. 3).

Convolutional neural networks (CNNs) is present high-tech model architecture for image classification tasks. CNN process a sequence of filters to the raw pixel records of

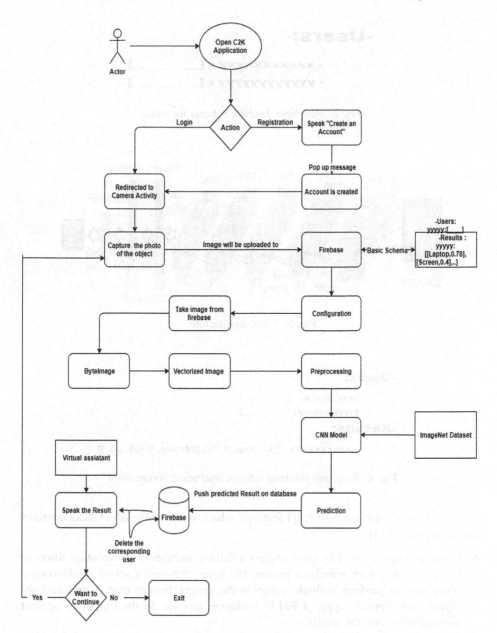

Fig. 1. Flowchart of proposed system

-Users:

- xxxxxxxxxxxx : [_____]
- yyyyyyyyyyyy : [_____]

Fig. 2. Real time database schema for users

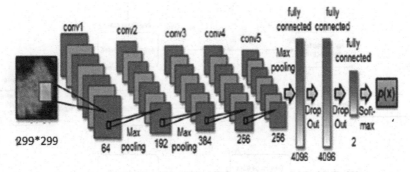

Fig. 3. CNN architecture

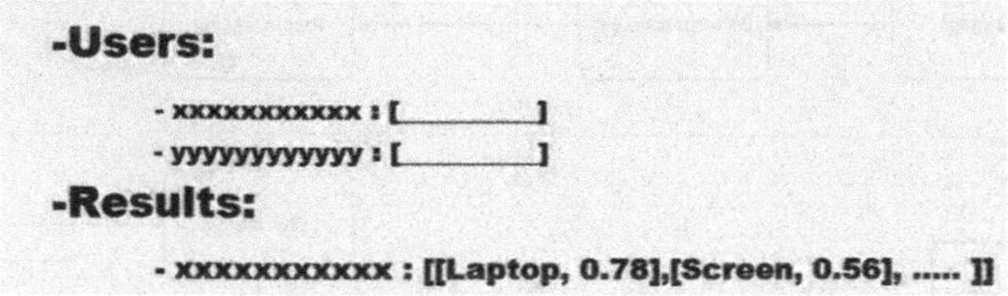

-Users:

- xxxxxxxxxxx : [_____]
- yyyyyyyyyyyy : [_____]

-Results:

- xxxxxxxxxxx : [[Laptop, 0.78],[Screen, 0.56], ...]]

Fig. 4. Real time database schema after result computation

an image to mine and lean high-level features, which is use to classify. CNNs contains three components [5]:

A. Convolutional layers: This layer applies a definite number of convolution filters to the image. For each substitute region, the layer executes a set of mathematical operations to produce a single value in the output feature map. Convolutional layers then typically apply a ReLU initiation function to the output to present nonlinearities into the model.

B. Pooling layers: Pooling layers depressed sample the image data mined by the convolutional layers to shrink the dimensionality of the feature map in order to decline processing time. A frequently used pooling algorithm is max pooling, which mines sub regions of the feature map (e.g., 2 × 2-pixel tiles), keeps their maximum value, and discards all remaining.

C. Dense (fully connected) layers: Dense layers accomplish classification on the features mined by the convolutional layers and down experimented by the pooling layers. In a dense layer, every node in the layer is linked to every node in the prior layer. After classification results will be in the form of like Fig. 4.

Here 0.78 means the probability of laptop being in the image is 0.78 in the similar fashion 0.56 conveys the probability of screen being in the image is 0.56 so on. The among the computed results the images having top 4 probabilities will be pushed onto the real time database within fraction of seconds for that specified user. Now the modified schema will look like in the Fig. 2, where xxx... is unique UID.

After computed results were pushed onto the real time database the android application will fetch it instantly. As result were fetched from real time data base there is no need to store them for future use. In order to facilitate fast and efficient working of real time database the user along with result are deleted. The android application will show the result of object recognition in audio as well as text format in real time.

4 Experimental Results

For implementing the system, ImageNet dataset is used. The dataset includes 14,197,122 total images in .jpeg form. These images consist of total 21841 non-empty synsets. Other details about the dataset are as [6]:

Total number of non-empty synsets: 21841. Total number of images: 14,197,122. Number of images with leaping box annotations: 1,034,908. Number of synsets with SIFT features: 1000. Number of images with SIFT features: 1.2 million.

The proposed approach for object recognition is based on Convolutional Neural Network which is able to recognize about 21841 objects (classes) in ideal case. The performance of the system is depicted in Table 1. For calculating confusion matrix, the application is tested 60 times. 26 times out of 30, the system recognizes object correctly (Here, correct recognition is considered as the predicted object is equivalent to the class of object in which it falls). It is important to note that the system is able to compute results correctly even when image consists of multiple objects.

Table 1. Confusion matrix for performance of system

		Actual	
		True	False
Predicted	True	26	7
	False	4	23

But when the accuracy of predictions of entire system is concerned, only 60 tests are not enough. According to the online documentation, the model used in this application is able to give accuracy as high as 94.4% in top five predictions. However top one accuracy is bit less i.e., 78.8% [7]. Thus the application flow goes like this. One has to create an account (By speaking "Create an Account") once when he/she installs the application for the first time. After successful account creation, the application intent redirected to the camera activity. Here user can capture images of surrounding.

The image captured is then converted into Bytes of integer and sent to the firebase database. Because of his/her account on the firebase (which is created earlier), user will get unique UserID/QueryID which is useful for differentiating user requests from each other. Now the continuously running python script fetch the Byte-converted image and convert it into .PNG format image. The converted image is then fed to the CNN model after doing some pre-processing as mentioned in the proposed system. The top-four predictions are computed and results are again pushed back to Android application via Firebase. Such fetched results are been shown on the screen along with voice guidance/pronunciation. Some sample results are shown in Fig. 5.

Fig. 5. Sample result

5 Future Scope

The work can be extended further with the various ideas like: Switching from CNN to Recurrent Neural Network (RNN) in order to reduce the computation time and accuracy. Deploying the Object Recognition (server-side scripting) over the wearable smart devices [8] like smart Google Glass, Smart Shoes or Smart Sticks which can result in more usefulness of the application for physically challenged people

Acknowledgements. We sincerely thank to all panel members for their guidance and encouragement in carrying out this work. We also highly indebted to Walchand college of Engineering Sangli for providing necessary information regarding this research and financial support to carry out this work.

References

1. Elisa, M., Giulia, P., Lorenzo, R., Lorenzo, N.: Interactive data collection for deep learning object detectors on humanoid robots. In: 2017 IEEE-RAS 17th International Conference on Humanoid Robotics, pp. 862–868 (2017)
2. Qi, M., Zong, Z.: A new method of moving targets detection and imaging for bistatic SAR. In: 2014 Seventh International Symposium on Computational Intelligence and Design, vol. 2, pp. 224–227 (2014)
3. Zainab, A.: Research blog about various libraries that can be used in deep learning. https://medium.com/mindorks/detection-on-android-using-tensorflow-a3f6fe423349. Accessed 27 July 2017
4. Raj, D., Gupta, A., Tanna, K., Garg, B., Rhee, F.C.H.: Principal component analysis approach in selecting type-1 and type-2 fuzzy membership functions for high-dimensional data
5. Fakhrulddin, A.H., Fei, X., Li, H.: Convolutional neural networks (CNN) based human fall detection on body sensor networks (BSN) sensor data, pp. 1461–1465 (2017)
6. Article showing detailed explanation of ImageNet dataset including various classes and their total counts, etc. http://image-net.org/about-stats
7. Document of keras library giving accuracies of various models. https://keras.io/applications/
8. Delrobaei, M.: Errata to "using wearable technology to generate objective Parkinson's disease dyskinesia severity score: possibilities for home monitoring". IEEE Trans. Neural Syst. Rehabil. Eng. 25(11), 2214 (2017)

Artificial Intelligence

Hot Spot Identification Using Kernel Density Estimation for Serial Crime Detection

S. Sivaranjani[✉], M. Aasha, and S. Sivakumari

Department of Computer Science and Engineering, Avinashilingam Institute for Home Science and Higher Education for Women, Coimbatore 641 108, India
sivaranjanicse@gmail.com

Abstract. A hot-spot mapping is an advanced crime detection technique which helps police personnel to identify high-crime areas and the best way to respond. However identification of crime location with less man power would be a more difficult process. In this work, Social crime data aware kernel density estimation based serial crime detection approach (SAKDESD) is implemented to group the serial and social crime data set in terms of more similarity. Social crime data set consists of various user comments about the crime happening in different locations which can provide the in-depth information about the serial crimes. The unstructured social crime data set is pre-processed to obtain meaningful structured format. This work also adapts the latent semantic approach for finding the similar topics present in the social crime data set which can lead to accurate prediction and efficient grouping of serial crimes. The experimental tests were conducted in matlab simulation environment which proves that the proposed approach SAKDESD provides a better result than the existing approach such as Modified graph cut clustering algorithm (MGCC).

Index Terms: Serial crime · Social crime data · Kernel density estimation Latent semantic

1 Introduction

Serial crimes are constant threats happening in different locations made by same person in similar manner. These crimes need to be identified and addressed for ensuring public safety. Investigation departments are responsible for assuring the social protection by finding the criminal who are responsible for the series threats happening in the world. Detection of serial crime hotspots and the persons who involved in that would be a tedious process which needs to be analysed and processed in efficient manner. Inadequate man power and the voluminous data about the crimes happening in multiple locations reduce the efficiency of the investigation process. To address these issues, several attempts were made by various researchers using data mining approaches to make ease of investigation processes [1].

In this work, we attempt to find and group the similar kind of crimes happening at various locations in terms of crime features by using the kernel density estimation approach which overcomes the issues like grouping and overlapping problem faced by

© Springer Nature Singapore Pte Ltd. 2018
I. Zelinka et al. (Eds.): ICSCS 2018, CCIS 837, pp. 253–265, 2018.
https://doi.org/10.1007/978-981-13-1936-5_28

modified graph cut clustering (MGCC) algorithm which was introduced in our previous work [2]. The social crime data is considered in this work along with the serial crime data set for the accurate prediction of the similar kind of crime features is done in terms of their characteristics. Social crime data would be in the unstructured format which must be pre-processed to handle it in an efficient manner. The latent semantic approach is employed to identify the most similar topics present in the crime data set. Hence this work can provide flexibility in the investigation processes involved in the serial crime detection.

2   Related Works

Susmita and Sharmistha [3] analysed the crimes that were happening against women at Tripura district using direct methods with multiple object. The fuzzy membership value was applied for the conversion of qualitative data into quantitative data. This work predicts the nature of crimes by gathering details from women in the Tripura district. Data mining approaches were utilized in this work for analysing the nature of crimes in terms of different risk factors. This research work concludes that the proposed methodology helps to deal with all type of data.

Clare et al. [4] analysed the risk factors of the serial killers in terms of their physiological behaviour. To do so, this approach implements the Neuro development methodology in which serial crimes would be identified in terms of interrelationship between different types of crimes which are located in different places. These interrelationships are identified based on the factors called the risk factors such as types of crime, level of crime effect, and so on. The neurological development approach leads to an improved finding of the serial crimes in terms of various risk factors associated with them.

Duygu and Murat [5] introduced the novel way of crime analysis process in the computer science student's point of view. This approach would lead to an improved finding of the various research methodologies. This analysis is conducted over the undergraduate students of the Trakya University where the finding of the work is demonstrated in the computer implemented programming language which was proved that the crime analysis was better than the manual analysis. This approach was employed to analyse the crime factors that are occurred in the environment in terms of the ethics and law of the crime factors.

Nabeela et al. [6] studied the socio-economic factors which are reason for the crime that are occurring in different locations of the Pakistan. The factors are which includes unemployment, educational qualification, poverty, and the economic growth. This study concludes that the government of Pakistan needs to concentrate on creating more job opportunities, alleviate poverty and promote education in order to reduce the rate of crime.

Omowunmi et al. [7] interrogate the crime situation and their behaviour by constructing the pattern model in terms of their location, time and the type of crime behaviour. The Author proposed an automatic threshold selection method based on quartile floor-ceiling functions. This approach is based on the pruning process which was done based on the Apriori techniques using which the unnecessary data samples would be eliminated from the data set, so that the crimes can be identified accurately. The author concluded with the result indicating that Revised Frequent Pattern Growth was more promising than Traditional FP-Growth model.

This section provides a detailed analysis about the different related works which has been conducted previously in terms of crimes that are happening in the different crime location in terms of their better provisioning of the resources. From these analyses, it is concluded that the serial crime is the most frequently occurred threat in the real world environment which need to be detected in order to provide the secured environment to the people.

3   Kernel Density Estimation Based Serial Crime Detection Approach

Crimes are the behaviour disorder which cost our society dearly in several ways. Those crimes needs to be spotted and the person committed must be identified to avoid further crime actions performed by same persons. The crime might happen in different form based on their structure and the characteristics. Two most important forms of crimes happening in different locations are "personal trait crime" and the "serial crime". Personal trait crimes are the one which would be done by individual for their personal reasons. Serial crimes are the one which are repetitive in nature done by the individual or group of people continuously in different places. Serial crimes are the most dangerous threat than the personal trait crime that needs to be identified for ensuring the security.

The serial crimes can be identified by finding the similar features present among the crimes characteristics that are happening in the different locations. The crime mapping will be effective if the crime locations are classified into different forms and that are shown in the separate regions in the visualization [8–10]. In our previous proposed research modified cut clustering approach was used for finding the serial crimes which would group the similar features that are present in the crime type of T. This approach works on structured serial crime data set but doesn't support well for unstructured data like social media data. This serial crime data set is gathered from the police department which will contain the police investigation parameters like crime details and behaviours which is not only enough for predicting the serial crimes in accurate manner.

In this proposed research work, Social crime data aware kernel density estimation based serial crime detection approach is introduced which would find the serial crimes by finding the similar features that reside in the crimes of type T happened in different spatial points p. Social media data's about crime behaviour are gathered from the multiple social media's which is used to predict the crime behaviour along with the serial crime data set for predicting the serial crime in an accurate manner. Social media data would be in the unstructured format which is pre-processed to eliminate the unnecessary words and represent the dataset in the structured format. After pre-processing, Latent Dirichlet Allocation (LDA) is applied to find the most important features present in the document based on their relevance. Then the kernel density would be estimated for the extracted features using LDA [11]. This kernel density is calculated by using the characteristics of the serial crime which can be identified from the serial crime data set.

Finally, the spatial point 'p' with more kernel density is taken as the hot spot location of the serial crime.

The Overall flow of the proposed research work is given in Fig. 1 as follows:

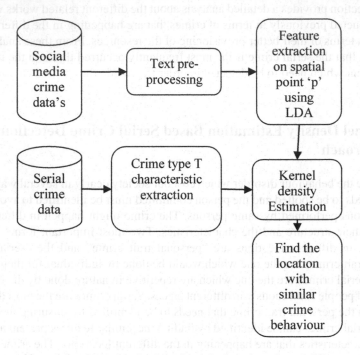

Fig. 1. Flow of kernel density estimation based serial crime detection of different spatial point 'p'.

The above flow diagram provides the steps in serial crime detection based on public comments on social media data in different locations. This process flow is explained in the following sub sections.

3.1   Data Collection

In this proposed research work, two types of data sets namely, Crime Data Set and Social Media Crime Data Set are considered for finding the serial crimes that are happening in different crime locations in terms of their similar characteristics.

Crime Data Set: The crime data considered in this work consists of attributes like different types of crimes for e.g., Murder, dacoity, robbery, theft and housing breaking etc., latitude and longitude values of crime which denotes the exact location where the crimes had happened. By using these information, the different types of crimes are analysed and finally the similar types of crimes are grouped together to predict the serial crimes [12–14].

Social Media Crime Data Set: The social media crime data sets are gathered from different public social media web sites such as Facebook, Twitter and so on which reveal the semantic meaning of the crime in terms of comments about different crimes at different locations. It would be in the unstructured format with various unnecessary tags and words. Pre-processing would be done to represent it in the structured format, so that semantic meaning of those data's can be obtained.

3.2 Text Pre-processing

Text pre-processing is the most essential process which is used to convert the unstructured data into structured format. It will also avoid the unnecessary words from the data set that are irrelevant to the concept. The social crime data's gathered from the online social media web sites would consists of the noises, html tags, advertisement messages and so on. This might cause the high dimensionality problem at the time of processing the social crime data set where every tag would be considered as one feature. The kernel density would be calculated for those features which are unnecessary, so that computation overhead would be increased. The text pre-processing would consist of the following steps:

Online Text Cleaning: Online text cleaning is the process of eliminating the online text characters such as html tags, scripting tags, and advertisement contents and so on. This step will remove all the contents that are irrelevant to the concepts.

White Space Removal: After performing online text cleaning white space removal would be done, so that the complete representation of the documents can be obtained. White space removal is used to provide the sentences in complete format, from which semantic meaning can be extracted in the flexible manner.

Expanding Abbreviations: In this step, abbreviation expansion would be done by parsing the sentences from start to end. By abbreviating the sentences, complete meaning of the document can be obtained.

Stemming: Stemming is the process of avoiding the derived words through which repetition is avoided. In this research work, porter stemming algorithm is used to perform stemming process which will reduce the inflected words from their base words.

Start and Stop Word Removal: Start and Stop word removal is the process of eliminating the Starting and ending words from the sentences from which meaning of the document cannot be extracted. Some of the start words are 'a, an, the' and so on. Some of the stop words that are considered in this work are 'the, is, at, which', which are removed before processing the social media crime data's.

3.3 Feature Detection Using Latent Dirichlet Allocation

Latent dirichlet allocation is the natural language processing based generative model which is used to observe the more similar parts of the documents which are given as input. LDA provides better performance in finding the similarity in two ways. Those are word distribution and the topic distribution. This would be identified by taking crime type T detail from the crime data set which is compared with the documents. Here documents are public comments about the crime data's gathered from the social websites.

Word distribution and Topic distribution are defined as the probability of corresponding crime type T belongs to the words and topics of document D respectively. The Algorithm is discussed as follows:

Algorithm 1:

   Input: Social crime data set gathered in particular time period, Crime types

   Output: Features

   1. Repeat

   2. Collect the input documents

   3. Initialize weight values of all documents as null

   4. Find the log likelihood to find the words that are most related

$$-2\ln \lambda = 2 \sum_i O_i \ln\left(\frac{O_i}{E_i}\right) \tag{1}$$

   5. Assign the temporary crime types for every word that has more log likelihood present in the user review

   5. a. If any word is repeated multiple times assign with different crime types

   6. Find the similarity of the word with the corresponding crime types

   7. Find the similarity of the crime type with the user review.

   8. Update the weight values of reviews based similarity

   9. until final solution obtained

where

-2ln λ → Log likelihood ratio

O_i → Observed Value

E_i → Expected Value

The above algorithm provides a pseudo code of the latent Dirichlet allocation procedure which is used to find the most similar and relevant words that are present in the social crime data set in terms of different crime types represented in the serial crime data set. This procedure will assign the labelling of crime type for every word present in the document. This process would be done for every crime types that are considered in this research methodology. Finally all the words that are labelled using corresponding crime types T would be considered as the features f(p) of the crime types at particular spatial point p. These features would represent the serial crimes that are happening in the different location of the world in similar manner. Based on these features, the location in which these crimes are happening more would be identified by calculating the kernel density of those particular features.

3.4   Kernel Density Estimation of Features at Spatial Point 'P'

Kernel density estimation is defined in terms of statistics as the non-parametric way which is used to identify the probability density distribution of the corresponding feature in a particular location. In this research work KDE is used to find the density level of the features f(p) in the corresponding spatial location 'p' in terms of crime type T. KDE is one of the most improved data smoothening approach which can find the density estimation of the corresponding feature on the particular spatial location by omitting the noises present in the location. This smoothening is achieved in this work by setting the bandwidth parameter value of 'h' as optimal. This 'h' value would indicate the surface of data which need to be processed.

KDE estimation would be done for all feature values in terms of every crime types T within a particular time period. The equation that is used to estimate the kernel density of features at the spatial point 'p' for the crime types are calculated by using the formulae as like follows:

$$f(p) = k(p, h) = \frac{1}{Ph} \sum_{j=1}^{P} K\left(\frac{\|p - p_j\|}{j} \right) \tag{2}$$

where,

p → spatial point in which density to be calculated

h → bandwidth of KDE used to smoothen the data processing problem

P → total number of crime types T that are considered

j → single crime location during the time period

K → density function

∥.∥ → Euclidean distance

Density function is calculated in this research work by using the probability density function procedure which will find the probability of likelihood of occurrence of particular feature in the spatial point 'p' in terms of different crime types T.

The overall working flow of this kernel density based serial crime detection for the social crime data is given in the following algorithm.

Algorithm 2:

   Input: Social crime data set, Crime types from serial crime data set, time window

   Output: Location in which serial crimes are happened

   1. Gather the serial crime data from different social web sites

   2. Pre-process the data's gathered from the social media web sites

   a. Remove the html tags, advertisement

   b. Remove the white spaces present in the data set

   c. Abbreviate the acronyms

   d. Remove the start and stop words from the data sets

   e. Apply porter stemmer algorithm to remove the stemming words

   3. Find the most relevant features that are related to the crime types present in the data set by using LDA

   4. For every features $f_i \in F$

   5. Find the kernel density function k of features f in spatial point p for every crime type T

   6. $$f(p) = k(p, h) = \frac{1}{Ph} \sum_{j=1}^{P} K \tag{3}$$

   7. End for

   8. Return spatial point p with more kernel density

The above algorithm provides kernel density estimation calculation for the features at different spatial point p in the particular time window. This approach provides a better findings of the locations in which serial crimes are happened most in an accurate manner. This proposed research work is implemented in the matlab simulation environment in terms of performance measure values which is evaluated and compared with the existing approach which is discussed in detail as follows.

4   Experimental Results

SAKDESD is adapted for analysing and predicting the different number of features from the social crime data set which is most similar with the crime type T. This methodology is implemented in the matlab simulation environment and compared with the existing approach MGCC. This performance evaluation is done based on the metrics called as the mantel index and the jaccard index. This analysis is graphically represented in the proceeding sections. The results reveal that the proposed methodology produces better results than the existing one.

4.1   Data Set

Social crime data set obtained from multiple social media web sites in terms of different crimes activities happening in the different crime locations and the data gathered from the police stations of Coimbatore city, India are considered in this work for predicting the locations in which serial crimes are happening mostly in the real world environment. Both data set would consists of the details like crime type, number of crimes happening in given time period, people opinion about those particular crime happened in different location. By using these information, the different types of crimes are analyzed and finally the similar types of crimes are grouped together to predict the serial crimes.

4.2   Mantel Index

The mantel index is defined as metric used to calculate the correlation between the different features that are located as similar crime type in the geographical area. Mantel index of the proposed research approach should be high than the existing research approach which represents the high data correlation. The mantel index is calculated as follows:

$$r = \frac{1}{(n-1)} \sum_{i=1}^{n} \sum_{j=1}^{n} \frac{(x_{ij} - \bar{x})}{s_x} \cdot \frac{(y_{ij} - \bar{y})}{s_y} \tag{4}$$

where,
x, y = variables measured at locations i, j
n = number of elements in the distance matrices
s_x, s_y = standard deviation of variable x and y
$\bar{x}, \bar{y}$ = mean value of variables x and y

The actual values obtained for the mantel index is given in the following Table 1.

Table 1. Mantel index values.

Number of data points	Mantel index	
	SAKDESD	MGCC
20	0.75	0.48
40	0.79	0.63
60	0.84	0.76
80	0.87	0.78
100	0.91	0.82
120	0.94	0.85
140	0.94	0.88
160	0.94	0.90
180	0.98	0.92
200	1	0.94

The graphical representation of the comparison of the proposed research work with the existing research work for the above mentioned actual values are depicted in the following Fig. 2.

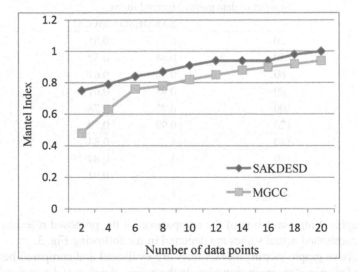

Fig. 2. Mantel index comparison.

In the above graph, mantel index values are evaluated and compared between the existing and proposed research scenarios. In the x axis, number of data points are taken where in the y axis mantel index values are taken. The analysis is done for varying number of data points which is taken in the range of 2 to 20. The mantel index value is increased linearly for increasing number of data points in both existing and proposed methodologies. From this graph it can be proved that the proposed research approach called SAKDESD provides better result than the existing approach called MGCC.

4.3 Jaccard Index

The jaccard index is used to represent the similarity between the data points. The jaccard index value is used to denote the number of crimes happened in different locations that are matched with each other. The jaccard coefficient measures similarity between finite sample sets, and is defined as the size of the intersection divided by the size of the union of the sample sets:

$$J(A, B) = \frac{|A \cap B|}{|A \cup B|} \tag{5}$$

where
    A, B = Data points

The actual values of Jaccard index obtained for both existing and proposed approach is indicated in the following Table 2.

Table 2. Jaccard index values.

Number of data points	Jaccard index	
	SAKDESD	MGCC
20	0.77	0.55
40	0.84	0.57
60	0.86	0.61
80	0.93	0.61
100	0.95	0.78
120	0.99	0.79
140	1	0.84
160	1	0.89
180	1	0.94
200	1	0.96

The graphical representation of the comparison of the proposed research work for the above mentioned actual values are depicted in the following Fig. 3.

In the above graph, jaccard index values are evaluated and compared between the existing and proposed research scenarios. In the x axis, number of data points are taken where in the y axis jaccard index values are taken. The analysis is done for varying number of data points which is taken in the range of 2 to 20. The Jaccard index value is increased linearly for increasing number of data points in both existing and proposed methodologies. From this graph it can be proved that the proposed research approach called SAKDESD provides better result than the existing approach called MGCC.

The GIS representation of clustered results of crimes which were happened in the various crime locations are depicted using MGCC and SAKDESD as follows:

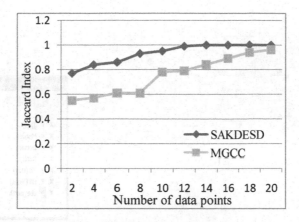

Fig. 3. Jaccard index comparison.

Figure 4 shows the clustering of crime spots performed using MGCC. The MGCC clusters crime spots effectively but the overlapping problem reduces the overall performance.

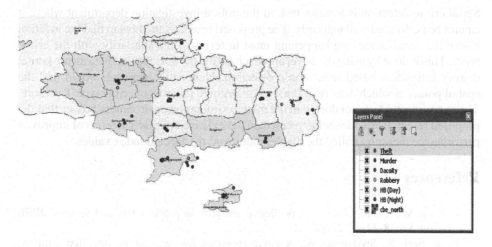

Fig. 4. Clustering crime spots using MGCC.

Figure 5 shows the clustering of crime spots performed using SAKDESD. The SAKDESD clusters crime spots effectively and has better performance than MGCC which is also proved in the graphical representation in terms of performance parameters.

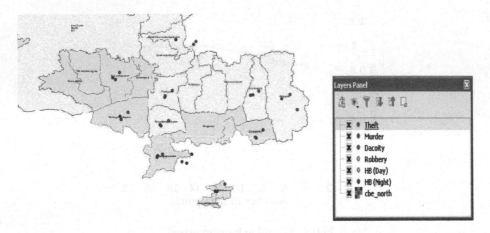

Fig. 5. Clustering crime spots using SAKDESD.

5   Conclusion

Serial crime detection is a major task in the police investigation department where it cannot be performed well manually. The proposed research attempts to find the location where the serial crimes are happening most in terms of the similarity with the crime types. This is done by using the novel approach called the Social crime data aware kernel density estimation based serial crime detection. Hence this approach would find the spatial points in which density of the similar features in terms of crime type T is more. The experimental tests conducted in the matlab simulation environment proves that the proposed research provides better result than the existing approach in terms of improved performance measures called the mantel index and the Jaccard index values.

References

1. Colleen, M.C.: Data mining and predictive analytics in public safety and security. IEEE Comput. Soc. **8**, 12–18 (2006)
2. Sivaranjani, S., Sivakumari, S.: A novel approach for serial crime detection with the consideration of class imbalance problem. Indian J. Sci. Technol. **8**, 1–9 (2015)
3. Susmita, R., Sharmistha, B.: Application of fuzzy-rough oscillation on the field of data mining (special attention to the crime against women at Tripura). Procedia Comput. Sci. **45**, 790–799 (2015). International Conference on Advanced Computing Technologies and Applications
4. Clare, S.A., Helen, M., Lucy, T., Philip, W., Christopher, G.: Neurodevelopmental and psychosocial risk factors in serial killers and mass murderers. Aggress. Violent Behav. **19**, 288–301 (2014)
5. Duygu, S., Murat, T.: The perception analysis of cyber crimes in view of computer science students. Procedia – Soc. Behav. Sci. **182**, 590–595 (2015)
6. Nabeela, K., Junaid, A., Muhammad, N., Khalid, Z.: The socio-economic determinants of crime in Pakistan: new evidence on an old debate. Arab Econ. Bus. J. **10**, 73–81 (2015)

7. Omowunmi, I., Antoine, B., Sonia, B.: A revised frequent pattern model for crime situation recognition based on floor-ceil quartile function. Procedia Comput. Sci. **55**, 251–260 (2015)
8. Tomoki, N., Keiji, Y.: Visualising crime clusters in a space-time cube: an exploratory data-analysis approach using space-time kernel density estimation and scan statistics. Trans. GIS **14**, 223–239 (2010)
9. Wang, D., Ding, W., Lo, H., Stepinski, T., Salazar, J., Morabito, M.: Crime hotspot mapping using the crime related factors-a spatial data mining approach. Appl. Intell. J. **4**, 772–781 (2013)
10. Devendra Kumar, T., Arti, J., Surbhi, A., Surbhi, A., Tushar, G., Nikhil, T.: Crime detection and criminal identification in India using data mining techniques. AI Soc. **30**, 117–127 (2015)
11. Matthew, S.G.: Predicting crime using Twitter and kernel density estimation. Decis. Support Syst. **61**, 115–125 (2014)
12. Christopher, R.H.: The dynamics of robbery and violence hot spots. Herrmann Crime Sci. **4**, 33 (2015)
13. Sivaranjani, S., Sivakumari, S.: Mitigating serial hot spots on crime data using interpolation method and graph measures. Int. J. Comput. Appl. **126**, 17–25 (2015)
14. Fitterer, J., Nelson, T.A., Nathoo, F.: Predictive crime mapping. Police Pract. Res. **16**(2), 121–135 (2015)

Automated Seed Points and Texture Based Back Propagation Neural Networks for Segmentation of Medical Images

Z. Faizal Khan[✉]

Department of Computer Science, College of Computing and IT, Shaqra University, Shaqraa, Kingdom of Saudi Arabia
faizalkhan@su.edu.sa

Abstract. In this paper, a combination of Pixel based Seed points and textural Back Propagation Neural Networks is proposed for segmenting the Region of Interest (ROI) from the medical images. Medical images such as Fundus and Skin images are used to test the proposed algorithm. To develop the proposed algorithm, Pixel based Seed points are combined with the Texture based Back Propagation Neural Network (TBP-NN) by a trained knowledge of textural properties for segmenting the medical images which can be used in early Diabetic Retinopathy (DR) detection and Skin lesion detection. The proposed algorithm is tested with a total of 200 fundus and skin images each which is stored in a database used for further testing. The medical images were processed such that a knowledge in form of texture features such as Energy. Homogeneity, Contrast and Correlation were automatically obtained. The proposed algorithms efficiency was compared with traditional BP-NN methods and Support Vector Machine for segmenting medical images. The results obtained from the proposed methodology reveals that the accuracy of proposed algorithm is higher. It indicates that the proposed algorithm could achieve a better result in medical image segmentation more effectively.

Keywords: Segmentation · Back-propagation neural network
Texture features · Skin lesions

1 Introduction

Segmenting medical image is an important task in image analysis. Segmentation of a medical imaging is used to extract the required information regarding the presence of different abnormalities present in the respective image. Artificial Neural network (ANN) is a machine learning model which is used by simulating the anatomic neuron connections present in it. Its effects have been validated in scientific studies for various applications, such as image processing, pattern recognition, system control, and medical image diagnosis [4–7, 13, 14, 16]. It has some limitations such as false segmentation and region growing faults, when it is applied in some type of medical images. In this paper, improvements in this Neural Networks has been obtained when a trained knowledge as features was incorporated into it which is used to train it. Therefore, we aimed to combine the trained knowledge as features from medical

© Springer Nature Singapore Pte Ltd. 2018
I. Zelinka et al. (Eds.): ICSCS 2018, CCIS 837, pp. 266–273, 2018.
https://doi.org/10.1007/978-981-13-1936-5_29

images with a doctor's trained knowledge in the segmentation of Retinal vessels and Skin lesions, for the application of early finding of diabetic retinopathy in eye fundus images and lesions in the Skin images. Important aim of this proposed methodology is to segment the region of interest in eye fundus images for finding the Diabetic Retinopathy and Skin lesions in medical images more effectively which is applied for training and testing as samples. Many methodologies have been given in the literature for the medical image segmentation. An automatic detection methodology was framed by Gardner et al. [8] for diabetic retinopathy using a neural network. The artifacts which are present in the medical images can be easily identified more effectively from the images of grey level. An automated back propagation type neural network is adopted to examine the eye fundus images. It does not work more effectively since the images are of low contrast.

Thresholding based technique proposed on eye fundus images by Sinthanayothin et al. in [9] The performance of their method is validated o using a 10×10 pixels instead of using the whole image. Usher et al. [10] endowed the various region of candidate exudates which are present in the eye fundus images by using a combination of Random Graph and intensity based adaptive thresholding methods. In their methodology, the regions of various candidate are removed and those are further used as an input for an artificial based neural network technology. Due to their algorithm the images with poor quality was affected and then all these images are extracted due to the bright and dark type of lesions.

Zheng et al. in [11] extracted the various types of artifacts which are present in the eye fundus images by combining the algorithms called as thresholding and region growing. The combination of Color normalization and local contrast enhancement methodology along with the clustering based on fuzzy C-means and neural network methodologies was proposed by Osareh et al. [12]. Their techniqua works more effectively only on Luv based colour space images even if it has an illumination of non-uniform in nature. Hence, the accuracy of detection less because of this dis advantage. Stoecker et al. [15] presented a new textural based segmentation methodology in skin images adopting the gray-level co-occurrence matrix which is a statistical based approach. Jeffrey et al. [1] adapted a novel methodology segmentation and classification of lesion present in the skin images. In their methodology, various lesions which are present in the skin images are segmented using a Distinctiveness based Joint Statistical Texture method. It results in an overall accuracy of 93%.

Menzies et al. in [5] proposed a novel algorithm based on a fusion of Semi-automatic and manual based methodologies A regression-based classifier is used for the purpose of classifying the further segmented results. They applied their algorithm. Their algorithm is applied for a set of 2430 lesion images of skin. A 65% average specificity and 91% average sensitivity is obtained to their algorithm.

2 Methodology

This section presents the overall process present in segmentation of region of interest in medical images. In this work, Skin and Retinal images are taken as testing datasets. Figure 1 gives the overall architecture of the proposed methodology for segmenting the medical images.

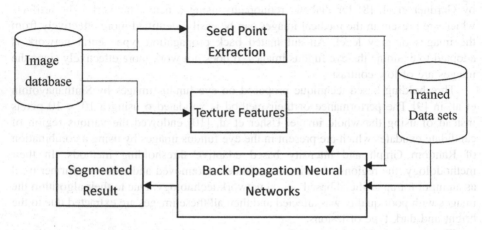

Fig. 1. Overall architecture of the proposed methodology

3 Image Pre-processing

All the obtained medical images were normalized for illumination, size, and color using an image processing process. In this work, the trained knowledge is defined as the texture features of the Region of Interest. This work is done after consulting with various experienced ophthalmologists and doctors for drawing the ground truth images.

A. Computerized seed points extraction

In the segmentation algorithm, usually all the seed points were formed by manual process. To remove each and every seed points in all the images is a big-time process. The seed points which are corresponding to each and every pixel present in the image should be given prior to the segmentation of entire image. The further segmented image should be different from the previously segmented image. For this process, all the regions present in the current image is merged with one another for the process of splitting and then merging. More number of seed points should be given since it should be merged with all the pixels present in the original image. Since the given put is a medical set of data, all the seed points should be merged with all the pixels present in the entire image so that the entire image is considered. The overall method is as follows: The medical image which is to be segmented is considered as s binary image.

Each pixel present in the original binary image which is considered as the target region is considered as 1, and all the others are considered as 0. The proposed algorithm is as follows

(1) Define the similarity threshold for color as a.
(2) Choose the pixel number manually which is present in the predefined target region, and also define them as an initial seed pixel as (x_0, y_0);
(3) Define the initial seed pixel in center, in order to get the eight pixels in neighborhood as $(x_i, y_i)(i = 0, 1, \ldots, 10)$
(4) Calculate the similarity in color between the predefined pixels (x_i, y_i) and (x_0, y_0)
(5) If the similarity in color in between the predefined pixel is larger than the set similarity threshold a, those pixels can be clustered towards a common region. Hence, the difference Pixel (x_i, y_i) should stored in a stack.
(6) Repeat the above process until the stack become empty.

B. **Seed points based TBPNN**

The seed point based Textural back propagation neural network (STBP-NN) [17, 18] is trained with pixel values as knowledge present in Fig. 2 has three layers such as an input layer, a hidden layer $H_{in}(j)$ and an output layer. Initial Seed point is assumed as X_i and final seed point is assumed as X2. the pixel values as seed points and a Trained knowledge in the form of textural features are given as the input towards the input layer.

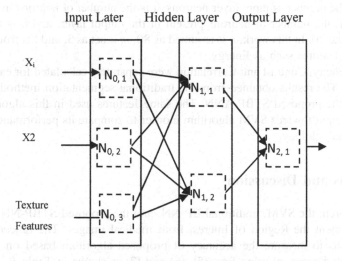

Fig. 2. Proposed TBPNN architecture

The framework of our STBPNN is as follows. The hidden layer input $H_{in}(j)$ was defined as:

$$H_{in}^n(j) = \sum_{i=1}^{M} \omega_{ij}^n x_i^n + d_j^n \tag{1}$$

where x_i is the input features, ω_{ij} is the weights between the neurons from the input layer and hidden layer, and a_j represents the threshold for hidden layer neurons.

The estimates of RBP-NN is as follows:

$$\hat{\beta} = \arg \min_{\beta} \left\| Y - \sum_{j=1}^{P} x_j \beta_j \right\|^2 + \lambda \sum_{j=1}^{P} |\beta_j| \tag{2}$$

where, λ is a non-negative regularization parameter, x is the Blood vessel width and blood vessel tortuosity of retinal images, Y is the average accuracy and βs is the regression coefficient.

The number of neurons in the hidden layer is as follows:

$$h < n - 1 \tag{3}$$

$$h < \sqrt{(m+n)} + a \tag{4}$$

where n is the number of input layer neurons, h is the number of neurons in the hidden layer, m was the number of neurons present in the output layer, and a is a threshold between 0 and 20. In this work, n is assumed as 80, m is set as 3, and his from 21 to 30. The texture features such as Energy.

Homogeneity, Contrast and Correlation were separately calculated for each training and testing. The results obtained from the traditional segmentation methods are compared with the proposed STBP-ANN. The same features used in this algorithm were used as the input towards SVM algorithm inorder to compare its performance with our proposed methodology.

4   Results and Discussion

In this research, the SVM, traditional BP-NN, and the proposed STBP-NN were each used to segment the Region of Interest from medical images. The objective of this comparison is to measure the accuracy of proposed algorithm based on sensitivity, specificity, and accuracy using Eqs. (5), (6) and (7) as shown in Table 1.

$$\text{Sensitivity} = \frac{TP}{TP + FN} \tag{5}$$

$$\text{Specificity} = \frac{TN}{TN + FP} \tag{6}$$

$$\text{Accuracy} = \frac{TP + TN}{TP + FN + TN + FP} \tag{7}$$

The images shown in Figs. 3 and 4 are the original. Ground truth and the segmented results of Skin and Retina images. The segmented images are very close to the ground truth image. In this approach, the ophthalmologists and doctors specified images are considered as ground truth images for the calculation of segmentation accuracy.

Table 1. Comparison of the results of methodologies for Retinal Images

Performance	SVM	Traditional BP-NN	Proposed STBP-NN
Specificity (%)	92.67	93.40	95.83
Sensitivity (%)	93.50	93.65	95.18
Accuracy (%)	94.39	94.69	95.71

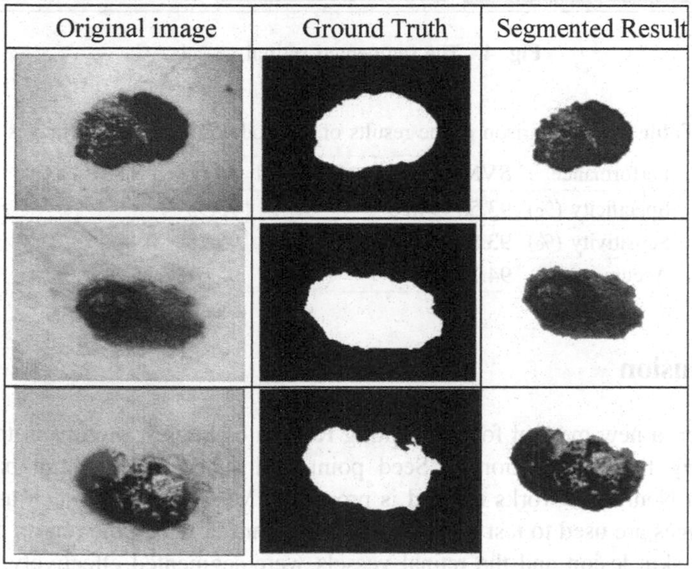

Fig. 3. The segmented skin lesions

The results showed that the accuracy of this proposed methodology is higher than the SVM, BP-NN and STBP-NN was 94.39%, 94.69% and 95.71% respectively as shown in Table 1 for retinal images. This indicates that the Proposed STBP-NN with a priori knowledge can achieve better segmentation results for retinal images.

Table 2 presents the accuracy of this proposed methodology is higher than the SVM, BP-NN and STBP-NN was 94.75%, 94.84% and 95.61% respectively for skin images indicating that the Proposed methodology can achieve better segmentation results for segmenting skin images.

Original image	Ground Truth	Segmented Result

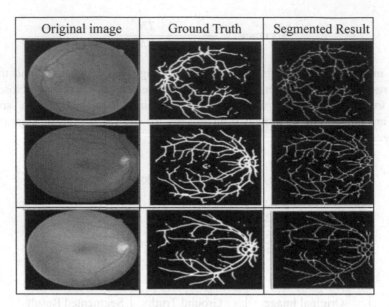

Fig. 4. The segmented retinal vessels.

Table 2. Comparison of the results of methodologies for Skin Images

Performance	SVM	Traditional BP-NN	Proposed STBP-NN
Specificity (%)	93.58	94.26	95.93
Sensitivity (%)	93.87	94.35	95.28
Accuracy (%)	94.75	94.84	95.61

5 Conclusion

In this paper, a new method for segmenting Region of Interest present in the medical images using the combination of Seed point and supervised Texture based Back Propagation Neural Networks method is proposed. Medical images such as Skin and Fundus images are used to test the proposed algorithm in which the region of interests such as the skin lesion and the retinal vessels were segmented effectively. The main purpose of this approach is to improve the DR in Retinal images and Lesion in skin image detection accuracy by segmenting the region of interest in it. In this method, the obtained seed points are given as input for training and testing the proposed TBP-ANN. Compared with other segmentation method, our method can better for processing the fundus image in its vessel and lesion in skin image segmentation. This STBP-ANN could segment the region of interest in medical images better than the traditional Neural Networks and SVM methods and could be a promising measure for early DR and Lesion detection more effectively.

References

1. Glaister, J., Wong, A., Clausi, D.A.: Segmentation of skin lesions from digital images using joint statistical texture distinctiveness. IEEE Trans. Biomed. Eng. **61**(4) (2014)
2. Stolz, W., Riemann, A., Cognetta, A.B., et al.: ABCD rule of dermatoscopy: a new practical method for early recognition of malignant melanoma. Eur. J. Dermatol. **4**, 521–527 (1994)
3. Nachbar, F., Stolz, W., Merkle, T., et al.: The ABCD rule of dermatoscopy high prospective value in the diagnosis of doubtful melanocytic skin lesions. J. Am. Acad. Dermatol. **30**, 551–559 (1994)
4. Lapuerta, P., L'Italien, G.J., Paul, S., et al.: Neural network assessment of perioperative cardiac risk in vascular surgery patients. Med. Decis. Making **18**, 70–75 (1998)
5. Menzies, S.W., Bischof, L., Talbot, H., Gutenev, A., Avramidis, M., Wong, L.: The performance of solarscan: an automated dermoscopy image analysis instrument for the diagnosis of primary melanoma. Arch. Dermatol. **141**(11), 1388–1396 (2005)
6. Argenziano, G., Soyer, H.P., De Giorgi, V., Piccolo, D., Carli, P., Delfino, M., et al.: Dermoscopy: a Tutorial. EDRA Medical Publishing & NewMedia, Milan (2002)
7. Salvi, M., Dazzi, D., Pellistri, I.: Classification and prediction of the progression of thyroid-associated ophthalmopathy by an artificial neural network. Ophthalmol. **109**, 1703–1708 (2002)
8. Gardner, G.G., Keating, D., Williamson, T.H., Elliott, A.T.: Automatic detection of diabetic retinopathy using an artificial neural network: a screening tool. Br. J. Ophthalmol. (1996)
9. Sinthanayothin, C., Boyce, J.F., Williamson, T.H., Cook, H.L., Mensah, E., Lal, S.: Automated detection of diabetic retinopathy on digital fundus image. J. Diabet. Med. **19**, 105–112 (2002)
10. Usher, D., Dumskyj, M., Himaga, M., Williamson, T.H., Nussey, S., Boyce, J.: Automated detection of diabetic retinopathy in digital retinal images: a tool for diabetic retinopathy screening. Diabet. Med. **21**, 84–90 (2004)
11. Liu, Z., Opas, C., Krishnan, S.M.: Automatic image analysis of fundus photograph. In: Proceedings of the International Conference on Engineering in Medicine and Biology, vol. 2, pp. 524–525 (1997)
12. Osareh, A., Mirmehdi, M., Thomas, B., Markham, R.: Automated identification of diabetic retinal exudates in digital colour images. Br. J. Ophthalmol. **87**, 1220–1223 (2003)
13. Mitra, S.K., Lee, T.-W., Goldbaum, M.: Bayesian network based sequential inference for diagnosis of diseases from retinal images. Pattern Recogn. Lett. **26**, 459–470 (2005)
14. Dupas, B., Walter, T., Erginay, A.: Evaluation of automated fundus photograph analysis algorithms for detecting microaneurysms haemorrhages, and exudates, and of a computer-assisted diagnostic system for grading diabetic retinopathy. Diabet. Metab. **36**, 213–220 (2010)
15. Stoecker, W.V., Chiang, C.-S., Moss, R.H.: Texture in skin images: comparison of three methods to determine smoothness. Comput. Med. Imag. Graph. **16**(3), 179–190 (1992)
16. Faizal Khan, Z., Nalini Priya, G., Anwar, M.K.: Texture based back propagation neural networks for segmentation of arteriole and venule in fundus images. In: IEEE International Conference on Power, Control, Signals and Instrumentation Engineering (ICPCSI), pp. 84–89 (2017)

ALICE: A Natural Language Question Answering System Using Dynamic Attention and Memory

Tushar Prakash[✉], Bala Krushna Tripathy, and K. Sharmila Banu

School of Computer Science and Engineering, VIT, Vellore, India
tusharprk@yahoo.com, {tripathybk,sharmilabanu.k}@vit.ac.in

Abstract. With the growing amount of textual information in recent years, it has become quite challenging to keep up with content produced by humans. Many models have been proposed that can perform reading comprehension on a variety of texts; however, past models either excel at information retrieval on complex texts or inference on simple texts. In this paper, we propose a model called ALICE that can perform information retrieval as well as inference tasks on any text. It is scalable to any document size and can be used to aid professionals in quickly finding answers to their problems using natural language queries. We will explore how ALICE achieves this and test it on some common datasets.

Keywords: Language · Processing · Reading · Comprehension · Attention Memory · Question · Answer

1 Introduction

In recent years, many models have been developed that can perform reading comprehension. Most recent models utilize deep learning techniques augmented with attention and memory mechanisms [1, 2] to decode answers from the encoded questions and documents [1, 3]. Many models either use information retrieval or inference techniques to find answers. However, current state-of-the-art models still face a decline in accuracy when processing larger and more complex text documents [3–6]. In this paper, we introduce a new approach towards text comprehension by utilizing predictive word embeddings, matched attention and external memory. Our proposed model can handle any document size due to its expandable architecture. Additionally, we predict the semantic relationship of out-of-vocabulary (OOV) and rare words, thereby, allowing us to generate significantly more accurate vector based word embeddings. End to end training is also possible with this model because the entire model is differentiable. The general architecture of this model can be expanded to applications beyond reading comprehension, such as, sentiment classification, image captioning and machine translation.

© Springer Nature Singapore Pte Ltd. 2018
I. Zelinka et al. (Eds.): ICSCS 2018, CCIS 837, pp. 274–282, 2018.
https://doi.org/10.1007/978-981-13-1936-5_30

2 Related Work

Most natural language processing models use recurrent neural networks (RNNs) for encoding and decoding operations as they can handle sequential data very easily. More recently, LSTM and GRU [7] model variations have been adopted because they can encode larger contexts [6, 7]. The recent addition of attention mechanisms over gated RNNs has also enabled a plethora of applications including, but not limited to, sentiment analysis, neural machine translation [1, 3], image caption generation [8] and question answering [2]. Furthermore, external memory manipulation also involves attention as seen in [4]. In this paper, we use a similar approach to understand the context of a document. However, unlike previous approaches [6, 9], our model combines the benefits of inference capability in augmented memory models with the scalability of sequential gated attention based models. This allows our model to perform a variety of tasks that were previously limited to only certain types of models. In recent years vector embedding models such as Word2Vec [10] and GloVe [11] have become a common method for encoding words because they can capture some of the semantic relationships between words. However, a major drawback of these models is that they can't generate representations of words or phrases that were not part of their training set. So, rare or out-of-vocabulary (OOV) words either have poor representations or don't have any representation at all. This results in diminished accuracy because the attention based models can't effectively determine how the poorly represented word relates to the context. Methods to mitigate this issue have been described in [12]. In this paper we propose a new method for embedding words that takes care of the underlying issues with vector based word embedding models, thereby significantly boosting their accuracy. To test our model we use common datasets like Facebook's bAbI dataset [13]; Stanford's SQuAD dataset [14] and Microsoft's MS MARCO dataset [15]. The bAbI dataset tests inference tasks over simple sentences, whereas, The SQuAD and MS MARCO datasets test more complex inference and information retrieval aspects of reading comprehension.

3 Model and Methods

In this section we provide a brief overview of the proposed model. Subsequently, we describe each component of the model in detail and give the intuition for its creation. We begin the task of reading comprehension by sequentially encoding the question and document using two bidirectional GRUs [7, 16]. A paired matching matrix then associates each word in the question with words in the given document [17]. The relevance of each sentence is determined using a soft attention mechanism on the matching matrix. Subsequently, a temporal controller writes the weighted encoding of the word into memory using a method similar to the one described in [18]. The encoded question and the word pair matching matrix is then passed as input to the read controller which selects the weighted encodings of the words (memory vector) that are relevant to the question.

Each selection is fed back to the read controller along with the question to find more evidence for supporting an answer (Fig. 1).

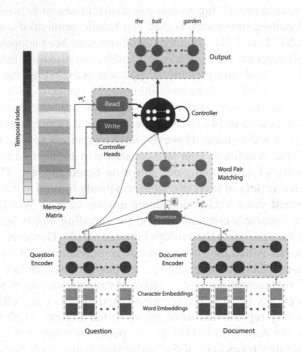

Fig. 1. A high level view of ALICE's architecture showing one time step

The weighted results are given to a bidirectional GRU which decodes the answer [1].

3.1 Embedding and Encoding

We propose a new method to handle rare or OOV words in vector based word embeddings [10, 11] by guessing the context of the words. A bidirectional LSTM (BiLSTM) is trained separately to predict the next word in a sentence. For each rare or OOV word, the BiLSTM generates 'n' candidate results to replace that word. We average the 'n' candidate vectors and insert then result into our word embedding model. We train the BiLSTM on texts used to train the word embedding model to ensure we don't encounter any OOV words. To encode the question Q and the document D, we generate their corresponding word embeddings $\left[w_i^Q\right]_{i=1}^m$ and $\left[w_i^D\right]_{i=1}^k$ along with their character embeddings $\left[c_i^Q\right]_{i=1}^m$ and $\left[c_i^D\right]_{i=1}^m$, where 'm' is the number of words in the question Q and 'k' is the number of words in the document D. For word embedding we use pre-trained GloVe embeddings [11] and for character embedding we use an LSTM-charCNN [19]. We then combine the word and character embeddings using bidirectional GRUs [16] to form encodings $\left[e_i^Q\right]_{i=1}^m$ and $\left[e_i^D\right]_{i=1}^k$ for all words in the question and document respectively.

$$e_i^Q = BiGRU\left(e_{i-1}^Q, \left[w_i^Q, c_i^Q\right]\right) \tag{1}$$

$$e_i^D = BiGRU\left(e_{i-1}^D, \left[w_i^D, c_i^D\right]\right) \tag{2}$$

By combining the enhanced word vector representations with character embeddings we get an encoding that effectively handles rare and OOV words.

3.2   Word to Word Relevance

To determine the importance of a word in answering the given question, we follow the suggestions outlined in [17] to generate word pair representations. For a given question encoding $\left[e_i^Q\right]_{i=1}^m$ and a document encoding $\left[e_i^D\right]_{i=1}^k$, the word pair representation $\left[p_i^D\right]_{i=1}^k$ is calculated using soft-alignment of words:

$$p_i^D = RNN\left(p_{i-1}^D, c_i\right) \tag{3}$$

where c_i is a context vector formed by merging all word pair attention vectors with question's encoding e_i^Q.

$$c_i = \sum_{i=1}^m a_i e_i^Q \tag{4}$$

$$s_i = \frac{\exp\left(a_j\right)}{\sum_{j=1}^m \exp\left(a_j\right)} \tag{5}$$

$$a_i = w^T tanh\left(W^Q e_i^Q + W^D e_i^D + W^p p_{i-1}^D\right) \tag{6}$$

Here, a_i is an attention vector over the individual question-document word pairs and s_i is the softmax over a_i. w is a learned weight vector parameter whose transpose is w^T. In [20], they add e_i^D as another input to the recurrent network used in $\left[p_i^D\right]$ (Fig. 2):

$$p_i^D = RNN\left(p_{i-1}^D, \left[e_i^D, c_i\right]\right) \tag{7}$$

Since the document may be very large, we introduce a gate g_i over the input $\left[e_i^D, c_i\right]$. This allows us to find parts of the document that are relevant to the question.

$$g_i = sigmoid\left(W_g\left[e_i^D, c_i\right]\right) \tag{8}$$

$$\left[e_i^D, c_i\right]' = g_i \odot \left[e_i^D, c_i\right] \tag{9}$$

The gate g_i is a learned value over time. This gate filters out the irrelevant words when g_i is closer to zero and gives importance to words when g_i is closer to one. The gate g_i learns different values for W_g over various time steps. Thus modeling a mechanism to effectively select parts of the document relevant to the question. In our model we use a BiGRU in place of an RNN.

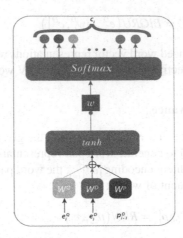

Fig. 2. A visual depiction of the attention mechanism used

3.3 Memory Controller

The memory architecture is similar to, but less complex than the one described in [18]. Memory is stored in an $N \times M$ matrix where N is the number of memory locations and M is the vector size at each location. Our first step is to write the weighted word vectors in to memory. This is achieved by using an LSTM for the controller network defined by:

$$i_t^l = sigmoid\left(W_i^l\left[x_t, h_{t-1}^l, h_t^l - 1 + b_i^l\right]\right) \tag{10}$$

$$f_t^l = sigmoid\left(W_f^l\left[x_t, h_{t-1}^l, h_t^l - 1 + b_f^l\right]\right) \tag{11}$$

$$s_t^l = f_t^l s_{t-1}^l + i_t^l tanh\left(W_s^l\left[x_t, h_{t-1}^l, h_t^l - 1 + b_s^l\right]\right) \tag{12}$$

$$o_t^l = sigmoid\left(W_o^l\left[x_t, h_{t-1}^l, h_t^l - 1 + b_o^l\right]\right) \tag{13}$$

$$h_t^l = o_t^l tanh\left(s_t^l\right) \tag{14}$$

where l denotes the layer of the LSTM and sigmoid is the logistic sigmoid function defined as:

$$sigmoid(x) = \frac{1}{1 + e^{-x}} \tag{15}$$

$i_t^l, f_t^l, s_t^l, o_t^l,$ and h_t^l are the input, forget, state, output and hidden gates, respectively, at layer l and time t. The input vector x_t is supplied to the controller at each time-step t. Since we want to maintain the order of sentences occurring in the document, we concatenate an increasing time index i to p_i^D from the word matching step:

$$p_i^D = RNN\left(p_{i-1}^D, \left[e_i^D, c_i\right]\right) \tag{16}$$

$$x_t = \left[p_i^D, i\right] \tag{17}$$

To read the memory vectors, we use the weighted averages across all locations:

$$r_t^i = M_t^T w_t^r \tag{18}$$

Here, M_t denotes the memory matrix at time t. The read vectors are appended to the controller input after each time-step. For the first time step, the question's encoding e_Q^i is supplied as input. The read weighting w_t^r determines the importance of a vector for answering the question. It is defined as:

$$w_t^r = f_t[1](i-1) + f_t[2]C(M,k) + f_t[3](i+1) \tag{19}$$

Here, f_t is a read mode decided by applying a softmax over a collection of 3 states: move backward, find similar vectors, move forward. The read weight w_t^r is applied over all memory locations, allowing the model to select important facts relevant to the given question. To find evidence supporting an answer, we use content-based addressing [18] on the read head to perform lookups over the memory:

$$C(M,k) = \frac{\exp\left(D(k, M[i, \cdot])\right)}{\sum_j \exp\left(D\left(k, M\left[j, \cdot\right]\right)\right)} \tag{20}$$

Here, $k \in R$ is the key or address of a memory location and D is the cosine similarity function. In our case, the temporal index is also used as the key.

3.4   Output

The output vectors are fed into a BiGRU which decodes the answer. Once the output is decoded, we calculate the loss using a standard cross-entropy loss function. We use this to minimize the sum of log probabilities when comparing the decoded answer with the actual one:

$$L(y, z) = -\sum_{\{i=0\}}^{m} z \cdot \log\left(P(y|z)\right) \tag{21}$$

Here, m is the vocabulary size, z is the actual answer and y is the prediction given by the model. The error is calculated and back-propagated until it is minimized.

4   Results

The bAbI dataset [13] consists of 20 tasks for testing a model's ability to reason over text. From the results listed in Table 1, we observe that ALICE performs significantly better than the DNC on basic induction tasks (task 16), which significantly contributes to a higher mean accuracy for ALICE. We also observe that the DMN [4] performs better than ALICE on basic induction tasks, yet, ALICE performs better on all other tasks. The accuracy of ALICE converges towards 96.8% over 10 training sessions. We only report the best results obtained from one of the 10 sessions.

Table 1. Comparison of results obtained on the bAbI dataset

Task	DMN [4]	DNC [18]	ALICE
1. Single supporting fact	100	100	100
2. Two supporting facts	98.2	99.6	98.6
3. Three supporting facts	95.2	98.2	97.2
4. Two argument relations	100	100	99.9
5. Three argument relations	99.3	99.5	99.4
6. Yes/no questions	100	100	100
7. Counting	96.9	99.8	99.7
8. List/sets	96.5	99.9	99.4
9. Simple negation	100	100	100
10. Indefinite knowledge	97.5	99.8	98.1
11. Basic co-reference	99.9	100	100
12. Conjunction	100	100	100
13. Compound co-reference	99.8	100	100
14. Time reasoning	100	99.7	99.8
15. Basic deduction	100	100	100
16. Basic induction	99.4	47.6	73.44
17. Positional reasoning	59.6	88	82.6
18. Size reasoning	95.3	99.2	97.5
19. Path finding	34.5	99.9	98.3
20. Agent motivations	100	100	100
Average accuracy	93.605	96.56	97.209
Number tasks failed (accuracy < 95%)	2	2	2

We further test our model on the SQuAD [14] and MS MARCO datasets [15]. The SQuAD dataset [14] contains question-answer pairs derived from 536 Wikipedia articles. SQuAD uses exact match (EM) and F1 score metrics to measure the performance of a given model. We report the best results obtained from 1 of 10 training sessions in Table 2.

Table 2. A comparison of results obtained on the SquAD dataset

Model	EM	F1
SAN ensemble model [23]	79.608	86.496
Reinforced mnemonic reader ensemble model [22]	77.7	84.9
ReasoNet ensemble model [23]	73.4	81.8
ALICE	78.024	85.212

While ALICE outperforms competitive models like ReasoNet [23] and Reinforced Mnemonic Reader [22]; the Stochastic Attention Network (SAN ensemble model) [21] still beats ALICE by a small margin. We attribute this to SAN's ensemble nature. To test our model (ALICE) on larger texts we use the MS MARCO dataset [15]. It contains multiple passages extracted from anonymized Bing search engine queries and the answers may not be exactly worded in those passages. The metrics used for evaluating a model for the MS MARCO dataset are BLEU and ROUGE-L scores.

From the results in Table 3, we see that ReasoNet [23] marginally outperforms ALICE on the MS MARCO dataset. Note that the results for ReasoNet are obtained by Microsoft AI and Research group after the paper was published. From our tests, we can clearly see that ALICE performs similar to, or better than some competitive models on the bAbI [13], SQuAD [14] and MS MARCO [15] datasets.

Table 3. A comparison of results obtained on the SquAD dataset

Model	BLEU	ROUGE-L
ReasoNet [23] [Microsoft AI and research results]	**38.81**	**39.86**
ALICE	38.43	38.67

5 Conclusion

In this paper, we propose a novel model, ALICE, aimed at the task of reading comprehension and question answering. We simplify an existing state-of-the-art architecture and combine it with a matching layer, to attend over a question and document. We also provide a method to improve the accuracy of similar models by using a bidirectional LSTM to generate contextual word embeddings for out-of-vocabulary words. Results for our model show that it is scalable in size and complexity. Our model achieves results that are close to the state-of-the-art and similar to, or better than some competitive models. Future work includes simplifying the current model and applying this model to generate captions for images.

References

1. Bahdanau, D., Cho, K., Bengio, Y.: Neural machine translation by jointly learning to align and translate (2014). arXiv preprint arXiv:1409.0473
2. Sukhbaatar, S., Weston, J., Fergus, R., et al.: End-to-end memory networks. In: Advances in Neural Information Processing Systems, pp. 2440–2448 (2015)
3. Luong, M.T., Pham, H., Manning, C.D.: Effective approaches to attention-based neural machine translation (2015). arXiv preprint arXiv:1508.04025
4. Kumar, A., et al.: Ask me anything: dynamic memory networks for natural language processing. In: International Conference on Machine Learning, pp. 1378–1387 (2016)
5. Gong, Y., Bowman, S.R.: Ruminating reader: reasoning with gated multi-hop attention (2017). arXiv preprint arXiv:1704.07415
6. Sordoni, A., Bachman, P., Trischler, A., Bengio, Y.: Iterative alternating neural attention for machine reading (2016). arXiv preprint arXiv:1606.02245

7. Chung, J., Gulcehre, C., Cho, K., Bengio, Y.: Empirical evaluation of gated recurrent neural networks on sequence modeling (2014). arXiv preprint arXiv:1412.3555
8. Vinyals, O., Toshev, A., Bengio, S., Erhan, D.: Show and tell: a neural image caption generator. In: 2015 IEEE Conference on Computer Vision and Pattern Recognition (CVPR), pp. 3156–3164. IEEE (2015)
9. Iyyer, M., Boyd-Graber, J., Claudino, L., Socher, R., Daumé III, H.: A neural network for factoid question answering over paragraphs. In: Proceedings of the 2014 Conference on Empirical Methods in Natural Language Processing (EMNLP), pp. 633–644 (2014)
10. Mikolov, T., Sutskever, I., Chen, K., Corrado, G.S., Dean, J.: Distributed representations of words and phrases and their compositionality. In: Advances in Neural Information Processing Systems, pp. 3111–3119 (2013)
11. Pennington, J., Socher, R., Manning, C.: Glove: global vectors for word representation. In: Proceedings of the 2014 Conference on Empirical Methods in Natural Language Processing (EMNLP), pp. 1532–1543 (2014)
12. Luong, M.-T., Sutskever, I., Le, Q.V., Vinyals, O., Zaremba, W.: Addressing the rare word problem in neural machine translation (2014). arXiv preprint arXiv:1410.8206
13. Weston, J., et al.: Towards AI-complete question answering: a set of prerequisite toy tasks (2015). arXiv preprint arXiv:1502.05698
14. Rajpurkar, P., Zhang, J., Lopyrev, K., Liang, P.: Squad: 100,000 + questions for machine comprehension of text (2016). arXiv preprint arXiv:1606.05250
15. Nguyen, T., et al.: MS MARCO: a human generated machine reading comprehension dataset (2016). arXiv preprint arXiv:1611.09268
16. Cho, K., et al.: Learning phrase representations using RNN encoder–decoder for statistical machine translation. In: Proceedings of the 2014 Conference on Empirical Methods in Natural Language Processing (EMNLP), pp. 1724–1734 (2014)
17. Rocktäschel, T., Grefenstette, E., Hermann, K.M., Kočiský, T., Blunsom, P.: Reasoning about entailment with neural attention (2015). arXiv preprint arXiv:1509.06664
18. Graves, A., et al.: Hybrid computing using a neural network with dynamic external memory. Nature 538(7626), 471 (2016)
19. Kim, Y., Jernite, Y., Sontag, D., Rush, A.M.: Character-aware neural language models. In: AAAI, pp. 2741–2749 (2016)
20. Wang, S., Jiang, J.: Learning natural language inference with LSTM. In: Proceedings of NAACL-HLT, pp. 1442–1451 (2016)
21. Liu, X., Shen, Y., Duh, K., Gao, J.: Stochastic answer networks for machine reading comprehension (2017). arXiv preprint arXiv:1712.03556
22. Hu, M., Peng, Y., Qiu, X.: Mnemonic reader for machine comprehension (2017). CoRR, abs/1705.02798 http://arxiv.org/abs/1705.02798
23. Shen, Y., Huang, P.-S., Gao, J., Chen, W.: ReasoNet: learning to stop reading in machine comprehension. In: Proceedings of the 23rd ACM SIGKDD International Conference on Knowledge Discovery and Data Mining, pp. 1047–1055. ACM (2017)

An Improved Differential Neural Computer Model Using Multiplicative LSTM

Khushmeet S. Shergill[(✉)], K. Sharmila Banu, and B. K. Tripathy

School of Computer Science and Engineering, VIT University, Vellore, India
{khushmeetsingh1996,sharmila.k,tripathybk}@vit.ac.in

Abstract. Artificial neural networks excel at doing specific tasks like image recognition, sequence learning, machine translation. But, the direction of research has moved towards the creation of more general purpose neural network architectures. Recently, DeepMind introduced Differentiable Neural Computer (DNC), with an external memory system that is capable of working on complex data structures. DNC can infer from graph problem, solve block puzzle using reinforcement learning and so on. DNC uses LSTM as controller network that manipulates the memory matrix. In this paper, we introduce a change to DNC architecture by replacing LSTM network with multiplicative LSTM and measure the performance of the improved model by training it on three different tasks; namely question answering task using bAbI dataset, character level modelling using harry potter text and planning search using air cargo problem. We compared the performance of the previous model to determine its behavior.

Keywords: LSTM · Memory · Planning search · Recurrent neural networks

1 Introduction

Differentiable Neural Computer (DNC) is a powerful neural network model coupled with external memory matrix. The model itself consists of two main components, controller and memory. Controller is recurrent neural network that acts like the CPU of the model, whereas memory is just an $N \times W$ matrix which can be thought of as the RAM of the model. The operations of the manipulating the memory is learned through gradient descent. The system is end to end differentiable.

Memory consists of weights, which represents the degree to which the particular location is involved in read or write operation. System has read and write heads that are involved in memory manipulation.

Differentiable Neural Computer can be thought of as a LSTM network, in a sense that they both try to remember things. The only difference is that DNC is given an external memory to remember things where as LSTM uses its hidden states for learning long term dependencies. But LSTM fall short in this regard.

DNC has taken inspiration from mammalian hippocampus. For e.g. humans recall the memories in the order they were remembered. Same happens in DNC with temporal link matrix, that retrieves memory in the same sequence they were written. Another is

© Springer Nature Singapore Pte Ltd. 2018
I. Zelinka et al. (Eds.): ICSCS 2018, CCIS 837, pp. 283–290, 2018.
https://doi.org/10.1007/978-981-13-1936-5_31

memory allocation is DNC, which is similar to dentate gyrus region of the hippocampus that performs neurogenesis.

2   Related Works

Artificial Neural Networks were created with the goal of trying to replicate thinking and remembering, which are the hallmark of human brain. With some setbacks ANNs started to perform well on many tasks, where brain excelled, such as image and speech recognition, machine translation and so on. But these neural networks still weren't good at doing inferencing and answering complex queries. This changed when Recurrent Neural Networks (RNN) were introduced. They have this feedback loop, that let them remember input through their hidden states. RNNs take sequence as vector $[x_1, x_2, x_3 \ldots x_n]$, $x_t \in \mathbb{R}^X$ and each output is a prediction sequence given as probability distribution with the help of softmax function. The hidden state $h_t \in \mathbb{R}^H$ is updates as a linear combination of input x_t and hidden state in the previous time step h_{t-1}. Later Bengio et al. [1] showed that RNNs suffer from vanishing gradient problem, where the model cannot store long term dependencies and training start to suffer.

In 1997 Schmidhuber and Hochreiter [2] introduced Long Short-Term Memory Network which solved vanishing gradients problem by introducing forget gate and new memory gate, that allowed hidden state to either pass through time (remembered) or not (forgot). This enabled the modelling of sequence with much greater length than before.

This was a significant win over RNNs and are very popular in sequence modelling, but still fall short when it comes to very long input sequence (limit of LSTMs). It can be seen in tasks like question answering.

To solve this, neural networks were given a separate memory component that can remember inputs and recall those inputs at later time steps. There are several networks that have memory component. Dynamic Memory Networks (DNM) from Kumar et al. [3] was built for question answering. It has semantic memory module that had GloVe vectors that are used to create sequence of word embeddings. The input module processes the text into a set of vectors. The episodic memory consists of 2 GRU modules. Outer GRU generates the memory vector from episodes, which are generated by inner GRU by moving over the facts from input modules.

Neural Turing Machine (NTM) (Graves et al.) [4] is a predecessor to DNC. It is different from memory networks in a sense that instead of just working on a specific task like question answering, they are built to perform algorithmic tasks (generalize to perform various tasks). The model is similar to DNC, meaning that it has same architecture as DNC, but differs in access mechanism. NTM uses location-based addressing, restricting how memory is being written (contiguous blocks). NTMs has no way to guarantee non-overlapping of memory locations. NTMs cannot free used memory locations.

3   High Level Overview of DNC

Model works by giving an input to the controller $x_t \in \mathbb{R}^X$ at every time step and it then emits an output vector $y_t \in \mathbb{R}^Y$. Along with input vector x_t controller receives a set of read vectors from the previous time step. Controller then emits an interface vector, that defines the way the controller will interact with the memory matrix denoted formally as $M \in \mathbb{R}^{N \times W}$. Both the input vector and read, write vectors are concatenated to obtain single vector with is then input into the controller.

$$X = \left[x_t, r^1 \ldots r^n \right]$$

After one iteration, model emits and output vector and an interface vector (δ) consisting of parameters for interacting with memory.

Reading memory is done by using read weightings to compute weighted average of the content of location, which are called read vectors, which are then sent to the controller network, therefore giving it access to memory content.

For write operation, a single write weighting $\left(w_t^w \right)$ calculated from memory allocation mechanism, erase vector and write vector (both taken from interface vector) is used to modify memory content.

DNC uses three mechanisms for concentrating on specific memory location while reading and writing. This increases the effectiveness of the model in storing and recalling. DNC uses content bases attention mechanism, which uses similarity between memory vector and the key vector. This is helpful when recalling some fact or finding a similar pattern. Second is memory allocation. DNC reuses memory location by defining usage vector u_t (for complete information refer to original paper by Graves et al.). Third is temporal linkage. This allows the DNC to recall facts in the order they were presented, simulating how human brain recall facts.

4   Multiplicative LSTM as Controller Network

Multiplicative LSTM (mLSTM) (Krause et al.) [5] is a hybrid architecture that combines LSTM and Multiplicative RNN (Sutskever et al.) [6]. Multiplicative RNN modifies Tensor RNN (Weston et al.) [7] to have hidden weight matrix for every input.

$$h_t = \tanh\left(W_{hx} x_t + W_{hh}^{(x_t)} h_{t-1} + b_h \right)$$
$$o_t = W_{oh} h_t + b_o$$

The hidden matrix used for given input is

$$W_{hh}^{(x_t)} = \sum_{n=1}^{N} W_{hh}^{(n)} x_t^{(n)}$$

Where $x_t^{(n)}$ is a one hot encoding of the input, and N is the dimensionality of the input. W_{hh} is a tensor here, which can be seen as a list of hidden matrices. With increasing dimensionality of x_t, W_{hh} becomes immensely huge to work with. Therefore W_{hh} is factorized as follows.

$$W_{hh}^{(x_t)} = W_{hf} \cdot diag\left(W_{fx}x_t\right) \cdot W_{fh}$$

To get mLSTM, hidden state from mRNN is plugged into gating units of LSTM.

$$m_t = \left(W_{fx}x_t\right) \odot \left(W_{fh}h_{t-1}\right)$$
$$h_t = W_{hx}x_t + W_{hm}m_t$$
$$i_t = \sigma\left(W_{ix}x_t + W_{im}m_t\right)$$
$$o_t = \sigma\left(W_{ox}x_t + W_{om}m_t\right)$$
$$f_t = \sigma\left(W_{fx}x_t + W_{fm}m_t\right)$$

5   Experimental Results and Analysis

The first experiment is done on *bAbI dataset* (Weston et al.) [8]. bAbI contains set of tasks and for each task there are 1000 questions for training and 1000 for testing. These tasks are designed to see whether network is able to answer query where deduction is required or can it count or can it answer questions in yes/no style etc. For full details on bAbI task, please refer to the paper by Weston et al.

In preprocessing phase, inputs and targets are encoded as vectors of length 100 by constructing a lexicon of unique words and padded zeros at the beginning if sequence length is small and is serialized to be used later during training phase.

During training, input and output vectors are taken one at a time, since batch size is 1 and are then converted to one hot vectors which are then fed to the model.

Two different DNC models are used DNC1 with LSTM as controller network and DNC2 with mLSTM variant both using 256 cells. Rest of the parameters are same as of DeepMind DNC. Both models are run for 300,000 iterations.

Following table compares DeepMind DNC with our model (Table 1).

Model performed poorly on Path finding and Basic Induction, but performed really well on Agent Motivation, Compound Coreference and Basic Coreference. Model if trained further could lower the error rate on every task. Model is further compared with DeepMind's DNC with their results taken from the paper.

Above Fig. 1 displays loss for 300,000 iterations but is showing only last 200,000 iterations because the training was later continued from a checkpoint. Loss fluctuation is high, but model was able to perform moderately accurate on bAbI tasks.

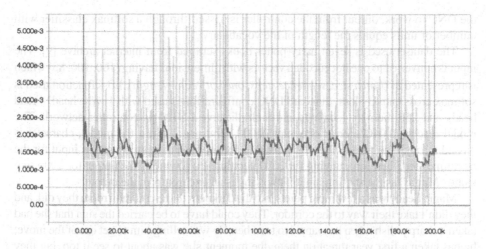

Fig. 1. Loss vs Iterations (bAbI task)

Table 1. Results for bAbI task

Task	DeepMind DNC	DNC
Positional reasoning	24.1	43.75
Two argument relation	0.0	5.69
Agent motivation	0.0	1.10
Counting	0.2	9.00
Single supporting fact	0.0	13.30
Path finding	0.1	90.23
Basic deduction	0.0	53.19
Three supporting fact	2.4	30.70
Indefinite knowledge	0.2	27.04
Three argument relation	0.5	13.20
Negation	0.0	4.40
Conjunction	0.1	8.40
Basic coreference	0.0	4.14
Time reasoning	0.3	27.98
Size reasoning	4.0	8.43
Compound coreference	0.0	1.05
Lists	0.1	7.70
Basic induction	52.4	63.19
Two supporting fact	1.3	30.37
Yes/No	0.0	5.80

The second experiment is *character level language modelling*. This means that DNC tries to model probability distribution of the next character is the given sequence. For the DNC to work, output from the controller is passed through a softmax classifier with number of units equals the length of the vocabulary.

The dataset used is seven harry potter books scraped from internet archive [9]. The input is converted to one hot encoded vectors with dimensions having (100, 100, 83) which is represented as (*batch size, sequence steps, one hot vector length*). Loss function used is softmax cross entropy loss. Standard DNC configuration is used with 1 write head, 4 read heads and 256 memory locations and multiplicative LSTM as controller network. The model reaches 0.04076 error rate after 26000 iterations. To test the model, character is passed through the model, the output for that corresponding input is used as input for the next time step, therefore running for 2000 iteration. Following is the output for 2000 step using "Magic" as the starting input. Following is an excerpt from the output.

Magic Leady was a few more points into the darkness to the common room they did, and they didn't take their way to the corridor. They could have to be carried the sign that she had taken a sharp polish when he wanted to trust her. He was telling him to get out of the move; he has taken a first year threat on them, the moment she was about to see it too, but they were tenting to have a good stone way, and then he started at him. "It's be there to do what te was stopped, you said to his stomach," said Dumbledore, starting to strain to Harry into her hand. "Well, if you come back there. The Muggle birth was a bulging second time. … They didn't look like it has the bar in the silence and take it."

Observing the above text, it can be seen that individual words generated were meaningful but the sentence structure did not make any sense. Some of the words that is particular to the input text, for example "Dumbledore" or "Muggle" were correctly generated, but later in the sentence, Dumbledore was referred as her, concluding that words as a collection does not possess any meaning. But overall grammar of the text is correct.

The model was able to converge quite nicely at 26,000 iterations and text resembles with the input given. Various words unique to this text, were generated accurately by the model. Loss vs Iteration of model is present in Fig. 2.

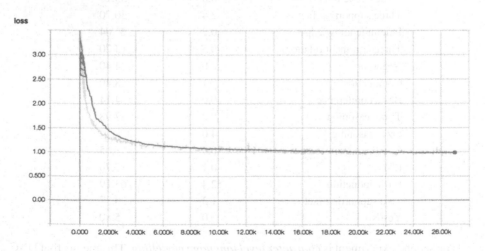

Fig. 2. Loss vs Iteration

The third experiment is *planning problem*. Air cargo planning problem is chosen from [8] Russel and Norvig. Given initial conditions, goal and list of actions that can be performed, model's task is to find optimal number of steps to reach that goal.

For example, given an initial condition and Goal and an action schema (Fig. 3).

Init(At(C1, SFO) ∧ At(C2, JFK) ∧ At(P1, SFO) ∧ At(P2, JFK) ∧ Cargo(C1) ∧ Cargo(C2)
        ∧ Plane(P1) ∧ Plane(P2)∧ Airport(JFK) ∧ Airport(SFO))
Goal(At(C1, JFK) ∧ At(C2, SFO))
Action(Load(c, p, a), PRECOND: At(c, a) ∧ At(p, a) ∧ Cargo(c) ∧ Plane(p) ∧ Airport(a)
        EFFECT: ¬ At(c, a) ∧ In(c, p))
Action(Unload(c, p, a), PRECOND: In(c, p) ∧ At(p, a) ∧ Cargo(c) ∧ Plane(p) ∧ Airport(a)
        EFFECT: At(c, a) ∧ ¬ In(c, p))
Action(Fly(p, from, to), PRECOND: At(p, from) ∧ Plane(p) ∧ Airport(from) ∧
Airport(to)
        EFFECT: ¬ At(p, from) ∧ At(p, to))

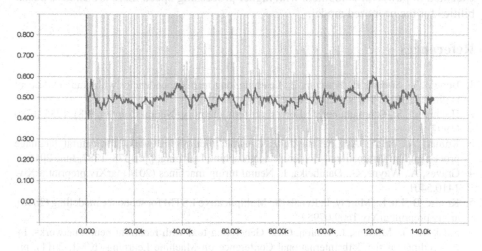

Fig. 3. Loss vs Iterations (Planning problem)

Following this data, the goal will be achieved in 6 steps, by following the plan given below.

Load (C1, P1, SFO)
Load (C2, P2, JFK)
Fly (P1, SFO, JFK)
Fly (P2, JFK, SFO)
Unload (C1, P1, JFK)
Unload (C2, P2, SFO)

DNC's task is to figure out the plan in minimum number of steps.

DNC has to figure out the correct order of the item in the tuple, like the arguments given to the action $(C1, P1, SFO)$ and to determine which action to use, Load or Fly or Unload.

The experiment is conducted with 2 cargos, 2 planes and 2 airports. After training for 2 days for 140,000 iterations on NVidia K80 12 GB GPU, it gave 62.5% accuracy on 8 planning problems. Graph below shows model reaches 0.2 loss.

Model is taking long time to converge, with another 100,000 iterations required to reach loss below 0.1.

6   Conclusion

In this paper we proposed a new DNC model with the LSTM being replaced with multiplicative LSTM. The proposed model has a slightly larger training time in comparison to the previous model. But, being a generalised model it is applicable to larger class of problems. Moreover, the new DNC model tries to emulate algorithmic tasks. When executed in powerful computers with higher processing speed there are endless possibilities for the proposed model.

References

1. Bengio, Y., Simard, P., Frasconi, P.: Learning long-term dependencies with gradient descent is difficult. IEEE Trans. Neural Netw. **5**, 157–166 (1994)
2. Hochreiter, S., Schmidhuber, J.: Long short-term memory. Neural Comput. **9**(8), 1735–1780 (1997)
3. Kumar, A., et al.: Ask me anything: dynamic memory networks for natural language processing. In: International Conference on Machine Learning, pp. 1378–1387, June 2016
4. Graves, A., Wayne, G., Danihelka, I.: Neural turing machines (2014). arXiv preprint arXiv: 1410.5401
5. Krause, B., Lu, L., Murray, I., Renals, S.: Multiplicative LSTM for sequence modelling (2016). arXiv preprint arXiv:1609.07959
6. Sutskever, I., Martens, J., Hinton, G.E.: Generating text with recurrent neural networks. In: Proceedings of the 28th International Conference on Machine Learning (ICML 2011), pp. 1017–1024 (2011)
7. Weston, J., et al.: Towards AI-complete question answering: a set of prerequisite toy tasks (2015). arXiv preprint
8. Russell, S.J., Norvig, P., Canny, J.F., Malik, J.M., Edwards, D.D.: Artificial Intelligence: A Modern Approach, vol. 2, no. 9. Prentice Hall, Upper Saddle River (2003)
9. Rowling, J.K.: Internet Archive. https://archive.org/details/HarryPotterCompleteCollection

Abnormal Activity Recognition Using Saliency and Spatio-Temporal Interest Point Detector

Smriti H. Bhandari$^{(\boxtimes)}$ and Navnee S. Babar

Department of Computer Science and Engineering,
Walchand College of Engineering, Sangli, India
smriti_bhandari@yahoo.com, navni010190@gmail.com

Abstract. Detecting abnormal activities is a crucial research topic nowadays because of its wide variety of applications in such as security monitoring, video surveillance and healthcare applications. The proposed method is used to distinguish between normal and abnormal human activities. Two-dimensional visual saliency map is created from color video sequences and used for further processing. Selective spatio-temporal interest point (STIP) detector is used to extract interest point features from saliency. 3D Image gradients are calculated using intensity patches to describe STIPs and feature vector is computed by quantizing them. The activities are described finally using bag-of-features representation. Support Vector Machine is used as a classifier to distinguish between normal and abnormal activities. The performance of the system is evaluated using UR fall Dataset and dataset S provided by Le2i CNRS that shows significant accuracy.

Keywords: Abnormal activity recognition · Spatio-temporal interest points
Visual saliency

1 Introduction

Recognizing behavior of a human from videos is a challenging problem in many application areas, such as video surveillance and retrieval, security and health care involving computer vision and machine learning applications. To moniter the behaviour, daily activities and any other information of elderly, a close observation is necessary to protect them [1]. With increased population of older adults in society, there is an urgent need for assistive technologies in the home. Older adults face many difficulties while undergoing their daily activities because of age-related changes. Thus, to take care of older adults, it is very important to know if any unusual activities are there.

Abnormal activities of human are still difficult to recognize because they are not able to predict it prior and those types of strange events does not occur frequently [2]. It is important to identify an emerging medical condition prior to it gets critical. So, it becomes necessary to observe the activities of daily livings (ADLs) and seek for abnormal behaviour in daily life [3]. In this paper, the work is focused on to detect unusual or abnormal activities instead of considering regular activity recognition. "Abnormal activities" can be defined as "activities which are infrequent and not predicted in advance" [4].

© Springer Nature Singapore Pte Ltd. 2018
I. Zelinka et al. (Eds.): ICSCS 2018, CCIS 837, pp. 291–301, 2018.
https://doi.org/10.1007/978-981-13-1936-5_32

In this work, a system is proposed to distinguish between normal and abnormal activities which uses 2D visual saliency map from color video sequences. Two datasets such as UR fall detection [5] and Dataset S by Le2i CNRS [6] are used to evaluate the performance. Selective STIP detector is used to find the interest points which are robust to the complex and moving background [7]. To extract relevent features is crucial step to detect and recognize the activity. Hence STIP detector is used which focuses on local spatio-temporal information and the performance is improved by retaining most repeatative, balanced and distinguishable STIPs for human subjects after elimination of undesired STIPs in background. Image gradients are computed after extracting selective interest points and then the gradients are quantized using spherical co-ordinate method to form a vector of final features. With the use of bag-of-features (BoFs) representation and support vector machine (SVM) performance is evaluated.

The rest of the paper is organized as follows. In Sect. 2 work related to abnormal activity detection is discussed. The proposed methodology is in Sect. 3. In Sect. 4 experimental results are presented with all the necessary details and discussion. Finally, the conclusion is provided in Sect. 5.

2 Related Work

Human activity recognition has given much attention especially from those who work in the field of machine learning and computer vision. Human activities are classified into normal and abnormal activities. To detect abnormal activities is main concern nowadays and lots of research is focused on identifying abnormalities in video surveillance. Abnormality detection belongs to video analysis which includes human activity detection and recognition. These systems are mainly classified into single-layered and hierarchical approaches as shown in Fig. 1. Single layered techniques are used to represent the activity directly based on image sequences and further classified into methods such as space-time and sequential approaches. Space-time approaches consider the activity in 3D volume and represent it in space-time features from given video sequences. These approaches are again divided based on features used from 3D

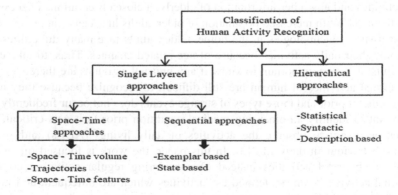

Fig. 1. Classification of human activity recognition system

volume [8]. Sequential approaches describe the activity in sequence of observations and can be further classified depending on the technique they use for recognition such as Exemplar or state based recognition [9].

Hierarchical techniques used for recognizing complex activities by analyzing the video into various feature descriptors [10]. Hierarchical approaches are categorized into statistical, syntactic and description based methodologies. Statistical strategies used to represent the high-level human activities by concatenating state based models hierarchically. Hidden Markov Model (HMM) is an example of such type of approach. Syntactic models use grammar based syntax and model the activities as strings of symbols [11]. Description based models describe the sub-event of activities to represent human activities. Abnormality or anomaly can be realized by using normal activities and depends on the approaches used to classify them. Three broad categories such as supervised, semi-supervised and un-supervised are used to construct the model [12–14].

The training data of normal and abnormal behavior is provided in case of supervised approaches to detect anomalous or unusual data. Semi-supervised approaches use only normal behavior to train the model and detect abnormality either automatically or through the training process. Un-supervised approaches do not need any kind of training [15]. These approaches work on some rule base or the conditions describing distinction between classes.

3 Proposed Methodology

This work proposes the method for abnormal activity detection in the home environment. Here we consider normal activities as daily activities of the person such as sitting on a chair, lying on the bed, reading, writing, picking up the fallen object, tightening shoelace, sweeping, cleaning, etc. Abnormal activities include forward fall, backward fall, fall from standing position, fall from a chair, etc. The overall methodology is depicted in Fig. 2.

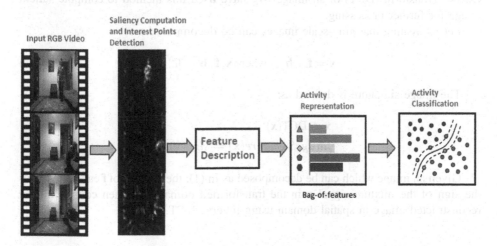

Fig. 2. Block diagram for activity recognition

Input videos are converted into frames. Further, we detect salient regions from each frame. The saliency of each frame is computed using DCT-based image signatures [16]. Once we find the salient image, next task is to find interest points those effectively contribute in describing the motion of the object. We use selective Spatio-Temporal Interest Point (STIP) detection method proposed by Chakraborty et al. [7]. Authors have claimed that performance is improved by retaining most repeatative, balanced and distinguishable STIPs for human subjects after removing undesired STIPs in background. However, in our experiments, we observed few unwanted background STIPs when the frames are processed directly for feature description. To remove or minimize unwanted background STIPs, we introduced intermediate step of saliency computation before interest point detection. Further, based on interest points, feature descriptor is formed for each frame in a video. The activity in the video is represented using a feature vector after application of bag-of-features technique. Finally, SVM classifier is used for classification of activities into two classes: normal and abnormal.

The detailed methodology is described in the following text.

The color visual data (i.e., color videos) are denoted as a sequence of 2D frames $\{I_1, \ldots, I_T\}$. The 2D frame at time instance t is denoted by $I_t = (x, y, i, t)$, $\forall t \in [1, T]$, where, x and y denote the spatial co-ordinates of pixel in the image frame; and i is the intensity value computed from its respective RGB values.

3.1 Saliency Computation

A part of an image that catches the attention of a viewer as it stands out from its neighborhood is called salient region. An object or a pixel in an image is referred as salient depending on the measure or quality by which it is distinguished from its surrounding. With respect to the application under consideration, we need to detect salient object; that is the moving foreground object in video under consideration. Hou et al. [16] used a binary, and holistic image descriptor called the "image signature" to highlight salient regions in the image. It is defined as the sign function of the Discrete Cosine Transform (DCT) of an image. We have used this method to compute salient image for further processing.

Let us assume that gray-scale images can be decomposed as:

$$\mathbf{x} = \mathbf{f} + \mathbf{b}, \quad \text{where } \mathbf{x}, \mathbf{f}, \mathbf{b} \in \Re^N \tag{1}$$

The image signature is defined as:

$$\hat{\mathbf{x}} = \text{DCT}(\mathbf{x})$$
$$\text{Im} ageSignature(\mathbf{x}) = sign(\hat{\mathbf{x}}) \tag{2}$$

Given an image which can be decomposed as in (1), the support of $\mathbf{f}$ can be taken as the sign of the mixture signal $\mathbf{x}$ in the transformed domain and then computing the reconstructed image in spatial domain using inverse DCT.

$$\bar{\mathbf{x}} = IDCT(sign(\hat{\mathbf{x}}))$$ (3)

Foreground of an image is assumed to be visually aparent and discernible with respect to its background, then we can form a saliency map $\mathbf{m}$ [17] by smoothing the squared reconstructed image as in (4)

$$\mathbf{m} = g * (\bar{\mathbf{x}} \circ \bar{\mathbf{x}})$$ (4)

where g is the Gaussian kernel. '*' is convolution operator and 'o' is Hadamard (entrywise) product operator.

3.2 Interest Points Detection

STIP-based methods avoid temporal alignment problem. Also these methods exhibit invariance to geometric transformations. We have adopted Selective STIP detection method [7], in which surround suppression mask is used to remove undesired points in the background considering local and temporal constraints. This formulation makes the system robust to camera motion as well as background clutter.

3.3 Feature Description

For each interest point detected in a frame, we construct a patch $i_p(x, y, t)$ of size m x m. Then, spatio-temporal gradients are computed for intensity patch sequence along x, y, and t dimensions as:

$$\nabla i_p = \left(\frac{\partial i_p}{\partial x}, \frac{\partial i_p}{\partial y}, \frac{\partial i_p}{\partial t} \right)$$ (5)

Here, we use 3D sobel operator [18] to compute gradient along each dimension.

The gradients of image patch sequence are quantized using a spherical coordinate-based scheme [19]. The azimuth angle $\theta(\nabla i_p)$ and elevation angle $\phi(\nabla i_p)$ are computed for each gradient vector obtained as per (5). This characterizes 3D orientations of image patch sequence in xyt space, as:

$$\theta(\nabla i_p) = \arctan\left(\frac{\partial i_p}{\partial y} \middle/ \frac{\partial i_p}{\partial x} \right), \qquad \phi(\nabla i_p) = \arctan\left(\frac{\partial i_p}{\partial t} \middle/ \sqrt{\frac{\partial^2 i_p}{\partial x} + \frac{\partial^2 i_p}{\partial y}} \right)$$ (6)

Feature descriptor is computed based on image gradient orientations in xyt space. Interest points provide visual cue to be used as features. As, orientation does not depend on its magnitude, it is not affected by changes in the illumination as well as noise. Thus, orientation quantization is a robust way for describing features as depicted in Fig. 3(b). Figure 3(a) shows orientation computation of azimuth and elevation angles. Further as shown in Fig. 3(b), these angles are quantized in bins. As an example, Fig. 3(b) shows subdivision of azimuth angle and elevation angle into six

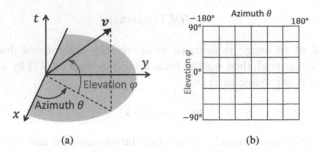

Fig. 3. 3D feature description using spherical coordinates. (a) Orientation computation. (b) Orientation quantization [19]

different bins, resulting in 36 bins of 1D histogram. By normalizing histogram of intensity gradient orientations h_i, final feature vector h is computed as given in (7).

$$h = \frac{h_i}{N_i} \tag{7}$$

where N_i is the total number of gradient orientations in h_i.

3.4   Activity Representation

We apply the standard BoFs representation to encode visual cues representing human activities. The BoF representation is based on a codebook or visual vocabulary. To construct vocabulary k-means clustering is used. Each cluster has a cluster center and is indexed by a visual word. Euclidean distance is used to assign every feature vector to its nearest visual word. Then, a video consisting of a sequence of color frames can be encoded as a histogram of visual word occurrences.

4   Results and Discussion

4.1   Dataset

As the real dataset containing abnormal activities of elderly people is difficult to acquire, two benchmark datasets such as Dataset UR-fall detection (URFD) and Dataset S provided by Le2i CNRS are used to estimate the performance of the proposed system to detect abnormal activity. Detail description of datasets are given below.

UR fall dataset [5]: The dataset consists of 70 (30 fall activities + 40 activities of daily living) RGB and depth videos. The proposed system uses only RGB videos for processing. Here, fall is considered as abnormal activity and activities of daily living are considered as normal activities to evaluate the performance. Normal activities such as walk, sit down, bend, lye on a bed and abnormal activities such as fall while sitting on a chair, fall while walking has been performed by five subjects.

Dataset S by Le2i CNRS [6]: The entire dataset contains total 221 videos out of which 126 are of abnormal activities and 95 are of normal activities. The original resolution of the frames in the dataset (640 × 480) is resized to 320 × 240 pixels for the analysis. Nine different subjects have performed various The dataset includes normal activities such as walk, sit down, stand up, bend, house keeping, move chair and abnormal activities as forward falls, falls when inappropriate sitting-down, loss of balance, stroke, etc.

4.2 Results

Experiments are performed with UR Fall dataset and Dataset S by Le2i CNRS. The results of intermediate steps as per methodology are shown in Figs. 4 and 5. SVM classifier is used for classification. Holdout method is used for dataset partitioning with 50% samples under both categories are used for training and remaining 50% are used for testing. Results are obtained by undergoing the methodology as explained in Sect. 3 and reported as with saliency. As mentioned in feature description we have computed orientation histograms to describe interest points. A group of angle bins controls the granularity of orientation histograms. The elevation angle ϕ is divided into 9 bins; whereas azimuth angle θ is divided into 18 cells. Thus, the resulting feature vector contains 162 elements. Size of the vocabulary for BoF activity representation is kept as 100. This parameter determines the size of final feature vector that describes the activity by encoding the original features obtained by STIPs.

The experiments are also carried out by omitting the step of saliency computation. For UR Fall dataset the accuracy is 100% for both normal as well as abnormal categories when STIPs are detected after saliency computation. However, it is observed that the accuracy obtained for correct classification is 86.67% for abnormal class when STIPs are computed by omitting saliency detection. The results are depicted in Fig. 6. The further experimentation of the results pointed out the possible reason for reduced accuracy. As shown in Fig. 4, for few of the frames in the video the interest points are detected in the background too which may mislead training as well as testing.

For Dataset S, the results are comparatively low than the results reported in [6]. In [6], the authors have reported the results separately for the dataset with different backgrounds. In the experiments, we have collectively used the videos to train and test the dataset and obtained the results irrespective of the room (background) used for acquiring the dataset. Further, in case of results with saliency for dataset S, though the system provides false alarms, those can be accepted to some extent as we are much concerned about correctly detecting abnormal activities. For dataset S, the accuracy for correct classification of abnormal activities is 96.83%, i.e. out of 63 abnormal activities tested, 61 are correctly recognized as abnormal and 2 activities are wrongly reported as normal. The results carried out in experimentation are depicted graphically in Fig. 6.

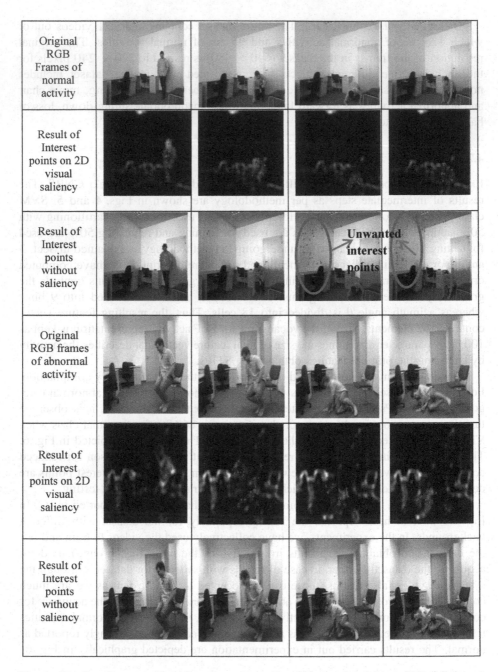

Fig. 4. Results of interest points detection on normal and abnormal activities of UR fall dataset

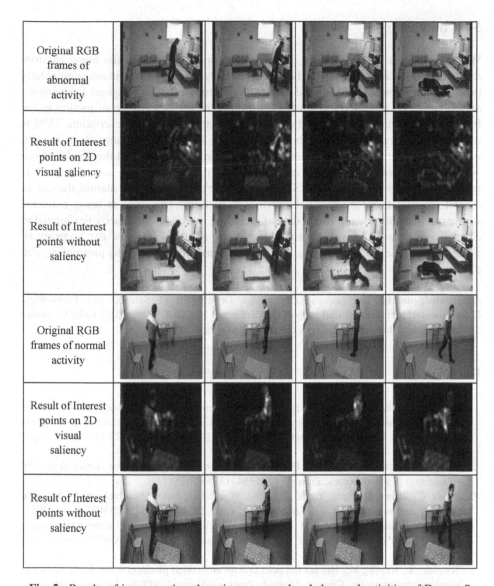

Fig. 5. Results of interest points detection on normal and abnormal activities of Dataset S

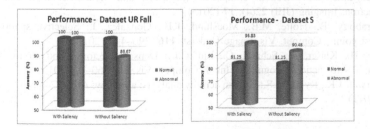

Fig. 6. Performance of proposed system for activity recognition

5 Conclusion

We have presented a method for detecting abnormal activities in the home environment. This is an attempt towards the aim of building a support system assisting elderly people living alone. The proposed system first computes salient regions from frames and then detects Spatio-Temporal Interest Points for the description of motion in the frame. BoF technique is used for building vocabulary for activity description. SVM is used as a classifier to recognize whether the activity is normal or abnormal. As our aim is to report abnormal activity of elderly people correctly, we found the results of our system encouraging. For UR Fall dataset the system is giving 100% accuracy whereas in case of Dataset S by Le2i CNRS, though a system is giving false alarms, the rate of correct classification for abnormal activities is 96.83%. The work is being extended with experimentation on our own dataset including more abnormal activities related to elderly health issues. Also, the attempts will be made to improve the accuracy of abnormal activities as well as reducing the false alarms by varying the parameters, size of codebook etc.

Acknowledgements. The authors acknowledge Department of Science and Technology, Government of India for financial support vide Reference No. SR/CSRI/84/2015 under Cognitive Science Research Initiative (CSRI) to carry out this work.

References

1. Mathur, G., Bundele, M.: Research on intelligent video surveillance techniques for suspicious activity detection critical review. In: IEEE International Conference on Recent Advances and Innovations in Engineering 2016 (ICRAIE-2016), 23–25 December, Jaipur (2016)
2. Tong, Y., Chen, R., Gao, J.: Hidden state conditional random field for abnormal activity recognition in smart homes. Entropy 17, 1358–1378 (2015)
3. Wang, C., Zheng, Q., Peng, Y., De, D., Song, W.-Z.: Distributed abnormal activity detection in smart environments. Int. J. Distrib. Sensor Netw. 2014, Article ID 283197, 1–15 (2014)
4. Hua, D.H., Zhang, X.-X., Yinc, J., Zhenga, V.W., Yang, Q.: Abnormal activity recognition based on HDP-HMM models. In: International Joint Conference on Artificial Intelligence (2009)
5. Kwolek, B., Kepski, M.: Human fall detection on embedded platform using depth maps and wireless accelerometer. Comput. Methods Programs Biomed. 117(3), 489–501 (2014)
6. Charfi, I., Mitéran, J., Dubois, J., Atri, M., Tourki, R.: Optimised spatio-temporal descriptors for real-time fall detection: comparison of SVM and Adaboost based classification. J. Electron. Imaging (JEI) 22(4), 17 (2013)
7. Chakraborty, B., Holte, M.B., Moeslund, T.B., Gonzàlez, J.: Selective spatio-temporal interest points. Comput. Vis. Image Underst. 116, 396–410 (2012)
8. Wang, H., Kläser, A., Schmid, C., Liu, C.L.: Dense trajectories and motion boundary descriptors for action recognition. Int. J. Comput. Vis. 103, 60–79 (2013)
9. Aggarwal, J.K., Ryoo, M.S.: Human activity analysis: a review. ACM Comput. Surv. 43, 1–43 (2011)

10. Jhuang, H., Serre, T., Wolf, L., Poggio, T.: A biologically inspired system for action recognition. In: Proceedings of IEEE International Conference on Computer Vision, Rio de Janeiro, pp. 1–8 (2007)
11. Tsai, W., Fu, K.S.: Attributed grammar-a tool for combining syntactic and statistical approaches to pattern recognition. SMC **10**, 873–885 (1980)
12. Brax, C., Niklasson, L., Smedberg, M.: Finding behavioural anomalies in public areas using video surveillance data. In: Proceedings of 11th International Conference on Information Fusion (2008)
13. Zhang, D., Gatica-Perez, D., Bengio, S., McCowan, I.: Semi-supervised adapted HMMs for unusual event detection. In: Proceedings of IEEE Conference on Computer Vision and Pattern Recognition (CVPR), pp. 611–618 (2005)
14. Xiang, T., Gong, S.: Video behavior profiling for anomaly detection. IEEE Pattern Anal. Mach. Intell. **30**, 893–908 (2008)
15. Beddiar, D.R., Nini, B.: Vision based abnormal human activities recognition: an overview. In: 8th International Conference on Information Technology (2017)
16. Hou, X., Harel, J., Koch, C.: Image signature: highlighting sparse salient regions. IEEE Trans. Pattern Anal. Mach. Intell. **34**(1), 194–201 (2012)
17. Itti, L., Koch, C., Niebur, E.: A model of saliency-based visual attention for rapid scene analysis. IEEE Trans. Pattern Anal. Mach. Intell. **20**(11), 1254–1259 (1998)
18. Sun, B., Sang, N., Wang, Y., Zheng, Q.: Motion detection based on biological correlation model. In: Zhang, L., Lu, B.-L., Kwok, J. (eds.) ISNN 2010. LNCS, vol. 6064, pp. 214–221. Springer, Heidelberg (2010). https://doi.org/10.1007/978-3-642-13318-3_28
19. Zhang, H., Parker, L.E.: CoDe4D: color-depth local spatio-temporal features for human activity recognition from RGB-D videos. IEEE Trans. Circuits Syst. Video Technol. **26**(3), 541–555 (2016)

An Improved ALS Recommendation Model Based on Apache Spark

Mohammed Fadhel Aljunid[✉] and D. H. Manjaiah

Department of Computer Science, Mangalore University, Mangalore, India
Ngm505@yahoo.com, drmdh2014@gmail.com

Abstract. Recommender Systems (RS) have become very imperative in several fields such as e-commerce, and social media networking. In recommender systems, there is a problem of filtering information, which is considered one of the complex challenges in building these systems. Recently, there are many algorithms available to address RS's challenges, one of the most common algorithms is collaborative filtering recommendation. This algorithm is based on Alternating Least Squares (ALS) is one of the widespread algorithms using matrix factorization method of recommendation system which is using to address that intricate challenges. In this paper, we suggest an approach called improved ALS to improve the performance of conventional ALS model on two different datasets, Movielene and Book-crossing using Apache Spark. The model evaluation is done using Root Mean Squared Error (RMSE) metrics, as well as compared with the conventional ALS Model.

Keywords: Recommendation system · Collaborative filtering · Alternating least squares · Apache Spark

1 Introduction

Recently, the building of Recommender Systems (RS) becomes a significant research area that attractive several scientists and researchers across the world. The Recommender Systems (RS) are used in many areas including music, movies, books, news, search queries, and commercial products.

There are different methods for building a recommender system, such as, user-based, content-based, or collaborative filtering. Collaborative filtering calculates recommendations based on similarities between users and products. For example, collaborative filtering assumes that users who give the similar ratings on the same movies will also have similar opinions on movies that they haven't seen. The alternating least squares (ALS) algorithm provides collaborative filtering between users and products to find products that the customers might like, based on their previous ratings. In this case, the ALS algorithm will create a matrix of all users versus all movies. Most cells in the matrix will be empty. An empty cell means the user hasn't reviewed the movie yet. The ALS algorithm will fill in the probable (predicted) ratings, based on similarities between user ratings. The algorithm uses the least squares computation to minimize the estimation errors, and alternates between solving for movie factors and solving for user factors [1, 2].

© Springer Nature Singapore Pte Ltd. 2018
I. Zelinka et al. (Eds.): ICSCS 2018, CCIS 837, pp. 302–311, 2018.
https://doi.org/10.1007/978-981-13-1936-5_33

The following trivial example gives us an idea of the problem to solve. However, keep in mind that the general problem is much harder because the matrix often has far more missing values.

User\Product	1	2	3	4	5
1	3.5	1	3	2	?
2	1.5	4.5	?	4	5
3	1	?	?	4.5	?
4	0.5	4	2	4	?

This paper introduces an efficient ALS recommendation model based on apache spark in big data analytics. We proposed the method to efficient the performance of ALS model. The performance analysis and evaluation of proposed approach are done on two different datasets, MovieLens and Book- crossing.

The structure of this paper is organized as follows. Related work is stated in Sect. 2. Proposed system is introduced in Sect. 3. Experimental and result is given in Sect. 4. Finally, a conclusion is given in Sect. 5.

2 Related Work

Recent time, many researchers presented and research introduced in the fields of recommendation systems.

Zhou, et al. [4], they proposed a parallel algorithm for large-scale collaborative filtering. This model was designed to be scalable to very large datasets.

By Wang, Yuan, Sun [5], they introduced a model based on item combination demographic information and feature, and it focuses on searching for a set of neighboring users shared with the same interest, which supports to enhance system scalability. The use of genetic algorithm to learn the weight features in the user model and significantly improves the accuracy of the recommender system.

An explicit trust and distrust clustering based collaborative filtering recommendation method were proposed by Ma, et al. [6]. A SVD signs-based clustering method was proposed to cluster the trust and distrust relationships. A sparse rating complement algorithm was proposed to generate dense user rating profiles which alleviates the sparsity and cold start problems to a very large extent. The model should be exploring more trust inference metrics and validate their method on some other datasets.

Kumar, et al. [7], introduced a new technique of building a hierarchical two-class structure of binary matrix factorization to manage matrix completion of ordinal rating matrix. The proposed technique verifies to be a more accurate matrix completion process.

Kumar, et al. [8] proposed a new idea of matrix factorization for multi-level ordinal rating matrix to handle the overfitting problem. They draw inspiration from a current work on proximal support vector machines wherein two parallel hyperplanes are used for binary classification and points are classified by assigning them to the class corresponding to the closest of two parallel hyperplanes.

A confidence-weighted bias model (CWBM) for online collaborative filtering (OCF) was provided by Zhou, et al. [9] to realize real-time updates of recommendation results, and to enhance the stability and accuracy of OCF. They applied the proposed approach on two real-world data sets, Movie-Lens100 K and MovieLens1 M. The results showed that the proposed approach not only achieve lower RMSE values but also simultaneously more stable than the other baseline approaches.

3 Proposed System

The proposed approach utilizes on two different datasets, 20 M benchmark dataset of Movie Lens consisting of 100,000 ratings, users can tag to a movie or also rate to a movie on a range of 1 to 5 [10]. while 30 M Book-Crossings consisting of 1.1 million ratings of 270,000 books by 90,000 users. The ratings are on a scale from 1 to 10 [11].

The Recommender System has two main modules, ALS Module and improved of ALS Module as shown in Figs. 1 and 2 respectively.

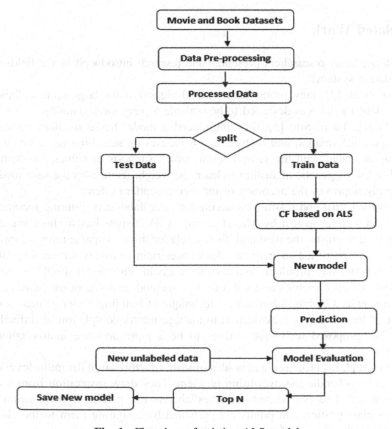

Fig. 1. Flowchart of existing ALS model.

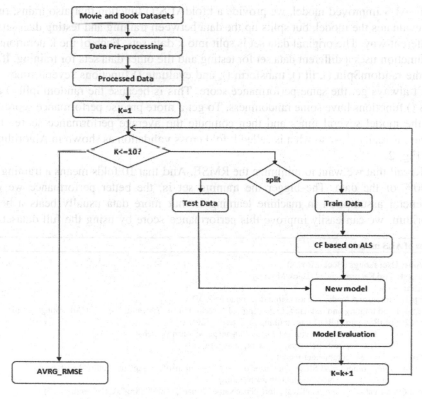

Fig. 2. Flowchart of improved ALS model.

In exiting ALS model, the first thing we'll do is import requisites packages. The regression evaluator will allow us to measure the performance of ALS model. The train validation split and param grid builder allow us to cross validate and fine-tune the hyper parameters of our model and of course we'll want to import the ALS algorithm. From here we'll split our data into training and test sets and then build the ALS model specifying the user, item and rating columns. We set cold start strategy to drop and the non-negative argument to true because we don't want it to return any negative predictions. After this we'll build our parameter grid using/am grilling can fine- tune or model, the hyper parameters we can tune include the U and P metrics max iteration which tell spark how many times to alternate between U and P when minimizing the error and the regularization parameter to prevent ALS from overfitting to the data well set our evaluator to our RMSE tell spark which column contains our labels and then what we wanted to name the output production column, then we'll build out the cross validator and tell it to use our ALS model the parameter grid and the evaluate which we just built once that's complete. We'll fit model through the training data once it finds the best combination of parameter from our parameter it will have spark take that model which is called the best model.

In ALS improved model, we provide a kfoldALS () function that also trains, runs, and evaluates the model, but splits up the data between training and testing data sets in a different way. The original data set is split into k data sets. Each of the k iterations of the function uses a different data set for testing and the other data sets for training. If we run the randomSplit (), fit (), transform (), and evaluate () functions several times, we won't always get the same performance score. This is because the randomSplit () and ALS () functions have some randomness. To get a more precise performance score, we run the model several times and then compute the average performance score. This process is really close to what is called k-fold cross validation as shown in Algorithm 2 and Fig. 2.

Recall that we want to minimize the RMSE. And that 10 folds means a training set of 90% of the data. The bigger the training set is, the better performance we get. A general assumption in machine learning is that more data usually beats a better algorithm. we can easily improve this performance score by using the full dataset.

Steps of ALS model

Input: User ratings (Book/Movies)
Output: Top Recommended (Book/Movies).
Begin:
Step 1: Parsing and loading datasets into dataframe or RDD
Step 4: Load training and test dataset into (int(Col "userId"), int (Col "itemId"), float(Col "rating")) tuple
Step 2: SplitRandom () RDD into (training [80], testing[20])
          training = sc.trainingfile('*https://*').map(parse_rating).cache()
          test = sc.testfile(' *https://* ').map(parse_rating)
Step 5: Train the recommender model.
          New_model = ALS(userCol="userId", itemCol="itemId", ratingCol="rating").fit(ratings)
          predictions = New_model.transform(ratings)
Step 6: Evaluator = RegressionEvaluator(metricName="rmse", Col="rating", Col="prediction")
          print ("RMSE for new model is: " str(evaluator.evaluate(predictions.na.drop())))
Step7: Adding new user ratings
Step8: Display top N recommended(Book/movies).
Step9: Save New_model

Algoritm1: Existing ALS Model

Steps of Improved ALS model

Input: User ratings (Book/Movies)
Output: Top Recommended (Book/Movies).
Steps:
Step1: Parsing and loading datasets into RDD
Step2: def kfoldALS

```
            k = 10
            evaluations = []
            weights = [1.0] * k
            splits = data.randomSplit(weights)
            for k = 1 to i [i is the number of flods]
                testingSet = splits[n]
                trainingSet = spark. createDataFrame(sc.emptyRDD(), data. schema)
                    for k = 1 to j [j is the number of flods]
                        if i == j:
                            continue
                        else:
                            trainingSet = trainingSet.union(splits[j])
                    als= ALS (userCol=userCol, itemCol=itemCol, ratingCol=ratingCol)
                    New_model = als.fit(trainingSet)
                    predictions = New_model. transform(testingSet)
                    evaluator = RegressionEvaluator (metricName=metricName, labelCol="rating", predictionCol="prediction")
                    evaluation = evaluator.Evaluate(predictions.na.drop())
                    print ("Loop " + str(i+1) + ": " + metricName + " = " + str(evaluation))
                    evaluations. Append(evaluation)
            return sum(evaluations)/float(len(evaluations))
```

Step3: Adding new user ratings
Step4: Display top N recommended (Book/movies).
Step5: Save New_model

Algoritm2: Improved ALS Model

4 Experiments and Result

All the experiments were performed on Ubuntu 16.04 operating system running on Intel® Core™ i5-2400 CPU @ 3.10 GHz × 4 processor as well as a hard disk of 500 GB. We used the latest released Apache Spark.2.3.0, Python 3.5 for the purposes of all the experiments.

The subsections of Experiments and Result are organized as follows: Apache spark introduced in Sect. 4.1. Evaluation is given in Sect. 4.2. Finally, a result is given in Sect. 4.3.

4.1 Apache Spark

Spark is an Apache software foundation open source project, it's a flexible in-memory framework that allows handling batch and real-time analytics and data processing workloads. Apache Spark is a fast and general- purpose cluster computing system for large-scale data processing. The main idea behind Spark is to provide a memory abstraction which allows us to efficiently share data across the different stages of a map-reduce job or provide in-memory data sharing.

At a high level, every Spark application consists of a driver program that runs the user's main function and executes various parallel operations on the worker or processing nodes of the cluster. The main memory abstraction that Spark provides is of a Resilient distributed dataset (RDD), which is a collection of elements partitioned across the nodes of the cluster that can be operated on in parallel. RDDs can be created from a file in the file system, or an existing collection in the driver program, and transforming it. So as the name suggests, the data from a file in the file system or from an existing collection in the driver program that forms a RDD is partitioned and distributed across the worker or processing nodes in the cluster, thereby forming a distributed dataset. The important point to remember is that RDDs are immutable distributed datasets across the cluster and are generated using the coarse-grained operations i.e. operations applied to the entire dataset at once. We can persist an RDD in-memory, allowing it to be reused efficiently across parallel operations or different stages of a map-reduce job. So, the reason why Spark works so well for iterative machine learning algorithms and interactive queries is that, instead of sharing data across different stages of the job by writing it to the disk (which involves disk I/O and replication across nodes) it caches the data to be shared in-memory which allows faster access to the same data [15] (Fig. 3).

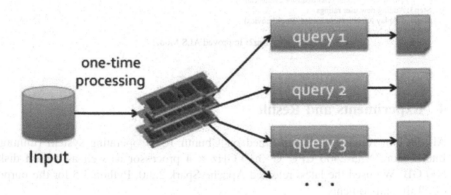

Fig. 3. Function of apache spark [15]

For fault tolerance, Spark automatically records how the RDD is created i.e. the series of transformations applied to the base RDD to form a new RDD. So, when the data is lost, it reapplies the steps from transformations graph to rebuilt the RDD or lost data. Generally, only a piece of data is lost when a machine fails, and so RDD tracks the transformations at machine level and recomputes only the required operations or a part of transformations on the previous data to perform recovery.

Its high- level APLs in Java, Scala, Python and R. Spark ecosystem consist of some components including Spark Core, Spark SQL, Spark Streaming, Spark Machine learning and GraphX, as shown in Fig. 4.

Spark core is the based engine for large-scale parallel and distributed data processing. it is responsible for memory management and fault recovery, scheduling, distributing and monitoring jobs on cluster as well as it responsible for interacting with storage system. Spark SQL used for structure data. It can run unmodified hive queries on existing Hadoop deployment. Spark streaming enable to analysis and interactive apps for live streaming data. Machine learning libraries begin built on top of spark. Machine learning classified into two classes of algorithms: supervised algorithm used labeled data and output are provides to algorithm, and unsupervised algorithm don't have the output in advance. Graph computation engine (similar to graph) combines data-parallel and graph-parallel concepts [3, 14].

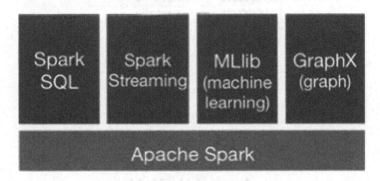

Fig. 4. Spark ecosystem

4.2 Evaluation Metrics

The most popular used metric in recommender system is Root Mean Squared Error (RMSE). In this paper we have used RMSE to measure the accuracy of the result as Eq. (1)

$$RMSE = \sqrt{\frac{1}{n} \sum_{\{i,j\}} (u_{i,j} - r_{i,j})^2} \tag{1}$$

Where $u_{i,j}$ is the predicted rating for user i on item j, $r_{i,j}$ is the actual rating again, and n is the total number of ratings overall users.

4.3 Results

In this paper, the improved ALS model is applied on two different datasets for various values of Kfolds and the RMSE as shown in Table 1 and Fig. 5. The improved ALs model is compared with conventional ALS model based on Average RMSE. From the results, we conclude that the proposed improved ALS model gives better performance in compared with conventional ALS.

Table 1. Compute the average performance score for 10 folds of RMSE of improved ALS model for book and movielens datasets

kfolds	RMSE	
	Movielens dataset	Book crossing dataset
1	0.90415955	4.355162
2	0.90831975	4.386369
3	0.89108102	4.406535
4	0.90200572	4.376952
5	0.90527143	4.413179
6	0.90132191	4.402714
7	0.88968362	4.425357
8	0.91184825	4.380932
9	0.90564973	4.40496
10	0.91061087	4.38408
Avrag	**0.902995185**	**4.393623848**

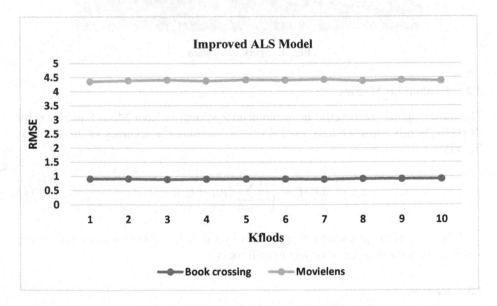

Fig. 5. Improved ALS model for book and movielens datasets.

Table 2. Comparison between ALS and improved ALS based on RMSE.

Type of dataset	ALS model	Improved ALS model
Movielens	0.956353411	0.902995185
Book crossing	6.65212705	4.393623848

5   Conclusion

Recently, Recommender Systems (RS) have become very important research topic. These system categories into three main types; Collaborative filtering, Content-based filtering, and Hybrid filtering. The techniques that are widely used in recommender systems are collaborative filtering. The main problems of collaborative filtering are sparsity of data, scalability, and cold start. Recently, Big Data Analytics has Effectiveness to resolve those problems. In this article, we proposed a new method that applied to two different datasets, Movielene and Book-crossing on big data platform which called apache spark to improve the performance of ALS model. we compared it with the existing model as shown in Tables 1 and 2. We evaluated the model by using Root Mean Squared Error (RMSE) metrics.

References

1. Cacheda, F., Carneiro, V., Fernández, D., Formoso, V.: Comparison of collaborative filtering algorithms: limitations of current techniques and proposals for scalable, high-performance recommender systems. ACM Trans. Web (TWEB) 5(1), 2 (2011)
2. Verma, J.P., Patel, B., Patel, A.: Big data analysis: recommendation system with Hadoop framework. In: 2015 IEEE International Conference on Computational Intelligence & Communication Technology (CICT) (2015)
3. https://spark.apache.org/docs/latest/. Accessed 10 Mar 2018
4. Zhou, Y., Wilkinson, D., Schreiber, R., Pan, R.: Large-scale parallel collaborative filtering for the netflix prize. In: Fleischer, R., Xu, J. (eds.) AAIM 2008. LNCS, vol. 5034, pp. 337–348. Springer, Heidelberg (2008). https://doi.org/10.1007/978-3-540-68880-8_32
5. Wang, Q., Yuan, X., Sun, M.: Collaborative filtering recommendation algorithm based on hybrid user model. In: 2010 Seventh International Conference on Fuzzy Systems and Knowledge Discovery (FSKD), vol. 4. IEEE (2010)
6. Ma, X., et al.: An explicit trust and distrust clustering based collaborative filtering recommendation approach. Electron. Comm. Res. Appl. 25, 29–39 (2017)
7. Kumar, V., Pujari, A.K., Sahu, S.K., Kagita, V.R., Padmanabhan, V.: Collaborative filtering using multiple binary maximum margin matrix factorizations. Inf. Sci. 380, 1–11 (2017)
8. Kumar, V., Pujari, A.K., Sahu, S.K., Kagita, V.R., Padmanabhan, V.: Proximal maximum margin matrix factorization for collaborative filtering. Pattern Recognit. Lett. 86, 62–67 (2017)
9. Zhou, X., Shu, W., Lin, F., Wang, B.: Confidence-weighted biasmodel for online collaborative filtering. Applied Soft Computing (2017)
10. https://grouplens.org/datasets/movielens/. Accessed 10 Mar 2018
11. http://www2.informatik.uni-freiburg.de/~cziegler/BX/. Accessed 10 Mar 2018
12. Herlocker, J.L., Konstan, J.A., Terveen, L.G., Riedl, J.T.: Evaluating collaborative filtering recommender systems. ACM Trans. Inf. Syst. (TOIS) 22(1), 5–53 (2004)
13. https://spark.apache.org/docs/2.2.0/ml-tuning.htm
14. Xie, L., Zhou, W., Li, Y.: Application of improved recommendation system based on spark platform in big data analysis. Cybern. Inf. Technol. 16(6), 245–255 (2016)
15. https://www.kdnuggets.com/2015/06/introduction-big-data-apache-spark.html. Accessed 10 Apr 2018

Big Data Analytics

Privacy Preserving and Auto Regeneration of Data in Cloud Servers Using Seed Block Algorithm

Aansu Nirupama Jacob[✉], B. Radhakrishnan, S. Deepa Rajan,
and Padma Suresh Lekshmi Kanthan

Department of Computer Science, Baselios Mathews II College of Engineering,
Kollam, India
aansunirupamajacob94@gmail.com

Abstract. Cloud storage is nowadays trending in case of storage and the cloud service providers (CSP) provides varieties of offers for organisations and made the resources available online. Data integrity checking and regeneration of failed servers nowadays becomes critical and we need to safeguard the delegated data in cloud storage against corruptions. It will become the major delinquent towards the cloud servers. Surviving paper works offers only private auditing and requires data owners to always available and handle auditing, and repairing which may sometimes be unrealistic. Our work-scheme focuses with regeneration and for that cloud user's first produce their secret keys and the TPA audits the files, in case of file corruption then forwards to a proxy agent by sharing partial private key and regenerates the corrupted files. Here we uses the seed block algorithm for the purpose of regeneration and also introduces an additional agent for auto regeneration, which reduces the downloading time related issues.

Keywords: Cloud storage · Regeneration · Seed-block · Privacy preserving

1 Introduction

Cloud computing is believed to be an ultra-fascinating storage of meta-datas with incredible features where data is kept in virtualised lochs of storage, typically laid on by an arbitrator. The incredible features with cloud fascinates the customers to utilize the cloud and to store their information's with the profits of cost saving, mobility and more over scalable services as it is a prototype of networked enterprise storage.

Cloud storage is nowadays trending in case of storage and the cloud service providers (CSP) provides varieties of offers for organisations and made the resources available online. It provides the customers a flexible data subcontracting services with the profits of relief on online burden, scalable access of data with independent location and evasion of capital outflow on hardware-software as well as on personal maintenance, etc., However these sometimes causes new security threats towards the user data and feels hesitant with the cloud storage. The top cloud security threats includes data spill, data loss, insider harm, denial of service attacks, untrusted APIs, etc.

© Springer Nature Singapore Pte Ltd. 2018
I. Zelinka et al. (Eds.): ICSCS 2018, CCIS 837, pp. 315–325, 2018.
https://doi.org/10.1007/978-981-13-1936-5_34

According to the enquiry of cloud security alliance from 10 years of record, data loss has been scaled from 25% to 68%. Some more technologies are there for backups and recoveries for failed servers. Sometimes the servers itself freezes and lost their backup copies. In other situations the main data server crashes and the information contained within is lost data and can't be recovered, so the cloud is not free from technology failure or human error. The most common aspects for the data loss in cloud includes

- Accidental error or human error: It will happens accidently or from the side of a careless user. It may be intentionally or unintentionally.
- Over-writing of data: It is possible for users or applications to post incorrectly. Software-as-a-Service (SaaS) applications may cause massive data loss. These apps store large data sets that are constantly updated or built-in. This new information has the ability to overwrite the old information, resulting in data sessions for partially overwriting the process.
- Malicious Performance: Most cloud storage providers, their networks, and databases are trying to maximize the strength and security of data, but not all attacks can be prevented.

The existing remote testing methods for the regeneration-coded data provide private auditing and for that data owners need to be online so that the auditing can be done. If any issues arises, it should have to be resolved. The study focuses mostly on PDP (Provable Data Possession) model and POR (proof of retrievability) model for single-server consequences. They are designed for a private audit and the holder of the data owner is allowed to check the integrity and correct the errors. The current system was slow and having some problems with:

- There is no method or procedure to check whether the file is corrupted or not.
- Whenever a user downloads a file, there might be a chance for the system to hang due to the file scan or in case of non-availability of file. There is no way to overcome it.
- Files that are normally uploaded directly into the cloud on the current system are not using any cryptographic strategies for security.
- Auditing and settlement in the cloud are both outstanding and cost-effective to the customers when it comes to extra size of outsourcing data and user's limited resource capabilities.

As cloud computing is a fastest growing technology, it is used by many major applications across the Internet. We inspire public auditing of data storage security in cloud computing and provide the Privacy-Maintenance Auditing Protocol, which means an external auditor supports the scan to audit user's outrage data in the cloud without consulting the data content. Here we uses Seed Block Algorithm (SBA) for the file regeneration process with an advanced encryption scheme and for better performance we introduces a new agent for the purpose of auto-regeneration. The new agent automatically regenerates the missing files in the cloud and hence reduces the downloading time issues.

2 Related Works

We shortly discuss the recent and closely related works here. Considering that files are barred and repeatedly stored over multi-servers or multi-clouds. [1] Chen proposes a framework for checking the integrity of regenerating-coded data tenuously and then outfits data integrity protection (DIP) method. DIP plays an ordinary role in the cloud server managers which uses for checking the data integrity. The protection of regenerated data will be must and outcome is provided using DIP. [2] Bowers announces about a framework called HAIL (High-Availability and Integrity Layer), a distributed storage. This framework scheme licenses the servers for proving to a client that a warehoused metadata file is intact and retrievable in many other complications. It mainly evaluates the performance of how when and why, the HAIL advances on the security and proficiency of existing tools, like PORs prearranged on each and every servers. [3] Zou, manners a work about data handling and data storage security. It mainly concentrates on privacy preserving with highly secured data. Data handling can be done for preserving the data or be managed privately. Finally we deliberate the currently used regeneration methods in cloud server domain. There are collections of procedures presently such as HSDRT, PCS, ERGOT, Linux Box, etc. for the file recovery process. They are currently doing an important part of data recovery from the failed servers. [4] Parity Cloud Service Technique-(PCS) is a frank and straight. It is simple to exploit and more helpful for information recuperation. Sometimes it will get corrupted with errored parity bits. It is the one of the method of regeneration of files and uses the parity bit information. This framework is more reliable and efficient in storage cost. PCS is forwarded with the concept of parity bit information. [5] Efficient Routing Grounded on Taxonomy (ERGOT) provides an efficient way for data retrieval in distributed computing. It fails to focus on time complexity. It is highly efficient in precise matching retrieving policy. This work performs efficiently but downs in time related matters. It provides the incredible way of data retrieval from the failed servers (Fig. 1).

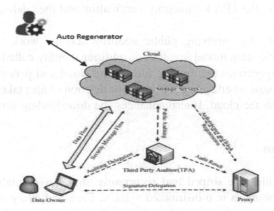

Fig. 1. Architecture of the public auditing model with auto regenerator in cloud computing

3 Problem Formulation

3.1 Architectural Model

Let us consider the privacy preserving public auditing scheme with auto regenerator in cloud computing. It encompasses mainly six entities:

- **Cloud Server**: It provides storage services which are managed by the cloud service providers (CSP) or cloud authorities. They promises to have a significant computational properties.
- **Data Owner**: Person who owns large amount of metadata files, to be stored in the cloud server. After the metadata file upload procedure, the data owner may become off-line. There is no serious issue in it.
- **Third Party Auditor (TPA)**: Familiarity and capability to conduct public audits on the coded data (perfectly and carefully encrypted) in the cloud. TPA is a stable body and its audit results is impartial for both the entities i.e., data owners and cloud servers.
- **Client**: Client is the one who downloads or shares or uses the information in the cloud with access permissions granted by the data owner. Data Owner can grant or block the access permission for data retrieval.
- **Proxy Agent**: An agent who is partially reliable and acts on behalf of the data owner for regenerating the missed files or corrupted metadata files on the failed servers during the repair procedure. The proxy, who would always be online. So that it will be much more authoritative than the data owner but less than the cloud servers in terms of computation and memory capacity.
- **Auto Regenerator**: An entity which automatically regenerates the missing files in the cloud and periodically checks the cloud in order to know whether any corruption or deletion of files takes place and hence acts as a checker in the cloud. It also enhances the downloading time issues.

To safeguard the metadata files or to secure the metadata files from the online burden hypothetically brought by the periodic auditing and deliberate repairing. The data owners remedy the TPA for integrity verification and then delegate the reparation to the proxy.

Compared with the surviving public auditing scheme work with regeneration system model, our system model involves an additional entity called auto regenerator. It automatically regenerates the missing files in the cloud and periodically checks the cloud in order to know whether any corruption or deletion of files takes place and hence acts as a checker in the cloud. It also enhances the downloading time issues.

4 Formulation

Considering that files are striped and recurrently stored over multi-clouds i.e., 'n' servers or clouds same as in a distributed system. Each and every server is a storage location and is self-governing over other servers. Stripping of data files and then finally storing in a distributed way improves the security of data stored in the cloud. No leakage happens as it is handled carefully by masking with a PRF (Pseudorandom Function). The method of masking with a pseudorandom function is a light-weight process and also preserves the data privacy.

Here we uses the seed block algorithm for the regeneration purpose. The framework approach recouples the stripped data over 'n' servers, to retrieve the data effectively in case of document erasure or if the cloud gets crashed because of any reason. SBA uses exclusive OR (XOR) operation for computation. For example, Consider two data files a_1 and a_2, $a_1 + a_2$ produces F. When a file may be destroyed or deleted and if we want to retrieve that file, it can be done using X-OR of file F and a_2.

$$\text{i.e. } a_1 = F \text{ (XOR) } a_2 \tag{1}$$

In the framework, cloud user gets an exclusive id as well as a set of arbitrary numbers in the cloud while registering on cloud. Whenever the data owner registers in the cloud, the exclusive owner id and set of arbitrary numbers produced in the cloud get X-ORed and then generates the seed block for that particular client. The generated seed block correspond to each data owner is stored in servers.

When the data owner uploads, the file it will be stored in the main cloud. When it gets stored in main cloud, that file being X-ORed with seed block of that particular owner. The X-ORed file is stored in servers. In case, if file corrupts in main cloud, user can get original file by X-ORing that file with seed block of particular owner to get the original uploaded file.

Seed Block Algorithm
Initialization: Main Cloud: C;
Remote Cloud: S;
Clients of Main Cloud: Ci;
Files: f1 and f2
Seed block: Si;
Random Number: r;
Owner's ID: Owner-id;
Input: f1 created by Ci;
          r is generated at C;
Output: Recovered file f1 after deletion at C;
Given: Authenticated clients can do uploading,
          Downloading and modification on its
          Own files.
Step 1: Generate a random number as;
          Int r = rand ();
Step 2: Create a seed Block Si for each Ci and
          Store
          Si at S; Si = r $\oplus$ Owner-id (For all
          Owner)
Step 3: If Ci/Admin create/modify f1 and
          Stores at C, Then f1 create as f1 = f2 $\oplus$ Si;
Step 4: Store f1 at S;

Step 5: *If server crashes f1 deleted from C;*
  Then, (original file) f1 = f2 ⊕ Si;
Step 6: *Return f1 to C1;*
Step 7: *END*

4.1 Encryption and Decryption Using Advanced Encryption Standard (AES) Algorithm

Advanced Encryption Standard (AES) Algorithm contains three cryptographs such as AES-128, AES-192 and AES-256. Symmetric ciphers uses the same key for encrypting and decrypting. A round consists of several processing steps that include substitution, exchange and mixing of the input plaintext and transform it into the final output of ciphertext. Each cryptogram encrypts and decrypts the metadata in block of 128 bits using cryptographic secrets of 128, 192,256 bits respectively.

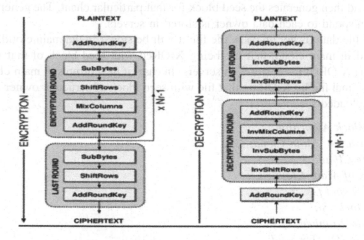

Fig. 2. AES encryption and decryption

4.2 Basic Operations in Formulation

Our goalmouth is to amplify the basic operations that takes place in the scheme for the process of auto-regeneration of files in the cloud such as file uploading, file verification, file downloading, file regeneration and an additional operation of Auto regeneration.

Upload Operation: We first describe how we upload a file F to the servers.

Step 1: Generate the Secrets Keys and Split the File into Equal Chunks: Before uploading the file F, we produce secrets-keys i.e., K_{ENC}, K_{PRF}, K_{Hash}, etc. Splitting considers the idea of maximum distance separable (MDS) codes. An MDS code is well-defined by the limits (n, k), where k < n always. Each chunk has size of

$$\text{i.e., } |F|/k \, (n - k) \tag{2}$$

with a file of size |F|.

Step 2: Encrypt Each File Chunks and Compute the Hash Value of Each File Chunks: Consider the code chunk and apply advanced encryption standard (AES) to the chunks for security. Consider code chunks and then we computes the hash value using Secure Hash Algorithm (SHA), its enhanced version called SHA-1 (Fig. 2) method. It has the ability to compress a fairly lengthy message and create a short message abstract in response.

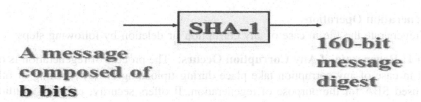

Fig. 3. Hashing of data using SHA-1

We stores the hash values of the uploaded file in the main server. Another hashing takes place in the cloud space. Both hash values get compared for checking the integrity of files received in the cloud.

Step 3: Update the File Using Seed Block Algorithm and Upload: We uses seed block algorithm in this section for the regeneration purpose. When the owner registers in main cloud an owner id and random number will be created initially, get XORed to generate seed block for that particular owner and stored in remote servers. Then we upload the encrypted code chunks to their respective servers.

Verification Operation: In the verification operation, we verify the rows of bytes of the currently stored chunks in the cloud servers.

Step 1: Verify the Metadata File and Check the Integrity of Each File Chunks: Download a copy of the encrypted metadata from each and every server and check that if all copies are alike. Integrity checking can be done by comparing the hash values using a temporary storage like list, dictionary etc. It works normally and compares the hash value of files in the cloud and files in the main server. If any inconsistence on comparing occurs there might be chance for data corruption.

Step 2: Error Localisation and Trigger the Regeneration Process: On comparing the hash values and gets inconsistent results then we know that some are get corrupted. Then localize the corruption and goes for regeneration process. If a server has more than a user specified number of bytes marked as corrupted, we consider it a failed server and trigger the Regeneration operation.

Download Operation
We download a file F from the servers as follows:

Step 1: Verify and Decrypt the Metadata File: Here the auditor verifies the file to check whether it promises data integrity. During verification, auditor compares the hash values and verifies the integrity. After verification, it decrypts the files using AES algorithm for decryption.

Step 2: Download the File: To rebuild the file |F|, we download k (n − k) chunks from any k servers and verifies the file integrity. Then recombines the file chunks using the function StringBuilder

$$i.e.,\ F\ (n,\ k) = [k\ (n - k)]\ StrBr \tag{3}$$

Regeneration Operation
We regenerate the file in case of any corruption or deletion by following steps:

Step 1: Regenerate, if Any Corruption Occurs: The process of regeneration is only used in case of any corruption take place during uploading or downloading or other. We used SBA for the purpose of regeneration. It offers security, integrity, confidentiality, trustworthiness, and cost efficiency.

Step 2: Download the Regenerated File F': In case of any corruption in files, we use the data recovery process and verifies the regenerated file and then downloads the file.
  For regeneration process,

$$i.e.,\ F' = Regen\ (|F|)\ such\ that\ f_1 = F\ (XOR)\ f_2 \tag{4}$$

Auto Regeneration Operation
This operation performs periodically with an additional trusted agent and regenerates the missed files and hence reduces the downloading time issues.

Step 1: Auto Regenerate the Missing Files in the Cloud: Auto regeneration means regenerating the file automatically with an additional trusted agent which reduces the downloading time between the normal regeneration process and the failures in cloud.

Step 2: Verify Periodically the Cloud Server: Verification in auto regenerator takes place periodically with offline auditor and data-owner. So that it promises high data security and integrity.

4.3 Cloud Storage Security Protocol

In addition with the cryptographic primitive, we needs a secure cloud storage protocol for the purpose of credentials exchange and keeps them secure in an untrusted network. We suggests Secure Socket Layer Protocol that helps to avoid the security challenges when exposed directly on the internet.

5 Evaluation

We evaluate the storage cost analysis and Mean Time To Failure (MTTF) during a particular time period overhead of using seed block algorithm for regeneration of codes and an additional entity of auto regeneration. The goal of our scheme work is to understand the file overhead of our regeneration method over the existing regeneration methods in the cloud, which ordinarily is based on seed block construction.

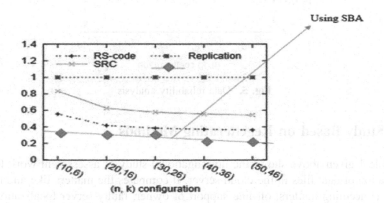

Fig. 4. Analysing the download-waiting time

5.1 Download-Wait Time Issue Analysis

Figure 3 above shows as that the waiting time of both the RS-code, the SRC and using SBA decreases as n grows in case (n − k) is stable. The stabilised budget of SBA is 0.68, the SRC is 0.54 and that of RS-code is 0.36 when (n, k) raises to (60, 56). If we use larger values of f, the cost of SBA's will auxiliary reduce, but at the cost of slower repair. When (n, k) raises to (60, 56), the stabilised budget of SRC is 0.54, SBA is 0.68 and that of RS-code is 0.36.

5.2 Data Reliability Analysis

Figure 4 above shows that the data reliability analysis. It mainly shows how it depends with other systems. This is more consistent and can be observed that the reliability of SBA's is much higher. Even for the high rate (60, 56) circumstance, SBA's are several orders of magnitude and become more reliable. The high repair speed of SBA's is highly beneficial and helpful for cloud users. Our work- scheme is highly reliable when compared to other methods (Fig. 5).

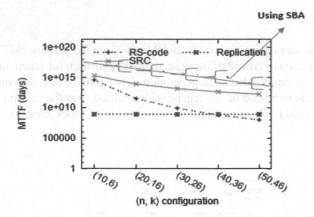

Fig. 5. Data reliability analysis

6   A Study Based on Regenerating Methods

The Table 1 given above shows the investigational study of assessment work for auto regeneration of data files in the cloud server. It compares the matters like auditability, privacy concerning matters, off-line support of owner, faulty server localization time, regeneration security, data reliability and storage cost. The table effectively shows that our scheme is better than our existing schemes in case of regeneration methods.

Table 1.   An investigational study

Comparison of different audit schemes for regeneration code based cloud storage				
Items	B. Chen	Henry Chen	Jian Liu	Ours
Public auditability	No	No	Yes	Yes
Privacy preserving	Yes	Yes	Yes	Yes
Owners off-line support	No	No	Yes	Yes
Faulty server localization time	0(1)	$O(C_n(n - k) * I)$	0(1)	0(1)
Regeneration data security	No	No	No	Yes
File downloading and regeneration time	High	High	High	Less
Storage cost	High	High	Little high	Affordable

7   Conclusion

Our work focuses mainly on regeneration methods with public auditability and preserving the privacy in the cloud, where original data can be protected by splitting, encrypting and masking with PRF's and then stored in a distributed way. To better the regenerating-code-scenario, we uses seed block algorithm for the repairing of missed files and an auto regenerator for enhancing the downloading time. The wide range investigation shows that our scheme is provable confident and the evaluation experiment shows our work scheme is highly efficient and can be attainably integrated into a regenerating system with code-based cloud storage.

References

1. Chen, H., Lee, P.: Enabling data integrity protection in regenerating coding-based cloud storage: theory and implementation. IEEE Trans. Parallel Distrib. Syst. **25**(2), 407–416 (2014)
2. Bowers, K.D., Juels, A., Oprea, A.: Hail: a high-availability and integrity layer for cloud storage. In: Proceedings of the 16th ACM Conference on Computer and Communications Security. ACM, pp. 187–198 (2009)
3. Zou, D., Xiang, Y., Min, G.: Privacy preserving in cloud computing environment. Secur. Commun. Netw. **9**(15), 2752–2753 (2016)
4. Song, C.W., Park, S., Kim, D.W., Kang, S.: Parity cloud service: a privacy-protected personal data recovery service. In: International Joint Conference of IEEE TrustCom-11/IEEE ICESS-11/FCST-11 (2011)
5. Pirro, G., Trunfio, P., Talia, D., Missier, P., Goble, C.: ERGOT: a semantic-based system for service discovery in distributed infrastructures. In: 10th IEEE/ACM International Conference on Cluster, Cloud and Grid Computing (2010)

Secure Data Deduplication and Efficient Storage Utilization in Cloud Servers Using Encryption, Compression and Integrity Auditing

Arya S. Nair$^{(\boxtimes)}$, B. Radhakrishnan$^{(\boxtimes)}$, R. P. Jayakrishnan$^{(\boxtimes)}$, and Padma Suresh Lekshmi Kanthan$^{(\boxtimes)}$

Department of CSE, Baselios Mathews II College of Engineering,
Sasthamkotta, India
aryanair12@gmail.com, radhak77@gmail.com,
jayakrishnanrp@gmail.com, suresh_lps@yahoo.com

Abstract. A burning issue of recent times is the concern regarding the management of huge volume of data. Since the local devices has its limits in storage, there was an urgent demand for a new technology The emergence of cloud-based storage satisfied all these needs of user. Cloud-based storage safely handles these huge data. But even cloud has its own demerits. Clients do not yet have the ability to test the integrity of cloud files. Another problem is that duplicate files takes up a great deal of storage space in cloud. Our works focusses on eliminating duplicates and auditing the integrity of files. We use SecCloud+ that offers an added advantage ie encrypting previous to uploading. We also modify the system by introducing a compression procedure following encryption. As a result, cloud space is saved to a great extent.

Keywords: Deduplication · Encryption · Integrity auditing · Compression

1 Introduction

Cloud based storage is judged to be one among the widely appreciated technique for storing data at recent times. This facility can remotely store massive units of data. This data is handled by third party devices. This feature knocks out the need of purchasing and developing individual infrastructures for storage, thus avoiding the installation and maintenance cost involved with it. Although numerous features are provided through backup services online, cloud based storage still fails to meet some urgent needs of users. Cloud clients do not yet have the ability to test the integrity of cloud files. Another problem is that duplicate files takes a great deal of storage space in cloud. Among these remotely stored data, most of them are repeated: according to a last review by EMC, 75% of current digital data is duplicated copies. Such diplomas and dudiclines are still an unsolved problem. Data Integrity is also an indispensable feature of cloud based storage. As data is carried via internet and stored in remote location, it is susceptible to many varieties of security threats like intruders placing backdoor on storage; altering permissions, editing files while leaving the server set up to function

© Springer Nature Singapore Pte Ltd. 2018
I. Zelinka et al. (Eds.): ICSCS 2018, CCIS 837, pp. 326–334, 2018.
https://doi.org/10.1007/978-981-13-1936-5_35

normally without any defect. By setting up periodic scan and auditing, the user can be notified within a short span time span about any corruption in the files that is stored in the cloud. This helps to reduce the anxieties of user about the integrity of data that is stored in the cloud.

Data saved in cloud hold a large fraction of duplicates. In this scenario, the need for reducing cloud data emphasizes the importance of minimizing storage and cost. Lowering data volume is possible by deduplication. When the new data arrives in cloud, it will be compared with the existing information and discarded if it is previously present in cloud, thus avoiding all non-unique blocks.

The method consists of several steps like partitioning input into blocks, computing the hash of each individual block, observing if an already stored block contain same value of hash and substituting duplicate data that denotes the object stored in the database with its reference. An index is created after dividing the input to blocks. Only a particular event of every kind is protected and saved in the index. This paper discusses the two security techniques SecCloud+ and SecCloud, aimed at obtaining deduplicated content along with data integrity in cloud.

SecCloud includes three protocols

- File uploading protocol
- Integrity auditing protocol
- Proof of ownership protocol

SecCloud+ also uses the same protocols but the protocol used for uploading file in SecCloud+ needs an additional step for communicating between cloud client and server ie for obtaining the convergent key to encrypt the uploading file. SecCloud+ offers an added advantage ie encrypting previous to uploading. Confidentiality is maintained through this method. To bring down the expenses of storage of large quantity of data, we put forward a compression procedure following encryption.

Apart from reducing the expenses, the compression results in minimal usage of cloud space. After dividing the file into partitions of equal size a randomly key is generated to encrypt these partitions. Once encrypted, compress the content prior to uploading. The retrieval process comprises decompression followed by decryption. The decrypted blocks are properly united with the help of MapReduce cloud in order to get the original file.

2 Related Work

In this section we briefly discuss the existing integrity auditing methods along with the deduplication. In 2007 Ateniese designed first framework for public auditing which is called Provable Data Possession (PDP). The PDP apply homographic tags in RSA algorithm. This method does not support privacy preserving and batch auditing. Moreover, there is considerable overhead associated with communication. So later Ateniese enhanced the model to a partially dynamic one. This partially dynamic PDP support dynamic auditing. The next work supporting integrity of data is proof of retrievability (POR) which aims at assuring the full recovery of the mentioned file in addition to confirming its presence in cloud server. The recovery option makes is more competent as compared to PDP.

Wang et al. enhanced the POR schema by introducing a Merkle hash tree structure that facilitates block tag-based authentication. This tree structure is used for deduplicating text in the client side. Xu and chang further enhanced the POR framework by commitment of polynomial which evidently reduced the cost involved in communication. In order to minimize the computational overhead.

Li et al. proposed a new architecture for cloud storage that comprises of two independent cloud servers. Assaurate et al. [3] developed new POR protocol by combining Privacy preserving word search algorithm with the insertion in the data segments of the randomly generated short bit sequence. Li et al. [4] introduced a key-disperse model that was used to solve a complex number of keys in convergent encryption. Private data duplication was first proposed by Ng et al. The new idea complemented the works of Halevi et al. [5] on public data duplication.

Bellare introduced message locked encryption which was widely appreciated as it guaranteed space efficient storage. Future works focused on the deduplication of encrypted data. keelveedhi et al developed a dupLESS system in which a client make use of a pseudorandom protocol. All the previously stated works solves the issues related with either deduplication or integrity auditing. This paper simultaneously solves both the problems.

3   System Model and Problem Formulation

3.1   System Model

The four main modules in the system are

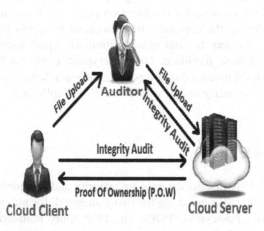

Fig. 1. System architecture for deduplication and integrity auditing

- Administrator - has the ultimate authority to access the system. The administrator regulates the work of cloud server, cloud client as well as the auditor.

- Cloud client - They are either individual or commercial organization. Prior to uploading they perform duplicate check ie checking if such a file is already present in cloud storage area. If there is duplicated content, another protocol called proof of ownership will be run between the client and cloud store-age server.
- Auditor - he or she helps clients upload and audit their outsourced data, maintains a MapReduce cloud and acts like a certificate authority. The auditing entity make use of a set of public and private keys. The public key can be accessed by all other entities. verification associated with deduplication is another function of the auditor.
- Cloud server - The cloud server offers the pools of storage for the clients (Fig. 1).

The users lend storage space from cloud server for storing their individual data or organizational data. The cloud server initializes the proof of ownership protocol along with the file uploading protocol. The proof of ownership protocol is used for proving that the claimed file is exactly owned by the user. Here client is the prover and cloud server has the role of verifier.

The three mainly protocols used in this setup are

(1) FILE UPLOADING PROTOCOL:

The different steps are

- After computing hash value of each chunks of file not yet uploaded, an examination is done to verify if an alike file ie file with identical hash is existing in cloud space just before uploading.
- If found, then proof of ownership protocol is called. It involves cloud server as well as client.
- Auditor receives file chunks from client.
- Auditor create tags for the file chunks
- File chunks are then encrypted and compressed.
- Finally, cloud receives the file chunks and tags.

(2) INTEGRITY AUDITING PROTOCOL

This protocol is used in verifying the integrity and is invoked by anyone other than the cloud server. Here verification is done by auditor (or client) and cloud server functions as a prover. Verification process include following steps

- Challenges felt by the verifier ie auditor or client is passed to prover ie cloud server
- Prover generate the proof and pass it to the verifier after examining the files as well as tags stored in cloud server
- Verifier outputs positive results (or true value) if integrity test produces satisfactory results.

(3) PROOF OF OWNERSHIP PROTOCOL

The main features of this protocol are:

- To prove that the claimed file is exactly in the ownership of the client, a POW protocol is activated simultaneously with the file uploading protocol+.

- Drainage of side channel data is minimized.
- Here client functions as prover and cloud server as verifier
- Sharing of secret key of uploaded file results in sharing of ownership among users.

3.2 Merkle Hash Tree Based Integrity Auditing

Suppose F is the file under consideration. we split the file into 8 equal sized chunks using function called splitfile. SHA-1 algorithm generate hash of each chunk. Leaf node has hash of each individual chunk. Concatenated hashes is contained in nodes which are in the internal positions. The verifier receives only the hashes of those nodes which are present in the path of authentication. Consider if the receiver needs to verify the integrity of chunk 2 then h(1), h(3, 4) and h(5, 8) are only transferred to the receiver. The receiver can calculate the h(2) from data block 2. h(1, 2) can then be calculated by using the received h(1) and calculated h(2). In the same way, h(1, 4) can be calculated and then h(1, 8). The receiver then can compare the calculated h(1, 8) with the already shared h′(1, 8) and if both the hashes match then the integrity of chunk 2 is confirmed (Fig. 2).

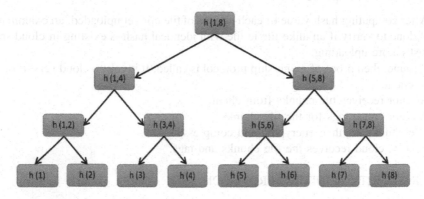

Fig. 2. Merkle hash tree for calculating the concatenated hashes from individual hashes

3.3 Algorithms Used

The general algorithm for the setup is

Data: File content
Result: Hash
Step1: For each (File content in File content []) do
    Hash<- calculatehash (File content)
Step2: If (Hash == alreadyhash [ ])
    */*do not save this File content*/*
    Give a pointer to the already saved file
    End
    Else

/save the File content after encryption and compression/
    End
Step 3: *Ownership proving (duplicate, hash)*
    End

File retrieval algorithm
Data: *Filename*
Result: *String*
Step1: *Begin Downloading from cloud*
Step2: *RequestedForFile(filename)*
Step3: *GetfileFromCloud(filename)*
Step4: *String content = DownloadStringFrom*
    Cloud(filename)
Step 5: *Receive(StringFromCloud)*
Step 6: *Decrypt (StringFromCloud, secretkey)*
Step 7: *Decompress (StringFromCloud, secretkey)*
Step 8: *File write decompressed content.*

3.4   Hashing Using SHA-1

Splitting produces eight equal parts of input file. The next step is hashing. SHA-1 is selected for hashing. It is more reliable since it eliminates brute force attack. In the beginning zeros and ones (bits of padding) are added to input to prepare it for following operations. Next 64- bit length is concatenated to data (message) obtained after padding. Then it is subjected to 80 preprocessing methods and constants. SHA-1 operates on 160 bits or 5 buffers. The message digest is shown below (Fig. 3).

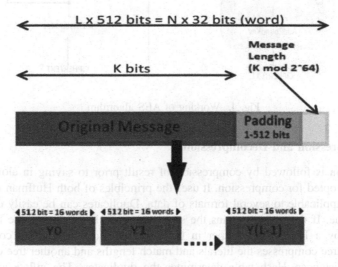

Fig. 3. Hashing computation using SHA-1

3.5   Encryption and Decryption Using AES

Input data in AES undergoes to a series of changes. Initially data is saved in a 4 * 4 column major matrix. Key length determines the count of rounds requires for conversion. The main transformation operation are substitution and permutation. This encryption protects system from extensive key search attack. The working is demonstrated below (Fig. 4).

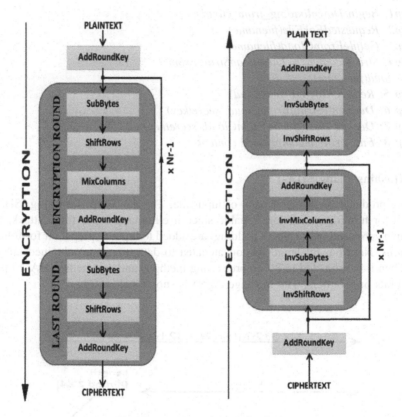

Fig. 4. Working of AES algorithm

3.6   Compression and Decompression

Deduplication is followed by compression of result prior to saving in cloud. Deflate algorithm is opted for compression. It uses the principles of both Huffman coding and LZ77. It is applicable to several formats of data. Duplicates can be easily detected by this technique. If the input contains the same string (ie bytes) twice, the later one is substituted by a link to the former in the form of a distance-length combination. A Huffman tree compresses the literals and match lengths and another tree compresses the match distances. Hash table determines the duplicates. The inflate algorithm is opted for decompression. This algorithm decodes smaller codes first and then the longer codes. This speedup the system.

4 Experimental Result

The table projects the variation in space usage resulting from deduplication as well as compression. By discarding duplicates, deduplication minimizes cloud space utilization Compression again saves 30% of cloud space (Table 1).

Table 1. Comparison of file size

Test run	No of files	File size before deduplication	File size after deduplication	File size after compression
1	13	261 MB	201.3 MB	140.91 MB
2	11	85.8 MB	23 MB	16.1 MB
3	8	37.4 MB	37.2 MB	26.04 MB
4	16	74.8 MB	37.2 MB	26.04 MB

Similarly, bandwidth usage is minimized by giving only unique file chunks when client asks for an item. The table shows the difference in bandwidth (Table 2).

Table 2. Comparison of bandwidth

File size (in MB)	Bandwidth used before deduplication (in MB)	Bandwidth used after deduplication (in MB)
2200	3.775	2.9
4500	3.925	2.55
9000	4.025	2.7
60000	2.2	1.475
125000	2.6	1.53

5 Conclusion

Cloud is a very fascinating technology. Providing storage space is its greatest advantage. Through our works we have proposed efficient methods for integrity auditing as well as deduplication. Our main focus was in the area of efficient space utilization of cloud. Deduplication along with compression helped in achieving this goal.

References

1. Ateniese, G., et al.: Provable data possession at untrusted stores. In: Proceedings of the 14th ACM Conference on Computer and Communications Security, Service. CCS 2007. ACM, New York, pp. 598–609 (2007)
2. Ateniese, G., et al.: Remote data checking using provable data possession. ACM Trans. Inf. Syst. Secur. 14(1), 12:1–12:34 (2011)
3. Azraoui, M., Elkhiyaoui, K., Molva, R., Önen, M.: StealthGuard: proofs of retrievability with hidden watchdogs. In: Kutyłowski, M., Vaidya, J. (eds.) ESORICS 2014. LNCS, vol. 8712, pp. 239–256. Springer, Cham (2014). https://doi.org/10.1007/978-3-319-11203-9_14
4. Li, J., Chen, X., Li, M., Li, J., Lee, P., Lou, W.: Secure deduplication with efficient and reliable convergent key management. IEEE Trans. Parallel Distrib. Syst. 25(6), 1615–1625 (2014)
5. Halevi, S., Harnik, D., Pinkas, B., Shulman-Peleg, A.: Proofs of ownership in remote storage systems. In: Proceedings of the 18th ACM Conference on Computer and Communications Security, pp. 491–500 (2011)
6. Armbrust, M., et al.: A view of cloud computing. Commun. ACM 53(4), 50–58 (2010)
7. Yuan, J., Yu, S.: Secure and constant cost public cloud storage auditing with deduplication. In: Proceedings of IEEE Conference on Communications and Network Security, pp. 145–153 (2013)

Secure Data Sharing in Multiple Cloud Servers Using Forward and Backward Secrecy

L. Gopika[⊠], V. K. Kavitha[⊠], B. Radhakrishnan[⊠],
and Padma Suresh Lekshmi Kanthan[⊠]

Department of Computer Science and Engineering,
Baselios Mathews II College of Engineering, Sasthamcotta,
Kollam, Kerala, India
lgopika93@gmail.com, kavithavk86@gmail.com,
radhak77@rediff.com, suresh_lps@yahoo.com

Abstract. In this era, where technologies are in its pace the amount of data generated is huge. These data's are usually stockpiled into cloud with an instinct of cost reduction and easy access of data. Thus there arises a need of securing the shared data. In this paper this disadvantage is overcome using revocable repository identity base encryption. RS-IBE provide both forward security and backward security of cipher text by introducing user functionalities and updating of cipher text. The enciphered facts are then uploaded into multiple cloud servers thereby providing more stability to the hoarded facts.

Keywords: Data sharing · Forward secrecy · Backward secrecy
RSIBE · Multiple cloud

1 Introduction

With concoct of cloud, millions of problems confronted by most of the business organizations were resolved. One such was the reduced cost of data repository in cloud. In cloud, facts are hoarded in lonesome user's which is controlled by repository servers. It makes use of virtual repository architecture such that retrieval and hoarding of facts in cloud is made easy for the users [6]. Although cloud repository is built with multiple servers it acts as though it is made from one single server. Some of the specialities of cloud repostiory are:

- Durability: The same data uploaded by different users are saved as different versions.
- Fault Tolerant: Fault tolerance is obtained in cloud by redundancy checking.
- Cost: Data repostiory in cloud enables the users to carry out green business by downsizing energy intake up to 70%. This is made feasible because the users need to pay only operating cost of repository.
- Web service interface: By using web service interface cloud users are able to use applications provided by other organization.
- Back up service mechanism: Cloud protects the data by providing back up mechanism.

I. Zelinka et al. (Eds.): ICSCS 2018, CCIS 837, pp. 335–342, 2018.
https://doi.org/10.1007/978-981-13-1936-5_36

In order to overcome the expense of storing data most of the business organization outsource their data in cloud repostiory. But the security and privacy concerns are one of the major issues that have to be faced by them. In cloud the data is stored in more than one location whereby increasing the possibility of unauthorized accesses [5]. Use of encipherment and decipherment may limit this to some extent. But again the need of repostiory area for storing these key is another great issue. The performance of hoarding data in cloud gets degraded due to increase usage of bandwidth. When choosing a cloud repository provider usually take into consideration factors like security, price, and ease of access. The data stored in cloud is considered to be less secure because the cloud is also a third party. Thus there arise a critical issue of data leakage. If we share the same credentials used for one with another set of customers then it will open a pathway for data leakage since access to facts are based on credentials used.

2   Related Work

[1] Prof Vishal More proposed a method in which data is securely stored by using key aggregate cryptosystem. This paper also provides an insight of various cryptographic techniques. [2] Lawanya Shri proposed a method that make use of perfect forward secrecy for enhancing efficiency of existing ring based security. [3] Patil proposes a scheme called multi-authority attribute based access control to reduce key management complexity. The user accesses files on a criteria formulated from given set of attributes. [4] Awadh has proposed a method using steganography for secure data sharing in cloud and efficient sharing of data. Secure data sharing in cloud is carried in two phases: embedding and extracting. [5] Subramanian proposes a scheme that prevents an anonymous user from accessing any files from multi cloud repository. [6] Anjali proposed a cloud prototype that discards the use of key management and file encipherment and decipherment. It aids dynamic enrollment of users and users can access data whenever needed. [7] Mohan Kumar proposed a method in which the user can select among the consignee with whom the facts are to be shared. AES algorithm is made used in this paper. [8] Chandankere recommended a method that make use of passel signature and vehement simulcast encryption techniques for storing facts. This paper uses role base encipherment techniques.

3   Methodology

Today, when the technologies are at its height, organizations are in search for a better way of storing their facts. The solution to this is cloud repository. It is a rostrum where users can share their facts and execute various applications. The model of repository is scrutinized as a place where facts are accessed by the end-user after paying for the same via internet. In this rostrum the sharing of facts take place in an open environment that is less secure. The catastrophe occurs when the cloud itself outsource the data to an unbiased observer for its sole profit. The totality of these obstacles can be visibly moved by bestowing cryptanalysis. Cryptanalysis involves shielding data by deploying secret keys. On shielding data with secret keys the data is initially converted to an intermediate form called cipher text and send to receiver where its reconverted to original information utilizing decipherment.

3.1 Architecture

The architecture incorporates three personnel's: fact provider, end-user, key arbiter. The fact provider decides between the end-users among whom the facts are to be allotted. He then enciphers the facts on basis of traits provided by each end-user. The encrypted text is uploaded to multiple cloud servers. This ensures more stability and security to data. The multi cloud repository ensures that the data is made available to end-user even if one of the cloud repositories is fallacious. The uploaded facts are then deciphered by the equivalent end-user. The RS-IBE techniques used here provide both backward and forward secrecy, preventing anonymous personnel's from accessing facts. If end-user authorization has expired, the fact is downloaded by fact provider and re-enciphered using a new cipher key and reloaded by fact provider. The revocation mechanism is carried out by Key arbiter. The end user who need to access facts must possess a secret key. The secret key is provided to each end user by fact provider after verifying the revocation list (Fig. 1).

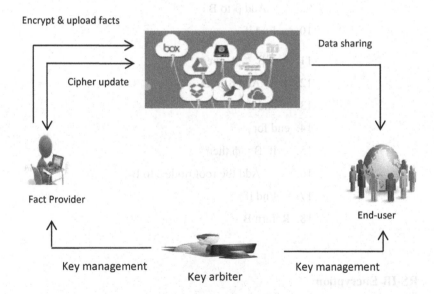

Fig. 1. Proposed architecture

3.2 Ku Node Algorithm

The RS-IBE scheme makes use of the KU node algorithm for versatile revocation. The various terms used are root node of binary tree (BTree) denoted by α and path from root node to leaf node by γ. A non-leaf node is denoted by β where β_l denote left node and β_r denote right node. The revocation list represented by $Revoc(\gamma_i, T_i)$ indicate node γ_i which was revoked at time period T_i. The KU node outputs a small subset of nodes B such that B contains nodes that are not revoked before time period. The revocation mechanism can be represented as a tree in which all the predecessor of revoked nodes are marked revoked. The algorithm then outputs all non-revoked nodes (Fig. 2).

ALGORITHM OF KU NODE

KU NODE (BTree, Revoc, T)

1.   AB $\longleftarrow$ ϕ

2.   For all$(\gamma_i, T_i) \in$ Revoc do

3.   If $T_i \leq T$ then

4.   Add path(γ_i) to A

5.   End if

6.   End for

7.   For all $\beta \in$ A do

8.   If $\beta_l \notin$ A then

9.   Add β_l to B

10.   End if

11.   If $\beta_r \notin$ A then

12.   Add β_r to B

13.   end if

14.   end for

15.   if B = ϕ then

16.   Add the root node ε to B

17.   End if

18.   Return B

3.3   RS-IB Encryption

RS-IBE includes message space, identity space, total time period and seven polynomial time algorithm:

- **PPGen(1^η, T, n):** *It accepts security parameter η, time bound T and entire system users n as input and outputs public parameter PP and master secret key MS associated with initial revocation list Revoc = ϕ and state S.*

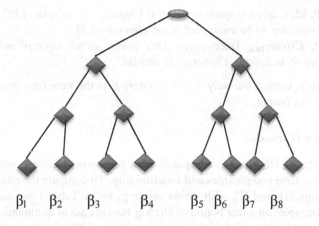

a. No node is revoked

⬭ Node in the set B

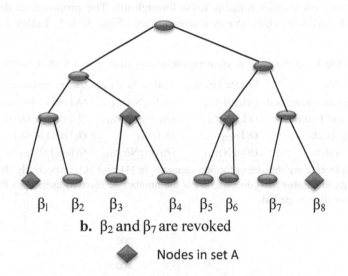

b. β_2 and β_7 are revoked

◆ Nodes in set A

Fig. 2. Illustration of KUnode

- **PrivGen(PP, MS, and Id):** *It generates private key PK_{id} and update state S by taking PP, MS, and Id.*
- **KeyUP(PP, MS, Revoc_l, t, S):** *Outputs key update KU_t for the current revocation list $Revoc_l$, state s, and for time period $t \leq T$.*
- **CipherTextUP(PP, Cipher_Id,t, t'):** *Output $Cipher_{Id,t'}$ for a new time slot $t' > T$, PP, $Cipher_{Id,t}$.*
- **DecrypGen(PP, S_Id, KeyUP_t):** *This generates $Decryp_{Id,t}$ for Id with time period t or a symbol Ψ to illustrate Id which was previously revoked.*

- **EncryptM(PP, Id, t, m):** *It outputs cipher text Cipher$_{Id,t}$ by In-take of PP, Id where t ≤ T and m message to be encrypted is an element of M.*
- **DecryptM(PP, Cipher$_{Id,t}$, Decryp$_{Id,t}$):** *This phase either decrypt message or generate symbol Ψ indicating Cipher$_{Id,t}$ is invalid.*

The decryption is carried out only if t ≥ t' where t' is the new time period else a symbol Ψ is given as output.

3.4 Performance Discussion

The proposed scheme's efficiency is compared with prior works using computational cost, repository cost, time complexities and functionality. To compare the efficiency we consider two groups H1 and H2 with prime order p. From Table 1 we can see the proposed system occupies an upper bound of $O(r \log N/r)$ in case of communication and storage cost compared to prior works which are constant size. This is obtained by use of binary data structure. The cipher text size for current and prior scheme remains same. The proposed model has a storage cost of $O(T)(\tau H1)$ Compared to prior models thanks to key arbiter that hoards end users secret key for entire time in a table. The decipher and encipher key complexity remains same throughout. The proposed model is built on decisional *l*-dBDHE which makes it more secure (Figs. 3, 4, 5, Tables 2 and 3).

Table 1. Comparison of communication and storage cost with previous works

Schemes	Private key size	Update key size	Cipher text size
Libert and Vergnaud	$O(\log N)\tau_{H1}$	$O(r\log N/r)_{H1}$	$O(1)\tau H1 + O(1)\tau_{H2}$
Seo and Emura	$O(\log N)\tau_{H1}$	$O(r \log N/r)\tau_{H1}$	$O(1)\tau H1 + O(1)\tau_{H2}$
Liang et al.	$O(1)\tau_{H1}$	$O(1)\tau_{H1}$	$O(1)\tau H1 + O(1)\tau_{H2}$
Our system	$O(\log N)\tau_{H1}$	$O(r \log N/r)\tau_{H1}$	$O(\log(T)2)\tau_{H1} + O(1)\tau_{H2}$

τH1 and τH2 are the sizes of group elements in H1 and H2, respectively. N is the maximum number of system users. r is the number of revoked users. T is the total number of time periods.

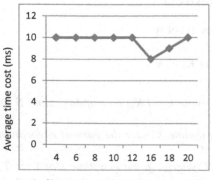
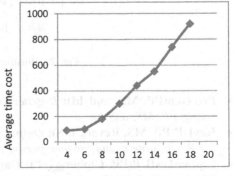
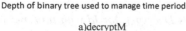

a)decryptM   b) CipherUP

Fig. 3. Time cost of algorithm DecryptM and CipherUP

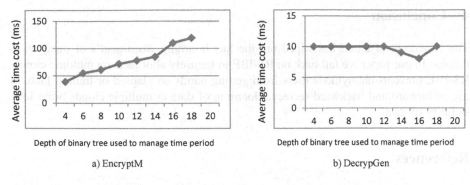

a) EncryptM   b) DecrypGen

Fig. 4. The time cost of DecrypGen and EncryptM

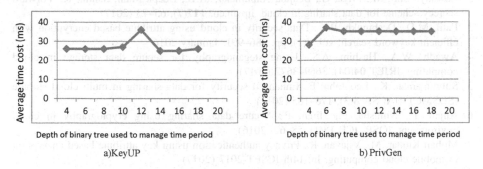

a) KeyUP   b) PrivGen

Fig. 5. The time cost of KeyUP and PrivGen

Table 2. Comparison of time complexity with previous works

Schemes	Encipherment	Decipherment	CTUpdate
Libert and vergnaud	$O(1)c + O(1)p_r$	$O(1)p_r$	0
Seo and Emura	$O(1)c + O(1)p_r$	$O(1)p_r$	0
Liang et al.	$O(1)c + O(1)p_r$	$O(1)p_r$	$(O(N))c + O(1)p_r$
Our system	$O(\log T)c + O(1)p_r$	$O(1)p_r$	$O(\log(T)2)c + O(1)p_r$

P_r and c indicate the cost of performing a bilinear pairing and exponentiation.

Table 3. Comparison of communication and storage cost with previous works

Schemes	Model	Assumption	PKU	PCU	CA	DKE	FS	BS
Libert and vergnaud	Adaptive	DBDH	Y	N	Y	N	Y	N
Seo and Emura	Adaptive	DBDH	Y	N	N	Y	Y	N
Liang et al.	Adaptive	DBDH	N	N	N	Y	Y	Y
Our system	Adaptive	ℓ-dBDHE	Y	Y	Y	Y	Y	Y

4 Conclusion

The concoction of cloud has wiped out the fact hoarding worriment's of many institutions. In our paper we fall back on RS-IBE to securely allot facts on multiple clouds. RS-IBE prevents anonymous users from getting hands on elapsed or trailing facts by use of forward and backward secrecy. Hoarding of data in multiple clouds bring forth stability of data.

References

1. More, V., Singh, A.K.: Key aggregate cryptosystem for scalable data sharing in cloud storage. In: NCRIET, pp. 182–187 (2015)
2. Lawanya Shri, M., Priya, G., Benjula Anbumalar, M.B., Deepa Mani, Santhi, K.: Forward secrecy scheme for data sharing in cloud. ijpam.eu, 117(7), 65–73 (2017)
3. Patil, M.D., Deshmukh, P.K.: Data security in cloud using attribute based encryption with efficient keyword search, vol. 7(1), pp. 986–991, January 2016
4. Awadh, W.A., Hashim, A.S.: Using steganography for secure data storage in cloud computing. IRJET 04(04), 3669–3672 (2017)
5. Subramanian, K., Leo John, F.: Enhanced security for data sharing in multi cloud storage (SDSMC). IJACSA 8(3), 176–185 (2017)
6. Anjali, P., Nimisha, P., Hiren, P.: Secure data sharing using cryptography in cloud environment. IOSR_JCE 18(1), 58–62 (2016)
7. Mohan Kumar, M., Vijayan, R.: Privacy authentication using key attribute based encryption in mobile cloud computing. In: 14th ICSET-2017 (2017)
8. Chandankere, R., Begum, M.: Secure data sharing in an untrusted cloud. Int. J. Eng. Res. Appl. 1, 49–54 (2015)

Privacy Preserving in Audit Free Cloud Storage by Deniable Encryption

L. Nayana$^{(\boxtimes)}$, P. G. Raji$^{(\boxtimes)}$, B. Radhakrishnan$^{(\boxtimes)}$,
and Padma Suresh Lekshmi Kanthan$^{(\boxtimes)}$

Department of Computer Science and Engineering, Baselios Mathews II College of Engineering,
Sasthamcotta, Kollam, Kerala, India
nayanamadhukumar2@gmail.com, rajiipg1985@gmail.com,
radhak77@rediff.com, suresh_lps@yahoo.co.in

Abstract. Cloud storage is a prominent storage plan and is widely accepted. Data's from cloud can be accessed at any time and from anywhere. Most of us believes that data in cloud cant be hacked. But in reality the data can be grabbed by illegitimate users. In this work we have created fake documents to misguide intruders to prevent illegal access. We setup an encryption method that convinces the intruders. A superior encryption method such as Cipher text Policy Attribute Based Encryption method is used and it helps to promote user's privacy. Only right user can attain the right data by using this technique. We also modify the system by introducing a new method such as once the file is deleted from cloud the key authority serve public key for restore again the file only based on some resembling conditions connected the attributes with that file.

Keywords: Deniable encryption · Cloud storage · Bilinear pairing
Attribute based encryption

1 Introduction

Cloud computing is a latest style of computing in which all resources are stored without any harm, accessed easily and work simply etc. Cloud computing is a skilled technology for all fields and it provides a sophisticated option for data storage. Cloud computing consist of a provider and subscriber. The internal groups of a particular company or trusted third party or the aggregation of both is called Provider. One who uses the cloud computing service is called Subscriber. Cloud provides some advantages such as low cost, simple and better storage of data. There are three common type of cloud: private, pubic and hybrid. If the data is only accessible by limited organizations is known as Private cloud. In public cloud data is accessed by public domain in public network. Hybrid cloud is the combination of both private and public cloud. Cloud computing offers various advantages such as scalability, low cost, simple in repair, faster development and large scale testing. The idea of attribute based encryption scheme was originated from the concept of deniable encryption. In Deniable encryption the sender sends fake documents along with valid files to the receiver, so that they can misguide and satisfy outside coercers. The outside coercers cannot detect whether the obtained

I. Zelinka et al. (Eds.): ICSCS 2018, CCIS 837, pp. 343–351, 2018.
https://doi.org/10.1007/978-981-13-1936-5_37

document is true or not and he believes that it might be the original document. So it provides high security for user's data (Fig. 1).

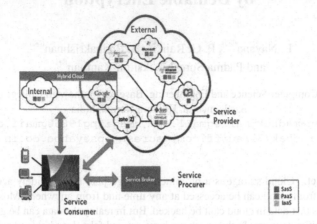

Fig. 1. Three types of cloud computing model

Cloud storage is the type of data storage in which a data pool can store all type of user's data. Any user's can easily access this data from anywhere at any time. Cloud storage Providers are responsible for this on user's behalf. For protecting the privacy of user's data, various encryption schemes are used for storage in cloud. But the currently existing technologies consumes more time for data uploading and retrieval and occupies more space in cloud. To resolve this problem we have used 'Cipher Text policy Attribute based encryption' scheme for encrypting data before storing in cloud. ABE is a public key encryption, in which user's secret key and cipher text are tied up to some attributes [3]. At the time of decryption, key attribute and user's secret key are compared and the decryption will be performed if a match occurs. Attribute based encryption scheme is a relevant encryption scheme for cloud computing and it provides preserved data transfer and difficult to hack.

2 Related Works

[1] Dürmuth suggested Deniable Encryption with Negligible Detection Probability. In this a public key bit encryption scheme with dense cipher text is used with single encryption algorithm, which was used as the basic concept of this paper. [2] Bi-Deniable Public-Key Encryption Protocol was proposed by Moldovyan and this intensifies the encryption scheme which is based on RSA cryptosystem. For refusing the coercive attack some cryptographic protocols are used. [3] Bethencourt suggested a Cipher Text Policy Attribute Based Encryption technique for recognizing difficult access control of encrypted data. The new policy is called CPABE. [4] Hohenberger introduced 'Attribute-based Encryption with Fast Decryption' that enhances fast decryption and encryption based on users attributes. [5] Goyal proposed Attribute Based Encryption for Fine Grained Access Control of Encrypted Data that establish a new technique called

KP-ABE (Key Policy Attribute Based Encryption) that promotes elegant sharing of encrypted data [6] Sahai and Waters suggested Attribute Based Encryption Scheme with Constant Sized Cipher Text. This technique upgrades encryption based on constant sized cipher text.

3 System Architecture and Methods

3.1 System Model

This system consists of four modules and they are cloud storage provider, users, owners, key authorities etc.

- Data Owner
  An authorized user is who outsources the data to store in cloud is called data owner. At first the data owner registers himself and the login to the system. On login the owner receives a key from KDC for authentication process. The owner will upload the files after the validation process.
- Data User
  The data user or client who wish to download a particular file should have to register and login to the system after receiving the key from authorised agency. Data user is allowed to access the file uploaded by the owner only after the authentication process.
- Key Authority
  Key authority is responsible for distributing keys to users and data owners. For validation process KA compares the key from KDC and key entered by user. If a match occurs the user is authorized to view the file, otherwise permission will be denied (Fig. 2).

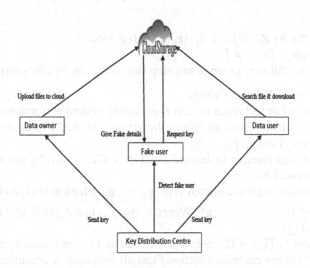

Fig. 2. Proposed system architecture

User can also search file using file id and name. If wrong file id and name is used, then file access is denied. By using the secret key the system decrypts the file and plain text is provided to the valid user. Key distribution authority can provide keys for owners according to their request. We modify our system as if a file is accessed only based on the permission granted by key centres. Once the file is erased it cannot recover anyone from the cloud. It is only accessible by requesting the public key to key centres. The file recovered only based on verification of key and verification performed only by one who knows the related attributed connected to that file. So only right user can access the right file.

3.2 Bilinear Groups

For to enhance the performance of our system we use prime order group rather than composite order group. Bilinear prime order group is much faster as compared to composite order group, it results as progress in computational performance of our system. The data's are encrypted in a multidimensional space and for establishing multidimensional space, bilinear composite order group is used. Most of the algorithms have many decryption problem. Our proposed algorithm is deterministic and have a powerful decryption, that helps to reduce decryption errors.

Definition 1: A bilinear mapping is denoted as e: $U \times V \to M$ Where U is left unitary, V is right unitary and M id module. If $V = W$ says that e is bilinear form of module V and V is also a metric structure.

Prime Order Bilinear Groups

Let R and R_V be two cyclic groups of prime order p, there is a map function f: $R \times R \to R_v$. Let r be a generator of R. Bilinear group R has the following properties such as,

- Bilinearity: $\forall g, h \varepsilon R$ and $b, c \varepsilon Z$, $e(g_b, h_c) = f(g, h)bc$
- Non-degeneracy: $f(r, r) \neq 1$
- Computability: Bilinear group R and map function f can be efficiently measured.

Composite Order Bilinear Groups

A most appropriate technique used to evaluate the system performance is composite order bilinear grouping. Consider two cyclic groups G1 & G2 of composite order N. That is the primes N are $p_1, p_2, \ldots, p_n$.

The bilinear map function is denoted as f: G1 $\times$ G2 $\to$ Gv. G_p has a subgroup for each prime p of order N.

Then the generators are considered as $g_1, g_2, \ldots, g_m$. Element in G can be represented in the form of $g_1^{b1}, g_2^{b2}, \ldots \ldots \ldots, g_t^{bt}$. Where $b_1, b_2, \ldots, b_t \varepsilon Z_n$. If $b_i \equiv 0$ modulo P_i the element has no G_{pi}.

If the element is $\prod_i \varepsilon S G_{pi}$, S is a subset that is {1....m}, then $b_i \equiv 0$ modulo p_i. Orthogonality between mapping function f and all subgroup is considered as the main property of composite order group.

Which defines that if $r \, \varepsilon \, G_{pi}$, $s \, \varepsilon \, G_{pj}$ i.e., $i \neq j$ and $f(r, s) = 1$ where 1 is an identity element. The subgroup decision assumption is considered as the complexity assumption in composite group without applying any orthognality testing method. The general form of this assumption is defined as follows;

Definition 2 (General subgroup decision assumption):

Consider some non empty subsets $T_0, T_1, \ldots \ldots, T_K$ of 1 ... .k such that for each $2 \leq u \leq v$, $T_u \cap T_0 = \emptyset = T_u \cap T_1$ or $T_u \cap T_0 \neq \emptyset \neq T_u \cap T_1$. For some group generator G we describe following definition

$$PP: = \{N = p1p2 \ldots pm, G, GT, f\} \leftarrow g$$
$$Zi \leftarrow GSi\forall i \, \varepsilon \, \{1 \ldots k\}$$
$$D: = \{PP, Z2, \ldots, Zk\}.$$

Adv G, A: $= |P[A(D, Z0) = 1] - P[A(D, Z1) = 1]|$ is negligible.

This assumption is very hard to distinguish the outputs of the bilinear map function from other elements when they contain at least one subgroup.

3.3 Algorithm Used

The steps of more advanced CP-ABE Algorithm is defined as follows,

Step 1: Setup $(1\mu) \rightarrow (PUP, MK)$
It generates public parameter (PUP) and master key (MK) by taking security parameter μ as input.

Step 2: Key generation $(MK, A) \rightarrow PR$
It gives private key (PR) as output by receiving set of attributes A and Master key (MK).

Step 3: Encrypt $(PUP, M, AS) \rightarrow C$
It can encrypt message (M) and output as a cipher text (C). It receives input as PUP, M and an access structure $AS = (M, p)$ over set of attributes. It can be decrypted by using the attribute which satisfies the access structure.

Step 4: Decrypt $(PUP, PR, C) \rightarrow \{M, \emptyset\}$
It can decrypt the cipher text and returns the original message. Decryption is performed by receiving public parameter, secret key and cipher text with access structure S. If A satisfies S it returns the original message M otherwise $\emptyset$.

Step 5: OpenEn $(PUP, C, M) \rightarrow EP$
It generates an encryption proof for the sender. Where EP is the encryption proof for (M, C).

Step 6: OpenDe $(PUP, PR, C, M) \rightarrow DP$
It generates a decryption proof for the receiver.

Step 7: Verify $(PUP, C, M, EP, DP) \rightarrow \{True, False\}$
It checks the exactness of encryption and decryption proof.

Step 8: DenSetup $(1^\mu) \rightarrow (PUP, MK, PR)$
It takes input as a security parameter μ and outputs as PUP, MK and PR. All system users knows the PR and kept private to the outsiders.

Step 9: Denkey generation $(MK, A) \rightarrow (PK, FK)$
It receives input as attribute A and master key MK. It then returns public key PK and FK is used for creating fake proof.

Step 10: DenEn $(PUP, PK, M, M', A) \rightarrow C'$
Deniable encryption algorithm takes input as PK Public key and fake message M'. It outputs cipher text which is not identical with output of Encrypt().

Step 11: DenOpenEn $(PUP, C', M') \rightarrow EP'$.
It can generate encryption proof EP' for fake message M. The output is not identical with result of OpenEn() and must pass for verify algorithm.

Step 12: DenOpenDe $(PUP, PK, FK, C', M') \rightarrow DP'$
It outputs decryption proof DP' for fake message M'. The output is not equivalent with result of OpenDe().

The Fig. 3 shows the implementation of CP-ABE Algorithm. Setup phase involves receiving implicit security parameter as input and generates MK (master key). During key generation phase it accepts set of attributes (S) and master key as input and output the secret key (SK). During encryption it receives input as message and outputs the cipher text. In decryption cipher text is decrypted and outputs the original message.

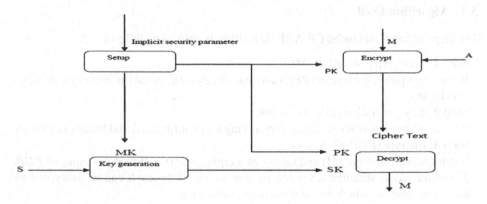

Fig. 3. CP-ABE algorithm implementation

3.4 ABE Construction

Let R_1, R_2 are the bilinear group of order p(prime) and r1 is the generator group of R1. Bilinear mapping $e: R_1 \times R_2 \rightarrow R_2$, d is threshold value.

- **Setup (1) -> (PU, MK),** A bilinear group R of order $N = p_1 p_2 p_3$ is generated and where p_1, p_2, p_3 are primes of order N with bilinear map function $e: R \times R \rightarrow Rt$ of order N. The three orthogonal subgroups is created is denoted as Rp_1, Rp_2, Rp_3. This algorithm selects generators $r1 \varepsilon Rp1$, r3 ε Rp3, and randomly select α ε Z_N. Select hash function H and Public parameter

$$PU = \left\{ \left\{ R, e, H_1, r_1 r_3, (r_1 r_3)^\alpha e(r_1 r_3, r_1 r_3)^\alpha \right\} \right\} \text{ and the master key } MK = (r_1 r_3)^\alpha.$$

- **Keygen**(*MK, A*) -> **PR**, where *A* is set of attributes and $t \, \varepsilon \, Z_N$ and output the private key as $PR = \{(r1r3)^{\alpha+at}, (r1r3)^t, \{H_1(x^t)\} \forall x \, \varepsilon \, A\} = \{k, m, \{k_x\} \forall x \, \varepsilon \, A\}$
- **Encrypt** *(PU, M, S = (M, p))* -> *C*: Where *M* is message and an access structure *(M, p)*. *M* be a $l \times n$ matrix, *Mi* denotes i[th] row of *M*. The algorithm first chooses two random vectors *u* and *v* and algorithm then calculates $\lambda i = vMi$. Algorithm setup a one way hash function *H* and it may be different for each transaction. The cipher text will be $C = \{A0, A1, B, (C1, D1), \ldots, (Cl, Dl), H, t0, t1, V\}$

$$A_{b0} = M \cdot e(r_1 r_3, r_1 r_3)^{\alpha s}$$

$$B = (r_1 r_3)^s$$

$$V = H(M, t_{b1})$$

- **Decrypt**(*PU, PR, C*) -> *M*. The algorithm decrypt cipher text *C* and return original message *M*.

  After decryption it then compute encryption proof and decryption proof, then verifies the exactness of proof (Fig. 4).

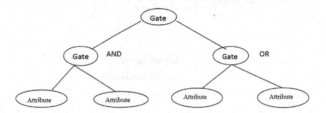

Fig. 4. Monotonic access tree used in CP-ABE algorithm

4   Result and Discussion

By comparing two concepts such as bilinear composite order and prime order group we can compute the performance of our system. This is performed by using our privacy preserving CP-ABE algorithm. In this paper we focused on the performance of encryption and decryption. From the experiments we find that Figs. 5 and 6 shows the encryption and decryption time grows linearly based on number of attributes. More time is utilized by using composite order group. So it cant be used in practical applications. By using prime order group rather than composite will reduce the overall time taken to finish both the process ie. less time is required for fetching files from the cloud. Our scheme strengthens the privacy of users and reduces the security threats. CP-ABE algorithm is deterministic so it reduces decryption errors (Table 1).

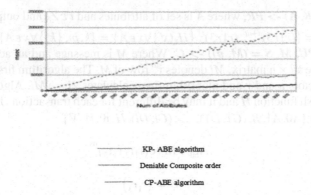

KP- ABE algorithm

Deniable Composite order

CP-ABE algorithm

Fig. 5. Encryption process

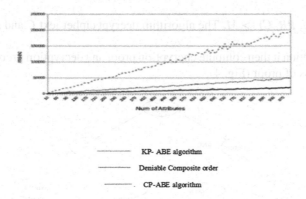

KP- ABE algorithm

Deniable Composite order

CP-ABE algorithm

Fig. 6. Decryption process

Table 1. Comparison of CP-ABE with other schemes

Scheme	Properties			
	Security	Efficiency	Decryption time	Fine grained access control
ABE	Average	Average	High	Low
KP-ABE	Average	Average	Average	Low
CP-ABE	High	High	Less	Average

5 Conclusion

In this work we suggested CP-ABE Algorithm for secure data sharing in cloud. Privacy conserving and preventing the illegal access by illegitimate users is the important advantage of this system. We have proposed a technique that generate fake documents for convincing coercers, and resist outside auditing. System performance is improved by using prime order group along with ABE algorithm. From the modification our system achieves an advantage that only right user can access the right file. The key centres perform authentication based on attribute or any other information's related to that file. So unauthorized users cannot recover the erased file.

References

1. Dürmuth, M., Freeman, D.M.: Deniable encryption with negligible detection probability: an interactive construction. In: Paterson, K.G. (ed.) EUROCRYPT 2011. LNCS, vol. 6632, pp. 610–626. Springer, Heidelberg (2011). https://doi.org/10.1007/978-3-642-20465-4_33
2. Moldovyan, A.A., Moldovyan, N.A., Shcherbacov, V.A.: Bi-deniable public-key encryption protocol, 23–29 (2014). ISSN 1024–7696
3. Bethencourt, J., Sahai, A., Waters, B.: Ciphertext-policy attribute-based encryption. In: IEEE Symposium on Security and Privacy, pp. 321–334 (2007)
4. Hohenberger, S., Waters, B.: Attribute-based encryption with fast decryption. In: Kurosawa, K., Hanaoka, G. (eds.) PKC 2013. LNCS, vol. 7778, pp. 162–179. Springer, Heidelberg (2013). https://doi.org/10.1007/978-3-642-36362-7_11
5. Goyal, V., Pandey, O., Sahai, A., Waters, B.: Attribute-based encryption for fine-grained access control of encrypted data (2006)
6. Sahai, A., Water, B.: Attribute based encryption scheme with constant sized cipher text. Comput. Sci. **422**, 15–38 (2012)

Data Mining

Cyclic Shuffled Frog Leaping Algorithm Inspired Data Clustering

Veni Devi Gopal[1(✉)] and Angelina Geetha[2]

[1] iNurtute Education Solutions, Bangalore, India
venidevig@gmail.com
[2] Department of Computer Science and Engineering,
BSAR Crescent Institute of Science and Technology, Chennai, India
angelina@bsauniv.ac.in

Abstract. The era of internet has been filling our globe with tremendous high volume of data. These data have become the main raw material for various researches, business, etc. As the data volume is huge, categorizing the data will help in faster and quality data analysis. Clustering is one way of categorizing the data. As the digital data that is generated by any transaction is unpredictable, clustering can be the best option for categorizing it. Numerous clustering algorithms are at our disposal available. This paper focuses on adding modifications to the existing Shuffled Frog Leaping Algorithm and cluster the data. The proposed algorithm aims at enhancing the clustering, by taking the outliers into consideration and thereby improving the speed and quality of clusters formed.

Keywords: Clustering · Cyclic SFLA modified SFLA · Outlier detection
Shuffled frog leap algorithm

1 Introduction

Clustering is the process of grouping the data based on the similarity among them. This helps to represent the data with limited groups and helps in the simplification of the data. Clustering is a process of minimizing the distance within the clusters and maximizing the distance in-between the clusters. Clustering segregates the data into unknown groups. As clustering is unsupervised, it can be applied to domain independent scenarios. Pattern analysis is the most widely used technique in machine learning. Clustering forms the foundation for pattern analysis. When a proper clustering is not possible with the available data, there is a possibility for natural clustering. Memetic algorithms are one of the nature inspired algorithms used for optimization. The Memetic algorithms are combined with the evolutionary algorithms for finding the local best or global best solutions depending upon the problem they are used for. There are many optimization algorithms and one such is Shuffled Frog Leap Algorithm (SFLA). This paper explores the modifications that can be made to the Shuffled Frog Leap Algorithm (SFLA) and the effects on the performance of the algorithm.

The literature review is given in Sect. 2, the steps in the original Shuffled Frog Leap Algorithm are discussed in Sect. 3. The modifications that can be made in the original

© Springer Nature Singapore Pte Ltd. 2018
I. Zelinka et al. (Eds.): ICSCS 2018, CCIS 837, pp. 355–363, 2018.
https://doi.org/10.1007/978-981-13-1936-5_38

SFLA and the steps to be followed are discussed in Sect. 4. The conclusion along with the suggestions for future direction is presented in the last section.

2 Literature Review

With an aim to improve the quality of clusters Arun Prabha et al. [1] proposed an improved SFLA based on K-Means Algorithm by altering the attribute values into their precise range before clustering. By exploring the normalization for supervised clustering the proposed algorithm helped to find the best fitness function along with good convergence and local optima. Ling et al. [2] considered the social behavior and proposed a modified shuffled frog leap algorithm that suitably adjusted the leaping step size for optimizing the result. Karakoyun et al. [3] developed an algorithm for clustering data according to optimum centers by using SFLA algorithm. This work used partitional data clustering and it was found to be effective when compared with standard classification algorithms. Vehicle routing problems were addressed by Luo et al. in implementing the algorithm of Shuffled frog Leap in an improved manner [4]. The local search ability had been increased and the convergence speed had been increased by adding Power Law External Optimization Neighborhood Search (PLEONS) which readjusted the position of all the frogs to form new clusters and then analyzed the clusters to get the best solution. In the proposed SFLK algorithm by [5] more than the initial and individual solutions, the global solution was achieved by exchanging the information received from the solutions of the individual clusters. Effective clustering as addressed by Jose and Pandi [6] introduced a parameter for the acceleration of the process of searching in the traditional SFLA. The position of the frog was changed randomly in order to accelerate the global search. This extended SFLA was compared with SFLA and other stochastic algorithms and it was proved that ESFLA was better in terms of performance despite the time required being more. It was also suggested that by making changes in the ranking and evolution process, there could be chances to improve the execution time. The LSFLA algorithm [7] had been used for data clustering by including chaos and combination operator in the local search as well as entropy in the fitness function. The results showed increased efficiency and less error rate when compared with k-means, GA, PSO, and CPSO.

SFLA was to be the best when compared with the other algorithms for optimization while reducing the total harmonic distortion and improving the power factor in power systems according to [8]. In proposed work [9], a Multivariable Quantum SFLA used quantum codes to represent the position of frogs. The mutation probability was used to avoid the solutions from the locally optimal instead of globally optimal. The results showed that the convergence rate and the accuracy were improved. The same algorithm was applied in the telecom field and the results were effective.

3 Overview of Shuffled Frog Leap Algorithm

The Darwinian principles of evolution and Dawkin's idea of a meme were the key factor to the Memetic Algorithm (MA). It was in 1989, Moscato introduced it to the world. He found that the hybrid genetic algorithms, when added with methodical upgradation of knowledge, resembled the hybrid-genetic algorithms. Memetic Algorithms were are able to maintain a balance between the symbolic Darwinian evolution and the local search heuristics of the memes. Having these two phases made the Memetic Algorithms be a special case of Dual-phase evolution [11]. The memes are the transmittable information pattern. The pattern can be transmitted to another animal/human being and change their behavior. These patterns will be alive forever because of their parasitic nature. Though the contents in the meme are similar to gene, the memes can be alive only if they are transmitted. While gene can be transmitted only between the parent and the offspring, the memes can be transmitted between any individuals. The number of individuals having the gene is limited to the number of offspring produced by a parent, whereas there is no such restriction in the case of memes. Similarly, the taken to process the memes is much less than that taken for genes [13]. Inspired by all these above concepts, Eusuff, Lansey and Pasha came up with an optimization algorithm in 2006. This was called "Shuffled Frog Leaping Algorithm (SFLA)". Basically, frogs have a tendency to search for food in groups. This is the main idea employed in SFLA. The frogs form groups among themselves while searching for food. Each group is known as memeplex. The frogs in each memeplex will have different culture. The frog which is at the greatest distance from the food changes its place based on the information it receives from the others frogs in its own memeplex and also from the other memeplexes. Within each memeplex the frogs communicate among themselves to come with an idea which can contribute towards the global solution.

3.1 Steps in SFLA

The steps of SFLA are given below:
    It is pictorially represented in Fig. 1.

Step 1: Assume a group of 'p' frogs as the initial population.
Step 2: Compute for each frog the fitness using pre-defined fitness function.
Step 3: Frogs are arranged in sorted order (descending order) based on the computed fitness value.
Step 4: Form 'm' memeplexes each with the capability to hold 'n' frogs.
Step 5: The frogs are arranged in the respective memeplexes based on their order of ranking.
Step 6: Based on the fitness function the best frog and the worst frog are identified. By considering the best frog in each group, a global best frog is identified. This helps in changing the current position of the worst frog in each group based on the information from fellow frogs in the same memeplex.
Step 7: The convergence criterion is checked and steps 2–6 are repeated until the criterion is satisfied.

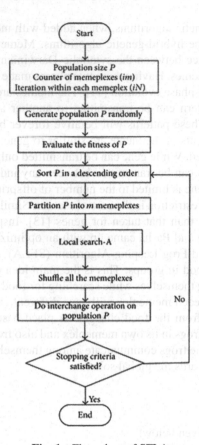

Fig. 1. Flow chart of SFLA

The sequence to be followed in the algorithm is given in the flowchart. Each block in the figure represents each step in SFLA. The algorithm ends when the stopping criteria are attained. The fitness of the frogs in each memeplex improves with every iteration and once they reach a constant value the algorithm is stopped.

4   Overview of Shuffled Frog Leap Algorithm

Though the SFLA has been proved to be more effective than many of the optimization algorithms, it suffers from a few disadvantages:

1. It takes too much time to converge.
2. The processing time is more.
3. There are no clear convergence criteria.
4. The number of iterations given does not guarantee convergence.

The Cyclic Shuffle Frog Leap Algorithm takes all these disadvantages into account while clustering the data given. Most of the research works based on SFLA have only the iteration count as the convergence criterion. In Cyclic SFLA, the step size is included which is used as a convergence criterion. The step size is the number steps a frog can be away from the best frog in a memeplex. Any frog with step size above the given value will be taken as the worst frog.

Once all the frogs inside have the step size less than or equal to the step size, then the memeplex is considered to be stable and once all the memplexes become stable, the algorithm can be terminated. In this case there is no need to mention the number of iterations in the beginning itself.

The main objective of the proposed work is to cluster the data; the sorting process can be carried out in the end once the memeplex is stabilized. This reduces the overhead to the processor. Instead of randomly shuffling the worst frogs in each stage, for every memeplex a worst frog is identified and exchanged with the other memeplexes in rotation, till all the worst frogs fit into a proper memeplex. This reduces the number of iterations needed to cluster the data.

4.1 Steps in SFLA

An algorithm for CSFLA is given in Fig. 2. This is also pictorially represented in Fig. 3.

Step 1:   Start.
Step 2:   Assume an initial frog population with size 'p'.
Step 3:   Decide the step size of the frogs within the memeplex 's'.
Step 4:   Randomly group the frogs to form 'm' memeplexes to hold 'n' frogs.
Step 5:   For every frog inside the memeplex, compute the fitness as:
          Fitness=Distance of each from the mean value
Step 6:   Check if all the frogs in the memeplex have step size less than or equal to the value specified in step 3.
Step 7:   If yes, go to Step: 12.
Step 8:   Else, find the frog at maximum distance and mark it as the worst frog of that memeplex.
Step 9:   Repeat steps 3 and 4 for all the 'm' memeplexes.
Step 10: The worst frog in memeplex 1 is moved to the position of the worst frog in memeplex 2, the worst frog in memeplex 2 to memeplex 3, and so on. Swap the worst frog from 'm' memeplex to the worst frog in memeplex 1.
Step 11: Repeat the steps from 4 to 9.
Step 12: Check if all the 'm' memplexes are stable.
Step 13: If yes, Sort the memplexes.
Step 14: Stop

Fig. 2. Algorithm for Cyclic SFLA

The sequence to be followed to cluster the data is given. Each block represents each step in the proposed algorithm. When all the memeplexes have frogs at step size less than or equal to s, the algorithm terminates.

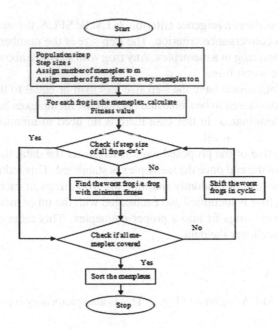

Fig. 3. Flow chart of Cyclic SFLA.

5   Results and Observation

The proposed CSFLA algorithm was implemented and its performance was evaluated based on the number of iterations and the processing time. The CSFLA algorithm was

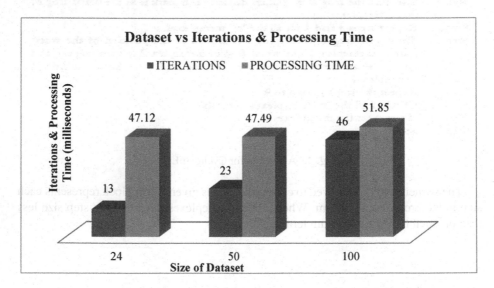

Fig. 4. Comparison of iteration count with processing time with varying sizes of dataset.

provided with univariate sequential data. Initially 24 set data was given as input, the data was clustered, and the number of iterations and the processing time were recorded. The same procedure was repeated for 50 and 100 dataset. The comparison between the various sizes of dataset with the iteration count and performance is shown in Fig. 4.

With a 24 data, the number of clusters was set as 3, the number of data in each cluster was set as 8, and the step size was set as 5. The data was clustered in 13 iterations with a processing time of 47.12 ms (Figs. 5, 6, 7 and 8).

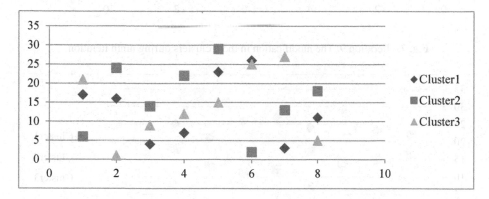

Fig. 5. Iteration 1. The initial arrangement of data in 3 clusters for dataset size 24.

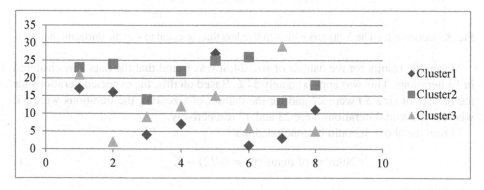

Fig. 6. Iteration 4. The rearrangement of data in 3 clusters for dataset size 24 during the fourth iteration.

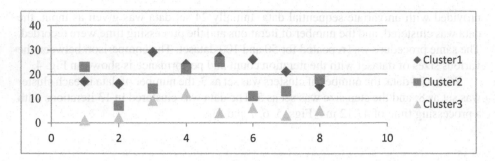

Fig. 7. Iteration 9. The modification in the 3 clusters during ninth iteration.

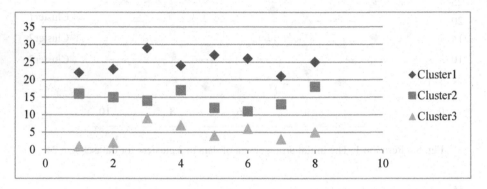

Fig. 8. Iteration 13. The 3 clusters with step size less than or equal to 4 in the thirteenth iteration.

From the results for the dataset of size 24, it was found that the data was clustered in 13 iterations. This was approximately 54%. Based on this, the expected iterations for the dataset of size 50 were 27 and for the dataset of size 100, the iterations were 54, whereas the actual iterations were 23 and 47 respectively.

From the above it could be concluded that

$$\text{Number of iterations} = (n/2) - 2, \tag{1}$$

approximately,

where 'n' is the size of the dataset.

Similarly, the processing times were 47.12, 47.49, and 51.85 ms for the datasets of size 24, 50, and 100 respectively. From this result, it could be inferred that as the data size increased, the processing size increased marginally and thereby it was established that our proposed algorithm was more suited to datasets of larger size.

6   Conclusion

The Cyclic SFLA algorithm has been proposed by modifying the SFLA for clustering the data. Since this algorithm focuses on fixing the worst data in each cluster first, it is claimed that the proposed method possesses properties like reduced number of iterations, condition for convergence and sorting of data only once. In the proposed work, data are grouped, worst data in each group are identified and then the identified data are shuffled with the other clusters in cyclic order. The process is repeated till all the data are placed in the proper cluster and the step size of the elements in the cluster is less than or equal to the given step size. The implementation of the proposed algorithm shows that the performance of the algorithm improves with the increase in the data size.

References

1. Arun Prabha, K., Karthikeyani Visalakshi, N.: Improved shuffled frog-leaping algorithm based k-means clustering. In: 4th National Conference on Advanced Computing, Applications and Technologies, May 2014. ISSN 2320-0790, Special Issue
2. Ling, J.-M., Khuong, A.-S.: Modified shuffled frog-leaping algorithm on optimal planning for a stand-alone photovoltaic system. Appl. Mech. Mater. **145**, 574–578 (2012)
3. Karakoyun, M., Babalik, A.: Data clustering with shuffled leaping frog algorithm (SFLA) for classification. In: International Conference on Intelligent Computing, Electronics Systems and Information Technology (ICESIT 2015), 25–26 August 2015, Kuala Lumpur, Malaysia (2015)
4. Luo, J., Chen, M.-R.: Improved shuffled frog leaping algorithm and its multi-phase model for multi-depot vehicle routing problem. Expert Syst. Appl. **41**, 2535–2545 (2014)
5. Amiri, B., Fathian, M., Maroosi, A.: Application of shuffled frog-leaping algorithm on clustering. Int. J. Adv. Manufact. Technol. **45**, 199–209 (2009). https://doi.org/10.1007/s00170-009-1958-2
6. Jose, A., Pandi, M.: An efficient shuffled frog leaping algorithm for clustering of gene expression data. In: International conference on Innovations in Information, Embedded and Communication Systems (ICIIECS 2014) (2014). Int. J. Comput. Appl. ISSN 0975-8887
7. Poor Ramezani Kalashami, S., Seyyed Mahdavi Chabok, S.J.: Use of the improved frog-leaping algorithm in data clustering. J. Comput. Robot. **9**(2), 19–26 (2016)
8. Darvishi, A., Alimardani, A., Vahidi, B., Hosseinian, S.H.: Shuffled frog-leaping algorithm for control of selective and total harmonic distortion. J. Appl. Res. Technol. **12**, 111–121 (2017)
9. Cheng, C., et al.: A novel cluster algorithm for telecom customer segmentation. In: 16th International Symposium on Communications and Information Technologies (ISCIT), pp. 324–329 (2016). https://doi.org/10.1109/iscit.2016.7751644
10. https://en.wikipedia.org/wiki/Memetic_algorithm
11. Muzaffar, E., Kevin, L., Fayzul, P.: Shuffled frog-leaping algorithm: a memetic meta heuristic for discrete optimization. Eng. Optim. **38**(2), 129–154 (2006)

Performance Analysis of Clustering Algorithm in Data Mining in R Language

Avulapalli Jayaram Reddy[1(✉)], Balakrushna Tripathy[2],
Seema Nimje[1], Gopalam Sree Ganga[1], and Kamireddy Varnasree[1]

[1] School of Information Technology and Engineering, VIT, Vellore, India
ajayaramreddy@vit.ac.in, seemadnimje@gmail.com
[2] School of Computer Science and Engineering, VIT, Vellore, India

Abstract. Data mining is the extraction of different data of intriguing as such (constructive, relevant, constructive, previously unexplored and considerably valuable) patterns or information from very large stack of data or different dataset. In other words, it is the experimental exploration of associations, links, and mainly the overall patterns that prevails in large datasets but is hidden or unknown. So, to explore the performance analysis using different clustering techniques we used R Language. This R language is a tool, which allows the user to analyse the data from various and different perspective and angles, in order to get a proper experimental results and in order to derive a meaningful relationships. In this paper, we are studying, analysing and comparing various algorithms and their techniques used for cluster analysis using R language. Our aim in this paper, is to present the comparison of 5 different clustering algorithms and validating those algorithms in terms of internal and external validation such as Silhouette plot, dunn index, Connectivity and much more. Finally as per the basics of the results that obtained we analyzed and compared, validated the efficiency of many different algorithms with respect to one another.

1 Introduction

R utilizes accumulations of bundles to perform diverse capacities. CRAN venture sees give various bundles to various clients as per their taste. R bundle contain diverse capacities for information mining approaches. This paper looks at different bunching calculations on Hepatitis dataset utilizing R. These grouping calculations give diverse outcome as indicated by the conditions. Some grouping methods are better for huge informational index and a few gives great come about for discovering bunch with subjective shapes. This paper is wanted to learn and relates different information mining grouping calculations. Calculations which are under investigation as takes after: K-Means calculation, K-Medoids, Hierarchical grouping algorithm, Fuzzy bunching and cross breed bunching. This paper contrasted all these grouping calculations agreeing with the many elements. After examination of these grouping calculations we depict what bunching calculations ought to be utilized as a part of various conditions for getting the best outcome.

I. Zelinka et al. (Eds.): ICSCS 2018, CCIS 837, pp. 364–372, 2018.
https://doi.org/10.1007/978-981-13-1936-5_39

2 Related Work

Few of the researches have worked on different algorithms and implemented few of them, as per that while others have worked on the existing algorithm few have implemented the new one's. applied various indices to determine the performance of various clustering techniques and validating the clustering algorithms.

3 Clustering Analysis Using R Language

Data mining is not performed exclusively by the application of expensive tools and software, here, we have used R language. R is a language and it's a platform for statistical computing and graphics. The clustering techniques which we used here are of basically four types, Partitioning methods, Hierarchal methods, Model based methods, Hybrid Clustering. Here hepatitis dataset is used to validate the results.

4 Clustering Concepts

Clustering analysis is the task of grouping a set of objects or very similar data in such a way that objects in same group or cluster are very similar to each other than to those in another groups or clusters. It is an unsupervised learning technique, which offers different views to inherent structure of a given dataset by dividing it into a many number of overlapping or disjoint groups. The different algorithm that we used in this paper to perform the cluster analysis of a particular given dataset is listed below.

4.1 Partition Based Clustering

It is based on the concept of iterative relocations of the data points from the given dataset between the clusters.

4.1.1 K-Means

The aim of this algorithm is to reduce objective function. Hers, the objective function that is considered is Square error function.

$$J = \sum_{j=1}^{k} \sum_{i=1}^{n} \left\| x_i - c_j \right\|^2$$

Where $\left\| x_i - c_j \right\|^2$ is the distance between the data point xi, and even the cluster points centroid cj.

Algorithm Steps:

- Consider a hepatitis dataset/data frame, load and pre-process the data
- Keep K points into the workspace as presented by the objects that has to be clustered. These are called the initial or starting group centroids.
- Here, the number of clusters is considered as 3.

- Closest centroid being identified and each object has been assigned to it.
- When all objects been assigned, the centroids is recalculated back again.
- Repetition is being done with Steps 2 and 3, till the centroids have no longer move.
- This gives out a separation of the objects into the groups from where the metric to be minimized should be calculated (Fig. 1).

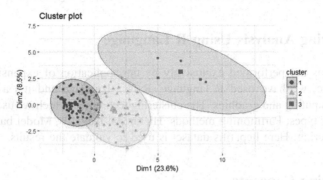

Fig. 1. K-means technique performed on Hepatitis data set in R studio.

4.1.2   K-Mediods (Partitioning Around Mediods)

K-Medoids algorithm is one of the partitioning clustering method that has been modified slightly from the K-Means algorithm. Both these algorithms, are particular meant to minimize the squared – error but the K-medoids is very much strong than the K-mean algorithm.

Here the data points are chosen as such to be medoids.

Algorithm steps:

- Load the dataset and pre-process the data
- Select k random points that considered as medoids from the given n data points of the Hepatitis dataset.
- Find the optimal number of clusters.
- Assign the data points as such to the closest medoid by using a distance matrix and visualize it using fviz function (Fig. 2).

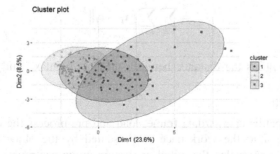

Fig. 2. K-medoids technique performed on hepatitis dataset using R studio

4.2   Hierarchy Based Clustering

This clustering basically deals with hierarchy of objects. Here we need not to pre-specify the number of clusters in this Clustering technique, like K-means approach. This clustering technique has been divided into two major types.

4.2.1   Agglomerative Clustering
This clustering technique is also known as AGNES, which is none other than Agglomerative Nesting. This clustering works as in bottom-up manner (Fig. 3).

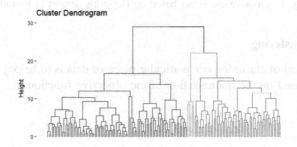

Fig. 3. Agglomerative clustering based on Hepatitis dataset in R studio

Algorithm Steps:

- Load and pre-process dataset then load the factoectra, nbclust, fpc packages..
- Assign each data object to a formed clusters such a way, that each object is assigned to one particular cluster.
- Find nearest pair of such clusters and combine them to form a new node, such that those are left out with N-1 clusters.
- Calculate distance between old and new clusters.
- Perform previous two steps till all the clusters have been clustered as one size.
- As we have N data objects, N clusters to be formed.
- At last, the data is visualized as a tree known as dendogram.

$$d = \sum_{i=1}^{n} |x_i - y_i| \qquad d = \sqrt{\sum_{i=1}^{n} (x_i - y_i)^2}$$
$$\text{Manhattan Formula} \qquad \text{Euclidean Formula}$$

4.2.2   Divisive Clustering
Divisive Clustering is just the opposite is Agglomerative algorithm. Divisive Clustering Approach is also known as Diana [3] (Fig. 4).

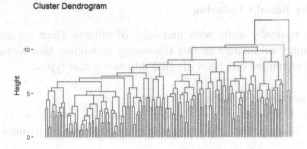

Fig. 4. Divisive clustering based on Hepatitis dataset in R studio

4.3 Fuzzy Clustering

Fuzzy is a method of clustering one particular piece of data is to belong to one or more clusters. It is based on the minimization of the objective function.

$$J_m = \sum_{I=1}^{N} \sum_{j=1}^{C} u_{ij}^m \|x_i - c_j\|^2 \quad 1 \leq m < \alpha$$

Where m is a real number which is greater than 1, u is a degree of membership of xi, in the ith dimensional data, cj is the centre dimension of the cluster.

Algorithm Steps:

- Load the dataset.
- Load the fanny function.
- At k – steps: Calculate the centres of the vectors c(k) = c(j) with U(k).

$$c_j = \frac{\sum_{i=1}^{N} u_{ij}^m . x_j}{\sum_{i=1}^{N} u_{ij}^m}$$

- Update the values of U(K), U(K + 1)

$$u_{ij} = \frac{1}{\sum_{k=1}^{C} \left(\frac{\|x_i - c_j\|}{\|x_i - c_k\|} \right)^{\frac{2}{m+1}}}$$

- If the U(K + 1) – U(K) < E then stop of the function, otherwise return back to Step 3.
- Visualize the data in the clustered format (Fig. 5).

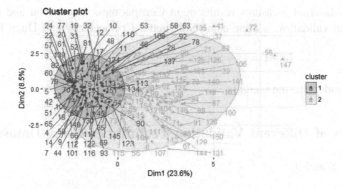

Fig. 5. Fuzzy clustering based on Hepatitis dataset in R studio

4.4 Model Based Clustering

The data will considered here is a mixture of two or more clusters.
  Algorithm Steps:

- Load and pre-process the Hepatitis dataset.
- Install Mass, ggpubr, factoextra, mclust packages in library in R studio.
- Apply mclust function to cluster the data. Then visualize the data (Fig. 6).

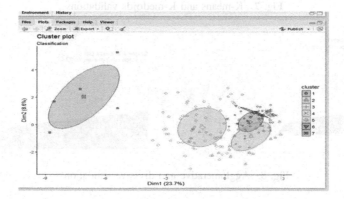

Fig. 6. Model based clustering based on Hepatitis dataset in R studio

5 Performance Analysis

5.1 Cluster Validation

Here the term of cluster validation is used here to evaluate and compare the goodness
and accuracy of different clustering algorithms results. This Internal Cluster Validation,
basically uses the internal information of all the clustering process to find out the
effectiveness and goodness of a cluster structure without knowing the external

information. Internal measures results upon Compactness, separation and connectedness. Internal validation is done using Silhouette, Connectivity and Dunn Index.

$$Index = (x * Separation)/(y * Compactness)$$

Here x and y are the weights.

6   Results of Different Validation Techniques Using Dataset

See Figs. 7, 8 and 9.

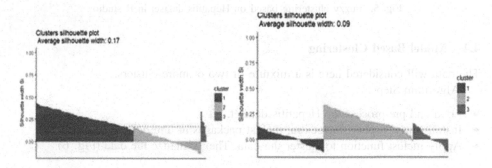

Fig. 7. K-means and K-medoids validations

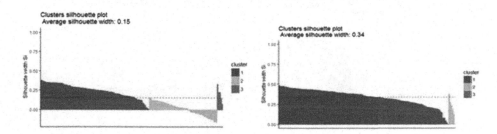

Fig. 8. Agglomerative and divisive validations

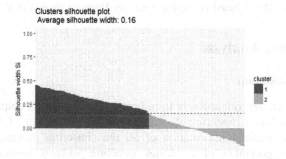

Fig. 9. Fuzzy validation

7 Choosing the Best Algorithm

Internal Validation of different clustering techniques results are listed here (Table 1).

Table 1. Comparision of clustering algorithms

K-means	Connectivity	8.0996	59.2643
	Dunn	0.6354	0.2907
	Silhouette	0.4395	0.1727
K-medoids	Connectivity	8.0996	11.0286
	Dunn	0.6354	0.6283
	Silhouette	0.4395	0.3358
Diana	Connectivity	8.0996	11.0286
	Dunn	0.6354	0.6283
	Silhouette	0.4395	0.3358
Pam	Connectivity	75.3393	102.3099
	Dunn	0.1061	0.2053
	Silhouette	0.1643	0.0942
Fanny	Connectivity	49.1099	NA
	Dunn	0.2004	NA
	Silhouette	0.1633	NA
Model	Connectivity	60.4056	NA
	Dunn	0.2138	0.14
	Silhouette	0.1298	−0.0289

8 Conclusion

This paper deals with defining few algorithms, and all those algorithms have been implemented and visualized in R studio. The clustering is done on hepatitis dataset. All the algorithms have been validated using internal measures and results have been displayed in the tabular format in terms of connectivity, Dunn, silhouette index. The measure has been considered for every algorithm and then compared overall to find out the best algorithm. As, per this we conclude that, the K-means is used for the large datasets and large number of clusters, Fuzzy clustering is not well suitable for the large number of clusters and also K-means have maximum dunn and silhouette index values when compare to all other algorithms.

References

1. Smith, T.F., Waterman, M.S.: Identification of Common Molecular Subsequences. J. Mol. Biol. **147**, 195–197 (1981)
2. May, P., Ehrlich, H.-C., Steinke, T.: ZIB structure prediction pipeline: composing a complex biological workflow through web services. In: Nagel, W.E., Walter, W.V., Lehner, W. (eds.) Euro-Par 2006. LNCS, vol. 4128, pp. 1148–1158. Springer, Heidelberg (2006). https://doi.org/10.1007/11823285_121
3. Foster, I., Kesselman, C.: The Grid: Blueprint for a New Computing Infrastructure. Morgan Kaufmann, San Francisco (1999)
4. Czajkowski, K., Fitzgerald, S., Foster, I., Kesselman, C.: Grid information services for distributed resource sharing. In: 10th IEEE International Symposium on High Performance Distributed Computing, pp. 181–184. IEEE Press, New York (2001)
5. Foster, I., Kesselman, C., Nick, J., Tuecke, S.: The physiology of the grid: an open grid services architecture for distributed systems integration. Technical report, Global Grid Forum (2002)
6. National Center for Biotechnology Information. http://www.ncbi.nlm.nih.gov
7. Kanungo, T., Mount, D.M., Netanyahu, N.S., Piatko, C.D., Silverman, R., Wu, A.Y.: An efficient k-means clustering algorithm: analysis and implementation. IEEE Trans. Pattern Anal. Mach. Intell. **24**(7), 881–892 (2002)
8. Liu, Y., Li, Z., Xiong, H., Gao, X., Wu, J.: Understanding of internal clustering validation measures. In: 2010 IEEE 10th International Conference on Data Mining (ICDM), pp. 911–916. IEEE, December 2010
9. Liu, Y., Li, Z., Xiong, H., Gao, X., Wu, J., Wu, S.: Understanding and enhancement of internal clustering validation measures. IEEE Trans. Cybern. **43**(3), 982–994 (2013)

Efficient Mining of Positive and Negative Itemsets Using K-Means Clustering to Access the Risk of Cancer Patients

Pandian Asha[(⊠)], J. Albert Mayan, and Aroul Canessane

Department of Computer Science and Engineering,
Sathyabama Institute of Science and Technology, Chennai, India
ashapandian225@gmail.com, albertmayan@gmail.com,
aroulcanessane@gmail.com

Abstract. Application of Data Mining tasks over health care has gained much importance nowadays. Most of the Association Rule Mining techniques attempts to extract only the positive recurrent itemsets and pay less attention towards the negative items. The paper is all about medical assistance, which concentrates on retrieving both positive and negative recurrent itemsets in a efficient way by compressing the overall data available. Stemming methods help in this compression of data to half of its size in order to reduce and save memory space. To analyze data, the clustering technique is applied, especially the k-means clustering is used, as it is found to be more effective, easy and less time consuming method when compared to other clustering flavours.

Keywords: Clustering · Positive itemsets · Negative itemsets · Stemmer

1 Introduction

People all over are worried about their health conditions and to screen those diseases, they undergo some online applications, download softwares, search in google, etc. Thus the first verification step is processed by the information available online or some dedicated mobile applications. Only if this initial verification step provides no better solution to heal the disease or if the situation is not controlled, then they adapt for consulting a doctor. Our work also concentrates on such an online information as a helpline for patients to screen their disease and to provide a detailed description about that disease such as cure, prevention, side effects.

Cancer is the dreadful disease and it is the most increasing disease too. But most of the people is not aware of the information regarding cancer like, what are its symptoms? What all cancers are there? What are its treatments and side effects? Awareness about the disease is less as there are more types of cancer, such as lung, eye, kidney, liver, stomach, etc. Eventhough cancer is a dreadful disease, there is a cure when it is recognized at an early stage. The paper discusses all about the early stage cancer assistance to help patients by providing them both the positive itemsets and negative itemsets, inorder to provide them suggestions to consult for further references i.e. a doctor. Thus,

I. Zelinka et al. (Eds.): ICSCS 2018, CCIS 837, pp. 373–382, 2018.
https://doi.org/10.1007/978-981-13-1936-5_40

in this work partition of data collected is carefully done and is analyzed for better performance.

2 Existing System

The existing method aims to mine and discover risk factors from an electronic medical record which is a large dataset. Mining association rules [1] from this large set of data is a bit complicated and more time consuming. Inorder to save time, different summarization techniques are being discussed and the best suitable method is found. Thus the key contribution in this method is identifying rules with high significant risk from a summarized set of data. Four summarization techniques [2–4] such as APRX-Collection, RPGlobal, TopK, BUSwere discussed from where the most suitable and best method is found. Bottom Up Summarization method is selected as it provides a better summarized quality dataset as it reduces redundancy more in number. In this method, only positive itemsets are focused and concentrated more on summarizing the dataset.

PNAR calculation was proposed by Zhu et al. [5] that mines legitimate guidelines snappier through relationship coefficient measure and pruning methodologies. In the first place, positive and negative standards are removed from the regular and rare itemsets. Utilizing a pruning system, fascinating positive principles are mined that fulfills both least backing and certainty measure alongside a relationship coefficient keeping in mind the end goal to evacuate repudiating rules. At that point intriguing negative guidelines [6] are mined as positive tenets with the exception of that, the base backing and certainty is distinctive. Along these lines, all legitimate affiliation tenets are found.

Swesi et al. [7] Integrated two calculations, for example, Positive Negative Association Rules (PNAR) and Interesting Multiple Level Minimum Support's (IMLMS) to another methodology called PNAR_IMLMS. The unique IMLMS methodology is marginally adjusted at prune step in order to evacuate insignificant tenets, this produces fascinating incessant and occasional itemsets. At that point relationship and Valid Association Rule in light of Correlation Coefficient and Confidence (VARCC) measures are utilized to mine positive guidelines from regular itemsets and negative principles from both successive and rare itemsets [8, 9]. Accordingly, legitimate positive and negative affiliation principles are come about abstaining from uninteresting guidelines. Shang et al. [10] described PNAR_MDB in P_S measure algorithm to mine association rules from multiple databases. From a large company, multiple database along with its weightage is retrieved. Thus support count is calculated with slight changes by including weight factor, but confidence remained same. Then itemsets are pruned with correlation measure, the survived rules are then passed on to undergo P–S measure for mining more interesting rules. The number of rules gets decreased with more interestingness [10] that avoids knowledge conflicts within the database while mining association rules simultaneously.

Shen et al. [11] introduces a new Interest_support_confidence approach which overcomes traditional Support_confidence that misleads association rules in 2009. The new mining method initially checked whether minimum interest has met and then correlation measure with the support measure is determined. This evaluation finds positive, negative and independent rules. After that the positive and negative rules are checked whether it satisfies minimum support and confidence. The only difficulty is, support and confidence [12–14] for negative rules cannot be found directly as it includes absence of itemsets. Still the method generates a reduced set of positive association rules with more meaningful negative association rules. Negative Association Rule (NAR) was effectively focused in the work [15], which uses Apriori to recover positive itemsets at first. At that point from the tenets recovered, k negative itemsets are extracted. Later applicant era and pruning is done to locate the legitimate positive and adversely related standards. Therefore, this methodology delivers contrarily related standards from the absolutely related principles decreasing an additional output to the database.

3 Proposed System

3.1 Proposed Method

The overall concept is about rule mining from existing information available about cancer using stemmer, clustering analysis and ranking based on a weightage provided to each rules.

Initially user query is nothing but the symptoms they undergo. The query then undergoes stemming and the stemmed query is passed on to the database server where all further information about cancer is partitioned and stored in it (Fig. 1). Then the stemmed input is associated with the dataset inorder to retrieve all the associated rules. All rules are ranked and the most prioritized one is retrieved and submitted to the user.

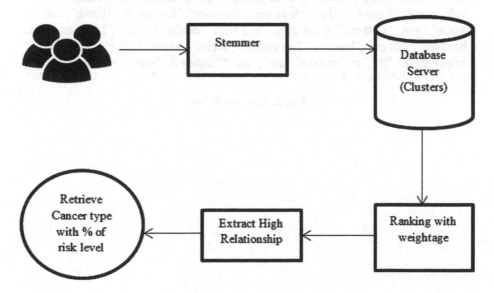

Fig. 1. Proposed architecture

3.2 Stemmer

Stemming is a process of reducing the words to its root word by stripping or replacing the prefix or suffix or even both. Here we use the affix stemming method to reduce the word as it stripes both the suffix and prefix. But while applying the stripping method alone may result in meaningless words. Thus we include affix replacement along with stripping where ever it is needed. Affix stemmer is the best and fastest method as it does not maintain a separate lookup table thus saving memory space. Before we start with stemming, stop words should be filtered, in order to provide better mining of accurate result. Stop words are nothing but the common words to interconnect between terms to produce a meaningful sentence (Fig. 2). Thus absence of stop words will not fulfil a completeness in the sentence.

```
// Removewords start

String stop[]={"a","able","about","above","abroad","according","accord
("allow","allows","almost","alone","along","alongside","already","also","a
"anybody","anyhow","anyone","anything","anyway","anyways","anywl
"asking","associated","at","available","away","awfully","b","back","bacl
"begin","behind","being","believe","below","beside","besides","best","be
"cause","causes","certain","certainly","changes","clearly","c'mon","co",
"contains","corresponding","could","couldn't","course","c's","currently",
"doing","done","don't","down","downwards","during","e","each","edu","
"ever","evermore","every","everybody","everyone","everything","ever
"following","follows","for","forever","former","formerly","forth","forwa
"gone","got","gotten","greetings","h","had","hadn't","half","happens","ha
"hereby","herein","here's","hereupon","hers","herself","he's","hi","him",
"immediate","in","inasmuch","inc","inc.","indeed","indicate","indicated",
```

Fig. 2. Stop words list

Algorithm Affix-Stemming:

1. *Initialize a list of stop words S to be filtered.*
2. *For each keyword K in input query I,*

   I is compared with the list of stop words.
   Matched keywords from I is removed.
3. *Affix_Stemmer()*
   a. *step1(word)*

   if K ends with "at" -> "ate", else
   if K ends with "bl" -> "ble", else
   if K ends with "iz" -> "ize".

   b. *step2(stem)*

   if K ends with "y" -> "i".

   c. *step3(stem)*

   if K ends with " ational ","ation","ator" -> " ate "
   if K ends with " tional " -> " tion "
   if K ends with " anci " -> " ance "
   if K ends with " izer ","ization" -> " ize "
   if K ends with " iveness "-> " ive"

   d. *step4(stem)*

   if K ends with " icate ","iciti","ical"-> " ic "
   if K ends with " ative ","ful","ness"-> " NULL "
   if K ends with " alize "-> " al "

   e. *step5(stem)*

   if K ends with "al","ance","ence","er","ic","able","ible",
   "ant","ment","ent"-> no change

   f. *step6(stem)*

   if K ends with "sses" -> ss, elseif K ends with "ies" -> i

Applying the affix stripping stemmer to the input we get,

Muscle-cramps → *Muscle-cramp*
Irregularities → *regular,*
Depression → *Depress*

Applying the affix stripping and replacement stemmer to the input we get,

Decreased → *Decreas* → *Decrease*
Troubling → *Troubl* → *Trouble*
Frequent urination → *Frequent urinat* → *Frequent urinate*

3.3 Cluster Formation

Clustering does grouping of similar objects(data) from the dataset D and thus forming different clusters(C). Every cluster characteristics is dissimilar to all other cluster. The objects which donot belong to any of the clusters is the outlier. In our work, frequent itemsets are extracted from the cluster formed as infrequency lies with analyzing the outliers also. So every different cancer types are grouped in different clusters. As we use partitioning method, every object must belong to atleast one cluster and every cluster must have atleast few objects belonging to it. K-means partitioning method is used to form clusters where clusters are selected randomly and applied distance formula to locate every objects to its minimum distant clusters and thus changes are made for k resultant clusters. K-means method is one of the best cluster partitioning method.

Algorithm_K-Means:

1. The initial number of clusters(c_l) are randomly choosed where $l = 1, 2, ...p$.
2. For each clusterc_l,
    Mean is calculated by

$$m_l = \left(\frac{1}{n}\right) \sum_{i=1}^{n} x_i$$

where x_l is the objects inc_l.
3. With m_l and x_l ,Compute the Euclidean distance between every mean of a cluster c_l for every object$x_i$present within and outside the cluster.
4. Find the minimum distance,
    For each x_i to c_l
        Find the minimum distance
        Replace the objects to minimum distant cluster where ever it is necessary.
5. Repeat the process till every x_i in D is covered.
6. Output the p resultant clusters.

3.4 Ranking

Ranking is used to prioritize the rules extracted from D by apriori algorithm which is a candidate generation algorithm. After extracting the associated rules from D, ranking is done to the extracted rules based upon the weightage given to every objects in the cluster. The weightage is based upon some analysis of term frequency where the term implies the symptoms that has been so frequent with the patients who have undergone that type of cancer.

3.5 Retrieval of Related Information

The output to the user is top ranked rules where they more frequently occur within the patients. And to alert the patients with its disease risk, the type of cancer along with the percentage(%) of possibility is provided as a result to the user (Fig. 3). The resultant cancer type can be viewed further in order to assist the user with its full description about the cure, prevention and treatment (Fig. 4). Another part is the negative rules, where the patients who have undergone the same type of symptom will not have resulted in cancer. Such part of the symptoms are also considered as outliers and there occurrence can be rare in some cases also. These itemsets can also be considered as a strength to users as they may also become one of such an outlier.

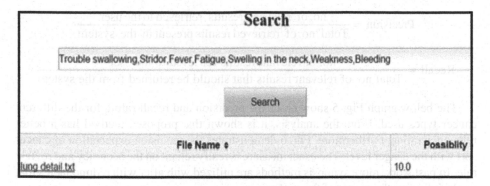

Fig. 3. Output with all possible cancer type and its percentage of probability

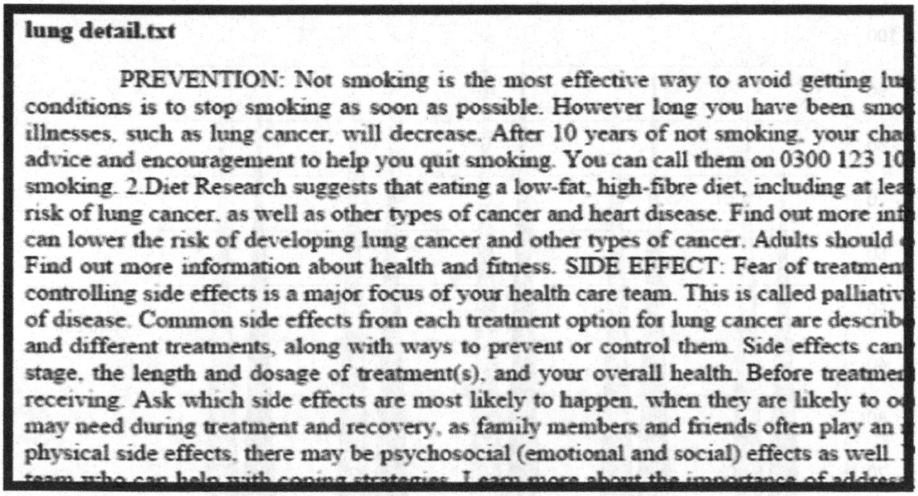

lung detail.txt

      PREVENTION: Not smoking is the most effective way to avoid getting lu
conditions is to stop smoking as soon as possible. However long you have been smo
illnesses, such as lung cancer, will decrease. After 10 years of not smoking, your cha
advice and encouragement to help you quit smoking. You can call them on 0300 123 10
smoking. 2.Diet Research suggests that eating a low-fat, high-fibre diet, including at lea
risk of lung cancer, as well as other types of cancer and heart disease. Find out more inf
can lower the risk of developing lung cancer and other types of cancer. Adults should
Find out more information about health and fitness. SIDE EFFECT: Fear of treatmen
controlling side effects is a major focus of your health care team. This is called palliativ
of disease. Common side effects from each treatment option for lung cancer are describ
and different treatments, along with ways to prevent or control them. Side effects can
stage, the length and dosage of treatment(s), and your overall health. Before treatmen
receiving. Ask which side effects are most likely to happen, when they are likely to o
may need during treatment and recovery, as family members and friends often play an
physical side effects, there may be psychosocial (emotional and social) effects as well.
team who can help with coping strategies. Learn more about the importance of addres

Fig. 4. Detailed guidance on prevention, side effect and treatment for every cancer type

4   Performance Evaluation

4.1   Evaluation Measures

Accuracy of the system is evaluated with Precision and Recall measures. The correctness of the system can be predicted with the accuracy measure. From the proposed model, the accuracy is the relevant result retrieved regarding the user query. Two terms such as 'relevant result' and 'retrieved result' is to be discussed to evaluate accuracy. Relevant result is when the system generates more relevant output to the input given and retrieved result is nothing but extracted output to the user input which may not be much relevant. Thus, 'Precision' and 'Recall' measures the quality and quantity of relevant results retrieved related to the query.

$$\text{Precision} = \frac{\text{no. of relevant results retrieved to the user}}{\text{Total no. of retrieved results present in the system}}$$

$$\text{Recall} = \frac{\text{no. of relevant results retrieved to the user}}{\text{Total no. of relevant results that should be returned from the system}}$$

The below graph Fig. 5 show cases the precision and recall rating for the different cancer types used. From the analysis, it is shown that proposed method has a better relevancy rating. Furthermore, Fig. 6 demonstrates the precision expectation of cancer sort with the danger level. The exact measure is really reliant on the accuracy and review rate. In past work, more synopsis methods are utilized with after with redundancies that might influence the nature of the outcome.

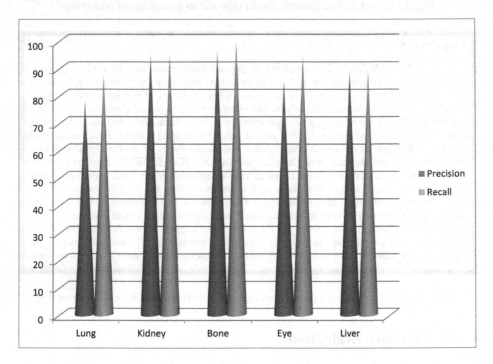

Fig. 5. Precision and recall analysis

In the proposed procedure, stemming strategy is utilized for packing the dataset which diminishes more excess standards than past work. In this manner, from the assessment come about, the proposed strategy creates a right and important result to the client question given. The measure is more, as it result in better exact framework giving a direction and consciousness of the clients who hunt down a help to cancer.

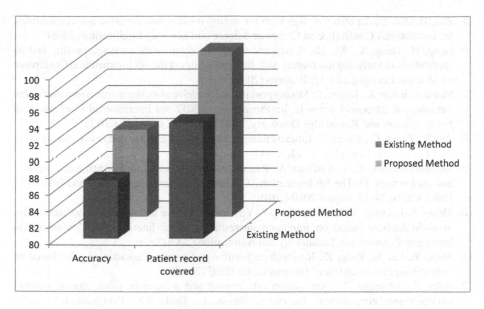

Fig. 6. Comparison of existing and proposed system

5 Conclusion and Future Enhancement

Overall, we conclude that the affix stemmer used for both removal and replacement enhances speed and correctness of stemming thus reducing the memory space. But it's limited to some irregular forms and compound words. And the k-means clustering result in a better cluster formation to mine the positive itemsets where outliers helped in negative ones. K-means is used for better computation time. Ranking the rules provided with an advantage of prioritized and relevant retrieval of output to the user in order to assist them with accurate and more relevant result. In future, more efficient stemmer can be used in order to handle the irregular and compound words. Patient record history of cancer can be included along with the cancer types so as to provide a better solution for the better treatment of user. Also, the work can be extended such that it can fit into the giant, Big data.

References

1. Agrawal, R., Srikant, R.: Fast algorithms for mining association rules. In: Proceedings of the 20th VLDB, Chile (1994)
2. Veleti, A., Nagalakshmi, T.: Web usage mining: an incremental positive and negative association rule mining approach. Int. J. Comput. Sci. Inf. Technol. **2**, 2862–2866 (2011)
3. Soltani, A., Akbarzadeh-T, M.-R.: Confabulation-inspired association rule mining for rare and frequent itemsets. IEEE Trans. Neural Netw. Learn. Syst. **25**, 2053–2064 (2014)
4. Simon, G.J., Caraballo, P.J., Therneau, T.M., Cha, S.S., Castro, M.R., Li, P.W.: Extending association rule summarization techniques to assess risk of diabetes mellitus. IEEE Trans. Knowl. Data Eng. **27**(1), 130–141 (2015)

5. Zhu, H., Xu, Z.: An effective algorithm for mining positive and negative association rules. In: International Conference on Computer Science and Software Engineering (2008)
6. Geng, H., Deng, X., Ali, H.: A new clustering algorithm using message passing and its applications in analyzing microarray data. In: Proceedings of the 4th International Conference on Machine Learning and Applications (2005)
7. Swesi, I., Bakar, A., Kadir, A.: Mining positive and negative association rules from interesting frequent and infrequent itemsets. In: Proceedings - 2012 9th International Conference on Fuzzy Systems and Knowledge Discovery, FSKD 2012, pp. 650–655 (2012)
8. Wu, X., Zhang, C., Zhang, S.: Efficient mining of both positive and negative association rules. ACM Trans. Inf. Syst. (TOIS), **22**(3), 381–405 (2004)
9. Ramasubbareddy, B., Govardhan, A., Ramamohanreddy, A.: Mining positive and negative association rules. In: The 5th International Conference on Computer Science and Education, Hefei, China, 24–27 August 2010 (2010)
10. Shang, S.-J., Dong, X.-J., Li, J., Zhao, Y.-Y.: Mining positive and negative association rules in multi-database based on minimum interestingness. In: International Conference on Intelligent Computation Technology and Automation (2008)
11. Shen, Y., Liu, J., Yang, Z.: Research on positive and negative association rules based on "interest-support-confidence" framework. In: IEEE (2009)
12. Asha, P., Jebarajan, T.: Association rule mining and refinement using shared memory multiprocessor environment. In: Padma Suresh, L., Dash, S.S., Panigrahi, B.K. (eds.) Artificial Intelligence and Evolutionary Algorithms in Engineering Systems. AISC, vol. 325, pp. 105–117. Springer, New Delhi (2015). https://doi.org/10.1007/978-81-322-2135-7_13
13. Asha, P., Jebarajan, T.: SOTARM: size of transaction based association rule mining agorithm. Turk. J. Electr. Eng. Comput. Sci. **25**(1), 278–291 (2017)
14. Asha, P., Srinivasan, S.: Analyzing the associations between infected genes using data mining techniques. Int. J. Data Min. Bioinform. **15**(3), 250–271 (2016)
15. Tseng, V.S., Cheng-Wei, W., Fournier-Viger, P., Yu, P.S.: Efficient algorithms for mining top-k high utility itemsets. IEEE Trans. Knowl. Data Eng. **28**(1), 54–67 (2016)

Machine Learning

Forecasting of Stock Market by Combining Machine Learning and Big Data Analytics

J. L. Joneston Dhas[1](✉), S. Maria Celestin Vigila[2], and C. Ezhil Star[3]

[1] Department of Computer Science and Engineering, Noorul Islam Centre for Higher Education, Kumaracoil 629180, Tamilnadu, India
joneston.jl@gmail.com
[2] Department of Information Technology, Noorul Islam Centre for Higher Education, Kumaracoil 629180, Tamilnadu, India
celesleon@yahoo.com
[3] Department of Computer Science and Engineering, Arunachala College of Engineering for Women, Manavilai 629203, Tamilnadu, India
ezhilstar@gmail.com

Abstract. Big data has large volume, velocity and variety. The data is taken from social network, sensor, internet etc. It has stream of information and has both structured and unstructured and one petabyte of information is generated daily and the data's is not in an order. Stock market analysis is not an easy task and prediction is created based on several parameters. The financial markets are fast and complex and the market participants face difficult to manage the overloaded information. The sentiment analysis is useful to process the textual content and the results are filtered and give the meaningful and relevant information. The technical analysis is to predict future value based on the past value. In this research the combination of technical analysis using machine learning and big data analytics is implemented and an accurate prediction is generated in the stock market.

Keywords: Big data · Stock market · Market hypothesis · Random walk
Data science

1 Introduction

Big data has several types of data such as text (structured, unstructured or semi-structured), multimedia data (audio, images, video). Dobre and Xhafa [1] reports 2.5 quintillion bytes of information are produced in the world every day and out of these 90% of the data are unstructured. As it has huge amount of information and has both valuable and unwanted information. So from this large amount of information short and valuable needed information will be taken and it is used to analysis for the future prediction and is used to improve the business.

Big data analytics is used in all areas like improving the performance of networks, business, stock market prediction etc. Chin-lin et al. [2] proposed big data analytics improving the performance of network and improve the performance. Thaduri et al. [3] implemented the analytics method to improve the railway management. In this paper

I. Zelinka et al. (Eds.): ICSCS 2018, CCIS 837, pp. 385–395, 2018.
https://doi.org/10.1007/978-981-13-1936-5_41

they take several parameters to improve it. Biag and Jabeen [4] create the data analytics to monitor the student's behaviour. So the big data is used in all the areas to improve the efficiency of the business.

In this research article the big data is used to predict for the stock market. The stock market is one of the main businesses in all over the world and millions of people involved in it and they do trading or investment. Prediction is very important in the stock market and accurate prediction will give good profit. One way of identify the economy status of the country is by the stock market. It is not isolated in one country and it depends on global economics. The data in the stock market varies on every second. In the social network many websites are available and from all the websites the sample data is taken and many people give many commands about the company and it is the sentiment data. Sometimes the people may be biased and it will not give the good prediction.

The stock market will move up or down depending upon the selling or buying habit of the customers and it gives the volume of the company. Some of the people lose their valuable money because of wrong prediction. So prediction is very important in the share market to make profit. The smarter person than others can able to make money in the stock market. So each and every day the data will be changed and depending on the data the prediction will be varied daily.

The stock value is changed due to many reasons like, profit or loss, profit booking, new order booking, agreement with other company or government, economy crisis of the same country, economy crisis of the other country and election result of the country. The result impact of the stock value of the same owner's other company etc. So investing money in the share market is not a little easy job. Before investing the money in the share market a big analysis must be done to book the profit.

The analysis is not done before buying the share and is should done before selling the share. The person does not know when the share value will be decreased. So the analysis must be done daily and the person will sell the stock in correct time and make profit from that stock. In the stock market there are thousands of company are involved and lot of channels, web sites and many analysis are done by many people and the refer some of the company stock to buy or sell. So the analysis will be done for a 360 degree view of the company to make a profit. So a manual calculation is not able to calculate all the values and find the buying pressure or selling pressure. Saul and Roweis [5] implemented unsupervised learning of the data and it will be done in large amount of data.

1.1 Big Data Analytics Model

Big data collects the information from various sources and take only the useful information for the people, government and business people. This information is used for the business people to analysis the data and to improve their business and it is used to analysis the stock market to make the profit. Kolaczyk [6] proposed a statistical analytical method of network data. In this the data is collected from the network and the data is analysis. Also he proposed an efficient model and method for analysis the data. Skretting and Engan [7] proposed a dictionary learning algorithm which analysis the information from the internet and social media. Nowadays the stock analysis is done using the big data. Uhr et al. [8] proposed a sentiment analysis for stock market. Here the data is collected from various sources and the useful information related to stock analysis is taken and it is analysed to predict the future value.

Current technologies will not be able to analyse the data from large amount information. In the case of data warehouse architecture it analyse the small data and also it delivers the product and not the value. It is a batch process and not the real value and the implementation is done by programming. So the new technology big data is able to analyse the information from the social media, sensor network etc. It analyse the real time information and implementation is done by orchestration. The task of big data analyser has to understand the detailed technical knowledge and choose the correct platform and software for analysing the data. Before analysing the data many challenges has to be addressed. The data challenge is the challenge that depends on the data characterisation i.e., volume, velocity, visualization, value, volatility, veracity, variety and discovery. Process challenge depends on the techniques used for analysis, i.e., how to get, integrate, transform the data, data acquisition, cleansing, choose the correct model for analytics and providing the result and the management challenge denotes the challenges that will be faced by the management for analysing big data. It includes security, privacy, data ownership, information sharing, data governance, operational expenditures. To analyse the information in big data it has three layer structures which is used to create the expected result with more accurate and is shown in the Fig. 1.

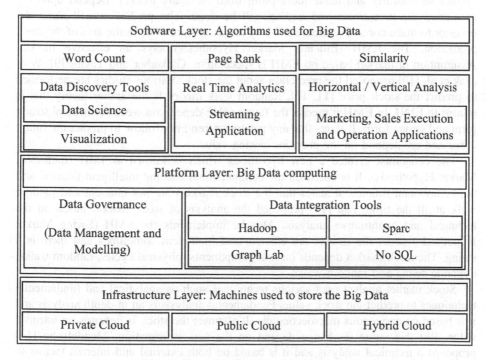

Fig. 1. Layer model of big data analytics.

The analytic process is divided as three layers. The first layer is the infrastructure layer and it is used for setting up communication, storage and computing. The analysis is performed in the platform layer. Many internet companies like amazon, Yahoo twitter and Face book and traditional analysis company use hadoop for analysis the data. But it is not used for all the analysis. So depending on the analysis the different platform was used. In software layer the different program abstractions are available. The user will have to choose the software depending on their needs.

The selection of analytical method is important to extract the sense of the data. The descriptive analysis method helps to identify the current status of the business. Inquisitive analysis method which will validate or reject the business hypothesis and give the answer to the questions why this moment happened in the past. Predictive analysis method estimates the future outcome of the supply chain management and used to identify the opportunities or risks in future. Prescriptive analysis method helps responding like how it is now and when decision is changed and how it will be in future and Pre-emptive analysis helps to recommend and to take precautionary action.

2 Related Work

Many researchers proposed various models to predict the future value. Nowadays the stock market forecasting gives more attention because it will guide the investors to predict successfully and make more profit from the share market. Depend upon the prediction the investment and trading will be done. The prediction will make the investor to make corrective measure. Many researches are done in the area of the stock prediction. The EMH (Efficient Market Hypothesis) says an effective market. Assumption of market based on EMH is speculative. Gallagher and Taylor [9], Walczak et al. [10] proposed the prediction about the stock. Various studies are performed to predict the stock price [11, 12]. Random walk theory is the anther theory that is related to EMH. In EMH it predict the future value depend on weak form, semi strong form and strong form. It states that any pattern or trend not follow to predict the future value and are depend on the previous closing value.

The economist created a new hypothesis which is known as IMH (Inefficient Market Hypothesis). It is based on the computational and the intelligent finance, and the behavioural finance. It states that the stock markets are not efficient and random walk at all the time. Pan [13] proposed the analysis of stock market based on the technical and quantitative analysis. Also he implements the SMH (Swing Market Hypothesis) states the market are efficient and inefficient sometimes and there is a swing. The stock market depends on four components: physical cycles, random walks, dynamic swing and random walks.

Stock market analyst uses various techniques such as analytical and fundamental techniques to predict the stock value. Fundamental analysis is an in-depth analysis and it is based on exogenous macroeconomic. It assumes the stock is depends on intrinsic value. But this value is change depend on the new information. Mendelsohn [14] proposed a technical analysis and it is based on both external and internal factors to predict the stock value. It uses the statistical chart, open price, closing price, high, low and volume to predict the future value. The stock market analysis can also be done based on Fuzzy Cognitive Maps (FCM) [15]. Here the authors proposed the FCM based on dynamic domination theory. Koulouriotis et al. [16] proposed a Fuzzy Cognitive Map-based Stock Market Model and it is the powerful tool to forecast the stock market. Senthamarai Kannan et al. [17] proposed stock market forecasting using the data mining and it predict the value using the global and other issues.

Schumaker and chen [18] implements textual analysis for the prediction of stock market and they take the information from the financial news and depending on the news they forecast the stock market. Alkhatib et al. [19] proposed a K-Nearest Neighbour (KNN) Algorithm for the prediction of stock market. They predict the value based on the closing value of the current day. Joneston Dhas et al. [20] proposed a framework to securely store the big data. Maria Celestin Vigila and Muneeswaran [21] proposed a security method to store the data. Maheswaran and Helen Sulochana [22] proposed the bandwidth allocation to transfer the data to the cloud.

ANN is a supervised learning method and it automatically trains the data and the output will be generated automatically. In this case many several artificial neurons are interconnected and produce the output. In the feed forward neural network several neurons are interconnected in the form of layers. The neurons process the data and provide the output. It has input set [Xi], where $i = 1, 2, \ldots a$, and it produces output [Zi], where $i = 1, 2, 3 \ldots p$. The input signal gives the input to the neuron and transmitted through the connection which multiplies the strength by weight W and forms the product WX. The bias b is added with weighted input and passed through the transfer function and the output is generated. The bias b and weight w are adjusted so a desired behaviour is exhibited by the neuron. The Fig. 2 explains the artificial neural network.

Input Layer   Hidden Layers   Output Layer

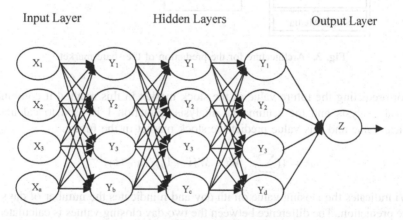

Fig. 2. Artificial neural network

In the case of Back Propagation Neural Network there are input layer, hidden layer and output layer. The input is given in the input layer and much process is done in the hidden layer. The output of one neuron will be given as the input to the next neuron. The output layer generates the output. It has a inputs, b, c, d number of neurons in first second and third hidden layer respectively and in the output layer it has p neurons.

3 Proposed Architecture

In the proposed architecture it combines the Artificial Neural Network and big data to predict the accurate stock value. In artificial neural network it will analyse the data based on the historical value and it predict the value automatically. Based on the historical value it generates the output and technically the result will be accurate. The neural network which will be trained is an expert in that particular area and the output generated by it is very accurate. But this technical calculation is not predicting the accurate value. Because in the share market the value depends on many factors like Half yearly or Quarterly result, National or International level dealing, Combined with other company, Anti-dumping duty of their product, National or International eco-nomical factor etc. In this paper the technical result and the sentimental analysis will be combined and the sentimental analysis is done using the big data analytics. So the combination of machine learning and big data analytics will give the accurate pre-diction about the share market value. The basic architecture of the analysis of stock market is shown in Fig. 3.

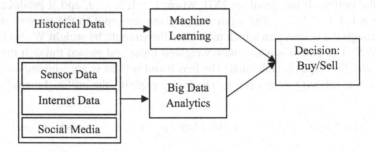

Fig. 3. Architecture for the prediction of the share market.

For predicting the future value of the stock market in this method it combines the technical analysis and the sentimental analysis. Equation 1 is used to calculate the technical value and this value predict the stock market in the future.

$$A = \left(\left(\sum\nolimits_{i=1}^{n} Wi - W(i-1)\right)/(n-1)\right)^{1/2} \tag{1}$$

Wi indicates the closing value on ith day and n indicates the number of days taken for the prediction. The difference between the two day closing values is calculated upto n days and the average is the final value for the final prediction and shown in Eq. 2.

$$B = \sum\nolimits_{i=0}^{n} Vi/n \tag{2}$$

Vi indicates the volume of the ith day and the value is calculated for n number of days and the average volume for n days is in Eq. 3.

$$C = (\sum\nolimits_{i=0}^{n} Vi/n)/U \qquad (3)$$

U indicates the average volume of the month and the final prediction value is calculated by the Eq. 4.

$$D = ((\sum\nolimits_{i=1}^{n} Wi - W(i-1))/(n-1))^{1/2} * ((\sum\nolimits_{i=0}^{n} Vi/n)/U) \qquad (4)$$

In this case A refers the average technical analytic value for the n trading days. B indicates the average volume of n trading days. C is the average volume of n days to the average volume of one month. If D is positive then it indicates a buying pressure and the share value will be increased in near future and if D is negative a selling pressure. If D is near to zero the share can be hold by the investor. Only this technical calculation will not predict the correct value. Some other factor like half yearly or Quarterly result, National or International level dealing, combined with other company, Anti-dumping duty of their product, National/International economy crisis will also play a major role in the stock market and it will be analysed by the big data. Depending upon the big data analysis result the prediction will be accurate.

In big data the analysis involves three steps:

Step 1: Capture – Collect the information from the social media, sensor and internet. In the case of stock market different websites, channels are available and it provides the new valuable information about the stock market. The data is streamed in HDFS (Hadoop Distributed File System).

The company data is taken from the internet and the recent activities of the company is taken from the news and tweet. News article is taken by Mozenda web crawler and tweet information is taken by twitter search API. The data will be streamed in HDFS and analyse the positive and negative of each company. In this method the data is taken from NSE (National Stock Exchange), Financial Express, Economic Times and Money Control websites and also the tweet information is taken from money control and from this websites different information is taken from the expert's overview.

Step 2: Analyse – Analysis the data using Hadoop. Here the news article is collected from different sources and it will be analysed. The collected information is processed before analysis will take place. Processing includes removing stop words, URL and duplicates. In sentiment analysis the processed data is analysed using HDFS. Hive is used for collecting sentiment of tweets and news.

Algorithm: Sentiment analysis
Input   : Data from various sources in the internet with Keyword.
Output: All the words with the keyword.
Begin:
    Sentiment[R] = 0
    For row 1 to n
        Compare word in dictionary for all rows R and apply Sentiment Word.
        Sentiment[R] = Sentiment[R] + 1
End

The news article and tweet information are aggregated and to give all sentiment about the company. The sentiment indicates the positive or negative impact about the company. The obtained result are observed in the form of graph using R and Hadoop.

Step 3: Result – Provide the valuable and summarized result to the user.

4 Performance Analysis

The performance metrics is calculated for the technical analysis and then combined with the sentimental analysis so any false positive and false negative can be identified and the stock forecasting depending on international economics or any other external factor can be predicted easily. In the technical analysis based on the previous data like volume, closing value the future value is predicted. The following Fig. 4 gives the prediction result of technical analysis.

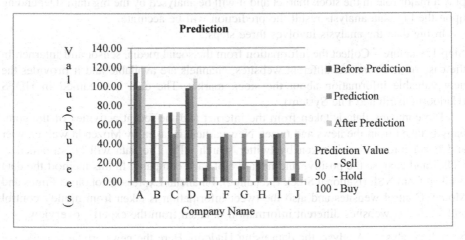

Fig. 4. Prediction using technical analysis

The precision recall is used to find the accuracy of the predicted result and precision is the average value of the probability in relevant value. Recall is the average value of the probability in complete value. Precision and recall is defined in Eqs. 5 and 6.

$$\text{Precision} = \frac{tp}{tp + fp} \tag{5}$$

$$\text{Recall} = \frac{tp}{tp + fn} \tag{6}$$

The accuracy is calculated by the Eq. 7.

$$\text{Accuracy} = \frac{tp + tn}{tp + tn + fp + fn} \tag{7}$$

Where tp is true positive, tn is true negative, fp is false positive and fn is false negative and all this parameter is used to find the accuracy of the predicted value. In this method it has 87% of precision, 89% of recall and 89% of accuracy.

By using the big data analytics the sentiment analysis is created and it analysis the stock prediction by the news channel, tweet information and internet. In this the positive and negative values are separately identified and by this the stock prediction is done. The Fig. 5 gives the prediction of the company.

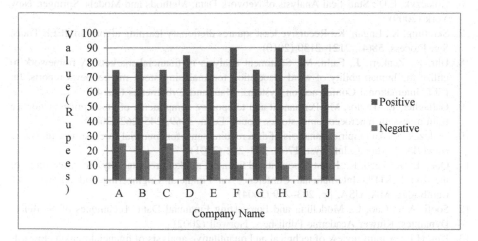

Fig. 5. Prediction using sentiment analysis

So by combine the technical analysis and sentiment analysis the prediction result will be more accurate and help the investor to make profit in the stock market.

5 Conclusion

In this research the stock market prediction is forecast using the combination of technical calculation and sentimental analysis. Machine learning and sentiment analysis predict the stock market. It shows the future prediction of stock market is also changed due to political news, economic and social media. Big data analytics predict the stock market in real time. The sentiment analysis algorithm gives the summative assessment in the tweet and news article and it is in real time. So the combination of technical analysis and sentiment analysis improve the stock prediction.

References

1. Dobre, C., Sc Xhafa, F.: Intelligent sendees for big data science. Future Gener. Comput. Syst. **37**, 267–281 (2014)
2. Chih-Lin, I., Liu, Y., Han, S., Wang, S., Liu, G.: On big data analytics for greener and softer RAN. IEEE Access **3**, 3068–3075 (2015)
3. Thaduri, A., Galar, D., Kumar, U.: Railway assets: a potential domain for big data analytics. Procedia Comput. Sci. **53**, 457–467 (2015)
4. Baiga, A.R., Jabeen, H.: Big data analytics for behaviour monitoring of students. Procedia Comput. Sci. **82**, 43–48 (2016)
5. Saul, L.K., Roweis, S.T.: Think globally, fit locally: unsupervised learning of low dimensional manifolds. J. Mach. Learn. Res. **4**, 119–155 (2003)
6. Kolaczyk, E.D.: Statistical Analysis of Network Data: Methods and Models. Springer, New York (2009)
7. Skretting, K., Engan, K.: Recursive least squares dictionary learning algorithm. IEEE Trans. Sig. Process. **58**(4), 2121–2130 (2010)
8. Uhr, P., Zenkert, J., Fathi, M.: Sentiment analysis in financial markets. A framework to utilize the human ability of word association for analysing stock market news reports. In: IEEE International Conference on Systems, Man, and Cybernetics (2014)
9. Gallagher, L., Taylor, M.: Permanent and temporary components of stock prices: evidence from assessing macroeconomic stocks. South. Econ. J. **69**, 245–262 (2002)
10. Walczak, S.: An empirical analysis of data requirements for financial forecasting with neural networks. J. Manag. Inf. Syst. **17**(4), 203–222 (2001)
11. Qian, B., Rasheed, K.: Hurst exponent and financial market predictability. In: Proceedings of the 2nd IASTED International Conference on Financial Engineering and Applications, Cambridge, MA, USA, pp. 203–209 (2004)
12. Soofi, A.S., Cao, L.: Modelling and Forecasting Financial Data: Techniques of Nonlinear Dynamics. Kluwer Academic Publishers, Norwell (2002)
13. Pan, H.P.: A joint review of technical and quantitative analysis of financial markets towards a unified science of intelligent finance. In: Paper for the 2003 Hawaii International Conference on Statistics and Related Fields (2003)
14. Mendelsohn, L.B.: Trend Forecasting with Technical Analysis: Unleashing the Hidden Power of Intermarket Analysis to Beat the Market. Marketplace Books, Columbia (2000)
15. Zhang, J.Y., Liu, Z.-Q.: Dynamic domination for fuzzy cognitive maps, pp. 145–149. IEEE (2002)
16. Koulouriotis, D.E., Diakoulakis, I.E., Emiris, D.M.: A fuzzy cognitive map-based stock market model: synthesis, analysis and experimental results. In: IEEE International Fuzzy Systems Conference, pp. 465–468 (2001)
17. Senthamarai Kannan, K., Sailapathi Sekar, P., Mohamed Sathik, M., Arumugam, P.: Financial stock market forecast using data mining techniques. In: Proceedings of the International MultiConference of Engineers and Computer Scientists, IMECS 2010, vol. 1 (2010)
18. Schumaker, R.P., Chen, H.: Textual analysis of stock market prediction using financial breaking news: the AZFin text system. ACM Trans. Inf. Syst. **27**, 1–19 (2009)
19. Alkhatib, K., Najadat, H., Hmeidi, I., Ali Shatnawi, M.K.: Stock price prediction using k-nearest neighbour (KNN) algorithm. Int. J. Bus. Humanit. Technol. **3**, 32–44 (2013)

20. Joneston Dhas, S., Maria Celestin Vigila, S., Ezhil Star, C.: A framework on security and privacy-preserving for storage of health information using big data. Int. J. Control. Theory Appl. **10**(10), 91–100 (2017)
21. Maria Celestin Vigila, S., Muneeswaran, K.: A new elliptic curve cryptosystem for securing sensitive data applications. Int. J. Electron. Secur. Digit. Forensics **5**(1), 11–24 (2013)
22. Maheswaran, C.P., Helen Sulochana, C.: Utilizing EEM approach to tackle bandwidth allocation with respect to heterogeneous wireless networks. ICT Express **2**, 80–86 (2016)

Implementation of SRRT in Four Wheeled Mobile Robot

K. R. Jayasree[✉], A. Vivek[✉], and P. R. Jayasree[✉]

Department of EEE, Amrita Vishwa Vidyapeetham, Amritapuri, Kollam, India
krjayasree0@gmail.com, vivekkrishna65@gmail.com,
sraavanam@gmail.com

Abstract. A mobile robot shall efficiently plan a path from its starting point or current location to a desired target location. This is rather easy in a static environment. However, the operational environment of the robot is generally dynamic and as a result, it has many moving obstacles or a moving target. One or many, of these unpredictable moving obstacles may be encountered by the robot. The robot will have to decide how to proceed when there are obstructions in its path. How to make the mobile robot proceed in dynamic environment using SRRT technique in hardware is presented here. Using the proposed technique, the robot will modify its current plan when there is an obstruction due to an unknown obstacle and will move towards the target. The hardware model of four wheeled mobile robot and target are developed. The experimental platform is developed and control of the system is obtained using an Arduino UNO and Arduino Mega platforms.

Keywords: Smoothed rapidly exploring random tree (SRRT)
Car-like mobile robot (CLMR) · Autonomous mobile robot (AMR)
Non-holonomic constraints · Smoothed RRT

1 Introduction

Path planning is one of the most researched problems in the area of robotics. The primary goal of any path planning algorithm is to provide a collision free path from a starting point till the end, within the configuration space of the robot. Probabilistic planning algorithms, such as the Probabilistic Roadmap Method (PRM) [1] and the Rapidly-exploring Random Tree (RRT) [2], provide a quick solution with the help of optimality. The RRT algorithm has been one of the most popular probabilistic planning algorithms since its introduction. The RRT is a fast, simple technique which incrementally generates a tree in the configuration space until the goal is reached. The RRT has a significant limitation in finding an asymptotically optimal path, and has been shown as never converging with an asymptotically optimal solution [3, 4]. There are wide researches happened to improve the performance of the RRT. Simple improvements like the Bi-Directional RRT and the Rapidly-exploring Random Forest (RRF) improve the search coverage and speed at which a single-query solution is found. The SRRT algorithm provides a significant improvement in the optimality of the RRT and has been shown to

© Springer Nature Singapore Pte Ltd. 2018
I. Zelinka et al. (Eds.): ICSCS 2018, CCIS 837, pp. 396–408, 2018.
https://doi.org/10.1007/978-981-13-1936-5_42

provide a smoothed path and so the time taken is also less [5]. Visual tracking is the most commonly used technique. In this research paper, an SRRT technique is used for tracking; distance between source and target is a main criteria for proper tracking. The distance is computed from the Received Signal Strength Indicator value obtained from target [12].

The following sections of the paper are structured as follows: In Sect. 2 description of the path planning technique is presented. Section 3 describes hardware model design used. Obstacle avoidance is discussed in Sect. 4. The tracking of target is discussed in the Sect. 5. Hardware test results are explained in Sects. 6 and 7 concludes the paper.

2   Path Planning Technique

For the motion of robot towards the target by avoiding obstacles, a path is to be planned. Smoothed RRT technique (SRRT) is used for path planning. The flowchart for SRRT shown in Fig. 1 describes how SRRT can be implemented in hardware of four wheeled robot. Beginning from the initial robot position, path is planned. If signal is available from sender, the vehicle moves. It reads the received signal strength which is then converted from hexadecimal to dBm.

As SRRT technique is used, tracking can be done only using the distance between source and target. The distance can be obtained in indoor environments only using RSSI. Hence, distance is obtained using (1). Every time, signal is sent from transmitter, the angle at which the sender moves is sent to receiver side and the receiver turns by the same angle on giving signal to servomotor attached to front wheels. At the same time, speed control is done at back wheels attached to DC motor while it moves forward. Then checking is done for obstacle avoidance. If an obstacle is detected, obstacle avoidance loop is called, which is explained later. If obstacle is not detected, the following sequence of operations happens.

Whenever the mobile robot moves forward, if the distance between the source and target is less than some threshold distance from the goal position, algorithm checks if the goal can be reached in a straight line from the current position of the robot. If the target is in a reachable distance, vehicle stops. At this juncture, the path planning is complete. If the goal position is still not reachable, the mobile robot proceeds further by avoiding obstacles. If obstacle is detected, in obstacle avoidance loop, it checks if left sensor detected the obstacle. If so, then turning is done to $+20°$ to the right direction away from obstacle. Else, if right sensor detected obstacle, then turning is done $-20°$ to the left, away from obstacle. Then it checks for signal from sender and steer towards the target according to target's angle received.

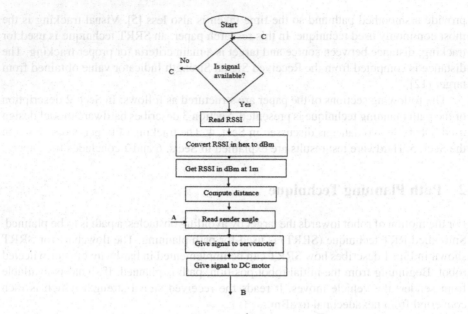

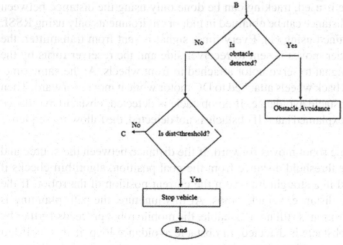

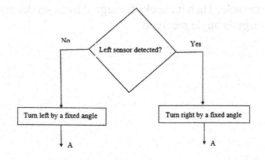

Fig. 1. Flow chart

3 Four Wheeled Robot Development

The robot models developed by different manufacturers for tracking is mostly done using camera. Here, a car-like mobile robot that tracks a target using distance between the robot position and target is developed. The robot should track the moving target by avoiding the obstacles. A single, efficient path planning technique that takes into account of target tracking and obstacle avoidance is adopted, instead of old techniques that use separate algorithms for target tracking and obstacle avoidance. To design such a mobile robot, the system has to be studied in detail. The system description for hardware model is shown in Fig. 2(a) and (b).

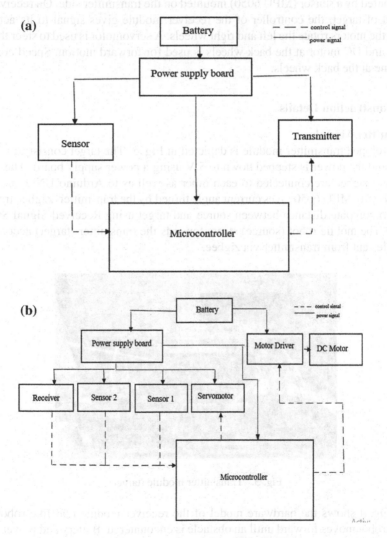

Fig. 2. (a) Block diagram: transmitter module. (b) Block diagram: receiver module

The hardware model requires a car-like robot as receiver module and target as transmitter module respectively. The microcontroller used for mobile robot (receiver module) is Arduino Mega 2560 and for target (transmitter module), Arduino UNO is used. Sensor 1 and sensor 2 are the ultrasonic sensors used for obstacle avoidance.

On the transmitter side, a transmitter, zigbee of Series-1 interfaced to the Arduino UNO is used to continuously transmit data. At the receiver side, a receiver, zigbee of Series-2 is used to receive the data. The required path for robot to move is planned using Smoothed RRT (Rapidly exploring Random Tree) algorithm. It computes the path and re-plan it on the spot or online based on the obstacles that comes in route. The actual path to be followed by mobile robot is according to the current position of target which is computed by a sensor (MPU 6050) mounted on the transmitter side. On receiving the position of target, the controller of the receiver module gives signal to its actuators. Hence, the motors rotate the left and right wheels. A servomotor is used to steer the front wheels and DC motor at the back wheels is used for forward motion. Speed control is also done at the back wheels.

3.1   Construction Details

Transmitter Module
The developed transmitter module is depicted in Fig. 3. The target consist of a battery of 6 V and the power is stepped down to 5 V using a power supply board. The sensor, MPU and zigbee are connected to each other as well as to Arduino UNO and power supply board. MPU 6050 gives current angle turned by the transmitter. Zigbee transmits signal to compute distance between source and target using Received Signal Strength (RSSI). The mobile robot (source) steers towards the transmitter (target) according to the angle sent from transmitter via zigbee.

Fig. 3.   Transmitter module (target)

In Fig. 4 shows the hardware model of the receiver module (car-like robot). The mobile robot moves forward until an obstacle is encountered. Battery and power supply board are used to give power signals. Two ultrasonic sensors having a maximum range of 400 cm are mounted on the front left and right of the car chassis to avoid obstacles.

Fig. 4. Hardware model of car-like robot (receiver side)

Zigbee is supported on a base which is the USB zigbee adapter (an inbuilt 3.3 V regulator). As Arduino Mega has 54 I/O pins and three serial ports, it is used to process signals.

Receiver Module
The S2 series zigbee receives the signal from transmitter and make movement with the front servomotor connected. The DC motor at the back wheels is connected to L293N motor driver. As motor driver can take up voltage higher than 5 V, it is directly connected to battery whereas the servomotor used operates at 5 V, hence it is connected to battery via power supply board.

4 Obstacle Avoidance

Ultrasonic sensors can be used to solve even the most complex tasks involving object detection or level measurement with millimeter precision, because their measuring method works reliably under almost all conditions. Infrared sensors too, find applications in many day to day products. Their low power requirements, their simple circuitry and their portable features make them desirable.

The ultrasonic sensor transmits sound waves and receives sound reflected from an object. When ultrasonic waves are incident on an object, diffused reflection of the energy takes place over a wide solid angle which might be as high as 180°. Thus some fraction of the incident energy is reflected back to the transducer in the form of echoes as shown in Fig. 5. If the object is very close to the sensor, the sound waves returns quickly, but if the object is far away from the sensor, the sound waves take more time to return. But if objects are too far away from the sensor, the signal is so weak when it comes back that the receiver cannot detect it [7].

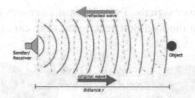

Fig. 5. Working principle

In order to determine the distance of an object, the sensor depends on the time it takes for the sound to come back from the object in the front. The distance to the object (s) can then be calculated with the help of speed of ultrasonic waves (v) in the medium by the relation where 't' is the time taken by the wave to reach back to the sensor. If the object is in motion, instruments based on Doppler shift are used. The ultrasonic sensor can measure distances in centimeters and inches. It can measure from 0 to 2.5 m, with a precision of 3 cm. The distance calculation is as follows:

$$Speed\ of\ sound\ v = 340\,\text{m/s}$$
$$= 0.034\,\text{cm/}\mu\text{s}$$

$$Time = distance/speed$$

$$Time\ t = s/v = 10/0.034$$
$$= 294\,\mu\text{s}$$

$$Distance\ s = t * (0.034/2)$$

In the research work presented, there are two motors at the front and back. The front wheels steer towards left and right directions using the servomotor whereas the DC motor at the back is used to move forward and backward. Hence, once the distance between the source and obstacle is calculated using the above given equation, if that distance is less than 30 cm, then the back wheels reduce the speed (using the enable pin) and front wheels turns right or left (based on if output is from left sensor and right sensor respectively) in order to evade the obstacle. Angles are directly given to servomotor to make a turn away from obstacle. HIGH and LOW signals are applied to DC motor pins in order to move back wheels forward.

5 Target Tracking

To track the moving target, target sends signal continuously. After receiving the signal from target, the mobile robot moves towards it accordingly. Based on the received signal strength (RSSI), the distance between mobile robot and target is found with which the robot tracks it.

RSSI stands for Received Signal Strength Indicator. It is the strength of the sender's signal as seen on the receiving device. RSSI provides an approximate result for the received signal strength. Digi radio modems send weak signals from a distant transmitter. The signal strength is obtained using a function, pulseIn() in Arduino IDE. The datasheet for Xbee RF module contains the description about the conversion from hexadecimal to dBm value and that is used in the project to convert the RSSI (in hex) to dBm. The RSSI (dBm) is used to get distance between the sender and receiver.

Here, it is given that, if RSSI is −40 dBm, then the hex value of which is 0×28 (decimal = 40) is returned. Hence, to convert the obtained RSSI to dBm, first convert the obtained RSSI (in hex) to decimal value and then take negative of it. Measured Power is a factory-calibrated, read-only constant which indicates the expected RSSI at a distance of 1 m to the sender. Combined with RSSI, it allows to estimate the distance between the device and the sender. Then the distance is calculated using (1).

$$Distance = 10 \char`\^ ((Measured\ Power - RSSI)/(10 * N)) \tag{1}$$

where measured power is also known as the 1 m RSSI. N is a constant that depends on the environmental factor, ranging from 2 to 4 m. A threshold value is set which is the minimum distance required to reach the target. In order to track a moving target, whenever the target steer at an angle, the mobile robot should steer at the same angle and in a specific direction so as to reach the target.

The desired position of the servomotor is send in the form of a *PWM* signal by the microcontroller. A PWM signal is an electrical signal of which the voltage periodically generates pulses. The width of these pulses determines the servo position. So when the width of the pulses change, the position of the servo gets changed. According to the angle sent from target using MPU 6050, the mobile robot will steer towards it, using servomotor to control its front wheels. Hence, proper tracking of mobile robot towards target by considering both range as well as orientation occurs. Servomotor itself has a potentiometer inside to make sure the correct angle is maintained. Hence servomotor is used to make the front wheels steer towards the target. The wheels of the receiver module, which are connected to servomotor turns according to the motion of transmitter.

6 Implementation Results

After the development of four wheeled mobile robot and target, it is tested for obstacle avoidance and target tracking, the results are described in this part of the research paper. The angles given by MPU according to transmitter movement is as shown in Fig. 6.

The servomotor movement is restricted to 90°. The wheels are at centre when the servomotor angle is at 50°. The left maximum angle is 0° and right maximum is 90°. Turning is done between 30 to 70°. From Fig. 6, when the transmitter is kept horizontal, the angle obtained is approximately 52° where 'ax' represents the angle along X-axis. When it is inclined at angle i.e., when transmitter is in an inclined position, the angle became 39°. When transmitter is tilted in the opposite direction, the angle became approx. 58°.

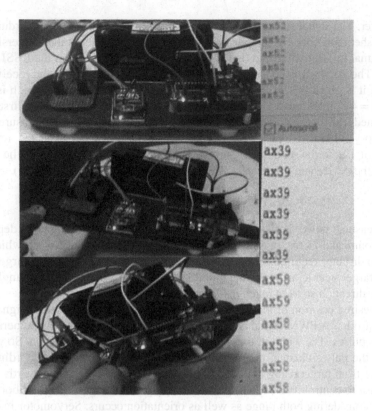

Fig. 6. Transmitter movement and angles

6.1 System Concept and Implementation

The car like robot that follows a path planned using SRRT technique is developed. At each step, according to the algorithm, the mobile robot checks if any signal is received from target and compute the distance between them. Also, it checks if there is any obstacle in its path from source to target. If not, it continues towards the target else if there is any obstacle, it turns away from the obstacle in the direction of target. Finally if the distance become less than threshold, vehicle stops as it has almost reached the goal, where threshold is some minimum distance from the target.

If any signal from transmitter is received at receiver, then its corresponding RSSI value is computed. In SRRT technique, using RSSI value, distance from source to target is found using formula given in (1). Then it checks for any obstacle in its path. The ultrasonic sensor used for obstacle detection was calibrated and distance to reach obstacle was found. It has a maximum range of 400 cm. If the distance between mobile robot and obstacle is less than 30 cm, obstacle is detected and has to be avoided. If the left sensor detects an obstacle, then servomotor turns −20° (which is right direction) else it turns 20° if right sensor detects an obstacle. If there is no obstacle, the angle is set as 50° which means wheels are at centre. At the same time, back wheels which are controlled by DC motors reduces their speed using enable pin. Then turning away from obstacle occurs. While turning, the servomotor steer at the same angle received as that of the transmitter and move towards target. If the distance become less than a threshold value, it stops as it has neared the goal.

The MPU present in transmitter transmits the angle via zigbee towards the receiver module and receiver turns accordingly. Every time, this checking for transmitter signal and moving towards it occurs. Hence, even if there is any obstacle encountered dynamically or target movement occurs robot is able to track the target. The obstacle avoidance using left sensor is shown in Fig. 7.

Fig. 7. Obstacle avoidance using left sensor

Initially from its position it moves forward and if any obstacle is detected using left sensor, it turns away from it towards right direction. The obstacle avoidance using right sensor is shown in Fig. 8.

Fig. 8. Obstacle avoidance using right sensor

On encountering an obstacle at the right side, the sensor detects it and turns away from it and moves in left direction. The obstacle avoidance using obstacles which are encountered dynamically is shown in Fig. 9. Initially on starting there was no obstacle in its exact path. But when obstacles are kept on its path, immediately the mobile robot turns away from it. Using both the ultrasonic sensors, system avoids obstacles on its path. Hence it works well with dynamic environments. The angle turned by receiver according to target angle is shown in Fig. 10.

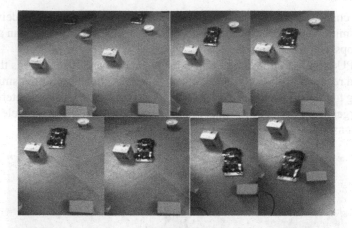

Fig. 9. Obstacle avoidance encountering dynamic obstacle

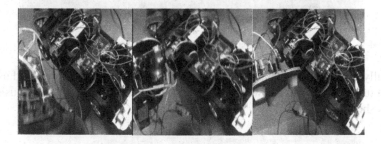

Fig. 10. The angle turned by receiver

The transmitter sends angle according to its movement via zigbee. The angle is received by receiver module using zigbee and it steers in that received angle. When servomotor is at 50°, wheels are at the center. From Fig. 10, it can be seen that wheel turned its angle to the right at around −20° when transmitter moved downwards.

Wheels came to center when transmitter was along X-axis or horizontal. Wheels moved to left maximum or to around 20° when transmitter moved upwards. The movement of transmitter is demonstrated in this way as the sensor is kept in opposite direction while setting up the hardware. The tracking of target is as shown in Fig. 11.

Fig. 11. Target tracking

On receiving the signal from sender, the mobile robot compute the distance and as there is no obstacle in its path, it directly tracks the target and stopped when that distance is less than the threshold value.

7   Conclusion

The wireless RF module, zigbee was used at the transmitter and receiver side. The receiver module turns the same angle as that of transmitter. As it tracks using SRRT algorithm, distance is needed which is computed using the RSSI value. This paper proposes obstacle avoidance and target tracking of a four wheeled mobile robot using SRRT path planning technique. The tracking is done by computing the distance between robot and target position using RSSI value obtained from target and the angle computed. Due to external factors influencing radio waves such as absorption, interference, or diffraction RSSI tends to fluctuate. Inside a room, GPS gives almost the same value everywhere. Hence, in this research work, RSSI was used instead of GPS. In future scope, if tracking is to be done in outdoor environment, Global Positioning System (GPS) can be used to get coordinates of mobile robot position and goal position and hence can be used to find distance.

References

1. Kavraki, L.E., Svestka, P., Latombe, J.C., Overmars, M.H.: Probabilistic roadmaps for path planning in high-dimensional configuration spaces. IEEE Trans. Robot. Autom. **12**(4), 566–580 (1996)
2. Lavalle, S.M.: Rapidly-exploring random trees: a new tool for path planning. Technical report (1998)
3. Karaman, S., Frazzoli, E.: Optimal kinodynamic motion planning using incremental sampling-based methods. In: 49th IEEE Conference on Decision and Control (CDC), pp. 7681–7687, December 2010
4. Karaman, S., Walter, M.R., Perez, A., Frazzoli, E., Teller, S.: Anytime motion planning using the RRT*. In: 2011 IEEE International Conference on Robotics and Automation, pp. 1478–1483, May 2011
5. Jayasree, K.R., Jayasree, P.R., Vivek, A.: Dynamic target tracking using a four wheeled mobile robot with optimal path planning technique. In: 2017 International Conference on Circuit, Power and Computing Technologies (ICCPCT), Kollam, pp. 1–6 (2017)
6. Ferguson, D., Stentz, A.: Anytime RRTs. In: 2006 IEEE/RSJ International Conference on Intelligent Robots and Systems, pp. 5369–5375, October 2006
7. Parrilla, M., Anaya, J.J., Fritsch, C.: Digital signal processing techniques for high accuracy ultrasonic range measurements. IEEE Trans. Instrum. Meas. **40**, 759–763 (1991)
8. Gammell, J.D., Srinivasa, S.S., Barfoot, T.D.: Informed RRT*: optimal sampling-based path planning focused via direct sampling of an admissible ellipsoidal heuristic. In: 2014 IEEE/RSJ International Conference on Intelligent Robots and Systems, pp. 2997–3004, September 2014
9. Salzman, O., Halperin, D.: Asymptotically near-optimal RRT for fast, high-quality, motion planning. In: 2014 IEEE International Conference on Robotics and Automation (ICRA), pp. 4680–4685, May 2014

10. Islam, F., Nasir, J., Malik, U., Ayaz, Y., Hasan, O.: RRT*-smart: rapid convergence implementation of RRT*; towards optimal solution. In: 2012 IEEE International Conference on Mechatronics and Automation, pp. 1651–1656, August 2012
11. Bruce, J., Veloso, M.: Real-time randomized path planning for robot navigation. In: IEEE/RSJ International Conference on Intelligent Robots and Systems, vol. 3, pp. 2383–2388 (2002). J. Robot. Res. **35**, 797–822 (2016)
12. Benkic, K., Malajner, M., Planinsic, P., Cucej, Z.: On line measurements and visualization of distances in WSN with RSSI parameter. In: 16th International Conference on Systems Signals and Image Processing 2009, IWSSIP 2009, pp. 1–4 (2009)
13. Ma, L., Xue, J., et al.: Efficient sampling-based motion planning for on-road autonomous driving. IEEE Trans. Intell. Transp. Syst. **16**(4), 1961–1976 (2015)

Personality-Based User Similarity List and Reranking for Tag Recommendation in Social Tagging Systems

Priyanka Radja[✉]

Delft University of Technology, Mekelweg 2, 2628 CD Delft, The Netherlands
radja.priyanka@gmail.com

Abstract. This paper is a proposal for efficient tag recommendation to a target user in social tagging systems by generation of a user similarity list and reranking of the list. The methodology involves a user similarity list generated for every target user based on the shared personality traits or the Big5 values obtained from a mandatory one-time questionnaire during the profile creation in the social tagging system. Different users are added to the neighborhood of similar users for the target user based on the Euclidean distance between the big5 values of these users and that of the target user. The User-Item matrix is replaced by a User-Item-Tag matrix with tags for the items used by the different users forming the 3rd dimension. The tags from the top k neighbors from the similar user neighborhood of a target user for a particular resource will be recommended to the target user. The idea is to maintain a ranked list of neighbors based on their similarity score (Euclidean distance) where the position of the neighbors in the list denotes the level of similarity the neighbor shares with the target user. It is essential to maintain the ranked lists of neighbors and perform any reranking when necessary as a fairly dissimilar user at the bottom of the neighborhood list may still respond in the same way to a context and use the same tag as the target user in more than one instance. This requires revising the rank of this fairly dissimilar user up the neighborhood list to reflect the change. This paper suggests an efficient method to perform such reranking based on logarithmic and exponential scale.

Keywords: Tag recommendation · Social tagging systems · User similarity
Reranking · Personality traits · Big5 values · Euclidean distance
User-Item-Tag

1 Introduction

Identification of similar users in social tagging systems is essential to recommend tags or even resources. Existing user similarity metrics under user-based collaborative filtering [4] consider two users as similar based on their previous history of ratings provided to the same resources. A more reasonable metric to identify similar users was proposed by [1]. This similarity metric identifies similar users based on their personality

© Springer Nature Singapore Pte Ltd. 2018
I. Zelinka et al. (Eds.): ICSCS 2018, CCIS 837, pp. 409–415, 2018.
https://doi.org/10.1007/978-981-13-1936-5_43

traits. The personality traits of the users are identified by determining the Big5 values - Extraversion (E), Agreeableness (A), Conscientiousness (C), Neuroticism (N) and Openness (O) [2], in the form of a 50 item questionnaire by IPIP [3].

According to [1], once the Big5 values of individual users are determined, similar users to the target user can be identified by calculating the Euclidean distance between these 5 values. Therefore, similar users to a target user are computed only once and not each time a user likes an item thus saving a lot of resources and computations [1]. However, a user with fairly dissimilar personality as the target user may choose the same tags for many resources as that of the target user. In the above case, this fairly dissimilar user is a potential neighbor who must be added to the target user's similar neighbors list so the tags used by the said dissimilar user can be recommended to the target user in the future. This case is not accounted for by [1]. Moreover, whenever the target user selects one of the tags recommended to him, the relationship of the target user with that of his neighbor whose tag he just chose must be coupled strongly with an increasing number of such tag matches. Therefore, the order of relevance of the neighbors to the target user must be recorded with the neighbor on top of the similarity list having the most similarity to the target user and the neighbor at the bottom of the list having the least similarity. So an ordered list of similar users must be maintained as neighbors to the target user and the tags used by top k neighbors for an item i must be recommended to the target user for the said item i. Note that k is an integer whose value is very crucial to the successful recommendation of tags. A very high k value results in a noisy neighborhood with unreliable neighbors and a low k value results in insufficient neighbors for a successful recommendation.

2 Previous Work

The user-based collaborative filtering [4] uses the rating data in recommender systems to identify neighboring users with similar rating patterns. The same cannot be applied to social tagging systems for tag recommendation as the tags do not have a definite scale like the Likert scale for ratings. The values of the tags are numerous and are thus added as a 3rd dimension to the user-item matrix which is referred to as the user-item-tag matrix [5] henceforth.

Personality based user similarity measure [1] has already been proposed employing the Big5 values [2] and the 50 item IPIP questionnaire [3] as proposed in this paper. The Big5 values of Extraversion (E), Agreeableness (A), Conscientiousness (C), Neuroticism (N) and Openness (O) were calculated for each user by the inputs provided by them for the 50 item questionnaire in which each of the 5 factors were covered by 10 items in the questionnaire [1]. Thus, users sharing similar personality traits to a target user were found by computing the Euclidean distance between the big5 values. The top k neighbors were the similar users to the target user who influenced the item/tag recommendation. The said method however ignores the capacity of influence of each of the neighbors on the target user. A ranked list of neighbors is not maintained where the neighbor ranked on top influences the tag recommendation to the target user more than the neighbor ranked at the bottom of the list. Therefore, the neighbors should be reordered each time there is a tag match i.e. a tag suggested by one of the neighbors is used by the target user to reflect the close relationship in terms of personality and tag selection given any context between the target user and the said neighbor. The research in [1] ignored the question of the existence of users with no personality match in terms of Big5 values with the target user or those who missed the top k neighborhood list by a tiny difference who however chose the same tags for the same resources in multiple instances as that of the target user. In the scenario mentioned above, these users must be added to the target user's neighborhood as the increasing number of same tags used by both the target user and these users for the same resources implicitly manifests a behavior match between the users in question.

3   Proposed Solution

The proposed solution involves reordering of the similar neighbor list to show the order of influence the neighbors have on the target user while recommending tags. When tags from top k similar neighbors are recommended to the target user, the user can select one of the recommended tags or insist on using a new tag. In the former case, the neighbor whose tag was selected by the target user is moved up the similar neighbors list. A count of the number of tags suggested by each neighbor that were selected by the target user is maintained in a variable #count for that neighbor in the list. When a tag recommended by a neighbor is used by the target user, the neighbor's count value is incremented and the neighbor's position is increased exponentially by 2 raised to the power of the count value starting from his original position in the list.

$$\text{new position}_i = 2^{\#count,i} + \text{old position}_i \tag{1}$$

For all i neighbors of the target user, new position$_i$ denotes the neighbor's new rank, old position$_i$ denotes the neighbor's old rank before the reranking is performed and #count denotes the number of tags suggested by each neighbor that were selected by the target user in (1).

When the target user uses a new tag instead of the ones recommended to him, the new tag and its synonyms are searched in the user-item-tag matrix to check for other users not in the similar neighbors list who may have used the same tag for the same resource. Such users are added to the bottom of the target user's similar neighbor list. The count value is incremented for these users in the same way as for the similar neighbors with matching personality traits but the position of these users is incremented by only 1 position until these users cross the kth position in the list or the similarity score i.e. the Euclidean distance decremented by log of the count value for each tag match falls below the threshold τ, after which their position is incremented exponentially by 2 raised to the power of the respective count value. This method accounts for the scenario where a user with very little personality match to the target user is allowed to influence the tags recommended to the target user, if the said dissimilar user crosses the kth position with an increasing number of tag matches with the target user. Therefore, the proposed solution takes into account the importance of ranking the user similarity list and also the fact that a dissimilar user may share some similarity with respect to behavior or response to the current context with the target user if not the personality which is implicitly manifested by their choice of same tags for resources.

$$\text{new Euclidean Distance}_i = \text{old Euclidean Distance}_i - \log(\# \text{count}, i) \qquad (2)$$

Equation (2) denotes how new Euclidean distance of each neighbor i is calculated as a decrement of log of count value for each tag match denoted by #count from the old Euclidean distance of neighbor i.

$$\text{new position}_i = 1 + \text{old position}_i \qquad (3)$$

Only for cases when the neighbor position is below k − 1, the new position is calculated as given in (3) for every tag match given in #count. The Euclidean Distance is calculated by (2).

Note in cases when the new tag used by the target user and its synonymous tags identified using a dictionary analysis have never been used before by any other user in the system, the new tag is simply added to the tag dimension of the User-Item-Tag matrix. If successful, the proposed project will remove bias on users that now exists in terms of similarity metric. Also note that the exponential and logarithmic scales are chosen for incrementing the position of the top k neighbors and for reducing the similarity score for the remaining users from k − 1 to bottom of the list until it falls below τ respectively, because the top k neighbors share personality traits and behave the same way in a given context by choosing the same tags as the target user. Hence, the position is increased drastically whenever there is a tag match in the exponential scale. But, the users from position k − 1 to bottom of the similar neighbors list do not have very similar personality traits to the target user yet choose the same tags hence their position in the similarity list is increased by only 1 position until the similarity measure (Euclidean distance) falls below threshold τ with decrease in its value each time there is a tag match by the log of the current number of tag matches (log #count). This is also the case until the neighbor's position crosses kth position after which the neighbor is treated like a top k similar neighbor and its position is incremented with 2 raised to power of #count henceforth.

4 Scientific Challenges and Objectives

The mandatory one-time questionnaire to be filled by each user upon their account creation in the social tagging system is a drawback as stated in [1] as such questionnaires are usually perceived as annoying by the users. Moreover, the users may not be diligent while filling in the questionnaire and may enter values blindly just to complete the questionnaire and to proceed further. Since the proposed solution does not rely fully on the matching personality trait for the tag recommendation but also includes the possibility of adding dissimilar users to a target user's similar neighborhood if a number of tags used by the target user match the tags used by these dissimilar users.

Another challenge would be storage and computational complexity of the proposed method. Since the similar users are reordered with each tag selection, the proposed solution is computationally expensive when compared to [1]. Storage is another challenge as even users with dissimilar personality traits to the target user are added to the neighborhood list when there is a tag match leading to an ever growing neighborhood of similar users for the target user. Although only the top k similar users influence the tag recommendation, the remaining similar users are not discarded as the next reranking may resurface some users from the bottom to become a part of the top k similar users. Users with the same similarity measure or the same rank in the list are stored as a linked list of values at the same position in the similar neighbors list. Selection of a proper k or a threshold value τ to select the similar neighbors whose tags are to be recommended to the target user is challenging. Too big of a k or τ value will result in many unreliable neighbors becoming a part of the noisy neighborhood and too small of a k or a τ value will result in insufficient neighbors and thus insufficient tags to be recommended.

5 Methodology

The first task in realizing the project involves making the users of the social tagging system to fill in the one-time questionnaire to determine the E, A, C, N, O values of the Big 5 personality model. The questionnaire can be made mandate to be filled by the users during their profile creation in the social tagging system. The IPIP questionnaire [1] was used in [2] to determine these 5 values for the different users as illustrated in Table 1.

Table 1. Big5 values of users from taken from [2]

Big5 values	Extraversion (E)	Agreeableness (A)	Conscientiousness (C)	Neuroticism (N)	Openness (O)
U_1	3.2	2.7	2.9	3.5	2.9
U_2	2.1	3.5	3.1	3.4	3.6
U_3	3.2	3.0	2.8	3.2	3.1
..	..	..	..	..	..
U_i	3.3	3.0	3.4	3.9	3.2

From the values in Table 1, the k similar users to a target user Ui can be determined by calculating the Euclidean distance between the respective Big5 values. For easy computations, the sum of the absolute difference between the E, A, C, N, O values of the different users with that of the target user are computed and divided by 5 to obtain the similarity metric on a scale of 0 to 5. Note 0 denotes two users with perfect match in personality as there is no difference in the Big5 values and 5 denote extremely contradictory personalities. Both the extreme values of 0 and 5 are highly unlikely to occur. A threshold τ is set between 0 and 5 to select the users falling with similarity metric between 0 and τ as the target user's neighbors.

The selection of an appropriate value of τ is crucial as a τ value closer to 5 may result in many users being selected as neighbors leading to a noisy neighborhood with unreliable neighbors. In the contrary, a τ value very close to 0 leads to insufficient neighbors to recommend tags. Once the neighbors of each user are determined, they are maintained in a database along with their similarity score to the target user value computed as either the Euclidean distance or the summation over the absolute difference between the Big5 values as mentioned above. Note that these neighbors are ranked in ascending order with the user with lowest similarity measure being ranked first. Two neighbors with the same similarity measure to a target user are entered at the same rank in the similarity list in the form of a linked list. In addition to the similarity value, a count of the number of tags suggested by the neighbor that were selected by the target user is stored in the database.

When the user selects an item to tag, a list of tags already used by his top k similar neighbors for that particular item are suggested to the user. The user can now choose to use one of the suggested tags or enter a new tag for the item. In the former case, the user similarity list is reordered to reflect the choice of the user which implicitly denotes a higher similarity between the target user and the user whose tag the target user chose. In the latter case, the new tag is added to the tag dimension of the User-Item-Tag matrix if even the dissimilar users in the social tagging system have never used the said tag or its synonyms before. In case the new tag the target user insisted on using for the item was used by another user not in the target user's similar neighbors list, the user will be added to the bottom of the target user's similar neighbors list.

Note that the top k similar neighbors alone influence the tags to be recommended to the user. For each tag match, if the neighbor is in the top k positions, his position is incremented exponentially by 2 raised to power of count of current tag matches(the tags suggested by him that the target user chose for a resource) from his original position and the count variable is incremented by 1. If the neighbor lies below the top k position, his similarity score is reduced by log of the count of tag matches (log #count) for each tag match and his position is incremented by 1 until his position crosses the kth position or his similarity score falls below the threshold τ. After either of the two cases occurs, the neighbor's position will be incremented exponentially.

6   Conclusion and Future Work

A new method for generating user similarity list for tag recommendation to target user was proposed in this paper. An efficient method for reranking of this ranked list of similar users and update on their similarity score to maintain a true, exact list of similar neighbors to the target user was also proposed. By employing this reranking of the ranked list of neighbors, the target user can benefit from relevant tag recommendations. As future work, the methodology will be employed to users in social tagging websites like Pinterest, Flickr etc. to evaluate the efficiency of tag recommendations to a target user through this methodology.

References

1. Tkalcic, M., et al.: Personality based user similarity measure for a collaborative recommender system. In: Proceedings of the 5th Workshop on Emotion in Human-Computer Interaction-Real World Challenges (2009)
2. McCrae, R.R., John, O.P.: An introduction to the five-factor model and its applications. J. Pers. **60**(2), 175–215 (1992)

3. Administering IPIP Measures, with a 50-item Sample Questionnaire, June 2009. http://ipip.ori.org/New_IPIP-50-item-scale.htm
4. Balabanović, M., Shoham, Y.: Fab: content-based, collaborative recommendation. Commun. ACM **40**(3), 66–72 (1997)
5. Kim, H.-N., et al.: Collaborative filtering based on collaborative tagging for enhancing the quality of recommendation. Electr. Commer. Res. Appl. **9**(1), 73–83 (2010)

A $21\text{nV}/\sqrt{Hz}$ 73 dB Folded Cascode OTA for Electroencephalograph Activity

Sarin Vijay Mythry[✉] and D. Jackuline Moni

Center for Excellence in VLSI and Nanoelectronics, School of Electrical Sciences,
Karunya University, Coimbatore, India
sarinmythry@gmail.com

Abstract. The electroencephalography signals are electrical signals with weak amplitudes and low frequencies recorded and displayed on screen from scalp or brain in the range of millihertz to kilohertz, created a tremendous demand amongst neuroscience researchers and clinicians. This paper presents a design analysis of single stage folded cascode (FC) OTA used for EEG activity recording applications. The FC OTA with 73.89 dB gain, $21.78\text{n V}/\sqrt{Hz}$ input referred noise and $4.5\,\mu\text{W}$ power is designed in 90 nm CMOS process. The Wilson current mirror technique is used in designing 1 V powered FC OTA for EEG signal Amplification.

Keywords: Folded cascode · EEG · BMI · Biopotentials
Human physiological signal · OTA

1 Introduction

Very low power consumption is essentially required in medical diagnosis devices. These devices are to be sub-micrometer designed to get inculcated on a single integrated circuit, requires channel length modulation, smaller area and supply voltage scaling. The biomedical and electrophysiological designs are leading towards the era of portability, demands low power for longer time to monitor the patient physiologically. Novel techniques are to be employed to design low power, maintenance free, light weight biomedical long monitoring and recording systems. The low power designs for low frequency bio signals like electroencephalograph (EEG) are to be operated in subthreshold [1]. This subthreshold region provides less distortion and high transconductance. The drawbacks of the subthreshold are large drain current mismatch and bandwidth reduction, are eliminated to some extent by proper offset compensation technique.

Biomedical systems like EEG systems and neural recording systems have signal with low frequency and low amplitude ranging from 100 Hz to few KHz (<20 kHz) in frequency and few microvolts to few millivolts in amplitudes [2]. EEG captures the physiological activities from scalp for diagnosing the epilepsy, brain dead patients etc. The long time EEG monitoring makes patient's life difficult because of bulky electrodes connected on patient's scalp, restricting their mobility [3]. Therefore, the demand for portable low power, ambulatory bio potential systems are growing day by day [4]. The

© Springer Nature Singapore Pte Ltd. 2018
I. Zelinka et al. (Eds.): ICSCS 2018, CCIS 837, pp. 416–424, 2018.
https://doi.org/10.1007/978-981-13-1936-5_44

recording of physiological signal requires large amount of brain signal information. The electroencephalogram signals are electrical signals with weak amplitudes and low frequencies recorded and displayed on screen from scalp or brain. EEG signals are of six main categories with delta activity having low frequencies and gamma waves has higher frequencies [5]. The information about attention, reasoning and focusing is depicted by beta activity. Alpha activity renders the blood flow, mental relaxation and stability information. The mental stress wave activity with amplitudes greater than 20 µV and 6 to 7 Hz frequency range is captured as theta activity. Mu Activity is the mesial cadence of the alpha activity. The normal amplitude of the EEG is 50 µV, with some activities having very low voltage amplitudes of 10 µV. So, EEG signal amplifier should be designed to process 0.5 Hz–100 Hz frequencies with 10 µV–150 µV amplitudes. Table 1 gives information about different biomedical signals amplitudes and bandwidth. Table 2 depicts source areas and applications of different biomedical signals.

Table 1. Amplitudes and bandwidths of biomedical signals [6].

Nerve signal	Amplitude	Bandwidth
EEG	1 to 10 mV	1 MHz–200 Hz
LFP	0.5 to 5 mV	1 MHz–200 Hz
EAPs	50 to 500 µV	100 Hz–10 kHz
IAPs	10 to70 mV	100 Hz–10 kHz
Ionic currents	1 to 10 nA	1 MHz–10 kHz
Redox currents	100 f to10 µA	1 MHz–100 Hz

Table 2. Source areas and applications of biomedical signals [10].

Bioelectrical signal	Source area	Applications (Recording)
EEG	Brain	Encephalography/Scalp
ECG	Heart	Cardiac
ECoG	Brain	Cerebral cortex
EOG	Eye dipole	Eye movement
EMG	Muscle	Muscle activity

In this research article a high gain, low noise and very low power consumed folded cascode (FC) OTA for EEG signal amplification is designed. This article organization is done as follows: Sect. 2 depicts folded cascode OTA for EEG signal amplification. Section 3 presents the small signal analysis of FC OTA. Section 4 presents the noise analysis of FC OTA. Section 5 presents the results and discussions of FC OTA. Section 6 concludes this article summarizing the results of the FC-OTA for EEG recording systems.

2 Folded Cascode OTA for EEG Signal Amplification

Operational transconductance (g_m) amplifier (OTA) is the very important and integral part of the biomedical signal amplifier design. To achieve precision analog signal processing, OTA requires high input impedance design with negative feedback. Basically, OTA is a voltage controlled current source (VCCS) where the differential input voltage is converted into output current. The electronically tuning ability of OTA, makes it most suitable to design precisioned analog devices. The variation in the transconductance that is varying the input voltage with controlled bias current (I_{bias}) defines the tuning ability of OTA. The telescopic configuration is used to construct the nMOS input signal triggered folded cascode (FC) OTA, which acts as a very useful configuration to record the biomedical signals. In telescopic amplifier; Mosfets are stacked to achieve high gain. The pMOS signal triggering is used to achieve the noise suppression and improved phase margins whereas nMOS triggered input signal provides high transconductance, high gain and high unity gain bandwidth (UGB). The designed folded cascode OTA is a modified structure from telescopic operational amplifier. It is called as folded cascode due to its folding of transistors. It folds p-channel cascade active loads of differential pair and replaces the input Mosfet devices with n-channel devices (M1 and M2). The input transistors (M1 and M2) are used to charge Wilson current mirrors (M7 and M8) in output stage. This topology ensures good output swing at an expense of power consumption. Folded cascode configuration is the synergy of common gate and common source stages. N-channel input signal triggered FC OTA topology is a better replacement to conventional operational amplifier due to greater mobility, high transconductance and larger output gain obtained than in p-channel driven folding topology.

This paper presents a FC-OTA design to amplify the desired low amplitude and low frequency EEG signals. The designed OTA is simulated in 90 nm CMOS technology using folded cascade architecture shown in Fig. 1. A single stage FC OTA powered by 1 V is designed instead of using more stages because of power hungry characteristics of more than one stage amplifiers. Transistor sizing is the vital approach in designing a functionality-based FC-OTA for EEG signal capturing applications. The FC OTA is a differential input and single ended output architecture with a N- channel MOSFETs (M_1 and M_2) which acts as the inputs. These transistors are folded to PMOS devices M_5 and M_6. The transistors M_3 and M_4 are appropriately biased at gate terminals. M_5 and M_6 acts as the PMOS cascaded stage whereas M_7 and M_8 acts the NMOS mirror load and M_9 and M_{10} are the NMOS cascade stage. MOS transistors M_{11} and M_{12} are mirror devices biased by 1.08 µA current. High gain is achieved by these reasons: To achieve high g_m of the MOS devices which lowers the noise, aspect ratios of the input transistors are increased and the lengths of the n-channel MOSFETs M_7 and M_8 are increased and by maintaining the magnitudes and amplitudes of input signal voltages in the range of few nano volts to 60 µvolts.

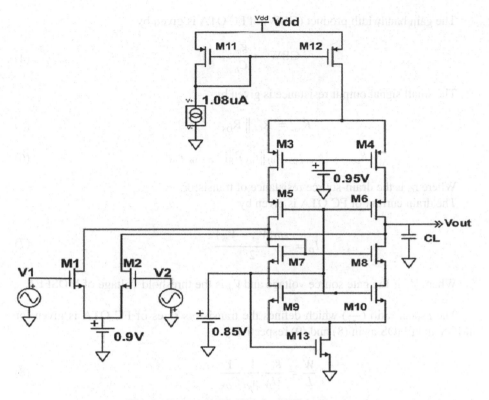

Fig. 1. Proposed FC OTA for EEG applications

3   Small Signal Analysis of FC OTA

The small signal analysis is used to approximate the behavior of the amplifiers containing non-linear devices with linear equations near bias points as linear.

The output current of an OTA is given by

$$V_{out} = g_m [V_1 - V_2] \qquad (1)$$

The transconductance of an FC OTA is given by

$$g_m = \frac{I_{bias}}{2V_t} \qquad (2)$$

Where V_t is the thermal voltage (26 mV) at room temperature.
The gain (A_v) of the FC OTA is given by

$$A_v = g_m \times R_{out} \qquad (3)$$

Where g_m is the transconductance of FC OTA.

The gain bandwidth product (GBW) of FC OTA is given by

$$GBW = \frac{g_m}{C_L} \qquad (4)$$

The small signal output resistance is given by

$$R_{out} = R_{O6} \| R_{O8} \qquad (5)$$

$$R_{out} = [g_{m6}.r_{05}.(r_{01}\|r_{05})]\|[g_{m8}.r_{08}.r_{10}] \qquad (6)$$

Where r_O is the drain-source resistance of transistor.
The drain current of FC OTA is given by

$$I_D = \frac{gm(V_{gs} - V_{th})}{2} \qquad (7)$$

Where V_{gs} is the gate source voltage and V_{th} is the threshold voltage of MOSFET.

The aspect ratio $(\frac{W}{L})$ which defines the transistors sizes of FC OTA is given for nMOS and pMOS as in (8) and (9) respectively.

$$\frac{W}{L} = \frac{g_m^2}{2I_D}.\frac{1}{\mu_n}.\frac{1}{C_{OX}} \qquad (8)$$

$$\frac{W}{L} = \frac{g_m^2}{2I_D}.\frac{1}{\mu_p}.\frac{1}{C_{OX}} \qquad (9)$$

Where μ_n and μ_p are the channel mobility of n-type and p-type devices respectively and C_{OX} is the oxide capacitance.

4 Noise Analysis of Folded Cascode OTA

The input referred noise is calculated with following considerations to find the major freelancers to noise level in FC OTA [7].

1. The voltage source (V_n) of the input Mosfets which represents input voltage noise is connected in series with gate terminal of MOS transistors. Thus assuming, transistor itself as a noise free. Neglect the current noise correlation with voltage noise [8].

2. The total output voltage noise power ($V_{n_{out}}^2$) of whole circuit is calculated by super position technique.

3. $V_{n_{out}}^2$ is referred to input triggered signal to obtain $V_{n_{eq}}^2$. Where $V_{n_{eq}}^2$ is the total equivalent noise.

The total output noise power of FC OTA is expressed as

$$V_{n_{out}}^2 = i_{n_{out}}^2 \cdot R_{out}^2 \tag{10}$$

$$R_{out}^2 = [[g_{m6} \cdot r_{05} \cdot (r_{01} \| r_{05})] \| [g_{m8} \cdot r_{08} \cdot r_{10}]]^2 \tag{11}$$

The input referred noise can be expressed as

$$V_{n_{eq}}^2 = \frac{V_{n_{out}}^2}{A_V^2} \tag{12}$$

The flicker noise is the dominant source of noise generation for low frequency designs, which is expressed as

$$V_n^2(f) \alpha \frac{K}{WL\,C_{OX}f} \tag{13}$$

5   Simulation Results of Folded Cascode OTA

The folded cascode OTA for EEG applications is designed and simulated in Cadence Virtuoso 90 nm CMOS process, powered by 1 V voltage supply. Figure 2 shows the results of FC OTA gain. The FC OTA achieved 73.89 dB gain for the bandwidth of 50 MHz to 25 kHz. Figure 3 shows the input referred noise analysis of FC OTA that is recorded as $21.78\,nV/\sqrt{Hz}$. Figure 4 shows that FC OTA achieved above gain at expense of 4.5 µW of power. Figure 5 gives the information about Monte Carlo simulation for gain with load capacitance varied from 0.5 fF to 5 pF. Table 3 gives the comparison performance of proposed FC OTA for EEG applications.

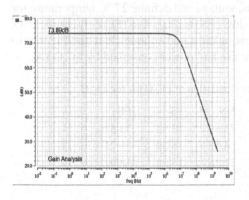

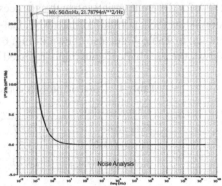

Fig. 2. Gain of FC OTA.   **Fig. 3.** Noise analysis of FC OTA.

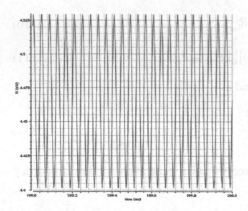

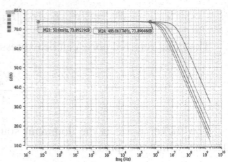

Fig. 4. Power analysis of FC OTA.   **Fig. 5.** Monte Carlo simulation for gain with CL varying from 0.5 fF to 5 pF.

Table 3. Comparison performance of proposed FC OTA for EEG

Parameters	[1]	[2]	[3]	[9]	This work
Year	2017	2017	2012	2017	2018
Technology	180 nm	65 nm	180 nm	130 nm	90 nm
Voltage (V)	0.5	1.2	1.8	1.8	1
Gain (dB)	58	36.34	54.2–72.3	64.5	73.89
Noise (V/$\sqrt{Hz}$)	1.15 μ	376.3 μ	2.2 μ	N/A	21.78 n
Power (W)	620 n	126.2 μ	1.26 μ	140 μ	4.5 μ

The process corner simulation is carried out to check the robustness of folded cascode OTA for different variations such as process, voltage and definite 27 °C temperature for slow-slow (SS), normal-normal (NN), fast-slow (FS) and fast-fast (FF) corner processes shown in Table 4 and also shown in Fig. 6 using clustered column chart and Fig. 6 also shows the pie chart of gain, input referred noise and power consumed by FC OTA.

Table 4. Different process corner simulation of FC-OTA for 27 °C

Characteristics	SS	NN	FS	FF
Gain (dB)	73.97	73.89	73.82	73.826
Power (μW)	3.2	4.53	6.30	6.21
Noise (nV/$\sqrt{Hz}$)	28.70	21.78	18.05	18.02

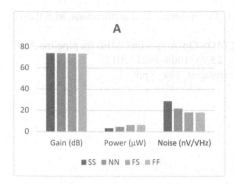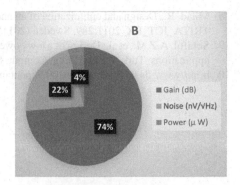

Fig. 6. A. Clustered column chart comparison of performance parameters for different process corners. B. Pie Chart displaying gain, input referred noise and power consumed by FC OTA.

6 Conclusion

High gain single stage FC-OTA is proposed for EEG activity capture, consumes low micro power and exhibiting low input referred noise characteristics is designed in 90 nm CMOS technology. The amplifier achieved 73.89 dB gain by consuming 4.5 µW power. The input referred noise is 21 nV/$\sqrt{Hz}$ for FC-OTA in the bandwidth of 50 MHz and 25 kHz. The presented design is best suited for recording EEG activities and other neural signals which are in the bandwidth of 50 MHz and 25 kHz.

Acknowledgment. We feel obliged to take this opportunity to thank Centre for Excellence in VLSI and Nanoelectronics, Karunya University, Coimbatore and CDAC, Hyderabad for taking keen interest and providing base, encouragement to our research work.

References

1. D. Hari P, A.S.C.S.S: A 0.5V low power single stage FC amplifier for bio-signals. ARPN J. Eng. Appl. Sci. (2017)
2. Laskar, N.M., Guha, K.: A low noise, high gain OTA for low frequency applications. In: 4th International Conference on "Microelectronics, Circuits and Systems" (2017)
3. Liu, H., Tang, K.-T.: A digitally trimmed low noise power analog front -end for EEG signal acquisition. In: IEEE-EMBS International Conference on Biomedical and Health Informatics, China, January 2012 (2012)
4. Refet Firat, Y., Patrick, M.: A 60 µW 60nV/$\sqrt{Hz}$ read out front end for portable biopotential acquisition systems. IEEE J. SSC **42**(5), 1100–1110 (2017)
5. Bautista-Delgado, A.F.: Design of an ultra-low voltage analog front-end for an electroencephalography system. Micro and nanotechnologies Microelectronics. UJF Grenoble I (2009). English. <tel-00418802>
6. Gosselin, B.: Recent advances in neural recording microsystems. Sensors **11**, 4572–4597 (2011). https://doi.org/10.3390/s110504572
7. Allen, P.E., Douglas, R.H.: CMOS Analog Circuit Design. Oxford University Press, New York (2002)

S. V. Mythry and D. Jackuline Moni

8. Milad, R.: Design and optimization of an analog FE for biomedical applications, M.S thesis, TRITA_ICT_EX_2011:289, Sweden (2011)
9. Sohiful, A.Z.M., et al.: Design of a low power CMOS Op-Amp with CMFB for pipeline ADC applications. Turk. J. Electr. Eng. Comput. Sci. **25**(3), 1908–1921 (2017)
10. http://www.fis.uc.pt/data/20062007/apontamentos/apnt_134_5.pdf

House Price Prediction Using Machine Learning Algorithms

Naalla Vineeth, Maturi Ayyappa, and B. Bharathi[✉]

Computer Science and Engineering, Sathyabama Institute of Science and Technology,
Chennai, India
naallavineeth6@gmail.com, maturiayyappa999@gmail.com,
bharathivaradhu@gmail.com

Abstract. Due to increase in urbanization, there is an increase in demand for renting houses and purchasing houses. Therefore, to determine a more effective way to calculate house price that accurately reflects the market price becomes a hot topic. The paper focuses on finding the house price accurately by using machine learning algorithms like simple linear regression (SLR), Multiple linear regression (MLR), Neural Networks (NN). The algorithm which has the lower Mean Square Error (MSE) is chosen as the best algorithm for predicting the house price. This will be helpful for both the sellers and buyers for finding the best price for the house.

Keywords: House price index · Simple linear regression (SLR)
Multiple linear regression (MLR) · Neural networks (NN)
Mean square error (MSE)

1 Introduction

The housing can be a shelter to fulfill the fundamental need of the individual, and it can also be a form of investment. Most of the people use the internet to schedule their life, such as finding a point of interest, looking for a nice restaurant, renting a good hotel and even letting out their own houses. Input parameters that are considered for predicting are price, bedrooms, bathrooms, sqft_living, area, year build, grade, waterfront, number of floors. The objective of this study is to find the accurate price by comparing the error between the algorithms.

The rest of the paper is as follows: Sect. 2 describes the Literature survey of previous papers, Sect. 3 explains the proposed model for the paper and the dataset, it's processing, Sect. 4 describes the methodology used and the two machine learning algorithms studied, Sect. 5 presents the results and discussion for house price prediction lastly Sect. 6 concludes the findings.

2 Related Work

Feng Juan & Gong Tingting in their paper they discuss the house price among small cities, by using yearly data calculation, works out Huludao's yearly house price index, whose

© Springer Nature Singapore Pte Ltd. 2018
I. Zelinka et al. (Eds.): ICSCS 2018, CCIS 837, pp. 425–433, 2018.
https://doi.org/10.1007/978-981-13-1936-5_45

statistics results show the goodness of fit. Also, by testing each item, it is completely feasible to apply Repeated Sales Model to calculate the house price index of small and medium-sized cities, which proves to have the potential prospect for development.

Mohd Fairuz and Hamid & Nor azuana Ramli in their research analyzed and mitigated the energy efficiency by using the artificial neural network as the most accurate method compared to the traditional method which is Linear Regression.

Akash Dutt Dubey in his research paper, used three different models has been developed for the gold price prediction. The models used are support vector regression, ANFIS-GP, ANFIS-SC. it was concluded that support vector regression method had better prediction ability for this purpose while the ANFIS-GP had a slight improvement over ANFIS-SC method.

3   Proposed Model

The dataset used for this paper contains house sales prices for king county, USA. It includes homes sold between May 2014 and May 2015.

The dataset contains 19 house features plus the price and the id column, along with 21613 observations.

The data processing and algorithm application follow the below block diagram (Fig. 1).

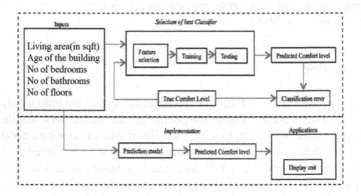

Fig. 1. Overview of proposed approach

The input parameters considered for the prediction are the price as the dependent variable or target variable and sqft_living, bedrooms, bathrooms, year build, grade, no of floors are taken as independent variables.

The observations in the dataset are divided into two parts for applying the algorithm. 75% of the observations in the dataset is taken as 1st part for training the algorithm. The remaining 25% of the observations in the dataset is taken as 2nd part for testing the algorithm.

The process of dividing the observations in the ratio 75:25 helps to predict the house price accurately.

The dividing ratio depends on the dataset and for predicting more accurate values training set data must at least 70% observations of the dataset.

Training Set:- For the training set (75% of the data) algorithm is applied to find the values of the dependent variable.

Testing Set:- The remaining data (25% of the data) will be given to the testing set. In this testing data, there will be only the independent variables and training set algorithm is applied to testing data. Now after applying training algorithm to testing data dependent variables (predicted) are obtained and compared with original values. The algorithm which gives the minimal error that will be considered as the best algorithm.

After applying algorithm to the testing dataset and error is calculated between the actuals and predicted values. If the error rate is high, then algorithm need to be modified such as choosing the columns into the algorithm are changed.

Data Preprocessing
In this process, the data need to check all completely. Initially, the data need to check for the blank spaces (it means no data available in that column). If any data is missing it can be processed using two types by imputing the data or removing the entire column. It can be chosen on basis of the dataset. In the dataset for this paper, there are no missing data. Next data to be checked for the outliers (outlier is the observation point that is distant from other observations). If any outliers are present it can either be removed or it can be brought to the range of other data or if there are very few outliers in our data, they can be ignored (Fig. 2).

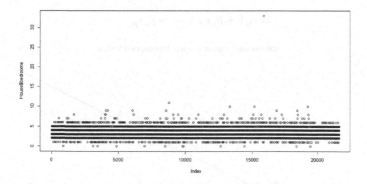

Fig. 2. Checking for outliers in bedrooms (one of the independent variable)

4   Model Implementations

4.1   Simple Linear Regression

Simple linear regression is a statistical method that allows us to summarize and study relationships between two continuous (quantitative) variables: One variable, denoted x, is regarded as the predictor, explanatory, or independent variable (Fig. 3).

$$y = \alpha + \beta x, \tag{1}$$

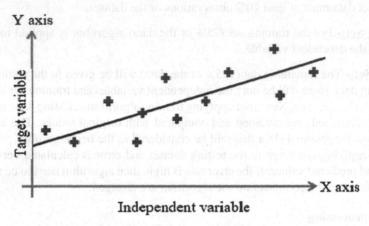

Fig. 3. Graphical representation of the simple linear regression

4.2 Multiple Linear Regression

Multiple linear regression is the most common form of linear regression analysis. As a predictive analysis, the multiple linear regression is used to explain the relationship between one continuous dependent variable and two or more independent variables (Fig. 4).

$$y_i = \beta_0 1 + \beta_1 x_{i1} + \cdots + \beta_p x_{ip} \tag{2}$$

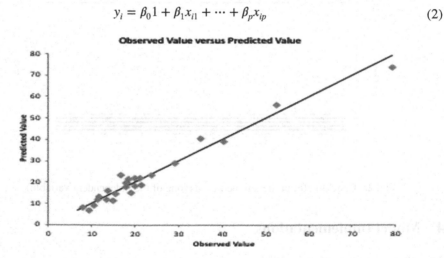

Fig. 4. Graphical representation of the simple linear regression

4.3   Neural Networks

An artificial neural network is an interconnected group of nodes, akin to the vast network of neurons in a brain. Here, each a circular node represents an artificial neuron and an arrow represents a connection from the output of one artificial neuron to the input of another (Fig. 5).

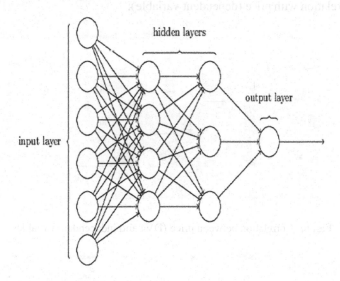

Fig. 5.   Graphical representation of a neural network

5   Results and Discussions

The purpose of this project has been mentioned in the introduction earlier which is to predict the House price of a building by using simple linear regression and multiple linear regression. This method is tested by using the same set of data with Linear Regression method in order to compare the error and accuracy. In simple linear regression, the independent variable which is more correlated with target variable is considered. For the dataset taken sqft_living has more co-relation with the target variable (price).

For applying simple regression algorithm the independent data column to be checked with the column which is mostly correlated with the target variable (dependent variable).

In the dataset chosen for applying simple regression sqft_living and bathrooms have more correlation than other variables.

So simple regression is applied to both sqft_living data column and bathrooms column.

Correlation:- Correlation is a statistical technique that can show whether and how strongly pairs of variables are related.

5.1   Output Graph for Actuals vs Predict Values in Simple Linear Regression

In simple linear regression, only two variables are taken independent and dependent variables.

Here independent variable is chosen by finding the correlation between the dependent and independent variable. From Figs. 6 and 7 it is clear that sqft_living has the best correlation with price (dependent variable).

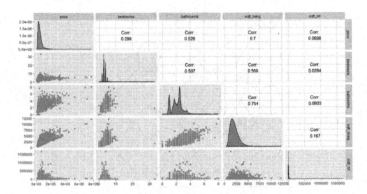

Fig. 6. Correlation between price (DV) and independent variables

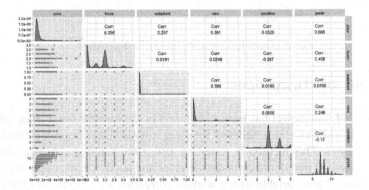

Fig. 7. Correlation between price (DV) and independent variables

The above graph Fig. 8 is plotted between the actuals (red lines) and predicted (blue lines) and the algorithm is said to be working good if most of the blue lines are coincide with the red lines. Where ever the lines are not coinciding we will calculate the error.

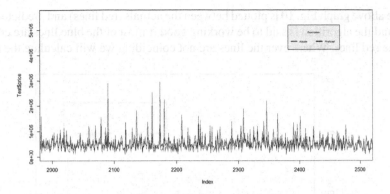

Fig. 8. Comparison of actuals and predicted values in simple linear regression. X-axis rows in test data and Y axis value of price in test data. (Color figure online)

5.2 Output Graph for Actuals vs Predict Values in Multiple Linear Regression

In multiple linear regression, two or more variables are taken as independent variables and one as dependent variable.

The above graph Fig. 9 is plotted between the actuals (red lines) and predicted (blue lines) and the algorithm is said to be working good if most of the blue lines are coincide with the red lines. Where ever the lines are not coinciding we will calculate the error.

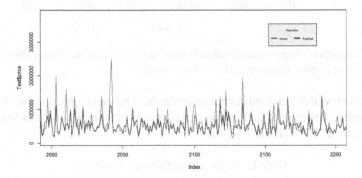

Fig. 9. Comparison of actuals and predicted values in multiple linear regression. X-axis rows in test data and Y axis value of price in test data. (Color figure online)

5.3 Output Graph for Actuals vs Predict Values in Neural Networks

In neural networks, two or more variables are taken as independent variables and one as dependent variable.

Here the data is brought into a specific range or into 0 to 1 using methods like normalization, standardization, etc.

Here it is not necessary that data to be normalized. But if data is normalized then it will give better output.

The above graph Fig. 10 is plotted between the actuals (red lines) and predicted (blue lines) and the algorithm is said to be working good if most of the blue lines are coincide with the red lines. Where ever the lines are not coinciding we will calculate the error.

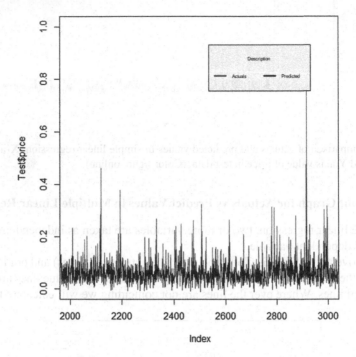

Fig. 10. Comparison of actuals and predicted values in neural networks. X-axis rows in test data and Y axis value of price in test data. (Color figure online)

The accuracy of the method cannot be determined based on observation of the graph only. Therefore, the error of each method is compared by using Root Mean Square Error (RMSE) as tabulated in Table 1 below.

Table 1. Performance of prediction models

Algorithm	RMSE
Simple linear regression	8.0337
Multi linear regression	5.4731
Neural networks	2.1905

It is clear from the above table neural networks has less error than remaining algorithms. Therefore, the most accurate method that can be used for prediction of house price is Neural Networks as it has the lowest error rate for the data.

6 Conclusion

In this research paper, four different models have been used for house price prediction. The models used are simple linear regression, multi linear regression, neural networks. The data samples used for the study ranges from May 2014 to May 2015 in king county of USA. From the results obtained, it was concluded that the neural networks method has better prediction ability.

References

1. Feng, Y., Gong, T.: The application of repeated sales model to calculating house price index among small cities. In: 6th International Conference on Information Management, Innovation Management, and Industrial Engineering. IEEE (2013). 978-1-4799-0245-3/13/$31.00 ©2013
2. Dubey, A.D.: Gold price prediction using support vector regression and ANFIS models. In: International Conference on Computer Communication and Informatics. IEEE (2016). 978-1-4673-6680-9/16/$31.00 ©2016
3. Hamid, M.F.A., Richard, H.G.A., Ramli, N.A.: An analysis on energy consumption of two different commercial buildings in Malaysia. In: 6th International Conference on Information Management. IEEE (2016). 978-1-5090-2547-3/16/$31.00 ©2016
4. Sudhakar, M., Albert Mayan, J., Srinivasan, N.: Intelligent data prediction system using data mining and neural networks. In: Suresh, L., Panigrahi, B. (eds.) Proceedings of the International Conference on Soft Computing Systems. Advances in Intelligent Systems and Computing, pp. 489–500, vol. 398. Springer, New Delhi (2016). https://doi.org/10.1007/978-81-322-2674-1_45
5. Li, L., Chu, K.-H.: Prediction of real estate price variation based on economic parameters. In: IEEE International Conference on Applied System Innovation (2017). ISBN 978-1-5090-4897-787

Content-Based Image Retrieval Using FAST Machine Learning Approach in Cloud Computing

N. Sharmi[(⊠)], P. Mohamed Shameem[(⊠)], and R. Parvathy[(⊠)]

Department of CSE, TKM Institute of Technology, Kollam, Kerala, India
mailtosharmi2017@gmail.com, toparvathyr@gmail.com,
pms.tkmit@yahoo.in

Abstract. The images are having significant role in our daily life. Images consume more storage space when compared to text documents. For preserving privacy of images, before deploying it to cloud storage images are encrypted. A scheme supporting CBIR (content-based image retrieval) from encrypted images is proposed in this paper. The features are identified from the outsourced images and by applying locality-sensitive hashing pre-filter tables are generated for increasing the efficiency of searching. The features of the outsourced images are represented by using interest points and are encrypted by using a stream cipher. A machine learning algorithm using FAST method identifies the interest point on image contour, which helps in retrieving most similar images from cloud. Besides these, for avoiding illegal distribution of retrieved images by query, the cloud server embedds a unique watermark to the encrypted images by using a watermark based protocol. The average search time and precision of the proposed system can be inferred from performance evaluations.

Keywords: Cloud computing · Content-based image retrieval (CBIR)
Contour-based shape descriptor · Machine learning · Watermark
Features from accelerated segment test (FAST)

1 Introduction

The advancement of the imaging technology increases the importance of images in this digital world. For both efficient storage and retrieval purposes large-scales databases are used. In many real-world applications CBIR is used. For example, CBIR helps in retrieving identical cases of patients which helps in clinical decision-making techniques.

Since huge numbers of images are stored on image database, CBIR is having high storage and computation complexity. The on-demand accesses to high computational and storages are provided by cloud computing, that makes cloud storage as a perfect choice for the storing images. Using CBIR services the images are deployed to the cloud server, which helps the image owner itself relieved from keeping the local image database and connecting with the database users online.

Among these astounding benefits of CBIR, image privacy is becoming the main concern of image outsourcing. For example, the medical reports of patients are not disclosed to any others other than to the concern doctor using CBIR medical applications.

© Springer Nature Singapore Pte Ltd. 2018
I. Zelinka et al. (Eds.): ICSCS 2018, CCIS 837, pp. 434–444, 2018.
https://doi.org/10.1007/978-981-13-1936-5_46

Shapes are used for object identification since it is provides as a unique identifier for object recognition which gives information about their identity. Human identifies the features of all objects based on their shape. So, this makes the shape of that object distinguishable from other object features. For objects retrieval and indexing, shape features are used by many applications. For example, in a video security system, sometimes the identity of intruder is determined by using object shape. By using these properties, finest retrieval performance is achieved.

The interest points in images can easily be identified by FAST method proposed by Rosten and Drummond [9]. An interest point is defined as a pixel which is detected and represented in an image. They are having high value of local information content and are repeated ideally over various images [12]. The main applications of Interest point identification are matching, recognition, and tracking of images.

Contribution: The proposed system preserve the privacy of outsourced images in applications using CBIR and retrieves most similar image in response to the query image with high search efficiency.

The following are the main contributions:

(1) A two layer index is generated these feature vectors are encrypted.
(2) The existing searching mechanism retrieve all the similar images based on the color descriptors defined by MPEG-7. So in order, achieve a better and finest search efficiency the search results are optimized by using FAST machine learning algorithm. By using the machine learning algorithm using FAST (Features Accelerated from Segment Test) method the searching is done. FAST method enables to retrieve most similar images based on the identified interest points on the contour of the queried image.
(3) In order to prevent unauthorized distributions of image retrieved by query user i.e., before distributing images to query users, cloud server embeds a unique watermark to the encrypted images in search result by a watermark-based protocol. The watermarked images which is in encrypted form is received by the query user are needed to be decrypted and the decryption could not cause any change to the watermark in the images.

This paper is described as follows. The related works is introduced in Sect. 2. The system model of the proposed scheme and the preliminaries used are illustrated in Sect. 3. The proposed scheme is explained in Sect. 4. Section 5 illustrates the performance evaluations inferred and the conclusions and future work are described by Sect. 6.

2 Related Work

The Searchable encryption (SE) methodology helps the users to perform searching on encrypted database in cloud server [2]. Xia, Wang, Zhang, and Qin [1] proposed a CBIR system, in which the privacy on retrieval of outsourced images based on feature vectors are preserved. The privacy of deployed images in cloud are protected by using four MPEG-7 color descriptors. The four color descriptors used are Scalable color descriptor, Color layout descriptor, Color structure descriptor and Edge histogram descriptor. Based on these descriptor values the features vectors are represented for each outsourced images. From the color descriptors used, colour structure descriptor retrieves all the images having similar color features within less time.

A watermark based protocol is used for preventing the unauthorized distribution of images by the unauthorised query user. For that, the image owner embeds a watermark on the image, before sending to the query user. If any query user has distributed the copies of the watermarked images, the distributed unauthorized user is identified by extracting the watermark from the doubtable images. This watermarking technique can prevent such illegal distribution to a great extent. This CBIR over encrypted images make good effort by the aforementioned work. However, the main challenging issue is the retrieval of the similar category outsourced images based on the features which gives a better search efficiency.

3   Problem Formulation

3.1   System Model

Four entities are representing the proposed system, which is illustrated using Fig. 1.

Image Owner is authorized user who can deploy his images i.e., n images $M = \{m_1, m_2, \ldots, m_n\}$ to the cloud server in an encrypted form $C = \{c_1, c_2, \ldots, c_n\}$. The values of features of an image, F are obtained from images M, $F = \{f_1, f_2, \ldots, f_n\}$ and I, a secure index is generated using F. The collection of encrypted images C, encrypted index I and the identity of the image owner are sent to the server for storage.

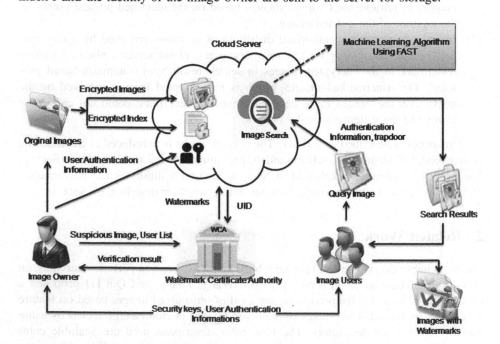

Fig. 1. Framework of the CBIR from cloud computing using FAST Machine Learning Algorithm.

Image users can retrieves required images from cloud server by using query image. The retrieval of images is achieved by using a trapdoor *(TD)*. After receiving requested image, a secret key is shared by the image owner to the user for decryption.

Cloud server consist of the encrypted images *(C)* and the encrypted index *(I)* for each image outsourced by image owners. The query requests from users are solved and besides these the cloud server takes the duty of watermarking the requested images. *WCA* is a trusted authority who helps in generating watermarks for the image users by using watermarking algorithm [1].

3.2 Preliminaries

Locality-Sensitive Hashing (LSH). Locality-sensitive hashing is defined as an algorithm used to perform searching in high dimensional spaces [4]. It represents similarities between objects using probability distributions over hash functions. The input items are hashed by LSH, such that the items are mapped to similar "buckets" having high probability. It aims at maximizing the probability of "collisions" for similar items. Some applications are data clustering and nearest neighbour search.

An LSH family can be defined over a metric space $M = (M, d)$, with $R > 0$ (threshold) and c >1 (approximation factor). $\{h : M \rightarrow S\}$, is defined as a function that maps elements in metric space to bucket $(s \in S)$. For two points $(p, q) \in M$, with function hε F, the following conditions are satisfied by LSH:

(i) If $d(p, q) \leq R$, then $h(p) = h(q)$ (i.e., p and q will collide each other) of probability at least P_1

(ii) If $d(p, q) \geq cR$, then $h(p) = h(q)$ of probability at most P_2.

A family can be defined as interesting when $P_1 > P_2$, and then F is called *(R, cR, P1, P2)-sensitive*. Each vector is assigned by a hash value, and for a fixed (a, b) the hash function $h_{a,b}$ is defined by,

$$h_{a,b}(v) = \left\lfloor \frac{a.v + b}{r} \right\rfloor \tag{1}$$

Hash function family is locality sensitive, if two vectors (v_1, v_2) are enough close (small $\|v1 - v2\|$) then they will collide with high probability and if two vectors (v_1, v_2) are far each other they collide with small probability [4].

Watermark Protocol. When the image owner deploys the image collection M to cloud server, the Image Encryption algorithm [1] is used to generate an encrypted image set C [1]. While a query request is received, the cloud server generates R, the temporary encrypted search results according to TD. Then, by using Watermark Embedding algorithm [1], cloud server will embeds the watermarks generated using Watermark Generation algorithm of image owner for the requested images R' [5]. While receiving R', the query user can decrypts the encrypted images using Image Decryption algorithm for retrieving the watermarked images [1]. If an illegal copy of an image m_t is found by image owner, then image owner initiate a checking by submitting both the suspicious copy (m_t) and original image (m_0) to WCA. Then the watermark w_t

is extracted by WCA using Watermark Extraction algorithm [1]. Finally, w_t, the extracted watermark will helps to identify the unauthorized user who had distributed the images for their benefits (Fig. 2).

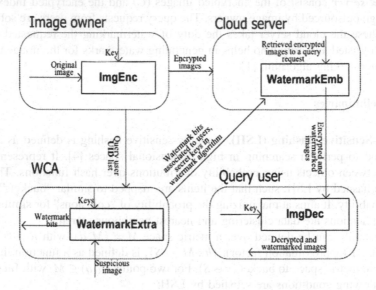

Fig. 2. Framework of watermark-based protocol.

Features from Accelerated Segment Test (FAST). FAST is an interest point identification method using machine learning approach [8, 9, 11]. An interest point is defined as a pixel which is detected and represented in an image robustly. They are having high local information content and are repeated ideally between various images.

Feature Detection Algorithm using FAST:

1. Select a pixel 'p' in the input image and I_p be the intensity of the pixel p.

2. Set a threshold intensity value, t.

3. A circle having 16 pixels around the pixel p is considered. (eg: Bresenham circle of radius 3)

4. A pixel is detected as interest point, if n pixels of the 16 pixels must be either above or below I_P by a threshold value t.

5. First step is done by comparing the intensity of pixels 1, 5, 9 and 13 points of the circle with associated with p. (i.e, at least three of these four pixels must satisfy the above threshold criterion Fig.3).

6. If any of the three pixel values out of the four pixels (I_1, I_5, I_9 I_{13}) are not below or above the pixel intensity I_P + t, then p is not considered as an interest point.

   Else if at least three of the pixels are above or below I_p+ t, then perform checking for all other 12 pixels of 16 pixels and if 12 contiguous pixels fall in the criterion then p can be considered as an interest point.

7. Repeat the above steps for all the pixels in the image.

Machine Learning Approach:

1.	Select an image set for training.
2.	The FAST algorithm executed to learn all the interest points in the images.
3.	For every pixel p, the 16 pixels around it are stored in vector form. (Fig. 4). Thus vector P contains all the data needed for training.
4.	For all the pixels in the images step 2 and 3 are repeated.
5.	Each value (say x, is one of the 16 pixels) in the vector of p, can have three states. Darker, lighter or similar to p. Mathematically,

$$S_{p \to x} = \begin{cases} d, & I_{p \to x} \le I_p - t(darker) \\ s, & I_p - t < I_{p \to x} < I_p + t(similar) \\ b, & I_p + t \le I_{p \to x}(brighter) \end{cases} \quad (2)$$

$S_{p \to x}$ represents the state, $I_{p \to x}$ represents the intensity of the pixel x and t denote the threshold.

6.	Depending on the states, the entire vector P can be divided into three subsets, P_d, P_s, and P_b.
7.	Define K_p, a variable which is assigned true, if p is an interest point and false otherwise.
8.	ID3 algorithm (decision tree classifier) is used to query each subclass using the variable K_p for getting the knowledge about the decision of true class.
9.	The ID3 algorithm is working on the basis of entropy minimization principle. The 16 pixels are queried to find the true class (interest point or not) with less number of queries. i.e, select the pixel x, which has the most information about the pixel p. The entropy for a point P can be represented mathematically as,

$$H(P) = (c + \bar{c}) \log_2(c + \bar{c}) - c \log_2 c - \bar{c} \log_2 \bar{c} \quad (3)$$

where $c = |\{p \mid K_p \text{ is true}\}|$ (number of corners)
and $\bar{c} = |\{p \mid K_p \text{ is false}\}|$ (number of non − corners)

10.	Apply entropy minimization tothe three subsets of P recursively.
11.	Terminate the process when entropy of a subset becomes zero.

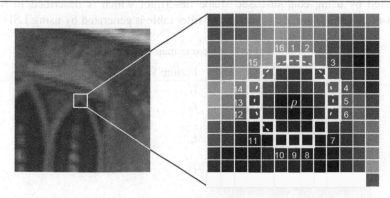

Fig. 3. Image shows an interest point p and the 16 pixels surrounding p.

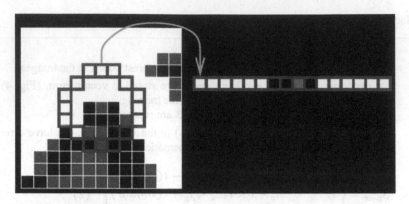

Fig. 4. The pixel p is stored in vector form having 16 values.

4 Proposed System

For a collection of images M, the secret keys set (K), the secured index (I), and the image collection in encrypted form (C) are generated using the following:

$K \leftarrow KeyGen(1^k)$ is an algorithm for key generation that uses the k as security parameter, and the secret keys set are returned as output.

$$K = \{S, M_1, M_2, \{g_j\}_{j=1}^{L}, \{k_j\}_{j=1}^{L}, k_{img}\}.$$

- S is a vector of $(l + 1)$ bits.
- M_1 and M_2 are two invertible matrices of size $(l + 1) \times (l + 1)$.
- $\{g_j\}_{j=1}^{L}$ defines set of LSH functions
- $\{K_j\}_{j=1}^{L}$ is a set of secret keys for bucket encryption
- k_{img} is the secret key used for the image encryption.

$I \leftarrow IndexGen(K, M)$ is an algorithm for index generation that is done by using the secret key set(K) and collection of images M as input, and index I is generated as ouput. Index generation is done by using two steps [1]. First step is encrypted index generation. In this step a one-to-one map index is generated by the image feature vectors represented by using contour-based shape descriptor which is described in Sect. 3. Then from the one-to-one map index a pre-filter table is generated by using LSH which

Table 1. One-one map index

Image identity	Feature vector
ID(m$_1$)	f$_1$
ID(m$_2$)	f$_2$
...	...
ID(m$_i$)	f$_i$
.....	...
ID(m$_n$)	f$_n$

is described in Sect. 3. From the generated pre-filter table, a cluster based on the similarity of images can be obtained from the buckets that are having same values (Tables 1 and 2).

Table 2. The j-th pre-filter table

Bucket value	Image identities
$Bkt_{j,1}$	$ID(m_3)$, $ID(m_9)$, $ID(m_{21})$, $ID(m_{53})$, $ID(m_{108})$
$Bkt_{j,2}$	$ID(m_{16})$, $ID(m_{66})$, $ID(m_{132})$
...	....
$Bkt_{j,Nj}$	$ID(m_{24})$, $ID(m_{243})$, $ID(m_{10})$, $ID(m_{150})$,
....	....

Second step is the encryption of index, in which the pre-filter tables is encrypted by using a one-way hash function. Since the bucket values will disclose the information regarding the features of the images, the pre-filter table cannot be outsourced to cloud storage directly. For ensuring the security, the pre-filter table having bucket values are encrypted before outsourcing. Finally, the index I in encrypted form, with the one-to-one map index of image features and the pre-filter tables generated using LSH, is uploaded to cloud server. Then image owner uploads the collection of encrypted images C, index I in encrypted form and authentication information to cloud server and also sends user identity $\{UID\}$ to WCA for generating watermarks. After receiving $\{UID\}$ of particular image user, a unique watermark w_i is generated by using Watermark Generation algorithm. Upon receiving the request for image uploading, the cloud server find the watermark w_i according to the UID of the image owner and embed the watermark using Watermark Embedding algorithm [1].

When an image user request for an image from cloud server, a query image is send for retrieving similar images from the outsourced image collection by using trapdoor TD. Trapdoor Generation algorithm is used by the query user to generate trapdoor TD [1]. The trapdoor (TD) and authentication information are sent to cloud server for performing searching. While receiving the request for searching, UID and authentication key of the query user is verified by the cloud server. If the verification is successful, then the cloud server allows to perform search by using Search (I, C, TD) for retrieving temporary result set R′ [6]. The R′ includes the top-k similar images containing the common interest points as that of query image, obtained by using machine learning approach based on FAST which is described in Sect. 3. Finally, the query user receives the watermarked images. After receiving the encrypted watermarked images query user can obtain the decrypted images by using Image Decryption algorithm [1].

If an illegal copy of an image m_t is found by image owner, then image owner initiate a checking by submitting both the suspicious copy (m_t) and original image (m_0) to WCA. WCA then exacts watermark w_t by Watermark Extraction algorithm [1]. Finally, w_t, the extracted watermark will helps to identify the unauthorized user who had distributed the images for their benefits.

5 Performance Evaluation

This section illustrates performances evaluating from proposed scheme on a Medical image dataset. The entire scheme is implemented in .NET language on Windows 10 (Intel(R) i5 2.70 GHz). Precision of a query is defined by,

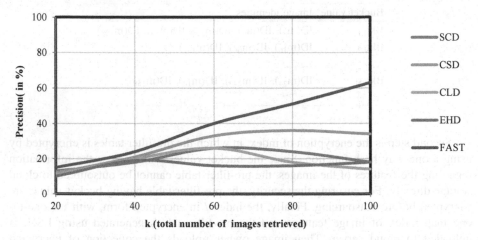

Fig. 5. Average search precision for SCD, CSD, CLD, EHD and FAST.

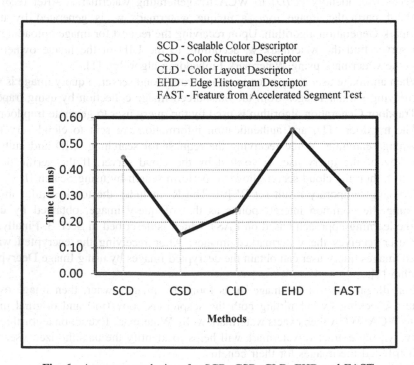

Fig. 6. Average search time for SCD, CSD, CLD, EHD and FAST.

$$P_k = \frac{k'}{k} \qquad (4)$$

Where, k' is the count of most similar images and k is the total images retrieved. The precision value is not affected by encryption of images. Average precisions of the FAST method with the four color descriptors are represented in Fig. 5. Based on the performances of the descriptors used the average search precisions can be evaluated.

The average search time of FAST with four color descriptors (SCD, CSD, CLD, EHD) are shown in Fig. 6. The search result obtained by using FAST based machine learning method is having better average search time. The FAST acquires images having similar features as the search result based on the interest points learned from the contour of input image.

6   Conclusions and Future Works

In this paper, a CBIR scheme in cloud computing scenario using machine learning algorithm based on FAST is presented. The image features are represented by using interest points identified by FAST. The locality sensitive hashing is utilized to group images having similar feature values which improve the search efficiency. Then, the machine learning algorithm based on FAST is applied over the outsourced images for identifying the similar images. Based on these identified interest point values the similarity score is obtained and the cloud server rank images without much effort. Since FAST algorithm is implemented by identifying the interest points on the detected contour, query user can easily retrieve the most similar images having common features with a better search efficiency.

Even though FAST takes less time, it is not robust in high level of noise and it is dependent on a threshold. Since, the features detected by FAST is less, for detecting more features and provides better results the MinEigen, an optimal feature detection algorithm can be used in future works.

References

1. Xia, Z., Wang, X., Zhang, L., Qin, Z.: A privacy-preserving and copy-deterrence content-based image retrieval scheme in cloud computing. IEEE Trans. Inf. Forensics Secur. **11**(11), 2594–2608 (2016)
2. Lu, W., Swaminathan, A., Varna, A.L., Wu, M.: Enabling search over encrypted multimedia databases. In: Proceedings of SPIE, vol. 7254, p. 725418, February 2009
3. Manjunath, B.S., Ohm, J.-R., Vasudevan, V.V., Yamada, A.: Color and texture descriptors. IEEE Trans. Circ. Syst. Video Technol. **11**(6), 703–715 (2001)
4. Datar, M., Immorlica, N., Indyk, P., Mirrokni, V.S.: Locality-sensitive hashing scheme based on p-stable distributions. In: Proceedings of 20th Annual Symposium on Computer Geometry, pp. 253–262 (2004)
5. Lian, S., Liu, Z., Zhen, R., Wang, H.: Commutative watermarking and encryption for media data. Opt. Eng. **45**(8), 080510 (2006)

6. Wong, W.K., Cheung, D.W.-L., Kao, B., Mamoulis, N.: Secure kNN computation on encrypted databases. In: Proceedings of ACM SIGMOD International Conference on Management Data, pp. 139–152 (2009)
7. Bober, M.: MPEG-7 visual shape descriptors. IEEE Trans. Circ. Syst. Video Technol. **11**, 716–719 (2001)
8. Viswanathan, D.G.: Feature from Accelerated Segment Test. http://homepages.inf.ed.ac.uk/rbf/CVonline/AV1FeaturefromAcceleratedSegmentTest.pdf
9. Rosten, F., Drummond, T.: Machine learning for high-speed corner detection. In: Leonardis, A., Bischof, H., Pinz, A. (eds.) ECCV 2006. LNCS, vol. 3951, pp. 430–443. Springer, Heidelberg (2006). https://doi.org/10.1007/11744023_34
10. Sai Anand, C., Tamilarasan, M., Arjun, P.: A study on curvature scale space. Int. J. Innov. Res. Comput. Commun. Eng. **2** (2014)
11. Rosten, E., Porter, R., Drummond, T.: FASTER and better: a machine learning approach to corner detection. IEEE Trans. Pattern Anal. Mach. Intell. **32**, 105–119 (2010)
12. CornerDetection. http://en.wikipedia.org/wiki/Corner_detection
13. Shokhan, M.H.: An efficient Approach for improving canny edge detection algorithm. Int. J. Adv. Eng. Technol. (2014)
14. Ullman, S.: High Level Vision. MIT Press, Cambridge (1997)
15. The MPEG-7 Visual Part of the XM 4.0, ISO/IEC MPEG99/W3068, December 1999

Panoramic Surveillance Using a Stitched Image Sequence

Chakravartula Raghavachari[1(✉)] and G. A. Shanmugha Sundaram[2]

[1] Centre for Excellence in Computational Engineering and Networking,
Amrita Vishwa Vidyapeetham, Coimbatore, India
raghavachari07@gmail.com
[2] Department of Electronics and Communication Engineering,
Amrita Vishwa Vidyapeetham, Coimbatore, India
ga_ssundaram@cb.amrita.edu

Abstract. Security threats have always been a primary concern all over the world. The basic need for surveillance is to track or detect objects of interest over a scene. In most of the fields, computers are replacing humans. One such field where computers play a great role is surveillance. Typically, computer based surveillance is achieved by computer vision that replicates human vision. Here, sequences of panoramic images are created and moving objects are detected over a particular time. Moving objects are being detected using various background subtraction methods like Frame Differencing, Approximate Median and Mixture of Gaussians. In real-time applications like surveillance, the time that takes to make a decision is critical. Hence, a comparison is made between these methods in terms of elapsed time.

Keywords: Surveillance · Computer vision · Image stitching
Moving object detection

1 Introduction

Security threats have always been a primary concern all over the world. Computer vision-based surveillance plays a major role in averting these threats. Here, we are developing a surveillance system that constitutes of image stitching and moving object detection. Image stitching helps in creating a panoramic mosaic of a scene and any objects that are moving over that scene can be detected using moving object detection. Image stitching is a process of stitching different images of a scene with an overlapping region between them. The result of an image stitching process would be a seamless photo mosaic. In order to view the scene completely about 360°, images are captured at every 20°. Using stitching algorithm [1–3], a single panoramic image is created by aligning and then compositing the acquired images. The correspondence and seam's visibility [4] between the adjacent images decide the quality of the stitching process. In computer vision, moving object detection is typically a feature extraction problem. For example, an image of both humans and non humans can be separated as a set of two features, one as humans and other as non humans. Similarly, in our case all moving objects comes

© Springer Nature Singapore Pte Ltd. 2018
I. Zelinka et al. (Eds.): ICSCS 2018, CCIS 837, pp. 445–452, 2018.
https://doi.org/10.1007/978-981-13-1936-5_47

under one set and all non moving objects into another. Several algorithms [5, 6] have been developed for this purpose. In [5], an optical flow based system is developed, in which, moving objects are detected and tracked for traffic surveillance. For moving object detection, a comparison is made between background subtraction and segmentation algorithm in [6]. The paper is organized as follows. The stitching algorithm is explained in the Sect. 2. Section 3 briefs about the different background subtraction methods used for detecting moving objects. The obtained results are shown in Sect. 4 and ends with conclusion in Sect. 5.

2 Image Stitching

Image stitching is a process of stitching images with a minimum amount of overlapping region between them. The following are the different stages of the image stitching process.

A. Image Acquisition

In an image acquisition stage, the focus of the camera is rotated for every 20°. Thus, covering an entire scene about 360°. Camera's center is fixed and a rotation is made with an angle of 20° about its center. This provides 50% of overlapping region between the adjacent images, which enables stitching process effortless.

In this work, a camera (Canon EOS 600D), placed on a tripod, is used for acquiring images. This setup helps in rotating the camera around its axis at any angle. For the reason, the rotation is made at every 20°, we get 18 images. These images are equally spaced with an overlapping region of more than 50% between the adjacent images. If θ_1, is the angle of rotation of the camera, then the number of images i, required to cover one complete rotation of 360° is given by,

$$i = 360°/\theta_1 \tag{1}$$

Here, the rotation angle is 20°. Therefore, the number of images required will be

$$i = 360°/20° = 18 \tag{2}$$

Hence, 18 images are required to cover 360° view. If the horizontal field of view (HFOV) of the camera used is θ_2, then the region of overlapping between adjacent images r is given by

$$r = \theta_2 - \theta_1 \tag{3}$$

The camera used has a HFOV of about 65°. Therefore, the region of overlapping in this case in terms of degrees is

$$r = 65° - 20° = 45° \tag{4}$$

The percentage of 45° of out 65° is about 70%, which is the overlapping percentage between the adjacent images, in our case.

B. Warping images onto cylindrical coordinates

For proper stitching, the images have to be warped to the same coordinates. We used a cylindrical projection for this purpose. Other projective layouts include rectilinear, spherical and stereographic. This projection results in the limited vertical view and complete horizontal view of the stitched image. Figure 1 shows how an image has warped into cylindrical coordinates. In particular, Fig. 1(a) depicts the original image. Figure 1(b) represents the warped image with actual focal length and Fig. 1(c) is the warped image with low focal length. The planar projection of the camera is converted to cylindrical projective layout by warping.

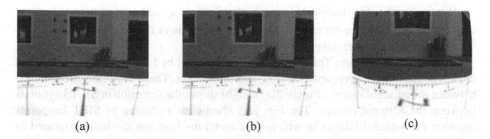

(a) (b) (c)

Fig. 1. Warping image onto cylindrical coordinates. (a) The original image. (b) The warped image. (c) The Warped image with low focal length.

The extent of warping can be changed by changing the focal length. After projection, the horizontal lines in the image appear as curves and the vertical line remains straight. This effect can be clearly noticed in Fig. 1(c). As shown in Fig. 1(a), every image of a scene is warped onto cylindrical coordinates. Converting a 3D point (X, Y, Z) to cylindrical image coordinates involves three steps.

Step-1: Map 3D point onto cylinder coordinates

$$(x, y, z) = \frac{1}{\sqrt{x^2 + y^2}}(X, Y, Z) \tag{5}$$

Step-2: Convert to cylindrical coordinates

$$(\sin \theta, h, \cos \theta) = (x, y, z) \tag{6}$$

Step-3: Convert to cylindrical image coordinates

$$(x_1, y_1) = (f\theta, fh) + (x_c, y_c) \tag{7}$$

Where, (x_c, y_c) is the unwrapped cylinder coordinates

C. Correcting radial distortion

Radial distortion must be removed to produce a perfect seamless panoramic image. In general, distortion can be caused by either the position of the camera with respect to the subject or the characteristic of the lens. In this work, to maintain the same focal length between images, the camera is set to manual mode. For correcting radial distortion, calibration toolbox in MATLAB is used for obtaining the focal length and radial distortion coefficients of the camera. An approximation for radial distortion is explained in the following equations.

$$r = x^2 + y^2 \tag{8}$$

$$x_d = x\left(1 + k_1 r^2 + k_2 r^4\right) \tag{9}$$

$$y_d = y\left(1 + k_1 r^2 + k_2 r^4\right) \tag{10}$$

Where x and y are undistorted image coordinates, x_d and y_d are distorted image coordinates. k_1 and k_2 are the radial distortion coefficients of the camera used.

D. Detection and matching of SIFT points

Scale Invariant Feature Transform (SIFT) developed by Lowe [7], is used to detect the keypoints that are invariant to scaling and orientation. These keypoints are matched between the adjacent images. Figure 2(a) and (b) shows the detection of SIFT keypoints between the adjacent images. The Fig. 2(c) shows the matching of SIFT keypoints between the adjacent images in which false matching between the images termed as outliers can also be seen.

E. Finding homographies by RANSAC

Images captured by rotating the camera are related by using homography. The matching keypoints (inliers) between the images can be found automatically by using RANSAC algorithm [8]. The adjacency between the images is explained as follows.

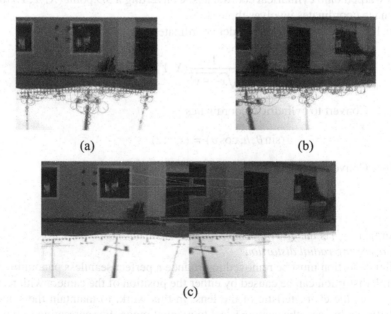

(a)   (b)

(c)

Fig. 2. Detection and Matching of SIFT keypoints. (a) and (b) SIFT keypoints detection in the adjacent images. (c) Matching of SIFT keypoints between the images.

Let us consider, $p = (x, y, 1)$ is a point in one image and $p^l = (x^l + y^l + 1)$ is the corresponding point in the adjacent image, then pixel coordinates of the two images are related by $p = Hp^l$, where H is homography matrix.

$$\begin{pmatrix} x \\ y \\ 1 \end{pmatrix} = \begin{pmatrix} H_{11} & H_{12} & H_{13} \\ H_{21} & H_{22} & H_{23} \\ H_{31} & H_{32} & H_{33} \end{pmatrix} \begin{pmatrix} x^l \\ y^l \\ 1 \end{pmatrix}$$ (11)

F. Image transform and stitching

The images are spatially transformed to align properly. This transformation would help in creating a proper mosaic corresponding to the scene. After aligning the images, they are blended to produce a seamless mosaic. For stitching, a minimum of about 50% overlapping region is maintained between the images. The images may not be in spatial form as shown in Fig. 3. These images have to be aligned properly before stitching to match with the scene.

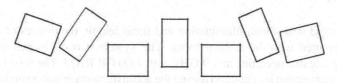

Fig. 3. Images with different spatial forms

Images with different spatial forms taken for stitching shown in Fig. 3 are aligned to match with the scene like in Fig. 4. After which they are stitched together into a seamless composite, as shown in Fig. 5.

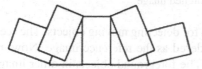   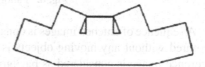

Fig. 4. Images aligned properly   **Fig. 5.** Seamless stitched image

3   Moving Object Detection

In this section, various methods used to detect moving objects are explained. There are several methods exists for detecting moving objects. Out of all, background sub-traction methods are the most widely used. When compared to other methods they are effective in terms of time and space complexities. Hence, different background subtraction methods like Frame Differencing, Approximate Median and Mixture of Gaussians (MoG) are used to detect moving objects over a particular scene.

A. Frame Differencing

This method carries pixel wise differences between two different image frames to extract the moving object. It is an efficient and most reliable method as the computational complexity required for this method is minimal when compared to other methods.

B. Approximate Median

In this method, the previous image frames are stored in a buffer and the background is calculated as the median of all the frames in the buffer. Then similar to that of the frame differencing, the background is subtracted from the current frame. If the absolute difference in pixel values for a given pixel position in both the images is greater than the threshold value, then those pixels are considered as the foreground.

C. Mixture of Gaussians (MoG)

MoG is parametric. The model parameters can be adaptively updated without keeping a large buffer of images. MoG maintains a density function for each pixel, making it capable of handling multimodal background distributions.

4   Results

In order to avoid issues with illumination and focal length, the images of a particular scene are captured with the manual focus. The system was developed on an AMD A8-4500M processor operating at 1.9 GHz with 4.00 GB RAM. The result of stitching the eighteen images that are captured during the acquisition stage into a single panoramic image is shown in Fig. 6.

Fig. 6.   Panoramic stitched image

A sequence of stitched images is considered for detecting moving objects. The scene captured without any moving objects is considered as the reference image. Now, the reference image is considered as background. The foreground objects in other images are detected with respect to the background of the reference image. In frame differencing, a reference image as shown in Fig. 7(a) is taken. This image is subtracted from an input image (Fig. 7(b)) to detect moving objects. Figure 7(c) is the resultant image in which only the foreground objects are highlighted.

(a)   (b)   (c)

Fig. 7. (a) The reference image, (b) The current image, (c) The result of Frame Differencing

In approximate median, the previous images are stored in a buffer. The back-ground of the current image is determined by the background median of previous images. The resultant of this method is shown in Fig. 8(b).

(a)   (b)

Fig. 8. (a) The current image, (b) The result of Approximate Median

Unlike approximate median, MoG can adaptively update the parameters that determine the background by using a density function for each pixel. The result obtained using MoG is shown in Fig. 9(b). The efficiency of the background subtraction methods with respect to response time is given in Table 1.

(a)   (b)

Fig. 9. (a) The current image, (b) The result of Mixture of Gaussians

Table 1. Response time for different background subtraction methods

Background subtraction method	Response time (approx.) in seconds
Frame differencing	14.779822
Approximate median	16.467282
Mixture of Gaussians	69.720255

Frame differencing is the most computationally efficient method while the MoG is the most accurate and complex method of all. The approximate median method is computationally very less complex when compared to MoG, but almost similar to that of frame differencing.

5   Conclusion

In this paper, a surveillance system for detecting moving objects over an entire scene is developed. A single panoramic image is created by stitching the sequence of images in a scene. In applications such as surveillance, stitching helps in monitoring the entire scene (360°). Further, moving object detection would enhance the surveillance. The detection of moving objects should be faster for real time applications (surveillance). Hence, background subtraction methods are used for detecting moving objects. Out of which Frame differencing is the most computationally efficient method while MoG is the most accurate and complex method of all.

References

1. Szeliski, R.: Video mosaics for virtual environments. IEEE Comput. Graph. Appl. **16**, 22–30 (1996)
2. Chen, S.E.: QuickTime VR: an image-based approach to virtual environment navigation. In: Proceedings of the 22nd Annual Conference on Computer Graphics and Interactive Techniques, pp. 29–38, September 1995
3. Brown, M., Brown, D.G.: Automatic panoramic image stitching using invariant features. Int. J. Comput. Vis. **74**(1), 59–77 (2007)
4. Levin, A., Zomet, A., Peleg, S., Weiss, Y.: Seamless image stitching in the gradient domain. In: Pajdla, T., Matas, J. (eds.) ECCV 2004. LNCS, vol. 3024, pp. 377–389. Springer, Heidelberg (2004). https://doi.org/10.1007/978-3-540-24673-2_31
5. Aslani, S., Mahdavi-Nasab, H.: Optical flow based moving object detection and tracking for traffic surveillance. Int. J. Electr. Electr. Sci. Eng. **07**(09), 1252–1256 (2013)
6. Mohan, A.S., Resmi, R.: Video image processing for moving object detection and segmentation using background subtraction. In: IEEE International Conference on Computational Systems and Communications (ICCSC), vol. 01, no. 01, pp. 288–292, 17–18 December 2014
7. Lowe, D.G.: Object recognition from local scale-invariant features. In: International Conference on Computer Vision, pp. 1150–1157, September 1999
8. Fischler, M.A., Bolles, R.C.: Random sample consensus: a paradigm for model fitting with applications to image analysis and automated cartography. Commun. ACM **24**(6), 381–395 (1981)

Epileptic Seizure Prediction Using Weighted Visibility Graph

T. Ebenezer Rajadurai and C. Valliyammai[(✉)]

Department of Computer Technology,
Anna University (MIT Campus), Chennai, India
ebenezerrajadurai5@gmail.com, cva@mitindia.edu

Abstract. Electroencephalogram (EEG) is commonly used for analyzing numerous psychological states of the brain. However, epileptic seizure prediction from EEG signals is quite challenging since it has more fluctuating information about the behaviour of the brain. Analyzing such long-term EEG signals to discriminate between interictal versus ictal regions is a difficult task. Also, EEG signals can be affected by noises from different sources. The proposed work presents an efficient approach based on Weighted Visibility Graph (WVG) for seizure prediction. In this work, the EEG signals are filtered to remove the artifacts due to power supply noise and then the filtered EEG time series data is segmented. The segmented time series data is converted into a complex network called WVG. This WVG inherits the dynamic characteristics of the EEG signal from which it is created. Features like mean degree, mean weighted degree and mean entropy are extracted from the WVG. These features are used to derive the essential characteristics of EEG from the WVG. Finally, classification is done using Support Vector Machine (SVM). The experiments show that the proposed system provides better performance than the existing methods in prediction of seizure in ictal as well as interictal states of EEG over the benchmark dataset.

Keywords: Electroencephalogram (EEG) · Epilepsy · Weighted Visibility
Graph (WVG) · Seizure prediction · SVM

1 Introduction

Electrical activity of the brain can be examined by Electroencephalogram (EEG). It is a cost-effective and preeminent technique used in clinical studies. It is commonly used for the diagnosis of Epilepsy. The Epileptic seizure also known as the epileptic fit is a neurological problem that is characterized by recurrent seizures in the brain. A Seizure is a sudden and uncontrolled change in electrical activity of the neurons in the brain. During a seizure, a person experiences abnormal behaviour, symptoms, and sensations, sometimes may lead to loss of consciousness.

The EEG signals of epileptic patient is classified into four states namely ictal, preictal, and postictal periods and interictal. The ictal period denotes the seizure activity. It may persist for a few seconds to 5 min. Interictal is the period between the seizures. Epilepsy affects 1% of world's population. About 10 million persons with

© Springer Nature Singapore Pte Ltd. 2018
I. Zelinka et al. (Eds.): ICSCS 2018, CCIS 837, pp. 453–461, 2018.
https://doi.org/10.1007/978-981-13-1936-5_48

epilepsy are there in India [1]. Seizure prediction from EEG signals is a challenging task. It needs long-term EEG and may have more artifacts. It also requires a patient-specific approach as the EEG signals are non-stationary in nature.

2 Literature Survey

The correlation based method for seizure prediction using dog iEEG (intracranial EEG) was previously used for EEG [2]. SVM based prediction mechanism was used with three features for classification which needs at least 5–7 seizures in order to achieve good prediction performance. Researchers proposed a patient-specific approach for seizure prediction [3]. Two features based on correlation were used for classification and high computation burden was minimized through least square support vector machine (LS-SVM).

A seizure prognosis mechanism based on Discrete Wavelet Transform (DWT) was presented by [4]. Both linear and non-linear classifiers were used for classification. The EEG method was primarily based on stationary wavelet transform and focused on separation of artifacts from EEG signals to assist seizure prediction [5]. A new feature extraction method using rational discrete short-time Fourier transform (DSTFT) [6] was presented. Rational functions were used for simple time-frequency representation of EEG signals. The proposed method was based on weighted Extreme Learning Machine (ELM) [7]. Wavelet packet transform was used for feature extraction and pattern match regularity statistic (PMRS) was used for quantifying the complexity of EEG time series data which has high event-based sensitivity. But, the influences on EEG signals may lead to false detections in this method.

A seizure detection method [8] based on Partial Directed Coherence (PDC) analysis was discussed in regard with EEG. Multivariate Autoregressive (AR) model was used and the Fourier Transform was applied. It only reflects the change of causal relationship between brain areas before and after a seizure. Three features [9] based on spectral power were extracted was adopted in our proposed work. The seizure prediction of EEG signals from minimum number of channels reduces the complexity, but its performance was degraded for scalp EEG.

The largest Lyapunov exponent was modified by [10]. The chaotic dynamics of EEG signals were obtained in fractional Fourier transform domain. Energy features were computed and the artificial neural network was used for classification. This method has higher accuracy when compared with original Lyapunov exponent. The key points are identified by finding the pyramid of the difference of Gaussian filtered signals [11]. Features were extracted by computing Local binary patterns (LBP) at the identified key points. Finally, SVM was used for classification and it is computationally simple and has high accuracy than the conventional Local Binary Pattern. Recently, deep learning classifiers like Convolutional Neural Networks (CNN) are used in seizure prediction [12–14].

3 Proposed Work

The proposed work presents a novel approach to predict epileptic seizures at the interictal stage using Weighted Visibility Graph method and it is shown in Fig. 1.

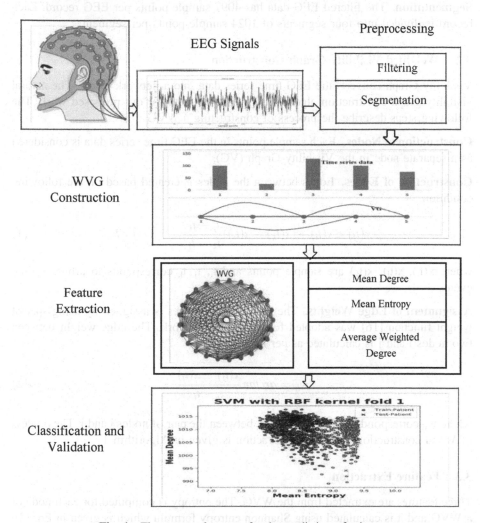

Fig. 1. The proposed system for the prediction of seizure

3.1 Data Preprocessing

The raw EEG data has so many artifacts. The artifacts may be due to eye blinking, body movements and other electrical equipment used in the recording room. These noises must be removed from the EEG in order to get the accurate result.

Filtering. EEG signals are commonly affected by power supply noise during collection of EEG signals. To remove this noise, notch filter of 50 Hz or 60 Hz is used. The proposed work uses a notch filter with cut off frequency of 50 Hz to remove the power supply noise.

Segmentation. The filtered EEG data has 4097 sample points per EEG record. Each record is divided into four segments of 1024 sample points per segment.

3.2 Weighted Visibility Graph Construction

Visibility Graph converts the EEG time series data into a network graph. The natural visibility graph construction algorithm [15] was adopted for our proposed work. The following steps describe the process of constructing a WVG.

Construction of Nodes. Each sample points in the EEG time series data is considered as a separate node in the Visibility Graph (VG).

Construction of Edges. Edges between the nodes is created based on the following condition.

$$x(t_j) < x(t_i) + (x(t_k) - x(t_i))\frac{t_j - t_i}{t_k - t_i}, \qquad i < j < k \qquad (1)$$

where $x(t_i)$, $x(t_j)$, $x(t_k)$ are sample points and t_i, t_j, t_k corresponds to arbitrary time events.

Assignment of Edge Weights. The weight of the edges is assigned by using special weight function [16] was adopted for our proposed work. The edge weight between two nodes i and j is calculated as per Eq. (2).

$$w_{ij} = arctan\frac{x(t_j) - x(t_i)}{t_j - t_i} \qquad (2)$$

where w_{ij} corresponds to the edge weight between the pair of nodes i and j. The process of WVG construction and feature extraction is given in Algorithm 1.

3.3 Feature Extraction

Three features are extracted from the WVG. The entropy is computed for each node of a WVG and it is calculated using Shannon entropy formula which is given in Eq. (3). Then mean entropy of the WVG is then calculated.

$$E(i) = -\sum_{j=1}^{m} p(i,j)log_2((p(i,j))) \qquad (3)$$

where,

$$p(i,j) = \frac{w_{ij}}{\sum_{k=1}^{m} w_{ik}} \tag{4}$$

w_{ij} is the edge weight between the nodes i and j.

Algorithm 1: WVG Construction and Feature Extraction

//**Input:**   Filtered segments of time series data
//**Output:** Mean Entropy, Mean Degree, Mean weighted degree
begin
        Initialize, {x} = EEG time series data points
        find number of data points
        for each data point i in {x}
                create separate node i in WVG
        end for
        for each pair of data points $x(t_i)$, $x(t_j)$, intermediate points $x(t_k)$; i < j < k
            if ($x(t_j) < x(t_i) + (x(t_k) - x(t_i))\frac{t_j - t_i}{t_k - t_i}$) **then**
                    create an edge between node i and node j
                    calculate weight w_{ij} using (2)
                    add edge weight to the edge E_{ij}
            end if
        end for
        for each node i in WVG
                calculate entropy(i) using (3)
        end for
        calculate the mean entropy of WVG
        total degree = 2 * number of edges   //Since undirected graph
        calculate the mean degree of WVG
        calculate the mean weighted degree of WVG
    end

$$E(i) = -\sum_{j=1}^{m} p(i,j) log_2((p(i,j)) \tag{5}$$

where,

$$p(i,j) = \frac{w_{ij}}{\sum_{k=1}^{m} w_{ik}} \tag{6}$$

w_{ij} is the weight of the edge between the nodes i and j.

The degree of a node is defined as the number of edges incident on a vertex. For each WVG average degree of the graph is calculated.

The sum of the weights of all the edges from a node gives the weighted degree (WD) of the node. The WD of the node a is given in Eq. (5).

$$WD_a = \sum_{b \in N(a)} w_{ab} \tag{7}$$

where, W_{ab} denotes the edge weight between the nodes a and b and N(a) represents the neighbor set of the node a. The mean weighted degree is calculated for the graph by finding the average of the WD of all the nodes of the graph.

3.4 Classification and Validation

The proposed work uses Support Vector Machine (SVM) for classification. SVM is a powerful classifier for classifying observed data into two classes. It uses optimal hyperplane to split the input data into two class namely normal and seizure. The proposed work uses radial basis function (RBF) kernel. For validation, 10-fold cross validation is performed. The performance of seizure prediction framework is measured using metrices like precision, recall, and accuracy.

4 Experiments and Results

The data is collected from open source EEG database of the Bonn University, Germany [17]. The EEG dataset consists of five sets namely A, B, C, D and E. Each set has 100 single channel EEG time series data in text file. The sets A and B have EEG data of normal persons. Set C and set D corresponds to the interictal EEG of the patients. The set E has the ictal EEG which recorded during seizure activity of the patients.

The implementation is done using Python language and runs on 3.60 GHz DELL CPU with 24 GB RAM. For the filtering process, the SciPy package is used. A notch filter of cutoff frequency 50 Hz is implemented to filter the power supply noise. Then each filtered EEG time series data is divided into four equal segments. For each segment, WVG is constructed. Figure 2 shows WVGs corresponding to normal, interictal and ictal EEGs with 25 sample data points. WVGs corresponding to normal and interictal EEGs have more edges when compared with the WVG corresponding to ictal EEG due to the sudden fluctuation of amplitude which acts as an obstacle between two time-series data points.

Three features namely mean degree, mean entropy, and mean WD are extracted from each WVG. Figure 3 shows the box plots of the extracted features for all the sets of EEG data. The features mean entropy and averages weighted degree shows a significant variation of ictal EEG with normal EEG. But, the feature mean degree shows a variation of both interictal and ictal EEG from normal EEG.

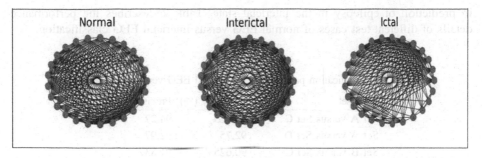

Fig. 2. WVGs corresponding to normal, interictal and ictal EEGs with 25 sample data points

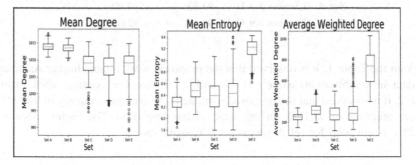

Fig. 3. Box plot of mean degree, mean entropy and mean weighted degree

These features are then used by SVM with RBF kernel for classification of EEG data into two categories namely normal versus ictal and normal versus interictal. The features Mean Entropy and Mean weighted degree are able to classify the seizure (ictal) EEG from normal EEG with improved accuracy than the existing method which is given in Table 1. The proposed work also improves the execution time of the seizure prediction. Modularity feature used by [16] takes 24.9 s while the features in the proposed work take only 0.43 s for computation from single WVG.

Table 1. Accuracy comparison of normal versus ictal seizure prediction

Data set	Accuracy (%)	
	Modularity method [16]	Proposed work
Set A versus Set E	100	100
Set B versus Set E	96.5	97.5
Set C versus Set E	98.5	97.75
Set D versus Set E	93	94.375

The existing works focused only on prediction epilepsy in the ictal state of the EEG. But, the proposed work also achieves state-of-the-art classification performance in prediction of epilepsy in the interictal state. Table 2 describes the performance details of different test cases of normal EEG versus interictal EEG classification.

Table 2. Classification performance of normal EEG versus interictal EEG

Data set	Accuracy (%)	Precision (%)
Set A versus Set C	94.125	94.27
Set A versus Set D	92.75	92.97
Set B versus Set C	90.625	91.837
Set B versus Set D	93.25	93.71
Set A, B versus Set C	92.75	93.15
Set A, B versus Set D	93.83	94.09
Set A, B versus Set C, D	91.43	92.09

From the Table 1, it is observed that the proposed work attains higher accuracy for two data set combinations namely B – E and D – E than the existing work. From the Table 2, it is observed that the proposed work delivers average accuracy of 92.68% and average precision of 93.15% for interictal seizure prediction. The highest accuracy of 94.125% is obtained for set A versus set C classification.

5 Conclusion and Future Work

The proposed system presents a patient-specific approach for epileptic seizure prediction. During the preprocessing stage, the system uses a notch filter to remove the power supply noise from the signal. The filtered signal is then segmented to reduce the computation complexity. The segmented data is converted into Weighted Visibility Graph. Three features namely, entropy, mean degree, and mean weighted degree are extracted from the constructed WVG. SVM classifier with RBF kernel is used for classification.

The proposed work delivers the highest accuracy of 100% for A – E data set combination. In other combinations, the proposed work gives higher accuracy than the existing works. It also delivers the highest accuracy of 94.125% in seizure prediction at the interictal state of EEG. In future, the interictal prediction accuracy can be further improved by adding additional features.

References

1. Santhosh, N.S., Sinha, S., Satishchandra, P.: Epilepsy: Indian perspective. Ann. Indian Acad. Neurol. **17**(Suppl. 1), S3 (2014)
2. Shiao, H.T., et al.: SVM-based system for prediction of epileptic seizures from iEEG signal. IEEE Trans. Biomed. Eng. **64**(5), 1011–1022 (2017)
3. Parvez, M.Z., Paul, M.: Seizure prediction using undulated global and local features. IEEE Trans. Biomed. Eng. **64**(1), 208–217 (2017)

4. Sharmila, A., Geethanjali, P.: DWT based detection of epileptic seizure from EEG signals using naive Bayes and k-NN classifiers. IEEE Access **4**, 7716–7727 (2016)
5. Islam, M.K., Rastegarnia, A., Yang, Z.: A wavelet-based artifact reduction from scalp EEG for epileptic seizure detection. IEEE J. Biomed. Health Inform. **20**(5), 1321–1332 (2016)
6. Samiee, K., Kovacs, P., Gabbouj, M.: Epileptic seizure classification of EEG time-series using rational discrete short-time Fourier transform. IEEE Trans. Biomed. Eng. **62**(2), 541–552 (2015)
7. Yuan, Q., et al.: Epileptic seizure detection based on imbalanced classification and wavelet packet transform. Seizure-Eur. J. Epilepsy **50**, 99–108 (2017)
8. Wang, G., Sun, Z., Tao, R., Li, K., Bao, G., Yan, X.: Epileptic seizure detection based on partial directed coherence analysis. IEEE J. Biomed. Health Inform. **20**(3), 873–879 (2016)
9. Zhang, Z., Parhi, K.K.: Low-complexity seizure prediction from IEEG/SEEG using spectral power and ratios of spectral power. IEEE Trans. Biomed. Circuits Syst. **10**(3), 693–706 (2016)
10. Fei, K., Wang, W., Yang, Q., Tang, S.: Chaos feature study in fractional Fourier domain for preictal prediction of epileptic seizure. Neurocomputing **249**, 290–298 (2017)
11. Tiwari, A.K., Pachori, R.B., Kanhangad, V., Panigrahi, B.K.: Automated diagnosis of epilepsy using key-point-based local binary pattern of EEG signals. IEEE J. Biomed. Health Inform. **21**(4), 888–896 (2017)
12. Antoniades, A., et al.: Detection of interictal discharges with convolutional neural networks using discrete ordered multichannel intracranial EEG. IEEE Trans. Neural Syst. Rehabil. Eng. **25**(12), 2285–2294 (2017)
13. Hosseini, M.P., Pompili, D., Elisevich, K., Soltanian-Zadeh, H.: Optimized deep learning for EEG big data and seizure prediction BCI via internet of things. IEEE Trans. Big Data **3**(4), 392–404 (2017)
14. Kiral-Kornek, I., et al.: Epileptic seizure prediction using big data and deep learning: toward a mobile system. EBioMedicine **27**, 103–111 (2018)
15. Lacasa, L., Luque, B., Ballesteros, F., Luque, J., Nuno, J.C.: From time series to complex networks: the visibility graph. Proc. Natl. Acad. Sci. **105**(13), 4972–4975 (2008)
16. Supriya, S., Siuly, S., Wang, H., Cao, J., Zhang, Y.: Weighted visibility graph with complex network features in the detection of epilepsy. IEEE Access **4**, 6554–6566 (2016)
17. EEG Database from the University of Bonn. http://www.epileptologiebonn.de. Accessed 16 July 2017

Comprehensive Behaviour of Malware Detection Using the Machine Learning Classifier

P. Asha$^{(\boxtimes)}$, T. Lahari, and B. Kavya

Department of Computer Science and Engineering,
Sathyabama Institute of Science and Technology, Chennai, India
ashapandian225@gmail.com, laharil2397@gmail.com,
kavyaboyina@gmail.com

Abstract. Everyone is using mobile phone and android markets like Google play and the model they offer to certain apps make the Google play market for their false and malware. Some developers use different techniques to increase their rank, increasing popularity through fake reviews, installation accounts and introduce malware to mobile phones. Application developers use various advertising campaigns showing their popularity as the highest ranking application. They manipulate ranking on the chart. In the past they worked on application permission and authorization. In this we propose a fair play - a novel framework that uses traces left to find rank misrepresentation and applications subjected to malware. Fair play uses semantic and behavioural signs gathered from Google play information.

Keywords: Google play · Fair play · Fake reviews · Malware

1 Introduction

Smart phone has rapidly become an important platform. Android in particular has seen an impressive growth in recent years. Due to the growth there are also cases of malware. Due to its open platform it is overtaking others competing platforms. Recently android malware has come with new advanced technology that makes difficult for us to identify the malware. Malware on a Smartphone can make unstable there by stealing the private information or affect the information and may behave abnormal. In this paper we use machine learning classifiers for detection of malware. Using malware samples a method is developed which uses combined methods to detect rank misrepresentation and thus detecting fraud applications and malware.

2 Existing System

The methods used by the android market to detect malware are not successful. The Google play uses a Google bouncer to remove malware. It is scanning software to scan the malicious software. It will scan the current, new applications; developer accounts

I. Zelinka et al. (Eds.): ICSCS 2018, CCIS 837, pp. 462–469, 2018.
https://doi.org/10.1007/978-981-13-1936-5_49

and detects red flags. It runs every application and looks for hidden and malware behaviour and also analysis developer accounts and prevent malicious developers from coming back. Android provides a permission system to understand the capabilities of the applications and thus deciding to install an application or not. Analysis revealed that the malware evolves quickly through antivirus tools.

Machine learning based system for detecting malware in android applications. They used different approaches such as machine learning and data mining. They extract data from android applications. The motive is to classify data as positive and negative sentiments. The features include various permission from the application that access various devices like camera, the microphone, reading contacts and divide permission as standard in-build applications and non-standard applications. They showed it has a very low negative rate and they are a number of positive improvements that can be made in the future [1]. Wang et al. [2] used crowd sourcing systems that can make users to do a certain type of things. Malicious activity is passed easily when attacks are generated by users working in the crowd sourcing systems. They described to study crowd sourcing systems. Extracting data and analysing their behaviour and campaigns offered and performed in the systems they analysed this using a micro blogging site similar to the twitter and extracted data from the mobile version where no of tweets and retweets, followers can be viewed. They concentrated on the worker's behaviour like submissions per worker and frequency. They get compensated to do this type of work. Results suggested that it might be an online threat in the future.

Pang et al. [3] tried to classify that the review is positive or negative. They take the information provided online and review sites and they classify according to the subject. The author worked with the reviews that are expressed with the star or some numerical value and these are converted into positive negative or neutral. They compared the text using standard text coordination problem using the number of positive and negative words in the text document. They created a list of positive and negative words. They use machine learning algorithms to examine sentiments. Positive and negative words are classified and divided into equal size folds maintaining balanced class distributions. They used three algorithms such as naive, entropy and support vector machines. They concluded that the machine learning algorithms needs to be improved. Ye et al. [4] tried to evaluate a product or service online every customer tries to see the online reviews. There are more fraudulent and fake reviews to mislead users. The attackers are organised as a group of spammers. The author proposed two methods to identify the group of spammers. First one is the network footprint score that is a graph based system and they use two observations. First one is neighbour diversity that observes the varying behaviour and levels of activity and other self-similarity. The second one is group strainer to cluster spammers on the graph basis using datasets with different domains, a large number of products. They applied these methods in real world datasets and showed case analysis, and it detected many user groups.

Shabtai et al. [5] presents a anomaly to identify malware in android devices. The system observe various features and applies the machine learning methods and classify data. They developed malicious applications and evaluated anomaly ability to detect malware based on known samples. They proposed a light weight malware detection and helps users in detecting suspicious activities on their headsets. They collected this data pre-processing and analysis of data. These are sent through various pre-processing to detect malware and generate threat assessment. Virus threats [6] are generated alert and if it is matched a notification is sent to the user. They evaluated several combinations of anomaly features in order to find the combination that is the best in detecting the malware. Android being the open platform mobile malware is increasing at an alarming rate. New advanced malware has been generated with much advanced capabilities which are more difficult to detect the malware. They proposed a parallel machine based models for early detection of malware. Using real malware samples, applications and combining different classifiers [7]. The authors [8] proposed a risk ranking information system to analyse Android application to improve risk communication and used different probabilistic generative models for effective rank scoring. Google has the permission system which when the user installs the application. They show the permissions that are required for the application thus decreasing risk communication. This is not an effective method so they introduced risk scoring function that assigns each application a score. If the application has high score then it is at high risk. The user knows the risk of different application based on the same functionality.

The growth of the android platform is the target for the malicious applications developers. An instance of the malware applications that track the personal data or applications investigates the application. They investigated using both the permissions [9]. They studied the permission that applications ask and observed that the malicious application requests more permission than the other applications. They designed a risk, signal that gives a warning. The data analysis is used for the effectiveness of the proposed system.

They proposed a novel system that deals with both the rank fraud and malware detection in applications. Behaviour and linguistic behaviour [10] is used. As the Google provides only some reviews, the data is collected from the Google play crawler. For searching rank fraud from the application the data is collected from freelancer, antivirus tools to get the malicious application detection and the last one is the mobile application recommendation [11]. They proposed the time efficient system for detecting fraud applications.

Most of us use android Mobile. Play store provides a large number of applications. Some applications may be fraud [10]. It damages the phone. So they proposed a web application which will process the information, comments and the reviews of the application with natural language processing to give results in the form of a graph. So it is easy to decide which application is fraud. Multiple applications can be processed at a time with the web application. Also, User cannot always get accurate reviews about the product on the internet. So we can check for more than 2 sites, for reviews of the same product. Reviews [12] and comments are fetched separately and analysed for positive and negative reviews. Rating will be combined with an average to give the final rating of the product. They proposed ranking fraud for the mobile application. The present concept of spam city to measure how likely a page is spam. Ranking based evidences finding fraud evidences [13] and check for historical ranking records. They classified into two categories ranking spam in web, spam in online reviews [14–16].

3 Comparative Study on Existing Methodologies

See Table 1.

Table 1. Comparison of existing system

Methodology	Advantages
Pseudo clique finder (PCF) algorithm	Correlates review activities using language and behaviour signs from app data
k-Means algorithm and natural language processing	Uses data aggregation based on the framework and uses two different websites for a single product and analyse them as positive and negative
Signed Inference Algorithm (SIA)	Scalable to large datasets and successfully reveals fraud in large datasets
Support Vector Machines (SVM)	Evaluate permission risk signals using dataset
Probabilistic generative model (Naïve Bayes)	Developed risk scoring for android applications based on permission. Assigns an application a score so that apps with high risk having high score
Max entropy classification, support vector machines (SVM)	Determining whether the review is positive or negative (sentiment analysis)

4 Proposed System

It is a machine learning approach to detect malware and fraud detection. We use two algorithms: De-duplication and time variance algorithm. De-duplication System decreases the amount of data by eliminating redundant information and observing

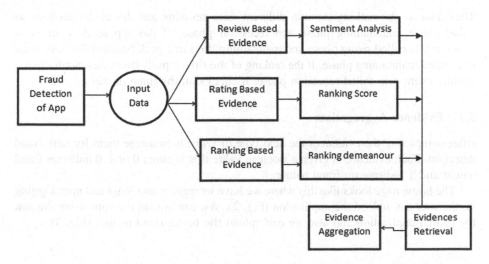

Fig. 1. Architecture diagram

whether it is stored before or not there by reducing fake reviews. The Time variant system measures the estimated and the actual time taken and output characteristics that depend on time or not (Fig. 1).

5 Methodologies Used

5.1 Rating Based Evidence

After downloading an application users generally rate the application. The rating given by the user is one of the most important factors for the popularity of the app. An application having higher rating always attracts more number of users to download an application. It is naturally ranked higher in the chart rankings. Hence, in ranking fraud of applications, rating based evidences is also an important feature so they are needs to be considered.

5.2 Review Based Evidence

Along with rating users are allowed to write their reviews about the app. Such reviews are showing the personalized experiences of usage for particular mobile Apps. The review given by the user is one of the most important factors for the popularity of the app. As the reviews are given in natural language so pre-processing of reviews and then sentiment analysis in pre-processed reviews is performed. The system will find sentiment of the review which can be positive or negative. The Positive review adds plus one to positive score, if negative it will add one to negative score. In this way it will find out the score of each of the reviews and determine whether app is fraud or not on the basis of the review based evidences.

5.3 Ranking Based Evidence

They analyse the ranking through different time sessions and divide the sessions as rising phase, maintaining phase and recession phase. If the app reaches the peak position it is called rising phase and maintaining the same peak position for some time it is called maintaining phase. If the ranking of the time rapidly decreases rapidly in the leading event it is called recession phase. It checks all the three phases.

5.4 Evidence Aggregation:

After completing the evidences the next type of work is to merge them for rank fraud detection. Each evidence is given a Boolean value that is either 0 or 1. 0 indicates fraud nature and 1 indicate no fraud nature.

The home page looks like this where we have to register and login and after logging in, the user can upload the application (Fig. 2). We can upload the apps using the apk file of that application and also we can upload the background picture (Fig. 3).

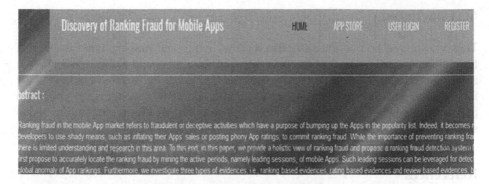

Fig. 2. Application upload

Fig. 3. Locate apps

After successfully uploading the application then the output looks like this. We can download the rating given by the user and can know whether it is fake or not and can also update the fake ranking (Fig. 4).

Fig. 4. Fake identification

6 Conclusion

Thus we showed how to classify false reviews, rating and ranking fraud using data from different applications and thus can detect ranking fraud and malware detection in applications. Rating, review and Ranking based approaches helps a lot in retrieving the fake reviews and ranking, thereby the real good products are saved. Else to promote a poor quality product, fake reviews may be upload in order to spoil the familiarity and sales of good products.

References

1. Sahs, J., Khan, L.: A machine learning approach to android malware detection. In: Proceedings of European Intelligence and Security Informatics Conference, pp. 141–147 (2012)
2. Wang, G., et al.: Serf and turf: crowdturfing for fun and profit. In: Proceedings of ACM WWW (2012). http://doi.acm.org/10.1145/2187836.2187928
3. Pang, B., Lee, L., Vaithyanathan, S.: Thumbs up? sentiment classification using machine learning techniques. In: Proceedings of ACL-02 Conference on Empirical Methods Natural Language Processing, pp. 76–86 (2002)
4. Ye, J., Akoglu, L.: Discovering opinion spammer groups by network footprints. In: Appice, A., Rodrigues, P.P., Santos Costa, V., Soares, C., Gama, J., Jorge, A. (eds.) ECML PKDD 2015. LNCS (LNAI), vol. 9284, pp. 267–282. Springer, Cham (2015). https://doi.org/10.1007/978-3-319-23528-8_17
5. Shabtai, A., Kanonov, U., Elovici, Y., Glezer, C., Weiss, Y.: Andromaly: a behavioral malware detection framework for android devices. Intell. Inform. Syst. **38**(1), 161–190 (2012)
6. Sarma, P., Li, N., Gates, C., Potharaju, R., Nita-Rotaru, C., Molloy, I.: Android permissions: a perspective combining risks and benefits. In: Proceedings of 17th ACM Symposium on Access Control Models Technology, pp. 13–22 (2012)
7. Yerima, S., Sezer, S., Muttik, I.: Android malware detection using parallel machine learning classifiers. In: Proceedings of NGMAST, pp. 37–42, September 2014
8. Peng, H., et al.: Using probabilistic generative models for ranking risks of android apps. In: Proceedings of ACM Conference on Computer and Communications Security, pp. 241–252 (2012)
9. Sanz, B., Santos, I., Laorden, C., Ugarte-Pedrero, X., Bringas, P.G., Álvarez, G.: PUMA: permission usage to detect malware in android. In: Herrero, Á., et al. (eds.) International Joint Conference CISIS'12-ICEUTE'12-SOCO'12 Special Sessions, vol. 189. Springer, Heidelberg (2013). https://doi.org/10.1007/978-3-642-33018-6_30
10. Zhou, Y., Jiang, X.: Dissecting android malware: characterization and evolution. In: Proceedings of IEEE Symposium on Security and Privacy, pp. 95–109 (2012)
11. Akoglu, L., Chandy, R., Faloutsos, C.: Opinion fraud detection in online reviews by network effects. In: Proceedings of 7th International AAAI Conference Weblogs and Social Media, pp. 2–11 (2013)
12. Burguera, I., Zurutuza, U., Nadjm-Tehrani, S.: Crowdroid: behavior-based malware detection system for android. In: Proceedings of ACM SPSM, pp. 15–26 (2011)
13. Oberheide, J., Miller, C.: Dissecting the android bouncer. In: Presented at the SummerCon2012, New York, NY, USA (2012)

14. Asha, P., Sridhar, R., Jose, R.R.P.: Click jacking prevention in websites using iframe detection and IP scan techniques. ARPN J. Eng. Appl. Sci. **11**(15), 9166–9170 (2016)
15. Asha, P., Jebarajan, T.: SOTARM: size of transaction based association rule mining algorithm. Turk. J. Electr. Eng. Comput. Sci. **25**(1), 278–291 (2017)
16. Asha, P., Srinivasan, S.: Analyzing the associations between infected genes using data mining techniques. Int. J. Data Min. Bioinform. **15**(3), 250–271 (2016). Inderscience Publishers

VLSI

Impact of VLSI Design Techniques on Implementation of Parallel Prefix Adders

Kunjan D. Shinde[(✉)], K. Amit Kumar[(✉)], and C. N. Shilpa

Department of Electronics and Communication Engineering,
PESITM, Shivamogga, India
Kunjan18m@gmail.com, amitkaller@gmail.com,
Shilpanandeesh07@gmail.com

Abstract. Adder in general is a digital block used to perform addition operation of given data and generates the results as sum and carry_out. This block is used in various platform for addition/subtraction/multiplication applications. There are several approaches to design and verify the functionality of the adder, based on which they may be classified on type of data it uses for addition, precession of the adder, algorithm used to implementation the adder structure. In this paper we are concentrating on the algorithm/method used to implement an adder structure while keeping the precision constant and considering the binary data for verification of the design. Use of conventional adders like ripple carry adder, carry save adder and carry look ahead adder are not used/implemented for industry and research applications, on the other hand the parallel prefix adders became popular with their fast carry generation network. The presented work gives a detailed analysis on the impact of various VLSI Design techniques like CMOS, GDI, PTL, and modified GDI techniques to implement the parallel prefix adders like Kogge Stone Adder (KSA), Brent Kung Adder (BKA) and Lander Fischer Adder with precession of 4bits, 8bits and 16bits. To measure the performance (in terms of Number of Transistors required, Power Consumed, and Speed) and verify the functionality of these adders we have used Cadence Design Suite 6.1.6 tool with GPDK 180 nm MOS technology, from the results and comparative analysis we can observe that the CMOS technique consumes less power and more transistors to implement a logic, whereas the GDI technique consumes slightly more power than CMOS and implements the logic with less number of transistors. In this paper we also present a simple approach to get the best of both techniques by new technique as modified GDI technique, using this we have optimized the design both in terms of power and transistors used.

Keywords: VLSI design techniques · Parallel prefix adders
Kogge Stone Adder · Brent Kung Adder · Ladner Fischer Adder
CMOS design · GDI design · Modified-GDI design · CADENCE
180 nm technology · Area · Power · Delay

1 Introduction

Addition is the most common arithmetic operation used in various digital blocks and binary adders are widely used to perform operations like addition/Subtraction/ Multiplication and in ALU (Arithmetic and Logical Unit). As the adder is most

© Springer Nature Singapore Pte Ltd. 2018
I. Zelinka et al. (Eds.): ICSCS 2018, CCIS 837, pp. 473–482, 2018.
https://doi.org/10.1007/978-981-13-1936-5_50

fundamental block in digital system the performance adder block plays a vital role in the design of other digital systems and hence the performance of the adders has to be improved. In VLSI system, the requirements of adder should be fast in performing the operation, low power consumption, and less area. The performance in digital system also depends on the algorithm/architecture used to implement adder. The major issue in the binary addition is propagation delay in carry generation stage, as the number of input stages increases the propagation delay also increases with reduction in the speed of operation. To overcome this problem, Parallel Prefix Adders (PPAs) are used and as they are effective, reliable and fast, hence they are better suited in the modern digital systems.

In this paper we have consider the three parallel prefix adder, which are Kogge Stone Adder, Brent Kung Adder and Lander Fischer Adder. The design and implementation of parallel prefix adders are performed using VLSI Techniques like CMOS design, GDI design and Modified-GDI design. Digital circuit design is an important phase, as most of the processing in todays chip are digital and the circuit that performs this operations should consume low power, occupy less area and compute in small delay. These are the main performance parameters and issues in the VLSI design and implementation. Several VLSI design techniques are proposed to implement digital circuits, among those the popular and most often used is CMOS technique, when compared with GDI design style, the GDI technique consumes less number of transistors for designing the digital circuits and consume more power when compared to CMOS technique. Some issues with GDI design style may be driving multiple load and it suffer from the swing degradation at the output signals, limitations of this design techniques are overcome by introducing the new design technique called as modified GDI technique. The modified GDI technique for a given digital circuit can be performed by drawing the given circuit in the form of layers i.e. vertical and horizontal layers, without altering the functionality of the circuit the each odd layer is designed using CMOS technique and the remaining even layers are designed using GDI technique, and at the last stage the design is made using CMOS technique in order to retain the full swing output. With such a combination of both the techniques used as an intermediate and optimal solution for digital circuit which provides good results in terms of accuracy in output, speed in computation and low power consumption [5, 7–9].

2   Literature Survey

The following are some papers that we have referred to design and implementation of parallel prefix adders using different design techniques. In [1], the authors have designed and compared various 8-bit different adders using Verilog HDL coding for conventional adder and parallel prefix adder. From [2], the basic design of parallel prefix adders like Kogge Stone Adder and Brent Kung Adder have been explained using different design techniques. In [3], a brief introduction about carry tree structure and working principle of KSA, BKA and LFA adders have been explained, a comparative analysis is coated based on area, delay and power consumption. In [4] the authors have focused on the design of high speed carry select adder (CSL) for replacing ripple carry adders (RCA) structure in conventional design of Ladner-Fischer Parallel Prefix Adders (LFA), with this replacement the authors have reduced the delay in generating the result. In [5], the GDI technique is applied for digital circuits and its performance is measured. In [6], the authors have implemented various parallel prefix adders and created a comparative analysis. In [7, 9] the authors have verified the functionality of the parallel prefix adders on various platforms like FPGA. and In [8], a modified GDI logic is used to design the system and verified its behaviors.

3 Design of Parallel Prefix Adder

The Parallel Prefix Adder has advanced architecture over the Carry Look ahead Adder (CLA) which is due to the Carry Network of the adder. In VLSI implementations, parallel-prefix adders are known to have the best performance, and widely used in industry for high performance Arithmetic Logic Units digital circuit operation. Compared to the other conventional adders the PPA performs high speed addition operation achieved with the help of its advanced carry generation network, reduce the delay and power consumption. In Parallel Prefix Adder, the execution of partial and final result is performed in parallel and the current stage outcome of the execution is dependent upon the initial input bits at that stage [3].

The following is the general structure of Parallel Prefix adder which involves three steps in process to generate the final results; the steps are explained with reference to Kogge Stone Adder architecture for better understanding (Fig. 1).

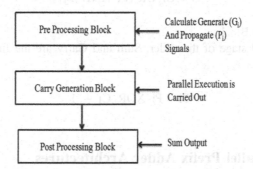

Fig. 1. Architecture of parallel prefix adder

A. Pre-Processing Block:
   The initial stage of the Parallel Prefix Adder is Pre-Processing, two signals are produced in this stage which are termed as generate signal (Gi) and the propagate

signal (Pi). The generated and propagate signal are computed for every ith stage of the input signal and its operation is represented using following equations.

$$Pi = Ai \ XOR \ Bi \qquad (1)$$

$$Gi = Ai \ AND \ Bi \qquad (2)$$

B. Carry Generation Block:

Carry generation stage is a most important block in Parallel Prefix Adder, as the carries are computed before the final result is available using a carry graph. Each adder has different carry graph and based on this the carries are computed. The carry graph consists of two components known as Black Cell and Gray Cell. Black Cell is used to produce the Generated signal and Propagated signal, needed to the calculation of the next stage. Gray Cell is used to produce only Generated signal and these signals are produced based on the earlier inputs received [1].

i. Black Cell: The black cell operator receives two set of generate and propagate signals (Gi, Pi) and (Gj, Pj) compute one set of generate and propagate signals (G, P).

$$G = Gi \ OR \ (Pi \ AND \ Pj) \qquad (3)$$

$$P = Pi \ AND \ Pj \qquad (4)$$

ii. Gray Cell: The Gray operator receives two set of generate and propagate signals (Gi, Pi) and (Gj, Pj) compute one set of generate signals (G).

$$G = Gi \ OR \ (Pi \ AND \ Pj) \qquad (5)$$

C. Post Processing Block:

This is the final stage of the adder; Sum and Carry are the final outcome of the adder.

$$Si = Pi \ XOR \ Ci - 1 \qquad (6)$$

4 Various Parallel Prefix Adder Architectures

The general structure of the parallel prefix adder is understood from the Sect. 3, these Parallel prefix adders differ from each other is by the method of generating carry from the carry generation stage of the adders, The following are the adders we have considered for analysis.

A. Kogge Stone Adders:

The Kogge Stone Adder is one of the most important Parallel Prefix Adders. It generates the carry signal in O (Log$_2$ N) time. This adder is widely used in the industry and considered as the fastest adder design. Carries are generated fast by computing them in parallel, speed of operation is very high due to the low depth of node and operation done in parallel and main important factor is the outcome of the adder is depend upon the initial inputs. Figure 2 gives the schematic of KSA [1].

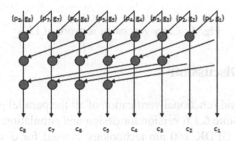

Fig. 2. Carry generation network of KSA

B. Brent Kung Adder:

Figure 3 shows the schematic of BKA. It is one of `the Parallel prefix adder's forms of the carry look ached adder. BKA prefix adder prefix tree is a bit complex to build the design because it has the most logic levels and it have a gate level depth of O(log$_2$n). Construction of design consumed less number of transistor count and it takes less area and speed of operation compare to other prefix adders. BKA structure reduced the delay without compromising the power performance of adder [6].

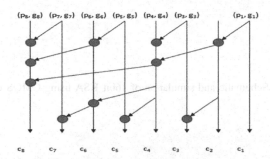

Fig. 3. Carry generation network of BKA

C. Ladner Fischer Adder:

This prefix tree structure shown in Fig. 4. The structure has the minimum logic dept and the number of logic level of (log$_2$n) is always the minimum in this scheme for an n-bit adder. Limits performance of the structure because of complex area by increasing the delay and consumed more power due to large drive cells [2–4].

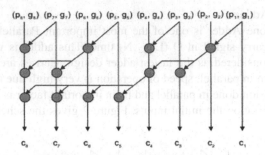

Fig. 4. Carry generation network of LFA

5 Results and Discussion

The implementation and functional verification of all the parallel prefix is performed on the Cadence Design Suite 6.1.6 version for design and simulation using Analog Design Environment (ADE), GPDK 180 nm technology is used for designing digital blocks using n-MOS and p-MOS transistor.

D. Simulation Results

The following are the simulation results of the Parallel Prefix adder used in this paper. The simulation results with schematic are shown only for adders designed using 16bit precession, the schematic and simulation of 4bit and 8bit precession is not show in this paper (Figs. 5, 6, 7, 8, 9, 10, 11, 12 and 13).

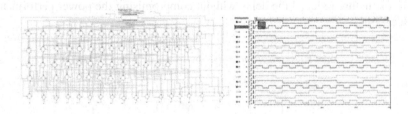

Fig. 5. Schematic and simulation of 16bit KSA using CMOS design.

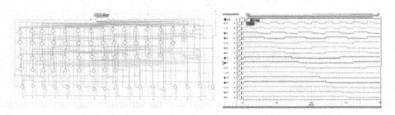

Fig. 6. Schematic and simulation of 16bit KSA using GDI design.

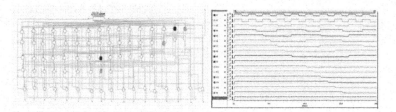

Fig. 7. Schematic and simulation of 16bit KSA using m-GDI design.

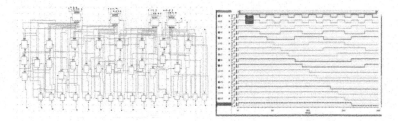

Fig. 8. Schematic and simulation of 16bit BKA using CMOS design.

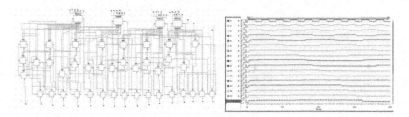

Fig. 9. Schematic and simulation of 16bit BKA using GDI design.

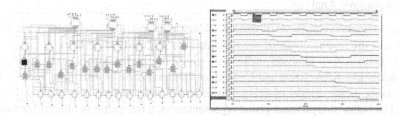

Fig. 10. Schematic and simulation of 16bit BKA using m-GDI design.

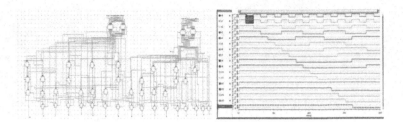

Fig. 11. Schematic and simulation of 16bit LFA using CMOS design.

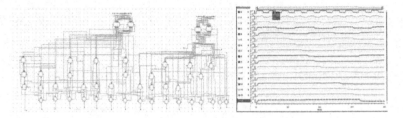

Fig. 12. Schematic and simulation of 16bit LFA using GDI design.

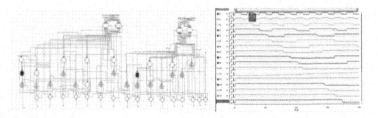

Fig. 13. Schematic and simulation of 16bit LFA using m-GDI design.

E. Comparative Analysis

The comparative analysis of various adders with 4bit, 8bit and 16 bit precession are shown and the results obtained are tabulated for comparison in Table 1.

The comparative analysis in Table 1 gives the performance analysis of parallel prefix adder like KSA, BKA and LFA adder with precision of 4bit, 8bit, and 16bit. The performance metric consist of delay, Power and number of transistor used to design the adder. For better analysis and visual representation, a bar graph is plotted for transistor used and delay of various adders.

From the comparative analysis it is clear that, the number of transistor required to design a parallel prefix adder using GDI design style is about 30% of transistors required to design the same adder using CMOS design style and 60% of transistor are required for GDI design style when compared to modified GDI design style. Power consumed by modified GDI design is higher than the GDI and CMOS design style, if the power is major issue and prime focus on selecting adder for application then CMOS design is the best choice. When compared with delay associated with different design

Table 1. Comparative analysis of parallel prefix adder

			Delay in s	Power in W	Transistors
Kogge Stone Adder	CMOS Design Style	4bit	41.43E-9	1.35E-5	240
		8bit	71.66E-9	2.7E-7	570
		16bit	161.1E-9	8.95E-9	1362
	GDI Design Style	4bit	90.95E-9	3.36E-7	68
		8bit	83.82E-9	3.99E-7	190
		16bit	181.0E-9	8.95E-9	454
	M-GDI Design Style	4bit	92.05E-9	0.0010520	74
		8bit	83.82E-9	0.0010870	248
		16bit	192.3E-9	0.0001083	772
Brunt Kung Adder	CMOS Design Style	4bit	41.67E-9	3.5E-7	192
		8bit	81.43E-9	4.746e-7	414
		16bit	161.4E-9	5.71E-7	804
	GDI Design Style	4bit	42.67E-9	1.05E-7	64
		8bit	82.59E-9	1.30e-7	122
		16bit	162.8E-9	1.54E-7	268
	M-GDI Design Style	4bit	52.73E-9	0.0010870	82
		8bit	93.33E-9	0.0004173	212
		16bit	173.4E-9	0.0010856	388
Lander Fisher Adder	CMOS Design Style	4bit	41.32E-9	4.66E-9	174
		8bit	81.43E-9	3.675e-7	414
		16bit	161.4E-9	5.54E-7	756
	GDI Design Style	4bit	42.15E-9	1.97E-9	100
		8bit	82.58E-9	1.274e-7	122
		16bit	162.4E-9	1.23E-7	252
	M-GDI Design Style	4bit	53.75E-9	7.47E-9	62
		8bit	89.49E-9	0.000568	246
		16bit	172.7E-9	0.0011088	410

styles, CMOS design style produces fast results (less delay in generating result) while consume more number of transistors.

Note: In results and comparative analysis, we have performed simulation for different types of adder like Kogge Stone Adder, Brunt Kung Adder and Lander Fisher Adder. We are not comparing the performance of the various adder architecture, but we are

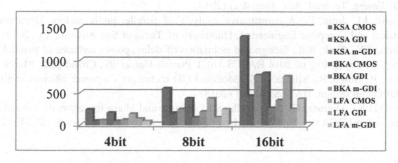

Fig. 14. Transistor count in VLSI technology for various parallel prefix adders

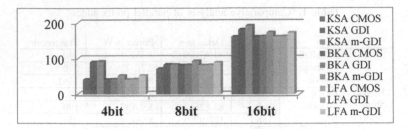

Fig. 15. Delay in ns versus VLSI technology for various parallel prefix adders

trying to measure the impact of designing the adder using different design style on various adders with existing architectures (Figs. 14 and 15).

6 Conclusion

With the presented work in this paper, the CMOS design style uses more number of transistor while generating faster results and consuming less power, GDI design style uses about 30% of transistors compared to CMOS style and does not provides a full swing output, using the modified GDI logic the adders consume moderate number of transistors with full swing output and generates results with an increased delay.

References

1. Shinde, K.D., Jayashree, C.N.: Modeling, design and performance analysis of various 8- bit adders for embedded approach. In: Second International Conference on ELSEVIRE, ERCIC 2014 (2014). ISBN 9789351072621
2. Brent, R.P., Kung, H.T.: A regular layout for parallel adders. IEEE Trans. **C-31**(3), 260–264 (1982)
3. Naganathan, V.: A comparative analysis of parallel prefix adders in 32 nm and 45 nm static CMOS technology. The University of Texas at Austin (2015)
4. Chakali, P., Patnala, M.K.: Design of high speed ladner-fischer based carry select adders. Int. J. Soft Comput. Eng. (IJSCE) **3**(1), 173–176 (2013). ISSN 2231-2307
5. Verma, P., Manchanda, R.: Review of various GDI technique for low power digital circuits. Int. J. Emerg. Technol. Adv. Eng. **4**(2) (2014)
6. Talsania, M., John, E.: A comparative analysis of parallel prefix adders. Department of Electrical and Computer Engineering, University of Taxas at San Antonio, Tx (2013)
7. Yezerla, S.K., Naik, B.R.: Design and estimation of delay, power and area of parallel prefix adders. In: Proceeding of 2014 RAECS UIET Punjab University, Chandigarh, March 2014
8. Verma, P., Singh, R., Mishra, Y.K.: Modified GD technique - a power efficient method for digital circuit design. IJATES. **10**(10) (2013)
9. Hoe, D.H.K., Matinez, C., Vandavalli, S.J.: Design and characterization of parallel prefix adders using FPGA. In: 2011 IEEE Hard South System on System Theory (SSST) (2011)

VLSI Implementation of FIR Filter Using Different Addition and Multiplication Techniques

N. Udaya Kumar[✉], U. Subbalakshmi, B. Surya Priya, and K. Bala Sindhuri

Sagi Ramakrishnam Raju Engineering College, Bhimavaram, India
n_uk2010@yahoo.com, sugunau52@gmail.com,
suryabattula97@gmail.com, k.b.sindhuri@gmail.com

Abstract. Today, in the modernized digital scenario, speed and area are the crucial design parameters in any digital system design. Most of the DSP applications such as FIR and IIR filters demand high speed adders and multipliers for its arithmetic operations. The structural adders, truncated multipliers, delay elements used in FIR filter implementation consume more area, delay and power. So, in this work by using efficient adders and compressed multipliers, different MAC units are designed and these MAC units are placed in FIR filter architecture to identify the best one structures of FIR filter by evaluating its performance with respect to slices, LUT's, and combinational delay. The coding is not in Verilog HDL and Simulation is carried by Modelsim 6.3 g. Finally, the design is implemented with Xilinx ISE 12.2 software on Spartan 3E kit.

Keywords: Dadda multiplier · Modified carry select AN-ta (MCSLA) · FIR filter
Vedic multiplier · Wallace tree (WT) multiplier

1 Introduction

The conception beyond digital communication in today's era is to satisfy the demand to send enormous data. Transmitting digital signals over analog signals will allow greater efficiency in signal processing. But when signal is transmitted digitally there is a greater scope of noise in the acquired signal which leads to efficient filter designing. Filters are basic building blocks of digital communication system [1]. Filters are hardly used for two reasons. First, signal separation, allowing an input signal, eliminating pre-defined frequency elements and transmitting the real signal with subtraction of noise components to output. Second, Signal restoration, this is employed when signal is distorted. However digital filters are preferred than analog filters because of its features like programmability, repeatability, ease of designing, testing, implementing [2] and the ability of digital filters to attain better SNR than the analog filters. Digital filters are classified into two types, FIR and IIR filters. Digital FIR filters are found substantial applications in communication systems and Software Defined Radio [3] systems. The basis for SDR system is exchanging the analog processors with digital processors in transmitters to furnish the flexibility along reconfiguration. The channelizer in SDR system desires a coherent filter structures to employ at greater sampling rate [4]. The delay required to transmit the input signal depends on the executing time needed for the

© Springer Nature Singapore Pte Ltd. 2018
I. Zelinka et al. (Eds.): ICSCS 2018, CCIS 837, pp. 483–490, 2018.
https://doi.org/10.1007/978-981-13-1936-5_51

multipliers and adders. However, multipliers are realized with shift and add operation [5]. As the order of the filter increments, complexity also increases [6]. So far, numerous investigations are made for designing efficient digital FIR filters i.e., by implementing effective multipliers and adders [7]. Normally, the multiplier takes input samples and filter coefficients, which are constant, and perform constant multiplication. The complexity level of the filter is defined by the number of adders used in the multiplier [8]. So, to minimize the complexity level, research work is still in progress to reduce the number of adders in the coefficient multiplier and for designing FIR filter with efficient multipliers [9].

The remaining paper is structured as follows. Section 2 describes about Theoretical concept of Digital FIR filter design. Section 3 presents various MAC techniques used in FIR filter design. Sections 4 and 5 contains the simulation results and comparisons. Finally, Sect. 6 deals with conclusion.

2 Design of FIR Filter

The filter with finite duration due to finite number of samples of impulse response is called FIR filter. Multipliers, delay elements and adders are the fundamental elements in FIR filter design. This FIR filters are broadly used in many DSP applications because of its linear phase and non-feedback characteristics. FIR filter architecture for transposed form is shown in Fig. 1.

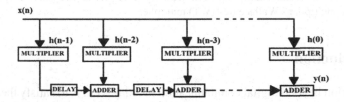

Fig. 1. Transposed direct form architecture for N-tap FIR filter

The output of FIR filter is the convolution of input sequence and the coefficients of filter.

$$Y[n] = X[n] * H[n] \tag{1}$$

For N^{th} order FIR filter, resultant signal of the filter is weighted function of the latter values of the input signal.

$$Y(n) = \sum_{p=0}^{N-1} h(p)x(n-p) = \sum_{p=0}^{N-1} x(p)h(n-p) \tag{2}$$

Here, $x(n)$ is the transmitted sequence, $h(n)$ shows the coefficients of digital FIR filter and $Y(n)$ is the obtained output of FIR filters [10]. Here N represents order of the filter. The multiplication process for 20-tap FIR filter is implemented by using 16-bit

Wallace tree, Dadda and Vedic algorithms. The adder circuit is designed by area and delay based CSLA and MCSLA.

3 MAC Techniques

Design of FIR filter by using MAC techniques is simple when analyze to window techniques and FIR filters typically need one MAC unit per tap. The work done by MAC unit in FIR filter design is, multiplication of filter coefficients with corresponding delayed input samples and add that result to an accumulator. To design high speed and area efficient FIR filters, the multipliers and adders preferred for MAC unit must consume less area and delay.

In this paper, various MAC units are designed by different multipliers and adders and eventually these MAC units are placed in FIR filter architecture to evaluate its performance. Figure 2 shows the design flow of FIR filter using different MAC techniques.

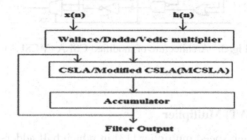

Fig. 2. FIR design flow using MAC unit

The combinations of Multiplier and Adder, used in MAC unit for scheming FIR filter architecture are

- WT Multiplier and CSLA
- WT Multiplier and MCSLA
- Dadda multiplier and CSLA
- Dadda multiplier and MCSLA
- Vedic Multiplier and CSLA
- Vedic Multiplier and MCSLA

Different kinds of Adders and Multipliers used in MAC unit are described below.

3.1 Adders

- **Carry Select Adder (CSLA):**

Among several adders, RCA is one of the easiest adders but it requires more delay [11] due to carry generation and propagation The problem in RCA is avoided by considering both the chances of input carry Cin i.e., '0' and '1'. The concept beyond CSLA is the sum and carry values are generated in advance for both the values of Cin. Further, by knowing the exact values of Cin, the corresponding results of sum and carry are selected by using 2X1 multiplexer. So, CSLA need less delay than RCA i.e., by manipulating the carry signal in before depend on input signal.

- **Modified Carry Select Adder (MCSLA):**

The traditional approach of CSLA is more area consuming because it requires two N-bit ripple carry adders and a multiplexer for choosing the sum. So, to avoid the specified problems in conventional CSLA, gate level optimization of the CSLA architecture for 1bee is proposed [12] by examining the accuracy in boolean expression for sum and carry outputs. So, the necessity of EX-OR gate to produce the half sum in conventional structure is avoided in each level. The 1-bit MCSLA architecture is as shown below (Fig. 3).

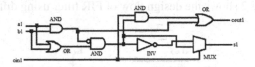

Fig. 3. Architecture of modified CSLA (MCSLA)

3.2   Multipliers

A.   Wallace Tree (WT) Multiplier

WT multiplier is a high speed multiplier [13] in which half adders and full adders are used to multiply two numbers in three steps:

I. Each bit of the n-bit multiplicand is multiplied with every bit in n-bit multiplier to produce n^2 result. Depending on the position of generated bits each bit carry different weights.
II. Afterwards, partial products are reduced with full adders and half adders. This process is sustained up to there are only two layers of partial products.
III. These two final layers are added by using traditional adder.

For scheming out the 20-tap FIR filter, 16×16 WT multiplier is employed to multiply the 16 bit input sequence with the coefficients of filter to produce final output.

B.   Dadda Multiplier

Dadda multiplier is same as Wallace tree multiplier but it is somewhat faster and diminishes the number of logic gates used. Dadda multiplier need N^2 AND gates for the generation of partial products. Moreover, the partial product matrix is diminished to two layers of full adders and half adders using (3, 2) and (2, 2) counters. The flow chart [14] of 16×16 Dadda multiplier is shown in Fig. 4, where the multiplier needs six stages to generate the final product.

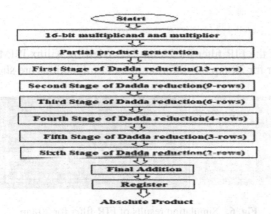

Fig. 4. Flow diagram of 16 × 16 dadda multiplier

At first the 16-bit multiplicand is multiplied with 16-bit multiplier to generate partial products by employing 256 AND gates and the number of rows existing at this stage are 16. Moreover, the number of rows is reduced by 13, 9, 6 and 4 in further stages and finally the last stage contains only 2 rows. Here, the height of transitional matrix does not exceed 1.5 times the height of its preceding stages.

C. Vedic Multiplier

Vedic Multiplier is one of the fastest multipliers used in various scientific and signal processing applications [15]. The 16-bit Vedic multiplier is used in Vedic-CSLA and Vedic-MCSLA based MAC technique. Figure 5 shows the block diagram of 16-bit Vedic Multiplier. It consists of four similar size 8-bit Vedic Multiplier blocks along with two 16-bit ripple carry adders. Urdhva-Tiryagbyam is pre-eminent technique which is relevant to all cases, compared to the remaining sutras in Vedic mathematics. The 16-bit input sequences are divided into two 8-bit sequences and are applied to four multiplier blocks according to this Vedic Sutra and partial products from each multiplier block are added by 16-bit Ripple Carry Adders.

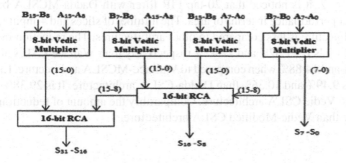

Fig. 5. Block diagram of 16-bit vedic multiplier

4   Results

Implementation of the FIR filter for 20-tap is carried by Xilinx ISIM tool. Simulation results of FIR filter for 20-tap for different input combinations are shown in Fig. 6.

Fig. 6.   Simulation results of FIR filter for 20-tap

5   Comparisons

Device utilization in terms of LUT's, slices and combinational delay for 20-tap FIR filter implementation with several combinations of multiplier and adder are shown in Table 1.

Table 1.   Analysis report for 20-tap FIR filter

FIR filtering using (multiple-adder)	No. of slices	No. of 4 input LUT'S	Combination delay (ns)
Wallace-CSLA	9135	16073	80.770
Wallace-MCSLA	8399	14648	58.054
Dadda-CSLA	7921	13905	94.647
Dadda-MCSLA	7169	12468	77.606
Vedic-CSLA	10152	17782	83.868
Vedic-MCSLA	9406	16406	61.07

From Fig. 7, it is noticed that 20-tap FIR filter with Dadda-MCSLA based MAC shows better performance in terms of area. The number of slices and 4-input LUT's used for this architecture is less compared to other architectures. The reduction in number of slices and LUT's is 21.52% and 22.42% respectively than Wallace-CSLA architecture. It is 14.64% and 14.88% when compared to Wallace-MCSLA architecture. Likewise the reduction is 9.49% and 10.36% than Dadda-CSLA architecture. It is 29.38% and 29.9% compared to Vedic CSLA architecture. Comparably the amount of reduction is 23.78% and 24.02% than Vedic-Modified CSLA architecture.

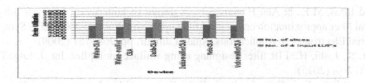

Fig. 7. Comparison of slices and LUT's for 20-tap FIR filter

Similarly from Fig. 8, it is also noticed that 20-tap FIR filter using Wallace-MCSLA based MAC unit shows reduction in delay than other architectures. Delay is reduced by 28.12% than FIR filter with Wallace-CSLA architecture and 38.66% than Dadda-CSLA architecture and 25.19% than Dadda-MCSLA architecture and 30.77% than Vedic-CSLA architecture and it is 4.93% than Vedic-MCSLA architecture.

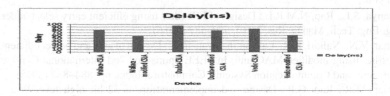

Fig. 8. Comparison of combinational delay for 20-tap FIR filter

6   Conclusion

In this paper, different MAC techniques are used in FIR filter design to identify the highly efficient FIR filter architecture. The result analysis shows that, area is less in terms of LUT's for FIR filter using Dadda–MCSLA based MAC when compared to all other architectures. Further, architecture of FIR filter using Wallace–MCSLA based MAC shows better performance in terms of combinational delay over other architectures. Finally, it can be concluded that FIR filter using Dadda-MCSLA based MAC is an area efficient architecture and FIR filter using Wallace–MCSLA based MAC is high speed architecture. In future Power analysis is also to be addressed for designing low power and high performance FIR filters.

References

1. Litwin, L.: FIR and IIR digital filters. IEEE Potentials 19(4), 28–31 (2000)
2. Mahesh, R.: New reconfigurable architectures for implementing FIR filters with low complexity. IEEE Trans. Comput.-Aided Des. Integr. Circuits Syst. 29(2), 275–288 (2010)
3. Vinod, A.P., Lai, E.: Low power and high-speed implementation of FIR filters for software defined radio receivers. IEEE Trans. Wirel. Commun. 5(7), 1669–1675 (2006)
4. Mittal, A., Nandi, A., Yadav, D.: Comparative study of 16-order FIR filter design using different multiplication techniques. IET Circuits Devices Syst. 11(3), 196–200 (2017)
5. Pridhini, T.S.: Efficient FIR filter design using Wallace tree compression. Int. J. Sci. Eng. Technol. Res. (IJSETR) 3(4) (2014). ISSN 2278

6. bin Md Idros, M.F., bt Abu Hassan, S.F.: A design of butterworth low pass filter's layout basideal filter approximation on the ideal filter approximation. In: 2009 IEEE Symposium on Industrial Electronics & Applications, Kuala Lumpur, pp. 754–757 (2009)

7. Chulet, S., Joshi, H.: FIR filter designing using wallace multiplier. Int. J. Eng. Tech. Res. (IJETR) 3(6) (2015)

8. Kesava, R.B.S., Rao, B.L., Sindhuri, K.B., Kumar, N.U.: Low power and area efficient Wallace tree multiplier using carry select adder with binary to excess-1 converter. In: 2016 Conference on Advances in Signal Processing (CASP), Pune, pp. 248–253 (2016)

9. AlJuffri, A.A., Badawi, A.S., BenSaleh, M.S., Obeid, A.M., Qasim, S.M.: FPGA implementation of scalable microprogrammed FIR filter architectures using Wallace tree and Vedic multipliers. In: Third Technological Advances in Electrical Electronics and Computer Engineering (TAEECE) (2015)

10. Hsiao, S.F., Jian, J.H.Z.: Low cost FIR filter designs based on faithfully rounded truncated multiple constant multiplications. IEEE Trans. Circuits Syst.-II Expr. Briefs 60(5), 287–291 (2013)

11. Sukanya, S.L., Rao, N.M.R.L.: Design of FIR filter using efficient carry select adder. Int. J. Mag. Eng. Tech. Manag. Res. 3(10), 580–587 (2016)

12. Kumar, V.N., Nalluri, K.R., Lakshminarayanan, G.: Design of area and power efficient digital FIR filter using modified MAC unit. In: IEEE Sponsored 2nd International Conference on Electronics and Communication Systems, Coimbatore, India, pp. 884–887 (2015)

13. Kumar, M.R., Rao, G.P.: Design and implementation of 32 bit high level Wallace tree multiplier. Int. J. Tech. Res. Appl. 1(4), 86–90 (2013). International Conference, pp. 159–162 (2015)

14. Ramesh, A.P.: Implementation of dadda and array multiplier architectures using tanner tool. Int. J. Comput. Sci. Eng. Tech. 2(2), 28–41 (2011)

15. Udaya Kumar, N., Bala Sindhuri, K., Subbalakshmi, U., Kiranmayi, P.: Performance evaluation of vedic multiplier using multiplexer based adders. In: International Conference on Micro-Electronics, Electro Magnetics and Telecommunications (ICMEET) (2018)

FPGA Performance Optimization Plan
for High Power Conversion

P. Muthukumar[1(✉)], Padma Suresh Lekshmi Kanthan[2], T. Baldwin
Immanuel[3], and K. Eswaramoorthy[4]

[1] Department of Electrical and Electronics Engineering,
PVP Siddhartha Institute of Technology, Vijayawada 520007, India
muthukumarvlsi@gmail.com
[2] Department of Electrical and Electronics Engineering,
Baselios Mathew II College of Engineering, Sasthamkotta 690521, Kerala, India
suresh_lps@yahoo.co.in
[3] Department of Electrical and Electronics Engineering,
AMET Deemed to Be University, Chennai 603112, India
bimmanuelt@gmail.com
[4] Department of Electrical and Electronics Engineering, Anna University,
Chennai 600025, India
keswaramurthi@gmail.com

Abstract. One of the major part of any power converter system is a blistering
implementation of PWM algorithm for high power conversion. It must fulfil
both requirements of power converter hardware topology and computing power
necessary for control algorithm implementation. The emergence of multi-
million-gate FPGAs with large on-chip RAMs and a processor cores sets a new
trend in the design of FPGAs which are exceedingly used to generate the PWM
in the area of power electronics. Of late, more and more large complex designs
are getting realized using FPGAs, because of less NRE cost and shorter
development time. The share of Programmable Logic Devices (PLD), especially
FPGAs, in the semiconductor logic market is tremendously growing year-on-
year. This calls for an increased controllability of designs, in terms of meeting
both area and timing performance, to really derive the perceived benefits. The
recent strides in FPGA technology favour the realization of large high-speed
designs, which were only possible in an ASIC, in FPGA now. However the
routing delay being still unpredictable and the pronounced nature of routing
delay over logic delay, in today's FPGAs impedes the goal of early timing
convergence. This paper introduces the few techniques for controlling the design
area/time right from architecture stage and the technique can adopt for any
FPGA based design applications including the high power conversion. This
paper also describes the trade off between Area, speed and power of the optimization techniques.

Keywords: Field Programmable Gate Array · Programmable Logic Devices
Application specific integrated circuits · Flip flops · Optimization
Register transfer logic · Block ram · Embedded array block
Configurable logic blocks

© Springer Nature Singapore Pte Ltd. 2018
I. Zelinka et al. (Eds.): ICSCS 2018, CCIS 837, pp. 491–502, 2018.
https://doi.org/10.1007/978-981-13-1936-5_52

1 Introduction

Field-Programmable Gate Arrays (FPGAs) have become one of the most popular implementation media for digital circuits, and since their introduction in 1984 FPGAs have become a multi-billion dollar industry. The key to the success of FPGAs is their programmability, which allows any circuit to be instantly realized by appropriately programming an FPGA. FPGAs have some compelling advantages over Standard Cells or Mask-Programmed Gate Arrays (MPGAs):

- **Accelerate time to market**—FPGA technology proffers flexibility and rapid prototyping proficiency in the countenance of increased time-to-market worries. The idea or concepts are tested and verified in hardware without going through the long fabrication process of custom ASIC design. The incremental changes and iterations are implemented on an FPGA design within hours instead of weeks. Commercial off-the-shelf hardware is also available with different types of I/O already connected to a user-programmable FPGA chip. The technological development of high-level software tools decreases the learning curve with layers of abstraction and often adduces valuable IP cores for modern control and signal processing.
- **Exploitation of FPGA**—Fetching boon of hardware parallelism, FPGAs surpass the computing power of digital signal processors by breaking the crux of sequential execution and performing more process per clock cycle. Controlling inputs and outputs at the hardware level affords faster response times and specialized functionality to closely match application demands.
- **Consistency**—FPGA circuitry is really a "hard" implementation of program execution whereas software tools provide the programming environment. Processor-based systems frequently involve several layers of abstraction to help schedule tasks and allocate resources among multiple processes. The driver layer controls hardware resources and the OS manages memory and processor bandwidth. For any given processor core, only one instruction can execute at a time, and processor-based systems are continually at risk of time-critical tasks preempting one another. FPGAs, which do not use OSs, minimize reliability concerns with true parallel execution and deterministic hardware dedicated to each task.
- **Long-term maintenance**—FPGA chips are field-upgradable and do not require the time and expense involved with ASIC redesign. Digital communication protocols, for example, have specifications that can change over time, and ASIC-based interfaces may cause maintenance and forward-compatibility challenges. Being reconfigurable, FPGA chips can keep up with future modifications that might be necessary. As a product or system matures, the functional enhancement of the design is promising without spending time redesigning hardware or modifying the board layout.
- **Cost**—The nonrecurring engineering cost of custom ASIC design far exceeds that of FPGA-based hardware solutions. The huge initial investment in ASICs is easy to justify for original equipment manufacturers shipping thousands of chips per year, but many end users demand custom hardware functionality for the tens to hundreds of systems in development. The very nature of programmable silicon means, no fabrication costs or long lead times for assembly. Because system demands

frequently change over time, the cost of making additive changes to FPGA designs is negligible when compared to the large expense of re-spinning an ASIC.

FPGAs are often used to reconfigure I/O module functionality. "For example, a digital input module can be used to simply read/write the true/false state of each digital line. Alternately, the same FPGA can be reconfigured to perform processing on the digital signals and measure pulse width, perform digital filtering, or even measure position and velocity from a quadrature encoder sensor," Thus, FPGA devices are very attractive for realizing modern, complex digital controller designs. Most real-time control systems, particularly those used in power electronics and ac motor drive applications, require fast processing, For example, a control algorithms executing at few 100 kHz importantly, the peripherals can be adapted to fit the algorithm." This is particularly true of high-speed A/D interfaces, resolvers and encoders.

Many researchers are used FPGAs for their research, especially in the area of high power conversion. In [1], different types of digital implementations are categorised for the implementation three phase sinusoidal pulse width modulation generation, which targets to control three phase induction motor. Assorted carrier variable frequency random pulse width modulation are implemented by using FPGA which targets to reduce the acoustic noise of the induction motor [2]. The potential of FPGA is highly used for realization of three different carrier waves which are, inverted sine carrier, sine carrier and triangle carrier for PWM generation. In [3, 4], proposes different configurations of SPWM techniques for harmonic reduction and improvement of fundamental peak voltage by using the FPGA implementation of third order harmonic injected SPWM. In [5], The high level calculation involved hybrid space vector pulse width modulation is implemented by using spartan3E FPGA device. The FPGA results are showing the FPGA adaptability of the industrial drives. The computational intensive direct torque control has been implemented by using FPGA [6].

IC designers today are facing continuous challenges in balancing design performance and power consumption. This task is becoming more critical as designs grow larger and more complex and process geometries shrink to 90-nm and below. FPGAs currently available provide performance and features that designers want, but suffer due to higher power consumption requirements. This growing need for maximizing performance while minimizing power consumption requires an increasingly efficient power optimization without sacrificing performance [7].

The trade off between the Optimization Techniques is augmented by a collection of area/time/power estimation guidelines. This paper presents the techniques to achieve area and time containment within the chosen FPGA device. FPGA Selection Guide ASIC Design Guide and HDL Coding Guide will lead to synergetic benefits in achieving design closure in time and with high quality [8]. This paper is generic enough to be applicable for all FPGA devices from various vendors such as Xilinx, Altera and etc.

The crux of these techniques lies in the "Design Level" processes, if when executed effectively and efficiently, will result in timing closure and Area closure at first level itself [9]. The weightage attached with various levels in this methodology is 50% for design level (strict adherence to design norms for timing closure), 40% for first level optimization (achieved by design/code change), 7% for second level optimization

(Changing the implementation options of tools) and 3% for fourth level optimization (applying more timing constraints to tools). This percentage distribution dictates where to concentrate more (indirectly amount of time spent) for achieving faster timing closure. In this paper describes more First level optimization.

During Design and code phase (First Level optimization), perform design and coding as per the established guidelines. It is encouraged to adopt the RTL coding guidelines of as well as from tool vendors. ASIC Design Guide [10–16] can also be referred for relevant sections. All tool vendors provide better coding styles for efficient implementation. Also any violations to norms and guidelines should be documented and analyzed for any impending risks. While coding, it is crucial to understand how your code will map into the logic blocks (LUTs) of the target FPGA [4]. This exercise, even though painful in the beginning but can be practiced, will lead to significant results later. Unlike ASICs, FPGAs provide too few library elements (Flip-Flops and 4-input LUT). Hence it is easier to visualize the implementation view of the code in terms of LUTs while coding, thereby judging the compliance to norms. In Sect. 2 introduces the various RTL Design methodologies to optimize the Area and Speed. In Sect. 3 Explain about the Discussed about the optimization techniques by using Soft and Hard FPGA Macros. In Sect. 4 Discussed about the optimization techniques by using FPGA system features. In Sect. 5 provides the information about the trade off between Speed/Area/Power optimization techniques.

2 Speed/Area Optimization Techniques

2.1 Reduce the Levels of Logic

Most FPGAs have only 4 input LUT architecture. This means that any four input equation, however complex it may be will take only one LUT. However a 5 input equation, however simple it may be (say simple ANDing), will take minimum two LUTs and two levels of logic. Note that two LUTs is only minimum requirement and depending on the complexity of the equation, it would require more than two LUTs and also would increase the levels of logic. The impact is both in area front and timing front if the number of terms in the equation increases beyond four. As a general suggestion, avoid wider decoding logic and also structure the logic in such a way that LUT utilization is minimized and also the levels of logic are kept optimum. Always be cautious of the number of terms in an equation (especially state machine design) during design/coding and see whether it can be reduced.

2.2 Affinity Flops

Some hard macros such as large embedded memory structures in the FPGA (Block-RAM, EAB etc.) have fixed locations in the FPGA. Often the associated interfacing logic would be placed far off in the chip. Hence all the interfacing signals to macro should always be driven from FF and interfacing signals from macro should be sampled directly into a FF without any combinatorial logic in between. This allows for the high routing delay on the signals to traverse from/to macro. This norm can be cautiously relaxed based on the placement of the associated interfacing logic with respect to the macro location and the frequency of operation. However the number of LUT/logic delays for the interfacing signals in that case shall be arrived at and complied with. It is recommended that the interfacing logic and the MACRO (if multiple macro elements are available in the FPGA) be placed closer to each other shown in Fig. 1. Because of the placement restrictions for macros in a FPGA, the routing delay for signals from the logic area to the macros such as memories could considerably affect the timing in very high-speed operations. In such cases, it is better to provide another proximity flop for all interface signals (Fig. 2) so that these flip-flops are placed closer to the macros to break the effect of routing delay.

Proximity flop needs to be provided even for I/O interface signals because of restriction on I/O pad placement. In this case, this proximity flop can be made located in the corresponding I/O pad itself.

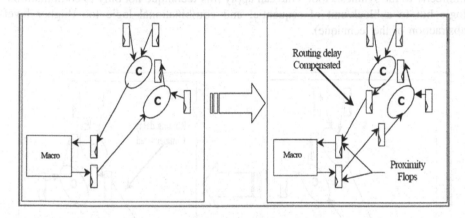

Fig. 1. Affinity flops with macros

2.3 Logic Structuring

Logic structuring technique helps in reducing the number of levels of logic experienced by particular signal(s), by rearranging the equation. The logic structuring deals with prioritizing the signals in an equation explicitly through design/code. The application of logic structuring technique is innumerable, left to the creativity of the designer. Some examples are given in this section for understanding.

One typical example of logic structuring is in grouping arithmetic functions. Instead of A1 + A2 + A3 + A4, which would produce three adders in cascade (as chain of adders), group it like (A1 + A2) + (A3 + A4). This gives a structured tree implementation as shown in Fig. 3. Same is the case with the parity tree. Instead of parity <= A1 ^ A2 ^ A3 ^ A4 ^ A5 ^ A6 ^ A7 ^ A8, group it like (A1 ^ A2 ^ A3 ^ A4) ^ (A5 ^ A6 ^ A7 ^ A8). Wherever possible, you should structure the equations in the code so that synthesis tool keeps the same structure, as you want. Another application is in prioritizing late-arriving signal by moving late-arriving source closer to the end-point. For example, assume that, signal A is a late arriving signal in the following equation.

if (A – B < 27)
Z <= C;
else
Z <= D;

The implementation of this equation is shown in left-hand side of Fig. 4. Instead, reduce the number of operations that have signal A in their fan-in cone.

if (A < 27 + B)
Z <= C;
else Z <= D;

Through logic structuring by design/code, you are conveying the signal priorities indirectly to the synthesis tool. One can apply this technique not only to combinational logic but to a block/unit of sequential and combinational logic too (higher level abstraction of the technique).

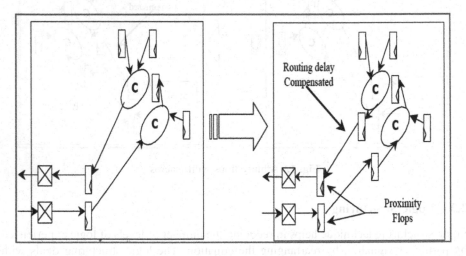

Fig. 2. Affinity flops with Interface ports

2.4   Logic Replication

Logic Replication technique is also referred as cloning technique. This is employed to reduce the fanout of a signal in order to control the routing delay. Figure 4 illustrates the concept of cloning. In this example, it shows a simplistic case of FF duplication. In some cases, one has to replicate (part of) the associated combinational logic too. However, while doing logic replication, care should be taken to ensure that load on the previous stage is not increased to problematic levels.

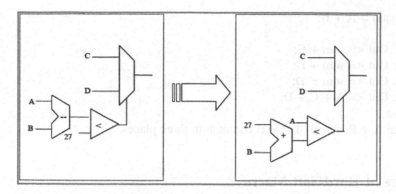

Fig. 3. Logic restructuring

It is a good practice to use macros for changing the number of signal copies (clones). The code piece below suggests a style that can be followed:

At posedge clk
for (I = 0; I < NO_OF_COPIES; I = I + 1)
enable[I] <= a & b & c;
At posedge clk
if (enable[DATA_ENB] == 1) then
data_out <= data_in;
if (enable[CTRL_ENB] == 1) then
ctrl_out <= ctrl_in;

By changing NO_OF_COPIES, DATA_ENB, CTRL_ENB values, one can easily do logic replication. Also the often ignored signal for fanout consideration is the enable signal. One has to consider data_out as N-bit wide flip-flop structure, thereby having N loads on enable signal. Accordingly, the fanout norms need to be applied for logic replication.

2.5 Code Restructuring

Structuring the RTL code to infer optimized logic is the key aspect in RTL coding. Some of the examples reflect the concept of "Resource sharing", wherein critical resources are reused while non-critical resources are duplicated. This point is illustrated through the following example.

Case 1
case (select)
2'b00 : Out <= A + B + C;
2'b01 : Out <= A + B + D;
2'b10 : Out <= A + B − D;
2'b11 : Out <= A + C + D;
end case;
Case 2

```
assign sum = A + B;
case
2'b00 : Out <= sum + C;
2'b01 : Out <= sum + D;
2'b10 : Out <= sum - D;
2'b11 : Out <= A + C + D;
end case;
```
Here A + B is reused as it is common in three places.

3 Use of Hard/Soft Macro

Most FPGAs provide hard/soft macros that are both area and timing efficient. For example, Xilinx's Look-Up Table (LUT) can be converted into a 16x1 Distributed RAM or ROM or 16-bit variable shift registers, reducing area consumption significantly. If you fear that the use of macros/cores make your design too target specific, then you can have wrappers for these cores so that you can code it in a generic HDL way if needed. It is suggested that frequently used logic functions (such as counters, adders etc.) be made as separate modules/entities and instantiated wherever needed. This helps in quickly trying out hard/soft macro option or going in for another way of implementation.

3.1 Use of BlockRAM/EAB

Most FPGA devices now provide large RAMs (BlockRAM in Xilinx and EAB in Altera). Often in most designs, you may not be utilizing all the RAMs available. In such cases, you can consider the following ways of utilizing the leftover RAMs

- Architectural resource sharing
- RAM based state machine design
- Wide high-speed binary counters, up/down counters (simple ROM lookup)
- Storing addition/multiplication tables, and performing simple lookup for complex addition/multiplication, thereby saving logic for these functions.
- Using it as conversion tables and lookup tables (basically as large ROM).

3.2 RAM Based State Machine Implementation

Large RAMs, if left unused, can be used to implement complex state machines. In most FPGA architectures, these large RAMs can be initialized with a specific data pattern. Hence when the write enable to the RAMs is permanently disabled, it becomes a ROM. This ROM can be used to implement wide state machines. It is pictorially depicted in Fig. 5. Address bus of the RAM is comprised of the State Machine inputs (branch control signals) and State Vector. And the Output Data bus gives us the next state and the output signals. (The next state value can be clocked by flip-flops if the RAM's data response is in the same clock period as that of address). All that is involved is programming the truth table values of the state machine as RAM INIT values. This RAM thus becomes a large lookup table.

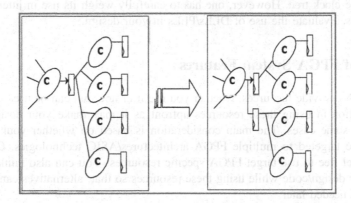

Fig. 4. Logic replication

This state machine is faster, running closer to the maximum clock frequency supported by the RAM and consumes zero logic area other than a RAM. One can realize large state machine with more number of states very easily. Most FPGA Architectures support Dual Ported RAMs. Each port can be independently used and hence these DPRAMs can be used to construct two independent complex state machines, which are timing and area efficient.

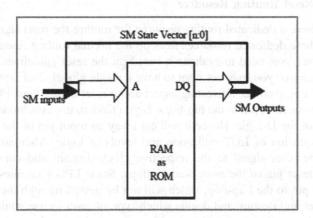

Fig. 5. Statemachine implementation by using ROM

3.3 Use of Multipliers

Some FPGA Devices have large multipliers. In most designs, these multipliers may go unutilized. It can be used as a shifter register.

3.4 Delay Locked Loop (DLL)

DLLs help in improving the I/O timing performance by compensating the insertion delay of the clock tree. However, one has to carefully weigh its use in jitter sensitive applications. Evaluate the use of DLLs/PLLs in your design.

4 Use of FPGA System Features

Most FPGAs provide resources, whether you use it or not, that can be used creatively. Apply caution in using these resources upfront as it may make your design target-specific in some cases. The main consideration is based on whether your design is likely to be targeted to multiple FPGA architectures/ASIC technologies. Otherwise, you can feel free to use target FPGA specific resources. You can also think of structuring your design/code while using these resources so that, alternatives can be easily explored if needed later.

4.1 Tri-state Buses

Most FPGAs provide tri-state buses on chip, which can be used to replace multiplexers (especially CPU I/F read path). However tri-state outputs would drive only specialized routing resources that will not be used for other time critical signals. Also on-chip tri-state buses are not preferred in ASICs. Hence weigh the various options before using tri-state buses.

4.2 Global Reset Routing Resource

Most FPGAs have a dedicated routing resource for routing the reset signal to all Flip-Flops. Using these dedicated resources frees up the normal routing resources for other signals. However, one need to evaluate it based on the reset guidelines. Also beware that in some designs, you may not want to have a single global reset signal for all flip-flops in the design. Another interesting aspect about reset is that most FPGAs provide only asynchronous reset pin to the flip-flops. If you have to use synchronous reset, then it will be part of the D-logic. Hence it will eat away an input pin of the 4-input LUT, resulting in explosion of LUT utilization and levels of logic. Alternatively, one can synchronize the reset signal to the respective clock domain and connect it to the asynchronous reset pin of the associated flipflops. Some FPGA families provide synchronous reset pin to the Flip-flop, which will not be passed through the lookup table. Study the target architecture and decide which type of reset to use while coding.

4.3 Slew Rate Control

The I/O pads of most FPGAs are slew-rate controllable. If the slew rate is high, it favours fast switching time. However it leads to ground bounce problems. Analyze from EMI front and if okay, you can enable fast switching of I/O pads to get better output delays. Also choose the proper current drive strength of the pads.

4.4 I/O Registers

Most FPGAs provide flip-flops for input, output and output enable on the I/O pad. Use these registers to improve I/O timing performance.

4.5 Programmable Delay on I/O Pads

Most FPGAs provide programmable delay element for the input on the I/O pad, which can be used to fix hold time violations. However this increases the setup time requirement. Apart from these resources, if you have any other unused resources in the FPGA, think of the creative use of such resources. Xilinx and Altera offer a broad selection of parts and compete in the same application cases.

While comparing the architectures of the Altera and Xilinx chips, it is more interesting fact that Altera chips are more amenable to subtle optimizations whereas Xilinx tends to be more technology-oriented and have better links to applications by offering more chips with custom circuits that implement specific functions. Table 1 depicts the conclusion table of the optimization techniques.

Table 1. Trade off between various optimization techniques

Optimization techniques		Speed	Area	Power
RTL design techniques	Reduce the levels of logic	Increase	Increase	Increase
	Affinity flip flops	Increase	Increase	Increase
	Logic structuring	Increase	No change	No change
	Logic replication	Increase	Increase	Increase-depends upon number of cloning FF
	Code restructuring	No change	Decrease	Decrease
Use of hard/soft macros	RAM based state machine implementation	Increase	Area will decrease, but usage of hard macros will increase	Increase
	Use of multipliers instead of using shift register	Increase	Usage of LUTs will decrease.	Increase
	Delay locked loop (DLL)	Increase	Usage of LUTs will decrease.	Increase
FPGA features	Synchronous reset	No change	Increase	Increase
	Asynchronous reset	No change	Decrease	Decrease
	Slew rate control	Increase	No change	Vary depends up on the slew rate and toggling rate
	I/O registers	Increases	Area slightly will increase	Increase
	Programmable delay on I/O pads	Depends up on setup time and hold time	Increase	Increase

5 Conclusion

This paper presented the techniques for area/time/power optimization with trade off. It is believed that the concepts discussed in this paper will help in achieving first-time-right design for implementing power conversion industry. There will always be newer challenges to ensure area/time closure for any FPGA design. However detecting the impact of such challenges early in the design cycle will help you in saving significant rework later, thereby favouring controllability of your design towards early area/time closure.

References

1. Jacob, T., Krishna, A., Suresh, L.P., Muthukumar, P.: A choice of FPGA design for three phase sinusoidal pulse width modulation. In: IEEE International Conference on Emerging Technological Trends (ICETT), Kollam (2016). https://doi.org/10.1109/icett.2016.7873768
2. Paramasivan, M., Paulraj, M.M., Balasubramanian, S.: Assorted carrier-variable frequency-random PWM scheme for voltage source inverter. IET Power Electron. **10**(14), 1993–2001 (2017)
3. Muthukumar, P., Mary, P.M., Deepaprincy, V., Monica, F.: Performance analysis of mixed carrier-pulse width modulation scheme. Res. J. Appl. Sci. Eng. Technol. **8**(23), 2356–2362 (2014)
4. Stephen, V., Suresh, L.P., Muthukumar, P.: Field programmable gate array based RF-THI pulse width modulation control for three phase inverter using matlab modelsim cosimulation. Am. J. Appl. Sci. **9**(11), 802–1812 (2012)
5. Muthukumar, P., Mary, P.M., Jeevananthan, S.: An improved hybrid space vector PWM technique for IM drives. Circuits Syst. **7**(09), 2120–2131 (2016)
6. Devipriya, D., Muthukumar, P., Selvarathinam, M., Deepika, T.: A novel FPGA based direct torque control for induction motor. Middle-East J. Sci. Res. **24**(23), 787–793 (2016)
7. Khan, M.: Power Optimization in FPGA Designs- SNUG San Jose 2006. Timing Closure by Whatcott, R. White paper. www.xilinx.com
8. AN 580: Achieving Timing Closure in Basic (PMA Direct) Functional Mode. Altera Corporation, June 2009. Altera.com
9. Marquardt, A., Betz, V., Rose, J.: Speed and area trade-offs in cluster-based FPGA architectures. Right Track CAD Corporation, Canada (2000)
10. Metzgen, P.: A high performance 32 bit ALU for programmable logic. In: Proceedings of the 2004 ACM-SIGDA 12th International Symposium on Field Programmable Gate Arrays, pp. 61–70 (2004)
11. FPGA Performance Benchmarken Methodology, White Paper. www.altera.com
12. FPGA Performance Benchmarken Methodology, White Paper. www.xilinx.com
13. Altera Corporation: The Stratix Device Hand Book, vol. 1. Altera Corporation, San Jose (2004)
14. Marshall, J.: RTL Coding and Optimization Guide for Use with Design Compiler. SNUG, San Jose (2002)
15. Cummings, C., Mills, D.: Synchronous Resets? Asynchronous Resets? I am so confused! How will I ever know which to use?. SNUG, San Jose (2002)
16. Lattice Semiconductor: HDL Synthesis Coding Guidelines for Series 4 ORCA Devices. Technical Note TN1008, March 2002

Cloud Computing

An Efficient Stream Cipher Based Secure and Dynamic Updation Method for Cloud Data Centre

Dharavath Ramesh$^{(\boxtimes)}$, Rahul Mishra, and Amitesh Kumar Pandit

Department of Computer Science and Engineering, Indian Institute of
Technology (ISM), Dhanbad 826004, Jharkhand, India
ramesh.d.in@ieee.org, mishrar93@yahoo.com,
amitesh90@gmail.com

Abstract. Cloud computing infrastructure – an environment of pay-as per- use model, enhance accessibility, advance adaptability, and also a adaptable computational model to provide rapid modification as per the user's requirement. At cloud data center, cloud service provider manages large database to attain these services with fast searching on low computation and introduces a concept of Virtualization-efficient partition of computing resources that reduces large number of servers. In spite of these excellence, cloud paradigm have some serious security and integrity concerns related to user's data stored in VM's disk at cloud data center. In this paper, we have proposed a scheme to handle security and integrity problems along with dynamic updation of user's data by introducing Merkle B+ Hash Tree (MBHT) with short signature without random oracle, which has worst case transmission cost- O (log n). The proposed methodology makes the updating criteria strong enough over existing methodologies.

Keywords: Cloud storage · MBHT · Stream cipher · Short signatures

1 Introduction

Cloud model, a platform of online storage business environment which provides service of data access and storage in pay-per-use manner. In the service of online storage business environment – users avail the service of on-demand provocation to access their data or account any-time-any-where model. In cloud model, there is rapid growth in internet services which increases the workload rapidly at cloud data center. In this new era of computing, internet enters in the phase of virtual paradigm. To make the environment computable, cloud model introduces concept of virtualization [1] at cloud data center to satisfy user requirements. To provide these services at data cloud center, cloud service provider act as controller and manager under an agreement protocol. Cloud paradigm [2] services are classified into the categories of – **SaaS** (software-as-a-service) platform managed by third party vendor and offers service like email, healthcare related applications (google apps, salesforcs etc.). **PaaS** (Platform-as-a-service) offers platform for development of customized applications where as cloud component to software (Apprenda – cloud PaaS for java in cloud model). **Iaas**

I. Zelinka et al. (Eds.): ICSCS 2018, CCIS 837, pp. 505–516, 2018.
https://doi.org/10.1007/978-981-13-1936-5_53

(Infrastructure-as-a-service) miniature for monitoring or casual management of cloud data center infrastructure i.e. virtualization concept, storage etc. (AWS, Joyent). Despite a long list of merits, cloud computing environment also have major security concern, virtualization related issues [3], secure VM migration, data security and data integrity/authentication related issues. Virtualization model have bulk number of virtual machines where each machine have excessive amount of storage for user's sensitive data at cloud data center. In assets of these benefits, virtualization environment have also some serious concerns like VM's disk data security, VM integrity, DOS/DDOS attacks etc. In the form virtualization manner, we access risk value related to VM and ensuring its security by administration to locate or find risk points, control, and reduce the chance of risks. In order to resolve this problem, cloud services user uses a technique of PORs (Proof of Retrievability) to achieve proper retrievability of their data in a good manner [4, 5]. But these models have been proposed only for static data. Now a day's users rapidly update their data in cloud model, so achieving retrievability in dynamic version of PORs model is a tricky task. In our proposed model, we discuss about sensitive data security and integrity issues for data stored at VM's disk at cloud data center with proper dynamic retrievability in cloud platform.

The roadmap of this paper is as follows:

First, we discuss about a stream cipher called (Cha-Cha20) encryption method to attain proper security by encrypting sensitive data. This stream cipher (Cha-Cha20) is much faster and secure than previously existing methods for securing crucial data. Further, we describe about a dynamic version of merkle hash tree - MBHT (Merkle B^+-Hash Tree) with Short Signature without random oracle to attain dynamic PoRs to maintain proper integrity of VM'disk at cloud data center. MBHT scheme has worst case complexity as $O(log\ n)$ instead of $O(n)$.

2 Related Works

Some researchers, Pearson [6], Gu and Cheung [7], and Siebenlist [8] have given brief description on virtualization in cloud paradigm. Takabi [9] proposed a work on VM's protection at cloud data center along with scheme based method on AES-128 bit encryption scheme for encryption to achieve security and Merkle Hash Tree with Merkle's one time signature scheme to attain integrity at cloud data center. In their method, they used SHA-1 hash function for hashing purpose of crucial data and arrange these hash values in Merkle hash tree.

Disadvantages: Now a day's cryptanalysis of AES by using access driven or timing attack is not so much hard. Some security related issues in cloud model related to virtual machine's disk discussed by Wang [10] and Hay [11]. A method called cloud visor [21] has been introduced to attain proper security and integrity for VM's disk data by using AES–CBC encryption technique with merkle hash tree to maintain hash values of VM's crucial data at data center respectively.

Disadvantages: Constraint of this model is side channel/access driven or time driven cryptanalysis of AES-128 bit with different cryptanalysis of MD5 hash function.

Shacham [12] formalized an advanced version of PoRs scheme which has meticulous security proofs based on BLS signature scheme which supports mutual authentication.

Disadvantages: Not suitable for dynamic environment where users rapidly update their data, so modification at one block node will affect adjacent blocks too. In dynamic environment, this scheme suffers from replay attack. After that Wang [13] updated the previous methods by introducing revised version of BLS scheme to enhance security aspects with classical version of MHT to attain integrity. But in this model, classical MHT with revised BLS scheme is used to manage hash values of large files blocks at their leaf level and maintain integrity by signing at root $Sign_{sup\_key}(Rt)$. Boneh and Boyen [14] introduces scheme of more secure short signature based on bilinear groups and computational Diffie-Hellman presumption on elliptic curve with low enclosing degree which has better performance over RSA based secure signature.

Disadvantages: This scheme is not much secure from chosen message attack.

3   Proposed Architecture

The objective of the proposed model is to bring transparent integrity and stability to VM's disk data at cloud data center. To attain proper privacy and for encryption, we use stream cipher (Cha-Cha20) [15] along with SHA 512 hash function to find hash values of VM's disk data and Merkle B$^+$-Hash Tree to arrange these large number of hash values of file blocks.

3.1   Stream Cipher (Cha-Cha20)

Basic of Cha-Cha20 has input state matrix and round functions. This stream cipher has the excellence of improved version of security to secure it from padding oracle attacks or timing attacks. It also contains an additional feature which is most effective for mobile and wearable devices due to less instructions CPU cycles, constant time computation, and key independent from cache driven attacks. Stream cipher have relevant characters to "column round" and "diagonal round", input state matrix of size - 16 words pattern as constant of size – 4 words, key-size of 8 words, block counter size - 2 words and nonce -size - 2 words (governed outside of chacha 20), 20 rounds (4 sub rounds to each so total 80 rounds) are executed on original input state matrix by altering column and diagonal matrix on every 16 words of new plaintext_data. In the first round, the quarter round function is performed as column wise operation and adjusting the first, fourth, third, second, first, fourth, third, second. In the second round, quarter function is being applied as diagonal wise operation and modifying first, fourth, third, second, first, fourth, third, second.

After that execute the sum operation on input state matrix with output of 20 rounds where this output is arranged in little endian format and XORed with 16 words of plaintext to generate ciphertext.

			Constant	Constant	Constant	Constant
a+=b	d^=a	d<<<=16	Key	key	key	key
c+=d	b^=c	b<<<=12	Key	key	key	key
a+=b	d^=a	d<<<=8	Counter	counter	nonce	nonce
c+=d	b^=c	b<<<=7				

1. Quarter round function   2. Input state matrix (Words) for Cha-Cha 20

3.2 MBHT (Merkle B$^+$-Hash Tree)

Users have outsourced their data at cloud data center and delete its provincial copies and store only a small part of data which provide a facility that whenever user asks the cloud server to provide proof of proper retrieval of its accurate data referred in PoRs manner. Users frequently update or modify their crucial data stored at cloud data center. It requires dynamic change in cloud storage to update the same. For this purpose, we introduced dynamic [16] version of MHT named; MBHT (merkle b$^+$-hash tree). This model have combination of dynamic MHT (use B$^+$-tree in place of simple binary tree) with short signature without random oracles. This combination has worst case complexity as $O(log\ n)$ instead of $O(n)$ for dynamic modification or updation. Figure 1 gives brief description of MBHT. MBHT stores hash values $H(M_1)$, H (M_2), ...$H(M_n)$ at leaf level. At the leaf level of MBHT, each node have 3 parts as left (t), right(t), middle(t) and rank(t) where rank represents number of dependent descents on any node. In Fig. 1, we have defined rank of any node on left side of node. For example rank of node $g_1 = 2$, $g_2 = 1$, $g_3 = 2$, and $g_4 = 1$ respectively.

$$p(t) = \begin{cases} 0, \text{if t has at most 2 child nodes} \\ 1, \text{ if t has more than 2 child nodes} \end{cases}.$$

u(elem) = h(H(M)), where H(M) defines SHA 512 hash value of message (M). So, compute $u(t) = h[u(left(t)) \parallel u(middle(t)) \parallel u(right(t)) \parallel p(t) \parallel rank(t)]$.

In MBHT short signature uses without random oracles method to attain proper authentication by signing [$Sign_{sup\_key}(w) = (u(w))$] at root of MBHT. Finally, user will be outsourced the data file block, signature (σ), dynamic tree (MBHT) and u(w) to cloud center. Our scheme has:

Key_generation (1^k) → (pub_key, priv_key): This function takes a security token as input and produces public and private keys (pub_key, priv_key) respectively.

Preparation (Priv_key, FB^1, FB$_{tags}$) → (σ, sign$_{priv\_key}$(u(w)), MBHT): This function create blocks of data files Block_ tag _set [FB$_{tags}$ = {$H(M_i)$}] in encoded form by using Reed −Solomon encoding method at leaf level of MBHT (M_i){ for $0 \le i \le n$} and produces sets of signatures (σ_i) as outputs.

Challenge_generation (l) → Qr: This function generates sets of queries Qr_k = (qr_1, qr_2, ...$qr_{k)}$ and a random value $r_i \in Z_p$ for each index (id_k) of block set at client side. These quires (Qr_i) consists collection of IDs (ID = id_1, id_2 ....id_k). Finally,

Client sends set of (Qr_i, r_i) to server. After receiving these IDs, server checks integrity of data file's block whose index number id $\in$ ID.

Proof_generation (Qr, MBHT, FB^1, FB$_{tags}$, σ) $\rightarrow$ Pr: Server generates proof (Pr_M) after taking (QR, MBHT, FB^1, FB$_{tags}$, σ) as inputs and authenticates integrity of queries(Qr_i, r_i) by producing proof (Pr_M) - $\eta = \sum_{i=1}^{i_k} r_i m_i \in Z_p$ and $v = \prod_{i=i_1}^{i_k} v_i r_i$. After completion of verification process cloud server sends Block_ tag _set [FB$_{tags}$ = {H (M_i)}] to the client or user.

Verification (priv_key, Qr, Pr, u(R)) $\rightarrow$ (T, F): User executes function Verification (priv_key, Qr, Pr, u(R)) to check integrity of blocks in Qr after receiving proof (Pr) and gives output TRUE(T) and FALSE(F).

Updation Process: For any dynamic updation, client runs **update_req ()** function which generates modification request. It takes **order_update $\in$ {insert, delete, modify}** and sends this modification request to server.

Update_r (FB^1, FB$_{tags}$, σ, R) $\rightarrow$ (Pr$_{new}$, Pr$_{old}$): After receiving modification request, server runs function Update_r(FB^1, FB$_{tags}$, σ, R) and generates the output (Pr$_{new}$, Pr$_{old}$). Server sends this output (**Pr$_{new}$, Pr$_{old}$**) to the client.

Update_verification (Pr$_{new}$, Pr$_{old}$) $\rightarrow$ (T, F): Client executes the function Update_ verification (Pr$_{new}$, Pr$_{old}$) to verify the complete updation process by producing the output TRUE(T) if server's action remain fair during updation process otherwise FALSE(F) output.

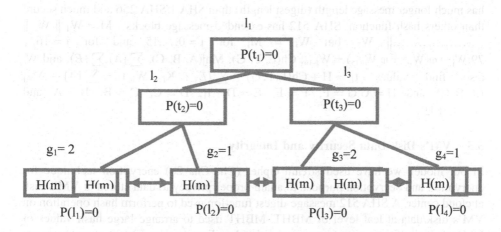

Fig. 1. Description of Merkle B$^+$- Hash Tree

3.3 Short Signature Without Random Oracles

In our proposed scheme, we use short signature without random oracles [17]. This secured short signature scheme is based on standard model of q-strong Diffie Hellman problem and bilinear map. This scheme requires only one pair computation of bilinear map. Let us consider cyclic group (having bilinear properties) $|G_1| = |G_2| = p$ and show isomorphism to each other i.e. $G * G \rightarrow G_T$ assume $p \leq 160$ at lest; as for the message space if the signature scheme is intended to be used directly for signing message then $|m| = 160$. SHA 512 function uses to hash any message of message length ≤ 160 bit. Short signature scheme has system tokens (G, G_T, e, p, g, $Z_p[+1]$) where $Z_p[+1]$ define as .{a $\in Z_p$ | a is a quadratic residue mod p}.

Short Signature without random oracle scheme has three parts:

(i) **Key_generation:** randomly select a, b $\in_R Z_p^*$ and compute $\mathbf{d = g^a, f = g^b}$ so public _key = (d, f) and Private_key (a, b).

(ii) **Signing:** Given secret key $(a, b) \in_R Z_p^*$ and compute $\alpha = g^{\sqrt{a + Mb + h}} \in G$; $\sqrt{a + Mb + h}$ computed over modulo p. Finally, generates output signature (α, h).

(iii) **Verification:** Given public key (G, G_T, p, g, d, f) and message M $\in Z_p[+1]$ and signature (α, h), finally verifies $e(\alpha, \alpha) = e(df^M g^h, g)$ for proper validation.

3.4 SHA 512

SHA 512 is a variant of SHA 256 which have 1024 bit message block with 512 bit intermediate hash value and stream cipher based encryption operation is performed on this intermediate value by using key (message block). First, perform padding to input message to attain result in 1024 bit multiple length after that apply parsing operation into 1024 bits message blocks i.e. $M_1, M_2 \ldots \ldots \ldots M_N$. Initially find hash values $H^{(0)}$ after that find other hash values in sequence:- $\mathbf{H^{(j)} = H^{(j-1)} + C_M^j H^{(j-1)}}$. SHA 512 has much longer message length (digest length) than SHA 1/SHA 256 and much secure than others hash function. SHA 512 has extended message blocks - $M = W_1 \parallel W_2 \parallel \ldots \ldots \ldots \ldots \parallel W_7$, but $W_i = M_i^j$ for j = 0.....15 and for j = 16... 79; $W_i \rightarrow \omega(W_{i-2} + \omega W_{i-15}) + W_{i-16}$; Ch(E, F, G), Maj(A, B, C), $\sum(A), \sum(E)$, and W_i also find values $t_1 = H + Ch(E, F, G) + \sum(E) + K_i + W_i, t_2 = \sum(A) + $ Maj (A, B, C) and H = G,G = F, F = E, E = D + t_1, D = C, C = B, B = A and A = $t_1 + t_2$.

3.5 VM's Disk Data Security and Integrity

In our model, we have used stream cipher (Cha-Cha 20) encryption technique for encryption and decryption purpose to attain proper privacy to data stored at VM's disk at cloud center. A SHA 512 message-digest function used to perform hash operation on VM's disk/data at leaf level of MBHT. MBHT used to arrange large hash values of VM's disk and for support dynamic updation by client or user. First it compute nonce or initial vector for all VM disks and computed hash values (by using SHA 512 Hash function) are arrange at leaf level of MBHT. After that perform encryption operation by using Cha-Cha 20 encryption technique. At the time of downloading operation for any stored file or message blocks, firstly, check the integrity of related VM's disk and after that user will able to download their requested file. SHA-512 with 96 bit IV/nonce arranges in 32-bit little endian format and IV/nonce stored at virtual machine monitor level. This instance has been described in Fig. 2 as a stepwise execution process. **Step 1:** user requests for any file or file blocks stored at cloud center, then on behalf of this request virtual device driver fetches related VM's disk. **Step 2:** attain all hash values which are initially stored at VMM level with initial vector (IV)/nonce. **Step 3:** after that find hash values for related VM's disk and check their authorization along with cached values at data storage level (leaf level) of MBHT. **Step 4:** compare hash values of requested VM's disk data with obtained hash value of VM's disk. **Step 5:** if successful, then virtual machine monitor provides secret key (initially stored at virtual machine monitor) for decryption process. **Step 6:** after the successful validation, data stored to I/O buffer of guest VM by virtual machine monitor.

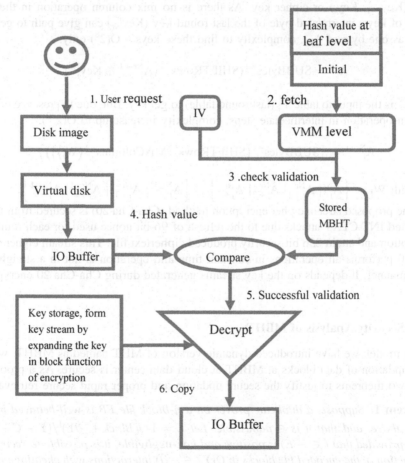

Fig. 2. Description of VM's disk data security and integrity

4 Security Analysis

In our proposed model, we have used stream cipher (Cha-Cha20) for encryption and decryption purpose. Due to major interest in advance method of cryptanalysis [18, 19], expansion of key length in AES doesn't provide required privacy to crucial data. AES–CBC 128 bit suffers from access driven attack and cache time driven. The access driven attack works based on two processes. Process one is crypto related operation by the user whereas another operation is by the secret agent to attain key by observing which lines of cache related to user's operation. But, in cache timing driven, secret agent detects variation during execution time of use's operation and tries to access secret key. In modern technique, only two round keys are sufficient to retrieve AES secret key. Let Key_{Nr} defined as last round key which is easily retrieved by using previous two round keys (Key_{i-1}, Key_i) or cipher key. As there is no mix column operation in the last round of keys, interrelated byte of the last round key (Key_{Nr}) can give path to get the value as one byte of A_{Nr}, complexity to find these keys – $O(2^8)$ only.

$$A_i^{Nr} = SUBBytes^{-1}(SHIFTRows^{-1}(A_j^{Nr+1} \oplus Key_j^{Nr}))$$

A_i^{Nr} is the input to table T_4 (last round table) to get key. But, due to presence of mix column operation in intermediate steps, complexity increase up to $O(2^{32})$.

$$A_i^{Nr-1} = \{SUBBytes^{-1}(SHIFTRows^1(MixColumns^{-1}(W_j)))\}$$

With $W_j = \left\{ A_j^{Nr} | A_{j+1}^{Nr} | A_{j+2}^{Nr} | A_{j+3}^{Nr} \right\} \oplus \left\{ A_j^{Nr-1} | A_{j+1}^{Nr-1} | A_{j+2}^{Nr-1} | A_{j+3}^{Nr-1} \right\}$.

The proposed stream cipher encryption method (Cha-Cha 20) is secured from IND-CPA and INT-CTXT attacks due to the refresh of 96-bit nonce used for each round of encryption and attain 256 bit security produced ciphertext file. This stream cipher Cha-Cha 20 performs all operations in constant time and operation is not in a straightforward manner. It depends on the key streams generated during Cha-Cha 20 encryption process.

4.1 Security Analysis of MBHT

In our model, we have introduced dynamic version of MHT named as MBHT, where each updation of data blocks at MBHT at cloud data center is secure. As a proof we state two theorems to justify the secure updation and proper rapid secure retrieval.

Theorem 1. *Suppose, a duplicate prover on a n-block file FB is well-behaved in the sense above, and that it is $\in$ admissible. Let $E = 1/\#Block + (\partial Y)^l/(Y - C + 1)^C$. Then, provided that ($\in - E$) is positive and non-negligible, it is possible to recover a ∂ - fraction of the encoded file blocks in $O(Y/ \in -\partial)$ interactions with cheating prover and in $O(Y^2 + (1 + \in Y^2)(Y)/(\in - E)$ time overall.*

Proof. Let us consider H(j) hash value at leaf level of MBHT and replace it with $H(M_j)$ i.e. every secure modification(insert, delete, update) can still be executed without involving the index information. As H(i) and $H(M_j)$ both have same value at leaf level in MBHT (which will reduce chance of potential attacks), on changing these values not affect any change in PoRs algorithm.

Theorem 2. *Given a fractional part of the Y-blocks of an encoded file FB, retrieval of complete original file FB with all but negligible probability is possible.*

Proof. As we know, file blocks are encoded using rate (∂) of Reed–Solomon code, as per the properties of Reed Solomon code all fractional part of file FB are easily retrievable, since any part of ∂ (encoded file blocks - FB) suffices from delete operation.

4.2 Security Analysis of SHA 512

SHA 512 is a truncated version of SHA 512. By adding extra 256 o's into output of SHA 256 to provide more resistance to both collision and pre image attacks, a small change to IV of SHA 512 prevents output of previous function related to next one. SHA 512 is secured from Dobberth's attack due to fast and strong diffusion of any message block having low weight difference. It is also secured from Chabaud's and joux's attack due to replacement of each addition operation with XOR operation. No more than 7 consecutive identical words may separate two consecutive differential collision patterns with the original message block difference having at least one non-zero word.

4.3 Security Analysis of Short Signature Without Random Oracle

Short signature scheme [20] is unforgeable in strong sense under chosen message attacks provided that the q-SDH assumption holds in (G_1, G_2). Security proof of this scheme is under these assumptions: Suppose that the (q, J^1, λ^1) SDH assumption holds in (G_1, G_2), then signature model (J, q_s, J) is secure against any forgery under chosen message attack provided by $q_s < q$, $\in \geq 2(\in^1 + q_s/p) \approx 2 \in^1$ and $J \leq J^1 - \theta(q^2 j)$, where j defines as maximum time for exponential operation in G1 and G_2.

5 Performance and Result Analysis

In that section, we have described the performance and graphically represent result analysis of our proposed scheme. On the basics of comparison between encryption and decryption scheme AES-128 bit and stream cipher based encryption scheme Cha-Cha20, we also compare static MHT and dynamic version MBHT which is used to attain proper integrity. Figure 3 shows the required time for encryption or decryption process of AES-128 bit and Cha-Cha20 with different file sizes. As an insertion, Fig. 4 describes about the performance of static MHT and dynamic MBHT on any data file/blocks (FB^1) on leaf level of hash tree and message transmission cost.

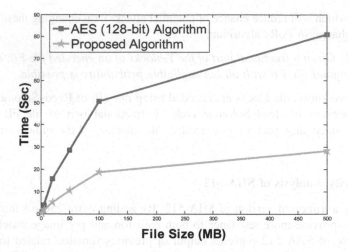

Fig. 3. AES-128 bit and proposed algorithm

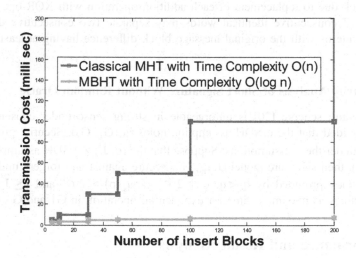

Fig. 4. Classical MHT and MBHT

6   Conclusions and Future Scope

As a proposed model, we have discussed about security concern related to VM's disk
in virtualization paradigm at cloud data center. It states a method to resolve problems
related to security and integrity issues by using stream cipher (Cha-Cha20) which is
very secure and fast encryption method and dynamic version of merkle hash tree called
MBHT. It works with SHA 512 hash function which provides proper dynamic modification to users in worst case complexity $O(\log n)$ and short signature without random
oracle of q-SDH model to remain unforgeable under chosen message attack. In future,
our system can be further enhanced by providing more efficient dynamic updation

method to attain better rapid modification in constant time and also provide recovery and backup to attain for cloud server.

References

1. Barham, P., et al.: Xen and the art of virtualization. ACM SIGOPS Oper. Syst. Rev. **37**(5), 164–177 (2003)
2. Mell, P., Grance, T.: The NIST definition of cloud computing (2011)
3. Lombardi, F., Di Pietro, R.: Secure virtualization for cloud computing. J. Netw. Comput. Appl. **34**(4), 1113–1122 (2011)
4. Ateniese, G., et al.: Scalable and efficient provable data possession. In: Proceedings of the 4th International Conference on Security and Privacy in Communication Netowrks. ACM (2008)
5. Juels, A., Kaliski Jr., B.S.: PORs: proofs of retrievability for large files. In: Proceedings of the 14th ACM Conference on Computer and Communications Security. ACM (2007)
6. Pearson, S.: Taking account of privacy when designing cloud computing services. In: Proceedings of the 2009 ICSE Workshop on Software Engineering Challenges of Cloud Computing. IEEE Computer Society (2009)
7. Gu, L., Cheung, S.-C.: Constructing and testing privacy-aware services in a cloud computing environment: challenges and opportunities. In: Proceedings of the First Asia-Pacific Symposium on Internetware. ACM (2009)
8. Siebenlist, F.: Challenges and opportunities for virtualized security in the clouds. In: Proceedings of the 14th ACM Symposium on Access Control Models and Technologies. ACM (2009)
9. Takabi, H., Joshi, J.B.D., Ahn, G.-J.: Security and privacy challenges in cloud computing environments. IEEE Secur. Priv. **8**(6), 24–31 (2010)
10. Liang, Q., Wang, Y.-Z., Zhang, Y.-H.: Resource virtualization model using hybrid-graph representation and converging algorithm for cloud computing. Int. J. Autom. Comput. **10**(6), 597–606 (2013)
11. Hay, B., Nance, K., Bishop, M.: Storm clouds rising: security challenges for IaaS cloud computing. In: 2011 44th Hawaii International Conference on System Sciences (HICSS). IEEE (2011)
12. Shacham, H., Waters, B.: Compact proofs of retrievability. In: Pieprzyk, J. (ed.) ASIACRYPT 2008. LNCS, vol. 5350, pp. 90–107. Springer, Heidelberg (2008). https://doi.org/10.1007/978-3-540-89255-7_7
13. Wang, Q., Wang, C., Li, J., Ren, K., Lou, W.: Enabling public verifiability and data dynamics for storage security in cloud computing. In: Backes, M., Ning, P. (eds.) ESORICS 2009. LNCS, vol. 5789, pp. 355–370. Springer, Heidelberg (2009). https://doi.org/10.1007/978-3-642-04444-1_22
14. Boneh, D., Lynn, B., Shacham, H.: Short signatures from the Weil pairing. In: Boyd, C. (ed.) ASIACRYPT 2001. LNCS, vol. 2248, pp. 514–532. Springer, Heidelberg (2001). https://doi.org/10.1007/3-540-45682-1_30
15. Nir, Y., Langley, A.: ChaCha20 and Poly1305 for IETF protocols. draft-nir-cfrg-chacha20-poly1305-01 (work in progress) (2014)
16. Zheng, Q., Xu, S.: Fair and dynamic proofs of retrievability. In: Proceedings of the First ACM Conference on Data and Application Security and Privacy. ACM (2011)

17. Boneh, D., Boyen, X.: Short signatures without random oracles. In: Cachin, C., Camenisch, J.L. (eds.) EUROCRYPT 2004. LNCS, vol. 3027, pp. 56–73. Springer, Heidelberg (2004). https://doi.org/10.1007/978-3-540-24676-3_4
18. Procter, G.: A Security Analysis of the Composition of ChaCha20 and Poly1305 (2014)
19. Neve, M., Tiri, K.: On the complexity of side-channel attacks on AES-256 (2007)
20. Zhang, F., et al.: A new short signature scheme without random oracles from bilinear pairings. IACR Cryptology ePrint Archive 2005, p. 386 (2005)
21. Zhang, F., Chen, J., Chen, H., Zang, B.: CloudVisor: retrofitting protection of virtual machines in multi-tenant cloud with nested virtualization. In: Proceedings of the Twenty-Third ACM Symposium on Operating Systems Principles, pp. 203–216. ACM (2011)

A Secure Cloud Data Sharing Scheme for Dynamic Groups with Revocation Mechanism

Anusree Radhakrishnan[✉] and Minu Lalitha Madha

Sree Buddha College of Engineering, Padanilam, Kerala, India
maloottyunnikkuttan@gmail.com, minulalitha@gmail.com

Abstract. Cloud is an extensive data sharing system in which it provides an IT paradigm for to access shared pools of resources for any type of organizations. It also provides higher level services with minimal management effort. Using the concept of cloud, only lower maintenance is needed for the group members. Use of cloud computing helps us to ensure the storage capacity and it can be shared between the group members. Since we are outsourcing the data there is the necessity of providing security to the data. Since the members are changing dynamically here it is needed to ensure the guarantee of the data security. Especially the collusion attack should be avoided. The prescribed paper describes about a privacy preserving data sharing technology that uses key distribution method along with the group manager.

Keywords: Encryption · Decryption · Authentication · Data security

1 Introduction

Cloud computing by the name is an IT paradigm in which it provides intrinsic data sharing and storage facilities. In cloud computing the organizations don't want to worry about the software and hardware spaces. All the storage concerns are relied on the cloud servers. So, the organizations don't want to concerns about the financial overhead. Since we outsource the data to cloud servers there are some issues in the cloud. In order to preserve the data privacy, the usual approach is to encrypt the data before the user uploading it into the cloud. But unfortunately, it is a common issue to design a privacy preserving scheme in order to ensure the security of data and the users. A cloud is initiated in an environment and more scalable resources need it. Cloud computing provides different services in different areas such as storage, servers, software development, etc., these services are provided through internet which is termed as "cloud". In all cloud computing vendors some of the cloud computing characteristics are common. They are

1. The hardware backend of the main application
2. The user pays the money on demand based on the type of service he uses
3. Services offered by cloud

Virtualization is a basic concept related to all the cloud computing paradigm. Depending on the type of service or the demand the organizations need to pay money.

© Springer Nature Singapore Pte Ltd. 2018
I. Zelinka et al. (Eds.): ICSCS 2018, CCIS 837, pp. 517–525, 2018.
https://doi.org/10.1007/978-981-13-1936-5_54

We can categorize the cloud computing services as follows. Infrastructure as a Service (IaaS), Platform as a Service (PaaS) or Software as a Service (SaaS). Mainly cloud computing services include servers, networking, database, hardware etc. Companies has to pay for the usage of the cloud on demand. The companies whom are offering these services are called the cloud providers. Cloud computing is actually a shift from traditional way of business. Some of the merits by using cloud computing is cost, speed, global scale, performance, productivity, reliability.

Cloud helps to the human to cope with the growing data. This will include the high definition audio, video, image etc. Cloud will help us to store, mine and access the crucial data. The field of cloud computing is now experiencing sudden growth in technology. It altered the normal business style and also shaken up the traditional structure of IT department. Cloud computing security refers to the set of policies and the control deployed on the computing security in order to protect the data. It can be considered as a sub domain of the information security. In cloud computing users data is stored and accessed with the help of third party. Different models of the cloud computing are used by the organizations in order to access the cloud services. Some of the deployment models of the cloud are public cloud, private cloud, hybrid cloud and community cloud. Since the data is highly sensitive in nature we have to design an algorithm in order to gather the concepts such as access control, data confidentiality, privacy etc. There should be proper password and authentication system.

Some security threats are already there in cloud computing they are illegal invasion, denial of service attack, some cloud security threats such as side channel attack, abuse of cloud service etc. In these situations, we need to limit the security threats with the help of some of the requirements. They are data confidentiality, access controllability, and integrity. Confidentiality ensures that data content is properly handled. And the data is not made available to any illegal users. Access controllability restricts the accessing power of other users with the data. Other users should not access the data without permission. Integrity somewhat other requirement factor. Integrity means ensuring the correctness and completeness of the data.

Whether the cloud is of private, public, hybrid nature we all know that the main benefits of the cloud system is the resource, hardware and software storage part. But our system makes use of public and private cloud. Most of the public clouds are used for quality assurance purpose. But in the private cloud is dedicated for a single client. There are some models available for public clouds. Namely utility model, shared hardware, self managed. Most of the public clouds are used for web servers and deployment systems. In those situations, the security is not a big concern. Private cloud computing by definition a single tenant environment. Security, compliance, customizability are the requirements of the private cloud.

In the literature review section [1] deals with the common cloud computing paradigm and its common issues related to it. By the analysis of the paper we can conclude that encryption is embedded in to it in order to provide security or the user data. [2] mentions key policy attribute-based encryption. It is mainly used for fine grained access control. Ownership and process history of data objects in the cloud is mentioned in [3]. [4] illustrates cipher text Based - Attribute Based Encryption (CB-ABE). It basically builds 5 frameworks for the cloud computing. [5] says about a key distribution system for data sharing among the dynamic members. And it proves that the data is free from collusion attack. [6, 7] says about the constant size cipher text. It illustrates hierarchical identity-basedencryption. Collusion free communication is focused here. [8] is based on a direct key distribution without correspondence channels. Roll based access control is discussed in [9]. A flexible key distribution system is discussed in [10]. It is specially for collaborative environment. [11] is basically a multi owner data sharing scheme especially for dynamic groups in the cloud. [12] is attribute based encryption for multi authority users. [9] is the base paper for the given work. It mainly focuses on dynamic clouds. It tells about a collusion free communication between the group members and group manager.

2 Problem Identification

From the survey of recent papers in cloud computing we will conclude the following findings, Each of the group members is owned by a secure secret key. This is authorized by the group manager. And every time a secure key is generated once the file is uploaded The group members can access the cloud with the key. The users in the group can access the file with the file key. This is quite unsafe because the secure key is generated once and users use the same secure key every time to download the file. Anybody who knows the secure key can access the file anywhere. This is one of the problem with the existing system. When we come closer to the problem it describes the security of data. How we will find the solution to the security issue in cloud computing this solution is the proposed scheme. More security is needed in the field of file uploading and downloading. There should not be any restriction to the file size that the user identifies. With the single encryption of user and file data the security can't be ensured. If we apply multiple encryption to the same data it will be more useful.

3 Proposed System

1. Methodology

System mainly runs on the domain of cloud. Group managers and actual users of the cloud. These are also called the modules of the system. Cloud mainly provides a storage space for all the data files. But it is not a trustworthy always. It can be easily accessed by unauthorised persons. It is possible to access the cloud data easily. The users are categorised in to groups according to the types of behaviour. The group manager takes the charge of account registration, key distribution, Revocation etc. Here we assume that the group manager is a trustworthy member. The group managers are registered users. Data is completely stored in the cloud and by the demand it is shared with others. Dynamic means the members can be added, deleted even can be modified.

In the above Fig. 1 illustrates the work flow of the system. Main aspects of the system are cloud, group manager, and group member. Registration and key distribution are the initialization part of the system. Group manager is in charge of distributing the keys. A unique key is generated for each user. Using these keys the users will access the cloud. Image, audio and video files are mainly handled. File key is visible to only those users in the group. The denial of the accessibility of file to other group members is possible here. Uploading is associated with encryption and downloading is associated with decryption.

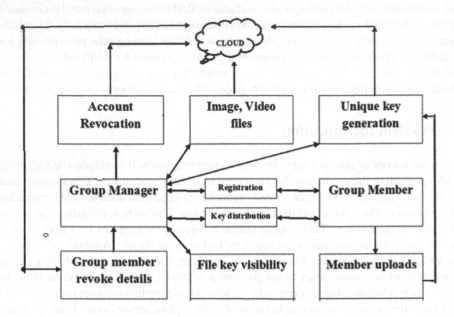

Fig. 1. Proposed system architecture

2. Cloud data sharing scheme

The secure key is generated each time while the file is uploaded. The secret key changes in every upload. This will avoid illegal access of data by unauthorized users. This Secret key will be visible only to the group admin. The modules are admin, user, public cloud private cloud. Main subsections coming under the system is listed below.

(1) User registration

User should have the facility to select the role of the user (ex: Director, project lead, engineer) during registration. The role should be approved or rejected by the admin in the private cloud. The user sends identity (ID), public key, and a random number. After the verification process the group manager will send the private key. This private key is known only to that particular user. After successful registration the user becomes the member of the cloud. The private key send by the manager will also be in encrypted format. The user need to decrypt the message using his public key. Once the registration process is completed the group manager will keep the details of the user in the local space.

(2) File tag creation

File tag is created to check whether this is an already uploaded one. The uniqueness of the file is maintained here. When a user need to upload a file, the user creates a file tag from the file using a cryptographic hash function (please make sure that the File tag is generating from the unique part of each file). The file tag will be sent to the private cloud for file token generation.

(3) Private cloud – administrator

Admin controls the private cloud. Process the user registration details and approve or reject the role selected by the user during registration. Each user role should have a privilege key associated with it and it is managed and stored in the private cloud by the admin.

(4) File upload

When a user need to upload and share a file to other users, the user sends the file tag to the private cloud and mention to which other users the file should be shared. The private cloud accepts the file tag sent by the user and generates file tokens using the file tag and the set of privileges [current user's privilege plus the privilege keys of the users to whom the user need to share the file]. The file tokens will be sent back to the user. For uploading the file, the user will send the identity and the time stamp first to the group manager. When each of the file is uploaded it will have the signature of the group manager. When each of the file is uploaded the group, manager will send the file list to the user. So that the user can check the freshness of the file. When the file is uploaded a secret key is generated for each of the files, But for uploading and downloading different keys are used for security purpose to the public cloud and the public cloud check whether the file tokens are already there in the public cloud or not. If any of the file token is already there, no need to upload the file again. Instead the remaining file tokens accepted from the user which are not in the public cloud will be updated to the database.

(5) Encryption and uploading

If any of the file token is not exist in the public cloud, the user module dual encrypt the file using the convergent encryption keys and upload the cipher text to the public cloud long with the file tokens. The encryption keys will be managed in the corresponding user's account in the private cloud.

(6) Interface in user account

After login, the file belongs to the user (files uploaded by the current user and shared by other users) should be listed with a download option. The file uploaded by the user should only have the permission to delete the file (proof of ownership protocol).

(7) Automatic account revocation

If the user tries to login with wrong password for more than three times the account will be automatically revoked by the group manager. And if the user wants to access the cloud he need to start from the initialization stage. And it is necessary to get the user private key after registration.

4 Result and Discussion

We made the performance evaluation with the WAPT tool in order to compare the existing system with the proposed one. WAPT Pro 4.7 is an Integrated performance testing solution. WAPT Pro can be considered as a load testing tool for running environment. It combines both efficiency and scalability and it is used in more than decades. We can use the tool in the following environment such as website, web server applications, mobile applications in order to test the load. It works on the following modules. For this evaluation we are using Windows XP. Some other options are Windows 2000, 2003, 2008 servers. We can use the following browsers such as Internet explorer, Firefox, chrome. Both group members and group managers processes are conducted on a laptop with Core 2 T5800 2.0 GHz, DDR2 800 2 G, Windows. The cloud process is implemented on a laptop with Core i7-3630.

Figure 2 illustrate the performance evaluation comparison of the proposed system with the current scheme. X axis shows the time period along with number of users on the y axis. Availability, with easy access to cloud services and the services are always available, performance will be increase. Number of users, if a data centre has a lot of users and this number is greater than that of the rated capacity, this will reduce performance of services. Location, data centres and their distance from a user's location are also an important factor that can be effective on performance from the users' view. Figure 3 illustrates the response evaluation. It has a better response behaviour than the current one. Figure 4 compares the bandwidth of current and proposed one. Network bandwidth, this factor can be effective on performance and can be a criterion for evaluations too.

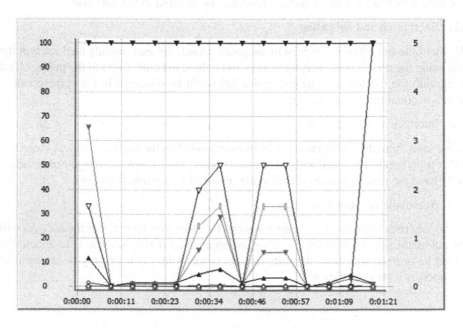

Fig. 2. Performance evaluation

For example, if the bandwidth is too low to provide service to customers, performance will be low too.

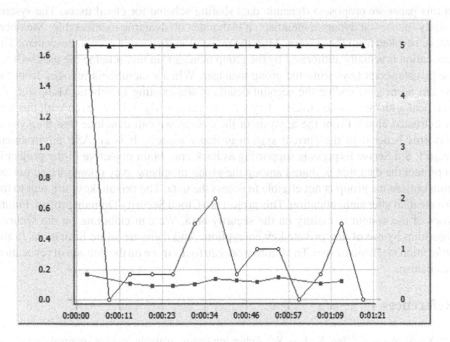

Fig. 3. Response evaluation

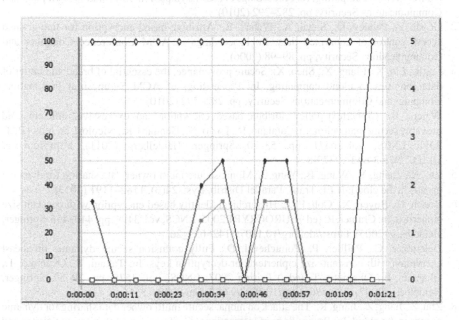

Fig. 4. Bandwidth evaluation

5 Conclusion and Future Work

In this paper we propose a dynamic data sharing scheme for cloud users. The system mainly focuses on dynamic members. It elaborates on dynamic membership. Members can be revoked from the system. After the revocation they can't access the system. The revocation is actually authorised by the group manager. In this scheme the uses will get the private secret keys from the group manager. When a member is revoked from the system no need to update the account details of the existing members. Moreover, we also deal with the revoked users. They will not get the original files even with the help of entrusted cloud. From the analysis of the system we can conclude that the system performs better than the current system in many aspects. It is an ASP environment project. Sql Server Express is supporting as back end. Main objective of the project is to protect the data that is shared among the group members. Any unauthorized person from outside the group is not eligible to access the data. The private keys are sent to the private mail after authentication. This project is a Cloud Security Domain project. Future work of the system is mainly on the security area. We can elaborate on the security algorithm by use of any private key encryption. And there are some limitations in the revocation mechanism also. The future work can focus more on the domain of revocation mechanism.

References

1. Yu, S., Wang, C., Ren, K., Lou, W.: Achieving secure, scalable, and fine-grained data access control in cloud computing. In: Proceedings of ACM Symposium Information, Computer and Communications Security, pp. 282–292 (2010)
2. Goyal, V., Pandey, O., Sahai, A., Waters, B.: Attribute-based encryption for fine-grained access control of encrypted data. In: Proceedings of ACM Conference Computer and Communications Security, pp. 89–98 (2006)
3. Lu, R., Lin, X., Liang, X., Shen, X.: Secure provenance: the essential of bread and butter of data forensics in cloud computing. In: Proceedings of ACM Symposium Information, Computer and Communications Security, pp. 282–292 (2010)
4. Waters, B.: Ciphertext-policy attribute-based encryption: an expressive, efficient, and provably secure realization. In: Catalano, D., Fazio, N., Gennaro, R., Nicolosi, A. (eds.) PKC 2011. LNCS, vol. 6571, pp. 53–70. Springer, Heidelberg (2011). https://doi.org/10.1007/978-3-642-19379-8_4
5. Liu, X., Zhang, Y., Wang, B., Yang, J.: Mona: secure multi owner data sharing for dynamic groups in the cloud. IEEE Trans. Parallel Distrib. Syst. 24(6), 1182–1191 (2013)
6. Boneh, D., Boyen, X., Goh, E.-J.: Hierarchical identity based encryption with constant size ciphertext. In: Cramer, R. (ed.) EUROCRYPT 2005. LNCS, vol. 3494, pp. 440–456. Springer, Heidelberg (2005). https://doi.org/10.1007/11426639_26
7. Delerablée, C., Paillier, P., Pointcheval, D.: Fully collusion secure dynamic broadcast encryption with constant-size ciphertexts or decryption keys. In: Takagi, T., Okamoto, T., Okamoto, E., Okamoto, T. (eds.) Pairing 2007. LNCS, vol. 4575, pp. 39–59. Springer, Heidelberg (2007). https://doi.org/10.1007/978-3-540-73489-5_4
8. Zhu, Z., Jiang, Z., Jiang, R.: The attack on mona: secure multi owner data sharing for dynamic groups in the cloud. In: Proceedings of International Conference on Information Science and Cloud Computing, pp. 185–189 (2013)

9. Zhou, L., Varadharajan, V., Hitchens, M.: Achieving secure role-based access control on encrypted data in cloud storage. IEEE Trans. Inf. Forensics Secur. **8**(12), 1947–1960 (2013)

10. Zou, X., Dai, Y.-S., Bertino, E.: A practical and flexible key management mechanism for trusted collaborative computing. In: Proceedings of IEEE Conference on Computer Communications, pp. 1211–1219 (2008)

11. Zhou, Z., Jiang, R.: A secure anti collusion data sharing scheme for dynamic groups in the cloud. IEEE Trans. Parallel Distrib. Syst. (2015). https://doi.org/10.1109/TPDS. 2015.2388446

12. Varun, I., Vamsee Mohan, B.: An efficient secure multi owner data sharing for dynamic groups in cloud computing. Int. J. Comput. Sci. Mob. Comput. **3**(6), 730–734 (2014)

Recovery of Altered Records in Cloud Storage Utilizing Seed Block Strategy

Radhakrishnan Parvathy$^{(\boxtimes)}$, P. Mohamed Shameem$^{(\boxtimes)}$, and N. Revathy$^{(\boxtimes)}$

Department of CSE, TKM Institute of Technology, Kollam, Kerala, India
parvathyrkv999@gmail.com, revathyram2006@gmail.com,
pms.tkmit@yahoo.in

Abstract. Cloud storage system provides service or assistance for file storage and sharing services for distributed clients. In order to address integrity, controllable outsourcing, file duplication and regeneration of outsourced files, a regeneration scheme is being proposed, which is equipped with seed block algorithm. Dedicated proxies upload data to the cloud storage server on her behalf, e.g., a file owner with a dedicated PC, otherwise proxies, uploads file to cloud server. Dedicated PCs are distinguished and qualified with their recognizable identities, which discards the scenario of file being altered at the time of upload. This scheme facilitates duplication checking at the client-side, i.e., it checks for presence of clone file in cloud, more importantly it checks for clone using SHA1 algorithm. Moreover it allows regeneration of manipulated file on request. i.e., when auditor identifies that a file is being hacked or missing, auditor could inform file owner about the need of file regeneration and file owner could send a regeneration request to the server for regenerating the corresponding file.

Keywords: Cloud storage · Data outsourcing · De-duplication
Remote integrity proof · Public auditing · Regeneration

1 Introduction

Cloud capacity is an online space that is utilized to store records. And too, it serves as a reinforcement for records on the physical capacity gadgets, which incorporates outside difficult drives or USB streak drives. Cloud capacity is more secure medium to store basic information as reinforcement. Online storage arrangements were a service for utilizing expansive arrangements of virtual servers that come with apparatuses for overseeing records and organizing virtual capacity space.

To essentially characterize advantages of cloud storage, foremost is the capacity of cloud, capacity makes sense happens when clients transfer records and files which are on their personal computers, laptops, portable gadgets, or it can be any handy electronic device to a web server by means of internet connection. The transferred records could serve as reinforcement in case the unique records are harmed or misplaced. The cloud server licenses the client to download records to any gadgets when required. It was guaranteed by encryption and was gotten to by the client with login qualifications and secret word.

© Springer Nature Singapore Pte Ltd. 2018
I. Zelinka et al. (Eds.): ICSCS 2018, CCIS 837, pp. 526–537, 2018.
https://doi.org/10.1007/978-981-13-1936-5_55

Cloud computing technology grants a global opportunity to approach the shared resources and services that would be right away handled with least attempt in management, which become frequently handy over the internet. The files are constantly accessible to the consumer if he/she has an internet connection to view or retrieve them. The benefit of using cloud storage was the synchronization [2] which was the access to user's digital content at any time, from anywhere using any device. Clients transfer records to cloud capacity and they are not being on edge almost how information capacity was taken care of at the cloud. Cloud computing was a service which was similar to a public utility, that depends on resource sharing to achieve consistency and ease of use.

Third-party clouds allow organizations to focus on their core activities and duties instead of disbursing resources on computer infrastructure and its preservation. Professionals say that cloud computing permits companies to minimize or evade IT infrastructure costs in advance. Cloud computing allows companies to get their works done with enhanced manageability and less cost for maintenance, and also it enables IT teams to adjust resources more swiftly, to meet shifting and unpredictable demands beforehand. Cloud services have on the way access, which allows files to be handled by the file owner from any place using a handy electronic gadget with internet connection.

Cloud service providers' serves clients agreeing to client needs utilizing diverse models, of which the three standard models are Infrastructure as a Service (IaaS), Platform as a Service (PaaS), and Software as a Service (SaaS). Out of the three administrations this work employs IaaS. Infrastructure as a service (IaaS) is a cloud computing benefit that aids in getting to computing assets which are said to be virtual over the web. i.e. it was the strategy of dispersing administrations like computing, storage, organizing and other capabilities through the Web. One of the fundamental highlights of IaaS is that, it grants companies to get to web-based working frameworks, applications and capacity without having, overseeing and supporting the concealed cloud foundation. The fastest-growing cloud portion was IaaS. The momentous improvement of IaaS was due to undertakings. They are unflinchingly moving from fact centers to the cloud.

Since there are so many cloud storage vendors available. Numerous elements have to be checked in advance. First and main was to make certain that the files stored at the cloud are secure. i.e. the information are in encrypted form or not. Second is the value of getting access to the cloud. It was determined by how much space is required for the motive. Many Cloud provider vendors offer either a time frame as free or a loose amount of storage as a provider to test out their services. It was not a wise idea to keep sensitive data during a trial period. And the remaining thing to recollect was the ease of use. The manner of report uploading and getting access to the cloud should be clear and easy to apprehend.

In this work, files would be stored at the cloud storage. Dedicated PCs are assigned for each users to upload files to cloud. This was done ensure security of file at the time of upload. Main reason for incorporating dedicated proxies was that a genuine file owner would use the facility to upload the specific type of file, which was the expected case. Despite that, an unauthorized user could take advantage of the facility by uploading malicious files. It would result in two scenarios. The first case involves the type of file uploaded. A wrongly uploaded file could overwrite another file that already exists in the cloud with similar name. If this were a vital file, the new file could cause the whole system to function improperly. In the second case a system could be hacked by a malicious file upload. The uploaded file could contain virus, Trojan or malware, which could be used to gain control.

Secondly, the need for ensuring absence of clone files in the cloud, which would help in memory management. The presence of duplicate files would lead to unnecessary wastage of memory space, which was wastage of resources. To ensure management of stored data in cloud computing, de-duplication has to be watched over. It was a technique that has become more popular recently. De-duplication was a useful technique, which reduce space required for storage and bandwidth required for file upload in cloud storage.

Lastly, the worst case scenario has to be considered so that there would be some backup plan. Integrity of the file, auditor does integrity auditing of the file in cloud. In this system, integrity would be checked and notifies file owner in case of any emergency. i.e. if the auditor come across to identify any sort of damage to the file in the cloud, the owner of the file would be notified about the incident and notifies the need of regenerating the file.

2   Related Works

The idea of information sharing was being discussed and was bolstered by most well-known cloud service providers (CSP) [2]. Record proprietor could share information with anybody and they get records from the cloud utilizing different sharing strategies. The one who gets to the record was basically trade accomplices, companions or colleagues. Sharing strategies are open sharing, secret-url sharing and private sharing. Clearly, the rudimentary way to share information was open sharing. This strategy was not pertinent to private data. For this reason, the private sharing was the most excellent and secure. The issue with private sharing was that everybody has to sign into the framework. It was not that simple to make everybody in the above-mentioned situation to enlist with the same CSP. The secret-url sharing jams a harmony between security and comfort. It permits the clients having the required URL to get to the shared information without a making.

Record ownership had to be guaranteed in the case of capacity of the record in the cloud [3]. The conspire of provable information ownership by Ateniese et al. [5] allowed to check the astuteness of an outsourced record by a reviewer. Astuteness is checked without recovering the whole record from the cloud server; at the same time, server does not require to get to the whole record for replying judgment questions. Provable Data Possession (PDP) permitted confirming whether the server has unique information (which was put away the client) at an untrusted server without recovering it. It gave rise to probabilistic proofs of ownership by assessing arbitrary sets of squares from the server. A steady sum of metadata had to be kept up by the client to confirm the verification. Network communication was minimized by transmitting a little, steady sum of information for challenge/response convention. This PDP model for remote information checking upheld expansive information sets in widely distributed frameworks. The building block for this PDP schemes was a concept called homomorphic verifiable tag.

A distinctive open examining instrument Panda [4] was for checking the keenness of shared information. It talks about the plot of intermediary re-signatures. Concurring to this, when a client in the bunch was disavowed, the pieces which are marked by the denied could be re-signed by the cloud with a key called re-signing key. This method essentially progresses the proficiency of client invalidating. Computation and communication belonging of existing clients were effectively moderated. In the interim, the cloud which was not in same trusted space of each and every client would be able to change over a signature of the denied client into a signature of an existing client on the same piece of data. And moreover it could not sign subjective pieces on sake of either the denied client or an existing client. In this intermediary re-signature plot, it was being able to check judgment of information shared without getting the whole information record from the cloud.

In Paterson and Schuldt's identity-based signature conspire [7] the designation was made as an identity based signature warrant, in this way the assignment could be freely approved in Audit protocol of IBDO framework [1]. SecCloud and SecCloud+ [6] accomplished both information judgment and de-duplication in the cloud. SecCloud presented an inspecting substance with the upkeep of MapReduce cloud, which made a different client to create information labels some time recently uploading records as well as to review the judgment of information put away in the cloud. In expansion, SecCloud helped secure de-duplication through the presentation of Proof of Ownership protocol and by anticipating the surge of side channel data in information de-duplication. In this paper, the computation by the client in the SecCloud was significantly diminished amid the record uploading and reviewing stages. SecCloud+ was an improvement propelled by the truth that clients continuously need to scramble their information some time before uploading and permits for integrity auditing.

Indeed and in spite of the fact that Cloud capacity was questionable and unsteady, Nishant and Ramkumar [8] proposes a combined approach for both information de-duplication and astuteness inspecting in the cloud. It was done by actualizing a cloud capacity which performs integrity auditing using hashing method and data de-duplication using block-Level de-duplication. Concurring to [9] data owner transfer records to the cloud with time and character in a scrambled arrange utilizing AES. When the record proprietor wishes to confirm the whether cloud record was hacked or debased by a Third Party Reviewer (TPA), at that point record proprietor sends Confirmation request to TPA, at that point TPA reviewer Send Confirmation reaction to information proprietor and Intermediary Server, the confirmation result signifies that record has been hacked, at that point intermediary recover file from intermediary cloud.

The calculation utilized to recuperate was Seed Block Algorithm [10]. It was well organized procedure to recoup the record in slightest time bound. And moreover it would keep up the information judgment. Concurring to the creators, it understands the trouble like fetched and the complexity of usage. It too stresses on security concept for the reinforcement records put away at the farther reinforcement server, without utilizing any encryption procedures. Seed Block Algorithm [11] was utilized for gathering data from inaccessible area and for recouping record in circumstances like record cancellation or record being devastated at the cloud. Concurring to this paper, Seed Block

Algorithm centers on integrity of the record which was put away at the farther server. It too diminishes the time required to recuperate the record.

A detailed comparative study had been conducted on recuperation procedures [12]. Agreeing with the authors, Seed block algorithm serves best for record recuperation as it was cheap and effectively implementable. An expansive amount of information in electronic shape was created in cloud. In order to preserve this information proficiently; the creator distinguished the need of information recuperation procedures. It was to cater this, a savvy remote information reinforcement algorithm, Seed Block Algorithm was proposed in this paper. The objective of proposed algorithm was twofold; firstly it helped the clients to gather data from any remote area in nonappearance of organize network and furthermore it helped to recuperate the records in case the record was being erased or in the event that the cloud gets crushed due to any reason. The time-related issues are too being illuminated by proposed SBA such that it would take the least time for the recuperation handle.

3 Proposed System

3.1 Proposed Architecture

The regenerator system architecture is shown in Fig. 1.

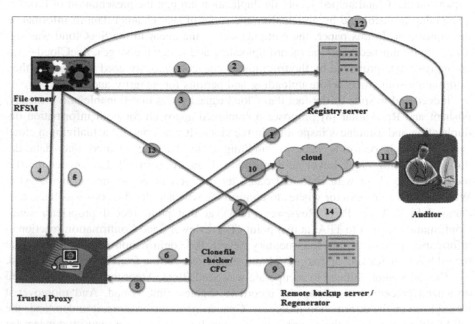

Fig. 1. Architecture of the system. 1- Register, 2- Original file, 3- Processed file, 4- Delegation, 5- Processed file, 6- Load file to CFC, 7- checks cloud for presence of clone, 8- Response from CFC, 9- File for Backup/XOR, 10- Uploads file to cloud(which is not a clone), 11- Auditing, 12- Regenerate message, 13- Regenerate Request, 14- Upload regenerated file

The system was composed of seven types of entities, which are, file owner, proxy, registry server, clone file checker, auditor, remote backup server and storage server. Cloud clients are the file owners, proxies and auditors. The registry server (RS) was a trusted party. RS does and responds for assembling of the system and the registration of clients'. RS allows the registered clients to store the public parameters of outsourced files. Storage services for storing outsourced files are provided to the registered clients by the cloud storage server. In real-world applications, an organization hires storage services from some CSP, and the IT department of the organization could performs the duty of the registry server. Thus, the file users (employees) could make use of the storage services.

The file-owner deploys files to the cloud server through the licensed proxies. Specifically, the authorized proxy processes the file on behalf of the owner. And they send the processed results to the Clone File Checker (CFC). CFC using the hash value of the file, checks for presence of duplicate file in cloud. After ensuring its absence, a copy of file was processed at remote backup server. Remote backup server was the regenerator which works on block seed algorithm and the copy of processed file was saved at there. Once it was done, the file was transferred to the cloud which was the storage server. The duty of the auditor was to check the integrity of outsourced files and also their origin-like general log information by interacting with the cloud storage server without retrieving the entire file. Upon the checking of integrity, if the auditor meets with any unusual activities like missing file, file content being altered; the auditor informs file owner about the need for regeneration of file. File owner then could regenerate his file from copy of file at the regenerator and was uploaded to the cloud.

- Registry Server (RS)

Registry server was a trusted server, which bargains with the enlistment of information proprietor and authorized intermediaries. The record proprietor, the subordinates (record clients) and the intermediaries (PCs) included has to enroll with the registry server. The record to be transferred has to be encrypted. The record would be encrypted utilizing RSA before uploading.

- Trusted Proxy (TP)

Trusted proxy was utilized to transfer record to the cloud. To include a PC as Trusted proxy, its MAC address was included to the database. For uploading the record, the admin chooses on which PC to be apportioned for the information proprietor from the list of trusted PCs, so that it was clear about which pc was to be utilized for record transfer.

- Clone File Checker (CFC)

Clone file checker checks for the presence of clone record in the cloud before uploading record. i.e., before uploading record to the cloud it would be sent to CFC to check presence of copy record in the cloud. For this, it uses Secure Hash Algorithm 1 (SHA1). Hash value of the record to be transferred would be calculated and it would be utilized to compare with the hash value of the records in the cloud. Comparative hash value signifies the presence of clone.

- Backup Server (BS)

The backup server stores the duplicate of record in the primary cloud. The primary cloud and remote reinforcement cloud are the central storage and the remote storage place respectively. When the primary cloud loses its information under any circumstances, which could be by a natural catastrophe, by human intercession or by erasing the substance by botch, at that point it uses the data from the remote storage. The fundamental objective of the remote backup was to facilitate the client to gather data from any remote area indeed when network was not accessible or on the off chance that information was not found on the fundamental cloud. As the title signifies, Backup Server was utilized for the reason of reinforcement. At BS, the reinforcement of record is put away in a distinctive format. It could be utilized to recover record in case of record manipulation. Record manipulation could be either record lost or a failure to demonstrate record astuteness. For this reason this framework employments Seed Block algorithm. This calculation stores XOR-ed form of unique record in BS. This XOR-ed file could be used to regenerate file on request.

3.1.1 Features of Backup Server

A secure backup server should have the following features:

(1) Integrity of Data

Information Integrity characterizes the server, which were total state and the entire structure of the server. It confirms that information perseveres unaltered amid broadcasting and received. It depicts the legitimacy and dependability of the information in the server.

(2) Security

It was the full assurance of the client's information which was the most extreme priority of the inaccessible server. Intentioned or inadvertently, it ought to not be gotten to by third party or any other users/client.

(3) Confidentiality of Data

Client's information records ought to be kept private such that on the off chance that numerous clients at the same time get to the cloud, at that point information records that are individual to a specific client ought to be kept covered up from other clients on the cloud.

(4) Truthfulness

The remote cloud must be reliable. Since the user/client stores, their secret information; the cloud and inaccessible reinforcement cloud ought to play a dependable part.

(5) Cost efficiency

The process of information recuperation ought to be cost-effective and minimum so that a most extreme number of company/clients could make utilize of the back-up and recuperation benefit.

3.2   Threats and Its Mitigations

There are two types of security attacks and four possible added features to overcome these flaws in this scheme. The security threats are impersonation and file modification. Impersonation by the cloud client, specifically, she may mimic a data owner or another authorized proxy, or abuses a delegation, so that she could operate a file and deploy it to the storage server. On the other hand, rarely accessed files could be modified or removed by a malicious storage server. It could be for saving storage space or for seizing up of hardware. Considering the above realistic attacks, a complete and secure system should satisfy the following requirements:

- Dedicated assignment: This designation process was issued by an information proprietor. It could be utilized by the particularly authorized intermediary to out-source specified records in the indicated way. Indeed the authorized proxy could not mishandle indicated records as it was scrambled before passing it to the proxy.
- Comprehensive auditing: The astuteness of the outsourced record, which incorporates log data as well as about the origin, type, and consistency of the outsourced records was approved by the reviewers. The astuteness checking guarantees that the outsourced records have been kept unharmed; the other common log data in reviewing guarantees that the record has been outsourced in the indicated way. With this point by point examining, this framework could be utilized to settle dispute since it could give persuading legal witnesses.
- De-duplication: Duplication check has to be done before record transfer as it would offer assistance a lot in lessening repetition, which would result in diminishing undesirable wastage of memory space. This framework as guaranteed was a total system that guarantees nonappearance of copy record at cloud before record transfer.
- Regeneration: To make a cloud storage system the best, it must have a backup plan to recover lost or damaged files. This system has a recovery system to regenerate missing and hacked files.

3.3   Seed Block Technique

As discussed, low implementation complexity, low cost, security and time related issues were still challenging in the field of cloud computing. To tackle these issues Seed Block Algorithm (SBA) is proposed in this section. Seed Block Algorithm was a simple algorithm which performs services like back-up and recovery. It uses the Exclusive-OR (XOR) concept in this algorithm. For instance: - Consider two data files: A and B. XOR operation was performed on files A and B it produces a new file which can be called as value Z. i.e. $Z = A \oplus B$. When data file A gets destroyed and we want to recover data file A, then we could recreate data file A by performing Exclusive-OR (XOR) operation on the data files B and Z.

Seed Block Algorithm consists of the Cloud storage and its clients and the Remote backup Server. In this algorithm random number was generated and unique client id granted for each client. Whenever a new client registers, client id was generated; then client id and random number was XOR-ed ($\oplus$) with each other to generate seed block of that specific client. This process was done before encryption. The client creates a file which is to be stored in the cloud. Before uploading, the file of the client was being XOR- ed with the Seed Block of the particular client. And that XOR-ed file was stored at the backup server as file A'. When the file in the main cloud gets tampered/damaged, then the user could recreate the original file back by XOR-ing backup file with the seed block of the corresponding client to produce the original file and return the original file back to the requested client. The regeneration algorithm works as discussed below.

Initialization Entities: Cloud - C; Cloud Client - C_1; Client's ID - $C_1\_Id_i$; Files - A and A'; Seed block - S_b; Random Number - r; Backup Server - B_S;

Input: File A was created by C_1; Random number generated was r;

Output: File A was recovered, which was previously deleted from the cloud.
             //File owner was allowed to perform operations such as download, upload and modify files owned by him.

Step1:   Trigger random number generator to put forward a random number.
             int r = random_no ();

Step2:   Create seed block S_b for each Client C_1 at client registration by XOR-ing with r and store S_b at B_S
             $r \oplus C_1\_Id_i = S_b$ (Repeat Step2 for all clients)

Step3:   When client C_1 creates a file A and stores at C, then file A' was created by XOR-ing file A and S_b
             $A' = A \oplus S_b$

Step4:   Store file A' at B_S.

Step5:   If server crashes or file A' was deleted from C, Then, perform XOR to recreate the original file A by XOR-ing file A' and S_b
             $A = A' \oplus S_b$

Step6:   Upload A to C_1.

Step7:   END

4 Performance Evaluation

4.1 Experimentation Results

In this segment, the experimentation and result examination of the Seed Block algorithm is talked about. Negligible prerequisite for the cloud server and reinforcement server considered for experimentation was portrayed in Table 1. From Table 1, memory necessary for the primary cloud's server and reinforcement server was kept 4 GB and 8 GB respectively, which could be expanded as per the need. It was apparent from Table 1, that the memory necessary for reinforcement server was more compared to the fundamental cloud's server since extra data like Seed Block of the corresponding client was spared onto the remote server.

Experimentation results show that the measure of unique information record put away at fundamental cloud was nearly comparable to the measure of Back-up record put away at Backup Server as portrayed in Table 2. In order to make this truth pleasant, this try was conducted on records of different sizes. Results arranged in Table 2 for the

Table 1. System environment

	Main cloud's server	Backup server
CPU	Core2 Quad Q6600 2.40 GHz	Core2 Quad Q6600 2.40 GHz
Memory	4 GB	8 GB
OS	Any OS	Any OS
HDD	SATA 250 GB or more	SATA 500 GB or more

Table 2. Performance analysis of files of various sizes

Type	Size of original file in cloud	Size of backup file in backup server	Size of recovered file after recovery
Text	450 KB	500 KB	450 KB
	800 KB	850 KB	800 KB
	2.5 MB	3 MB	2.5 MB
	5 MB	5.5 MB	5 MB

study shows that proposed Seed Block Algorithm was exceptionally much strong in keeping up the measure of recuperation record same as that the unique information record. It was apparent that proposed SBA recuperates the information record without any information loss.

Handling Time was added up to the time taken by the process when a client transfers a record at the fundamental cloud. It incorporates the gathering of information such as the irregular number from the main cloud, seed block generation of the corresponding client from the remote server for XOR-ing operation; after assembling, performing the XOR operation on the transferred record with the seed piece and at last putting away the XOR-ed record onto the remote server. It was watched that, as information measure increments, the handling time too increments.

4.1.1 CPU Utilization at Main Cloud and Backup Server
Main Cloud's CPU utilization begins with 0% and CPU utilization increments as the client transfers the record onto the cloud; cloud had to check whether the client was authenticated or not, at the same the time, Backup Server produces seed block for the corresponding client. Amid this period, load at Primary Cloud diminishes which in return comes about in the progressive decrease in CPU utilization at fundamental cloud. After seed generation, CPU utilization at Backup Server increments as it has to perform the XOR operation.

4.1.2 Recovery Analysis
The Fig. 2 shows the experimentation result of proposed SBA. As Fig. 2(a) shows the original record which was transferred by the client to the cloud. Figure 2(b) shows the XOR-ed record which was put away on the reinforcement server. It contains the secured XOR-ed substance of original record and seed block of the corresponding client. Figure 2 (c) shows the recuperated record; which was sent to record proprietor in case of the record erasure or in the event that the cloud gets devastated due to any reason.

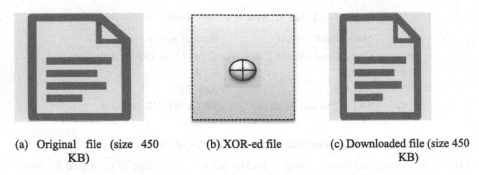

(a) Original file (size 450   (b) XOR-ed file   (c) Downloaded file (size 450
       KB)   KB)

Fig. 2. Sample output image of Seed Block Technique

5   Conclusions and Future Work

On these days, secure cloud storage was gaining its peak. In order to ensure security of file in cloud, it has to be secured from the moment of file creation and upload which was done at the desk file owner. File has to be uploaded to the cloud, only after a phase of encryption. In order to handle the case of avoiding preservation of multiple copies of file at cloud, a checking of duplicate files has to be done prior to upload. This would help in saving money as well as resources to a large extent. Finally, the most unpleasant thing to file owner would be the loss of integrity of file. In such cases the regenerator system would be helpful. It would help in regenerating the file to the fullest.

File owner would sign up/login to the system for file uploading. To ensure secure file upload on system, use dedicated proxies to upload file to the cloud. Prior to the file upload de-duplication was ensured. It was done by Clone file checker. SHA1 algorithm was used at this stage. Integrity of file was ensured and audited by auditor. Auditor runs SHA1 algorithm for this purpose. At the time of file upload, it was XOR-ed with random number at the regenerator and saved at the regenerator. In case of necessity, the XOR-ed file would be used for regeneration purpose.

Types of file: This system concentrated on text file, which was to be considered in upcoming works. A fully efficient uploading scheme should be able to handle any type of file irrespective of its file size. De-duplication: This scheme uses SHA1 for checking presence of duplicate files. Even though SHA1 fits for the requirement, more efficient algorithm could be used. Regenerator: In this work, regenerator was a separate entity which needs extra cost for its implementation. In further works, Regenerator has to be cost effective, either by clubbing its functionality with some other entity or by introducing a low cost scheme for the same.

References

1. Wang, Y., Wu, Q., Qin, B., Shi, W., Deng, R.H., Hu, J.: Identity-based data outsourcing with comprehensive auditing in clouds. IEEE Trans. Inf. Forensics Secur. **12**(4), 940–952 (2017)
2. Chu, C.-K., Zhu, W.-T., Han, J., Liu, J.K., Xu, J., Zhou, J.: Security concerns in popular cloud storage services. IEEE Pervasive Comput. **12**(4), 50–57 (2013)
3. Ranaware, N.S., Dalvi, R.: Regeneration of code based cloud storage. IJCAT – Int. J. Comput. Technol. **3**(5), 267–269 (2016)
4. Wang, B., Li, B., Li, H.: Panda: public auditing for shared data with efficient user revocation in the cloud. IEEE Trans. Serv. Comput. **8**(1), 92–106 (2015)
5. Ateniese, G,. di Pietro, R., Mancini, L.V., Tsudik, G.: Scalable and efficient provable data possession. In: Proceedings of the 4th International Conference on Security and Privacy in Communication Networks, Article no. 9, New York, NY, USA (2008)
6. Sahana Kumari, B., Rajni: Data deduplicating and auditing in cloud. Int. J. Emerg. Res. Manag. Technol. **5**(5) (2016)
7. Paterson, K.G., Schuldt, J.C.N.: Efficient identity-based signatures secure in the standard model. In: Batten, L.M., Safavi-Naini, R. (eds.) ACISP 2006. LNCS, vol. 4058, pp. 207–222. Springer, Heidelberg (2006). https://doi.org/10.1007/11780656_18
8. Nishant, S., Ramkumar, D.: Integrity auditing and deduplication of data in cloud. Int. J. Adv. Comput. Sci. Cloud Comput. **4**(2), 13–17 (2016). ISSN 2321-4058
9. Jadhav, S., Penchala, S.K.: Proxy oriented data uploading and auditing for regenerating-code-based cloud. Int. J. Innov. Res. Sci. Eng. Technol. **6**(6), 12217–12222 (2017)
10. Umesh, K., Mahesh, K., Ajay, M., Rathee, N.: Backup and recovery system using seed block algorithm. Int. J. Adv. Res. Comput. Eng. Technol. (IJARCET) **4**(1), 125–128 (2015)
11. Pophale, K., Patil, P., Shelake, R., Sapkal, S.: Seed block algorithm: remote smart data-backup technique for cloud computing. Int. J. Adv. Res. Comput. Commun. Eng. **4**(3) (2015)
12. Ranaware, N.S., Dalvi, R.: Regeneration of code based cloud storage. IJCAT – Int. J. Comput. Technol. **3**(5), 267–269 (2016)

Network Communication

Discrete Time vs Agent Based Techniques for Finding Optimal Radar Scan Rate - A Comparative Analysis

Ravindra V. Joshi$^{(\boxtimes)}$ and N. Chandrashekhar

Department of Computer Science and Engineering, Noorul-Islam-University,
KumaraKoil, TamilNadu, India
rvjoshi18@hotmail.com, drnshekhar@gmail.com

Abstract. Complexity Theory is relatively young discipline which has provided very powerful tools to analyse Complex Adaptive Dynamic Systems. Agent Based Modelling is one such technique which is widely practiced. In this paper, we study, why or how this ABM is different from conventional modelling techniques. In particular, we would like to understand ABM performance for the context of modern battle scenarios and systems. For this analysis we chose battle/confrontation between Radar and an Attack aircraft as a cases study. One of the distinguishing feature of modern battle systems like Radar is their extensive configurability. Detailed configuration must be made even after purchase and deployment of a given system to suit to specific mission scenarios. This configuration is knowledge intensive activity, where knowledge about resources of own side, rival side, environment and general science and mathematics becomes very important. In this paper we exam how suitable configuration parameters can be selected for a radar based on its mission needs. Purpose of analysis is to determine the scan rate with which Radar must scan the space. More specifically, planner must evaluate what should be the rate at which Radar should scan across azimuth and elevation. Agent based model helps to pick a suitable trade-off between the extremes of slow scanning and speeding through long hops and thus leaving holes in between. Discrete time-based simulation model is one of the oldest and conventional type of modelling. It is chosen as reference to compare and contrast with ABM performance. We simulate scenarios which by varying different target characteristics like approaching radial velocity and initial positions in both DTM and ABM. By evaluation of these scenarios, optimal scan rate value is determined. While determining the optimal value of the scan rate, we also examine comparative merits and demerits of using causative phenomenon (like discrete event simulations) vs complex, emergent phenomenon (agent-based simulation).

Keywords: Discrete Time-Based Simulation · Agent based simulation
Complexity Theory · Mono pulse radar · Scan rate

1 Introduction

Complexity Theory is an inter-disciplinary field which addresses the problem of dealing with Complex Adaptive Dynamic Systems [1]. It tries to answer questions in exploratory or "bottom-up" manner in contrast to conventional approaches. Agent

© Springer Nature Singapore Pte Ltd. 2018
I. Zelinka et al. (Eds.): ICSCS 2018, CCIS 837, pp. 541–547, 2018.
https://doi.org/10.1007/978-981-13-1936-5_56

Based Modelling is being adapted rapidly as a tool or method to analyse CADS. In this paper, we will study how ABM differs from conventional approaches, specifically, in the context of modern battle scenarios. Modern battles and battle systems are very complex. The differences in each type of battlefield have rich and wide variation in their functional and performance capabilities. [2–4]. This makes designing general solution virtually impossible. Usual solution is to design a generic component and customize it extensively as per the mission needs. Such customising of electronic systems to operate effectively in given environment is not a trivial task. Firstly, there is no clear analytical or mathematical models as these are multi-variable (multi-body) systems with non-linear interactions.

The case-study be formulated as answering following question. Can we determine a set of the radar configuration parameters that are optimized for a specific set of scenarios? To make the problem more concrete, can we determine optimal scan rate for a radar that is expecting a range of aircraft velocities and initial ranges?

To find the target Radar has to scan through the space in a particular Scan Pattern. The way space is scanned will have impact on efficacy of radar operation like how likely it is to detect radar and how early. Choosing appropriate scan pattern is an important and interesting design problem [5–8].

Even if we fixate scan pattern, determining scan rate is not trivial issue. If the whole space of interest could be scanned such that even fastest moving planes will be caught by its rotating beam at one place or other Radar can operate in fool-proof mode. Otherwise, it will be a trade-off between how fast you move, and how closely you move (i.e., without leaving any gaps between the steps).

Thus, taking too small step size will result in slow scanning of space, trying cover same by hopping may leave gaps/holes in between. A specific scan rate will be optimal for specific scenarios based on the initial approach angles of range, aircraft, velocity Determining Optimal Scan Rate will become interesting from this perspective.

Conventional approaches are difficult to model, analyse and interpret results. Even more difficult is to make changes and experiment. Many a times, gaining better insights calls for tweaking the model here and there. We have developed a develop a Discrete Time-Based Simulation Program called BSim. Here various scenarios like aircraft velocity, range are modelled and are run against various scan rates over required range. Though modelling power increases in this approach, it still lacks flexibility, capability to experiment. As shown, it is not very efficient to combine multiple scenarios and hence uncover hidden and emergent patterns.

2 Description of the Models and Simulations

2.1 Modelling Methods

There are multiple ways in which problem in complex problems can be solved. Some of the prominent techniques which have to be proven very successful are Analytical, Mathematical, Decision Theoretic (Operation Research), Simulation (discrete-time, discrete-event, Monte-Carlo). Agent Based Model, a Complexity Theory based approach, latest addition to this suite of tools which promises many new insights. Well as qualitative and also knowledge-based approaches. Detailed survey can be found at [9–14] for these methods.

2.2 Battle Scenario – Radar Optimal Scan Rate

Radar an Acronym for Radio Detection And Ranging is a time tested technology/ system for detection and tracking of targets in all weathers and mediums like air, ground, space. Here targets are detected by using Radio Waves. Essentially, Radio wave is transmitted and its echo from targets is analysed to determine target characteristics like Range, Direction, Velocity and other attributes [5].

To detect target with single pulse, radar's beam has to be directional. It means the attenuation echo experiences is precisely related to the angle between target and radar. Let us call the main direction at which radar is receiving as Main Beam Azimuth (MBAZ). Let Target's azimuth be TAZIM. Attenuation expressed by echo is directly proportional to |MBAZ – TBAZ|. (Refer Fig. 1). Figure 1 and Table 1 shows the relation between echo signal strength and difference between azimuths of target and radar main beam in quantitative way.

In this paper, for the purpose of simplification, all parameters are frozen and just what should be the optimal scan rate is explored. Radar Scan Rate, is the rate at which the whole search space has to be sampled. We also assume that, Radar scans only in azimuth in a fixed elevation. Thus, we are interested, to optimize, the step-size in azimuth, the Radar has to take to cover the whole search space. This is not trivial as trade-offs are involved. If step-size is too small, we may miss fast moving at farther end. On the other hand, hopping in larger steps will leave holes in between through which enemy can "sneak-in". Thus, an appropriate value which balances these two extremes should be chosen.

3 Model Description

3.1 Scenario

Optimal Scan Rate depends on various attributes of Radar as well as target. But modeling all of them will be computationally impractical. Including too many variables will make simulation go forever. Also, development cost will also go high. But more important fact is, with many factors participating, it will be very difficult to draw a coherent picture. Also, since this was an academic exercise, simplicity and comprehensibility took precedence of over sheer accuracy. The Model was simplified as follows.

a. Radar-Side (Refer Fig. 1)
    i. Radar scans only in Azimuth (fixed elevation)
    ii. Scan pattern from left to right and right to left (zig-zag manner)
    v. The earliest time (either indicated by Radar cycle no, or Range (as seen by radar can be seen to modle effectiveness of Radar.
b. Aircraft-Side
    i. Aircraft moves with radial velocity towards radar

ii. Its elevation of attack is same as elevation which Radar is looking at.

3.2   Discrete Time-Based Simulation

Battle Model and Simulator (Generic). Next step is to model this scenario in computer program and try to see how the behavior will be.
A Battle Engine is the central component that runs simulation thread (Fig. 2).

Customization to Current Scenario. In scenario outlined in 3.1 only two Battle Systems are there. A radar and an attack aircraft. If we apply the simplifications suggested 3.1, building radar and aircraft model would be fairly simple. We can assume Radar's static properties like Beam Shape, Transmission. Similarly, aircraft's if any. Now, in every iteration, Radar has to move its antenna a small step as per its scan strategy. Aircraft will move its position. Now, battle engine checks whether aircraft falls in the receptivity of zone of radar, and if so forms an echo matrix and writes back into the radar's reception buffer. Radar continues to process as if echo has come. Radar's echo strength is calculated by Radar Equation (Eq. 1) [5]. Simulations were run for following parameters Table 2. Let location of radar be denoted by letter O. If aircraft is A, OA indicates the direction at which Radar should be pointing. But at this point of time Main Beam Azimuth (MBA) will be pointing at some other direction. Results and Analysis is shown in next section DTM is shown in Fig. 3.

Radar Range Equation:

$$\sqrt{\frac{P_t\, G^2\, \lambda^2 \sigma}{(4\,\pi)^3 P_{min}}} \quad \begin{aligned} P_t &= Transmitted\ Power\ in\ dBm \\ G &= Gain\ of\ Radar\ Antenna\ Unitless\ Multiplication\ Factor \\ \lambda &= Wavelength\ of\ Electro\ Magnetic\ Wave\ in\ m \\ \sigma &= Radar\ Cross\ Section\ in\ m^2 \\ P_{min} &= Radar\ Sensitivity\ in\ dBm \end{aligned} \tag{1}$$

4   Agent Based Simulation

As an alternate candidate we chose Agent Based Modelling (ABM) to determine Optimal Scan Rate. ABM is recently emerging method based on Complexity Theory. The purpose of using this method was to gain insights into merits and demerits of using ABM vis a vis DTBM. ABM will be carried out in a world (usually toros based i.e., wrapped around). The world will be a grid consisting of many discrete cells. This world can have two types entities turtles and patches. While turtles move around and respond to world around them, patches are more passive entities. Once placed, they can't change their place, also they can't actively respond to events. Net Logo Model was built using following equations. NetLogo model is shown in Fig. 3.

Equation for Turtle (Agent) Movement:

$$BlueTurtle[x]_{n+1} = BlueTurtle[x]_n + blue_{scan_{rate}}$$
$$Red_{n+1} = Red_n + red_{velocity} \qquad (2)$$
$$S \quad dist = Red_x - Blue_x$$

Table 1. Gain pattern of radar

Theta (in degrees)	−45	−30	0	30	45
		0.3	0.7	0.9	0.7	0.3

Table 2. ABM simulation parameters

Red-turtle	Movements
Units	Patch/tick
Start-value	0.5
End-value	1.1
Step-size	With uniform distribution with 25% between 0.5 and 1.1
Blue-turtle	
Units	Patch/tick
Start-value	0.25
End-value	0.75
Step-size	0.05

5 Results

5.1 Result from DTBM

Plot above show the results from running Discrete Time Simulation Program. Two main conclusions can be drawn First, presence of blind spots i.e., 0 hits at large scan rates (step sizes). We can see that after 22.5°, the number of hits collapses to 0. Though it is possible to show hit for some values, probability becomes less and less gap it leaves becomes and larger and larger. Figures 4 and 5 capture this.

5.2 Aggregated Result of DTBM Vs ABM

Varying Initial Range -Varying Velocity and Scan Rang: Plots Fig. 6a and b aggregated results of DTM and ABM respectively. It can be seen that shape of the trend as function of scan rate is more or less same pattern in both simulations.

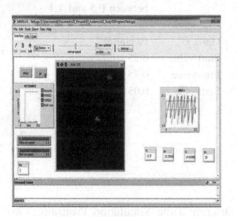

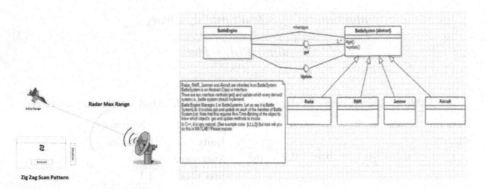

Fig. 1. Air defense model

Fig. 2. Battle model

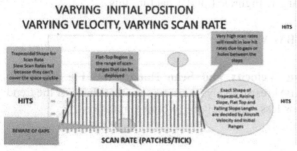

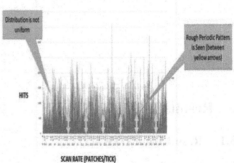

Fig. 3. ABM NetLogo model

Fig. 4. Model results

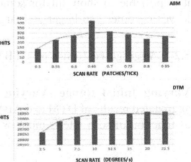

Fig. 5. Trapezoidal pattern

Fig. 6. DTM and ABM comparison

6 Conclusion

From discussion above and Fig. 6 this we can conclude that, an engineer irrespective of DTM or ABM is likely to chose Scan Rate having same median value. But, at a deeper level, there are differences between Discrete Time Battle Model and Agent Based Model.

a. We found the key difference between the two approaches lies in Battle Engine. Battle Engine makes the whole application centralized and it has to be empowered with whole Domain Knowledge. E.g., Target Echo Matrix Population

b. Whereas in ABM, there is no Battle-Engine. Though there may a tick-engine in background, it is Domain-Neutral. Whole knowledge is embedded in Agents and their interactions only. This helps truly de-centralized systems to emerge.

c. Without this key insight, handling promises of ABM like, heterogeneous Agents, Flexibility for Random behaviour and even bottom-up emergence would have been not possible.

References

1. Aghababa, M.P.: Adaptive control of complex systems with unknown dynamics and input constraint: applied to a chaotic elastic beam. Int. J. Adapt. Control Signal Process. **32**, 213–228 (2018)
2. Skalický, P., Palasiewicz, T.: Intelligence preparation of the battlefield as a part of knowledge development. In: International Conference on Knowledge-Based Organization, vol. 23, no. 1, pp. 276–280 (2017)
3. Fighting for victory on the learning battlefield. Dev. Learn. Org. **31**(4), 24–26 (2017)
4. Kress, M., Caulkins, J.P., Feichtinger, G., Grass, D., Seidl, A.: Lanchester model for three-way combat. Eur. J. Oper. Res. **264**(1), 46–54 (2018)
5. Skolnik, M.L.: Introduction to Radar Systems. 2nd edn. McGraw Hill Company International Edition (1981)
6. Yongqiang, G., Yumin, W., Hui, L.: Construction of waveform library in cognitive radar. Polish Marit. Res. **24**(s2), 22–29 (2017)
7. Han, J., et al.: The establishment of optimal ground-based radar datasets by comparison and correlation analyses with space-borne radar data. Meteorolog. Appl. **25**(1), 161–170 (2018)
8. Liu, H., Zhang, Y., Guo, Y., Wang, Q.: A two-stage space-time adaptive processing method for MIMO radar based on sparse reconstruction. Frequenz **71**(11–12), 581–589 (2017)
9. Lättilä, L., Hilletofth, P., Lin, B.: Hybrid simulation models–when, why, how? Expert Syst. Appl. **37**(12), 7969–7975 (2010)
10. Gonzalez, R.A.: Architecture of a discrete-event and response simulation model. Int. J. Adv. Intell. Paradig. 413–653 (2012)
11. Heard, D., Dent, G., Schifeling, T., Banks, D.: Agent-Based Models
12. Ramos, J., Lopes, R., Araújo, D.: What's next in complex networks? Capturing the concept of attacking play in invasive team sports. Sports Med. **48**(1), 17–28 (2017)
13. Wang, C., Koakutsu, S., Okamoto, T., Qian, F.: A collaborative learning automata team model for modeling multi-agent systems. Electron. Commun. Jpn. **101**(3), 28–37 (2018). Author, F.: Article title. Journal 2(5), 99–110
14. Wang, H., Olhofer, M., Jin, Y.: A mini-review on preference modeling and articulation in multi-objective optimization: current status and challenges. Complex Intell. Syst. **3**(4), 233–245 (2017)

Privacy Preserving Schemes for Secure Interactions in Online Social Networks

Devakunchari Ramalingam$^{(\boxtimes)}$, Valliyammai Chinnaiah,
and Abirami Jeyagobi

Department of Computer Technology, Anna University, Chennai, India
devakunchari.r@gmail.com, jabirami02@gmail.com,
cva@annauniv.edu

Abstract. Online Social networks (OSNs) play a major part in everyone's daily life. The Sybil uses the original identity of targeted or random victim to create an account on the social network. All these effects should be removed only by mitigating the creation of Sybil accounts. The existing techniques for Sybil detection employs reactive defence strategies which are initiated based on the user report, where by the time, the user becomes victim. In this paper, a novel Sybil detection framework is proposed to identify the attempt of account creation by the Sybil users during the admission phase and also making the defending pattern unpredictable to learn. The Sybil detection framework consists of a three level privacy scheme, which is capable of strongly suspecting the fake from the genuine. The extensively used open-source, distributed computing platform, Apache Hadoop with its ecosystem tools are utilized to implement the Sybil detection framework for effective processing of unstructured social data. The experiment is conducted with data collected from Twitter, Facebook and Google+. Experimental results show that proposed Sybil detection framework is well suited for mitigating the worst effects of fake users with efficient handling of large social data. The K-means clustering and Bayesian classification are used to detect the Sybil accounts with 94% accuracy and 92% precision using many public features.

Keywords: Social networks · Sybil detection · Apache hadoop
Attribute extraction

1 Introduction

The OSNs like Twitter, My space, LinkedIn, Instagram and Google+ have millions of registered users generating data at a massive extent. "Sybil's (alias Social bots)" create another form of themselves on the social network by means of fake profiles under the name of a high-profile user or random victim. Based on the report [1] filed by Facebook to Security exchanges and commission, there are 8.7% of 955 million of users are fakes. The threats, mainly target technical, non-savvy users, teens, females and kids. Most of the content and feature based approaches assume that fake users initiate only few attack edges with the real accounts. The fake user can evade these detection techniques by learning the defence patterns employed by the OSN providers. The huge

© Springer Nature Singapore Pte Ltd. 2018
I. Zelinka et al. (Eds.): ICSCS 2018, CCIS 837, pp. 548–557, 2018.
https://doi.org/10.1007/978-981-13-1936-5_57

amount of data shared in OSN is valuable for deciding the behaviour of any individual. The challenge also lies in analyzing the dynamic user generated information from heterogeneous sources and the restructuring of these collected data [2].

A reliable and proactive Sybil defence scheme which assists OSNs to identify the bogus accounts during initial account creation is designed. Apache Hadoop [3] and its ecosystem tools like Apache Flume, Apache Hive and Apache Mahout are utilized for processing the data available in social networks. The sections in this paper discuss about extracting the structured information from the unstructured social data and making use of those data for revealing the nature of an individual, thereby detecting the Sybil user in the social networks.

2 Related Work

Multi-stage level classification mechanism [4] was proposed for detection of sybils in Facebook and Twitter. A real time model was proposed for doing scalable processing over the social data resulting in classification and prediction [5]. Detection of fake profiles in LinkedIn [6] is experimented. An attribute based recommendation of friends for social relationships between strangers was developed in [7]. The complete community detection solution for large networks was focused using MapReduce [8]. Friend Book [9], a semantic based friend recommendation utilizes Latent Dirichlet Allocation algorithm for extracting the life styles out of daily activities with which similarity between the users are calculated. The empirical evaluation model exploits the features in existing techniques and added 9 new features with spatial and temporal correlations [10]. Zhou et al. stated that identifying users with screen name and profile images for large social networks [11] is tedious. Feizy et al. analyzed static and dynamic profile features to classify the fake profiles using Decision tree and Nearest Neighborhood approach [12].

3 Proposed Framework

The Sybil detection framework works under the supervision of the OSN service provider. Data collected from twitter is done using the python API and Twitter API using Apache Flume. Unstructured data are converted into the structured format using Apache Hive and stored into HDFS for processing. To ensure scalability, the framework is implemented on multi-node cluster setup to divide and distribute the workload. Sybil detection framework depicted in Fig. 1 consists of three levels, namely Verification Level 1 (VL1), Verification Level 2 (VL2) and Verification Level 3 (VL3) respectively. Each level has its own similarity metrics over the set of attribute values, followed by the classification to derive at the risk level of users being fake. The process flow of verification levels with its similarity measures is illustrated in Fig. 2. Algorithm 1 defines the overall Sybil detection process.

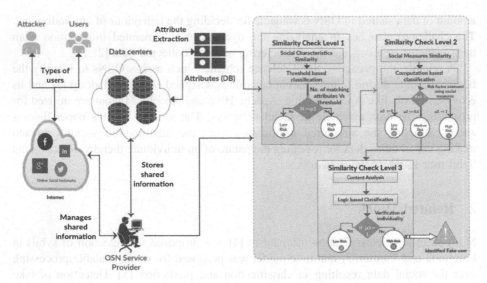

Fig. 1. Sybil detection framework

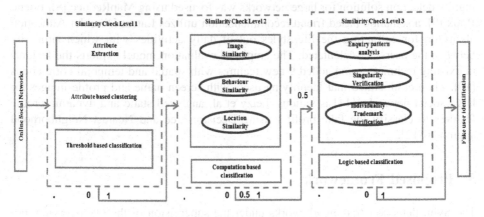

Fig. 2. Sybil detection process flow

3.1   Verification Level 1

The VL1 involves attribute extraction from the database and clustering the attributes using K-means. A threshold based classification is used for determining low and high risk users {0, 1}. The 20 low-cost profile features such as name, email, registration date, birthday, gender, current city, hometown, relationship status, education, current employment, employer, languages known, activities, Interests, groups, apps, alternate email, address, phone number and twitter screen name are considered for VL1.

Algorithm 1: Sybil Detection Algorithm

\# User Profile Attributes $Ua_i = \{Ua_1, Ua_2 \ldots Ua_n\}$, \# Users U_i, \#User Account UA_i,
\#Clusters $C_i = \{C_1, C_2 \ldots C_n\}$
\# μ - Threshold for number of matching attributes
\# t – time in seconds, \# T – Fixed time limit
begin
Detect_profile (U_i)
count = Sign_Up (U_i)
if (count > μ) **then**
\# Illegal account creation identified
\# Verify the original user U_1
Send email and verify proof
    if (t <=T) **then**
        Wait for response
            if (response (U_i)) **then**
            Create UA_i (U_i),
            Assign U_i.id = C.id
            end if
    else
Deny UA (U_i) creation
\# Denied bogus account creation
Alert user U_1 where U_1.id = C.id
Warn U_2.id = U_1.id
\# Warn the illegal attempt made by user U_2
    end if
else
 \# Not an existing user
 Create UA (U_i)
 Assign U_i.id = new id ()
end if
end

Algorithm 2 defines the clustering of user profile attributes existing in the social network. Each cluster consists of users who have many similar attributes. Attributes extracted from new user profile (Ua) are matched with clusters formed and finally the Sybil users are classified.

```
Algorithm 2: Clustering based on attributes
# User Profile Attributes Ua_i = {Ua_1, Ua_2.........Ua_n}, # Users U_i
# Clusters C_i = {C_1, C_2 ... C_n}
Input: User attributes of a profile
Output: Attribute based clusters
begin
U_i.id = new id ()
Ua [U_i] ←{ Ua_1, Ua_2 ... Ua_n}
C.id ← U_i.id
C_i ← {C_1, C_2... C_n}
    if ((Ua [U_i] =Ua [U_j]) && (U_j∈C_i)) then
        C_i (U_i) ← U_i
    end if
end
```

Algorithm 3 defines an attribute matching algorithm which further determines the existence of the new user in the social network. The user attributes are retrieved and matched with profile based clusters formed using Algorithm 2, when the new user gives out attribute values for creating the new account.

```
Algorithm 3: Attribute Matching algorithm
# User Profile Attributes Ua_i= {Ua_1, Ua_2 ...Ua_n}, # New User U_i,
# Clusters C_i = {C_1, C_2 ... C_n}, # μ - Threshold for number of matching attributes
Input: Attributes of new user, Attribute based clusters
Output: Number of matching attributes
begin
Sign Up (U_i)
# Get attributes of the user profile
Ua (U_i) ← {Ua_1, Ua_2...Ua_n}
U_i.id ← new id ()
for i = 1 to n
    if ((Ua_i == C_i) && (U_i.id! = C_i.id)) then
        Count [U_i] + = 1
    end if
end for
return Count;
```

The attribute values matched from the corresponding cluster should not have different ids. This condition is checked to avoid confusions related to many users with few different matching attributes. The count of number of matching attributes is taken to further decide the existence of the user in the social network. Equation (1) defines function which is capable of calculating the number of attributes between the same two persons.

$$AM(U_a, U_i) = \begin{cases} C + 1, & \text{if } ((Ua_i == C_i)\&\&(U_i.id! = C_i.id)) \\ C + 0, & \text{otherwise} \end{cases} \tag{1}$$

Where U_a is the attributes of the user to be verified, U_i is the attributes of the users in the database, AM (U_a, U_i) is the Attribute Matching function and C is the count of number of matching attributes. Equation (2) is an assignment function.

$$N = \sum AM(Ua_j, U_i) \tag{2}$$

The user attributes are matched with clusters formed based on the similarity of the profiles. The number of matching attributes exceeding the threshold ($\mu1$) will be of high risk user profiles. Otherwise, it is of low risk user profiles. The high risk user profiles of VL1 are taken to the next level VL2. The overall decision around everyone reveals the truth as mentioned in Eq. (3).

$$Dec(Ua)VL1 = \begin{cases} 0 & \text{Low risk, if } N < \mu1 \\ 1 & \text{Low risk, if } N \geq \mu1 \end{cases} \tag{3}$$

where N is the overall matching attributes and $\mu1$ is the threshold for matching attributes.

3.2 Verification Level 2

The VL2 takes the social measures like images, location and behavior which are not considered in previous works. The social measures such as Google Search Images (GSI), Google plus friends Circle Images (GCI), location of friends in friend list, number of friend requests and accepted requests (FR), number of status updates, number of photos uploaded, number of report abuses and likes, IP Trace back (IPT) and Privacy settings (PR) are considered for VL2. Equations (4)–(10) reveals the design behind how the attributes are mapped and used for detection at this level. This risk is classified as medium and high based on the metrics.

$$GSI\ (Pu) = \begin{cases} 0, \text{Same Person;} & \text{Same attribute} \\ 0.5, \text{Nothing appears;} & \text{Nothing} \\ 1, \text{Many persons; Different attributes} \end{cases} \tag{4}$$

$$GCI\ (Pu) = \begin{cases} 0, & \text{Images of friends appears} \\ 1, & \text{Images of friends not appears} \end{cases} \tag{5}$$

Location (Loc) of friends in the list and user's locations are verified for a normalized ratio. Number of Likes and Dislikes (LD), unfriend and report abuses gained, frequent activities in the network by the user is taken into account for analyzing the behaviour. IPT can reveal the log in attempts of the user from their residency area. These are calculated as ratios to be resulting in range of 0 to 1 and resulted as a decision as Eq. (11).

$$Loc(U) = \frac{\text{No. of related location friends in friend list}}{\text{Total no. of friends in the friends list}} \tag{6}$$

$$FR(U) = \frac{\text{No. of friend requests received}}{\text{No. of friend requests sent}} \tag{7}$$

$$LD(U) = \frac{\text{No. of likes and tagging}}{\text{No. of report abuses, blocks and untagging}} \tag{8}$$

$$IPT(U) = \begin{cases} 0, & \text{Usual location} \\ 1, & \text{Different Location} \end{cases} \tag{9}$$

$$PR(U) = \begin{cases} 1, & mostly\ public \\ 0, & \text{otherwise} \end{cases} \tag{10}$$

$$Dec(U)VL2 = \begin{cases} 0, & \text{social measures are undoubtful} \\ 0.5, & \text{social measures are undecidable} \\ 1, & \text{social measures are doubtful} \end{cases} \tag{11}$$

3.3   Verification Level 3

The VL3 verifies the identification of the individual person by enquiring about few possible known questions followed by verifying any national identity proofs and finally requests the handwritten signature for verification to reveal the trademark individuality of the user. Equations (12)–(15) reveal the flow of assignments, mapping and decision resulted at the end of VL3. The categories of information such as Questionnaire (Que), National ID verification and Signature authentication are considered for VL3.

$$Que\ (U_i, Q) = \begin{cases} 1, & \text{All questions answered} \\ 0, & \text{If any one question answered incorrectly} \end{cases} \tag{12}$$

$$Sig\ (U_i, DS\ (U_i)) = \begin{cases} 1, & \text{If signature matches} \\ 0, & \text{If signature not matches} \end{cases} \tag{13}$$

$$Nat\ (U_i, DI(U_i)) = \begin{cases} 1, & \text{Identity proved} \\ 0, & \text{Identity not proved} \end{cases} \tag{14}$$

$$Dec(U_i)VL3 = \begin{cases} 1, & \text{Low Risk,} & \text{All contents proved \& verified} \\ 0, & \text{High Risk,} & \text{Anyone content not verified} \end{cases} \tag{15}$$

After the positive responses from all the three steps in this level, user will be designated as genuine user. If any one of three steps in this level is negative, then the user is identified to be fake user. Thus, identification of fake user is achieved at the end of three verification levels.

4   Experimental Results

The Sybil detection framework is implemented on multi-node Hadoop cluster deployed with Mahout Machine learning library for making out the decision over the given profile in the dataset. Due to the current restrictions and blocking of social network APIs, only a limited number of datasets are collected from Twitter and Google+. The missing fields are simulated with synthetic datasets for experimentation. Apache Flume and Apache Hive are deployed upon Hadoop for data collection, aggregation and transforming the unstructured data into structured format. The results with respect to number of profiles in the dataset and the time taken for steps involved in machine learning techniques are comparatively analyzed for examining the efficient processing of the huge data from OSNs. The profile attributes are clustered using k-means, where each cluster contains similar profiles. The clustering is done by the following steps. The profiles are converted into sequence file format. The next step is to convert the sequence files into feature vectors which will be used for identifying the similarity. The final step is to cluster the points with respect to the feature vectors. The classification of new profile into the dataset is based on the stored feature vectors and clustered attributes of the profiles in the dataset. There are few steps involved in Naïve Bayes classification. Again, it starts with the sequence file generation and then the feature vectors are formed and analyzed. The dataset is split into two, namely training and test set using the split function on the hold set. The test set is evaluated using the trained classifier. The clustering of attribute values is done using k-means algorithm. Each cluster will have a set of possible attribute values of a similar user profiles. The performance analysis of the k-means clustering algorithm is tested by varying the number of profiles as 10, 50, 100 and 1000. The time taken by the single node and multi-node setup to form clusters based on profiles in k-means clustering is compared and analyzed. The Fig. 3 shows that time taken is gradually increasing when number of profiles are increased in both single node and multi-node setup. The classification of user profiles is done using a Naïve Bayes algorithm. The process of classification using Naïve Bayes involves training and testing followed by the resulting classified instances. The time taken by the single node and multi-node setup to form classes for training the model based on the number of profiles is shown in Fig. 4. The performance of classifier at various threshold settings is illustrated in Fig. 5, using Receiver Operating Characteristics (ROC) curve. The threshold value varies from the range of 0.0 to 0.2. The lower the threshold value only less number of sybils are obtained with less false positives. Accuracy, Precision and Recall acquired are given in Table 1.

Accuracy of a classification lies on the important factors namely Fp - False positive, Fn - False negative, Tp - True positive and Tn- True negative and is shown in Eq. (16). The number of instances correctly and wrongly classified over the total number of given instances is evaluated. Precision is a measure calculated to determine the number of required classification done correctly and is shown in Eq. (17). Recall which is given as Eq. (18) is a measure calculated to determine the number of required instances that are required as the result of classification.

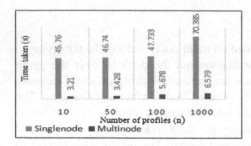

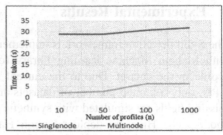

Fig. 3. No. of profiles vs Time taken for K-means clustering

Fig. 4. No. of profiles vs Time taken for training the dataset in Naïve Bayes

$$Accuracy = (Tp + Tn)/(Tp + Tn + Fp + Fn) \qquad (16)$$

$$Precision = Tp/(Tp + Fp) \qquad (17)$$

$$Recall = Tp/(Tp + Fn) \qquad (18)$$

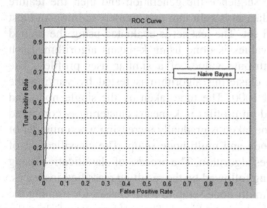

Fig. 5. ROC – FPR vs TPR

Table 1. Evaluation of classification results

Metrics	Value
Accuracy	0.94
Precision	0.92
Recall	0.91

5 Conclusion

The proposed Sybil detection framework uses three levels for identifying the fake users in OSN. The proposed framework is found to be memory efficient, computationally efficient and user friendly as it compares less number of profiles in the subsequent levels of the detection model. The Sybil detection framework identifies the suspected fake user by analyzing the attributes of the profile with the proposed privacy scheme. Scalable machine learning techniques are utilized for fake account detection. The performance also increases, when the number of training profiles is increased. In future, the image features will be taken along with text to identify sybils using deep learning.

Acknowledgments. The authors gratefully acknowledge DST, New Delhi for providing financial support to carry out this research work under PURSE scheme. I also thank Big Data Analytics Lab of MIT campus, Anna University for infrastructure and support.

References

1. Cluley, G.: Facebook: There are over 83 million fake accounts on our site [INFO-GRAPHIC]. Naked Security (2017). https://nakedsecurity.sophos.com/2012/08/02/fake-facebook-accounts/. Accessed 12 Jan 2017
2. Liu, S., Wang, S., Zhu, F.: Structured learning from heterogeneous behavior for social identity linkage. IEEE Trans. Knowl. Data Eng. **27**, 2005–2009 (2015)
3. HDFS Architecture Guide. https://hadoop.apache.org/docs/r1.2.1/hdfs_design.html. Accessed 12 Jan 2017
4. Gao, P.: Sybil frame: a defense-in depth framework for structure-based sybil detection. arXiv: 1503.02985v1 [cs.SI] Social and Information Networks (Cs.SI); Cryptography and Security (Cs.CR), arXiv: 1503.02985 [cs.SI] (2015)
5. Baldominos, A., Albacete, E., Saez, Y., Isasi, P.: A scalable machine learning online service for big data real-time analysis. In: IEEE Symposium on Computational Intelligence in Big Data (CIBD), pp. 1–8 (2014)
6. Adikari, S., Dutta, K.: Identifying fake profiles in LinkedIn. In: PACIS Proceedings, p. 278 (2014)
7. Shi, J., Xue, W., Wang, W., Zhang, Y., Yang, B., Li, J.: Scalable community detection in massive social networks using MapReduce. IBM J. Res. Dev. **57**, 12:1–12:14 (2013)
8. Boshmaf, Y., Logothetis, D., Siganos, G., Lería, J., Lorenzo, J., Ripeanu, M., et al.: Integro: leveraging victim prediction for robust fake account detection in large scale OSNs. Comput. Secur. **61**, 142–168 (2016)
9. Wang, Z., Liao, J., Cao, Q., Qi, H., Wang, Z.: Friendbook: a semantic-based friend recommendation system for social networks. IEEE Trans. Mob. Comput. **14**(3), 538–551 (2015). https://doi.org/10.1109/tmc.2014.2322373
10. Ji, Y., He, Y., Jiang, X., Cao, J., Li, Q.: Combating the evasion mechanisms of social bots. Comput. Secur. **58**, 230–249 (2016)
11. Zhou, X., Liang, X., Zhang, H., Ma, Y.: Cross-platform identification of anonymous identical users in multiple social media networks. IEEE Trans. Knowl. Data Eng. **28**, 411–424 (2016)
12. Feizy, R., Wakeman, I., Chalmers, D.: Are your friends who they say they are?: data mining online identities. Crossroads **16**, 19–23 (2009)

Design and Parameters Measurement of Tin-Can Antenna Using Software Defined Radio

R. Gandhiraj[1(✉)], K. P. Soman[1], Katkuri Sukesh[2], K. V. S. Kashyap[2], Karanki Yaswanth[2], and Kolla Haswanth[2]

[1] Center for Excellence in Computational Engineering and Networking,
Amrita School of Engineering, Amrita Vishwa Vidyapeetham, Coimbatore, India
r_gandhiraj@cb.amrita.edu
[2] Department of Electronics and Communication Engineering, Amrita School of Engineering,
Amrita Vishwa Vidyapeetham, Coimbatore, India
saisukeshredddy1010@gmail.com

Abstract. Antenna is a device which transmits and receives and electromagnetic signal. This paper suggests the making of a tin can antenna which is of low cost. The properties of the tin can antenna like radiation pattern, Half Power Beam width, Frequency response and also the linear polarization characteristics are measured in this paper using USRP with the association of a software interface, GNU Radio Companion. The first three properties were tested using a single tone signal and the polarization test was performed using a chirp signal.

Keywords: USRP · GNU radio companion · Can antenna

1 Introduction

An antenna is an electrical device which converts an electrical signal into radio waves and vice versa. When an alternating electric current flows through a conductor, it makes the antenna radiate electromagnetic waves, which is the transmitting characteristic. When an electromagnetic wave strikes a conductor, it induces an alternating current, which is the receiving characteristic. There are various types of antennas such as dipole antenna, aperture antenna, waveguide antenna, log periodic antenna, loop antenna, etc. A *can* antenna is one such type which can be designed using tin cans. It is a type of a circular waveguide antenna which can be designed as per the requirements.

Its characteristics can be found using Software Defined Radio (SDR) which enables implementing the components of signal processing like modulation, mixing and filtering using software on an embedded system or personal computer. The advantage of using this system is that it decreases the heavy dependence of hardware by providing all functionalities in SDRs [1]. The waveform is completely generated through software which is then transmitted. At the receiving end, the signal is subjected to few processes that are controlled by the software algorithms. An SDR system basically means a PC connected to a Digital-to-Analog Converter (DAC) and that is connected to an RF front end which in turn is connected to an antenna. This constitutes the transmission part. The reception part consists of an RF front end, which receives the signal through another or

I. Zelinka et al. (Eds.): ICSCS 2018, CCIS 837, pp. 558–568, 2018.
https://doi.org/10.1007/978-981-13-1936-5_58

the same antenna and is passed into the Analog-to-Digital Converter (ADC) and then to the PC. The signal processing and other related works are performed by the processor of PC. This kind of design produces a radio that receives and transmits widely large kinds of radio waveforms completely based on the software used.

The advantages of using SDR are many. First of all, it costs less than hardware. The open source software like GNU radio can be used. Second, if a change is required in the processing stacks or modulation schemes, an update in GNU radio would fix everything. Less mechanical parts are being used; hence there would be no wear and tear which might occur due to the ageing of the components and due to the environments in which they work. Third, new hardware components have to be replaced with the old ones, but here in GNU radio, just an update would suffice. Combined all together, it gives us a more efficient device in all terms.

Universal Software Radio Peripheral (USRP) is an FPGA based RF transceiver which is made up of Analog to Digital and Digital to Analog converters with RF front ends to transmit and receive signals. They are available from Ettus Research, a National Instruments Company which manufactures them [2]. It consists of one motherboard and can integrate with up to four daughter boards. In the motherboard, the analog to digital and digital to analog conversions happen. It uses USB or Ethernet connection to interact between the PC and USRP. Among them, Ethernet has the highest transfer of data. USRP that is used for this paper is USRP N210 which consists of Xilinx Spartan 3A DSP 3400 FPGA, an analog to digital converter that is used has a speed of 100 MS/s dual ADC, a digital to analog converter that has a speed of 400 MS/s dual DAC and an Ethernet interface to stream data between the processors. This USRP is designed to operate from DC up to 6 GHz signal.

The daughter board which was used is the CBX board. It is a full duplex, wide bandwidth transceiver that provides up to 100 mW of output power, and a typical noise figure of 5 dB. The local oscillators for receive and transmit chains operate independently, which allows dual-band and multi-frequency operation. It has 2 ports, one for TX/RX and the other for RX2 and provides 40 MHz of bandwidth. The frequency range is between 1.2 GHz and 6 GHz.

There is so many software which is available that could be used to interact with the USRPs like MATLAB, LabVIEW, etc. But GNU radio is known for its open source and is specially designed for working with an USRP [3]. Though it is not as user-friendly as MATLAB, it has many advantages over it. Firstly, it can be used for real-time analysis, while MATLAB is for offline. Secondly, its driver supports for USRP, while it is not that easy in MATLAB. Fourthly, the blocks are built using either Python or C++ code and they don't need to be compiled separately. Thirdly, it is a graphical tool, while MATLAB requires Simulink to work in graphical mode. Finally, GRC shows different colours for each of the links like float, complex etc., in graphical flow, while in Simulink, graphical flow shows all the links only in black colour. That is why GNU radio has a better hand over USRP than MATLAB.

2   Tin Can Antenna

With the integration of the software and the transceiver, the various parameters of the tin *can* antenna can be determined which are shown in this paper. The sweets which are sold by Haldiram's were used to design the antennas. They are available in almost any supermarket and each cost around INR 200. Along with the cans, SMA female connectors were utilised which are used to connect the antenna to the USRP via the SMA to SMA coaxial cable. Each cost around INR 120. In addition to that, 1 mm thickness of copper wire was used for the feed which costs around INR 40 for a length of 4 feet. So, all together, the antenna would cost a maximum of INR 350.

The dimensions of the antenna are,

Diameter of the *can* = D = 10 cm
Length of the *can* = L = 11.5 cm

In addition, the antenna is linearly polarised as the feed is a small copper strip which acts like a dipole antenna through an SMA connector.

2.1   Distance of Feed from Backwall

An electromagnetic wave has $1/r$ field attenuation and a phase shift as it transverses at a distance of r [4]. The formula for electric field is

$$E(r) = \frac{exp(-j * \beta * r)}{r} \tag{1}$$

Here $\beta = 2 * \pi */\lambda$ is the phase constant. If an electromagnetic wave travels one-quarter of a wavelength or $\lambda/4$, the phase shifts by $\pi/2$ radians or 90°. If the feed is placed at a distance of $\lambda/4$ from the back wall, an electromagnetic wave which is propagating towards the back wall will undergo a phase shift of 90° from the feed line to the back wall and another 90° phase shift after the wave is reflected.

From the back till the feed line, i.e., reflected wave adds with the direct wave, 90° + 180° + 90° = 360°. Thus, overall, it will undergo a phase shift of 360°, which will appear as if it has started from the feed itself.

Therefore, to enhance the radiation pattern in a particular direction, the feed is placed at a distance of a quarter of a wavelength from the back wall of the *can*. Since the electromagnetic wave first propagates within the waveguide, we have taken the guided wavelength into consideration for the calculation of this distance.

2.2   Length of the Feed

The length of the feed should be one-quarter the wavelength, as that is the wavelength for which there will be maximum radiation from the feed.

The resulted antenna is shown in Figs. 1 and 2. The former shows the side view and the latter shows the inside view.

Feed

Fig. 1. Side view of tin can antenna   **Fig. 2.** Inside view of tin can antenna

2.3   Caluclation

The frequency selected for the design of the antenna is in the range from 2.7 to 2.8 GHz. This range is selected because in that range, the disdrometer works and detects the size and velocity of raindrop [4]. Thus for designing the antenna, the frequency of operation was considered to be 2.75 GHz. The wavelength is given by,

$$\lambda = \frac{c}{f} \tag{2}$$

Which gives a value of 10.9 cm. The formula to find the cut-off wavelength is,

$$\lambda_c = 1.705 * D \tag{3}$$

This equates to 17.05 cm. Hence, the cut-off frequency is 1.759 GHz, given by the equation

$$f_c = \frac{c}{\lambda_c} \tag{4}$$

The guided wavelength is calculated using the formula,

$$\lambda_g = \frac{\lambda}{\sqrt{\left(1 - (\lambda/\lambda_c)^2\right)}} \tag{5}$$

It gives value of 14.194 cm. Hence, the length of the feed is 2.72 cm given by the formula,

$$l = \frac{\lambda}{4} \tag{6}$$

And the distance of feed from the back wall is 3.55 cm given by the formula,

$$d = \frac{\lambda_g}{4} \tag{7}$$

The dimensions of the *can* antenna are given in the Fig. 3.

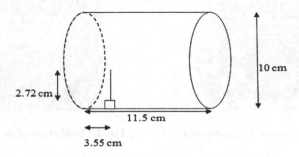

Fig. 3. Dimensions of antenna

2.4   Construction of Tincan Antenna

The antenna was constructed with a tin can, an SMA female connector, copper wire, cello tape and soldering kit. Firstly, the distance from the back wall is measured on the outside of the can, according to the calculations performed above, and marked. A hole is inserted into the marked point using a sharp object, which will be used to insert for the feed and then 4 other holes are inserted around the central hole for the SMA female connector to sit comfortably. Then, the required length of copper wire is cut, according to the calculations performed above and soldered into the central pin of the SMA female connector. This becomes the feed for the antenna. Cello tape is wound around the feed so that the tin can doesn't conduct by contact with the copper wire. It is then inserted into the hole made in the can and fixed rigidly using glue.

3   Experiments and Analysis

3.1   Radiation Pattern

The radiation pattern of the tin can antenna was found for both horizontal (H) and vertical (V) planes. It was performed by using an omnidirectional antenna to transmit a single tone signal using one USRP and received the signal using the tin can antenna [5].

The GNU Radio block diagrams used to transmit and receive the signal are shown in Figs. 4 and 5. The sampling rate of the entire experiment is set at 640 kHz. In the transmitting block diagram, a sine wave of 100 kHz frequency and 1 V amplitude is produced using the signal source block. This input signal is depicted in Fig. 6. It shows that it has a peak power of −5 dB. It is then sent to the UHD: USRP Sink block, which transmits the signal using a USRP at a frequency of 2.7 GHz and a gain of 30 dB. Simultaneously, the FFT of the signal is plotted using the WX GUI FFT Sink with a throttle block in between. The throttle limits the data throughput to the specified sampling rate, which is 640 kHz. It prevents GNU Radio from consuming all CPU resources when the graph is not being regulated by an external hardware. The amplitude, frequency of signal tone signal, and transmitter gain are varied using the WX GUI Slider.

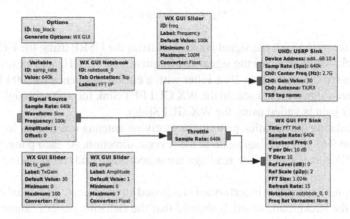

Fig. 4. Transmission flowgraph

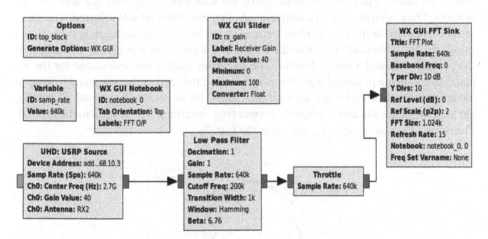

Fig. 5. Reception flowgraph

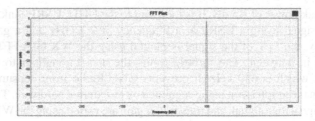

Fig. 6. Input signal

For the reception part, the signal is received from the USRP using the UHD: USRP Source block, which receives the signal at a frequency of 2.7 GHz and a gain of 40 dB. It is then passed through a Low Pass Filter with a cut-off frequency of 200 kHz

The filtered signal is then sent to the WX GUI FFT Sink for a plot through a throttle. The receiver gain is varied using the WX GUI Slider.

For calculating the radiation pattern, the receiving antenna was rotated for every 10° starting from 0 to 180 and then again in the reverse direction. At each point, the power value (in dB) is calculated. The readings are noted down and the radiation pattern is plotted.

Initially, the experiment is performed in a closed environment. The signal is received along with the background noise. It showed that the radiation pattern cannot be calculated accurately because of a lot of reflections which occur due to side walls, the human body, and other objects in the room. Also, the side lobes constitute the noise of the antenna. Thus, for a precise measurement, an open environment was selected. The same setup was placed at the centre of the terrace (to avoid side wall and ceiling reflections) using an adjustable stand such that the height of the antennas was greater the height of the parapet wall and a seated human. The radiation pattern was calculated for the H-plane and V-plane polarised waveforms by rotating the feed of the antenna. H-plane of the antenna is when the copper feed is horizontal with respect to the ground. Similarly, the V-plane of the antenna is when the copper feed is vertical with respect to the ground. The three radiation patterns are shown in the Fig. 7.

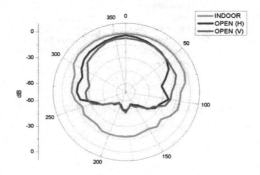

Fig. 7. Radiation patterns

This shows that the HPBW is around 60° which is the almost the same if calculated theoretically, which is given by,

$$HPBW = 58 * \lambda/D \tag{8}$$

which gives a value of 63.33°. This method can replace the conventional radiation pattern measurement techniques, where the accuracy of measurement is not necessary. The pattern's accuracy can be improved by connecting a motor and an electronic circuit to run it with the receiver antenna in order to change its angular position.

3.2 Frequency Response Measurement of Antenna

The frequency range of the antenna has also been experimented. It determines the working range the antenna, within the limitation of the frequency range of the daughter board used. The same setup and GNU Radio flow graph which was used for the radiation pattern has been utilized here. The only change in the flow graph is that that the WX GUI Slider used for frequency is now used to change the central frequency of the USRP Sink and Source.

It was observed that the antenna starts radiating at around 1.2 GHz, increasing till it becomes constant with maximum power at around 1.8 GHz to 3.5 GHz with minor deflections and then again starts dipping till it becomes negligible at 4.5 GHz. The frequency response is given as shown in the Fig. 8.

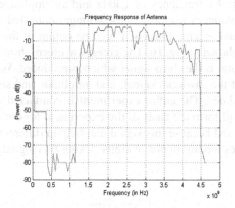

Fig. 8. Frequency response of antenna

3.3 Polarisation Test

The third experiment which was conducted was to find out whether the antennas show linear polarization as predicted. To find out, a linearly frequency modulated (LFM) or chirp waveform was used as the transmitted signal [5]. It is of the form

$$x(t) = \cos(\pi * \beta * t^2/\tau), 0 \le t \le \tau \tag{9}$$

Here the magnitude β represents the frequency span of the chirp, whose sign determines if the frequency of the signal is increasing (up chirp) or decreasing (down chirp) for the chirp duration of τ. This was implemented and checked using the following block diagram in Fig. 9.

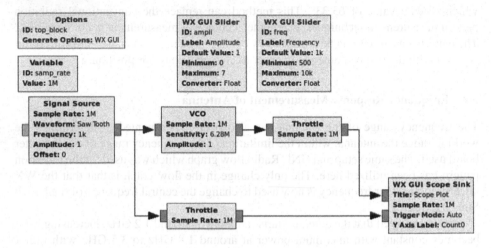

Fig. 9. Chirp signal generation flowgraph

A sawtooth signal of a frequency of 1 kHz and an amplitude of 1 V is generated which is used as the frequency response of the voltage controlled oscillator (VCO) with a sensitivity of 1 M and an amplitude of 1 V. The output of the VCO is the required chirp signal which is shown in Fig. 10. The green waveform shows the saw tooth signal and the purple waveform is the chirp signal which is produced. We can vary the amplitude and frequency of the sawtooth signal, and amplitude and sensitivity of the VCO using the WX GUI Slider. The entire experiment is conducted at a sampling rate of 1 MHz. This signal will form the basis of the experiment for the formation of the dual polarised radar. It is used because this signal can help in differentiating between chirp waveforms with slopes having opposite signs. Using this signal, the polarisation test

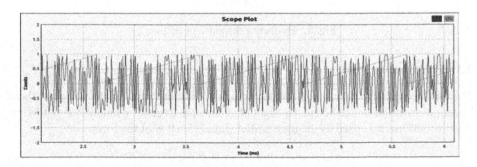

Fig. 10. Chirp signal (Color figure online)

was performed. The GNU Radio block diagram for this testing experiment is shown in (Fig. 11).

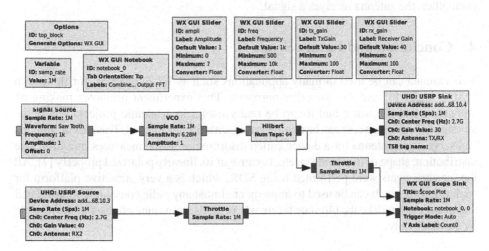

Fig. 11. Polarisation test flowgraph

The chirp signal is sent to USRP using the UHD: USRP Sink. This helps the antenna to transmit. It is also plotted with respect to time using the WX GUI Scope Sink through a Throttle block. Another USRP receives the signal via another *can* antenna. The received signal is plotted alongside the first plot by sending it to the same WX GUI Scope Sink as before through a Throttle block. The whole experiment is performed at the same sampling rate as before, that is, at 1 MHz. And thewhole setup for the experiment is performed at the same place as it was performed for the radiation pattern. The waveforms plotted are depicted in Figs. 12 and 13. Both graphs are comparison time plots between the transmitted signal and received signal. The former shows the when both the feeds are placed parallel to each other, that is, they are not polarised with respect to each other. The latter shows when one feed is perpendicular to another, that is, one antenna is orthogonally polarised with another. The green and blue represent the transmitted signal and the red and purple represent the received signal, which has almost zero amplitude.

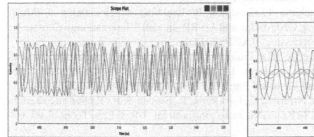
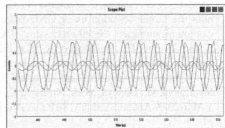

Fig. 12. Graph when both antennas have zero polarisation (Color figure online)

Fig. 13. Graph when both antennas are orthogonally polarised (Color figure online)

This shows that when the antennas are orthogonally polarised, the receiving antenna receives almost no input. And when they have zero polarisation angle with respect to each other, the antenna receives a signal.

4   Conclusion and Future Scope

This antenna can be used in many applications such as Wi-Fi antenna, for the use in FMCW radar [6] and various other purposes. This experiment promotes making an antenna out of the waste and it can be really useful in academic projects where the antenna of high range cost can be compensated with tin can antenna. This tin can antenna can serve as an antenna for a device called disdrometer which measures rain drop size distribution, shape of the raindrop, etc. because of its linearly polarized property [7]. All the measurements were performed using SDR, which is a very attractive platform for such experiments. It can be used to implement almost any radio communication-related technology and will be the ultimate future where all the communications will be possible by software only.

Acknowledgement. The authors would like to thank Mr. Vinod (CEN lab assistant) and Mr. Thomas (Fabrication lab) of Amrita Vishwa Vidyapeetham University for being so helpful throughout our project. We also thank our seniors NirmalAagash C, Madavan R V, Akshay T Shankar for their put forth effort in the project which helped us as a reference.

References

1. Fähnle, M.: Software-Defined Radio with GNU Radio and USRP/2 Hardware Frontend: Setup and FM/GSM Applications
2. USRP N210 Software Defined Radio (SDR)-Ettus Research. https://www.ettus.com/product/details/UN210-KIT
3. Seeder, B.: SDR Tutorials. https://files.ettus.com/tutorials/
4. Bringi, V.N., Seliga, T.A., Mueller, E.A.: First comparisons of rainfall rates derived from radar differential reflectivity and disdrometer measurements. IEEE Trans. Geosci. Remote Sens. **GE-20**(2), 201–204 (1982)
5. Gandhiraj, R., Soman, K.P.: Modern analog and digital communication systems development using GNU Radio with USRP. Telecommun. Syst. **56**(3), 367–381 (2014)
6. Sundaresan, S., Anjana, C., Zacharia, T., Gandhiraj, R.: Real time implementation of FMCW radar for target detection using GNU radio and USRP. In: 2015 International Conference Communications and Signal Processing, ICCSP (2015)
7. Ajitha, T., Joy, E., AnishJoyce, A., Gandhiraj, R.: Radiation pattern measurement of log-periodic antenna on GNU radio platform. In: 2014 International Conference on Green Computing Communication and Electrical Engineering (ICGCCEE), pp. 1–4 (2014)
8. Sreethivya, M., Dhanya, M.G., Nimisha, C., Gandhiraj, R., Soman, K.P.: Radiation pattern of YAGI-UDA Antenna Using USRP on GNU radio platform. Int. J. Res. Eng. Technol. **03**, 69–71
9. Archana, S., Sugatha Kumari, P.R., Vishakh, A., Gandhiraj, R., Soman, K.P.: Low cost RADAR design using GNU radio. In: International Conference on Singal and Speech Processing (ICSSP 2014), TKM College of Engineering, Kollam (2014)

Clustered Heed Based Cross Layer Routing Scheme for Performance Enhancement of Cognitive Radio Sensor Networks

S. Janani[1($\boxtimes$)], M. Ramaswamy[2], and J. Samuel Manoharan[3]

[1] Department of Electronics and Communication Engineering,
A.V.C College of Engineering, Mannampandal 609305, Tamil Nadu, India
jananiphdl5@gmail.com
[2] Department of Electrical Engineering,
Annamalai University, Annamalai Nagar 608002 Tamil Nadu, India
[3] Department of Electronics and Communication Engineering, Bharathiyar
College of Engineering and Technology, Karaikal 609 609, Tamil Nadu, India

Abstract. The process of data transfer owes allegiance to the dependence on the transmission power, fading and interference with the licensed users and thus invites alternative approaches for routing the data in a cognitive radio sensor network (CRSN). The paper develops a cross layer scheme in an attempt to inhibit the scarcity in the use of the spectrum and allows the formation of the clusters in the network and-engages the choice of the cluster head (CH) based on the CHEV value. Owing to the fact that the length of the packet plays a significant role, the formulation incites a dynamic packet size optimization procedure and follows it up with a forward error correction technique. The results obtained using NS2 simulation illustrate a superior QOS indices for the HEED protocol based routing techniques and thus claim a place for its use in real world applications.

Keywords: Cross layer routing · Cognitive radio sensor networks
Dynamic packet size optimization · Forward error correction

1 Introduction

The cognitive radio enabled wireless sensor network (WSN) appears to evolve as a new medium for the transmission of the data. It benefits to reduce the congestion and excessive packet loss which in turn increases the reliability in transmission [1]. However the challenges in the WSN increases the complexity of the spectrum controlling in cognitive radio sensor network (CRSN) and to tackle this problem, the sensor nodes empowered with cognitive radios speculatively access manifold alternative channels.

A CRSN forms a multi-channel system with two main differences from the conformist WSN. The first difference orients to the fact that while the WSN operates with a fixed number of channels, it varies in a CRSN. The other relates to the architecture where in the nodes occupy a set of channels in the WSN, whereas in CRSN, a set of channels surround each node.

© Springer Nature Singapore Pte Ltd. 2018
I. Zelinka et al. (Eds.): ICSCS 2018, CCIS 837, pp. 569–583, 2018.
https://doi.org/10.1007/978-981-13-1936-5_59

The innovative wireless products and services experience the delinquent of spectrum scarcity and static spectrum allocation to a particular application creates a room for unused spectrum in a licensed bands [2, 3]. The unused spectrum of the licensed band, called as the spectrum hole [4] belongs to a frequency band which may be allotted to a licensed user and remain unutilized at specific interval or position.

The federal communications commission (FCC) provides rights to the unlicensed or secondary users (SUs) to custom that spectrum holes in collaboration with that primary users (Pus) or licensed users without affecting the Pus spectrum [5, 6]. The dynamic spectrum access offers to be a promising technology which improves the overall spectrum usage effectively in both unlicensed and licensed crews [7]. It facilitates the cognitive radio allowed wireless devices to acclimate to the spectrum distribution so refining the spectrum usage.

The theory of clustering facilitates the routing methods to introduce self organized cell structures [8]. The advantages of clustering which moderates the communication overhead charge and growths the dependability of transmission. Besides the clustering allows a designed way to succeed the topology excellently and increases the system stability and capacity.

1.1   Related Works

A hefty number of MAC procedure have been used with the WSN completed the last epoch [5, 7–10] and the solutions seem to be unsuitable for the CR based WSN'S. The primary consumers in the CRSNs have been assigned the significance to access the spectrum in order to maintain the synchronization and warrant the detection of the occurrence of PU's.

A cluster based MAC procedure for CRAN has been introduced to form the clusters based on the environmental positions of nodes, obtainable channels and knowledgeable statistics to sustain the cluster permanency. The spectrum occupancy details have been stored in database to be useful in case of the neighbour discovery. The CH in each cluster has been assigned to broadcast a encouragement packet that covers the cluster regulator information. The channel access schedule has been defined for making the nodes that do not participate in communication to perform the neighbour discovery [9].

Cog-mesh has been known to be a cluster based MAC procedure for CRN that speculatively operates different spectrum holes for smooth earl to earl communication. The clusters have been formed with nodes that shares the local public channels called the master channel and invites the neighbouring nodes to join the cluster [12].

A schedule based KoN-MAC protocol used for multi-hop CRSN has been seen to select an interference free channel dynamically for making the adjacent clusters with different channels to mitigate the multi-channel hidden terminal [11].

A light load dynamic packet length control (DPLC) structure has been suggested for WSNs and its QOS parameters measured [14]. A packet length adaptation structure commissioning error approximating codes for low power sensor networks has been detailed [13].

Owing to reasons that energy efficient optimization scheme becomes essential for CRNs with the resource constrained sensor nodes, the efforts endeavour to benign suitable modifications in the process of packet delivery.

2 Problem Statement

The emphasis owes to form a network where it allows the selection of the CH based on CHEV and involve a dynamic packet size optimization procedure at the MAC layer along with an error correction method to enhance the throughput without affecting the energy efficiency. The efforts owe to examine its performance through simulation and relate the comparative merits of the routing method under study in terms of indices.

3 System Model

The exercise involves a CRSN with N node mentioned to as SU's, M data channels, primary users (PU's) and a sink node as seen from Fig. 1. Each fixed node K ε N identifies his position, represented as (x_k, y_k) and the procedure allows the communication to start between the two nodes only if their distance lies greater than the euclidean distance. It transfers the control information through a common regulator channel and the data complete the data channel.

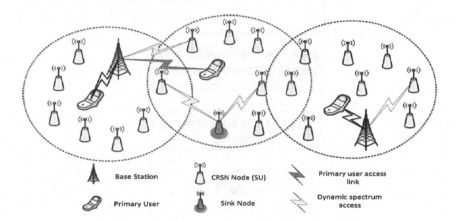

Fig. 1. System model of CRSN

4 Proposed Methodology

The attempt revolves around the formulation of a cluster constructed HEED strategy to route the data over a dynamic packet size optimization with forward error correction in a CRSN. The goal endeavours to maximize the throughput of the system with constraints on QoS and illustrate its merits for Dypktoec-HEED techniques.

The scheme assumes that a SU can procedure individual one channel at a time to transmit the data at a fixed packet transmission rate. Besides the SU can transmit on the channel only if it remains unoccupied and engages a variable packet size in each cycle depending on the status of the available channels.

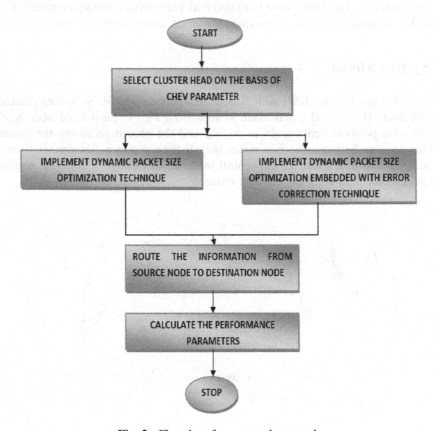

Fig. 2. Flowchart for proposed approach

The methodology explained through the flowchart in Fig. 2 forms the clusters based on availability of the information regarding the neighbouring nodes and thereafter reconstructs the clusters to form an interconnected network. The philosophy selects the CH selected based on CHEV parameter from among the available nodes and follows the Eqs. 1 and 2.

$$P_q^* = max_{Pq}(CHEV_{P_q}), 1 \leq P_q \leq Cm_q \qquad (1)$$

$$CHEV_{P_q} \propto N_{P_q}^{CC_{P_q}} \qquad (2)$$

Where

P_q indicates the node P in cluster q and

N_{P_q} is the total number of neighbouring nodes to node P in cluster q

CC_{P_q} is the total number of common channels for node P in cluster q

The CHEV parameter selects the node with the uppermost number of public channels and neighbours. It creates flexibility for the PU and the spectrum mobility allows reducing the big number of clusters in the network. Besides it ensures that the CH operates at a lower speed among the cluster participants in the network to circumvent normal re clustering in the network.

The Eqs. 3 and 4 define the constant of relationship

$$CHEV_{Pq} = nf_{Pq} x N_{P_q}^{CC_{P_q}} \tag{3}$$

$$nf_{Pq} = \frac{n_{Pq}}{\sum_{Pq} \gamma_{P_q}} \tag{4}$$

$$where \quad n_{Pq} = \frac{E_{P_q}^{\alpha}}{V_{P_q}^{\beta}} \alpha, \beta \, \varepsilon \, R \, and \, 0 < \alpha, \beta \leq 1$$

Where nf_{Pq} represents a normalization factor to indicate how powerful and static the node Pq exists relative to the additional nodes in the cluster n_{Pq}. It always holds a positive value and brings out the relationship between the node energy and speed. The design parameters α and β serve to prioritize the energy and speed constructed on the application condition. If energy assumes a higher priority, then α becomes larger than β and vice versa.

The process takes the log of CHEV to avoid a big CHEV value as a final collection metric of CHs. The Eq. 5 gives the optimized CH selection

$$m_n^* = \max_{m_n} (\log(nf_{m_n} x N_{m_n}^{CC_{m_n}})) \tag{5}$$

Hence a node with the uppermost log (CHEV) value methods the cluster and develops the CH. If the log (CHEV) value of the node CR_m falls to be smaller than that of its neighbour, then CR_m becomes the cluster member of the neighbour with large log (CHEV) value. The cluster member in the cluster arrange themselves built on log (CHEV) value for the selection of reserved cluster head (RCH). Whenever the current CH moves out, the RCH takes control of the cluster so that it avoids the possibility of re-clustering.

The succeeding define the cluster development process.

1. Start

2 . $m_n \, \varepsilon \, CM_n$ | m_n start broadcasting ACL_{mn};

3. m_n receives ACL_{pn}, where $m_{n \neq} P_n$;

4. m_n construct partitions and calculate $CHEV_{mn}$;

5. $m_n = \max_{m_n} (\log(nf_{m_n} \times N_{m_n}^{cc_{m_n}}))$;

6. m_n exchange $CHEV_{mn}$ with neighbours;

7. if $CHEV_{mn} > CHEV_{Pn}$ | $m_n \neq P_n$ then

8. $|CH_n = m_n$ /* m_n becomes cluster head*/

9. else

   $CM_n = m_n$; /* m_n develops cluster member*/

10.   if m_n collects any other ideal then

       $GW_n = m_n$; /*$m_n$becomes gateway*/

12.end

13.   end

14.END

  Where

  P, q $\rightarrow$ possible integer (1, 2, 3 ...)
  P_q $\rightarrow$ p member of cluster q
  N_{pq} $\rightarrow$ neighbour set of cluster q
  CM_q $\rightarrow$ Cluster Member of cluster q
  CH_q $\rightarrow$ Cluster head of cluster q
  $ACL_{pq(t)}$ $\rightarrow$ Accessible channel gradient of p_q at time t
  GW_q $\rightarrow$ Gateway of Cluster q

The CH receives the material from the assigns and nodes the node with the uppermost CHEV as a GW with every specific cluster to continue the inter-cluster communication.

4.1 Dypsoc Protocol Design

The node follows an asynchronous duty cycle approach where every cycle involves of sleeping and wakeful phases. The wakeful phase can be distributed into three sub phases namely channel negotiation phase, spectrum sensing phase and data transmission phase.

When a node requires transmitting data, it tests the spectrum, prepares a list of offered channels and sends a short introduction message on CCC. A introduction message covers terminus ID and the location $(x_k, y_k$ where $k \varepsilon N)$. The nodes on receiving the preamble messages allow the transmitter/receiver pair to enter the channel arbitration phase. It decides the packet size and channel for data transmission to be forwarded along with the acknowledgement (ACK) packet sent to the transmitter node, so that the transmission activates on the designated channel.

It lastly sends the end of data (ENDd) transmission concluded CCC for updating the channel. The ENDd identifier contains the modern sensing information found by the node through transmission. However in the event of the channel being un available, then the nodes sends an acknowledgement representing the no channel (ANC) obtainable on CCC and enables the transmission to be stretched till the next resulting cycle.

The packet transmission postponement communicated as a sum of waiting time of packet when the node remains sleeping, time expended terminated control channel, T_{CC}, time to spend over data channel T_{DC}. The Eq. 6 relates the delay

$$\int_0^{\zeta_0 T} (\zeta_0 T - t) \frac{1}{\Gamma} e^{-\mu dt} + E[T_{cc}] + \zeta_3 T \leq Delay_{max} \tag{6}$$

Where

ζ_3 indicates the immobile possibility of node being in transmit state.

$\zeta_3 T$ represents the average time expended in data transmission during every cycle

The energy efficiency, denoted as $R(P_S^K, CS_i)$ expresses the ratio of the real energy expended in transmitting one packet terminated the total energy spent in transmitting or receiving a packet as in Eq. 7

$$P_S = \left[P_S^{1,...P_S^{max}} \right] \rightarrow Different\ Packet\ Size\ Values$$

$$CS_i[CS_1, ...CS_M) \rightarrow Different\ set\ of\ channels$$

$$R(P_S^K, CS_i) = \frac{E_{SPKT}(\tau_3(P_S^K, CS_i)P_{STRANS})}{E_{SPKT} + E_{TCI}(P_S^K)} \tag{7}$$

Where

P_S^K denotes a function for packet size
CS_i shows the selected channel
E_{SPKT}, the energy consumed for efficacious transmission

P_{STRANS}, the prospect of efficacious transmission

$E_{TCI}\left(P_S^K\right)$, the energy spent for transmitting control information over channel

The problem engages the difficulty of the channel selection scheme and packet size optimization to be resolved by linear programming.

4.2 FEC Based Co-operative Identifying in a Noisy Cognitive Radio Sensor Network

The data in its passage to the fusion centre over the control channel [13] may experience noise interference and introduce errors in the received signal. It increases the error possibility which in turn lessens the detection possibility of the system. The forward error correction (FEC) encoder integrates termination to the message of every CR user earlier transmission.

It allows the use of convolution coding techniques for the channel encoder to diminish the noise possessions in the control channels and enables the data transmission from each cognitive radio user to the fusion centre using (n_{opb}, K_{ipb}, L_b) with the code rate $\left(\frac{K_{ipb}}{n_{opb}}\right)$ and constant length L_b

Where K_{ipb} is the number of input bits
n_{opb} is the number of output bits
The polynomial illustration of the code can be carved as in Eq. 8

$$[1 + b + b^2 + b^3 + b^6, 1 + b^2 + b^3 + b^5 + b^6] \tag{8}$$

The scheme uses quadrature amplitude modulation (4QAM) to map the coded categorization into points in the 2D plane. It unsamples the mapped categorization f(i) to decrease the effects of the inter symbol interference (ISI). The Eq. 9 gives the un sampled sequence g(i) as

$$g(i) = \begin{cases} f\left(\frac{1}{K_i}\right) & if \frac{1}{K_i} \text{ is an integer} \\ 0 & \text{Otherwise} \end{cases} \tag{9}$$

The finite impulse response root raised cosine filter (FIR RRC) with impulse reaction $h_i(t)$ in Eq. 10 shapes the pulse for the up sampled sequence [15] and transmits the filtered signal under AWGN with changed signal to noise fractions in the control channel (SNR-C).

$$h_i(t) = \frac{4\alpha}{\pi\sqrt{T_S}} \frac{\cos\left(\frac{(1+\alpha)\pi t}{T_S}\right) + \frac{T_S}{4\alpha t}\sin\left(\frac{(1-\alpha)\pi t}{T_S}\right)}{1 - \left(\frac{4\alpha t}{T_S}\right)^2} \tag{10}$$

Where
α is the roll-off factor
T_S is the symbol period

The viterbi technique decodes the convolution code and calculates the bit error rate (BER) at the fusion centre for the classification engendered with and without convolution coding using the formulation in Eqs. 11 and 12.

$$Q_{det-wc} = C_{d(Wgc)-\psi} - \theta_{b-wc} * \xi \tag{11}$$

$$Q_{det-woc} = C_{d(Wgc)-\psi} - \theta_{b-woc} * \xi \tag{12}$$

The above equation can be rearranged through Eqs. 13 and 14

$$Q_{det-wc} = C_{d(Wgc)-\varnothing} - \theta_{b-wc} * \xi \tag{13}$$

$$Q_{det-woc} = C_{d(Wgc)-\varnothing} - \theta_{b-woc} * \xi \tag{14}$$

Using cyclic mouth detection, the expressions can be expressed as in Eqs. 15 and 16

$$Q_{det-wc} = C_{d(Wgc)-\varnothing 1} - \theta_{b-wc} * \xi \tag{15}$$

$$Q_{det-woc} = C_{d(Wgc)-\varnothing 1} - \theta_{b-woc} * \xi \tag{16}$$

Where

$C_{d(Wgc)-\psi}$, $C_{d(Wgc)-\varnothing}$ and $C_{d(Wgc)-\varnothing 1}$ account the co-operative detection possibilities of entropy estimation and energy entropy using cyclic article detection.

θ_{b-wc} and θ_{b-woc} denote the bit error rates without and with coding ξ is the number of iterations.

5   Simulation Results and Discussions

The procedure simulates the model using NS2 simulator network with two hundred and fifty nodes distributed in a space of 1000 m × 1000 m and remains free from energy dissipation in the idle or carrier sensing mode. The nodes form the clusters and elect the CH based on the energy level of the nodes. It urges to carry the data between the chosen three source and target nodes in the network.

The Table 1 includes the simulation and parameters.

Table 1. Simulation parameters

Simulation parameters	Value
Sensor nodes (Numbers)	250
Communication range	250 m
Size of the Buffer	50 packets
Size of the Packet	1000 to 5000
Power (Transmission)	0.8 W
Power (Reception)	0.6 W

The exercise reflects the benefits of the dynamic packet size optimization and later with error correction in the process of routing the data over a packet size ranging from 1000 to 5000 for the HEED protocol. It explains an increase in the PDR and throughput values as seen from the entries in Tables 2 and 6 and the decrease in overhead, delay, packet loss and energy as observed from Tables 3, 4, 5 and 7.

Table 2. PDR (%) vs number of packets

Protocol/no. of pkts	1000	2000	3000	4000	5000
HEED-Dypsoc	99.83	98.72	96.40	93.76	92.06
HEED-Dypsoc &EC	100.00	99.15	97.43	94.96	92.64

Table 3. Overhead vs number of packets

Protocol/no. of pkts	1000	2000	3000	4000	5000
HEED-Dypsoc	0.93	2.76	3.64	4.81	5.09
HEED-Dypsoc &EC	0.43	1.76	2.64	4.21	4.29

Table 4. Delay (seconds) vs number of packets

Protocol/no. of pkts	1000	2000	3000	4000	5000
HEED-Dypsoc	5.28	6.75	7.77	8.34	9.29
HEED-Dypsoc&EC	4.57	6.44	6.76	7.33	8.29

Table 5. Packet loss (%) vs number of packets

Protocol/no. of pkts	1000	2000	3000	4000	5000
HEED-Dypsoc	0.17	1.28	3.60	6.24	7.94
HEED-Dypsoc&EC	0.00	0.85	2.57	5.54	7.36

Table 6. Throughput (Kbps) vs number packets

Protocol/no. of pkts	1000	2000	3000	4000	5000
HEED-Dypsoc	16.01	29.51	42.98	54.84	69.07
HEED-Dypsoc&EC	16.71	30.33	43.80	55.64	70.07

Table 7. Energy (J) vs number packets

Protocol/no. of pkts	1000	2000	3000	4000	5000
HEED-Dypsoc	1.62	5.67	11.89	13.72	17.20
HEED-Dypsoc&EC	0.62	4.67	10.89	12.72	16.20

The Fig. 3 compares the bar chart for the PDR, defined as the packets successfully transmitted between source and destination when the model transmits the packets under normal routing, with the dynamic packet size optimization and with error correction obtained through HEED routing to bring out the relative merits.

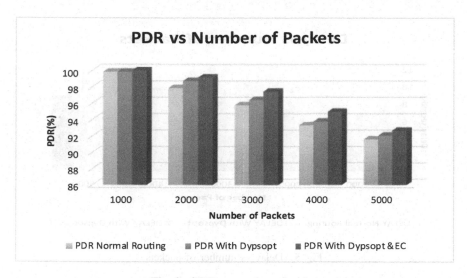

Fig. 3. PDR vs number of packet

The bar chart in Fig. 4 drawn for the overhead, defined as the number of control packets generated for signalling and packets received at destination with the number of packets for HEED displays the decrease in the number of control packets for the transfer with both packet size optimization and error correction when compared with

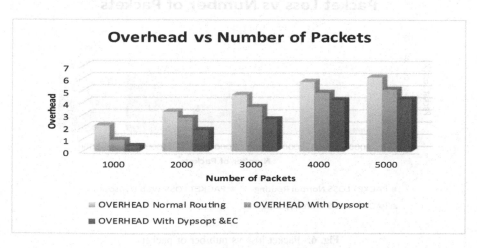

Fig. 4. Overhead vs number of packets

the other two approaches. The Fig. 5 shows the smaller Delay, defined as the ratio of sum of every packet delay to the number of packets received with the number of packets for HEED in the event of data transmission using dynamic packet size optimization with error correction.

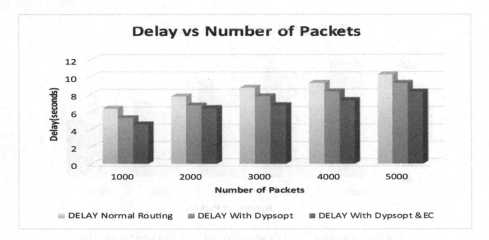

Fig. 5. Delay vs number of packets

The variation of Packet loss, defined as the packets not received by destination due to packet drop with the number of packets for HEED seen from the bar diagram in Fig. 6 displays a lower value for the case with packet size optimization along with error correction.

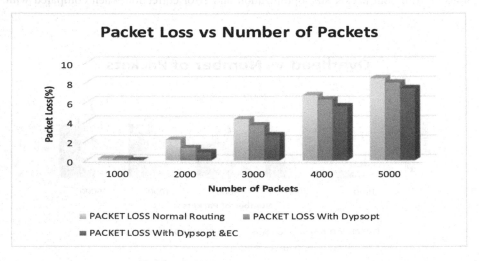

Fig. 6. Packet loss vs number of packets

The Fig. 7 depicts a higher throughput, defined as the ratio between the packet size transmitted in a particular range across a range of packet sizes for HEED for the case with dynamic packet size optimization and error correction. The energy, defined as the consumption of energy between transmission and reception of packets in a network related with the number of packets for HEED related in Fig. 8 exhibits a progressive decrease and claims to be the lowest for the formulation with packet size optimization and error correction.

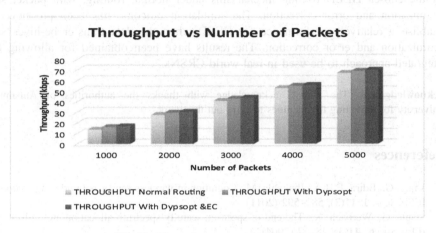

Fig. 7. Throughput vs number of packets

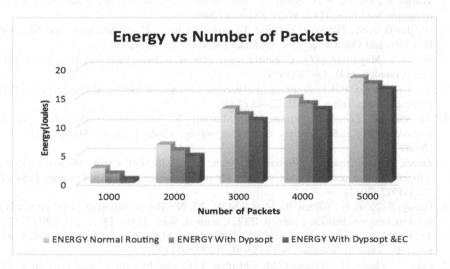

Fig. 8. Energy vs number of packets

582   S. Janani et al.

6 Conclusion

A cluster based routing scheme has been developed for a cross layered CRSN and the performance examined with three different methods. The formulation has been enticed with dynamically optimizing the packet size and further incorporated with an error correction technique. The methodology has been evaluated through simulation on a 250 node network model over a range of packet sizes. The indices have been computed for the chosen HEED routing mechanisms under normal routing, with packet size optimization and error correction. The simulation outcome has been presented to establish a relatively performance for HEED when the data transfer be-hives size optimization and error correction. The results have been obtained for allowing the integrated approach to be used in real world CRSNs.

Acknowledgment. The authors acknowledge with thanks the authorities of Annamalai University for providing the facilities to carry out this work.

References

1. Vijay, G., Bdira, E.B.A., Ibnkahla, M.: Cognition in wireless sensor networks: a perceptive. IEEE Sens. J. **11**(3), 582–592 (2011)
2. Staple, G., Werbach, K.: The end of spectrum scarcity[spectrum allocation and utilization]. IEEE Spectr. **41**(3), 48–52 (2004)
3. FCC spectrum policy task force, report of the spectrum efficiency working group (2002). http://transition.fcc.gov/sptf/files/SEWGfinalreport_1.pdf
4. Tandra, R., Mishra, S.M., Sahai, A.: What is the spectrum hole and what does it take to recognize one? Proc. IEEE **97**(5), 824–848 (2009)
5. Wyglinski, A.M., Nekovee, M., Hou, T.: Cognitive Radio Communications and Networks: Principles and Practice, p. 736. Academic Press, Cambridge (2009)
6. Mitola, J., MaguireJr, G.Q.: Cognitive radio: making software radios more personal. IEEE Pers. Commun. **6**(4), 13–18 (1999)
7. Yu, J.Y., Chong, P.H.J.: A survey of clustering schemes for mobile ad hoc networks. IEEE Commun. Surv. Tutor. **7**(1–4), 32–48 (2005)
8. Akyildiz, I.F., Lee, W.Y., Vuran, M.C., Mohanty, S.: Next generation/dynamic spectrum access/cognitive radio wireless networks:a survey. Comput. Netw. **50**(13), 2127–2159 (2006)
9. Zareei, M., Taghizadeh, A., Budiarto, R., Wan, T.C.: EMS-MAC energy efficient contention – based medium access control protocol for mobile sensor networks. Comput. J. **54**(12), 1963–1972 (2011)
10. Huang, P., Xiao, L., Soltani, S., Mutka, M.W., Xi, N.: The evolution of MAC protocols in wireless sensor networks: a survey. IEEE Commun. Surv. Tutor. **15**(1), 101–120 (2013)
11. Yigitel, M.A., Incel, O.D., Ersoy, C.: QOS aware MAC protocols for wireless sensor networks: a survey. Comput. Netw. **55**(8), 1982–2004 (2011)
12. Chen, T., Zhang, H., Maggio, G.M., Chlarntac, I.: Cogmesh: a cluster based cognitive radio network. In: 2nd IEEE International Symposium on New Frontiers in Dynamic Spectrum Access Networks. IEEE, pp. 168–178 (2007)

13. Li, X., Hu, F., Zhang, H.: A cluster – based MAC protocol for cognitive radio adhoc networks. Wirel. Pers. Commun. **69**(2), 937–955 (2013)
14. Dong, W., et al.: Dynamic packet length control in wireless sensor networks. IEEE Trans. Wirel. Commun. **13**(3), 1172–1181 (2014)
15. Xu, Y., Wu, C., He, C., Jiang, L.: A cluster based energy efficient MAC protocol for multi-hop cognitive radio sensor networks. In: IEEE Global Communications Conference (GLOBECOM), pp. 537–542 (2012)

Survey on Multiprocessor System on Chip with Propagation Antennas for Marine Applications

A. Benjamin Franklin[✉] and T. Sasilatha

Department of EEE, AMET Deemed to be University, Chennai, India
benfrank234@gmail.com, deaneeem@ametuniv.ac.in

Abstract. In this paper, our objective to design a system on chip with multiple numbers of antennas for high speed and efficient transmission of data in marine communication fields. This survey paper explains many in-depth problems to design a multiprocessor system on chip with antennas for to improve the communication and integration of electronics in marine based equipments. This paper analyze the performance of various design and it gives clear idea about how to design a low power, low frequency, noise less, low return loss and path loss chip antenna with high gain, high directivity and less complicated properties.

Keywords: Antennas · Directivity · Marine based equipments · Return loss
System on chip

1 Introduction

The communication problems can be resolved by using Network-on chip (NoC). The communication issue which is been found in the radio medium for on-chip communication are enabled using antennas and transceivers. The attenuation of the system introduced by the wireless channel since the magnetic force waves are propagated in lossy semiconducting material, the ability attributable to the wireless communication represents a vital contribution of the whole communication.

The larger area multiprocessor system on chip is not scalable on topology based design. These interns increase the latency, power consumption and hop count while transferring data. The limitation of this problem has been countered as to separate the topologies into the smaller subnets. To improve the performance and energy efficiency of the system especially for this case of problems the network on chip architecture introduces the on chip antennas [3]. The main benefits to use the propagation and bidirectional antennas are ability to transmit simultaneously across multiple wireless channels instead of single channel in the network.

Networks-on-Chip (NoCs) are throw out for viable result of on-chip communication problem. However, the two-dimensional (2D) architectures limit the performance of NoCs due to its property, restricted floor designing selection and world interconnects and this problem overcome by the three Dimensional network on-chips (3D - NoCs) were it emerged as sensible and power full infrastructure. They proffer specific development in performance by reducing average length and delay of interconnects.

© Springer Nature Singapore Pte Ltd. 2018
I. Zelinka et al. (Eds.): ICSCS 2018, CCIS 837, pp. 584–592, 2018.
https://doi.org/10.1007/978-981-13-1936-5_60

2 Related Work and Novel Contribution

This study of design deals with two major fields they are "Antennas and wave propagation" and "Wireless network on chip (WiNOC)". Normally, the antennas are very useful for long range distance communications and the performance of throughput cannot be degrade. An antenna is a key component in wireless communication and global positioning system. The miniaturization of the antenna has been accomplished by various forms which may include high dielectric constant substrate modifications. Wireless network on chip (WiNOC) improves the quality and standard of transmitting data with high energy efficiency.

This is needed in numerous applications like remote sensing, medicine application, mobile radio, satellite communication etc. It analyse the impact of a recent substrate propagation technique for on-chip communication on WiNOC with antennas.

An advance in recent on–chip technology makes it potential to utilize wireless antennas on a wafer die. So that many works are given in the study to use the efficient output of long ranged communication links by the sort of millimeter-wave over the standard wired Network-on-Chips architecture [4]. To enhance the result and performance of each normal and worldwide information can be forwarded by using hybrid wired-wireless NoCs.

The main problems of Wireless communications are known that the data forwarding reliableness of the broad wireless communication channel. To deal this problems and comfortableness of the on-chip network architecture the technology has been proposed by implementing a joint interference error correction technique and quadruple error detection technique with in the wired infrastructure links and performing code primarily with the product codes of WiNoC Links with carbon nano tubes (CNT) transceivers [4]. It had been incontestable that, the solidness of the wireless communication channel might be enhanced.

Some routing algorithmic programs enhance the on-chip communication needablity and histrionics by choosing less congested channels between the routers and those algorithm are contemplate for congestion management in three-dimensional on-Chip architectures [8]. That transmits the datum and packets with efficiency towards network area with reduced congestion to avoid network traffic. This issue can be implemented by few steps. Firstly, the algorithmic program chunks world congestion data in every layer of the communication network, after that, the algorithmic program needs the received data to mention a best channel for sending a packets by dimension-order routing algorithm. Global congestion ethics are transmitted by embedded congestion channel data in header zip of packets [2]. Moreover, flip models are based on deadlock-freedom in planned routing algorithmic program.

3 Various Performance Evolution

This section explains various architecture of network on chip properties with benefits and drawbacks of the design.

3.1   Power Transmission Technique

This power transmission technique presents the prospective transmitting power communication equipments that adaptatively find the optimum transmission power, supported
the packed destination address and underneath reliability constraints expressed in terms
of most allowed communication Bit Error Ratio (BER). The runtime adjustable power
transmission technique is especially for improving the energy efficiency of the data
transmission equipments and data links in WiNoC designs. Figure 2 shows as fundamental plan to tuning or adjusting the transmission power with the personal mandatory
point of the existing communication. mainly, based on the incoming packets of the
designation address, the communication transreceivers adjust its transmission power to
a low level, however high enough to achieve the terminal antenna while not exceptional
a particular bit error ratio [6]. This method is common and can be applied to all WiNoC
design. It is applicable on different representative WiNoC architectures leads to a median
energy reduction up to five hundredth with none impact on performance and with a

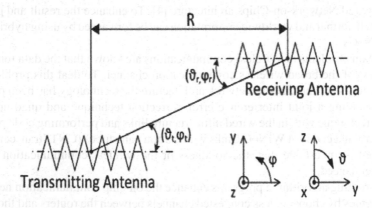

Fig. 1. Transmitting and receiving antennas geometrical orientation. By a spherical coordinate
system, φ is considered as azimuth angle. and θ is considered as the elevation angle, [6].

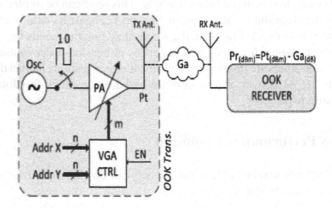

Fig. 2. Adaptive transmitting power transmitter

minimum expense in terms of semiconductor space. It explains clearly the way to improve the transreceiver energy efficiency in WiNoC design with the destination address of incoming packets and this method saves the ability whereas transmitting the packets. This technique saves the energy but it have the scalability problems and it saves only minimum of power (Fig. 1).

3.2 NoC's Using Antennas

This technique shows to enhance the performance of long range data transmission inside a NoCs and it introduces the innovative substrate propagation technique over a silicon area with some Antennas. The System on chip (SoC) transreceivers wanted mandatorily these WNoCs are often Uni-directional and bi-directional, broadcasting to each receiving antenna or directional solely during some special transmission of data, severally [5]. There are positives and negatives for each sort of transmitting equipments, though bi-directional antennas with none of interlink methods have the more advantages of reducing the likelihood of signal cross over of channels. This work is analyzed the metrics of a Wired NoC with bi-directional antennas that applicable for an proposed substrate propagation technique. This technique has been introduced as newly in system on chip architecture research and it demonstrated the capable of longer and wide range of communications compared to typical properties of on-chip antennas that can be using surface layer propagation method showed in Fig. 3. These design systems have used 8 numbers of antennas in the experimental setup without degradation of performance (Table 1).

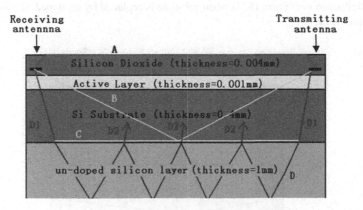

Fig. 3. Typical ray path of on-chip antenna propagation.

Guidance medium has combined with a integrated antennas to increase the system efficiency and throw out in experimental setup. The below table clearly shows the substrate chip architecture technique performance with results (Figs. 4 and 5).

Table 1. Performance chart of substrate propagation technique [5].

S. No	Parameter	Results
1	Resistivity	10–20 Ω-cm
2	Radiation efficiency	3% - Low with significant SNR
3	Transmission gain	−40 dB
4	Relative frequency	15 GHz
5	Throughput	±5%
6	Substrate used	Silicon substrate (10 cm)

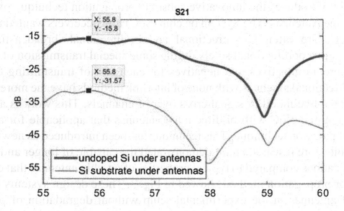

Fig. 4. Reflection coefficient (S21) when Si substrate is not replaced by un-doped Si (red, bottom); Reflection coefficient (S21) when substrate is replaced by un-doped Si (blue, top) [5]. (Color figure online)

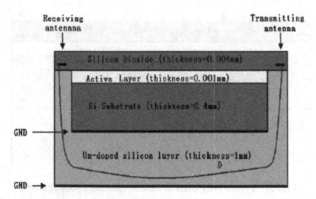

Fig. 5. Dominant ray path after propagating a ground layer between Si substrate and un-doped silicon layer [5].

3.3 2D Wave Guide Communication

The existing wireless design for WiNoCs among the kind of mm wave system depends the free space signal interference of very high power dissipation and it was degraded the rate at which the signal quality per range of transmission. Additionally the combined wired and wireless medium can be highly reduced the complete reliability of on-chip communication architecture over the lossy wireless medium [1]. Surface wave transmission technology has been implemented by another wireless technique for low power on-chip communication system. Moreover, the actual and correct design can concern the reliability and performance metric analysis of the surface wave radio medium are often elaborated. In this technique has intended a surface wave communication system for Wireless NoCs and also the designed system can able to match the dependable of existed wired NoCs.

A proper channel model has clearly explains that existing mm-Wave wireless NoCs suffers from free-space spreading loss (FSSL) and in addition molecular absorption attenuation (MAA), notably at high frequency, that decreases the reliability of the design. In the same way, this idea has tendency to use a strictly designed device and easily available in electronics market with low restive metal strip coated with minimum value dielectric material to induce surface wave signals with increasing range of transmitting gain. This design has been used for low power on-chip communication and it improves the transmission reliableness of wireless network but the foremost disadvantage of the mm wave WiNoc's suffer from Free space Spreading Loss (FSSL) and Molecular Absorption Loss (MAL) notably at higher frequency and It reduces the reliableness of the system [1] (Fig. 6).

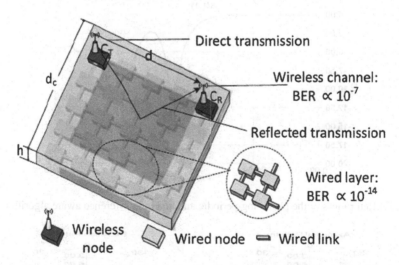

Fig. 6. Model of system with communication between two cores in existing hybrid wired and wireless Network-on-Chip [1].

3.4   Interference Algorithms in Network on Chip Architecture

Interference aware WiNoC to subsume the minimisation of NoC by optimum wireless interfaces (WIs) placement to maximise efficiency whereas reducing the signal cross over. It demonstrates an interference-aware WIs placement rule with routing strategy for wireless NoC style by using directional two-dimensional log-periodic antennas (PLPAs) and it ovver comes the realizable performance blessings because just single wireless structure can transreceiving at a time. It's in addition not sensible inside the immediate future to randomly proportion the number of non-overlapping channels by designing transceivers operating in disjoint frequency bands inside the millimeter-wave spectrum sometimes adopted for on-chip wireless interconnects. likewise, the directional antennas applicable to multiple wirelesses interconnect pairs can communicate at identical time and it have to coincident wireless communications could lead to signal cross over frequency [7]. This may be used for optimum wireless interfaces (WIs) placement to high performance wherever as reducing the signal fading. To manage this it proposes an interference-aware WI's placement algorithmic rule with routing strategy for WiNoC style by incorporating directional two-dimensional log-periodic antennas (PLPAs). The interference of channel can be reduced by connecting the point-to-point links between different wireless link and antennas or transreceiving devices at same time in directional wireless network-on-chip (DWiNoC) [7]. The most advantage is that this design permits point-to-point links between transceivers and therefore multiple wireless links will operate at identical time while not interference. The utilization of directional antennas whereby multiple wirelesses interconnect pairs will communicate at the same time (Figs. 7 and 8).

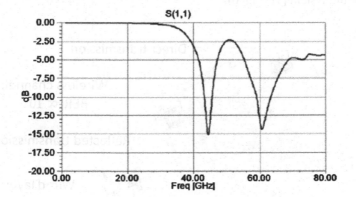

Fig. 7. Return loss of the planar log-periodic antenna in interference aware algorithm [4]

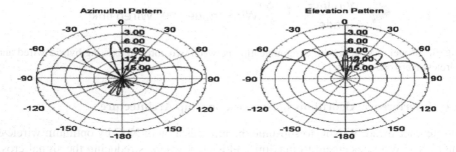

Fig. 8. Radiation pattern along with azimuth and elevation plane [4].

4 Applications

The main advantages of this survey of design have lightweight, smaller in size, mechanically robust when mounted on rigid surfaces. The proposed design would very useful in commercial applications such as delivery of data and packets in wireless long range communication systems. This massive level of integration makes modern multi-core chips. These modern multi core chips are applicable in many fields such as weather forecasting, astronomical data analysis, biological applications, consumer electronics, smart phones, etc. Especially in marine fields Ship to ship communications, Ship to port communications via high speed Wireless chip antennas, Digital service calling (DSC) in marine communication system can be improved and this will be very useful for Emergency Position Indicating Radio Beacons (EPIRB) boards.

5 Concluding Remarks

In this survey, we analyzed the various multiprocessor systems on chip design with different types of architectures and algorithms. This analysis can give clear idea about the entire network on chip properties with the advantages and drawbacks of existed design to address various challenges in multiple areas including device fabrication, circuit design and system architecture. From the above analysis antennas and other transreceiver has been played an important role with the characteristics of transmitting property and this will be used for the NoC designers has to note this criteria and limitation while they are designing the system using antenna with NoC's architecture. However, this analysis of survey is used to design a system on chip with multiple numbers of antennas for high speed and efficient transmission of data in marine communication fields and this review of network on chip explained many in-depth problems to design a multiprocessor system on chip with antennas for to improve the communication and integration of electronics in marine based equipments and other commercial systems and also it analyzed the performance of various design and it gives clear idea about how to design a low power, low frequency, noise less, low return loss and path loss chip antenna with high gain, high directivity and less complicated properties.

References

1. Agyeman, M.O., Vien, Q.-T., Ahmadinia, A., Yakovlev, A., Tong, K.-F., Mak, T.: A resilient 2-D waveguide communication fabric for hybrid wired-wireless NoC design. IEEE Trans. Parallel Distrib. Syst. **28**(2), 359–373 (2017)
2. Li, X., Duraisamy, K., Bogdan, P., Majumder, T., Pande, P.P.: Network-on-chip-enabled multicore platforms for parallel model predictive control. IEEE Trans. Very Large Scale Integr. (VLSI) **24**(9), 2837–2850 (2016)
3. Catania, V., Mineo, A., Monteleone, S., Palesiand, M., Patti, D.: Energy efficient transceiver in wireless network on chip. In: Architectures, 2016 Design, Automation and Test in Europe Conference and Exhibition (DATE), pp. 1321–1326 (2016). ISBN 978-3-9815370-7-9

4. Mondal, H.K., Shamim, S.: Interference-aware wireless network-on-chip architecture using directional antennas. IEEE Trans. Multi-Scale Comput. Syst. **3**, 193–205 (2016). ISSN 2332-7766
5. Pano, V., Liu, Y., Yilmaz, I., More, A., Taskin, B., Dandekar, K.: Wireless NoCs using directional and substrate propagation antennas. In: 2007 IEEE Computer Society Annual Symposium on VLSI (2007)
6. Mineo, A., Palesi, M., Ascia, G., Catania, V.: Runtime tunable transmitting power technique in mm-Wave WiNoC architectures. IEEE Trans. Very Large Scale Integr. (VLSI) Syst. **24**, 1535–1545 (2015). ISSN 1063-8210
7. Nosrati, N., Shahhoseini, H.S.: G-CARA: a Global Congestion-Aware Routing Algorithm for traffic management in 3D networks-on-chip. In: 25th Iranian Conference on Electrical Engineering (ICEE2017), pp. 2188–2193 (2017)
8. Jacob, T., Krishna, A., Suresh, L.P., Muthukumar, P.: A choice of FPGA design for there phase sinusoidal pulse width modulation. In: International Conference on Emerging Technology and Trends, pp. 1–6 (2016)

PRLE Based T – OCI Crossbar for On-Chip Communication

Ashly Thomas[✉] and Sukanya Sundresh

Department of ECE, T.K.M. Institute of Technology, Kollam, Kerala, India
ashlythomas1994@gmail.com

Abstract. One of the most important factor which affects the performance of system on chips are on chip interconnects. The most suitable interconnection method which is capable of addressing several high performance applications are Network on Chip. The widely used method to implement on-chip crossbars are Code Division Multiple Access (CDMA). Overloaded CDMA is used to intensify the capacity of the CDMA based Network on Chip and to overcome the problems due to Multiple Access Interference (MAI). In this paper, to reduce the excess area needed to store the message data and to provide security, a parallel compare and compress technique is used to the T – OCI crossbar. The Walsh spreading codes has a property that enables adding more number of non orthogonal spreading codes which increases the capacity of CDMA bus. Codec is the key component of the PaCC architecture, which effectively balances the area and performance. The Parallel Run Length Encoding (PRLE) scheme observes q bit in parallel. The results show that the PRLE based T – OCI attains greater bus capacity, less power consumption and efficient in area when compared with the conventional CDMA bus.

Keywords: System on chip · PRLE · Overloaded CDMA · On-chip interconnect
Crossbar · Network on Chip · Codec · Code Division Multiple Access

1 Introduction

System on Chip integrates several intellectual property (IP) blocks into a single chip. All of these IPs need to communicate in the Gbps range. So the on-chip communication requirements for these systems are very demanding. The IP blocks must comprise an interconnection architecture and several interfaces to connect the peripheral devices. The interconnection architecture includes many physical interfaces and communication mechanisms. On-chip data transfer affects the area, performance and the power utilization of the System on Chips. Developing an suitable high performance on-chip interconnect architecture has been of supreme significance while considering the high speed computing technologies. Network on Chips provide a way to prevail over the restrictions inherent in regular bus based interconnection schemes and offers several benefits like high throughput, lower energy dissipation, flexible scalability and design reusability.

In the case of Network on Chips, data from the routers are considered as several packets and on - chip processing elements are examined as network nodes which are

© Springer Nature Singapore Pte Ltd. 2018
I. Zelinka et al. (Eds.): ICSCS 2018, CCIS 837, pp. 593–601, 2018.
https://doi.org/10.1007/978-981-13-1936-5_61

interconnected via routers and switches. A crossbar is one of the most important compo-nent of the NoC physical layer. It is a shared communication medium which helps in the exchange of packets. Time Division Multiple Access (TDMA), Space Division Multiple Access (SDMA) and Code Division Multiple Access (CDMA) are the foremost resource allocating techniques utilized by the existing network on- chip crossbars. In the case of CDMA communication each transmit – receive pairs is assigned a distinctive bipolar spreading code. In the communication channel all the data from the transmitters are added. The ordinary CDMA systems use Walsh – Hadamard codes to facilitate the sharing of medium [8]. The spreading codes that used in normal CDMA communication systems are mostly orthogonal and it allows the CDMA receiver to accurately decode the sum from the channel. Multiple Access Interference (MAI) may occur if any addi-tional codes were added. The supreme number of users in the CDMA based communi-cation system may limited because of the Multiple Access Interference problem. Over-loaded CDMA can be used to intensify the number of users sharing the communication channel. The interconnect capacity of the on chip interconnects can be boosted with overloaded CDMA concept.

In this paper, we apply the idea of overloaded CDMA to the crossbar of Network on Chip to appreciably increase the capacity of the bus. Also, a parallel compare and compress (PaCC) architecture is used to diminish the surplus area needed to store the message data. Codec is an important component of the PaCC architecture and it effec-tively balances the area and performance. The Parallel Run Length Encoding scheme observes q bit in parallel.

2   Literature Review

The ordinary CDMA bus depend on orthogonal Walsh codes to validate the allocation of bus. Tatjana Nikolic, Mile Stojcev and Goran Djordjevic proposes a Code Division Multiple Access related bus structure in [2] for the reduction of the parallel data transfer lines to TDMA based buses. Combination of Code Division Multiple Access and Time Division Multiple Access in the CT bus communicates the data over the time domain as well as the code domain. A multilevel 2 bit Code Division Multiple Access proposed in [7] was mainly used as an input and output redesign scheme and also reduces the bus contention over Time Division Multiple Access. In [3] a comparison of based Network on Chip and a Point To Point duplex ring based Network on Chip is done. The outcome after simulation shows that the Code Division Multiple Access Network on Chips irre-versible data transfer latency is equal to the best case latency of the Point To Point of the identical channel width. The irreversible data transfer latency of the Code Division Multiple Access based Network on Chip is attributed to the co-occurent allocation of the channel by the nodes of the network. In [4], a wireless Code Division Multiple Access Network on Chip system was indicated to have notably lower energy consumption and larger bandwidth than a Time Division Multiple Access Network on Chip. Most of the associated work with the Code Division Multiple Access interconnect makes the enhancements in the architecture and the topology. Also the performance of the normal DS – Code Division Multiple Access communication scheme is evaluated.

In this paper, we find a solution to reduce the area and power consumption of the T-OCI crossbar which increases the bus capacity by applying overloaded Code Division Multiple Access to the existing on chip Code Division Multiple Access bus.

3   Overview of Conventional CDMA Crossbar

Figure 1 demonstrates the block diagram of the architecture of the normal CDMA NoC router. The classical CDMA crossbar as shown in the Fig. 2 consists of three sections. They are encoder section, channel section, and decoder section. In the encoder part, the spreading code generator module (Walsh spreading code) generates binary orthogonal code which has a chip length of C. After the XOR operation between the data and the spreading code the output is sent to the communication channel in a serial manner. It indicates that spreading of each single bit takes place in a time span of C clock cycles. Maximum number of IP core Transmit- Receive pairs that shares the channel bus is equal to B i.e., B > C. For the normal CDMA bus, which uses the Walsh spreading codes B = C. The Serial data streams from every transmitting IP cores that shares the CDMA bus system are added together. Then the resultant sum that obtained is denoted in the binary form and it is sent to the decoding unit which is connected to the receiving IP cores.

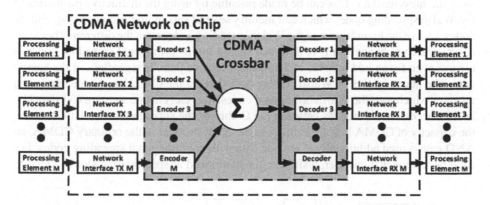

Fig. 1. Architectural diagram of normal CDMA based Network on Chip router.

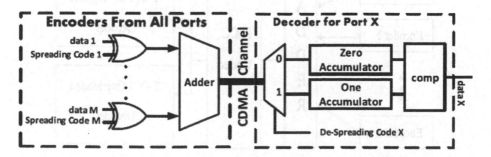

Fig. 2. CDMA crossbar.

It cross correlates the serialized data (sum) from the channel with the spreading code which is assigned for every transmit receive pair. The data despreading mainly require two operations. They are summultiplication by ± 1 and the accumulation. The data from the channel (bus) is passed to zero or one accumulator on the basis of the chip value. If the chip value is zero, then it is passed to the zero accumulator and if the value is one, then it is passed to one accumulator [1]. At the beginning of each decoding cycle the accumulators is reset to zero. Consequently, each of the accumulators adds half of the chip length of different inputs during the decoding cycle. This is possible due to the balancing nature of spreading codes. During the completion of each decoding cycle, if the content of the zero accumulator is greater than the content of one accumulator, the data bit transmitted is one otherwise it is zero.

4   PRLE Based T – OCI Crossbar

The main objective of this on-chip interconnect system is to increase the number of elements that shares the normal CDMA bus and to reduce the area and power consumption without changing the complexity. Fig 3 shows the PRLE based T-OCI architecture. It uses a simple encoding circuitry. But there are several changes in the accumulator based decoder. In addition to this, a PaCC codec is used to reduce the area needed to store the message data. This can be made possible by using the distinctive properties of the Walsh spreading code, which can identify several sets of non orthogonal spreading codes. One of the main feature of the Walsh spreading code is that the difference between any successive channel sums of data that spread by the orthogonal spreading codes is always even for an odd number of Transmit - Receive pairs B regardless of the spread data. The above feature of Walsh code reveals that for the C-1 pairs of transmitter and receiver which uses the Walsh orthogonal codes, user can encode an extra C-1 data bits in the successive differences between the C chip of the spreading code. This increase the capacity of CDMA bus. In addition to the XOR encoder in the ordinary CDMA, an AND gate is used additionally to encode data with nonorthogonal spreading codes. For each nonorthogonal encoder, a single chip summoned at specific time slot is added to

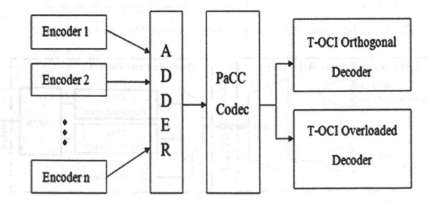

Fig. 3.   PRLE based T-OCI crossbar architecture.

the bus sum if the transmitted data is one. This causes the deviation of consecutive sum difference. Because the suggested codes are similar to TDMA signals. The above defined code family is known as T – OCI codes and the bus architecture is called TDMA overloaded on CDMA Interconnect (T – OCI).

4.1 Encoder Module

Figure 4 shows the encoder module. The encoder has the capability to encode both the orthogonal and nonorthogonal data. So it is known as hybrid encoder [1]. The XOR operation of the data bit and the Walsh spreading code produce the orthogonal code and AND operation of data bit and the Walsh spreading code to produce nonorthogonal spread data. A multiplexer is used to choose between the orthogonal and nonorthogonal inputs in accordance with the code type allotted [1].

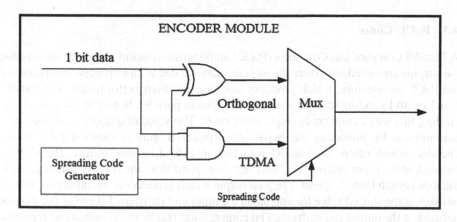

Fig. 4. Encoder module.

4.2 Crossbar Adder

If the spreading code has a length C, then number of crossbar Transmit- Receive ports is 2C-2. In the case of TDMA-OCI crossbar, when transmitting a "one" data chip to the adder or channel is mutually exclusive between the nonorthogonal transmit ports. This reveals that among the 2C-2 inputs, which is given to the channel as shown in Fig. 5, there are C-2 zeros, while the supreme number of "one" chips is C.

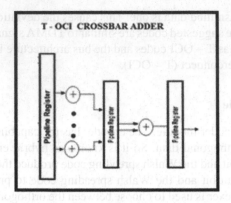

Fig. 5. T-OCI crossbar adder.

4.3 PaCC Codec

A Parallel Compare and Compress (PaCC) architecture is mainly used for diminishing the surplus area needed to store the message data. Codec is an important component of the PaCC architecture, which produces less area and high performance. The Parallel Run Length Encoding scheme observes the q bits in parallel. If they all are *zero* or *one* all the q bits are circumvent in single clock cycle. The value of q influence the speed of compression by impacting the bypass opportunity. Figure 6 shows a PaCC based encoder, which offers the Parallel Run Length Encoding mechanism. The shifting network at the input end in PaCC encoder shifts p bit from the register and is given to the Run Length Encoding unit. The p bit output which produces in the input end shifting network is the shifted value for upgrading the input end registers. Likewise, the shifting network at the output end shifts the j bit compression results to the output end registers. The all *zero or one* detector unit helps to perform q bit parallel monitoring and generates a bypass signal. The signal is fed to the RLE encoder unit and to the length controller [9].

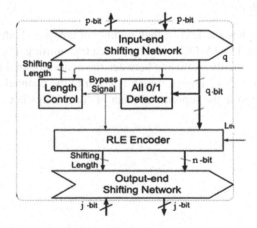

Fig. 6. PaCC encoder.

The function of the length controller is to provide the length of shifting to the shifting network at the input end in accordance to the bypass signal as well as Observation Window Width (OWW) q. The Run Length Encoder unit compresses the q bit input serially. This will perform only when the bypass signal is in disabled condition, otherwise bypasses the q bit input. Once the Run Length Encoder unit completes the compression, it sends a compressed segment having length n to the shifting network at the output end. The PaCC decoder is almost identical to the encoder. The PaCC decoder can reutilize the shifting networks at the two ends, since encoding and decoding are reverse operations. The input end and output end parts are interchanged and the direction of data flow is in reverse direction. The main dissimilarity in the PaCC decoder is that it comprises an Run Length decoding unit instead of the Run Length encoding unit. Because of the change in the shifting length, there are some challenges occurs while designing the shifting networks. Most commonly the barrel shifter is used for the shifting purposes, but it may consume much area. So that a novel area efficient shifting structure is used for the shifting network at the input and output end. The shifting process can be classified into two stages. The first stage shifting network is a coarse grained, which has a shifting length of X. Here the input is shifted by X. This is an area efficient shifter. In the second stage a barrel shifter of 2X bit is used. It takes the first 2X bits from the p bit input data and use as it's input. The detailed structure of the shifting network is shown in Fig. 7

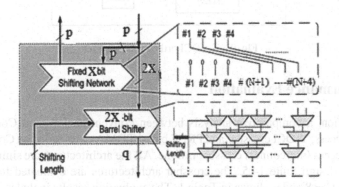

Fig. 7. Shifting network architecture.

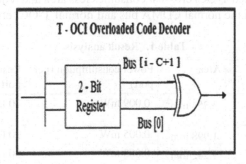

Fig. 8. T – OCI overloaded decoder.

4.4 Decoder

The orthogonal TDMA overloaded on CDMA Interconnect decoder is depicted in Fig. 9. As an alternative to the implementation of the two different an up- down accumulator is used and accumulated result is the difference between the zero and one accumulators of the normal CDMA decoder. The function of the accumulator is to perform the addition or subtraction of the data values in accordance with the despreading code chip. For every C clock cycle this will be reseted. The decoded data bit will be directly indicated by the sign bit of the accumulated data bit. The TDMA overloaded on CDMA Interconnect over-loaded decoder shown in the Fig. 8 is mainly made up of a register of 2 bit, which is used to store the Least Significant Bits of the two sum values. The output of the register are then given to the XOR gate, which produces the non orthogonal spread data.

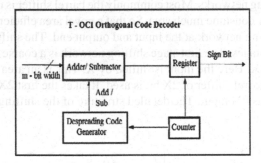

Fig. 9. T – OCI orthogonal decoder.

5   Performance Evaluation

In this portion, a comparative analysis between conventional/normal Code Division Multiple Access, T-OCI crossbar and PRLE based TDMA overloaded on Code Division Multiple Access Interconnect crossbar is done. All the architectures are simulated using Xilinx ISE design Suite 14.5. The crossbar architectures are evaluated for spreading code length C = 8 and is shown in Table 1. The evaluation results includes the resource utilization, area and power consumption. The below results show that the Parallel Run Length Encoding based T-OCI crossbar consumes less area and reduced usage of power when compared with the normal CDMA bus and normal T-OCI crossbar.

Table 1. Result analysis

Parameters Crossbar	Area	Power consumption (per port)	Resource utilization (per port)
Conventional CDMA crossbar	2.254 mm^2	0.009 mW	30 LUT 20FF
T – OCI crossbar	1.998 mm^2	0.005 mW	30 LUT 20 FF
PRLE Based T – OCI crossbar	1.242 mm^2	0.003 mW	30 LUT 25 FF

6   Conclusion and Future Work

İn this paper a PaCC Codec based overloaded CDMAcrossbar architecture is used. İt can enode both the orthogonal and the nonortogonal codes by using the same encoder module. By using sepertae decoder modules the orthogonal and the overloaded codes can be decoded seperately. This overloaded concept increases the number of users sharing the communication channel. The FPGA implementation of the normal T-OCI crossbar requires more area and consumes more power. But the novel PaCC based structure reduces the area needed to store the message data and enhances the capacity of the crossbar. The future scope includes more architectural improvements and addition of more number of codes having various properties to increase the interconnect capacity of the crossbar.

References

1. Ahmed, K.E., Farag, M.M.: Enhanced overloaded CDMA interconnect (OCI) bus architecture for on- chip communication. In: Proceedings of the IEEE 23rd Annual Symposium on High-Performance Interconnects (HOTI), pp. 78–87, August 2015
2. Nikolic, T., Stojecv, M., Djordjevic, G.: CDMA bus based on chip interconnect infrastructure. Microelectron. Reliab. **49**(4), 448–459 (2009)
3. Wang, X., Ahonen, T., Nurmi, J.: Applying CDMA technique to network on chip. IEEE Transactions Very Large Scale Integration (VLSI) Systems **15**(10), 1091–1100 (2007)
4. Vidapalapati, A., Vijayakumaran, V., Ganguly, A., Kwasinski, A.: NoC architectures with adaptive code division multiple access based wireless links. In: 2012IEEE International Symposium on Circuits and Systems (ISCAS), pp. 636–639, May 2012
5. Nikolic, T., Djordjevic, G., Stojcev, M.: Simultaneous data transfers over peripheral bus using CDMA technique. In: 26th International Conference on Microelectronics, MIEL 2008, pp. 437–440 (2008)
6. Kim, J., Verbauwhede, I., Chang, M.-C.F.: Design of an interconnect architecture and signaling technology for parallelism in communication. IEEE Trans. Very Large Scale Integr. (VLSI) Syst. **15**(8), 881–894 (2007)
7. Lai, B.-C.C., Schaumont, P., Verbauwhede, I.: CT – bus:a heterogeneous CDMA/TDMA bus for future SoC. In: Conference Record of the Thirty-Eigth Asilomar Conference on Signals, Systems and Computers, vol. 2, pp. 1868–1872, November 2004
8. Wang, J., Lu, Z., Li, Y.: A new CDMA encoding/decoding method for on chip communication network. IEEE Trans. Very Large Scale Integr. (VLSI) Syst. **24**(4), 1607–1611 (2016)
9. Wang, Y., Zhang, D., Chiang, M.-F.: PaCC: a parallel compare and compress codec for area reduction in Nonvolatile processors. IEEE Trans. Very Large Scale Integr. (VLSI) Syst. **22**(7), 1491–1505 (2015)

A Novel Approach to Design Braun Array Multiplier Using Parallel Prefix Adders for Parallel Processing Architectures

- A VLSI Based Approach

Kunjan D. Shinde$^{(\boxtimes)}$, K. Amit Kumar$^{(\boxtimes)}$, D. S. Rashmi,
R. Sadiya Rukhsar, H. R. Shilpa, and C. R. Vidyashree

Department of Electronics and Communication Engineering, PESITM,
Shivamogga, Karnataka, India
kunjan18m@gmail.com, amitkaller@gmail.com,
srvecpesitm@gmail.com

Abstract. Multipliers play a important role in current signal processing chips like DSP and general purpose processors and applications. In such high performance systems addition and multiplication operations are fundamental and most used arithmetic operations. Some case study shows that more than 70% of DSP algorithms and in microprocessor operations perform addition and multiplication. Hence these operations dominate the execution time. To meet the processing speed demand, the design of multipliers and adders plays a vital role. Low power consumption has consumption has became a major issue in design of multiplier. To reduce the power consumption, the components used in the design must be drastically reduced and in parallel it should not degrade the other performance metric. In this paper we are proposing a new architecture to design multiplier using parallel prefix adders. The Parallel Prefix Adder (PPA) have fast carry generation network and hence they are the fastest types of adder that had been created and developed. Most common types of parallel prefix adder are Brent Kung and Kogge Stone adders. The performances of these two adders in terms of worst case delay and transistor count and power consumption studied and an analysis of the same is presented. Utilizing the same to design and implement the multiplier using PPA is performed in this paper.

Keywords: Array multiplier · Braun multiplier · Parallel prefix adder
High speed adders · Parallel processing · Kogge Stone Adder (KSA)
Brunt Kung Adder (BKA) · Cadence design suite · GPDK 180 nm technology
CMOS design · GDI design

1 Introduction

A multiplierais one of the key hardware blocksain most digital signal processing (DSP) systems. Typical DSP applications, where aamultiplier plays an important role in digital filtering, digital communications and spectral analysis. In low power VLSI applications, it is desirable that theamultiplier should consume less power and less area

© Springer Nature Singapore Pte Ltd. 2018
I. Zelinka et al. (Eds.): ICSCS 2018, CCIS 837, pp. 602–614, 2018.
https://doi.org/10.1007/978-981-13-1936-5_62

in the design while achieving higher execution speed, combination of all these in a single multiplier is rarely found in several designs and hence a novel approach is proposed in this paper to meet the computational needs. The system's performanceais usually determined by the performance of the multiplier because the multiplier used in the design generally contributes significant amount of delay and consumes most of the area in the design. Statics shows that more than 70% instructions in microprocessor and most of DSP algorithms perform addition and multiplication. Power dissipation becomes one of the primary design constraints. Reducing the delay of a multiplier is an essential part of satisfying the overall design. Multiplication is very expensive and slows the overall operation. The performance of many computational problems is often dominated by the speed at which a multiplication operation can be executed. In order to achieve high execution speed and to meet performance demands in DSP applications, parallel array multipliers are widely used. One such widely used parallel array multiplier is the Braun Array Multiplier. The adderaunit is very important for designing any multiplier. Multiplication can be done serially or parallel. Theoretically multiplication can be done by repetitive addition.

In this paper we are proposing a new architecture and a novel approach to design braun multiplier using parallel prefix adder. The Parallel Prefix Adder (PPA) is one of the fastest types of adder that had been created and developed to achive a greater speed in the calculation of intermediate carries, carry and sum for the applied inputs. The two common types of parallel prefix adder are Brent Kung and Kogge Stone adder which will be used in this paper to increase the performance of the multiplier with the proposed architecture of the multiplier.

2 Literature Survey

The following are some of the references that we have gone thought for the design and analysis of various parallel prefix adders and multiplier design. The [8] is one of our paper which describes the modeling of parallel prefix adders used in this paper. In [1, 2] the authors have designed and simulated various adders like RCA, CSA, CLA and KSA using Cadence design suite at GPDK 45 nm technology and front end verification is performed on Xilinx ISE tool, these work is carried out for precision of 8-bit addition and have coated a comparative statement. In [2] the FPGA implementation of 8bit adders like RCA, CSA, CLA, and KSA is performed using Braun Multiplier. In [3] the detail investigation of the performances of different multipliers in terms of computational delay and design areas are studied. In [4] the authors have studied digital principles and designs. In [5] the authors have design and made a comparative analysis of various 8-bit adders for embedded application. In [8] we have analyzed the working and behaviors of various adders that could be used for the design of multipliers. From the analysis coated by us in [8] helps to select a adder in this paper.

3 Design of Array Multiplier

In this paper we are proposing a new architecture and a novel approach to design and implement Braun array multiplier using parallel prefix adders and the modeling of the same is performed. Several design styles were referred and few of them are used in the design, In this paper we have considered the braun multiplier with precession of 4-bit and 8-bit for design and implementation. It includes the proposed methodology of Braun's multiplier and various parallel prefix adders like Kogge Stone Adder, Brent kung Adder.

To design high performance Braun multiplier, the all stages of multiplier consisting of full adder array structure and last stage is also a set of full adder which seems like with ripple carry adder structure, most of the researchers and designers concentrate on the modification of the last stage of the multiplier structure and try to replaced with other conventional adder or some Parallel Prefix Adders like Kogge Stone Adder, Brent Kung Adder. Figure 2 shows the conventional Braun multiplier with array of full adders and the highlighted group of full adders form a ripple carry structure, this frames a worst case delay for the multiplier with theadelay incorporated from the upper full adder stages and as the precession of multiplier increases the delay increases proportionally, various other methods are framed to reduce this delay which involves the replacement of the last stage of ripple carry structure with carry look ahead adder or carry save adder or carry skip adders, or high speed adders, with those variations also the performance is having the limit, as the last stage computation is valid only when all the inputs are computed by stages above it, hence the proposed structure of Braun multiplier is shown in Fig. 3, here we are not only replacing the last stage of ripple carry structure with high speed parallel prefix adder but we are increasing the speed of the computation of stages above the last stage by observing the ripple stages and grouping those set of full adders into similar behavior of a ripple like structure and replacing those structures with the high speed adders as shown in the proposed architecture for the multiplier using parallel prefix adder, from the structure of the Braun array multiplier we can see the ripple structure excluding last stage which is indicated in Fig. 3 with dotted circles, these are now forming a ripple carry structure and limiting the performance of the multiplier, here in this paper we are replacing this indicated circle with any of high speed adder hence the new architecture and a novel approach to design the Braun multiplier with parallel prefix adder, the processing of the information from individual block to stages and then to several intermediate level give the new techniques to design and test its performance.

Figure 1 shows the general blockadiagram of Multiplier with internal component as adder, the adders of various sizes have been developed and these adders are used in different sized multipliers. To reduce the dynamic power dissipation and to increase the execution speed of multiplier. To build 4 × 4 Braun multiplier, a parallel prefix adders are designed and added at all the stages of the multiplier replacing 3 bit ripple carry adders. Similarly for 8 × 8 Braun multiplier we are replacing 7-bit Ripple Carry Adder by PPA. we are using Cadence Design Suite for designing and to perform functional verification of the adders and multiplier used in this paper. While designing we have set the following parameters.

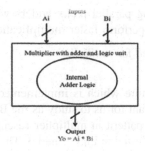

Fig. 1. General block diagram of multiplier.

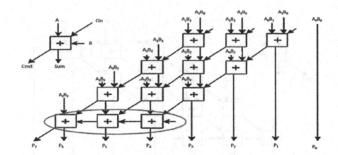

Fig. 2. Conventional Braun multiplier design.

- The precision of the multiplier is set, maximum up to the length of 8-bits.
- The design of multiplier will be performed in the conventional way and methods.
- Design of one conventional adder (like Ripple carry adder or Carry look ahead adder) and parallel prefix adder (Kogge Stone Adder, Brent Kung Adder).

Here we are using these adders for implement of multiplier as multiplication is done by successive addition. The performance can be measured and check for power consumption, computational delay and area utilized.

The propagation of signal in Braun multiplier is from one full adder to the other and hence the ripple like structure can be observed in the architecture of the Braun multiplier, this structure accounts to the delay in the result. And at the final stage also such structure can be observed. In conventional design of Braun multiplier, ripple carry adder is used at the last stage. The ripple carry adder at the all stage introduces significant amount of delay as the number of bits to be added increases.

The multiplier delay can be reduced by replacing the ripple carry adder with fast adder like parallel prefix adders. The proposed multiplier's design is shown in Fig. 3 where the Ripple Carry Adder at the last stage is replaced with parallel prefix adders like Kogge Stone Adder and Brent Kung adder at all the stages and we observed a similar ripple carry structure preceding to the final stage of the addition, in the proposed multiplier design these ripple stages are also replaced by the fast adders like Kogge Stone Adder and Brunt Kung Adder. To build 8 * 8 Braun multiplier, the bit size of inputs A and B is 8 bit wide and produces product of width 16 bit wide, partial products are generated at 7 stages using parallel prefix adders which are replaced in place of Rippleacarry structure and to perform faster multiplication of binary numbers.

Design of Various Adders used for multiplier

i. *Ripple Carry Adder*

It is the basic adder structure which is implemented by cascading full adders in series. This structure is important for us to study as the Braun multiplier uses 1-bit full adder in a specific patter, the pattern in multiplier resembles the ripple structure of a ripple carry adder which is indicated in Figs. 2 and 3. The drawback of such structure is that the performance of multiplier is degraded due to the ripple of result from first to last. [1, 2] (Figs. 4 and 5).

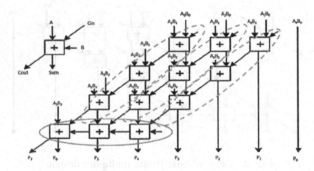

Fig. 3. Proposed Braun array multiplier design (Note: with reference to Fig. 3: A new architecture and novel approach, dotted circles form a ripple structure which are replaced by high speed adders like Kogge Stone or Brunt Kung Adder and also replacing the last stage of computation by high speed adder)

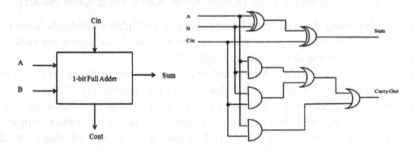

Fig. 4. Block diagram and schematic of 1-bit Full adder. [1, 2]

Equation of Ripple Carry Adder [8]

$$S_i = A_i \text{ XOR } B_i \text{ XOR } C_{i-1} \tag{1}$$

$$C_{i+1} = (A_i \text{ AND } B_i) \text{ OR } (B_i \text{ AND } C_i) \text{ OR } (C_i \text{ AND } A_i) \tag{2}$$

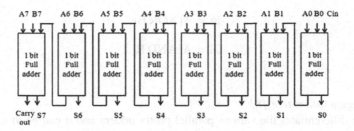

Fig. 5. Block diagram of 8-bit ripple carry adder

ii. *Parallel Prefix Adders*

Parallel prefix adders are most popular adder design used to speed up the binary addition. Parallel prefix adder, the acquirement of the parallel prefix adder carry bit differentiates the parallel prefix adder from other types of adders. In these adders, the carry bit is obtained in parallel form which makes the addition operation faster. Adders are derived from the family of Carry Look Ahead adder structure. The parallel prefix adders are represented in tree structure form to increase the speed of arithmetic operation. These adders are considered as the fastest adders and used in the high performance arithmetic circuits in the industries. Figure 6 shows the structured diagram of parallel prefix adder.

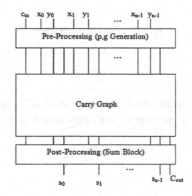

Fig. 6. Block diagram of parallel prefix adder structure. [1, 8]

Three stages involved in the design of parallel prefix adders are Pre-processing Stage, Carry generation and Post processing.

1. *Pre-processing:*
It computation a pair of signals in this phase namely generate and propagate signals, which corresponds to each ith state of A and B input sets. Propagate and generate signal are represent as shown below [8]:

$$Pi = Ai \, XOR \, Bi \tag{3}$$

$$Gi = Ai \, AND \, Bi \tag{4}$$

2. Carry generation tree [8]

Carry tree differentiates the various parallel prefix adders and it can be thought as the prime force for high performance in PPA Adders. In this phase the carries are computed much earlier using the two cells described below [8]:

Black Cell. It consumes two pairs of generate and propagate signals (Gi, Pi) and (Gj, Pj) as input and computes a pair of signals as generate and propagate signals (G, P) as output for the current stage [8].

$$G = Gi \, OR \, (Pi \, AND \, Pj) \tag{5}$$

$$P = Pi \, AND \, Pj \tag{6}$$

Grey Cell. It consumes two pairs of generate and propagate signals (Gi, Pi) and (Gj, Pj) as inputs and Computes a single generate signal G as output for the current stage [8]

$$G = Gi \, OR \, (Pi \, AND \, Pj) \tag{7}$$

3. Post processing

The last phase to compute sum which is common to all adders of PPA adder family. This stage involves the computation of sum bits which is given by [8]:

$$Si = pi \, XOR \, Ci - 1. \tag{8}$$

A. Kogge Stone Adder:

It is one of the parallel prefix adder, which has fastest carry tree to compute carries. It produces the carry signal in O (log n) time [1].

B. Brent Kung Adder:

Brent Kung Adder tree structure is similar to the Kogge Stone structure but the structure has a low fan-outs from each prefix cell and also has a large critical path which limits the high speed addition operation. Looking at the carry tree structure the design consumes less gates, uses less fan-outs and wiring compared to KS Adder [8].

4 Results and Analysis

In this section we have discuss the impact of different adders with various precision on the multiplier design. This section provides the simulation and comparative analysis with performance metric for various parallel prefix adder and conventional adder. We have also gives the impact of these adders, when used in the Braun Array Multiplier structure. The design of various blocks used in adders and multiplier is performed using CMOS technique and GDI technique, Cadence tool of version 6.1.6 to perform the schematic entry and Generic Process Design Kit of 180 nm technology to define the MOS device. In the first part schematic and simulation of various adder and Braun multiplier is depicted and in the second part a performance based comparative statement is presented (Figs. 7, 8, 9, 10, 11, 12, 13, 14, 15, 16).

A. *Multiplier Schematics and Simulations.*

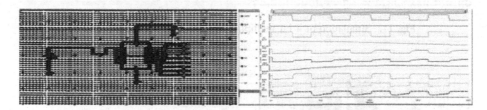

Fig. 7. Simulation of 4 bit multiplier in CMOS design using RC adder

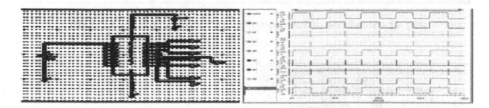

Fig. 8. Simulation of 4 bit multiplier in CMOS design using KS adder

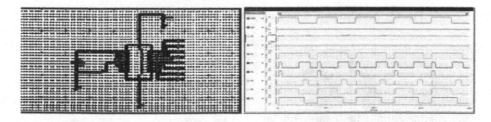

Fig. 9. Simulation of 4 bit multiplier in GDI design using KS adder

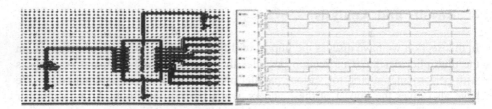

Fig. 10. Simulation of 4 bit multiplier in CMOS design using BK adder

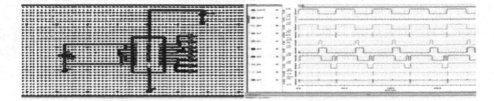

Fig. 11. Simulation of 4 bit multiplier in GDI design using BK adder

Fig. 12. Simulation of 8 bit multiplier in CMOS design using RC adder

Fig. 13. Simulation of 8 bit multiplier in CMOS design using KS adder

Fig. 14. Simulation of 8 bit multiplier in GDI design using KS adder

Fig. 15. Simulation of 8 bit multiplier in CMOS design using BK adder

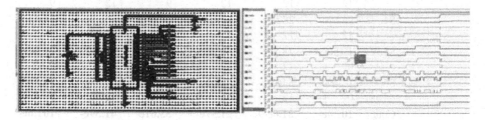

Fig. 16. Simulation of 8 bit multiplier in GDI design using BK adder

Table 1. Adder performance with 3-bit precession, CMOS and GDI technique [8]

Parameter		RCA CMOS	RCA GDI	KSA CMOS	KSA GDI	BKA CMOS	BKA GDI
Worst case delay	S0	11.36e-9	283.8e-12	11.24e-9	11.24e-9	11.32e-9	6.278e-9
	S1	21.38e-9	283.9e-12	21.46e-9	11.35e-9	11.32e-9	11.33e-9
	S2	41.38e-9	285.5e-12	41.46e-9	21.37e-9	31.46e-9	21.34e-9
	Cout	151.3e-9	21.06e-9	61.51e-9	31.42e-9	71.48e-9	71.24e-9
No. of transistors		144T	42T	156T	60T	106T	38T
Power		20.35e-6	55.97e-6	35.42e-6	198.4e-6	5.387e-6	181.1e-6

B. Comparative analysis

Multiplier designed using CMOS technique consumes low power in computing the result when compared with GDI Technique. Whereas the GDI technique utilizes less number of transistors to implement the given block and generates the results faster compared with CMOS technique. When a multiplier is to be designed using adder block, a care has to be taken in which adder itself is simplex and designed with optimal characteristics which is coated in this paper. In the initial stages the Gate Diffused Input

Table 2. Adder performance with 7-bit precession, CMOS and GDI technique [8]

Parameters		RCA CMOS	RCA GDI	KSA CMOS	KSA GDI	BKA CMOS	BKA GDI
Worst case delay	S0	392.9e-12	203.0e-12	556.8e-12	43.87e-9	518.9e-12	23.84e-9
	S1	392.9e-12	208.0e-12	556.7e-12	43.88e-9	518.9e-12	23.84e-9
	S2	593.0e-12	208.0e-12	556.9e-12	43.98e-9	518.9e-12	23.97e-9
	S3	779.0e-12	235.9e-12	565.2e-12	43.99e-9	518.8e-12	24.43e-9
	S4	964.0e-12	282.9e-12	649.2e-12	44.66e-9	566.9e-12	24.44e-9
	S5	1.150e-09	282.9e-12	653.8e-12	44.66e-9	681.2e-12	24.64e-9
	S6	1.340e-09	546.9e-12	671.1e-12	45.76e-9	695.8e-12	24.86e-9
	Cout	**101.4e-9**	**21.6e-9**	**21.57e-9**	**53.61e-9**	**41.39e-9**	**101.6e-9**
No. of transistors		**336T**	**98T**	**420T**	**164T**	**286T**	**106T**
Power		**15.4e-6**	**4.0447e-6**	**91.87e-6**	**293.1e-6**	**33.7e-6**	**360.0e-6**

Table 3. Multiplier performance with 4-bit precession, CMOS and GDI technique

Parameter		RC A CMOS	KS A CMOS	KS A GDI	BK A CMOS	BK A GDI
Worst case delay	P0	136.6E-12	136.6E-12	136.5E-12	136.6E-12	136.6E-12
	P1	596.0E-12	532.4E-12	408.6E-12	477.2E-12	393.7E-12
	P2	998.1E-12	580.1E-12	410.3E-12	555.6E-12	507.0E-12
	P3	1.051E-12	600.0E-12	777.3E-12	835.1E-12	717.5E-12
	P4	1.326E-9	771.6E-12	792.7E-12	965.1E-12	980.8E-12
	P5	1.50E-9	1.035E-9	797.3E-12	1.296E-9	1.171E-9
	P6	1.83E-9	1.481E-9	1.155E-9	21.36E-9	21.85E-9
	P7	**21.63E-9**	**21.33E-9**	**1.163E-9**	**21.64E-9**	**21.94E-9**
No. of transistors		**222**	**564**	**212**	**414**	**146**
Power		**8.22383E-5**	**9.2954E-5**	**0.002355**	**3.7741E-5**	**0.0016089E-6**

Table 4. Multiplier performance with 8-bit precession, CMOS and GDI technique

Parameter		RC A CMOS	KS A CMOS	KS A GDI	BK A CMOS	BK A GDI
Worst case delay	P0	136.6E-12	136.6E-12	136.5E-12	136.5E-12	136.5E-12
	P1	507.8E-12	136.6E-12	136.5E-12	136.5E-12	136.5E-12
	P2	507.8E-12	533.1E-12	400.1E-12	341.4E-12	341.4E-12
	P3	554.2E-12	533.1E-12	400.1E-12	341.4E-12	341.4E-12
	P4	507.8E-12	533.1E-12	400.1E-12	477.9E-12	477.9E-12
	P5	507.8E-12	533.1E-12	400.1E-12	477.9E-12	477.9E-12
	P6	920.2E-12	533.1E-12	400.1E-12	477.9E-12	477.9E-12
	P7	968.6E-12	533.1E-12	400.1E-12	481.2E-12	481.2E-12
	P8	968.6E-12	533.1E-12	1.069E-9	507.0E-12	507.0E-12
	P9	1.041E-9	905.8E-12	1.104E-9	866.5E-12	866.5E-12
	P10	1.015E-9	1.189E-9	1.105E-9	1.059E-9	1.059E-9
	P11	1.04E-9	1.259E-9	1.302E-9	1.088E-9	1.088E-9
	P12	1.11E-9	1.295E-9	1.151E-9	1.151E-9	1.151E-9
	P13	1.122E-9	1.295E-9	1.151E-9	1.194E-9	1.194E-9
	P14	1.122E-9	1.312E-9	1.315E-9	1.194E-9	1.194E-9
	P15	**1.978E-9**	**1.338E-9**	**1.519E-9**	**1.455E-9**	**1.445E-9**
No. of transistors		**678**	**3324**	**1276**	**2386**	**870**
Power		**7.43509E-05**	**0.000961**	**0.00116312**	**0.00011783**	**0.0084047**

technique is used to optimization of area & delay of adder blocks, whereas the CMOS technique can be used at the final stage to generate a full swing output (Tables 1, 2, 3 and 4).

5 Conclusion

Multipliers in Signal Processing and processors are core and important block. Hence the design of high speed multiplier is essential, In order to meet such needs the proposed architecture gives a novel approach to design and implement the multipliers using conventional adders and parallel prefix adders which is presented in paper and as the performance (like speed, area and delay) is prime concern in the design which is increased (reduction in number of transistors required to design, increasing speed and reducing delay). In this paper provided a new architecture and novel method to design a multiplier, and we have designed, modeled and implemented the same on cadence design suite and verified its behavior for the same. The multipliers designed are of 4-bit and 8-bit precession Braun multipliers using parallel prefix adders and coated comparative analysis obtained. From the analysis and comparative statement, we observe that the blocks and multiplier designed using Gate Diffused Input technique provides less delay in generating results and also uses few transistors to implement the structure; CMOS technique uses less power but uses more number of transistors to implement the logic. Designer can opt the proposed method where the initial stages are designed using GDI technique and final stage with CMOS technique, the circled up blocks in Fig. 3 are redefined with advance adder structure and thereby improving the result of multiplier.

References

1. Shinde, K.D., Jayashree, C.N.: Modeling, design and performance analysis of various 8-bit adders for embedded applications. In: Elsevier International Conference on Emerging Research in Computing, Information, Communication and Applications at NMIT, Bangalore, Karnataka, India, pp. 823–831, August 2014. ISBN 9789351072621
2. Shinde, K.D., Nidagundi, J.C.: Design of fast and efficient 1-bit full adder and its performance analysis. In: IEEE International Conference on "Control, Instrumentation, Communication and Computational Technologies (ICCICCT-2014)" at NIU, Tamilnadu, India, pp. 1275–1279, July 2014. ISBN 978-1-4799-4191-9
3. Shinde, K.D., Nidagundi, J.C.: Comparative analysis of 8-bit adders for embedded applications. In: IJERT in NC on "Real Time System". City Engineering College Bangalore, Karnataka, India, pp. 682–687, May 2014. ISSN 2278-0181
4. Zamhari, N., Voon, P., Kipli, K., et al.: Comparison of parallel prefix adder (PPA). In: Proceedings of the World Congress on Engineering 2012, WCE 2012, vol 2, 4–6 July 2012, London, U.K (2012)
5. Thakur, M., Ashraf, J.: Design of Braun multiplier with Kogge stone adder & it's implementation on FPGA. Int. J. Sci. Eng. Res. 3(10) (2012). ISSN 2229-5518
6. Prakash, M., Karthick, S.: Simulation and comparative analysis of different types of multipliers. IJAICT 1(7) (2014)
7. Givone, D.D.: Digital Principles and Design, chap. 5, pp. 231–240. Tata McGraw-Hill (2002)
8. Shinde, K.D., et al.: Modeling of adders using CMOS and GDI logic for multiplier application: a VLSI based approach. In: Proceedings of the IEEE International Conference on Circuit, Power and Computing Technologies (ICCPCT). https://doi.org/10.1109/iccpct.2016.7530213

An Avaricious Microwave Fiber-Optic Link with Hopped-up Bandwidth Proficiency and Jitter Cancelling Subsisting Intensity and Phase Modulation Along with Indirect Detection

Archa Chandrasenan[✉] and Joseph Zacharias

Department of Electronics and Communication Engineering,
Rajiv Gandhi Institute of Technology, Kottayam, India
archachandrasenan44@gmail.com, joseph.zacharias@rit.ac.in

Abstract. A extremely spectral potent, low cost microwave radio-over fiber link, which productively handle the transmitted power is proposed and experimentally demonstrated. To boost the bandwidth proficiency along with data security, ease of installation and signal integrity in the recommended approach, modulation of two microwave vector signal with have the same message frequency on a single optical carrier handling amplitude and phase modulation. At the target, coherent detection is favoured, one of the very promising techniques for maximizing power efficiency and thus improving the photonic link performance. The receiver enumerates decision variables established on the recovery of the full electric field, which contains both amplitude and phase information, which yields better performance in terms of dynamic range and linearity. The random fluctuations in the phase of a waveform, caused by time domain instabilities, caused due to Coherent detection is eradicated accurately and reliably using signal processing in Matlab, which has been described and verified by experiment. The transmission of 5 Gb/s QAM at 4 bits per symbol and 2.5 Gb/s QPSK information signals both at 550 MHz over 10 km OF is executed. The transmission accomplishment in terms of BER is figured out.

Keywords: Coherent detection · Phase modulation · Intensity modulation
Phase noise · Digital signal processing

1 Introduction

Hybrid fiber radio (HFR) technology is the effective combination of both optical and wireless access networks for the optical transmission of microwave signals from base station (BS) to central office (CO), which is an emerging tool for effective and faster communication [1]. The conventional uncomplicated system was adopted with intensity modulation and direct detection (IM + DD). In order to blow away the demerits of IM + DD [2] and make the link less sensitive to a variety of non linear effects, Intensity modulation in conjunction with Coherent Detection (IM + CD) came into action. It ensures admirable RF sensitivity. Also the BS can detect RF information signal and

© Springer Nature Singapore Pte Ltd. 2018
I. Zelinka et al. (Eds.): ICSCS 2018, CCIS 837, pp. 615–622, 2018.
https://doi.org/10.1007/978-981-13-1936-5_63

demodulate the full phase and amplitude information data even at lowest power level and is available for multiple users. In addition this approach can provide potential progress up to 20 dB in receiver sensitivity unlike DD based systems. For a given power budget [3], this would allow to expand the entire range of an optical link. Moreover higher transmission rates over extant optical links without lowering repeater spacing is achieved. Additionally CD system encourages economical utility of available frequencies, which allows to transmit simultaneously several carrier. Indeed channel spacing can be reduced to 1–10 GHz.

As the later stride, coherent photonic system proposes improved optical power utilization by photonic phase modulation (PM + CD) [4]. Aforementioned strategy attracts in view of the reliability in RF signal handling. Demodulating information coded on phase of photonic modulating signal stand in need of coherent optical receiver. Transporting signal encoded on phase allows the full signal to be transmitted to the receiver, without distortions induced by interferometric conversion to intensity modulation of IM + DD link. The indirect photonic links exhibits unprecedented performance which are critical for future generation reconfigurable networks in terms of their ease of implementation, monitoring and dynamic bandwidth provisioning.

The design is better than IM in terms of dynamic range and linearity but narrow bandwidth and desire low-set phase noise optical sources. Phase noise highly disgrace the execution of the receiver and act as a inhibit for improving the realization of MWP system. Though in the coherent systems information is impress on the optical beam using amplitude, frequency or phase modulation format, it fetch certain drawbacks [6] (1) Maximum detected bandwidth is few GHz, (2) Coherent receiver is electronically complicated, (3) Time delay between the signals leads to phase noise. The main limitation is the signal degradation due to phase noise. Hence intention while using CD is that to cancel the phase noise, that downgrade the information signal. Jitter cancellation edge to (1) Essentiality of high quality signals with superior phase noise, (2) Great speed data communication system, (3) Huge accurate connection. Phase noise highly disgrace the execution of the receiver and act as a inhibit for improving the realization of MWP system.

Phase noise highly disgrace the execution of the receiver and act as a inhibit for improving the realization of MWP system [7]. The distinct methods for controlling and cancelling phase noise [5] includes (1) Take up Optical phase locked loop to lock the phase of the two signals namely LO laser source at BS and CO source. But it led to difficulty in maintaining system stability. (2) Signal processing based carrier estimation technique, which make the system quite complex but can improve the overall stability. (3) Adoption of Dual-mode local light source or a very narrow line-width laser source and that maximize the system cost, even though it doesn't make the system complicated. To such a degree, option (2) is nominated. Signal processing increases productivity including coding efficiency. Still, these approaches suffer from ineffective optical carrier power usage and intolerances to transference losses. Aforementioned can be overthrown to an extend by Intensity and Phase modulating the optical carrier signal with two different messages. (IM + PM) [8] at the CO plus effectively revival of the informations at the BS using Coherent detection. The jitter is cancelled using the signal processing techniques, which are easy to realise and more accurate method than other choices.

2   Theory and Design

2.1   Transmitter Section and Double Modulation

The newly developed fundamental communication technology using high-level modulation methods on the optical side. Until now, just one modulation method was used, in order to support continued growth it is increasingly necessary to exploit all available multiplexing. In Fig. 1, utilising a single carrier, two informations are broadcasted using two modulations in the amplitude and phase parts of the carrier vector signal. The total carrier is being subdivided into two routes with power ratio 95:5. The signal in Route 1 bearing maximum amount of carrier power is being amplitude modulated in Electro-absorption modulator Mach-Zehnder Modulator (MZM). After MZM, the magnitude part is controlled by 16-QAM, applied via RF port [6]. For this reason MZM is biased at the Quadrature point.

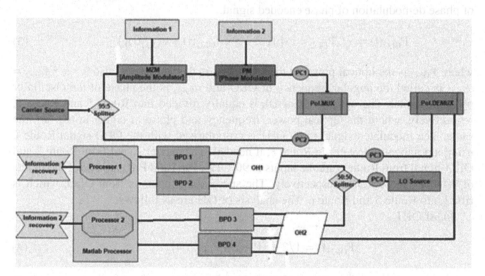

Fig. 1. Schematic of the proposed MWP link with double modulation and Matlab based Processing module [1]

The main allure of the system is to profit every measure of the modulating vector signal. Hence AM modulated wave is compelled to phase modulate (PM) by QPSK microwave signal, widely applied to short reach communication and long haul transmission. Route 2, carrying un-modulated carrier signal, is rotated to a polarization of 90^0OH.

$$P_{AM+PM}(t) \,\&= \sqrt{2P_{r1}} * \cos\left(\pi * E_{16\,QAM}(t)/2V_{IM}\right) \exp(j[\omega_{r1}(t) + \varphi_1(t) + [\pi * E_{QPSK}(t)/2V_{PM}) \quad (1)$$

where P_{r1} is the optical power through Route 1, $\omega_{r1}(t)$ is the angular frequency $\varphi_{R1}(t)$ is the phase of Route 01 signal. E_{QAM}, E_{QPSK} are the QAM and QPSK microwave vector signal applied to MZM and PM respectively, V_{PM} and V_{IM} is the half wave voltage of

PM and IM respectively. The signals coming on Route 1 and 2, are to be transmitted to the BS without conjoining. Thus polarization multiplexing technique is used to multiply user capacity and for multiplexing signals carried on EM waves allowing two channels of informations to be transported on the same carrier frequency. Thus routes 1 and 2 are using waves of rectangular polarization.

2.2   Receiver Section and Coherent Detection

$$P_3(t) = P_{AM+PM}(t) \text{ and } P_4(t) = P_{R2}(t) \tag{2}$$

The rectangularly polarised multiplexed signals are disseminated to the receiver through the optical fiber (OF), where it is demultiplexed using polarisation splitter to Route 3 and 4. Coherent optical techniques, which offer improved optical power utilisation by PM, demand Local Oscillator (LO) at the BS in Eq. 6, which assist linear demodulation of phase demodulation of phase encoded signal.

$$P_{OLO}(t) = \sqrt{2P_{OLO}} * \exp(j[2 * \pi * f_{OLO}(t) + \varphi_{OLO}(t)]) \tag{3}$$

where P_{OLO} is the optical power at the LO, f_{OLLO} is the frequency, also $2 * \pi * f_{OLO} = \omega_{OLO}$ is called the angular frequency of OLO and φ_{OLO} is the phase of the Oscillator photonic signal. The optical LO signal, is equally divided into Route 5 and Route 6, respectively, where the Optical power, frequency and phase of signal in two are the same. The modulated signal to 90^0OH1 is co-polarised with the OLO signal Route 6 using a polarization rotator in Route 3, if needed. The carrier signal from Route 2 and OLO signal from Route 6 are the inputs to 90^0OH2. $P_4(t)$ and $P_3(t)$ are one of the inputs of 90^0OH1 and 90^0OH2 respectively. The succeeding inputs is from OLO, which is rifted into Route 5 and Route 6. The analysis of OH are as follows.
   In 90^0OH1,

$$P_{OH1}(t) = 1/2 * (P_3(t) + P_{OLO1}(t)) \tag{4}$$

$$P_{OH2}(t) = 1/2 * (P_3(t) + P_{OLO1}(t)) \tag{5}$$

$$P_{OH3}(t) = 1/2 * (P_3(t) + P_{OLO1}(t)) \tag{6}$$

$$P_{OH4}(t) = 1/2 * (P_3(t) + P_{OLO1}(t)) \tag{7}$$

   Similarly, In 90^0OH2, identical operations are carried out using $P_{R2}(t)$ and $P_{OLO}(t)$ to obtain $P_{OH5}(t)$, $P_{OH6}(t)$, $P_{OH7}(t)$ and $P_{OH6}(t)$. Employing (P_{OH1}, P_{OH2}) and (P_{OH3}, P_{OH4}) to Balanced Photodetector 1 (BPD1) and BPD2 respectively, wherein two matched PDs can eliminate the need for lock-in amplifiers and can make all of the difference while detecting a small signal. It absorbs two laser beams and produces a voltage proportional to the difference in their intensities so as to outturn two photo-currents $I_1(t)$ and $I_2(t)$, given by

$$I_1(t) = 2R\sqrt{P_{r1}P_{OLO}} * \sin\left(\pi * E_{16\,QAM}(t)/2V\_IM\right)\sin\left(\Delta\omega_1 + \Delta\varphi_1 + \pi * E_{QPSK}(t)/2V_{PM}\right) \quad (8)$$

$$I_2(t) = 2R\sqrt{P_{r1}P_{OLO}} * \cos\left(\pi * E_{16\,QAM}(t)/2V\_IM\right)\sin\left(\Delta\omega_1 + \Delta\varphi_1 + \pi * E_{QPSK}(t)/2V_{PM}\right) \quad (9)$$

Out of the four BPDs, in the waiting two BPDs, BPD3 and BPD4 senses (P_{OH5}, P_{OH6}) and (P_{OH7}, P_{OH8}) and yields

$$I_3(t) = 2R\sqrt{\left(P_{r2}P_{OLO}\right)} * \sin\left(\Delta\omega_2 + \varphi_2\right) \quad (10)$$

$$I_4(t) = 2R\sqrt{\left(P_{r2}P_{OLO}\right)} * \cos\left(\Delta\omega_2 + \varphi_2\right) \quad (11)$$

From Eqs. 8, 9, 10 and 11, $\Delta\omega_1 = \omega_{R1} - \omega_{OLO}$ and $\Delta\omega_2 = \omega_{R2} - \omega_{OLO}$, where, $\Delta\omega_1$ and $\Delta\omega_2$ are the frequency difference between Transmitter laser source and OLO laser source in Route 1 and Route 2 respectively. The phase noise introduced by the Coherent detection corresponds to term φ_1 and φ_2, where $\varphi_1 = \varphi_{r1} - \varphi_{OLO}$ and $\varphi_2 = \varphi_{r2} - \varphi_{OLO}$, such that $\varphi_{OLO1} = \varphi_{OLO2} = \varphi_{OLO}$. From the four BPD currents, the two messages are to be separately sampled and fed to ADC Further more, both the QPSK and 16-QAM would be afflicted by the phase noise. Probably, it is impassable to recover and demodulate the two microwave signal precisely from the output of coherent receiver [7]. In order to demodulate 16-QAM and QPSK signals, Matlab signal processing $(\$MSP\$)$ schemes are employed to cancel the phase noise, eliminate the frequency shift and recover the signal from overlapped spectra. MSP takes repetitive tasks also operates analytical mathematical trends. The Amplitude modulated, 16-QAM is regained by squaring and adding $I_1(t)$ and $I_2(t)$

$$I_{16\,QAM}(t) = 4R^2 P_{r1}P_{OLO1}\left[1 - \sin\left(\Pi * E_{16\,QAM}(t)/V_{IM}\right)\right] \quad (12)$$

Thus, 16-QAM signal, free from phase noise is obtained for demodulation. It has to be noted that from Eq. 12, $I_{16\,QAM}(t)$, only 16-QAM is and no QPSK is procured after processing. From the four currents, the phase modulating QPSK signal is obtained according to the following digital processing of signal.

$$I_{QPSK}(t) = \operatorname{atan}\left(I_{2314}(t)/I_{1234}(t)\right) \quad (13)$$

where,

$$I_{2314}(t) = \sin\left(\pi * E_{QPSK}(t)/2V_{PM} + \varphi_1(t) - \varphi_2(t)\right) \quad (14)$$

$$I_{1234}(t) = \cos\left(\pi * E_{QPSK}(t)/2V_{PM} + \varphi_1(t) - \varphi_2(t)\right) \quad (15)$$

Again, substituting Eqs. 14 and 15 on Eq. 13,

$$I_{QPSK}(t) = \pi * E_{QPSK}(t)/2V_{PM} + \varphi_1(t) - \varphi_2(t) \quad (16)$$

where $\varphi(t) = \varphi_1(t) - \varphi_2(t)$, whose maximum frequency is determined by the width of the laser source.

3   Methodology

A research occupied on the setup shown in Fig. 1 is executed. A laser source of adjustable power running at frequency of 193.343 THz(λ = 1551.646 nm), with the width of the optical spectrum of proximate 1 MHz and an emitting power of 16 dBm, is employed at the carrier source at the transmitter, which is partitioned into 2 pathways, namely Route 1 and Route 2 with the power 15.2 dBm and .8 dBm respectively. Route 1 carrying supreme power in correlate with Route 2, Information 1 as 16-QAM at 550 MHz is first modulated with the carrier signal, using MZM. In MZM, the RF information 1 signal is fed to port 1 and 2 with and without 90 phase shift respectively (Fig. 2).

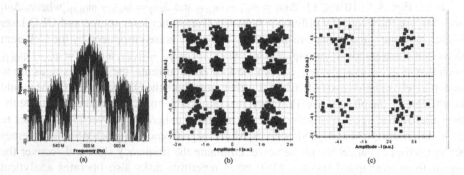

Fig. 2. Results obtained on simulation of the proposed system. (a) The recovered signal at 550 MHz, (b) The detected 16 QAM after PNC processing, (c) The detected QPSK after PNC processing

The intensity modulated signal at the output of MZM, is fed to phase modulate with information 2 with the same microwave frequency = 550 MHz which is the QPSK signal. Thus the vector carrier signal is forcefully exploited in its amplitude and phase to carry 2 separate information from CO to BS. The unmodulated carrier arriving through Route 2, is polarization shifted to 90^0 and the two routes are polarization multiplexed to Optical Fiber (OF) of 2 km to BS. The polarization controllers PC 1 and PC 2 can be used to restrain the state polarization in the two pathways, by fixing PC1 to 0^0 and PC2 to 90^0. At BS the two messages are to be coherently detected. Each of the de-multiplexed signal through Route 3 and Route 4 are one of the inputs to the two 90^0OH suitably. A second laser light source,as the local oscillator (OLO) for coherent detection, is at work at the BS. The OLO has frequency of 193.344 THz, λ = 1551.638 nm with the spectrum width of 10 MHz is assigned, having the power emiting of 9.3 dBm. 90^0OH1, the modulated signal, is co polarised with the OLO signal using PC3 and PC4. At 90^0OH2, the unmodulated carrier signal with 0.005 of the total power is beaten with better half the OLO signal.

The frequency difference between the two optical sources coincide to 1000 MHz, corresponding $\Delta\lambda$ = 0.008 nm. The four beated optical outputs from two 90^0OH are fed to the BPDs, to convert them to electrical signals correspondingly. $I_1(t)$ and $I_2(t)$ contain both the informations in an overlapped spectrum, whereas $I_3(t)$ and $I_4(t)$ act as the beating signal to recover QPSK and 16-QAM. The four PD currents are fed to Matlab processing, where the signals are segregated into 16-QAM and QPSK and the phase noise is excellently cancelled. Thus, the proposed signal processing scheme along with multiple information transmission is suitable for coherent system using any applicative laser sources that are designed for optical communication (Fig. 3).

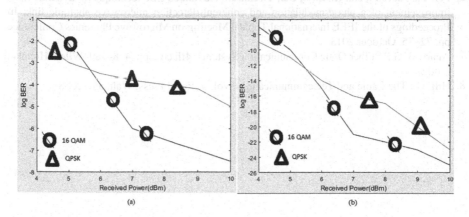

Fig. 3. BER analysis of 16 QAM and QPSK. (a) Conventional system, (b) Proposed system

4 Conclusion

A high bandwidth proficient, coherent jitter cancelling MPL supporting simultaneous intensity and phase modulation with signal processing in order to cancel phase noise was proposed and experimentally demonstrated. It is validated through the experiment that through advanced signal processing, the overlapped AM + PM signals would be recovered and the noise caused due to time delay between carrier laser source and LO source is cancelled. The proposed technique was evaluated experimentally. Two microwave vector signals, one being 16-QAM and other QPSK were transmitted over 10 km OF and recovered effectively by coherent technique at the receiver. The expected scope of the project is to employ additional messages on the uplink transmission and engage other effective PNC techniques. The BER for recovered QAM and QPSK were calculated to be −10 and −12 respectively, corresponding to the Received Optical power 5 dBm and 7 dBm respectively.

References

1. Chen, X.: A high spectral efficiency coherent microwave photonic link employing both amplitude and phase modulation with digital phase noise cancellation. J. Light. Technol. **33**(14), 3091–3097 (2015)
2. Chen, X.: A high spectral efficiency coherent RoF using OSSB modulation with low cost lasers and UDWDM-PONS. J. Light. Technol. **33**(11), 2789–2795 (2016)
3. Zhang, Y., Zhang, F.: Coherent OSSB modulation with tunable carrier to sideband ratio. IEEE Photonics Technol. Lett. **19**(16), 653–655 (2014)

4. Caballero, A., Zibar, D., Monroy, I.T.: Performance evaluation of digital coherent receivers for phase-modulated radio-over-fiber links. J. Light. Technol. **29**(21), 3282–3292 (2011)
5. Chen, Y., Shao, T., Wen, A., Yao, J.P.: Microwave vector signals transmission over an optical fiber based on IQ modulation and coherent detection. Opt. Lett. **39**(6), 1509–1512 (2014)
6. Pei, Y., Yao, J.P., Xu, K., Li, J., Dai, Y., Lin, J.,: Advanced DSP technique for dynamic range improvement of a phase-modulation and coherent-detection microwave photonic link. In: Proceedings of the IEEE International Topical Meeting on Microwave Photonics Conference, pp. 72–75, October 2013
7. Agarwal, G.P.: Fiber Optic Communication Systems, 4th edn, pp. 478–489. Wiley, Hoboken (2002)
8. Hill, G.: The Cable and Telecommunications, vol. 3. Focal Press, Waltham (2008)

Power Electronics

A Probabilistic Modeling Strategy
for Wind Power and System Demand

A. Y. Abdelaziz[1], M. M. Othman[1], M. Ezzat[1], A. M. Mahmoud[1],
and Neeraj Kanwar[2(✉)]

[1] Electrical Power and Machines, Faculty of Engineering,
Ain Shams University, Cairo, Egypt
[2] Electrical Engineering Department, Manipal University Jaipur, Jaipur, India
nkl2.mnit@gmail.com

Abstract. This paper presents a novel modeling technique for modeling wind based renewable energy sources considering their intermittent nature. This algorithm utilizes historical data of wind speeds and take into consideration wind power characteristics by determining the most closely cumulative distribution function. Monte Carlo Simulation is used for determining the most likelihood wind power at each hour at each season. Furthermore, the same modeling algorithm is applied for modeling system demand. The outcomes from the proposed technique and another probabilistic model are compared for showing the validity of the introduced technique. The introduced methodology is implemented using MATLAB program, Finally, the outcomes show that the introduced novel strategy maintains an appropriate result.

Keywords: Monte Carlo simulation · Renewable energy sources
Stochastic load and generation

1 Introduction

The disadvantages of traditional energy sources which depend on fossil fuels, especially their negative environmental impacts, lead to the mounting need of new sources of energy as wind and solar energy. Renewable energy sources (RES) output powers are uncontrollable and hard to be predicted due to dependability on temperature, time and season [1]. Among all RES, wind energy can be considered the lowest risk and most established technology due to the recent technical developments and financing options. However, the wind power is strongly dependent on the wind speed which has a stochastic nature. Therefore, stochastic modeling algorithms are required in order to model the wind power while addressing their uncertainties [2]. Moreover, the modeling strategy should depend on the uncertainties associated with the wind power as the dependence on only one single value of forecast make it not accurate in analyzing the system under study, Therefore, the distribution of the wind power depends on the stochastic nature of the wind speed, Moreover, the wind power distribution is characterized by a Weibull distribution [3].

Authors in [4] presented a new algorithm for evaluating power system capacity in the presence of wind energy. This methodology is dependent on Monte Carlo

© Springer Nature Singapore Pte Ltd. 2018
I. Zelinka et al. (Eds.): ICSCS 2018, CCIS 837, pp. 625–640, 2018.
https://doi.org/10.1007/978-981-13-1936-5_64

Simulation (MCS). Moreover, generating units are assumed to be located at single or more than one site which are considered to be independent sites. A sequential Monte Carlo simulation (SMCS) procedure was utilized in order to take the concept of combined generating units in consideration. In [5], the authors presented an algorithm for studying wind generation uncertainties effects on the power system operation by utilizing MCS for obtaining several wind generation scenarios. These scenarios were utilized for showing the effects of wind power on the prices of the electricity market and the capacity of the power system.

Deterministic power flows are usually utilized in order to analyze the electrical power system quantities; however, system demands variation and power fluctuation due to uncertainties in wind speed in case of wind generators cannot be considered with the deterministic power flows because of using certain power values. Probabilistic power flow (PPF) methods are used to overcome the problems of the deterministic power flows since they apply techniques for taking the uncertainty of input variables into consideration so leading to more accurate obtained results. The MCS methodology is usually considered the base for probabilistic power flows because it is easy to be developed. Probabilistic power flows which are dependent on MCS are also may be named as numerical methods which provide accurate results rather than the analytical methods [6]. Thus, numerical methods are much preferred rather than the analytical methods.

Authors in [7] presented an algorithm for stochastic generation of electrical power by using uncontrollable energy sources. This paper presented the principles of this modeling when MCS is utilized in order to check the dependence between the input powers that can be considered a unique choice as analytical method cannot be applied. Authors in [8] provided a Zhao's point estimate method (PEM). This methodology can treat with correlated input random variables (RVs) with either normal or non-normal probability distributions. Moreover, this method only needs the marginal distribution function data for each variable and their correlation coefficients rather than knowing the joint probability density function (PDF).

Reference [9] proposed a probabilistic load flow (PLF) algorithm in a high presence of renewable energy-based generation. The relation between the load demand and renewable energy resources generation was considered in this study. Reference [10] presented a novel for PLF for improving the inadequacy of the transmission system depending on both concepts of cumulants and Gram-Charlier expansion theory for having the PDFs of transmission line flows.

In [11], an application was presented of a two-point estimate method in order to take the uncertainties of optimal power flow problem into consideration. In this method probability distributions of locational marginal prices were calculated as a result instead of using computationally demanding methods. Authors in [12] introduced a technique that considers the wind power uncertainties by utilizing benders decomposition algorithm and MCS in order to study the effect of increasing wind power generation on transmission systems.

Reference [13] introduced an empirical method that considers incomplete wind data to generate a wind farm wind speed model. Monte Carlo filters and likelihood function maximization are considered in order to calibrate the obtained model and calculate the parameters. Therefore, stochastic modeling algorithms are required in order to model the wind power while addressing their uncertainties [2].

This paper introduced a novel modeling strategy for wind power generation. The introduced model depends on MCS technique and the accurate calculation of the most appropriate CDF. The modeling strategy utilizes the variations of wind power at each hour at each season. The outcomes of the introduced model and a probabilistic model presented in [14, 15] are compared. The rest of the paper is divided into four sections. Section 2 describes the proposed model. The modeling strategy presented in [14, 15] was based on the calculation of the most appropriate PDF. The rest of the paper is divided into four sections. Section 2 introduces the proposed methodology. Section 2.1 discusses the model in [14, 15]. System demand modeling strategy is described in Sect. 2.2. Results and conclusions are finally presented in Sects. 3 and 4, respectively. The proposed algorithm and literature algorithm are implemented in MATLAB environment. Results of both strategies are compared and the comparison shows the validity of the proposed algorithm in calculating the output wind powers at each hour.

2 Proposed Strategy

This part discusses the proposed strategy of wind power; the proposed technique depends on MCS. The available data is organized into 4-seasons data (i.e. each season data is separated). Moreover, each data of the 4-seasons data is divided into 24-h data. Therefore, there are 96-time segments for each year. The wind speeds nature is presented by utilizing the Weibull CDF. The modeling technique that calculates the hourly output wind power is minutely discussed in the following step by step algorithm

(1) Begin with the first hour of the 96-time segments.
(2) For each hour determine the Weibull CDF for the wind speed data.
(3) Make a vector that contains 10,000 random numbers with a uniform distribution and their values are between [0.1].
(4) Determine the corresponding wind speeds depending on the random number value from the inverse CDF function (CDF^{-1}).
(5) For the whole 96-time segments and at each simulated wind speeds value, calculate the output power by using (1). This output wind power depends mainly on the wind speed and the characteristics of the available wind turbine.

$$P(W) = \begin{cases} 0 & W \leq Wci \\ Prated\left(\frac{W-Wci}{Wr-Wci}\right) & Wci \leq W \leq Wr \\ Prated & Wr \leq W \leq Wco \\ 0 & W \geq Wco \end{cases} \tag{1}$$

where W can be defined as the wind speed (m/s), W_{ci} can be defined as the cut in speed, W_r can be defined as the rated speed, W_{co} can be defined as the cut out speed and P_{rated} can be defined as the maximum turbine power (MW).

(6) Calculate the most likelihood output wind power at each hour of the 96 time-segments at each season by getting the mean value of power for the 10000 simulated wind powers at each hour by utilizing MCS. This average value of the output wind power can be calculated by utilizing MC convergence. For the available 10000 simulations, The MC convergence is accurate. This most likelihood output wind power at each hour can be calculated by using (2).

$$P_{ave} = \frac{1}{NS} \sum_{J=1}^{NS} P(J) \tag{2}$$

Where P_{ave} is the average wind power value at each hour, P (j) is the power random variable at each hour and NS is the simulations number (i.e. Which is 10000 simulations in our study). The MCS method is utilized for all hours (i.e. 96 time-segments) in order to obtain the most likelihood wind power values at each hour at each season so that a typical day model is produced for each season. Thus, the representation output wind powers behavior during the intervals of the season is produced. Finally, the aforementioned steps for determining the most probabilistic output wind powers at each hour at each season is summarized in the next flowchart as presented in Fig. 1.

2.1 Methodology Presented in [14, 15]

This part discusses the algorithm presented in [14, 15] for the sake of comparison with the introduced model. This strategy is discussed by using the following steps:

1. Begin with the first hour of the 96 time-segments.
2. Determine the probability density function for each hour. This can be done by dividing the wind speeds into a specified number of states and each state have certain wind speed limits as shown in Table 1. The large number of states makes the analysis more difficult while small number of states can affect the accuracy of the solution. Thus, the number of states must be chosen carefully.
3. Determine each state power by using Eq. (1). Each state power is determined by using the intermediate speed of this state (i.e. if the boundaries of a certain state is between 5 and 6 m/s thus, the chosen wind speed is 5.5 m/s).
4. Determine the output power from the wind turbine for each hour by using Eq. (3).

$$P_{ave} = \sum (P\,state * (probabillity\,of\,each\,state)) \tag{3}$$

Where the probability of each state can be introduced as the division of the wind speeds number at each state by the total number of speeds.

2.2 System Demand Modeling Strategy

The proposed strategy depends on MCS. The available data is organized into 4-seasons data (i.e. each season data is separated). Moreover, each data of the 4-seasons data is divided into 24-h data. The Gaussian CDF is used to represent the system demand behavior. Then the data of system demand at each season is divided into several states

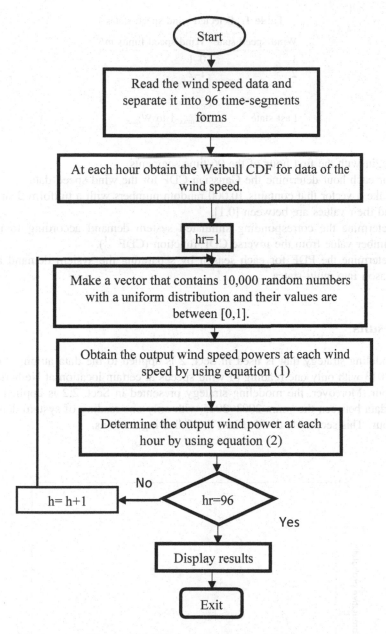

Fig. 1. Flow chart of the modeling strategy

(i.e. state 1: system demand from 0 to 10% of the peak load at this season, state 2: system demand from 10% to 20% of the peak load till reaching the final state), Thus the continuous probability density functions (PDFs) are calculated at each state. Finally, the following step by step algorithm is used for modeling the system demand with associated PDF at each state at each season.

Table 1. Selected wind speed states

Wind speed state	Wind speed limits m/s
1	0–1
2	1–2
.	.
.	.
.	.
Last state	$W_{max}-1$ to W_{max}

(1) Begin with the first hour of the 96-time segments
(2) For each hour determine the Gaussian CDF for the wind speed data.
(3) Make a vector that contains 10,000 random numbers with a uniform distribution and their values are between [0.1].
(4) Determine the corresponding simulated system demand according to random number value from the inverse CDF function (CDF $^{-1}$).
(5) Determine the PDF for each season by separating the system demand at each season into multi-states.

3   Results

The modeling strategy that is used in Sect. 2 is applied to the data among the years 1983–2013 with only one reading of wind speeds at certain location at Netherlands at each hour. Moreover, the modeling strategy presented in Sect. 2.2 is applied to historical data between the years 2003–2015 with only one reading of system demand at each hour. This section is divided into the following sections.

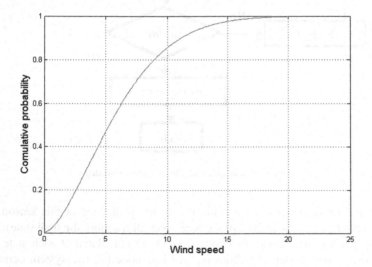

Fig. 2. CDF at h = 3 of the winter season.

3.1   Renewable Energy-Based Generator Model

3.1.1   CDF and PDF for the Historical Data

The Weibull CDFs for each hour at each season are obtained. The calculated CDF at hr = 3 is shown in Fig. 2.

From the aforementioned CDF it is noticed that at hr = 3. there are more than 60% of wind speeds is within cut-in and the cut-out wind speeds so a certain value of output wind power is generated. Almost 34% of wind speeds are below the cut in speed so it results in zero output power. Finally, about 1.08% of the wind speeds are above the cut-out speed so it results in zero output wind power. Figure 3 previews the calculated CDF at hr = 29 (i.e. 05:00 a.m. of the specified spring day). Figure 4 previews the calculated CDF at hr = 55. Figure 5 previews the calculated CDF at hr = 81.

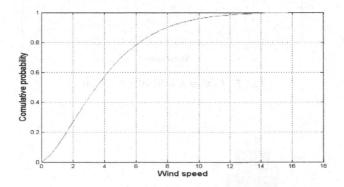

Fig. 3. CDF at h = 29 of spring season.

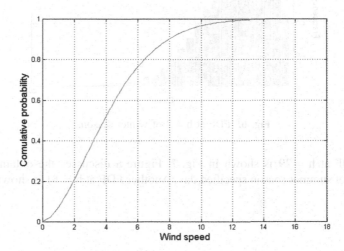

Fig. 4. CDF at h = 55 of the summer season.

The Hourly PDFs for each hour at each season are obtained. The calculated PDF at hr = 3 (i.e. 3:00 a.m., of the specified winter day) is shown in Fig. 6.

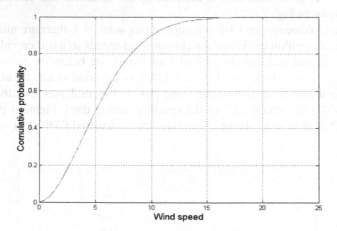

Fig. 5. CDF at h = 81 of fall season.

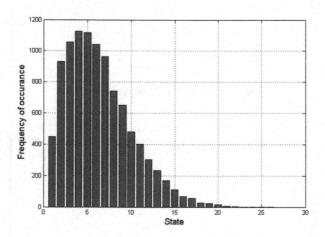

Fig. 6. PDF at h = 3 of winter season.

The PDF at h = 29 is shown in Fig. 7. Figure 8 discusses the obtained PDF at 07:00 a.m. of summer season specified day. Another PDF at h = 81 is shown in Fig. 9.

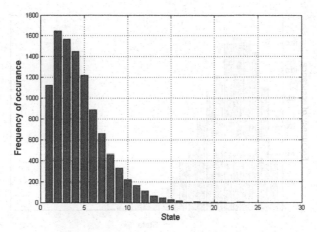

Fig. 7. PDF at h = 29 of spring season.

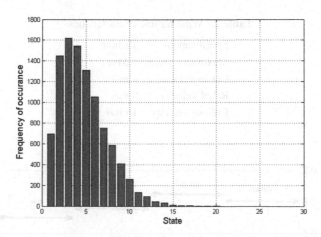

Fig. 8. PDF at h = 55 of summer season.

3.1.2 Outcomes of the Introduced Modeling Strategy

The results of the overall time segments are obtained by using a wind turbine which has the following specifications as shown in Table 2.

The obtained output wind power at each hour of the 96 time -segments of all season are shown in Fig. 10.

3.1.3 Validation of MCS Results

The graph between the average output wind power and the total number of iterations at hr = 3 of the winter season is shown in Fig. 11. Figure 12 previews the output power for the actual wind speed data of the winter season.

It can be deduced that from the Figs. 11 and 12, the obtained results from MCS is close to the actual data results which validates the introduced modeling strategy.

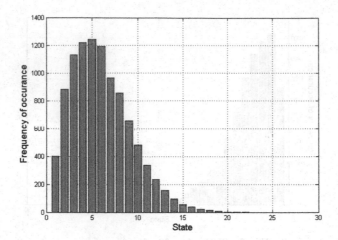

Fig. 9. PDF at h = 81 of fall season.

Table 2. Wind turbine characteristics

Specifications	Turbine
Maximum power	1 MW
Cut-in velocity	4 m/s
Rated velocity	8 m/s
Cut-out velocity	16 m/s

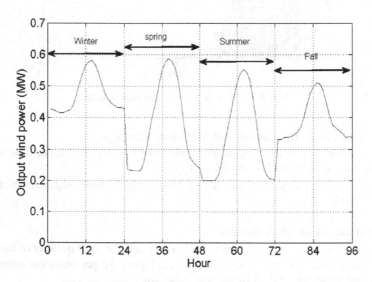

Fig. 10. Wind power for all hours of all seasons.

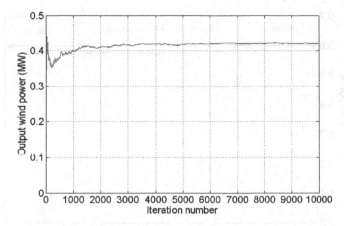

Fig. 11. MCS convergence for h = 3 of the winter season.

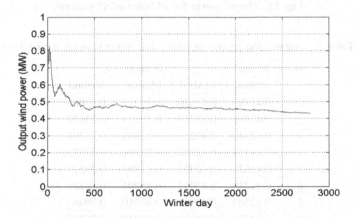

Fig. 12. Original data output power at hr = 3

3.1.4   Comparison with Previous Literatures

The graph between the output wind powers and all 96 h' time segments of the second algorithm is shown in Fig. 13 and it can be deduced that the output results from the second algorithm is very close to the obtained results of the introduced strategy. Figures 10 and 13 preview that the peak output wind power is occurred at the mid-hour of the chosen typical day as this mid-hour represents the peak system load.

Table 3 shows the output wind powers for winter and spring seasons for the proposed modeling strategy and the model in [14, 15].

Table 4 previews the output wind powers for summer and fall seasons for the proposed modeling strategy and the method in [14, 15].

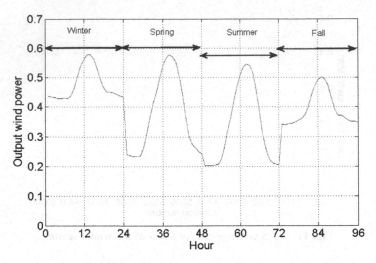

Fig. 13. Output power for all hours of all seasons.

Table 3. Output wind powers at winter and spring seasons at each hour

Hour	Winter		Spring	
	Method 1	Method 2	Method 1	Method 2
1	0.42787	0.43670	0.23316	0.23966
2	0.42582	0.43410	0.23034	0.23637
3	0.42040	0.43254	0.22889	0.23304
4	0.41559	0.42968	0.22897	0.23523
5	0.41628	0.42924	0.22863	0.23527
6	0.42239	0.43053	0.25201	0.26223
7	0.42509	0.42959	0.30310	0.30943
8	0.43628	0.44036	0.35940	0.36347
9	0.46696	0.46560	0.41197	0.41168
10	0.50553	0.50232	0.45931	0.45345
11	0.53809	0.53762	0.50937	0.50118
12	0.56991	0.56348	0.55309	0.54089
13	0.58040	0.57814	0.57616	0.56715
14	0.58406	0.57943	0.58800	0.57885
15	0.57117	0.56478	0.58049	0.57192
16	0.54342	0.53440	0.56022	0.55321
17	0.50591	0.49227	0.51656	0.50675
18	0.47562	0.46591	0.45189	0.44592
19	0.45928	0.45202	0.37011	0.36904
20	0.44619	0.45113	0.30211	0.30501
21	0.43985	0.44800	0.26948	0.27595
22	0.43356	0.44322	0.25358	0.26175
23	0.43331	0.43786	0.24305	0.25000
24	0.43001	0.43442	0.23617	0.24115

Table 4. Output wind powers at summer and fall seasons at each hour

Hour	Summer		Fall	
	Method 1	Method 2	Method 1	Method 2
1	0.19620	0.20249	0.33077	0.34159
2	0.19672	0.20293	0.33055	0.34039
3	0.19667	0.20297	0.33623	0.34358
4	0.19673	0.20367	0.33644	0.34558
5	0.19890	0.20643	0.33907	0.34753
6	0.21272	0.22261	0.34435	0.35453
7	0.25039	0.26056	0.35206	0.36051
8	0.31074	0.31623	0.36647	0.37429
9	0.36782	0.36921	0.39990	0.40052
10	0.41886	0.42273	0.44026	0.43770
11	0.47006	0.47107	0.47449	0.47226
12	0.51159	0.50903	0.49802	0.49211
13	0.53981	0.53682	0.51014	0.50261
14	0.55182	0.54663	0.50687	0.49544
15	0.54364	0.53914	0.48562	0.47155
16	0.51804	0.51166	0.44742	0.43123
17	0.46170	0.45306	0.40576	0.39396
18	0.39253	0.38372	0.37592	0.37221
19	0.31540	0.30834	0.36657	0.37154
20	0.25327	0.25276	0.35647	0.36375
21	0.22036	0.22344	0.34746	0.35524
22	0.20536	0.21130	0.33603	0.35231
23	0.20134	0.20735	0.33764	0.35134
24	0.19981	0.20499	0.33726	0.34793

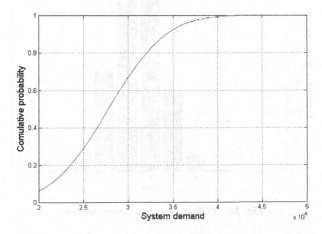

Fig. 14. CDF at hr = 3 of winter season.

3.2   System Demand Model

The proposed modeling strategy introduced in Sect. 2.2 is applied to the available data which contains only one certain reading of system demand value at each hour at each season. The Gaussian CDFs for each season are obtained based on the available system demand season, Fig. 14 shows the CDF of the available system data of the winter season at h = 3. Figure 15 previews the obtained CDF at hr = 29.

Figure 16 previews the calculated PDF at hr = 3 Fig. 17 previews the calculated PDF at hr = 29.

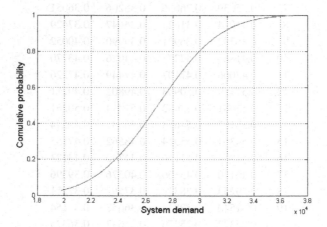

Fig. 15. CDF at h = 29 of spring season.

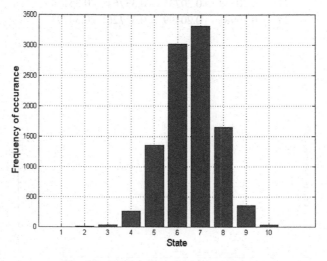

Fig. 16. PDF at hr = 3 of winter season.

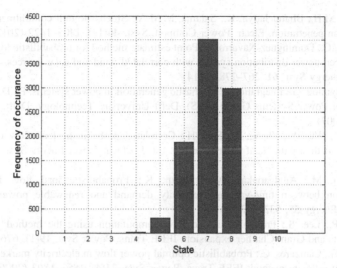

Fig. 17. PDF at h = 29 of spring season.

4 Conclusions

The introduced modeling algorithm for modeling the wind based renewable energy generators is presented based on the calculation of the accurate CDF of the wind speed. Moreover, it is compared with another modeling strategy presented in previous literature. Furthermore, the same modeling strategy is applied in order to get the system demand model.

The proposed modeling strategy depends on MCS technique for determining the hourly output wind power. There are two different techniques for characterizing the wind based renewable energy-based generators one of them is based on the calculation of the CDF and the other one is based on the calculation of the most appropriate PDF as presented in the paper.

The results are promising if it used in load management strategies by using energy storage systems (ESSs) in order to maximize the arbitrage benefit and minimize the net present value of energy losses cost.

References

1. Tsekouras, G., Koutsoyiannis, D.: Stochastic analysis and simulation of hydrometeorological processes associated with wind and solar energy. Renew. Energy **63**, 624–633 (2014)
2. Siano, P., Mokryani, G.: Probabilistic assessment of the impact of wind energy integration into distribution networks. IEEE Trans. Power Syst. **28**(4), 4209–4217 (2013)
3. Zhang, Z.-S., Sun, Y.-Z., Lin, J., Cheng, L., Li, G.-J.: Versatile distribution of wind power output for a given forecast value. In: 2012 IEEE Power and Energy Society General Meeting (2012)
4. Billinton, R., Bai, G.: Generating capacity adequacy associated with wind energy. IEEE Trans. Energy Convers. **19**(3), 641–646 (2004)

5. Ahmed, M.H., Bhattacharya, K., Salama, M.M.A.: Stochastic unit commitment with wind generation penetration. Electr. Power Compon. Syst. **40**(12), 1405–1422 (2012)
6. Delgado, C., Domínguez-Navarro, J.: Point estimate method for probabilistic load flow of an unbalanced power distribution system with correlated wind and solar sources. Int. J. Electr. Power Energy Syst. **61**, 267–278 (2014)
7. Papaefthymiou, G.: Integration of stochastic generation in power systems. Ph.D. dissertation, Electrical Power Systems Group (EPS), Delft University Technology, Delft, The Netherlands (2007)
8. Chen, C., Wu, W., Zhang, B., Sun, H.: Correlated probabilistic load flow using a point estimate method with Nataf transformation. Int. J. Electr. Power Energy Syst. **65**, 325–333 (2015)
9. Abdullah, M., Agalgaonkar, A., Muttaqi, K.: Probabilistic load flow incorporating correlation between time-varying electricity demand and renewable power generation. Renew. Energy **55**, 532–543 (2013)
10. Zhang, P., Lee, S.: Probabilistic load flow computation using the method of combined cumulants and Gram-Charlier expansion. IEEE Trans. Power Syst. **19**(1), 676–682 (2004)
11. Verbic, G., Canizares, C.: Probabilistic optimal power flow in electricity markets based on a two-point estimate method. IEEE Trans. Power Syst. **21**(4), 1883–1893 (2006)
12. Orfanos, G.A., Georgilakis, P.S., Hatziargyriou, N.D.: Transmission expansion planning of systems with increasing wind power integration. IEEE Trans. Power Syst. **28**(2), 1355–1362 (2013)
13. Hosseini, S.H., Tang, C.Y., Jiang, J.N.: Calibration of a wind farm wind speed model with incomplete wind data. IEEE Trans. Sustain. Energy **5**(1), 343–350 (2014)
14. Othman, M.M., Hegazy, Y.G., Abdelaziz, A.Y.: Electrical Energy Management in Power Delivery Systems: Virtual Power Plant Concept, LAP LAMBERT, December 2015
15. Abdelaziz, A., Hegazy, Y., El-Khattam, W., Othman, M.: Optimal allocation of stochastically dependent renewable energy based distributed generators in unbalanced distribution networks. Electr. Power Syst. Res. **119**, 34–44 (2015)

Performance Analysis of High Sensitive Microcantilever for Temperature Sensing

Balasoundirame Priyadarisshini[1], Dhanabalan Sindhanaiselvi[1],
and Thangavelu Shanmuganantham[2(✉)]

[1] Department of Electronics and Instrumentation Engineering,
Pondicherry Engineering College, Pondicherry 605014, India
priyasandra94@gmail.com, sindhanaiselvi@pec.edu
[2] Department of Electronics Engineering, Pondicherry Central University,
Pondicherry 605014, India
shanmugananthamster@gmail.com

Abstract. Temperature is one of the influential parameter in the environment for the production of irrigation and the progress of weather pattern. The objective of this paper is to design the MEMS micro cantilever based temperature sensor for the range of (0 °C to 100 °C). The micro cantilever is tend to deflect when applied with temperature. The performance of micro cantilever is investigated in respect of deflection and stress with different material as a sensing layer, varied length and width. Further investigations are carried out to improve the sensitivity by adding ribs and perforations.

Keywords: MEMS · Cantilever · Deflection · Perforations and ribs

1 Introduction

MEMS technology is the most widely held miniaturization technique. Nowadays, more research work is going on in implementing advanced MEMS based biosensor, inertial sensor, communication sensor, agricultural sensors, drug delivery system [3]. Temperature sensors is the frequently used parameter in a varied applications as of result monitoring soil science, metrology, agriculture, BioMEMS, chemical analysis and etc., [1, 2]. In the field of weather pattern and production of irrigation, temperature sensing is one of the key factor for the accurate measurement. Cantilever based sensing is a flexible approach since it is a simplified structure, affordable, highly sensitive, faster response. In this paper, a simple cantilever structure is constructed and investigated for performance with respect to material, length and width by measuring deflection and stress. Micro cantilever based temperature sensing is also tested for linearization. Further optimized structure is analyzed for enhancement of sensitivity by the addition of ribs and perforation. This study is analyzed for the temperature range of (0 °C to 100 °C).

I. Zelinka et al. (Eds.): ICSCS 2018, CCIS 837, pp. 641–648, 2018.
https://doi.org/10.1007/978-981-13-1936-5_65

2 Design of Micro Cantilever Structure

The simple micro cantilever structure for temperature sensing is shown in Fig. 1. The structure comprises of substrate, dielectric layer and sensing layer. The three different T-shaped structure namely i-shaped cantilever, cantilever with single T shaped sensing layer and cantilever with double T-shaped dielectric sensing layer is shown in Fig. 2. The temperature sensor is constructed using INTELLISUITE MEMS CAD tool.

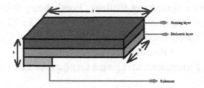

Fig. 1. Simple cantilever structure

Where l = length (μm),
w = width (μm),
h = height (μm)

In MODEL 1, i-shaped cantilever structure, the substrate is constructed with the dimension of 100 μm × 50 μm × 0.5 μm, dielectric layer with dimension of 50 μm × 100 μm × 0.5 μm and the sensing layer in the dimension of 50 μm × 100 μm × 0.5 μm.In MODEL 2, cantilever with single T-shaped sensing layer model having substrate with the dimension of 100 μm × 50 μm × 0.5 μm, dielectric layer with dimension of 50 μm × 100 μm × 0.5 μm and the sensing layer in the dimension of 50 μm × 100 μm × 0.5 μm. In MODEL 3, cantilever with double T-shaped sensing layer model having substrate with the dimension of 100 μm × 50 μm × 0.5 μm, dielectric layer with dimension of 50 μm × 100 μm × 0.5 μm and the sensing layer in the dimension of 50 μm × 100 μm × 0.5 μm.

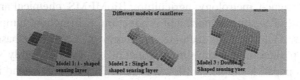

Fig. 2. Different models of cantilever based temperature sensing

2.1 Sensitivity Analysis with Material

The sensitivity of the temperature sensor is analysed with different sensing materials like gold (Au), Aluminium (Al), Platinum (Pt), Titanium (Ti), and Copper (Cu). In the above models, bottom layer is of silicon as substrate with thickness 0.5 μm and the middle layer is of silicon di oxide as dielectric layer and the top layer is a sensing layer coated with the various thin film temperature sensing material such as gold (Au), Aluminium (Al), Platinum (Pt), Titanium (Ti), Copper (Cu) and analysed for the deflection sensitivity. The materials and their properties used for the simulation is given in Table 1.

Table 1. Material properties for the material used in the simulation

Material	Thermal conductivity (At 300 K (W/m K))	Thermal expansion (At 300 K (10-6 Material/°C)	Density (ρ) (10-6 Ω × cm)	Young modulus (GPa)	Poisson ratio
Si	156	2.616	0.22	170	0.26
SiO₂	1.4	0.4–0.55	2.2	73	0.17
Al	236	25	2.7	70	0.36
Au	317	14	19.32	74.48	0.44
Pt	71.6	8.8	21.45	146.9	0.38
Cu	401	17	8.93	117	0.32
Ti	21.9	8.5–9	4.51	115	0.32

The deflections obtained for different sensing material at a temperature of 100 °C with the dimension of the micro cantilever 100 μm × 50 μm × 0.5 μm is compared for three different model is listed in Table 2.

Table 2. Sensitivity analysis with different material

Material			Deflection (μm)		
Substrate layer 1	Dielectric layer 2	Sensing layer 3	Model 1	Model 2	Model 3
Si	SiO₂	Pt	4.48867	2.44338	4.14933
Si	SiO₂	Au	7.73502	4.44485	7.21884
Si	SiO₂	Cu	9.32254	5.39952	8.73295
Si	**SiO₂**	**Al**	**14.1941**	**8.38154**	**13.3379**
Si	SiO₂	Ti	4.08763	2.10161	3.74263
Si	Pt	Au	4.94071	4.40689	5.59069
Si	Pt	Cu	6.5226	5.22337	6.97101
Si	**Pt**	**Al**	**11.3699**	**8.29302**	**11.4111**
Si	Au	Cu	4.70796	5.18804	6.01981
Si	Au	Al	9.9451	8.61472	11.1567
Si	Ti	Au	5.18964	4.42064	5.72642
Si	Ti	Al	11.7122	8.36421	11.7365
Si	Ti	Cu	6.76236	5.25494	7.12736
Si	Cu	Al	9.03137	8.57169	10.6163

From the results obtained in Table 2, MODEL 1 with Aluminium as a sensing layer and SiO$_2$, as dielectric layer gives the maximum deflection of 14.1941 μm. Similarly for MODEL 2 offers the maximum deflection of 8.38154 μm and MODEL 3 with 13.339 μm. The temperature sensing material which has the maximum deflection from Table 2 is analysed for linearity from 10 °C to 100 °C in the step of 10 °C. The temperature versus deflection is plotted in the Fig. 3.

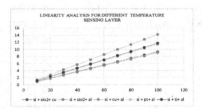

Fig. 3. Temperature (°C) versus deflection (μm)

From Fig. 3, it is found that micro cantilever results with linear performance which is suitable to convert deflection into electrical output using piezo resistor with wheat stone bridge configuration in future.

2.2   Sensitivity Analysis with Length

By varying the length of the structure keeping width and thickness as constant, the performance is analysed for maximum deflection and stress. The deflection sensitivity of micro cantilever MODEL 1, MODEL 2, MODEL 3 is carried out in this section. The width is 50 (μm) and thickness is 0.5 (μm) is kept constant for the MODELS. The length is varied from 100 μm to 500 μm in step of 50 μm. The displacement and stress (Sxx, Syy, Szz) at the maximum temperature of 100 °C is given in Table 3.

Table 3. Sensitivity analysis of MODEL 1 by varying length

Dimensions (μm)	Deflection (μm)	Sxx (Mpa)	Syy (Mpa)	Szz (Mpa)
100 × 50 × 0.5	14.1941	146.141	75.0286	25.1839
150 × 50 × 0.5	36.417	142.621	70.6123	24.6381
200 × 50 × 0.5	**62.1871**	**150.051**	**74.7256**	**26.1598**
250 × 50 × 0.5	122.882	35.9576	35.0288	4.17181
300 × 50 × 0.5	153.627	143.664	70.5585	24.5817
350 × 50 × 0.5	207.15	147.245	73.0934	25.5851
400 × 50 × 0.5	306.404	71.9328	70.069	8.33763
450 × 50 × 0.5	353.267	142.608	70.5461	24.8402
500 × 50 × 0.5	474.605	71.9963	70.0606	8.34438

From the result obtained for MODEL 1, though the dimension 500 μm × 50 μm × 0.5 μm gives maximum deflection = 474.605 μm but stress values Sxx = 71.9963 MPa, Syy = 70.0606 MPa and Szz = 8.34438 MPa is minimum. The dimension of 200 μm × 50 μm × 0.5 μm is chosen which gives 62.1871 μm deflection and the stress values Sxx = 150.051 MPa, Syy = 74.7256 MPa and Szz = 26.1598 MPa which is the maximum among the output obtained.

Similarly, the analysis is carried out for MODEL 2 and MODEL 3. It is found that Model 2 with dimension of 200 μm × 50 μm × 0.5 μm is chosen which gives 23.9524 μm deflection and the stress values Sxx = 152.989 MPa, Syy = 64.3502 MPa and Szz = 66.6146 MPa which is the maximum among the output obtained. For Model 3 with dimension of 100 μm × 50 μm × 0.5 μm is chosen which gives 13.3379 μm deflection and the stress values Sxx = 205.349 MPa, Syy = 77.9763 MPa and Szz = 14.7798 MPa which is the maximum among the output obtained.

2.3 Sensitivity Analysis with Width

The deflection sensitivity analysis of micro cantilever MODEL 1, MODEL 2, MODEL 3 with varying width is carried out in this section. The length (μm) and thickness (μm) is kept constant for the MODELS. The width is varied from 10 μm to 100 μm in step of 10 μm. The displacement and stress (Sxx, Syy, Szz) at the maximum temperature of 100 °C is shown in Table 4.

Table 4. MODEL 1 sensitivity analysis by varying width

Dimensions (μm)	Deflection (μm)	Sxx (Mpa)	Syy (Mpa)	Szz (Mpa)
200 × 10 × 0.5	78.4533	73.4614	70.7334	8.46291
200 × 20 × 0.5	77.1009	74.6797	70.7048	8.33932
200 × 30 × 0.5	78.4533	73.4614	70.7334	8.46291
200 × 40 × 0.5	79.1677	71.4263	70.062	84.9871
200 × 50 × 0.5	62.1871	150.051	74.7256	26.1598
200 × 60 × 0.5	81.0516	71.5458	69.9598	8.48661
200 × 70 × 0.5	81.9786	72.6274	69.8379	8.40265
200 × 80 × 0.5	81.9785	72.6272	69.8379	8.40266
200 × 90 × 0.5	84.1233	81.7203	69.3741	8.70932
200 × 100 × 0.5	**50.5893**	**254.065**	**67.0749**	**18.9623**

From the result obtained for MODEL 1, though the dimension 200 μm × 90 μm × 0.5 μm gives maximum deflection = 84.1233 μm but stress values Sxx = 81.7203 MPa, Syy = 69.3741 MPa and Szz = 8.70932 MPa is minimum. The dimension of 200 μm × 100 μm × 0.5 μm is chosen which yields 50.5893 μm deflection and the stress values Sxx = 254.065 MPa, Syy = 67.0749 MPa and Szz = 18.9623 MPa which is the maximum among the output obtained.

Similarly, the analysis is carried out for MODEL 2 and MODEL 3. It is found that Model 2 with the dimension of 200 μm × 70 μm × 0.5 μm is chosen which yields

29.8631 µm deflection and the stress values Sxx = 179.498 MPa, Syy = 86.8054 MPa and Szz = 59.4082 MPa which is the maximum among the output obtained. For Model 3 dimension of 100 µm × 50 µm × 0.5 µm is chosen which yields 13.3379 µm deflection and the stress values Sxx = 205.349 MPa, Syy = 77.9763 MPa and Szz = 14.7798 MPa which is the maximum among the output obtained.

3 Sensitivity Enhancement by Adding Ribs and Perforations in the Micro Cantilever

The sensitivity enhancement is further investigated by adding Ribs as revealed in Fig. 4. Square perforation as revealed in Fig. 5, and Rectangular perforation as revealed in Fig. 6. in the bending beam of micro cantilever [4]. The Ribs are nothing but solid supports added at the bottom. To detail the analysis number of Ribs at the bottom is varied for investigation. The perforation are holes added in the beam to reduce the thickness so as to increase the deflection. The perforations are added in different shape for investigation.

Fig. 4. Ribs added in the cantilever

Fig. 5. Square perforation added in the cantilever

The i-shaped cantilever with the dimension of 200 µm × 50 µm × 0.5 µm is chosen as optimized structure from previous section and investigated with its performance for enhanced sensitivity with the addition of ribs and perforations is listed in Table 5.

Fig. 6. Rectangular perforation added in the cantilever

From the result obtained in Table 5, for MODEL 1 with the dimension of 200 µm × 100 µm × 0.5 µm offers 54.9271 µm by adding perforation in corner near to the substrate. Also the stress is equally distributed at Sxx and Syy when compared with other cases.

Table 5. Sensitivity enhancement analysis for MODEL1

Structure name	Deflection (µm)	Sxx (Mpa)	Syy (Mpa)	Szz (Mpa)
Two base rib	41.7524	202.943	94.8153	36.9922
Three base rib	44.1022	164.364	124.535	29.0915
Single middle rib	33.3564	161.822	106.153	28.5707
Single base rib	35.0807	203.102	89.5711	37.0256
Corner perforation	**54.9271**	**193.719**	**197.005**	**35.1038**
Corner perforation 1	50.8656	163.287	117.036	28.8706
Corner rectangular perforation	49.257	162.83	110.077	28.7772
Single rectangular perforation	51.47	164.464	133.79	29.1118
Three perforation	51.4749	164.501	134.215	29.1194
Double rectangular perforation	51.0724	163.838	27.155	28.9837
Centre rectangular perforation	51.2434	163.989	128.26	29.0146
Double square perforation	51.2799	164.027	127.675	29.0224
Four square perforation	51.2386	163.961	126.863	29.0089

Comparison of deflection for cantilever with and without perforation is given in Table 6.

The relative sensitivity is enhanced by 4.4478 µm for the micro cantilever with the dimension 200 µm × 100 µm × 0.5 µm which shows 8% increase in deflection by adding corner perforations. The relative sensitivity is calculated for the cantilever without perforation.

Table 6. Comparison of sensitivity for MODEL1

Structure name	Dimension (μm)	Deflection (μm)	Relative reflection (μm)
Corner perforation	$200 \times 100 \times 0.5$	54.9271	4.4478
Without perforation	$200 \times 100 \times 0.5$	50.5893	1
Single corner perforation	$200 \times 50 \times 0.5$	62.9293	0.7422
Without perforation	$200 \times 50 \times 0.5$	62.1871	1

4 Conclusion

The micro cantilever is investigated with deflection analysis for temperature sensing application. The performance is analyzed with material, length and width and presented Aluminium material gives the highest deflection. In dimension 200 μm $\times$ 100 μm $\times$ 0.5 μm yields the maximum deflection of 50.5893 μm. This deflection sensitivity is improved to 54.9271 μm by adding corner perforation shows that deflection is improved by 8% than the ordinary micro cantilever model. This stress Sxx, Syy, Szz equally distributed than the ordinary model.

References

1. Ma, R.-H., Lee, C.-Y., Wang, Y.-H., Chen, H.-J.: Microcantilever-based weather station for temperature, humidity and flow rate measurement. Microsyst. Technol. **14,** 971–977 (2008). Springer
2. Chen, L.-T., Lee, C.-Y., Cheng, W.-H.: MEMS based humidity sensor with integrated temperature compensation mechanism. Sensors Actuators A: Phys. **147,** 522–528 (2008). Elseiver
3. Sindhanaiselvi, D., Shanmuganantham, T.: Double boss sculptured diaphragm employed piezoresistive mems pressure sensor with silicon-on-insulator (SOI). J. Eng. Sci. Technol. **12** (7), 1740–1754 (2017). School of Engineering, Taylor's University
4. Sindhanaiselvi, D., Shanmuganantham, T.: Design and analysis of MEMS based piezoresistive pressure sensor for sensitivity enhancement. J. Mater. Today Proc. 2214–7853 (2017, in press). Elsevier

SOS Algorithm Tuned PID/FuzzyPID Controller for Load Frequency Control with SMES

Priyambada Satapathy, Manoj Kumar Debnath$^{(\boxtimes)}$, Sankalpa Bohidar, and Pradeep Kumar Mohanty

Siksha 'O' Anusandhan University, Bhubaneswar, Odisha, India
lirasatapathy@gmail.com, mkd.odisha@gmail.com,
sankalpabohidar@gmail.com, pkmohanty68@rediffmail.com

Abstract. The article presents a newly advanced, novel and proficient symbiotic organism search optimization technique to resolve the stability problem of a power system. A two area reheat based thermal system is considered with the SOS tuned proportional-integral-derivative controller along with fuzzy-PID (FPID) controller separately. The supremacy of this designed power system is verified by introducing a disturbance on load of 0.15 p.u. in one control zone. The validation of the implemented SOS technique is analyzed by doing a comparison with the dynamic responses of the FPID controller over a two area reheat based thermal power system. Finally, a profound verification is achieved to analyze the robustness and non-linearity of the modeled controller by subjecting a 5% of generation rate constraints (GRCs) to the designed system with the existance of superconducting magnetic energy storage.

Keywords: Load Frequency Control · Symbiotic Optimization Search
Fuzzy control · PID control · SMES

1 Introduction

The key role of automatic generation control (AGC) is to retain limelight towards the stability of system and to uphold the scheduled interchange of power within the generating station at the most economical way [1]. The variations in demand of loads may disrupt the steadiness with in the total generation and consistent losses [2]. Researchers have been done lots of investigative research on the field of AGC in the past era. Lee et al. [3] executed PID controller over Simulink model of cascaded control system. The idea of 2DOF-PID controller is introduced in AGC and afterward a proportional-integral controller is added with the proportional derivative controller named as (PI-PD) cascaded controller by Dash et al. [4, 5]. A novel model predictive controller (MPC) has designed to achieve the preferred control of large scale networked power system constraints [6]. In 2015 a controller named PD-PID cascaded controller is injected over a multi area thermal system for LFC [7]. Firefly algorithm tuned fuzzy

© Springer Nature Singapore Pte Ltd. 2018
I. Zelinka et al. (Eds.): ICSCS 2018, CCIS 837, pp. 649–657, 2018.
https://doi.org/10.1007/978-981-13-1936-5_66

PI controller is established over a hybrid power system [8]. Yang and Deb invented meta-heuristic cuckoo search algorithm based upon aggressive reproduction strategy on automatic generation control [9]. In [10] presents an efficient QOGWO optimization technique to acquire the optimal parameters of PID over a two area intersects system to face all the challenges regarding automatic generation control. Laterally the same controller is executed over a four area power system to verify it's superiority [10]. Hybrid optimization technique named TLBO-DE implemented over fuzzy PI controller with generation rate constraints [11]. The performance analysis of teaching learning based optimization technique is examined over a multi area system [12]. Dated back in [13] various integral and proportional integral controller modes tuned by minority carrier inspired algorithm (MCI) are employed on a multi area hydro-thermal system with constant generation rate constraints (GRC). An investigation has been done on both integral and fuzzy logic controller over a hydro-thermal system under small load disturbances and fluctuation of sampling time period [14]. An advanced fractional order proportional-integral-differential (FOPID) is developed along with thyristor controlled series capacitor (TCSC) tuned by improved particle swarm optimization (IPSO) [15]. The article [16] refers to the fuzzy logic controller along with SMES for an automatic generation control. Two unequal areas have considered with multiple generating sources like thermal, hydro, wind and diesel power plants and introduced PSO, IPSO and bacteria Foraging algorithm tuned PID controller [17]. Optimal fuzzy-PID controller was tuned by a new and advanced differential evolution-grey wolf optimization algorithm over a multi- source intersect network for AGC in [18].

2 System Analyzed

In this article an interlocked two area dynamic model is considered with a reheat-thermal system in each area. In this research paper PID and fuzzy PID controller structure is implemented as secondary control loop to regain the system stability fully on the above mentioned power system along with SMES. Due to the slow responses of the governor to compensate the change in load was not preferred in the power system. SMES has the capability to handle that situation by controlling the active and reactive power simultaneously. The optimum values are obtained by tuning with symbiotic organism optimization method by taking ITAE as objective function. Figure 1 displays the structure of the observed system. The nominal parameters of this system are given in appendix. Area control error is known as input to the classical PID controller i.e. forcefully brings to zero to regain the stability.

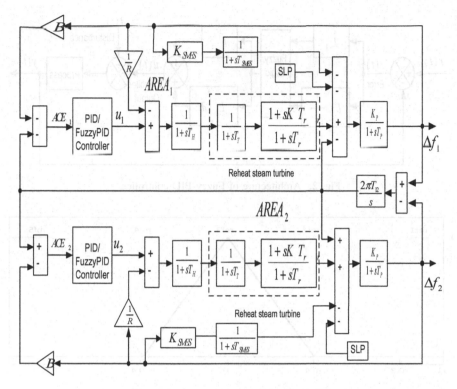

Fig. 1. Power system model with SMES

3 Proposed Controller

In this proposed system proportional Integral Derivative (PID) and FPID controller are implemented as secondary controller to diminish the instabilities in system frequency and tie-line power. A desirable level of control signal is developed by combining the silent feature of proportional, integral and derivative controller. Because of its flexibility and consistency characteristics, now-a-days the commands of PID controller are fed into a programmable logic controller. L_p, L_I and L_D are the gain constraints of this traditional PID controller which is optimize by symbiotic organism optimization approach. Even though PID controllers are able to afford satisfactory control mechanism for simple control systems, they are unable to pay compensation for undesired disturbances. The Fuzzy Logic controllers are used to improve the PID controller ability to handle disturbances. So in our current research purpose we mentioned a PID controller associated with fuzzy-logic accomplishment to standardize the frequency. Figure 2 signifies the design of proposed controller mechanism. In Fig. 3 five membership function are deliberated based upon the Mamdani based fuzzy rule. The fuzzy logic based rule base for this employed controller are represent in Table 1.

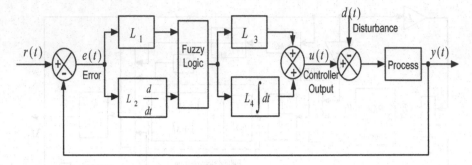

Fig. 2. Architecture of Fuzzy-PID controller.

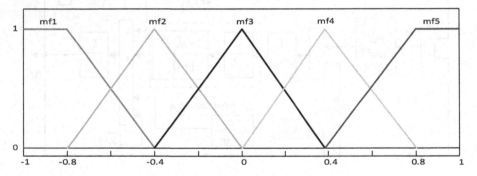

Fig. 3. Fuzzy membership functions

Table 1. Rule-base for the FPI controller

error	$\overset{\bullet}{error}$				
	mf1	mf2	mf3	mf4	mf5
mf1	mf1	mf1	mf2	mf2	mf3
mf2	mf1	mf2	mf2	mf3	mf4
mf3	mf2	mf2	mf3	mf4	mf4
mf4	mf2	mf3	mf4	mf4	mf5
mf5	mf3	mf4	mf4	mf5	mf5

4 SOS Algorithm

In 2014 an advanced stochastic nature stirred population based Symbiotic organism Search optimization process is discovered by Cheng and Prayogo. According to Greek word Symbiosis generally signifies 'living together' that describe the survival relationship between two different kinds of organisms in the ecosystem. The chances of alive are very rare for those species that rely on other species for living in this environment. Obligate and facultative are two kinds of symbiotic interactions within paired organisms. Dependency of two organisms on each other is known as obligate. If two species are not exactly in an essential relationship but cohabitate for mutual benefits is known as facultative. The fitness and survival duration of an organism are improved by four basic symbiotic relationships like Mutualism, commensalism and parasitism.

Mutualism Phase:
Mutualism is one type of relationship that two dissimilar species are profitable. Bees move here and there among flowers for collecting honey which is the perfect example of this mutualism phase. At first choose X_i as a species belonging to the ith participant of the ecology system and then it interacts with another arbitrarily select species X_j to develop the fitness value. The below equation is applied to get the improved solutions of X_i and X_j.

$$X_i^{new} = X_i^{old} + rand^*(X_{best} - MV * BF1);$$ (1)

$$X_j^{new} = X_j^{old} + rand^*(X_{best} - MV * BF2);$$ (2)

Here the rand number is computed with in the interval [0,1]. The mutual vector (MV) and the beneficial factor (BF) are calculated by implementing below procedure.

$$mutual\ vector = \frac{X_i + X_j}{2}$$

$$beneficial\ factor = round(1 + rand)$$

The major role of beneficial factor is to detect whether the interaction is fully or partially favorable for an organism. The term $X_{best} - MV * BF$ helps to improve the percentage of survival in ecosystem by repeating the mutualistic effort.

Commensalism Phase:
In this mentioned phase the interaction of two kinds of organism leads to benefit for one organism where as other species is unaffected by this collaboration. The bond between remora fish and shark is the flawless example of this phase. After collaborate itself with shark the remora fish take the food leftovers and the shark is affected by this action. The X_i is attached to the randomly selected species X_j as like as mutualism phase. But here X_j is neither gets profit nor agonize from this relation whereas X_i takes the benefits from X_j. To obtain the updated solution the below equation is implemented

$$X_i^{new} = X_i^{old} + rand(-1,1) * (X_{best} - X_j)$$ (3)

Parasitism Phase:
Parasitism symbiotic interaction occur among two dissimilar species where one organism get the benefit and the another species vigorously affected by this relationship. Plasmodium parasite describes the connection with the anopheles mosquito to enter into the hosts of human. In SOS X_i Plays alike character of anopheles mosquito over the development an artificial parasite vector (X_j). The X_j is selected randomly as a swarm to parasite vector. The parasite vector obtain the best fitness value then it kills the random parameter X_j otherwise parasite vector have to leave the system.

5 Simulation and Analysis

In the performed Matlab simulation, PID and FPID controller are experimented to sustain in stability in the described Fig. 1. The dynamic responses and performances of this robust system are tested with the step load perturbation of 0.15 p.u. in area 1 and GRC as 5%. The objective function taken in this case is the ITAE (Integral time absolute error).

$$J = \int_{0}^{t} (|\Delta f_1| + |\Delta f_2| + |\Delta p_{tie}|)dt. \tag{4}$$

5.1   Case-1 with Load Disturbances of 0.15 p.u. in Area 1

The system responses with the presence of PID and FPID controllers including frequency fluctuations in area 1 and area 2 ($\Delta f1$), ($\Delta f2$) and tie-line power exchange ($\Delta Ptie$) are represented in Figs. 4, 5 and 6. At the end of the optimization, the optimum gain values of PID and FPID controller and settling time of frequency and tie-line power fluctuations with and without GRC are provided in Tables 2 and 3 respectively. It is profound that the simulation outputs of modeled controller along with FPID controller exhibits better responses over PID controller.

5.2   Case-2 with the Presence 5% Generation Rate Constraint

GRC restricts the increase of power of the turbine of the system; once it is achieved its marginal upper bound of generation power. The GRC of 5% is considered to measure the non-linearity of system, the variations of frequency and interline power flow of the proposed reheat thermal system with the presence of both PID and FPID controller as per Fig. 1. The obtained results as in Figs. 7, 8 and 9 indicate that the implemented FPID controller confirms good performance in spite of load disrupts and intermediate 5% of generation rate constraints.

Table 2. The finest gain parameters of FPID/PID controller.

Fuzzy-PID controller							
Area 1				Area 2			
L_1	L_2	L_3	L_4	L_1	L_2	L_3	L_4
1.9065	0.9124	1.9798	1.989	1.9945	1.7894	1.2197	1.2998
PID controller							
L_P	L_I	L_D		L_P		L_I	L_D
2.051	9.4522	4.0121		0.710		0.6245	0.6741

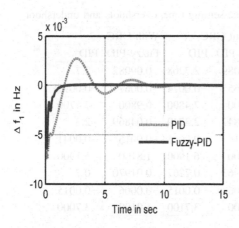

Fig. 4. Frequency change in area 1

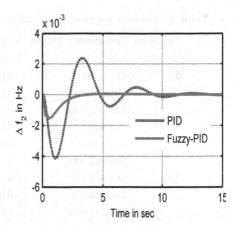

Fig. 5. Frequency change in area 2

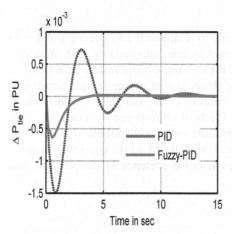

Fig. 6. Interline power change

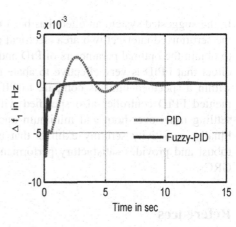

Fig. 7. Frequency change in area1 with GRC

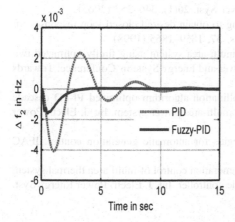

Fig. 8. Frequency change in area2 with GRC

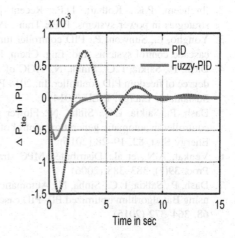

Fig. 9. Interline power change with GRC

Table 3. Dynamic response specifications like settling time, overshoots and undershoot

Variations	Performance indices	Without GRC		With GRC	
		Fuzzy-PID	PID	Fuzzy-PID	PID
Δf_1	$O_{sh} \times 10^{-3}$	0.09089	2.7208	0.09082	2.7
	Undershoot	−0.0065	−0.0074	−0.0067	−0.0075
	Settling time	0.9600	7.4800	0.9800	7.4700
Δf_2	$O_{sh} \times 10^{-3}$	0.04847	2.3749	0.04869	2.3
	Undershoot	−0.0016	−0.0041	−0.0016	−0.0041
	Settling time	1.7900	6.1900	1.8100	6.1900
ΔP_{tie}	$O_{sh} \times 10^{-3}$	0.01963	0.7267	0.01970	0.7
	Undershoot	−0.0006	−0.0015	−0.0006	−0.0015
	Settling time	1.0000	3.7100	1.0100	3.7000

6 Conclusion

In the suggested system an effect has been made to reinforce the frequency stability in the scrutinized intersect two area electrical power system optimized by SOS algorithm to obtain the optimal parameters of PID and FPID controller. The results of simulation direct that FPID is very effective to abate the deviations of power system constraints within a specified limit as compare to PID structures. The sturdiness of this implemented FPID controller also specified with the help of response specifications like settling time, overshoot and minimum undershoots under a disturbance of 0.15 p.u. Finally, sensitivity scrutiny exhibits that novel SOS tuned FPID controller is quite robust and provides satisfactory performances with the presence of disturbances and GRCs.

References

1. Kundur, P.: Power System Stability and Control. McGraw-Hill, New York (1994)
2. Ibraheem, P.K., Kothari, D.P.: Recent philosophies of automatic generation control strategies in power systems. IEEE Trans. Power Syst. **20**(1), 346–357 (2005)
3. Yongho, L., Sunwon, P.: PID controller tuning to obtain desired closed loop responses for cascade control systems. Ind. Eng. Chem. Res. **37**, 1859–1865 (1998)
4. Dash, P., Saikia, L.C., Sinha, N.: AGC of a multi-area system using firefly optimized two degree of freedom PID controller. In: 2014 Power and Energy Systems Conference: Towards Sustainable Energy. IEEE (2014)
5. Dash, P., Saikia, L.C., Sinha, N.: Flower pollination algorithm optimized PI-PD cascade controller in automatic generation control of a multi-area power system. Int. J. Electr. Power Energy Syst. **82**, 19–28 (2016)
6. Venkat, A.N., et al.: Distributed MPC strategies for automatic generation control. IFAC Proc. **39**(7), 383–388 (2006)
7. Dash, P., Saikia, L.C., Sinha, N.: Automatic generation control of multi area thermal system using Bat algorithm optimized PD–PID cascade controller. Int. J. Electr. Power Energy Syst. **68**, 364–372 (2015)

8. Pradhan, P.C., Sahu, R.K., Panda, S.: Firefly algorithm optimized fuzzy PID controller for AGC of multi-area multi-source power systems with UPFC and SMES. Eng. Sci. Technol. Int. J. **19**(1), 338–354 (2016)

9. Chaine, S., Tripathy, M.: Design of an optimal SMES for automatic generation control of two-area thermal power system using cuckoo search algorithm. J. Electr. Syst. Inf. Technol. **2**(1), 1–13 (2015)

10. Guha, D., Roy, P.K., Banerjee, S.: Load frequency control of large scale power system using quasi-oppositional grey wolf optimization algorithm. Eng. Sci. Technol. Int. J. **19**(4), 1693–1713 (2016)

11. Behera, A., Panigrahi, T.K., Sahoo, A.K., Ray, P.K.: Hybrid ITLBO-DE optimized fuzzy PI controller for multi-area automatic generation control with generation rate constraint. In: Satapathy, S.C., Bhateja, V., Das, S. (eds.) Smart Computing and Informatics. SIST, vol. 77, pp. 713–722. Springer, Singapore (2018). https://doi.org/10.1007/978-981-10-5544-7_70

12. Sahu, R.K., Gorripotu, T.S., Panda, S.: Automatic generation control of multi-area power systems with diverse energy sources using teaching learning based optimization algorithm. Eng. Sci. Technol. Int. J. **19**(1), 113–134 (2016)

13. Nanda, J., Sreedhar, M., Dasgupta, A.: A new technique in hydro thermal interconnected automatic generation control system by using minority charge carrier inspired algorithm. Int. J. Electr. Power Energy Syst. **68**, 259–268 (2015)

14. Chandrakala, K.R.M.V., Balamurugan, S., Sankaranarayanan, K.: Variable structure fuzzy gain scheduling based load frequency controller for multi source multi area hydro thermal system. Int. J. Electr. Power Energy Syst. **53**, 375–381 (2013)

15. Morsali, J., Zare, K., Hagh, M.T.: Applying fractional order PID to design TCSC-based damping controller in coordination with automatic generation control of interconnected multi-source power system. Eng. Sci. Technol. Int. J. **20**(1), 1–17 (2017)

16. Demiroren, A., Yesil, E.: Automatic generation control with fuzzy logic controllers in the power system including SMES units. Int. J. Electr. Power Energy Syst. **26**(4), 291–305 (2004)

17. Barisal, A.K., Mishra, S.: Improved PSO based automatic generation control of multi-source nonlinear power systems interconnected by AC/DC links. Cogent Eng. **5**(1), 1422228 (2018)

18. Debnath, M.K., Mallick, R.K., Sahu, B.K.: Application of hybrid differential evolution-grey wolf optimization algorithm for automatic generation control of a multi-source interconnected power system using optimal fuzzy–PID controller. Electr. Power Compon. Syst. (2018). https://doi.org/10.1080/15325008.2017.1402221

Location of Fault in a Transmission Line Using Travelling Wave

Basanta K. Panigrahi[1(✉)], Riti Parbani Nanda[1], Ritu Singh[2], and P. K. Rout[1]

[1] Department of EE, ITER, S'O'A University, Bhubaneswar, India
basanta1983@gmail.com, ritinanda96@gmail.com,
pravatrout@soa.ac.in
[2] Department of EE, Bhilai Institute of Technology, Durg, India
ritusingh02@gmail.com

Abstract. The Power system is mostly affected by the faults in the transmission line. Now a day it is very necessary to locate and detect the faults quickly in order to get better performance. The advantages of travelling wave fault location technique are it avoids complexity and minimizes the cost. In this strategy the faulted distances are calculated with the help of arrival time and propagation velocity. Single ended measurements have certain merits over multi ended measurements. In this work travelling wave technique is applied, which uses the travelling waves to locate the fault distance in transmission lines. Simulation models have been made to confirm the effects of the used methods which depend on the wavelets. Results show that the used method is able to locate the faults more accurately.

Keywords: Travelling Wave (TW) · Fourier Transform (FT)
Discrete Wavelet Transform (DWT)

1 Introduction

The electricity which is produced in the power plant is transmitted to the consumer and load centers by transmission towers and poles. When the system is operating under normal condition the system is in a balanced state. Abnormal conditions occur because of faults. Faults can be created in a power system because of many reasons; it may be because of natural events or may be because of mechanical failures of other equipment's connected in the system. Natural fault events occurred in the transmission line such as short circuiting of the individual conductor due to excessive wind, ice storm or felling of a tree. Analysis of power system can be done with the calculated values of system voltages and currents in balanced and unbalanced conditions. Occurrence of fault in a system leads to equipment damage because of large current. Because of fault currents voltage level of the system will change, it can affect the insulation of equipment. If the voltage will be lower than a particular value than equipment failure may occur. The purpose of this work is to give a general idea about travelling wave to calculate the fault location on an interconnected system. It is very necessary to mitigate the faults as quickly as possible for reliable services to the consumer in addition to

© Springer Nature Singapore Pte Ltd. 2018
I. Zelinka et al. (Eds.): ICSCS 2018, CCIS 837, pp. 658–666, 2018.
https://doi.org/10.1007/978-981-13-1936-5_67

restore the installation. We know that the most severe fault is three phase fault so in this study this type of fault is considered. In the nest section a brief literature review have discussed.

Casagrande et al. States that the location of the fault is needed for the consumer satisfaction and for the efficient operation of electrical components. For the past decades, there are several techniques for fault location methods such as travelling wave methods, Fourier analysis and line impedance method [1]. Yongli et al. states that in place of traditional methods, high frequency components are used [2]. According to Jung, Aburetal and some other researchers have discussed traveling wave mehod [3–9]. According to Panigrahiet al. the wavelet transform has the special property of time-frequency resolution, from which we can detect the fault. In this paper wavelet transform (WT) is used for determining the location of fault. All the signals are analyzed using the wavelet transform toolbox after selecting the suitable wavelet level. From the analyzed signal the pre fault and post fault coefficients are derived [10]. Jin et al. state that traveling wave fault location strategies are considered as the most correct strategies as compared to the Impedance based strategies and these are not affected by the source impedance, fault resistance and power flow [11].

2 Travelling Wave

This works focuses on determination of the faults using the travelling wave theory. Herefor analysis long and homogenous lines are considered for determination of location of the fault using travelling wave. The concepts of travelling waves are based on the high frequency signals of voltage and current on the buses. The faulted points were estimated by calculating the difference between the first two peaks at the bus. At the fault location, the two signals having equal amplitude but in opposite directions are superimposed. This technique is appropriate for the fault location at single end. In this technique the fault position is far from the primary travelling wave and hence a reflected wave arrives at the fault position which is proportional to the distance of the fault location. Therefore, this algorithm is not correct as it is difficult to find the spot of the fault as waves can be lost due to disturbances. In a transmission line, electromagnetic energy can be sent from one end to another [12]. For the analysis of transmission line let us assume dx is the line segment R is the resistance, G is the conductance, L is the inductance and C is the capacitance. All the parameters are in per unit system. Therefore the resistance conductance inductance and conductance for the line segment dx Rdx, Gdx, Ldx, and Cdx respectively. Electromagnetic wave generates electric flux and magnetic flux are Ψ and φ respectively and the instantaneous voltage and current are v(x,t) and i(x,t)

$$\partial \Psi(t) = v(x,t)Cdx \tag{1}$$

$$\partial \varphi(t) = i(x,t)Ldx \tag{2}$$

The voltage drop in positive x direction is calculated by

$$v(x,t) - v(x+dx,t) = -dv(x,t) = \frac{\partial v(x,t)}{\partial x} \partial x = \left(R + L\frac{\partial}{\partial t}\right) i(x,t)dx \qquad (3)$$

If dx cancels from both sides of (3), the voltage equation become

$$\frac{\partial v(x,t)}{\partial x} = -Ri(x,t) - L\frac{\partial i(x,t)}{\partial t} \qquad (4)$$

Applying Kirchhoff's current law, the current through conductance G and capacitor C while charging is given by,

$$i(x,t) - i(x+dx,t) = -di(x,t) = -\frac{\partial i(x,t)}{\partial x} = \left(G + C\frac{\partial}{\partial t}\right) v(x,t)dx \qquad (5)$$

If dx cancels from both sides of (5), the current equation becomes

$$\frac{\partial i(x,t)}{\partial x} = -C\frac{\partial v(x,t)}{\partial t} - Gv(x,t) \qquad (6)$$

The sign is negative in the above equations because of the voltage and current wave propagates in positive x direction, v(x,t) and i(x,t) is decreasing in amplitude when x is increasing. When one substitutes and differentiates once more with respect to x, we get the second order partial differential equations, $Z = R + \frac{\partial L(x,t)}{\partial t}$ and $Y = G + \frac{\partial C(x,t)}{\partial t}$ differentiate once more with respect to x, we get the second order partial differential equation

$$\frac{\partial^2 i(x,t)}{\partial x^2} = -\frac{Y\partial v(x,t)}{\partial t} = YZi(x,t) = \gamma^2 i(x,t) \qquad (7)$$

$$\frac{\partial^2 i(x,t)}{\partial x^2} = -\frac{Y\partial v(x,t)}{\partial t} = Yzi(x,t) = \gamma^2 i(x,t) \qquad (8)$$

In the above equations, γ is the propagation constant which is complex in nature and is given by

$$\gamma = \sqrt{ZY} = \alpha + j\beta \qquad (9)$$

Where, attenuation constant is α has effects on the amplitude of the travelling wave, and phase constant is β has effects on the phase shift of the travelling wave.

Equations (7) and (8) can be solved by transform or classical

$$v(x,t) = A_1(t)e^{\gamma t} + A_2(t)e^{-\gamma t} \qquad (10)$$

$$i(x,t) = -\frac{1}{Z}[A_1(t)e^{\gamma t} + A_2(t)e^{-\gamma t}] \qquad (11)$$

Where characteristics impedance is Z of the transmission line and is calculated as

$$Z = \sqrt{\frac{R+L\frac{\partial}{\partial t}}{G+C\frac{\partial}{\partial t}}} \qquad (12)$$

where arbitrary constants are A_1 and A_2, and it does not depend upon x.

TW technique needs communication link to urge data from the each ends, therefore the information is at a standard time base. While compared with the single-ended algorithm rule, this technique needs high priced and complicated. The frequency and time resolutions square measure each mounted. The above analysis is appropriate for the stationary signal which is periodic in nature and varies slowly.

3   Proposal Model

In this work an interconnected system is modelled in MATLAB. The system which is considered for this study is shown in the Fig. 1. It is paramount that the simulated system demonstrates to display an authentic system in all vital components. The model has to be homogeneous so that it transpires into an authentic situation. It is further described how this has been established. Figure 2 shows the flow chart is to calculate the fault location using travelling wave.

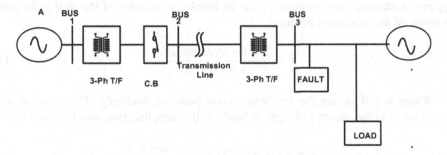

Fig. 1. Single line diagram of interconnected system

4   Results and Analysis

The V and I signal data are recorded by the simulation system. Those data's are in phase values. In order to change the phase value to a modal value, these modal values are converted to approximate coefficients and detail coefficients using DWTs with the db4 bus band. Then, the point where the maximum value of the detail coefficient

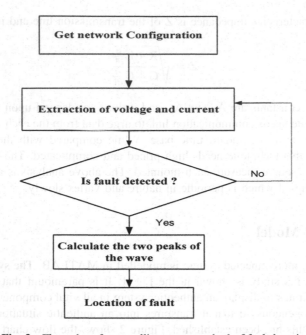

Fig. 2. Flow chart for travelling wave method of fault location

appears is the exact crest position so that the number of samples of the crest is the point position of the maximum modulus.

$$V = \frac{2 \times \text{the distance taken}}{\tau \times (a_1 - a_2)} \tag{13}$$

Where a_1 and a_2 are the first and second peaks respectively. The value of v is nearly equal to the velocity of light. Where τ is the sampling time and fault distance is

$$\text{Fault distance} \quad D = \frac{(a_1 - a_2) \times V}{2} \tag{14}$$

$$\text{Error} \quad e = \left| \frac{\text{actual distance} - \text{calculated distance}}{\text{total line length}} \right| \times 100\% \tag{15}$$

Figure 3 shows the wavelet modulus maxima for the current signals for distance = 10 km. It also shows the sample number of the first and second peaks of the travelling wave. The first peak value is 2567 and the value of second peak is 2651. Figure 4 shows the wavelet modulus maxima for the current signals for distance = 20 km. It also presents the sample number of the first and second peaks of the travelling wave. The first peak value is 2608 and the value of second peak is 2779.

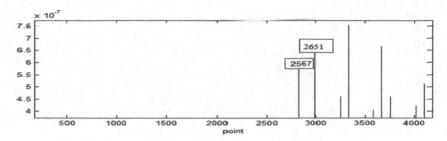

Fig. 3. Fault location for distance 10 km (three phases-ground fault)

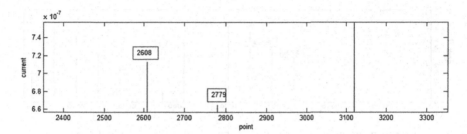

Fig. 4. Fault location for distance 20 km (three phase-ground fault)

Figure 5 shows the wavelet modulus maxima for the current signals for distance = 30 km. It also presents the sample number of the first and second peaks of the travelling wave. The first peak value is 2653 and the second peak is 2907. Figure 6 shows the wavelet modulus maxima for the current signals for distance = 40 km. It also presents the sample number of the first and second peaks of the travelling wave. The first peak value is 2693 and the second peak is 3035.

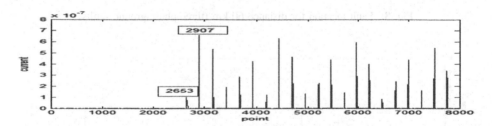

Fig. 5. Fault location for distance 30 km (three phase-ground faults)

Figure 7 shows the wavelet modulus maxima for the current signals for distance 50 km. It also presents the sample number of the first and second peaks of the travelling wave. The first peak value is 2736 and the second peak is 3163.

Figure 8 shows the wavelet modulus maxima for the current signals for distance 60 km. It also presents the sample number of the first and second peaks of the travelling wave. The first peak value is 2780 and the second peak is 3291. Figure 9 shows

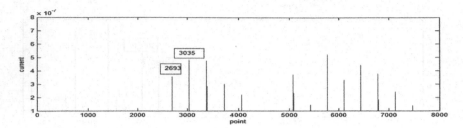

Fig. 6. Fault location for distance 40 km (three phase-ground fault)

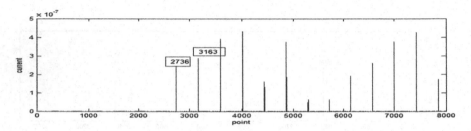

Fig. 7. Fault location for distance 50 km (three phase-ground fault)

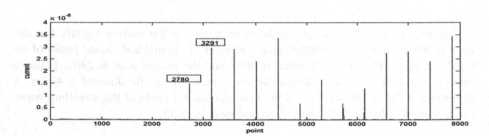

Fig. 8. Fault location for distance 60 km (three phase-ground faults)

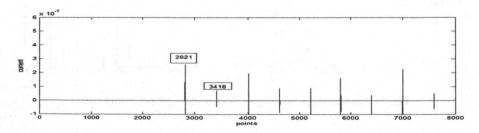

Fig. 9. Fault location for distance 70 km (three phase-ground faults)

the wavelet modulus maxima for the current signals for distance 70 km. It also presents the sample number of the first and second peaks of the travelling wave. The first peak value is 2821 and the second peak is 3418.

Figure 10 shows the wavelet modulus maxima for the current signals for distance 80 km. It also presents the sample number of the first and second peaks of the travelling wave. The first peak value is 2866 and the second peak is 3546. Figure 11 shows the wavelet modulus maxima for the current signals for distance 90 km. It also presents the sample number of the first and second peaks of the travelling wave. The first peak value is 2906 and the second peak is 3674.

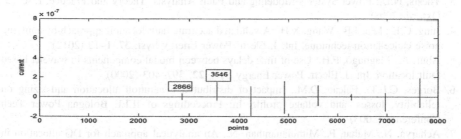

Fig. 10. Fault location for distance 80 km (three phase-ground faults)

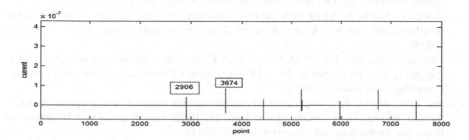

Fig. 11. Fault location for distance 90 km (three phase-ground faults)

5 Conclusion

Faults should be cleared in the transmission line as soon as possible. Smaller time for the fault clearing has many merits like less threat of damaging equipment's, reduces loss of income, less customer complaint. Faster restoration of supply is possible if accurate location of fault method is used.

The proposed scheme calculates the fault location of the fault point. The fault calculated at different locations. The fault location is estimated at different fault points. Faults are calculated at 10 km, 20 km, 30 km, 40 km, 50 km, 60 km, 70 km, 80 km and 90 km using Travelling wave method. When the fault is near the source, Estimation of fault location is better using travelling wave method. The error increases if the fault occurs away from the source. For a fault at a distance of 20 km the error value is only 0.43%, whereas for a fault at a distance of 60 km from the source the error increases to 1.06%. The percentage error due to travelling wave method was less. The error is less in travelling wave.

References

1. Casagrande, E., Woon, W.L., Zeineldin, H.H., Kan'an, N.H.: Data mining approach to fault detection for isolated inverter-based microgrids. IET Proc. Gener. Transm. Distrib. **7**(7), 745–754 (2013)
2. Ngu, E.E., Ramar, K.: A combined impedance and traveling wave based fault location method. Int. J. Electr. Power Energy Syst. **33**, 1767–1775 (2011)
3. Thesis, N.D.: Power Systems Modeling and Fault Analysis. Theory and Practice. Elsevier, Oxford (2008)
4. Jung, C.K., Lee, J.B., Wang, X.H.: A validated accurate fault location approach by applying noise cancellation technique. Int. J. Electr. Power Energy Syst. **37**, 1–12 (2012)
5. Abur, A., Magnago, F.H.: Use of time delays between modal components in wavelet based fault location. Int. J. Electr. Power Energy Syst. **22**, 397–403 (2000)
6. Borges, C.L.T., Falcao, D.M.: Impact of distributed generation allocation and sizing on reliability, losses, and voltage profile. In: Proceedings of IEEE Bologna Power Tech Conference (2003)
7. Acharya, N., Mahat, P., Mithulananthan, N.: An analytical approach for DG allocation in primary distribution network. Int. J. Electr. Power Energy Syst. **28**(10), 669–678 (2006)
8. Kunte, R., Gao, W.: Comparison and review of Islanding detection techniques for distributed energy resources. In: 40th North American Power Symposium, pp. 1–8 (2008)
9. Niazy, I., Sadeh, J.: Using fault clearing transients for fault location in combined line (overhead/cable) by wavelet transform. In: 24th International Power System Conference (2009)
10. Panigrahi, B.K., Ray, P.K., Rout, P.K., Sahu, S.K.: Detection and location of fault in a micro grid using wavelet transform. In: IEEE International Conference on circuits Power and Computing Technologies (2017)
11. Jin, Z.G., Zhang, P.J.: Application of wavelet neutral network and roughset theory to forecast mid-long-term electric power load. In: IEEE First International Workshop on Education Technology and Computer Science (2011)
12. Baseer, M.A.: Travelling waves for finding the fault location in transmission line. J. Electr. Electron. Eng. **1**(1), 1–19 (2013)

An Efficient Torque Ripple Reduction in Induction Motor Using Model Predictive Control Method

T. Dhanusha[✉] and Gayathri Vijayachandran

Electrical and Electronics, Sree Buddha College of Engineering, Pattoor, India
dhanu1446@gmail.com,
vijayachandrangayathri3@gmail.com

Abstract. Nowaday, model predictive based torque control is arise as one of the powerful control technique for the IM drives. The fast response and accuracy is the main features of predictive torque control technique. The control technique includes the predictive controller to obtain better dynamic response and PI controller to attain better steady state response. The main characteristics of Predictive Torque Controls (PTC) is by using the machine model for determining the future performance of the variables which is to be controlled. In Model Predictive Torque Control (MPTC) scheme, the command signals are indicated as cost function, which is to be reduced. It has increased resilience to use constraints that gives low computational complexity compared to simple vector controlled schemes. PTC offers increased dynamic behaviour and improved speed responses. A modified MPTC is suggested for the control of the torque ripple minimization. A portion of time interval is given to the non zero voltage vector, while the remaining time is given for a zero vector. The minimisation of torque ripple concept help to know the time period for individual vectors. The proposed method proves that it gives excellent steady state response by the reduction of the torque ripples.

Keywords: Direct torque controls · Induction motors
Model predictive controls · Predictive torque controls · Torque ripples

1 Introduction

In former days, DC machines were widely used for adjustable speed drive applications because of the decoupled management of torque and flux. DC drives has many merits like starting torque, speed variation, simple management and nonlinear performance. However due to the disadvantage of DC machine like the effect of commutator and brush assembly. In the industrial applications, DC machine drives are not used nowadays. AC motors are replaced by the DC motors due to their reduced price, excellent reliability, reduced heaviness, and maintenance requirement is less.

One of the better performance control strategy for the three phase ac electric drives is the Direct Torque Control (DTC) [1]. Accuracy and fast torque performance is the main features [2]. But, predefined switching tables and hysteresis comparators causes

© Springer Nature Singapore Pte Ltd. 2018
I. Zelinka et al. (Eds.): ICSCS 2018, CCIS 837, pp. 667–675, 2018.
https://doi.org/10.1007/978-981-13-1936-5_68

variable switching frequency and more torque ripple [3–6]. MPTC has developed as the best substitute to DTC by determining the future performance of the system [3–6].

2 Dynamic Equations of Induction Motor

The dynamic equations of induction motor are represented in stationary frame as

$$u_s = R_s i_s + \frac{d\Psi_s}{dt} \tag{1}$$

$$0 = R_r i_r + \frac{d\Psi_r}{dt} - j\omega_r \Psi_r \tag{2}$$

$$\Psi_s = L_s i_s + L_m i_r \tag{3}$$

$$\Psi_r = L_m i_s + L_r i_r \tag{4}$$

Where u_s, i_s, i_r, Ψ_s, Ψ_r, R_s, R_r, L_s, L_r, L_m are stator voltages and currents, rotor currents, stator and rotor flux linkages, stator and rotor resistances, stator and rotor inductances, mutual inductances respectively; ω_r is the rotor angular speed

$$i_s = \lambda(L_r \Psi_s - L_m \Psi_r) \tag{5}$$

$$i_r = \lambda(-L_m \Psi_s - s\Psi_r) \tag{6}$$

Where $\lambda = 1/(L_r L_s - L_m^2)$

Motors electromagnetic torque T_e is represented as

$$T_e = \frac{3p}{2} \text{Im}(\Psi_s i_s) \tag{7}$$

where p is the poles. Moreover, the motion equation of motor is as follows

$$\omega_r(t) = \frac{p}{2j} \int (T_e - T_L)dt \tag{8}$$

3 Model Predictive Control

Control with model predictive can be represented as an algorithm [12–15]. It can be used as a mathematical model in order to anticipate its future behavior. Predictive control has many merits: Idea is easy to understand and intuitive, it can be given to various conditions, multivariable situation can be included, and the controllers are easy to implement [16, 17]. The general concepts of Model predictive control are, a model of machine help to guess the future performance of controllable variables till a control range, a cost function that consider the appropriate performance of the systems and the proper actuations are selected by minimising cost function [8]. MPC is an optimization problem that includes minimization of the cost function g, for a known time range, subjects to the system limitations and the model of the system.

In existing MPTC, single voltage vector is chooses and which is given during single control period. Two-level inverters have restricted number of voltage vectors so as to decrease the torque ripples to a lower value are very difficult. So, the frequency of MPTC has to be large in order to attain proper steady state performance. In DTC [18], the zero vectors produce few torque variations. Hence, it is able to handle both active vector and non-active vector during single control span to obtain the decrease of torque ripple [4]. The proposed MPTC attempts to hold this principle by splitting the control span into two interims for both vectors. Figure 1 shows the whole control of the proposed MPTC, which consist of three sections: prediction and estimation of flux and torque, cost function minimisation or choice of vector, and reduction of torque ripple.

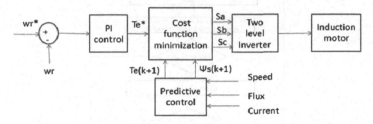

Fig. 1. Proposed system

3.1 Estimation and Prediction of Flux and Torque

With lesser sampling time, we can use first-order Euler equation to transform Eq. (7) into discrete form.

$$T_e(k+1) = \frac{3p}{2} \operatorname{Im}(\Psi_s(k+1) * i_s(k+1)) \qquad (9)$$

3.2 Vector Selection

The selection of vector is represented on the principle of minimisation of a cost function, which is a collection of series arrangement of torque and flux errors [19]. The cost function is

$$g = \left|T_e^{ref} - T_e^{k+1}\right| + A\left|\left|\Psi_s^{ref}\right| - \left|\Psi_s^{k+1}\right|\right| \qquad (10)$$

Where T_e^{ref} and Ψ_s^{ref} are the amplitude of reference torque and stator flux; A is the weighting factor of the stator flux. Eight discrete voltage vectors are available for Induction motor drives with two level inverter: $V_0, V_1, \ldots V_6, V_7$. For each individual voltage vector, we can find a value of T_e^{k+1} and Ψ_s^{ref}, and the one reducing (10) is taken as the voltage vector to be applied.

3.3 Torque Ripple Minimisation

The interval of nonzero vector must determine after the selection of active voltage vector based torque ripple minimisation principle.

Duration of the torque ripple in one control range can be represented as (Fig. 2):

$$\frac{1}{T_{sc}} \int_{kT}^{(k+1)T_{sc}} \left(T_e^{ref} - T_e\right)^2 dt \rightarrow min \tag{11}$$

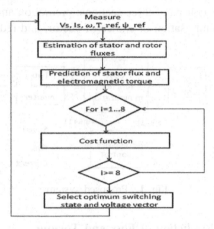

Fig. 2. MPTC algorithm

4 Simulation and Results

4.1 Modeling of Induction Motor

A 4 kW induction motor [8] is modeled with machine details specified in Table 1.

Table 1. Machine parameters

Rated power, P	4 Kw
Rated voltage, U	380 V
Rated frequency, f	50 Hz
Rated torque, T	10 Nm
Polepair	2
Stator resistance, Rs	3.126 Ω
Rotor resistance, Rr	1.879 Ω
Stator inductance, Ls	.230 H
Rotor inductance, Lr	.230 H
Mutual inductance, Lm	.221 H

Figure 3 shows the modeling of IM according to the Eqs. 1 to 8. The results obtained are as follows. Figure 4 shows the three phase input voltage given to induction motor. Figure 5 shows the waveforms of current, of an induction machine. A step time of 0.7 s is given to the machine, as a result up to 0.7 s the graphs shows the no load reading and after 0.7 s loaded condition is depicted here. The peak value of the current is about 10.2 A obtained. Figures 6 and 7 shows the electromagnetic torque and speed of induction motor respectively. At time 0.7 s, speed is decreased and torque is increased.

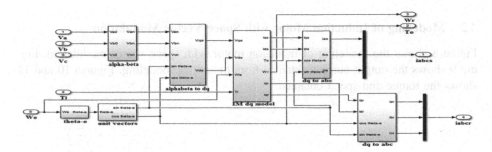

Fig. 3. MATLAB modeling of induction motor

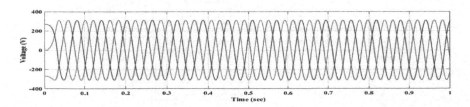

Fig. 4. Three phase voltage

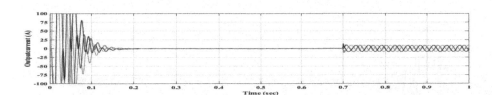

Fig. 5. Three phase output current

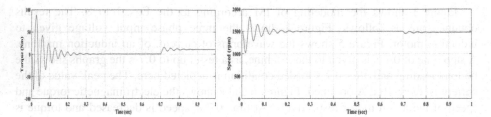

Fig. 6. Electromagnetic torque   **Fig. 7.** Speed of induction motor

4.2   Modelling of Induction Motor with Space Vector Modulation

Figure 8 shows the modeling of induction motor with space vector modulation. Figure 9 shows the output current obtained from the above modeling. Figures 10 and 11 shows the torque and speed obtained.

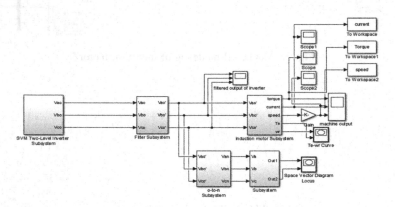

Fig. 8. Modeling of induction motor with SVM

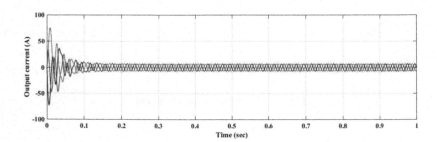

Fig. 9. Output current

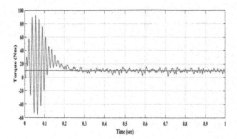

Fig. 10. Torque characteristics

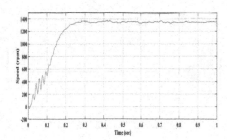

Fig. 11. Speed characteristics

4.3 Modelling of MPTC Control

Figure 12 shows the modeling of proposed model predictive torque control. Figure 13 shows the output current obtained from the above modeling. Figures 14 and 15 shows the torque and speed obtained.

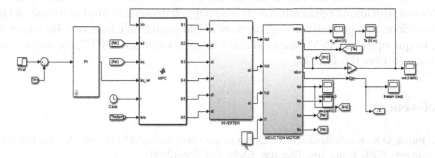

Fig. 12. Proposed MPTC

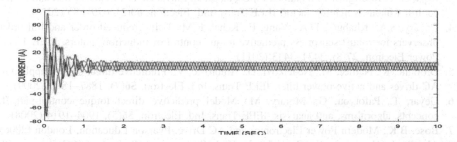

Fig. 13. Current waveform

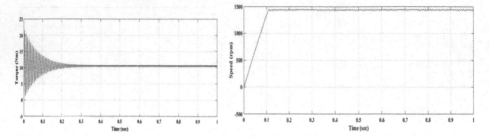

Fig. 14. Torque characteristics at full load

Fig. 15. Speed characteristics at full load

5 Conclusion

An efficient MPTC is used to achieve the reduction of torque ripple was proposed. For that purpose, the control range is splitted into two intervals. One is for appropriate null vector and other one for active vector. The control period of the nonzero vector is achieved with torque ripple minimisation strategy. Individual control of torque and flux can be done with good dynamic performances and steady state behavior. By comparing the torque ripples in SVM based induction motor control and MPTC, the torque ripple is reduced from 0.6 to 0.06.

References

1. Bujua, G.S., Kamierkoske, M.P.: Direct torque control of SPWM inverter-fed AC motors – a survey. IEEE Trans. Ind. Electron. **53**(4), 744–756 (2004)
2. Berten, J., Vervecken, J., Drisen, J.: Predictive direct torque controls for flux and torque ripple reduction. IEEE Trans. Ind. Electron. **58**(1), 404–422 (2010)
3. Roja, C.A., Rodriguis, J., Villaroel, F., Espinosa, J., Shiva, C.A.: Predictive flux and torque control without weighting factor. IEEE Trans. Ind. Electron. **60**(2), 679–685 (2013)
4. Davary, S.A., Khabury, D.A., Vang, F., Kenal, R.M.: Using reduced order and full order observers for robust sensorless predictive torque control of induction motors. IEEE Trans. Power Electron. **27**(8), 3431–3433 (2011)
5. Drobinic, K., Namec, M., Nedelkovi, D., Ambrosic, V.: Predictive direct control applied to AC drives and active power filter. IEEE Trans. Ind. Electron. **56**(7), 1887–1890 (2009)
6. Geyar, T., Pafotyou, G., Morary, M.: Model predictive direct torque control-part II: concepts, algorithms, and analysis. IEEE Trans. Ind. Electron. **56**(5), 1904–1914 (2008)
7. Bose, B.K.: Modern Power Electronics and AC Drive. Pyarson Education, London (2006)
8. Abu-Ruub, H., Iqbal, A., Guzinski, J.: High Performance Control of AC Drives with MATLAB/Simulink Model, 2nd edn. Wiley, Hoboken (2012)
9. Camacho, E.F., Bodon, C.: Model Predictive Control. Springer, London (1998)
10. Joashim, H.: Sensorless control of induction motor drive. In: Tutorial and Proceeding of IEEE-IECON, 28 November–1 December 2000 (2000)
11. Mirenda, H., Cortez, P., Yuz, J., Rodriguez, J.: Predictive torque control of induction machines based on state-space model. IEEE Trans. Ind. Electron. **56**(6), 1916–1924 (2009)

12. Duren, M.J., Prito, J., Barero, F., et al.: Predictive current controls of dual three-phase drive using restrained search technique. IEEE Trans. Ind. Electron. **56**(7), 3243–3260 (2010)
13. Barraero, F., Arahal, M.R., Gregor, R., et al.: A proof of concept study of predictive current controls for VSI-driven asymmetrical dual three-phase ac machine. IEEE Trans. Ind. Electron. **55**(6), 1927–1944 (2009)
14. Shyu, K.K., Lin, J.-K., Pham, V.-T., et al.: Global minimum torque ripples design for direct torque control of induction motor drive. IEEE Trans. Ind. Electron. **56**(8), 3150–3158 (2011)
15. Geyer, T.: A comparison of control and modulation scheme for medium-voltage drive: emerging predictive control concepts versus PWM-based schemes. IEEE Trans. Ind. Appl. **46**(1), 1370–1392 (2012)
16. Rodrigues, J., Kouro, S., et al.: Direct torque controls with imposed switching frequency in an 11-level cascaded inverters. IEEE Trans. Ind. Electron. **51**(4), 820–830 (2005)
17. Maeas, J., Melkebek, J.A.: Speed-sensorless direct torque controls of induction motor using an adaptive flux observers. IEEE Trans. Ind. Appl. **36**(3), 772–781 (2001)
18. Scoltocck, J., Madawaala, U.: A comparison of model predictive control scheme for mv induction motors drives. IEEE Trans. Ind. Inf. **10**(3), 908–916 (2012)
19. Geyer, T., Papafotiou, G., Morari, M.: Model predictive direct torque controls – part I: concepts, algorithms, and analysis. IEEE Trans. Ind. Electron. **57**(7), 1893–1900 (2010)

Modeling of an Automotive Grade LIDAR Sensor

Jihas Khan[✉], Jayakrishna Raj, and R. Pradeep

College of Engineering Trivandrum (CET), Thiruvananthapuram, Kerala, India
jihaskhan10@gmail.com

Abstract. Automobiles use LIDAR sensor to detect different objects around the vehicle. For system analysis and study, this paper is proposing a LIDAR sensor model, which takes into consideration the impact of the real world information. Impact of the environment on the LIDAR sensor is modeled and all the possible parameters of the LIDAR are modeled as configurable parameters. Option to import 3D objects of any shape, type or dimension via FBX file format is also incorporated. 3D objects in FBX format shall be converted to a set of triangles first, which approximate the surface mesh of the 3D object. Ray cast modeling is then used to detect whether in a vertical distribution of LIDAR beams, an intersection occurs between LIDAR beam and any of the triangular face. If there is a collision, the collision point shall be saved as the point cloud data. This will be repeated around the sensor, and all such point cloud data points shall be appended to the final point cloud data. These point cloud data is then subjected to segmentation and object detection using belief theory. Either the processed point cloud data in object information format or the unprocessed raw point cloud data can be produced as the output by the proposed LIDAR sensor model.

Keywords: LIDAR · Automotive · Point cloud data · FBX · Detection
Classification · Process data · Raw data · 3D objects

1 Introduction

LIght Detection And Ranging (LIDAR) makes use of LASER light pulses to scan the environment around it which allows it to create a map of the surrounding in the speed of light. As the electronics became more advanced, LIDAR sensor became more and more miniaturized. This enabled the use of LIDAR sensor in automotive applications. LIDAR works by first emitting a constant stream of LASER pulses. These LASER pulses travel outwards until contacting an object. When the LASER pulses fall on an object, they reflect back to the LIDAR system which receives and registers this as echo pulse. Based on the time of travel of the LASER pulse, LIDAR system calculates the distance of the target surface from the sensor system, which will then be converted to X, Y and Z coordinates. This process will be repeated for every horizontal and vertical positions, thus providing the target information, 360° around the LIDAR sensor system. X, Y, Z coordinates found out by each iteration will be appended to previous

© Springer Nature Singapore Pte Ltd. 2018
I. Zelinka et al. (Eds.): ICSCS 2018, CCIS 837, pp. 676–686, 2018.
https://doi.org/10.1007/978-981-13-1936-5_69

coordinate output. Collection of all the X, Y, Z coordinates output by LIDAR sensor system, is called Point Cloud data or PCD.

In automotive domain, these point cloud data will be used to implement autonomous driving and advanced driver assistance features. Since complete autonomous driving is safety critical which can affect human lives, functional safety with respect to ISO 26262 is very relevant here. To analyze the system, its working, how it behaves in different normal and extreme conditions, to analyze its performance, issues and to validate its robustness, it is very important to model the LIDAR sensor used in the modern day automobiles. Author used [1] to understand the basics of LIDAR especially automotive LIDAR.

[2] Explains an approach from algorithm level to detect pedestrian from an in vehicle LIDAR sensor. This give an idea, how the LIDAR data can be used for potential applications. Converting the point cloud data to category of the object detected, example – car, bike, truck, etc. requires lot of computations, which are explained in [3]. [4] gives an overview of using deep learning for detection of car from LIDAR sensor data. The concept of Optical Phase Array (OPA) is used by LIDAR sensor to steer the direction of LASER pulse streams, so as to achieve 360 degree coverage. This is explained in [5]. In prior literature, one can see, similar ventures for the simulation of the LIDAR sensors, [6] is such an attempt where stochastic volumetric model which better captures the complexities of real LIDAR data of vegetation is considered and it is far better suited for automatic modeling of scenes from field collected LIDAR data. [7] is another similar attempt for LIDAR simulation based on the modern computer graphics hardware making heavy use of recent technologies like vertex and fragment shading. [8] was used by the author to understand the existing simulation methods for LIDAR simulation and its analysis.

A combination of physical sensor modeling aspects and ray cast modeling methods is employed in the proposed LIDAR sensor model. Physical impact of the environment is modeled as equations and ray cast modeling is used for detection of point cloud data. FBX files are used for specifying the 3D objects. FBX files are first converted to surface composing triangular meshes. Each of these triangular mesh (a plane) is used to check for collision for the ray modeled LIDAR laser beam. Segmentation and object detection is also implemented on top of the generated point cloud data. Figure 1 gives an overall idea of the working of the proposed LIDAR sensor.

Rest of the paper is structured as follows. Section 2 is modeling the basic LIDAR equation while Sect. 3 discusses about importing different 3D objects in FBX format. Section 4 mainly focusses on the parameterization of the LIDAR sensor. Section 5 indulges in the design of core part of the sensor model. Section 6 explains about the logic to generate the point cloud data while Sect. 7 discusses about adding the impact of environment on the point cloud data. Section 8 discusses about the object detection and classification. Conclusion and Results of the simulation are summarized towards the end of the paper.

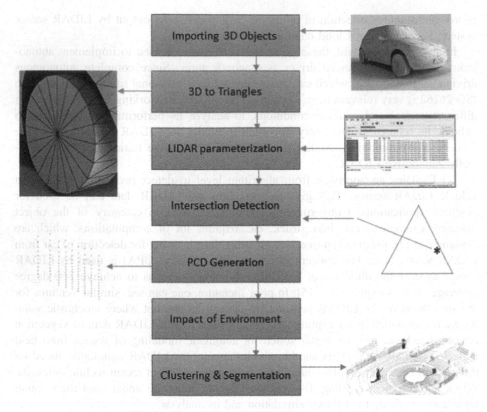

Fig. 1. Steps followed in LIDAR sensor model

2 LIDAR Equations

[9] Gives an idea about the equations used for modeling the LIDAR. Detected LIDAR signal can be represented by:

$$P(R) = KG(R)\beta(R)T(R) \tag{1}$$

Where P(R) denotes the power in Watts received from a distance R meters away, parameter K summarizes the performance of the LIDAR system, G(R) is the geometric factor depending on the range. Information about the environment and channel is contained in the parameters $\beta(R)$ and T(R). $\beta(R)$ is the backscatter coefficient at a distance of R meters and T(R) describes the transmission term which explains how much amount of light is lost during transmission and reception. The system factor K can be modeled as:

$$K = P_0(c\tau/2)A\eta \tag{2}$$

P0 is the average power of a single laser pulse, τ is the pulse duration in time domain in seconds, the factor ½ shows the effect of 'folding' the pulse due to back scattering. A is the area of the primary receiver optics responsible for the collection of backscattered light and overall LIDAR system efficiency is shown by η. G(R) can be modeled as:

$$G(R) = O(R)/R^2 \tag{3}$$

LASER beam receiver field of view overlap function is shown by O(R). The backscatter coefficient at a distance of R, β(R) is given by:

$$\beta(R, \lambda) = \sum N_j(R) \frac{d}{d\Omega}\sigma j(\pi, \lambda) \tag{4}$$

Nj be the concentration of scattering particles of kind j in the volume illuminated by the laser pulse, summation is over all j, and the differential term corresponds to the particles' differential scattering cross section for the backward direction at wavelength λ. Finally the transmission term T(R) which can take values between 0 and 1 is given by:

$$T(R, \lambda) = e^{-2\int_0^R \alpha(R,\lambda)dr} \tag{5}$$

The integral considers the path from the LIDAR sensor to distance R. The factor 2 stands for the two-way transmission path. The sum of all transmission losses is called light extinction, and α(R, λ) is the extinction coefficient. Thus Eq. 1, where each term in Eq. 1 are calculated by Eqs. 2 to 5, will be used to model the physical part of the LIDAR sensor model in this paper.

3 Importing 3D Objects

Primary output of a LIDAR sensor is the point cloud data of the targets in the range of the sensor. Therefore, a standard LIDAR sensor model should also produce the point cloud data of the virtual targets present in the range of the sensor model. Since the LIDAR sensor is capable to detect the three-dimensional properties of the targets, in the LIDAR sensor model also, there should be an option to incorporate three-dimensional objects. There are different formats available for the definition of three dimensional object models, like .FBX (FilmBoX File), .STL file (STereoLithography file), .OBJ (OBJect file) etc. In the proposed LIDAR sensor model, .FBX file format is used for importing information about the targets. Reasons for selecting the .FBX file as the standard for the LIDAR sensor model are mentioned below:

1. FBX has become a defacto standard choice of 3D model content and it provides interoperability between digital content creation applications
2. Efficient workflow, easier data exchange
3. Accommodates information about the geometry (example – Polygons, patches), Attributes (Normal, Transformation coordinates), camera information (focal length, aperture), Materials (emissive textures), Animation, Dynamics etc.
4. Entire animation scene information in one FBX File

We are more interested in only the surface geometry information of 3D objects, Why? Because any automotive sensor can only perceive the surface information of any 3D objects. This is because sound waves from an ultrasonic sensor, light waves from a camera, RF waves from a RADAR, LASER beams from LIDAR sensor falling on any 3D objects, gets reflected from the surface of the object. Thus the sensor's can 'SEE' only the surface information about the targets. Keeping this conclusion in mind, for the proposed LIDAR sensor model, it is logical to consider only the surface description of 3D objects. FBX file of the targets contain the surface description of the object under interest.

Surface is modeled in FBX file, as a mesh comprising of triangles (3 vertices) or quads (4 vertices). Increasing the number of triangles or quads will increase the resolution of 3D object, while also increasing processing time considerably and vice versa. Figure 2 gives an idea of how the surface of a 3D object is approximated using triangles. Thus in the proposed LIDAR sensor model, information of these triangles is made use of. Each of these triangles consist of 3 vertices, and each of the vertices shall have a X, Y, Z coordinate, expressed in float.

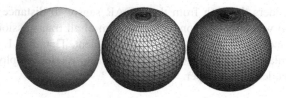

Fig. 2. A 3D sphere surface being approximated by triangles; increasing the number of triangles increases the accuracy and thus resolution

4 LIDAR Parameterization

A good sensor model should be as much as generic as possible. Parametrization should be there in the sensor model to make sure that any real world LIDAR sensor can be realized by parameterizing the base LIDAR sensor model. Identifying the sensor parameters is the primary step here. List of LIDAR parameters that are modifiable in the proposed LIDAR sensor model and its description are shown in Table 1.

5 Sensor Model

5.1 Importing 3D Objects

An input to the sensor will be the list of 3D objects present in the scenario. For each of the object, there will be a FBX file. Each of these FBX files will be parsed and converted to a set of triangles, saved in an array T. This T will be used for further processing of the LIDAR sensor model. Figure 3 shows an example of the imported 3D models, where the FBX file for a box and a pyramid are imported.

Table 1. LIDAR sensor parameters and its description

Parameter	Description
Number of layers N	Number of LIDAR laser beams sent in one vertical scanning
Range R	Distance in meters up to which LIDAR can 'see'
Horizontal resolution HR	Angle in degrees between one vertical scanning pattern and the subsequent vertical scanning pattern
Vertical resolution VR	Angle in degrees between N laser lines in one vertical scanning pattern
Scanning rate SR	Frequency in Hz, which is the inverse of the time taken to complete one horizontal scanning
Horizontal start HS	Angle in degrees, from which horizontal scanning will start
Horizontal stop HE	Angle in degrees, at which horizontal scanning will end
Vertical start VS	Angle in degrees, from which N lines will be created with a separation of VR angle between each line till VE
Vertical end VE	Angle in degrees, where the Nth line will be placed
System factor K	K summarizes the performance of the LIDAR system
Wavelength λ	Wavelength of LIDAR
Area of the primary receiver optics A	Area of the primary receiver optics responsible for the collection of backscattered light and
Efficiency η	Overall LIDAR system efficiency
Pulse duration τ	Pulse duration in time domain in seconds
P0	P0 is the average power of a single laser pulse
T(R)	T(R) describes the transmission term which explains how much amount of light is lost during transmission and reception
FBX Files	FBX Files for every targets in the scenario
LIDAR sensor position P	X, Y, Z parameters of the LIDAR sensor

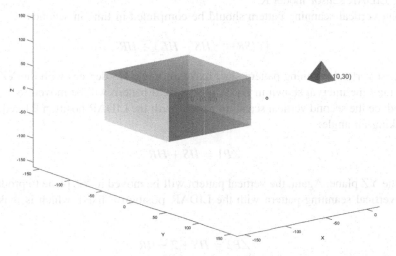

Fig. 3. 3D geometries imported into the sensor model

For each of the 3D object, its centroid location shall be mentioned, so that the 3D object can be uniquely linked to a specific location in the three dimensional coordinate system. Orientation of the 3D object with respect to X axis, Y axis and Z axis will also mentioned so as to make sure that the 3D object is having a unique angular position in three dimensional space.

5.2   Sensor Model

Line segments will be used to approximate transmitted LIDAR laser beams synonymous to ray casting methodologies. A set of line segments are modeled which starts from the LIDAR sensor position P and ending at another point E1. E1 is taken in such a way that the angle made by line segment PE1 is 'Vertical start VS' degrees with XY plane. Next line segment will start from P and end at E2, such that the angle made by the line segment PE2 with XY plane is:

$$\angle PE^2 = VS - VR \tag{6}$$

Next line segment will start from P and end at E3, such that the angle made by the line segment PE3 with XY plane is:

$$\angle PE^3 = VS - 2 * VR \tag{7}$$

This continues until the last line segment PEN which makes an angle:

$$\angle PE^N = VS - (N - 1) * VR \tag{8}$$

With the XY plane, where N can take any value. These set of lines comprises one vertical scanning pattern. Length of all the line segments PE1, PE2, PE3, PEN is range of the LIDAR sensor model R.

One vertical scanning pattern should be completed in time in seconds:

$$(1/SR) \div (HS - HE) \div HR \tag{9}$$

This first vertical scanning pattern shall make an angle HS degrees with the YZ plane. After the time interval shown in Eq. 9, this vertical pattern will be moved in YZ plane to produce the second vertical scanning pattern with the LIDAR position P fixed, which is making an angle:

$$\angle P1 = HS + HR \tag{10}$$

with the YZ plane. Again, the vertical pattern will be moved in YZ plane to produce the third vertical scanning pattern with the LIDAR position P fixed, which is making an angle

$$\angle P2 = HS + 2 * HR \tag{11}$$

with the YZ plane. This will be continued till the last vertical pattern again with the LIDAR position P fixed, which is making an angle

$$\angle Plast = HE \tag{12}$$

With the YZ plane. Thus the total number of horizontal scanning patterns will be:

$$NHS = (HE - HS)/HR \tag{13}$$

This replicates the real world physical LIDAR sensor which does a sequential horizontal scan, where at each horizontal scan, a set of equally vertically distributed LASER lines are placed at a specific horizontal angle. This allows the LIDAR sensor model to 'see' the full area around the sensor, given that HS = 0°, HE = 360°, VS = 0° and VE = 180°.

6 Point Cloud Generation

The processed data output of a conventional automotive grade LIDAR sensor is the Point Cloud File (PCD) which contains the X, Y, Z coordinates of the targets it detected. In real world, the transmitted LASER segments of length R will reflect off from the surface of targets which are within in the field of view of the sensor. These reflected LASER beams will be received by the sensor and the time taken for the round trip of the LASER beam is calculated. This round trip time is then converted to Euclidean distance of a point on the surface of the target. Since the sensor already know the exact horizontal and vertical angular location of the LIDAR beams, it can find the exact X, Y, Z coordinate of the surface of target from which each of these specific LASER beams were reflected.

In the proposed LIDAR sensor model, each of the LASER line segments mentioned in the previous section, shall be checked for its intersection with each of the triangles which comprises the surface of all the targets in the scenario under interest. To check whether a line segment intersects with a triangle in 3D coordinate system, Möller–Trumbore intersection algorithm is used. It is fast and a minimum storage ray triangle Intersection method is used. [10] is referred to make use of its equations in implementing ray casting method in the proposed LIDAR sensor model.

The algorithm translates the origin of the ray and then changes the base of that vector which yields a vector (t u v)T, where t is the distance to the plane in which the triangle lies and u, v represents the coordinates inside the triangle. A point on the triangle T(u,v) is given by:

$$T(u, v) = (1 - u - v)V_0 + uV_1 + vV_2 \tag{14}$$

where u, v are the barycentric coordinates which must fulfill the condition u >= 0, v <= 1 and u + v <= 1. Now, we need to find the intersection between the triangle T(u, v) and line segment R(t) given by:

$$R(t) = O + tD \tag{15}$$

Where O is the origin and D is the normalized direction, which is same as equating Eqs. 14 and 15, which gives:

$$O + tD = (1 - u - v)V_0 + uV_1 + vV_2 \tag{16}$$

Rearranging the terms results in a matrix form:

$$[-D, V1 - V0, V2 - V0] = \begin{bmatrix} t \\ u \\ v \end{bmatrix} = O - V0 \tag{17}$$

Where V0, V1 and V2 are the three vertices of the triangle, Cramer's rule is used to solve for t, u and v, where u and v gives the intersection coordinates and t the Euclidean distance of the intersection point from the LIDAR sensor position.

This entire operation will be repeated for each and every triangles in each and every 3D object's surface, for each and every LIDAR line segments in each and every vertical scanning patterns in each and every horizontal scans of the LIDAR sensor model. Each of the point cloud data vertices generated for each of the case above will be appended together to generate the final point cloud data in the current sampling time of LIDAR sensor operation.

7 Adding the Impact of Environment

It is imperative to add the impact of environment to the sensor model, so as to make the sensor similar to a real world sensor. Equation 1 gives the detected LIDAR signal power in watts, where the impact of environment, LIDAR efficiency, LIDAR geometrical properties etc. are already considered. An impact algorithm is designed so as to transfer the effect of real world information to the point cloud data where a calibrated and normalized value of detected power level P(R) will determine how much of point cloud data needs to be quantitatively transferred as the physical LIDAR sensor model output. It will also change the values of point cloud data, where the change made depends inversely on the normalized detected power level.

8 Object Detection and Classification

In certain LIDAR sensors, the point cloud data will be processed to detect the objects and classify them. [3] provides an overview about the automotive LIDAR objects detection and classification algorithm using the belief theory. Same concept is added to the proposed sensor model to detect the type of the object example: car, truck, pedestrian, traffic signs etc. from the point cloud data of the scene.

First step is to cluster the huge point cloud data, so that points from the same object can be grouped together. Clustering algorithm is simply based on the gathering of points, which are close to their neighbors using the Recursive Best Segment Split (RBSS) algorithm. Clustering also gives the 3D bounding box of each such clustered object. Next step is to time track the targets, where we can keep track of each object related to time. This allows in getting dynamic information about the targets, for example velocity. Next, we perform the object classification using the belief theory, where the truth of a hypothesis is validated using a degree of confidence. Degree of confidence of each object will be increased for a specific object type, based on a set of criterions, for example, width of bounding box is within the range allowed for a car, speed of the bounding box is within range allowed for a pedestrian etc. By looking at the confidence level, type of the object is decided.

9   Results

The full LIDAR sensor model is developed in CPP programming language to make sure that the memory usage is optimized. Point cloud data and object detection, classification is output via UDP protocol for verification. Simulation results were very close the real data got from physical LIDAR sensors. A sample output pint cloud data of the sensor model is shown in Fig. 4.

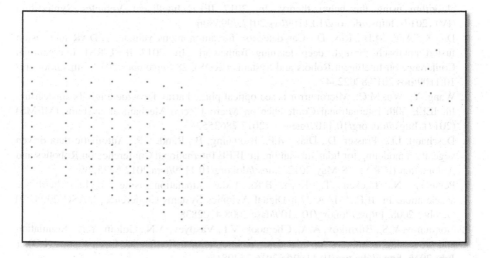

Fig. 4. Sample point cloud data output of the LIDAR sensor model

10   Conclusions and Future Works

A real world physical LIDAR sensor is modeled and simulated. Option to add any 3D objects into the scenario in FBX format, parse them and load triangles from its mesh surface was done. Logic for simulating the LIDAR beams and to check its intersection with the triangle faces was implemented using ray casting methodologies. Point cloud data thus generated will be subjected to the impact of environment to incorporate real world behavior of a LIDAR sensor. This point cloud data was then converted to object type for further processing. Author has completed the automotive grade RADAR sensor simulation, and thus the future work identified is sensor fusion of data from the RADAR sensor model and the proposed LIDAR sensor model.

Acknowledgements. The Author would like to acknowledge the support and guidance provided by Professor Jayakrishna Raj, Associate Professor, ECE Department, College of Engineering, Trivandrum and Professor Pradeep R, Associate Professor, ECE Department, College of Engineering, Trivandrum.

References

1. Kutila, M., Pyykönen, P., Ritter, W., Sawade, O., Schäufele, B.: Automotive LIDAR sensor development scenarios for harsh weather conditions. In: 2016 IEEE 19th International Conference on Intelligent Transportation Systems (ITSC) (2016). https://doi.org/10.1109/itsc.2016.7795565
2. Ogawa, T., Sakai, H., Suzuki, Y., Takagi, K., Morikawa, K.: Pedestrian detection and tracking using in-vehicle lidar for automotive application. In: 2011 IEEE Intelligent Vehicles Symposium (IV), 9 June 2011. https://doi.org/10.1109/ivs.2011.5940555
3. Magnier, V., Gruyer, D., Godelle, J.: Automotive LIDAR objects detection and classification algorithm using the belief theory. In: 2017 IEEE Intelligent Vehicles Symposium (IV) (2017). https://doi.org/10.1109/ivs.2017.7995806
4. Du, X., Ang, M.H., Rus, D.: Car detection for autonomous vehicle: LIDAR and vision fusion approach through deep learning framework. In: 2017 IEEE/RSJ International Conference on Intelligent Robots and Systems (IROS), 28 September 2017. https://doi.org/10.1109/iros.2017.8202234
5. Wang, Y., Wu, M.C.: Micromirror based optical phased array for wide-angle beamsteering. In: IEEE 30th International Conference on Micro Electro Mechanical Systems (MEMS) (2017). https://doi.org/10.1109/memsys.2017.7863553
6. Deschaud, J.E., Prasser, D., Dias, M.F., Browning, B., Rander, P.: Automatic data driven vegetation modeling for lidar simulation. In: IEEE International Conference on Robotics and Automation (ICRA), 18 May 2012. https://doi.org/10.1109/icra.2012.6225269
7. Peinecke, N., Lueken, T., Korn, B.R.: Lidar simulation using graphics hardware acceleration. In: IEEE/AIAA 27th Digital Avionics Systems Conference, DASC 2008, 30 October 2008. https://doi.org/10.1109/dasc.2008.4702838
8. Goryainov, V.S., Buznikov, A.A., Chernook, V.I., Vasilyev, A.N., Goldin, Y.A.: Simulation and processing techniques for lidar data. In: International Conference Laser Optics (LO), 1 July 2016. https://doi.org/10.1109/lo.2016.7549817
9. http://home.ustc.edu.cn/~522hyl/%B2%CE%BF%BC%CE%C4%CF%D7/lidar/intrduction%20to%20lidar1.pdf
10. https://cadxfem.org/inf/Fast%20MinimumStorage%20RayTriangle%20Intersection.pdf

A Solar Photovoltaic System by Using Buck Boost Integrated Z-Source Quasi Seven Level Cascaded H-Bridge Inverter for Grid Connection

R. Rahul(✉), A. Vivek, and Prathibha S. Babu

Department of EEE, Amrita Vishwa Vidyapeetham, Amritapuri, India
rahulrk0094@gmail.com,
{viveka, prathibhababu}@am.amrita.edu

Abstract. This paper brings up a grid-connected photovoltaic (PV) interface from the combined use of quasi-Z source inverter (qZSI) and cascaded H-bridge (CHB) inverter. Controlling and modeling of a Z-Source quasi CHB Seven level Inverter has been done. In one stage dc to ac conversion together with the buck/boost output is given by Quasi ZSI. PV applications requires best topology like this because of its numerous advantages such as reduced THD, high reliability and gain. Network for impedance has been drafted in this work for a seven level Z-source CHB Quasi inverter. To control switches of inverter shoot-through control along with Phase Shifted Inverted Sine carrier pulse width modulation is executed. In two stages power control scheme in closed loop condition is executed here. Controlling of each string voltage at the input side of photovoltaic is done by varying the inverter shoot-through type states with the help of individualistic Maximum Power Point Tracking control method. Each bridge DC link voltage was equalized with the help of DC link voltage control scheme. Controlling of grid transferred power is achieved from both of these controlling actions. A 2 Kilowatt photovoltaic inverter were designed simulated and executed with MATLAB simulation tool and also verified the open loop control.

Keywords: CHB inverter · Maximum power point tracking
Photovoltaic (PV) systems · Z source quasi inverter
Renewable energy · THD

1 Introduction

Energy from solar is becoming one of the most favorable renewable source of energy. Multilevel inverters are incorporated with PV systems because of the increase in power demand [1]. There are various types of multilevel inverters. They are capacitor clamped type, diode clamped type, and cascaded H-bridge (CHB) type inverter. CHB inverter is more advantageous when compared to other topologies. So it is generally used in Photovoltaic systems [2].

© Springer Nature Singapore Pte Ltd. 2018
I. Zelinka et al. (Eds.): ICSCS 2018, CCIS 837, pp. 687–697, 2018.
https://doi.org/10.1007/978-981-13-1936-5_70

The ZSI and the qZSI are extensively used in renewable power generation system because of its notable features. They can concurrently implement power conversion and voltage boost up in one step. The reliability can be improved due to the shoot-through technique without damaging the inverter. Researches are undergoing in the field of Photovoltaic embedded models depending on qZS-CHB multilevel type inverter [3]. With the introduction of qZS network to the CHB module, this system will bring up many benefits, such as boosting up of the PV string voltage, maximum power point tracking control of individual PV strings, therefore keeps same DC link voltages in all H-bridge type inverter modules [4, 5].

In grid tied PV generation systems string inverters, module inverters, and central inverters are commonly used. These types of inverters have many disadvantages such as the output voltage of inverter would be smaller compared to the DC input rail voltage and prevents switch conduction in identical leg simultaneously [6–8]. CHB multilevel inverter (MLI) is selected to PV systems for increasing the output voltage quality. For rising the input voltage in multi-level inverter, a DC-DC converter is needed in each stage that will cause an increase in the price and entanglement of the circuit. So work done here is proposing a Z-Source Quasi seven-level multilevel inverter (QZMLI) for photovoltaic enactment integrated to grid [9]. The Seven-level QZMLI has many features such as stepping up/stepping down the voltage and by providing a unique impedance network in all levels. Hence the device conduction takes place in identical leg which causes inversion of DC-AC.This happens in a single power stage [10]. The evolution of the QZMLI by integrating the merits of both the quasi network of impedance and H-bridge cascaded MLI. In future, this will be a propitious topology for PV applications.

The controlling of the power flow and output-side voltage into the grid is very difficult because of the frequent fluctuations in the solar power. Thus QZMLI employs an independent maximum power point tracking [11]. The generated power coming out of the photovoltaic arrays could be controlled by implementing maximum power point tracking method, controller of the total power injection and separate controller of the DC link voltages [12]. A power controlling method for PV based 1Φ grid tied QZMLI with PI controller is implemented in this paper. Each PV source employs MPPT based on Perturb and Observe (P & O) methodology. The shoot-through states can be varied by using maximum constant boost controlled MPPT and PWM will be controlled by method of DC link voltage control. On reducing stress on voltage across the devices and the total harmonic distortion, the power quality of the voltage in the grid can be enhanced by using sine carrier which is phase shifted and inverted [13].

Comparing to other multilevel inverters, cascaded multilevel inverter (CMI) is more advantageous. In conventional CMI, different modules have different DC link voltages due to the wide range variations in PV voltage. Even if the PV voltage is low, CMI will produces the rated voltage across output side [14]. After the photovoltaic array another DC-DC boost converter will be connected to the DC link of all modules to wipe out the demerits of conventional CMI based PV power systems. Thus each module consists of two-stage inverter. Additional inductors and power switches are needed for this topology. Thus the power circuit and the control implementation becomes more complex which leads the increase in cost of the system and power loss. The same purpose can be achieved in one stage power conversion by the qZS-CMI. Each leg of the bridge will allow the shoot through and an additional diode will be used instead of active switch. Comparing with traditional CMI based photovoltaic systems, one-third units can be saved with the qZS-CMI. Thus the cost can be reduced, the system efficiency and reliability are improved [15]. However, not much literature was happened in the control method and operating performance of three-phase z source quasi cascaded multilevel inverter based grid-tie photovoltaic embedded systems. In 3Φ systems, for MW scale power applications the qZS-CMI will be corelated with traditional CMI. This shows qZS-CMI is more advantageous with high efficiency, high reliability and low cost, but control method is not involved in it. For single-phase systems high efficiency can be achieved by connecting GaN devices to single-phase qZS-CMI based photovoltaic embedded system. The parameter designing of qZS H-bridge module was done. Here proposing a controlling method for 1Φ qZS-CMI photovoltaic embedded system which includes separate control for dc-link voltage, grid-tie control along with distributed maximum power point tracking control (MPPT). To achieve maximum output power tracking from photovoltaic cell it uses shoot-through type duty cycle and fixed peak DC-link voltage control is used for adjusting the modulation signal of the module to push every photovoltaic powers injected into the grid.

2   Quasi Z-Source Multilevel Type Inverter

A seven-level CHB - qZS Inverter with Quasi Network of Impedance connected to every DC link of photovoltaic unit is shown in Fig. 1. Three inductors namely L_1, L_2, L_3 are and three capacitors namely C_1, C_2, C_3 are in the impedance network of each stage of the inverter circuit. The working of the circuit is modified by the LC network that is linked to the inverter bridge, which allows the shoot through states. Thereby it protects the circuit from damage in case of short circuit. The quasi z source network increases the voltage at DC link by the effective utilization of this shoot-though state. The attractive features of QZSI in comparison to other Z-inverters are that the DC current drawn from source continuous and constant. The voltage obtained in the capacitor C2 is decreased tremendously. This topology is best suited for power conditioning of PV systems because of the steady and very fixed source side DC current drawing.

2.1   Circuit Working of Quasi Z-Source Inverter

QZSI have two different states of operation. They are known as shoot-through type state and the non-shoot through type state. Figures 2 and 3 represents the circuits of the shoot-through type state and the non-shoot through type state. Since the voltage in the dc link is zero, there will be no power transmission in shoot-through state. The power transmission into AC side from the DC side is occurring in non-shoot type through state. Based on the operating conditions, basic relationships of the parameters are given below.

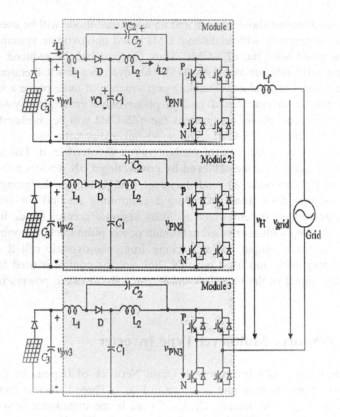

Fig. 1. Seven level cascaded QZSI

$$V_{ca} = \frac{1-D}{1-2D} V_{PhV} \tag{1}$$

$$V_{cb} = \frac{D}{1-2D} V_{PhV} \tag{2}$$

$$V_{dc} = \frac{1}{1-2D} V_{PhV} = B.V_{PhV} \tag{3}$$

$$V_{ac} = \frac{1}{2} * V_{PhV} * B * M = \frac{1}{2} * M * V_{dc} \tag{4}$$

$$B = \frac{1}{1-2D} \tag{5}$$

$$D = \frac{T_{sh}}{T_s} \tag{6}$$

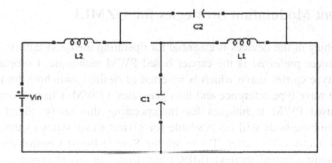

Fig. 2. Shoot state

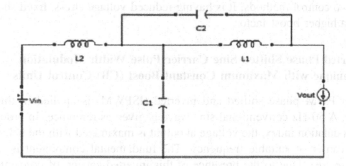

Fig. 3. Non-shoot state

2.2   Quasi Impedance Network Design

The impedance type network is designed on the basis of 2ω analysis of ripple scheme of the single phase QZSI. In single phase QZSI, there is a flow of second harmonic power, so this is a specified analysis. Size of impedance network becomes more complex due to the passive methods used for alleviating 2ω component. For finding the least QZS capacitance and inductance, here analyze 2ω power flow and proposes damping control of current ripple. Capacitors stores the 2ω Power Po-2ω and the capacitance of QZS is given by,

$$C = \frac{2V_{out}I_{out}}{\omega\varepsilon V_{PN}^2} \qquad (7)$$

To limit the inductance of switching frequency ripples of QZS,

$$L = \frac{D(1-D)T_S V_{DC}}{(1-2D).2.\Delta i} \qquad (8)$$

The capacitance and the inductance values can be find out from the Eqs. (7) and (8)

3 Different Modulation Strategies for QZMLI

Proper switching of the devices is essential for operating the qZSI faithfully. Therefore switching pattern preferred is the carrier based PWM technique. Compare the phase shifted sine type carrier wave which is inverted of desired switching frequency along with the sine wave type reference and thus generates a PWM. Changes are done in the commonly used PWM techniques for incorporating the newly added shoot state. Various control methods will be available for giving shoot states to the signals of Z source quasi multilevel inverter. Those are the Simple Boost Control (SBC) method, Maximum Boost Control method (MBC) and Constant Boost control (CB) method Depending upon the magnitude of shoot through line these strategies may differ. In the paper, CB technique is implemented because is more advantageous when comparing to the other two control methods. It is having reduced voltage stress, fixed shoot through ratios and a higher boost factor.

3.1 Inverted Phase Shifted Sine Carrier Pulse Width Modulation Technique with Maximum Constant Boost (CB) Control Units

Sine carrier PWM phase shifted and inverted (ISPWM) is applied in this proposed technology. A 50 Hz conventional sine wave is given as reference. In order to get the required modulation index, the voltage at output is maximized with the help of inverted sine wave carrier of suitable frequency. The fundamental component is increased at greater pulse area. Due to the presence of this inverted carrier, of conventional triangular PWM fundamental component becomes three times and the THD will be reduced. This PWM technique doesn't require any change in mode and has equal switching in each cycle along with enhanced fundamental component.

4 Control of Photovoltaic Based Z Source Quasi Multilevel Inverter for Grid Connected System

4.1 Total Power Controlling Scheme

Conventional methods convey independent controlling of the reactive and real inverter power. But a basic total power controlling method of the z source quasi multilevel inverter is established through this paper. It conjoins both Dc link voltage controlling technique and independent MPPT controlling scheme for each inverter bridge. Injected power into the grid will be the output-side power of the photovoltaic module. Let Pnx be the power reference. Then maximum grid current will be as follows

$$I_{grid} = \frac{2Pnx}{v_{gridd}} \tag{9}$$

To achieve unity power factor operating condition, for measuring the voltage phase at grid, phase locked loop will be implemented. After measuring the current at grid it will be fed backed into the current loop. Current loop gives the total output voltage to produce desired modulation index. The suitable gate pulse for QZSI can be generated by using this desired index for modulation obtained and MPPT provided shoot state duty ratio. Also DC link voltage controller generates power reference and by using this, can implement power control. Under DC link voltage control method, capacitor voltage will come in a comparison to the bridge DC link voltage obtained. With the help of a proportional integrator the error gets transferred to a proportional value of current. Now the grid voltage phase is measured using PLL and thus produces the current reference. For locking phase of grid voltage and grid current, the current reference should be compared with the grid current. PWM is modified by adding the shoot-state produced by MPPT with the help of Perturb & Observe method with the PWM which is inverted sine and phase shifted. Inverter power can be controlled by using this modified PWM which would control inverter power.

4.2   DC Link Voltage Control Simulation and MPPT

Modeling of grid tie photovoltaic system based on cascaded seven-level QZSI was done using MATLAB simulation tool and obtained outputs are mentioned here. The entire system simulation is depicted below in Fig. 4.

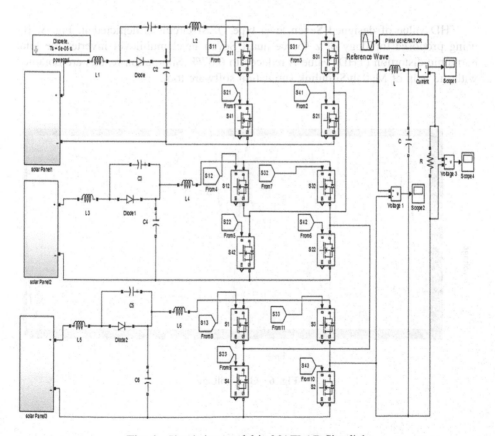

Fig. 4. Simulation model in MATLAB Simulink

The output voltage from the solar module obtained is depicted in Fig. 5. Figures 6 and 7 shown below depicts the voltage and current peak values at the grid. Grid voltage peak value is found as 415 Volts and peak value of grid current is found to be 4.5 A.

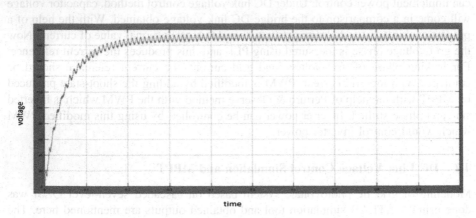

Fig. 5. Output voltage from the solar panel

THD value of designed Seven level type QZSI circuit is depicted in Fig. 8. By using proposed topology of z source quasi Seven level multilevel inverter the total harmonic distortion (THD) value is reduced to 6.87%. Simulation results are obtained with the help of Matlab/Simulink simulation software tool.

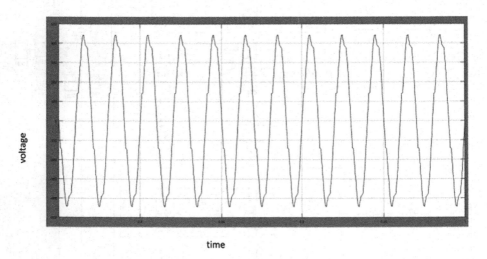

Fig. 6. Grid voltage

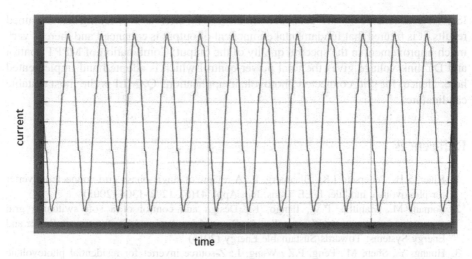

Fig. 7. Grid current

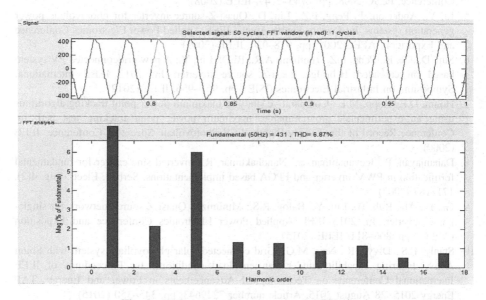

Fig. 8. THD value of the seven level QZSI circuit

5   Conclusion

A Z-Source Quasi Seven level grid connected photovoltaic system using cascaded type multilevel inverter is designed, modelled and controlled in the proposed paper. Using 2ω ripple analysis impedance network design was done as it proves and provides a minimized LC network. With use of single stage power conversion, proposed system can inject PV power to grid with lesser total harmonic distortion By implementing

IS-PSPWM scheme to the proposed multilevel inverter and checking the obtained results, it is figured that fundamental component of output is enhanced and there is very much improvement in the spectral quality of the output. Combination of MPPT control and DC link voltage gives the total power control which is adopted and implemented here. Hence, for grid connected photovoltaic applications QZMLI is the most suitable candidature.

References

1. Kjaer, S.B., Pedersen, J.K., Blaabjerg, F.: A review of single-phase grid-connected inverter for photovoltaic module. IEEE Trans. Ind. Appl. **41**(5), 1292–1306 (2005)
2. Soman, M., Manitha, P.V., Ilango, K.: Design and control of a soft switching grid connecting inverter using PI controller. In: Biennial International Conference on Power and Energy Systems: Towards Sustainable Energy (2016)
3. Huang, Y., Shen, M., Peng, F.Z., Wang, J.: Z-source inverter for residential photovoltaic system. IEEE Trans. Power Electron. **21**(6), 1776–1782 (2006)
4. Anderson, J., Peng, F.Z.: Four quasi-Z-source inverter. In: Power Electronics Specialists Conference, PESC 2008, pp. 2743–2749. IEEE (2008)
5. Li, Y., Anderson, J., Peng, F.Z., Liu, D.: Quasi-Z-source inverter for photovoltaic power generation systems. In: Twenty-Fourth Annual IEEE Applied Power Electronics Conference and Exposition, APEC 2009, pp. 918–924. IEEE (2009)
6. Sun, D., Ge, B., Peng, F.Z., Haitham, A.R., Bi,D., Liu, Y.: A new gridconnected PV system based on cascaded H-bridge quasi-Z source inverter. In: 2012 IEEE International Symposium on Industrial Electronics (ISIE), pp. 951–956. IEEE (2012)
7. Hohm, D.P., Ropp, M.E.: Comparative study of maximum power point tracking algorithms using an experimental, programmable, maximum power point tracking test bed. In: Conference Record of the Twenty-Eighth IEEE Photovoltaic Specialist Conference. IEEE (2000)
8. Dananjayan, P., Jeevananthan, S., Nandhakumar, R.: Inverted sine carrier for fundamental fortification in PWM ınverter and FPGA based ımplementations. Serb. J. Electr. Eng. **4**(2), 171–187 (2007)
9. Ge, B., Abu-Rub, H., Liu, Y., Balog, R.S.: Minimized Quasi Z source network for single-phase inverter. In: 2015 IEEE Applied Power Electronics Conference and Exposition (APEC), pp. 806–811. IEEE (2015)
10. Stanly, L.S., Divya, R., Nair, M.G.: Grid connected solar photovoltaic system with Shunt Active Filtering capability under transient load conditions. In: Proceedings of IEEE International Conference on Technological Advancements in Power and Energy, TAP Energy 2015, 28 August 2015, Article number 7229643, pp. 345–350 (2015)
11. Thangaprakash, S., Krishnan, A.: Comparative evaluation of modified pulse width modulation schemes of Z-source inverter for various application and demands. Int. J. Eng. Sci. Technol. **2**(1), 103–115 (2010)
12. Rakhi, K., Ilango, K., Nair, M.G.: Hardware implementation of solar photovoltaic system based half bridge series parallel resonant converter for battery charger. Int. J. Power Electron. Drive syst. **8**(4), 1622–1630 (2017)
13. Aiswarya, M., Ilango, K., Manitha, P.V., Nair, M.G.: Simulation analysis modified DC bus voltage control algorithm for power quality improvement on distributed generation systems. In: Power and Energy Systems Conference: Toward Sustainable Energy, PESTSE 2014 (2014)

14. Umarani, D., Seyezhai, R.: Design and simulation of cascaded h-bridge quasi z-source multilevel inverter for photovoltaic application. Int. J .Innov. Res. Electr. Electron. Instrum. Control Eng. 2(7) (2014)
15. Seyezhai, R., Mathur, B.L.: Performance evaluation of inverted sine PWM technique for an asymmetric cascaded multilevel inverter. J. Theor. Appl. Inf. Technol. 9(2), 91–98 (2009)

Modified Dickson Charge Pump and Control Algorithms for a Solar Powered Induction Motor with Open End Windings

Riya Anna Thomas[(✉)] and N. Reema

Electrical and Electronics, Sree Buddha College of Engineering, Pattoor, India
312riya94@gmail.com, n.reema3@gmail.com

Abstract. Induction motors are the widely used machines in many industrial applications. But the induction motors results in high starting current, occurrence of common mode voltages, difficult to control etc. This work proposes controlling of a dual inverter fed induction motor with open end stator windings. This system is functioned using an integrated control which includes Perturb and Observe maximum power point tracking algorithm and space vector pulse width modulation technique. A boosting converter with a high voltage gain is also proposed which requires no chemical storage elements such as batteries and make the system more effective. Space vector modulation is designed such that it excludes the common mode voltages generated. The modeling of dual inverter served induction motor with open end windings, PV system, MPPT algorithm, Dickson charge pump and switching of inverters using modulation by space vector has been done and the output waveforms obtained are analyzed.

Keywords: Common mode voltage · Open end winding induction motor
Space vector modulation

1 Introduction

One of the efficient solution for the utilization of resources is the use of systems delivered by photovoltaic solar energy. Among different machines, induction machines are most widely used due to many advantages such as low cost, ease of operation, durability, high starting torque etc. In this paper, a three level dual inverter fed open end winding induction motor (OEWIM) is proposed to overcome the shortcomings of conventional induction motor. Vast works were carried out for different inverter topologies [5–8] and different modulation schemes [9, 10] for OEWIM. In this machine stator windings are open and the supply is given from both sides and inverters are connected at either ends to obtain the multilevel output. Vast works has being carried out about the different modulation schemes used for open end winding induction motor [11, 12]. This works proposes a zero sequence elimination pulse width modulation scheme for the machine. The system is supplied by solar panel and MPPT algorithm is also used to path the maximum power and space vector modulation scheme is also used for switching of inverters efficiency etc. But the induction motors has certain disadvantages such as difficult to control, occurrence of common mode voltages, high

© Springer Nature Singapore Pte Ltd. 2018
I. Zelinka et al. (Eds.): ICSCS 2018, CCIS 837, pp. 698–706, 2018.
https://doi.org/10.1007/978-981-13-1936-5_71

starting current etc. Several multilevel converter topologies have also remained analysed and suggested during several periods [1–3]. To improve the overall performance neutral point clamped inverters was used which is beneficial over two level H bridge inverter [4]. The multilevel inverter is a feasible.

2   Control Algorithms of OEWIM

The proposed system includes control algorithms for an open end winding induction motor (OEWIM) besides modified Dickson charge pump. OEWIM is simply an induction motor with both the ends of the stator windings are opened. The Fig. 1 shows the block illustration of the proposed system.

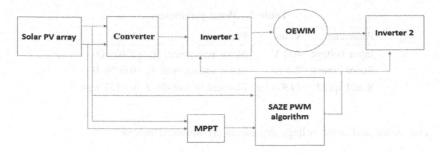

Fig. 1. Block illustration of control algorithms of OEWIM

Nowadays OEWIM received large attention in many applications and has many advantage over conventional induction machine. The proposed system is powered by solar photovoltaic panel and to track the maximum power MPPT algorithm is used. A great voltage gain DC- DC converter is also used to acquire a stable output. The proposed converter is Dickson charge pump voltage multiplier. An OEWIM contains dual inverters on either side of the machine to achieve a 3 level output voltage and supply is provided from both sides of the inverter. The machine can be used for variable speed applications, grid applications, hybrid vehicles etc. Figure 2 illustrates an induction machine with both ends of stator windings are open.

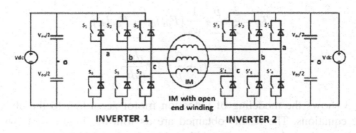

Fig. 2. OEWIM

Since both ends of stator windings are open the switches can be rated at half of the machine power rating and it removes the common mode voltage generated due to switching.

3   Simulation and Results

3.1   Modeling of Induction Motor

Modelling of OEWIM is similar to conventional induction motor. For simulating the proposed system, a 4 kW induction motor is modelled with the parameters specified in Table 1.

Table 1. Motor parameters

Parameter	Value	Parameter	Value
Input voltage	400 V	Stator resistance, R_s	1.405 Ω
Stator current	7.2 A	Stator inductance, L_s	0.0058 H
Rated speed	1430 rpm	Moment of inertia, J	0.0131 kg-m^2

The stator and rotor voltage equations are showed below.

$$V_{qs} = R_s i_{qs} + \frac{d\varphi_{qs}}{dt} + \omega_e \varphi_{ds} \tag{1}$$

$$V_{ds} = R_s i_{ds} + \frac{d\varphi_{ds}}{dt} + \omega_e \varphi_{qs} \tag{2}$$

$$V_{qr} = R_s i_{qr} + \frac{d\varphi_{qr}}{dt} + \omega_e \varphi_{dr} \tag{3}$$

$$V_{dr} = R_s i_{dr} + \frac{d\varphi_{dr}}{dt} + \omega_e \varphi_{qr} \tag{4}$$

The equation for determining torque and rotor speed of induction motor is given by

$$T_e = \frac{3}{2} * \frac{p}{2} * \frac{1}{w_b} \left(F_{ds} i_{qs} - F_{qs} i_{ds} \right) \tag{5}$$

$$T_e = T_l + \frac{2}{p} * J * \frac{dw_r}{dt} \tag{6}$$

Figure 3 shows the modeling of induction motor according to the above voltage and torque equations. The results obtained are as follows. Figure 4 shows the input voltage given to the induction motor with an amplitude of 325 V.

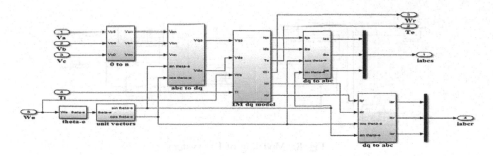

Fig. 3. Modeling of induction motor

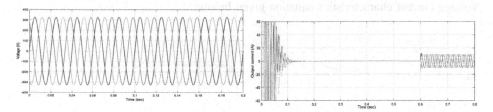

Fig. 4. Input voltage of induction motor   **Fig. 5.** Output current of induction motor

Figure 5 shows the waveforms of output current, of an induction machine. A step time of 0.6 s is given to the machine, as a result up to 0.6 s the graphs shows the no load reading and after 0.6 s loaded condition is depicted here. The topmost value of the current of about 10.2 A obtained (Figs. 6 and 7).

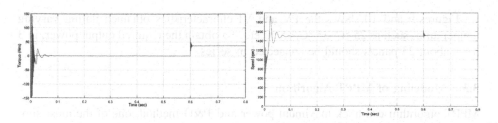

Fig. 6. Torque Vs Time graph   **Fig. 7.** Speed Vs Time graph

3.2 Modeling of PV System

Solar photovoltaic system is used to supply the machine and is modeled for running a 4 kW OEWIM. It is modeled according to the voltage current characteristic equation.

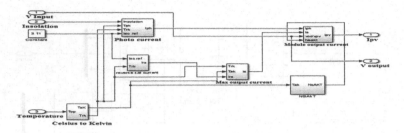

Fig. 8. Modeling of PV system

Figure 8 shows the modeling of photovoltaic system in MATLAB according to the voltage current characteristics equation given below:

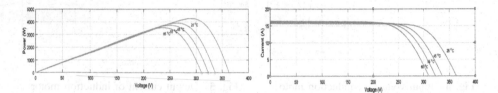

Fig. 9. PV characteristics   **Fig. 10.** VI characteristics

$$I = I_{ph} - I_s \left[\exp \left(\frac{q(V + IR_s)}{kT_cA} \right) - 1 \right] - \frac{V + IR_s}{R_{sh}} \tag{7}$$

Figures 9 and 10 shows the PV and VI characteristics obtained during varying temperatures such as 25 °C, 35 °C, 45 °C etc. To obtain the required output power, in 3 rows about 23 panels should be connected in series.

3.3   Modeling of MPPT Algorithm

MPPT algorithm can track maximum power and P&O method, one of the most simplest method is proposed here. Figure 11 shows the modeling of MPPT algorithm and the output obtained is used to trigger the converter switches.

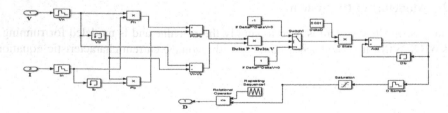

Fig. 11. Modeling of MPPT algorithm

3.4   Modeling of Converter

To obtain a stable output a converter is also proposed in this circuit. A Dickson charge pump voltage multiplier is used which had a high voltage gain and removes the need of chemical storages such as batteries (Table 2).

Table 2. Converter parameters

Parameter	Value	Parameter	Value
Input voltage	20 V	L_1	100 µH
Output voltage	400 A	L_2	100 µH
Load resistance	800 Ω	VM capacitors	60 µF

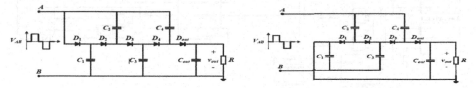

Fig. 12. Dickson charge pump voltage multiplier

Figure 12 shows the circuit of conventional and modified Dickson charge pump and Fig. 13 shows the different modes of operation of Dickson charge pump.

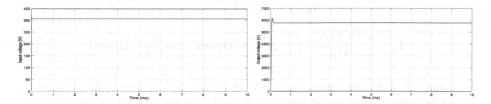

Fig. 13. Input and output of dickson charge pump voltage multiplier at 1000 W/m² and 25 °C

The circuit works by the charging and discharging of the capacitors in the circuit. It has three modes of operation, either two switches will be ON or either of the switches will be ON. The graph shows the input and output voltage at 1000 W/m² and 25 °C is shown in Fig. 13.

3.5   Modeling of OEWIM and Space Vector Modulation

Modeling of OEWIM is given by subtracting the difference of pole numbers of both the inverters which can be seen in Fig. 14. The equations below are used for converting abc to dq coordinate

$$V_d = \frac{2}{3}\left(V_a \sin\omega t + V_b \sin(\omega t - \frac{2\Pi}{3}) + V_c \sin(\omega t + \frac{2\Pi}{3})\right) \tag{8}$$

$$V_q = \frac{2}{3}\left(V_a \cos\omega t + V_b \cos(\omega t - \frac{2\Pi}{3}) + V_c \cos(\omega t + \frac{2\Pi}{3})\right) \tag{9}$$

$$V_0 = \frac{1}{3}(V_a + V_b + V_c) \tag{10}$$

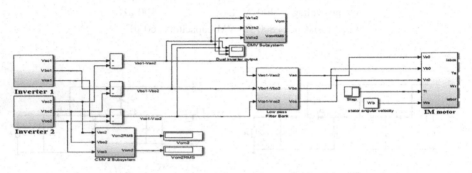

Fig. 14. Modeling of OEWIM

The below equations shows the determination of time duration

$$T_a = \frac{\sqrt{3} * T_Z * V_{ref}}{V_{dc}}\left(\sin\frac{n}{3}\Pi\cos\alpha - \cos\frac{n}{3}\Pi\sin\alpha\right) \tag{11}$$

$$T_b = \frac{\sqrt{3} * T_{Z*} V_{ref}}{V_{dc}}\left(-\sin\frac{n-1}{3}\Pi\cos\alpha + \cos\frac{n-1}{3}\Pi\sin\alpha\right) \tag{12}$$

Figure 15 shows the space vector model of the dual inverter. The inverters are pulsed by space vector modulation scheme and is modeled rendering to the above equations.

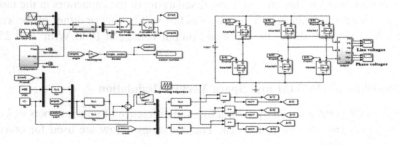

Fig. 15. Modeling inverter and space vector modulation scheme

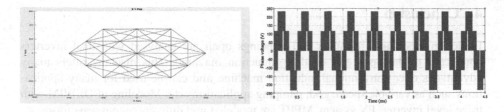

Fig. 16. Space vector of dual inverters   **Fig. 17.** Dual inverter voltage

The modulation scheme has a total of space vector locations of about 64 which can be depicted from the Fig. 16. Figure 17 shows the dual inverter output voltage obtained by space vector modulation.

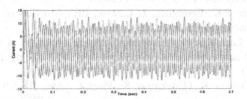

Fig. 18. Output current of OEWIM

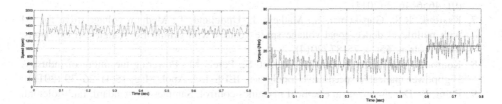

Fig. 19. Speed characteristics   **Fig. 20.** Torque characteristics

Figures 18, 19 and 20 shows the output current, speed and torque of an OEWIM. The dual inverter is modeled and the difference in voltages of inverters is given to induction motor to obtain the modeling of OEWIM. A 0.6 s a step load is applied and the variation during load can be shown in the figure. The use of OEWIM over conventional machine excludes the common mode voltages generated.

4 Conclusion

Induction machine with stator end windings open are analysed with dual inverters connected to the open end winding induction machine. This machine offers many advantages over conventional induction machine and can be used for many applications such as variable speed, grid, pumping applications etc. Modeling of OEWIM with three level inverter, PV system, MPPT are modeled and different output waveforms are analyzed.

References

1. Wheeler, P., Xu, L., Meng, Y., Empringham, L., Klumpner, C., Clare, J.: Analysis of multilevel matrix converter topologies. In: Proceedings of 4th IET Conference Power Electronics, pp. 286–293 (2008)
2. Rodriguez, J., JihSheng, L., Zheng, P.: Multilevel inverters, a survey of topologies, controls. IEEE Trans. Ind. Electron. 49(4), 719–724 (2003)
3. Bernet, S., Steimer, P.K., Rodriguez, J., Lizama, I.E.: A review on neutral point clamped inverters. IEEE Trans. Ind. Electron. 58(9), 229–2220 (2011)
4. Jiao, Y., Lu, S., Lee, F.C.: Switching performance optimization of a high power high frequency 3 level active neutral point clamped leg. IEEE Trans. Power Electron. 29(7), 3245–3266 (2014)
5. Rajeevan, P., Sivakumar, K., Gopakumar, K., Patel, C., Abu, H.: A 9 level inverter topology for medium voltage induction motor drive with oewim. IEEE Trans. Ind. Electron. 60(9), 3627–3636 (2013)
6. Somasekar, V., Vennugopal Reddy, B., Sivakumar, K.: A 4 level inversion scheme for a 6-*n*-pole OEWIM drive for an better-quality DC link utilization. IEEE Trans. Ind. Electron. 57(9), 4656–4672 (2014)
7. Kaarthik, R.S., Gopakumar, K., Mathew, J., Undeland, T.: Medium voltage drive for IM with multilevel dodecagonal voltage space vectors with symmetric triangles. IEEE Trans. Ind. Electron. 63(1), 75–87 (2014)
8. Kouro, S., Rodriguez, J., Bin, W., Bernet, S., Perez, M.: Powering the future of industry, Highpower malleable speed drive topologies. IEEE Ind. Appl. Mag. 15(4), 26–39 (2012)
9. Betin, F.: Developments in electrical machines control: Samples for standard, fault-tolerant, sensorless, techniques. IEEE Ind. Electron. Mag. 8(2), 43–55 (2014)
10. Meinguet, F., Sandulescu, P., Kestelyn, X., Semail, E.: Fault-tolerant operation of an OEW 5phase PMSM drive with inverter faults. In: Proceedings Conference IEEE Industrail Electronics, pp. 5191–5196 (2013)
11. Venugopal Reddy, B., Somasekhar, V.T., Kalyan, Y.: Decoupled space-vector PWM strategies for a four-level asymmetrical open-end winding induction motor drive with waveform symmetries. IEEE Trans. Ind. Electron. 58(11), 5130–5141 (2011)
12. Jacob, B., Baiju, M.R.: Vector-quantized space-vector-based spread spectrum modulation scheme for multilevel inverters using the principle of oversampling ADC. IEEE Trans. Ind. Electron. 60(8), 2969–2977 (2013)

The Torque and Current Ripple Minimization of BLDC Motor Using Novel Phase Voltage Method for High Speed Applications

Meera Murali$^{(\boxtimes)}$ and P. K. Sreekanth

Department of EEE, Sree Buddha College of Engineering, Pattoor, India
meeraprabha001@gmail.com,
Sreekanth.kallamvalli@gmail.com

Abstract. This paper introduce a novel sensorless control for the Brushless DC motor drive by phase voltage method for minimizing the torque and current ripple in high speed, high power applications. The commutation signals for the six switch inverter are generated from the phase voltage. In each phase commutation two switches conduct corresponding to the truth table logic. Delay occurring in speed is compensated by the phase voltage deviation. Selection of commutation logic in a parallel mode improves the efficiency of system by reducing the operation delay. Compared with the existing traditional control torque and current ripple of BLDC motor can be significantly reduced by using this proposed method.

Keywords: BLDC motor · Current ripple · Sensorless · Phase voltage

1 Introduction

In the present scenario, conventional Direct Current (DC) motors are replaced by Brushless Direct Current (BLDC) motors. Efficiency of onventional Brushed DC motors are high and their merits make the motors apt for use as servo motors. However, their main drawbacks are the presence of commutators and brushes,which require regular maintenance [1]. The Permanent Magnet Brushless DC (PMBLDC) motor is became popular in various applications because of its higher efficiency, improved power factor, high torque to weight ratio, simplicity in control and lower maintenance. The major demerits of PM motors are their high cost and relatively increased complexity introduced by the driver system because of the power electronic converter in it. BLDC motors do not have brushes for commutation, instead of mechanical commutation they use electronic commutation [2].

The control of BLDC motor is possible by sensor or sensorless method. The sensor based control is the basic control method. In this method the normally used sensor is hall sensor which determines the position of rotor for phase commutation. Using such position sensor has several drawbacks such as high cost, less reliability and some times it increases the criticality of the system [3]. To overcome the demerits of sensor based control, sensorless control methods are used where position sensor can be completely eliminated. In sensorless control it is possible to determine the position of rotor by

© Springer Nature Singapore Pte Ltd. 2018
I. Zelinka et al. (Eds.): ICSCS 2018, CCIS 837, pp. 707–714, 2018.
https://doi.org/10.1007/978-981-13-1936-5_72

sensing the back EMF. The back EMF sensing methods are direct detection method of back EMF and indirect detection method of back EMF. The direct detecting method of back EMF includes the Zero Crossing Detection (ZCD) or the terminal voltage sensing and indirect detecting method of back EMF includes the back EMF integration method, third harmonic voltage integration method and free wheeling diode conduction method [4]. Instead of back EMF phase voltages can be used for the sensorless operation of BLDC motor. Because for the high speed applications the outputs obtains with the back- EMF method contains more ripples. And this method can't be used for the wider speed range.

2 Dynamic Equations and Modelling of BLDC Motor

The circuit diagram of the BLDC motor drive system is shown in Fig. 1 below.

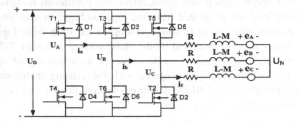

Fig. 1. BLDC drive system

The electrical and mechanical mathematical equations of BLDC are given in this section. The stator voltage equations are [5]:

$$U_A = Ri_a + (L - M)\frac{di_a}{dt} + e_A + U_N \tag{1}$$

$$U_B = Ri_b + (L - M)\frac{di_b}{dt} + e_B + U_N \tag{2}$$

$$U_C = Ri_C + (L - M)\frac{di_c}{dt} + e_C + U_N \tag{3}$$

The equations of back-EMF can be given as:

$$e_A = K_e\omega_m F(\theta_e) \tag{4}$$

$$e_C = K_e\omega_m F(\theta_e - 2\Pi/3) \tag{5}$$

$$e_C = K_e\omega_m F(\theta_e + 2\Pi/3) \tag{6}$$

The equations of electrical torques can be given as:

$$T_A = K_t i_a F(\theta_e) \tag{7}$$

$$T_B = K_t i_b F(\theta_e - 2\Pi/3) \tag{8}$$

$$T_C = K_t i_c F(\theta_e + 2\Pi/3) \tag{9}$$

$$T_e = T_A + T_B + T_C \tag{10}$$

$$T_e - T_l = J\frac{d^2\theta_m}{dt^2} + \beta\frac{d\theta_m}{dt} \tag{11}$$

Where

U_N is the neutral point voltage of motor
i_a, i_b, and i_c are the phase currents in A, B, C phases
U_A, U_B, and U_C are the three-phase voltages
e_A, e_B, and e_C are phase back EMFs.
R is the winding phase resistance
L is the self inductance of motor
M is the mutual inductance of motor
K_e is the back-EMF constant
K_t is the torque constant.

With these dynamic equations BLDC motor model was developed. Motor parameters used for modelling are shown in the Table 1.

Table 1. Motor Parameters

Parameters	Values
Input voltage	470 V
Rated current	220 A
Rated speed, N	20000 rpm
Rated torque, T	50 Nm
Winding phase resistance, R_S	1.31 mΩ
Phase inductance, L_S	0.052 mH

3 Phase Voltage Method

A novel self compensation method based on the phase voltages of the motor is used here to obtain better performance. A switching logic and a commutation algorithm for the inverter circuit are implemented to generate commutation signal from the phase voltage (Fig. 2).

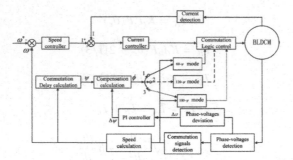

Fig. 2. Block diagram of proposed method

On comparing with the real time approach offline compensation method is very easily affected by small drift of sensors, unbalance in three phases and disturbance occur in the load etc. When suddenly the load is changed, the non real time compensation cannot easily respond to the disturbance occurs. Self tuning or auto regulation of Proportional Integral (PI) controllers in a feedback system is proposed based on the phase voltage deviations.

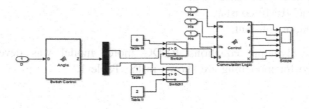

Fig. 3. Simulink model of new switching and commutation logic

Figure 3 shows the block diagram of closed loop control strategy. It contains a current controlling loop, a speed controlling loop, a commutation logic controller with a switching control, a commutation signal detection area, a compensation angle processing path, etc. Comparing the usual BLDCM speed control system; this self compensated sensorless commutation algorithm effectively increases the overall efficiency of the system. The commutation delay is calculated from the speed of motor. And this delay is compensated by the deviation in phase voltages. Corresponding to this compensation value the particular truth table will be selected and the control logic with that truth table will be actuated. There are three truth tables for the proper selection of commutation logic to the inverter.

4 Results

The output characteristics of the conventional back-EMF method and proposed self compensation phase voltage method are shown below. Simulation results of the phase current are also shown in figures. Figure 4 shows the phase current of the back-EMF method. And the phase current of the proposed system is shown in Figs. 5 and 6.

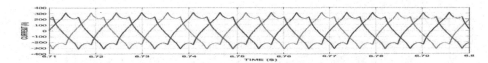

Fig. 4. Phase currents vs Time waveform of conventional system

Average value of the current obtained from the upper limit 300 and lower limit 200 is 250 A. Difference is 100 A. Ripples obtained from the conventional system are 0.4.

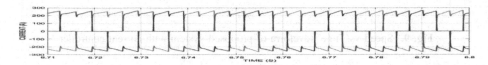

Fig. 5. Phase currents vs Time waveform of proposed system without current controller

Average value of the current obtained from the upper limit 280 and lower limit 200 is 250 A. Difference is 80 A. Ripples is reduced to 0.33 in proposed system without current controller.

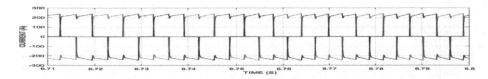

Fig. 6. Phase currents vs Time waveform of proposed system with current controller

When adding the current controller to the proposed control then the current ripple reduces to 0.19 as shown in Fig. 6. The current ripples are minimized effectively in the proposed method. Speed response of conventional system is shown in Figs. 7, 8 and 9 shows the speed characteristics of proposed method. In the proposed system rated speed 20000 rpm attained at very fast rate.

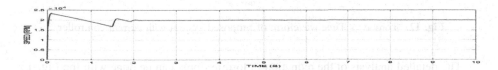

Fig. 7. Speed vs Time waveform of conventional system

Torque characteristics of the conventional and proposed system are shown in Figs. 10, 11 and 12.

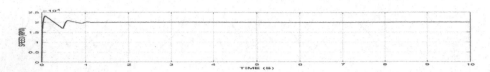

Fig. 8. Speed vs Time waveform of proposed system without current controller

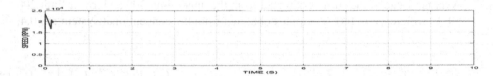

Fig. 9. Speed vs Time waveform of proposed system with current controller

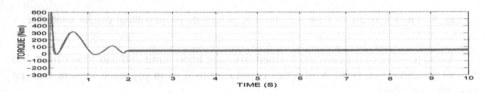

Fig. 10. Torque vs Time waveform of conventional system

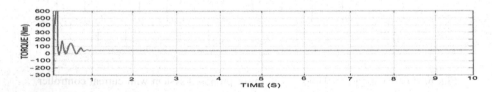

Fig. 11. Torque vs Time waveform of proposed system without current controller

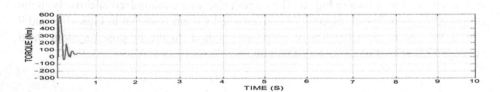

Fig. 12. Torque vs Time waveform of proposed system with current controller

The detailed analysis of the reduction in torque ripple can be done with the Fig. 13 below. Both of the results attain the rated torque 50 Nm. But the ripples are effectively minimizes in the proposed method. Torque ripple minimization is shown in Figs. 14 and 15.

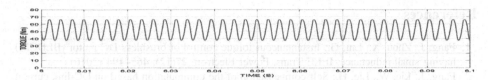

Fig. 13. Torque ripple in the conventional system

In conventional system torque ripples are 0.5384. In proposed system without current controller torque ripples are reduced to 0.48 as in Fig. 14.

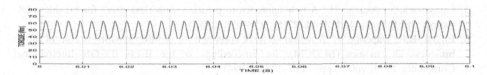

Fig. 14. Torque ripple in the proposed system without current controller

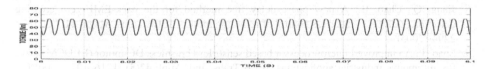

Fig. 15. Torque ripple in the proposed system with current controller

In proposed system with current controller torque ripples are reduced to 0.3921 as in Fig. 15. i.e., In proposed control method the torque ripple can be effectively reduce from 53.84% to 39.21%.

5 Conclusion

A new control method has been proposed to minimize the current and torque ripple of BLDC motor in this work. With the new switching logic and commutation algorithm simplicity in the control increases and the motor operates in wide speed range. From the simulation results it is found that with the proposed method we can obtains fast convergence speed, minimizing the current and torque ripples effectively. In addition it also improves the dynamic performance and is effective and can be easily implemented.

6 Future Scope

Six switch inverter can be replaced with 3 Phase 4 Switch (3P4S) inverter. 3P4S can reduce the cost of the system, since it has reduced number of components when compared to six switch inverter. The efficiency of the system can be improved since 3P4S inverter has less switching loss when compared to six switch inverter.

References

1. Fang, J., Zhou, X., Liu, G.: Instantaneous torque control of brushless DC motor (BLDCM) having small inductance. IEEE Trans. Power Electron. **27**(12), 465–494 (2016)
2. Fang, J., Lie, W., Lie, H.: Self compensation of the commutation based on DC-link current for high-speed brushless DC motors (BLDCM) with low inductance. IEEE Trans. Power Electron. **29**(1), 438–448 (2015)
3. Plunkett, A.B.: Back EMF sampling circuits for the phase locked loop (PLL) motor control. US Patent 4 928 043, May 2015
4. Ertugul, N., Acamley, P.: A new method for sensorless control of permanent magnet brushless DC motors. IEEE Trans. Ind. Appl. **30**(1), 127–133 (2004)
5. Ogesawara, S., Akaagi, H.: A new approach to position sensorless drive for brushless DC motors. IEEE Trans. Ind. Appl. **27**(5), 928–938 (2017)
6. Engtai, H.F., Dapeing, T.: A neural network approach for position sensorless control of the brushless DC motors (BLDCM). In: Proceedings of the IEEE IECON International Conference 1994, vol. 2, pp. 1277–1281 (1994)
7. Shao, J., Nolan, D., Tsisier, M.: A novel microcontroller based sensorless brushless DC motor (BLDCM) drive for automotive fuel pump applicaton. IEEE Trans. Ind. Appl. **39**(6), 1735–1739 (2003)
8. Jiang, Q., Chao, B.: A new phase delay free method for the back-EMF zero crossing detection (ZCD) for sensorless control of spindling motors. IEEE Trans. Magn. **41**(7), 2287–2294 (2005)
9. Shao, J.: An improved microprocessor based sensorless brushless DC motor (BLDCM) drive for automotive industrial applications. IEEE Trans. Ind. Appl. **42**(5), 1216–1221 (2006)
10. Mohamed Asiq, T.S., Govindaraju, C.V.: Design of two inductor boost converters for photovoltaic applications. Singap. J. Sci. Res. (SJSR) **8**(2), 12–19 (2016)
11. Liy, C.G., Moi, I.O.: Sensorless control BLDC motor using third harmonic back-EMF. Power Electron. **9**(3), 1132–1147 (2011)
12. Shen, J.X., Zhiu, Z.Q., Hove, D.: Indirect and direct methods of sensorless back EMF control for PMBLDC motors. IEEE Trans. Ind. Appl. 1629–1639 (2004)
13. Yan, Y.B., Zho, W.C.: Analysis of sensorless control of permanent magnet motor using third harmonic method. IEEE Trans. Ind. Appl. **3**(5), 726–738 (2008)
14. Zhao, Z., Howe, D.: Analysis of the material used for the permanent mangent in motors. IEEE Trans. Magn. **16**(4), 101–121 (2001)

Analysis of Switching Faults in DFIG Based Wind Turbine

Surya S. Kumar[✉] and N. Reema

Electrical and Electronics Engineering, Sree Buddha College of Engineering, Pattoor, India
suryasajikumar64@gmail.com, n.reema3@gmail.com

Abstract. The doubly fed induction generator (DFIG) is a standout amongst the most extensively used generators in wind turbines owing to its variable speed operation, power control, littler converter limit, and network tie possibility. The power electronic converters assume a key role in wind energy conversion system for controlling and conditioning of the output power from wind energy. These power electronic converters may undergo different fault conditions which may influence the performance of the system, so the fault in power electronics is considered as a noteworthy issue. The performance of doubly fed induction generator based wind energy conversion system in power electronic converter fault conditions has been analyzed in this paper. The impact of short circuit and open circuit fault in the switches of grid side and rotor side converters has been explore by means of MATLAB simulation. DFIG based wind energy conversion system in case of healthy condition and faulty condition were analyzed.

Keywords: DC link capacitor · Doubly fed induction generator · Open circuit Short circuit · Wind energy conversion system

1 Introduction

Nowadays the wind energy plays a significant role in the world owing to its environmental friendliness and availability. The wind turbines used in wind energy conversion system (WECS) are mainly constant speed & variable speed [1]. Among them variable speed wind turbine was more vitality productive since the speed of wind is varying in nature. Squirrel cage induction generator (SCIG), doubly fed induction generator (DFIG) & permanent magnet synchronous generator (PMSG) are the three types of generator used in wind energy conversion system, among this DFIG is most commonly used for WECS based on the advantages i.e.; improved efficiency, active & reactive power control, reduced losses, reduced converter cost and ability of power factor correction.

In DFIG based WECS, the stator windings are directly connected to the grid whereas the rotor windings are connected to the grid by means of power electronics converters such as grid side and rotor side converter. The DC link capacitor connected between grid side and rotor side converter act as DC voltage source. The converters used in DFIG based WECS need to process only a fraction of the generated output power so that the converters are intended to transmit about 30% of the full load power [2]. The active & reactive power from the stator to the grid of wind turbine was controlled by rotor side

© Springer Nature Singapore Pte Ltd. 2018
I. Zelinka et al. (Eds.): ICSCS 2018, CCIS 837, pp. 715–724, 2018.
https://doi.org/10.1007/978-981-13-1936-5_73

converter. The grid side converter is designed to control the dc link voltage & set aside the converter to generate or soak up reactive power [3, 4]. In DFIG the stator and rotor will supply power so that it is called as doubly fed.

In DFIG based WECS fault could originate internally from the machine or externally from the grid [5]. Internal fault may occur either in turbine, gearbox, generator and power electronic converters which include generator winding open circuit faults, winding insulation failure, semiconductor device failure, DC link capacitor failure etc. Over 50% of internal faults are attributed to the power electronic converters [6]. The main problems associated with the converters in DFIG based WECS include short circuit fault, open circuit fault, fire through, flashover within the voltage source converter switches etc. and the fault in DC link capacitor is also considered as a main problem [7]. The causes of these faults includes driver circuit break down, supplementary power supply failure, sudden change of voltage, intrinsic failure which may occur due to avalanche stress or overvoltage & temperature overshoot, opening or failure of the power switches in the converter [8–12]. This fault easily degrades the power quality and causes potential secondary faults in other components. Pulsating torque may also come out, which cause rigorous vibration on the mechanical system parts & possibly cause total malfunction of the system.

2   Wind Energy Conversion System

The block diagram of DFIG based WECS is shown in Fig. 1. In this DFIG based wind energy conversion system, the stator winding is connected directly to the grid but the rotor winding was connected to the grid through grid side and rotor side converters. The main area of this work includes analyzing switching fault in rotor side and grid side converter along with the DC link capacitor fault.

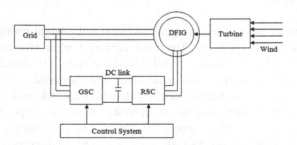

Fig. 1. Block diagram of DFIG based WECS

The MATLAB model of DFIG based WECS is shown in Fig. 2. The operation of DFIG based WECS is regulated by the help of a control system, which include the control of grid side & rotor side converter.

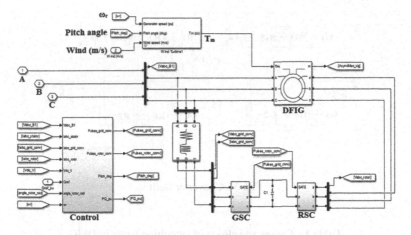

Fig. 2. SIMULINK model of WECS

3   Fault in DFIG

In DFIG based WECS fault might originate internally from the machine or externally from the grid.

3.1   Grid Fault

Many fault types can be appear in the DFIG when it connected to grid such as: single line to ground at input supply, line to line short circuit at the machine terminal, voltage dip, single line to ground fault at the machine. The grid fault can bring about substantial over-currents & over-voltages put the entire facility under stress.

3.2   Converter Switching Fault

Internal fault may occur either in turbine, gearbox, generator and power electronic converters which include generator winding open circuit faults, winding insulation failure, semiconductor device failure, DC link capacitor failure etc. Over 50% of internal faults are attributed to the power electronic converters.

The main problems related with the converters in DFIG based WECS include short circuit fault and open circuit fault within the switches & also fault in DC link capacitor. The various switching faults discussed in this work are grid side converter (GSC) open circuit fault (F1), GSC short circuit fault (F2), rotor side converter (RSC) open circuit fault (F3), RSC short circuit fault (F4), DC link capacitor open circuit fault (F5), DC link capacitor short circuit fault (F6) & DC link capacitor ground fault (F7) as displayed in Fig. 3 (Table 1).

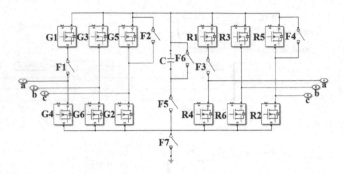

Fig. 3. Converter fault

Table 1. Causes and effects of switching faults in DFIG

SI No	Fault	Causes	Effects
1	GSC open circuit	• Thermal stress • Failure of gate driver	• Degrade the power quality • Causes secondary fault in other components
2	GSC short circuit	• Wrong voltage in IGBT gate • Inherent failure	• Oscillation in power torque results in rigorous vibration on the mechanical system part
3	RSC open circuit	• Thermal stress • Failure of gate driver	• Degrade the power quality • Causes secondary fault in other components
4	RSC short circuit	• Wrong voltage in IGBT gate • Inherent failure	• Oscillation in power torque results in rigorous vibration on the mechanical system part
5	Capacitor open circuit	• Mechanical stress	• Total shut down of system occur
6	Capacitor short circuit	• Due to effect of temperature	• Power decreases
7	Capacitor ground fault	• Damaged wiring and insulation • Thermal stress	• Power and rotor speed decreases

4   Simulation Analysis

DFIG based WECS is developed & simulated using MATLAB SIMULINK. The simulation results under normal condition and under different fault are analyzed. The stator current and rotor current under normal working condition is displayed in Figs. 4 and 5 correspondingly.

Fig. 4. Stator current under normal condition

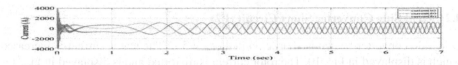

Fig. 5. Rotor current under normal condition

At starting both stator and rotor current is increased to 4000 A and then it take some time settle at constant value of 1255 A at t = 3 s. The dc link capacitor voltage was maintained at constant value 1150 V as displayed in Fig. 6.

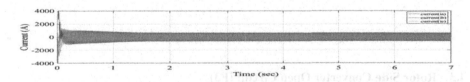

Fig. 6. DC link capacitor voltage under normal condition

4.1 Grid Side Converter Open Circuit (F1)

Grid side converter switch is open circuited at t = 5 s. The stator current, rotor current, rotor speed and DC link capacitor voltage is almost same as normal condition which is displayed in Figs. 7, 8 and 9 correspondingly.

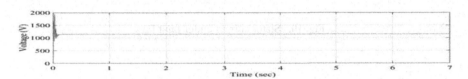

Fig. 7. Stator current under fault F1

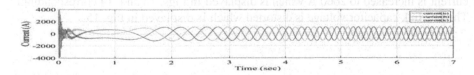

Fig. 8. Rotor current under fault F1

Fig. 9. DC link capacitor voltage under fault F1

4.2   Grid Side Converter Short Circuit (F2)

Grid side converter switch is short circuited at t = 5 s. The stator current is increased which is displayed in Fig. 10. The rotor current is distorted and is displayed in Fig. 11. The DC link capacitor voltage is decreased to 100 V as displayed in Fig. 12.

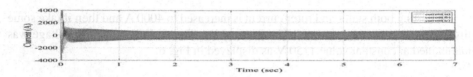

Fig. 10.   Stator current under fault F2

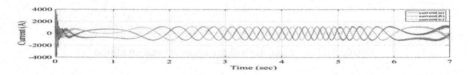

Fig. 11.   Rotor current under fault F2

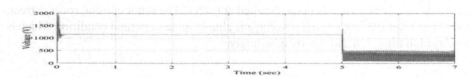

Fig. 12.   DC link capacitor voltage under fault F2

4.3   Rotor Side Converter Open Circuit (F3)

Rotor side converter switch is open circuited at t = 5 s. The stator current and rotor current is increased to 2600 A which is displayed in Figs. 13 and 14 correspondingly. The DC link capacitor voltage is distorted which is displayed in Fig. 15.

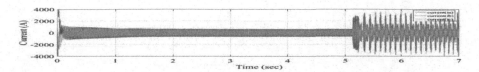

Fig. 13.   Stator current under fault F3

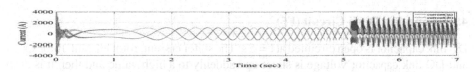

Fig. 14.   Rotor current under fault F3

Fig. 15.   DC link capacitor voltage under fault F3

4.4   Rotor Side Converter Short Circuit (F4)

Rotor side converter switch is short circuited at t = 5 s. The stator current is increased which is displayed in Fig. 16. The rotor current is distorted and is displayed in Fig. 17. The DC link capacitor voltage is decreased to 400 V as displayed in Fig. 18.

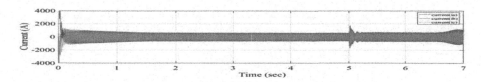

Fig. 16.   Stator current under fault F4

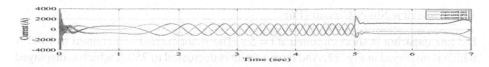

Fig. 17.   Rotor current under fault F4

Fig. 18. DC link capacitor voltage under fault F4

4.5   Capacitor Open Circuit (F5)

DC link capacitor is open circuited at $t = 5$ s. The stator current, rotor current, rotor speed and DC link capacitor voltage is shoot-up suddenly to a high value and then it is drop to zero which is displayed in Figs. 19, 20 and 21 correspondingly.

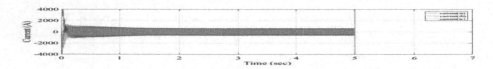

Fig. 19.   Stator current under fault F5

Fig. 20.   Rotor current under fault F5

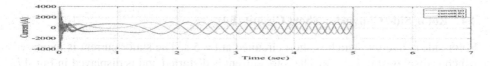

Fig. 21.   DC link capacitor voltage under fault F5

4.6   Capacitor Short Circuit (F6)

DC link capacitor is short circuited at $t = 5$ s. The stator current is increased to 1400 A which is displayed in Fig. 22. And rotor current is decreased to 750 A which is displayed in Fig. 23. The DC link capacitor voltage is decreased to 50 V as displayed in Fig. 24.

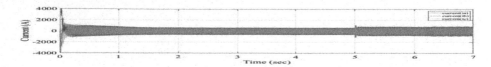

Fig. 22.   Stator current under fault F6

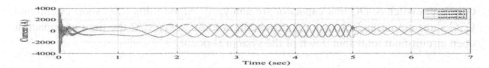

Fig. 23. Rotor current under fault F6

Fig. 24. DC link capacitor voltage under fault F6

4.7 Capacitor Ground Fault (F7)

DC link capacitor is grounded at t = 5 s. The stator current is increased which is displayed in Fig. 25. The rotor current is distorted and is displayed in Fig. 26. The DC link capacitor voltage is decreased to 200 V as displayed in Fig. 27.

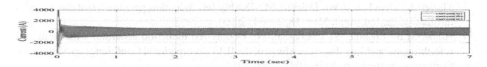

Fig. 25. Stator current under fault F7

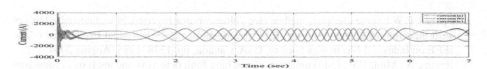

Fig. 26. Rotor current under fault F7

Fig. 27. DC link capacitor voltage under fault F7

5 Conclusion

DFIG based WECS is modeled & simulated by means of MATLAB SIMULINK. Switching fault in converter was introduced by breaker switch & various fault conditions were simulated. Short circuit & open circuit fault of grid side converter, rotor side converter switch & DC link capacitor are analyzed. Among this RSC short circuit and GSC short circuit and capacitor short circuit is more severe. At the time of capacitor open circuit fault total shut down of the system is occur.

Appendix A: Machine Parameters

Rated power = 1.5 MW
Rated voltage = 690 V
Rated speed = 1750 rpm
No of poles = 4
Rated current = 1255 A

References

1. Gantie, V., Sinh, B., Aggarwal, S.: DFIG based wind power conversion with grid power leveling for reduced gusts. IEEE Trans. Sustain. Energy **4**(2), 13–21 (2011)
2. Priya, B., Anitta, R.: Modelling simulation and analysis of doubly fed induction generator for wind turbines. J. Electr. Eng. **61**(3), 80–86 (2008)
3. Muller, S., Doncker, R.W., Deickie, M.: Doubly fed induction generator systems for wind turbines. IEEE Ind. Appl. Mag. **18**, 27–34 (2003)
4. Hilale, M.B., Erramie, Y., Cherkaoui, M.: Doubly fed induction generator wind farm fault ride-through capability. In: Proceedings of the IEEE International Conference on Multimedia Computing System, pp. 1078–1081, April 2012
5. Wei, Z., Li, J., Zheng, T.: Short circuit current analysis of DFIG with crowbar under unsymmetrical grid fault. In: Proceedings of the IEEE International Conference RPG, March 2013
6. Fuches, E.W.: Some diagnosis methods for voltage source inverters in variable speed drives with induction machines-A survey. In: Proceedings of the 29th Annual Conference of the IEEE on Industrial Electronics Society, USA, Canada, pp. 1378–1385, August 2004
7. Reema, N., Mini, V.P., Ushakumary, S.: Switching fault detection and analysis of induction motor drive system using fuzzy logic. In: Conf. Proc. ICAGE 2014 (2014)
8. Merabet, H., Bahi, T., Halem, N.: Condition Monitoring and Fault Detection in Wind Turbine Based on DFIG by the Fuzzy Logic. in science direct 2015 (2015)
9. Bouzekri, A., Allaoui, T., Denai, M.: Intelligent open switch fault detection for power converter in wind energy system. Appl. Artif. Intell., February 2017
10. Giaourakis, D.G., Safacas, A., Tsotoulidis, S.: Dynamic behaviour of a wind energy conversion system including doubly-fed induction generator in fault conditions. Int. J. Renew. Energy Res. **2**(2) (2012)
11. Zhao, H., Cheng, L.: Open-circuit faults diagnosis in back-to-back converters of DF wind turbine. J. IET Renew. Power Gener., September 2016
12. Abdoua, A.F., Abu-Siadab, A., Potaa, H.R.: Effect of intermittent voltage source converter faults on the overall performance of wind energy conversion system. Int. J. Sustain. Energ. **33**(3), 606–618 (2014)

A Novel Self Correction Torque and Commutation Ripples Reduction in BLDC Motor

Megha S. Pillai[✉] and K. Vijina

Electrical and Electronics, Sree Buddha College of Engineering, Pattoor, India
meghaspillai02@gmail.com, vijinasbce@gmail.com

Abstract. Brushless DC motor has permanent magnets in the rotor side that rotates around fixed armature. The current rotates in armature continuously as it is electronically commutated. The motor's efficiency will otherwise reduce due to commutation angle errors. The novelty of the work is to reduce commutation errors that are obtained based on commutation point phase shift and the dc-link current difference. Based on these relationship analysis is made under ideal, advanced and delayed commutations. The self-compensation method of commutation instant deviation delays the commutation angle by $10°$ thus reducing the impact caused by commutation ripple thereby improving dynamic response and control. The dc-link current difference decrease gradually and phase deviation converges to correct the commutation angle. The proposed correction method can achieve ideal commutation effect thereby attaining fast convergence speed, current and torque.

Keywords: Brushless DC motor · Buck converter
Commutation signal deviation corrections · Dc-link current · Hoist

1 Introduction

Brushless DC motors are used tremendously from household to automobiles and industries due to its high power density, efficiency and higher torque to inertia ratio. These motors are often selected for its high torque performance over long life working and also the rugged construction makes them suitable in extreme environments. BLDC motor commutation process does not use any brushes and is done electronically. The motors permanent magnet and electronic commutation cause BLDC to have merits over brushed DC motors and induction motors. The permanent magnets display higher efficiency that is thus used for industrial heavy load applications such as weight lifting. The BLDCM commutation ripples can be compensated using wide range of methods. The ripples cause substantial the motor to damage permanently as it generates temperatures at defect locations causing mechanical deformations. The kind of failures do not cause immediate breakdown, but deteriorates the operation of machine decreasing the performance of the machine. Voltage drop occurs while switching therefore a buck converter is used for maintaining the voltage and reducing the variations caused by it thus increasing current for the use in the windings of inverter. The self-compensation

© Springer Nature Singapore Pte Ltd. 2018
I. Zelinka et al. (Eds.): ICSCS 2018, CCIS 837, pp. 725–733, 2018.
https://doi.org/10.1007/978-981-13-1936-5_74

method resolves the commutation errors based on back EMF waveforms, which refers to switching current in phases to generate motion thereby thus adjusting the commutation angle for proper commutation. The position of the rotor can be estimated by sensing the Back-EMF from one of the motor terminal voltages. Sensing each terminals extract two commutation instants for determining the position. The method is obtained based on the relationship between commutation point phase shift and the difference of dc-link current.

This project proposes a BLDC motor that can be used in industrial applications such as in tractions or hoist's. Most of the commercial system uses induction motor where power drops at low loads and also starting torque is poor. When compared to induction motors BLDC motor has high efficiency and high energy saving with low start up current capacity, as it is powered by DC electric source and the switching power supply produces AC supply to drive the motor, hence BLDC motor has higher torque ratio for its application in industries and also positioning and controlling is possible which is more efficient than induction motors.

2 Proposed System

Rapid self-compensation method solves the commutation angle error as otherwise it would reduce the overall performance of the system. The commutation error is eliminated using a dc-link current sensor, the difference from that of the inverter and motor is estimated and corrected. The effects of commutation position deviation occurs due to certain machine aspects such as electric component delay, measurement noise etc. These commutation errors can be reduced by self-compensation method. The ideal com-mutation point is found to be the intersection of every two-phase back EMFs. The self-compensation method analyses the sampled current difference and commutation phase point shift (Fig. 1).

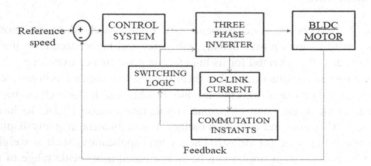

Fig. 1. Block diagram representing self-compensation technique

The self-compensating method is composed of a buck converter for increasing the current value that is needed to be extracted from the inverter for the dc link current sensor, velocity controller to regulate the motors speed and a current controller to restrain the motor current to below the maximum value. The combination of these processes produce high frequency ripple compensation by testing the current difference under certain instants to solve the problems related to commutation angle errors as otherwise it will reduce the overall performance and efficiency of the motor. By identifying the commutation position of a BLDC motor the change in dc-link current induced by commutation phase shift is analyzed for any current ripples. The difference of the dc-link current and phase currents are compared and analyzed for any ripples and the compensated current is driven to the switching circuit and back to the inverter circuit to that of the motor. Thereby the compensation of commutation time errors with the dc-link current obtains good steady state response and smooth operation of motor therefore can be applied for heavy load applications.

The advantage of this analysis compared to other existing methods is its simplicity to implement, and it is influenced by the current sampling from the source and phase currents of the motor not disturbed by any other sources of magnetic interferences present in industrial environment.

3 Estimating Commutation Position in Motor

The motor operates in three phase six states and transistor commutates at every 60°. The three phase armatures are symmetrical, i.e.; Ra = Rb = Rc = R and La = Lb = Lc = L. Then motor terminal voltage can be written as:-

$$U_x = Ri_x + (L - M)\frac{di_x}{dt} + e_x + U_N \tag{1}$$

BLDC motor phase back EMF waveform is in a trapezoidal form. The ideal commutation point is the intersection of every two-phase back EMFs. However, under effects of low pass filters, commutation delay, electric component delay, measurement noise, etc., the extracted commutation position deviation occurs. This effect's on commutation deviation can be minimized by using self compensation method. The current transfers from upper bridge of phase A to that of phase B leading to commutation position deviations. The law of dc-link current commutation ripple can be achieved by analyzing the current in non-commutation phases (Fig. 2).

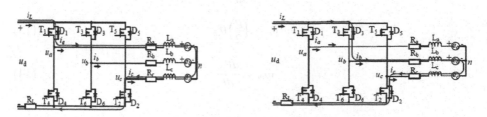

Fig. 2. Current flow during commutation from T1 to T3.

The electrical and mechanical mathematical equations for modelling of BLDCM are:

$$V_a = Ri_a + (L - M)\frac{di_a}{dt} + E_a \tag{2}$$

$$V_b = Ri_b + (L - M)\frac{di_b}{dt} + E_b \tag{3}$$

$$V_c = Ri_c + (L - M)\frac{di_c}{dt} + E_c \tag{4}$$

BACK EMF EQUATIONS:

$$E_a = K_e\omega_m f(\theta_e) \tag{5}$$

$$E_b = K_e\omega_m f\left(\theta_e - \frac{2\pi}{3}\right) \tag{6}$$

$$E_c = K_e\omega_m f\left(\theta_e + \frac{2\pi}{3}\right) \tag{7}$$

TORQUE EQUATIONS:

$$T_a = K_t i_a f(\theta_e) \tag{8}$$

$$T_b = K_t i_b f\left(\theta_e - \frac{2\pi}{3}\right) \tag{9}$$

$$T_c = K_t i_c f\left(\theta_e + \frac{2\pi}{3}\right) \tag{10}$$

$$T_e = T_a + T_b + T_c \tag{11}$$

MECHANICAL EQUATIONS:

$$T_e - T_l = J\frac{d^2\theta_m}{dt^2} + \beta\frac{d\theta_m}{dt} \tag{12}$$

$$\theta_e = \left(\frac{p}{2}\right)\theta_m \tag{13}$$

$$\omega_m = \frac{d\theta_m}{dt} \tag{14}$$

CURRENT EQUATIONS:

$$i_a = \int \left(\frac{1}{3L}\right)[2V_{ab} + V_{bc} - 2E_a + E_b + E_c - 3Ri_a] \tag{15}$$

$$i_b = \int \left(\frac{1}{3L}\right)[-V_{ab} + V_{bc} + E_a - 2E_b + E_c - 3Ri_b] \tag{16}$$

Where K:	a, b, c
V_a, V_b, V_c:	Three phase voltage applied from inverter to BLDC
I_a, I_b, I_c:	Three phase current
R:	Resistance
L:	Inductance
M:	Mutual inductance
E_a, E_b, E_c:	Three phase Back- EMF
T_a, T_b, T_c:	Electric torque produced in each phase
K_e:	Back EMF- constant
K_t:	Torque constant
ω_m:	Angular mechanical speed of rotor
Θ_m:	Mechanical angle of rotor position
Θ_e:	Electrical angle of rotor position
$F(\Theta_e)$:	Back- EMF reference as function of rotor position

4 Simulation and Results

4.1 Model of BLDCM

Modeling of BLDC Motor using commutation logic is been simulated and the output waveforms are attained using MATLAB. A PI controller has also been used for generating the rated speed. The Simulink representation of the overall system is shown below (Fig. 3).

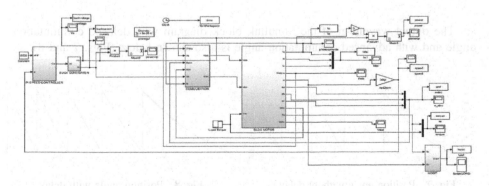

Fig. 3. Modelling of BLDC motor

For simulating the proposed system, a 1.5 kW motor is modeled and the input signals to that of the controller are attained through a buck converter.

4.2   Model of Buck Converter

The buck converter is implemented in front of the 3Φ inverter to weaken the influence caused by low inductance. Therefore the input signals given are 250 V and a 5 A current to the converter to attain the desired output that is been fed to the commutation circuit for generating voltage (Figs. 4, 5 and 6).

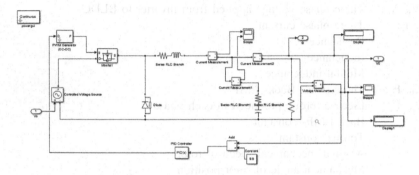

Fig. 4. Modelling of buck converter

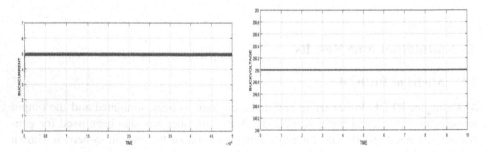

Fig. 5. Output current   **Fig. 6.** Output voltage

The output waveform of the Simulink block diagram with delayed commutation angle and with advanced commutation angle is also shown below (Figs. 7 and 8).

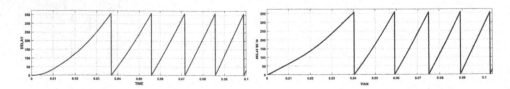

Fig. 7. Position angle without delay   **Fig. 8.** Position angle with delay

The dc-link current waveforms changes with the commutation point shift. If the commutation point is advanced, then the dc-link current sampled before commutation instant is smaller than that of the after value. Commutation phase delay leads to reducing the speed accordingly. Thereby also decreasing the commutation point errors. Simulation results of the current waveform of one of the phases and DC link current of the motor with non-ideal back EMFs is shown in figures below. The back EMF's are obtained depending on the rotor position of the motor forming ideal trapezoidal waveforms (Figs. 9 and 10).

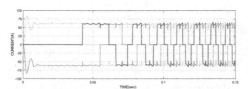

Fig. 9. Current Vs Time

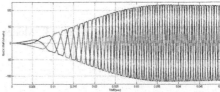

Fig. 10. Back EMF Vs Time

The rotor aligns with the rotating magnetic field thereby achieving proper commutation. When the input voltage is higher then the back EMF is high enough for the detection circuit, by the use of buck converter the sensorless commutation signals is at a desired level and the signals will be sent to the commutation table and the motor is changed to the self-commutation mode thereby correcting the position depending on the commutation. The speed response of the machine is shown in Fig. 12, the rated speed of the machine is 5000 rpm. The torque waveform is shown in Fig. 11 and it is analyzed that the variation of torque occurs between 0.95 Nm to 2.75 Nm.

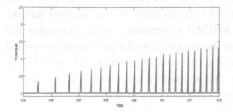

Fig. 11. Torque Vs Time

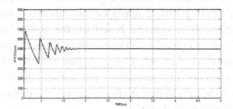

Fig. 12. Speed Vs Time

4.3   Model of Hoist System

The validation of the simulation model proves that the DC motors are easier to control in case of heavy loads than AC. The forward and reverse motoring action of the motor is done for lifting and lowering of the loads. Therefore in the forward motoring mode the current required depends on the load and during the lowering of the load i.e.; reverse motoring the motor functions as a generator, the Simulink model is shown below (Figs. 13, 14 and 15).

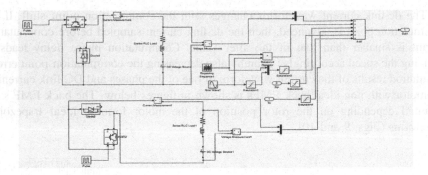

Fig. 13. Model of a hoist system

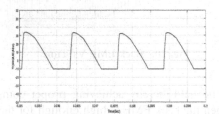

Fig. 14. Torque Vs Time of Hoist   **Fig. 15.** Speed Vs Time of Hoist

5 Conclusion

The unbalanced three-phase back EMFs and the commutation ripple would make the commutation error compensation disabled, a new compensation method of commutation instant deviation is proposed in this paper. Control techniques for minimizing the pulsating torque apply certain advanced method that depends on the machine parameters. It is derived based on the actual back EMF waveform, and it eliminates the impact caused by commutation ripples. From the results it is found that the proposed method start reliably with this method obtains fast convergence speed, avoids the freewheeling current of the non-conduction phase thereby eliminating commutation error. In addition it also improves the dynamic performance and is effective and can be easily implemented.

References

1. Niasar Alij, H., Vaheda, M., Moghbalina, H.: Position control for a four-switch, BLDC motor drive without the use of a phase shifter. IEEE Trans. Power Electron. **23**(8), 3078–3086 (2008)
2. Park, J.S., Jungshin, M., Ki, H.W., Youlani, M.J.: Position tracking observer power control for sensorless drive permanent magnet synchronous machine. IEEE Trans. Power Electron. **26**(7), 2585–2597 (2014)

3. Po-ngamnaig, S., Sanwinch, S.: Performance improvement of full-order observers for PMSM drive. IEEE Trans. Power Electron. **37**(7), 588–604 (2012)
4. Tiwari, V., Rani, B.L.: Torque ripple reducing technique in BLDC motor with unideal back EMF method. In: Proceedings of 2nd International Conference on Emerging Trends Engineering technology (ICETET), pp. 687–690 (2010)
5. Kangshi, S., Sungshi, K.: Torque control of BLDC with nonideal trapezoidal back EMF. IEEE Trans. Power Electron. **10**(2), 796–802 (1994)
6. Bhogineni, S., Rajagopal, K.: Position control in a brushless DC motor based on average line to line voltages. In: Proceedings of the IEEE International Conference On Power Electronics, Drives and Energy Systems, pp. 1–6 (2012)
7. Ogasahara, S., Akagini, H.: An approach to position sensorless drive for brushless DC motors. IEEE Trans. Ind. Appl. **27**(5), 928–933 (1992)
8. Urasaki, N., Senjyungthang, T., Uezato, K., Sungabasi, T.: Adaptive dead-time compensation strategy for a PMSM drive. IEEE Trans. Energy Convers. **22**(4), 271–280 (2008)
9. Hu, T., Jung, K., Ehsanigom, M.: An error analysis for ataining position estimation for BLDC motor drive. In: Proc. Conf. Rec. Ind. Appl. Conf., vol. 1, pp. 611–617 (2003)
10. Pellechangnou, G., Gugulung, P., Arnandoshi, E., et al.: A Self-Commissioning Algorithm for Compensating Inverter Nonlinearity in Sensorless Induction Motors. IEEE Trans. Ind. Appl. **47**(8), 1415–1424 (2012)

Reduction of Torque Ripples in PMSM Using a Proportional Resonant Controller Based Field Oriented Control

P. S. Bijimol[(✉)] and F. Sheleel

Electrical and Electronics, Sree Buddha College of Engineering, Pattoor, India
bijimolps2012@gmail.com, fsheleel@gmail.com

Abstract. Permanent magnet synchronous motors (PMSM) are extensively used in many applications including robotics, precision machining etc. because of their good features such as, high efficiency, light weight, better accuracy, and low maintenance requirements compared to induction motors. Because of the increasing demand for energy efficiency, PMSM replaces the traditional induction motors. The main problem with this motor is the formation of torque ripples at low-speed which may cause mechanical vibrations and induces oscillations in speed. So low-speed application of this motor have some limitations. Vector controlled PMSM drives can be used to supply lesser torque ripples and better dynamic response. Conventionally Proportional integral (PI) controllers are used for this. But the performance of the PI controllers are affected by disturbances in load, variations in speed and parameter variations due to its constant proportional gain and integral time constant. The novelty of this work is implementing a new control technique by implementing a proportional resonant controller by paralleling a variable frequency resonance controller with the traditional PI controller and the performance of the two controllers is compared with the help of simulation results.

Keywords: Permanent magnet synchronous motors
Proportional integral (PI) controllers · PI-resonance (PI-RES) controllers
Torque ripples · Field oriented control

1 Introduction

The permanent magnet synchronous motor (PMSM) drives are widely used for many industrial applications such as industrial servo applications, robotics etc. They have high efficiency, smaller parts, less weight, high torque density and small size [1]. Nowadays the induction motors used in compressors are gradually being replaced by PMSM due to the increasing demand for energy efficiency and variable-speed systems performance. The major drawback with this motor for some applications is the presence of torque ripples [2]. These torque ripple induces vibrations which may destroy the whole drive system and can generate serious noise problems. The extensive application of variable speed compressors have some limitations due to the speed fluctuations at low-speed range and it causes noise at low-frequencies and serious vibration related

© Springer Nature Singapore Pte Ltd. 2018
I. Zelinka et al. (Eds.): ICSCS 2018, CCIS 837, pp. 734–742, 2018.
https://doi.org/10.1007/978-981-13-1936-5_75

problems. To compensate these periodic torque pulsations additional control effort should be used [3].

In this work, the conventional PI speed controller and a variable frequency resonant controller are applied in parallel to form a proportional resonant speed controller. It eliminates the ripples by providing a reference torque current. The resonance controller generates a compensation torque current and the PI controller produces a main reference current [4]. The proposed controller combines both of this current to reduce the speed ripples. The performance comparison between the controllers are done and evaluated through simulation results.

2 Modelling of PMSM

The d-q model of PMSM on rotor reference frame without having damper winding has been developed. The stator and rotor mmf rotates at the same speed. The modelling follows these assumptions:

1. Rotor flux is concentrated along d axis.
2. The induced EMF is sinusoidal.
3. Hysteresis losses and eddy currents are negligible.
4. There are no field current dynamics.
5. The stator windings are balanced with sinusoidal distributed magneto-motive force (mmf).
6. The saturation and parameter changes are neglected.
7. Variations in rotor temperature with time is neglected.

The stator voltages in d-and q-axes are obtained as the sum of the resistive voltage drops and the derivative of the flux linkages in the corresponding windings [5]. The flux-linkage equation for stator are given by:

$$V_q = R_q i_q + P\lambda_q + \omega_r \lambda_d \tag{1}$$

$$V_d = R_d i_d + P\lambda_d - \omega_r \lambda_q \tag{2}$$

where, V_d and V_q are the voltages in the d-axis and q-axis windings, i_d and i_q are the stator currents in d-axis and q-axis, R_d and R_q are the stator resistance in d-axis and q-axis, λ_d and λ_q are the stator flux linkage in d-axis and q-axis, ω_r is the rotor speed of the machine. Flux Linkages in d and q axis is given by,

$$\lambda_q = L_q i_q \tag{3}$$

$$\lambda_d = L_d i_d + \lambda_f \tag{4}$$

According to the method of field-oriented control of PMSM, the d-axis current is usually taken to be zero. The developed motor torque is given by,

$$T_e = \frac{3P}{4}(\lambda_m i_q) = k_t i_q \tag{5}$$

where, P is the number of poles of the motor and k_t is the torque constant. Then the mechanical torque equation is,

$$T_e = T_L + B\omega_m + J\frac{d\omega_m}{dt} \tag{6}$$

The mechanical speed and position of the motor are represented as,

$$\omega_m = \int \frac{1}{J}(T_e - T_L - B\omega_m)dt \tag{7}$$

$$\omega_e = \frac{P}{2}\omega_m \tag{8}$$

$$\frac{d\theta_m}{d_t} = \omega_m \tag{9}$$

where, ω_m is the mechanical speed, θ_m is the mechanical position, J denotes the inertia, T_L is the external load and B represents the viscous coefficient.

3 PMSM with Compressor Load

PMSM motors are widely used to improve the efficiency of compressors used for air conditioning purpose. Compressors in refrigeration application also require better efficiency and torque performance at low speeds [6]. These requirements are achieved by PMSM motors due to their increased life time compared to DC motors, and high torque at low speeds. But PMSM motors produce speed ripples at low speeds. It may upset the performance of the refrigeration systems. Figure 1 shows the MATLAB model of PMSM with a compressor load. Since viscosity coefficient Bm is very small, it can be neglected. Differentiator s can be used instead of (d/dt), from (3), the plant transfer function between the motor speed and the torque is,

$$\omega_m(s) = \frac{\Delta T_m}{J_m s} \tag{10}$$

where,

$$\Delta T_m = T_e - T_L \tag{11}$$

At low operating speeds, the speed will oscillate at the same harmonic frequencies as those of the torque ripple, ΔT_m. It is essential to reduce the speed ripples, which are the major cause of these oscillations in speed [7]. For that, the error in torque ΔT_m must be reduced. In the case of a compressor, the load torque is position-dependent. So with the various positions of rotor, the torque differs. Also for the different rotor speeds, the torque ripple frequency varies.

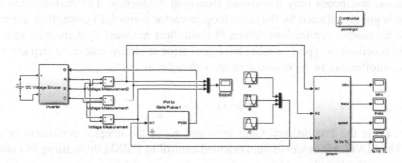

Fig. 1. MATLAB model of a PMSM with compressor load

4 Field Oriented Control

A PMSM can be operated with rapidly changing load in a broad range of speeds in adjustable speed drives applications by using Field oriented control (FOC). The torque and flux of the motor can be controlled in an efficient way using FOC. Irrespective of the machine parameters and variations in load parameters, FOC enables the motor to precisely follow the command trajectory. There are two input references or two constants for a field orientated controlled machine. First one is the component of torque aligned along the q-coordinate and the other is the component of flux aligned along the d co-ordinate [8]. This allows perfect control in both steady state and transient working operation and it is independent of the mathematical model with limited bandwidth. Torque control can be obtained by adjusting the orientation of the stator current vector with respect to the rotor field. When the angle between rotor flux and stator current is 90 degree, torque production will be maximum [9]. Also FOC can maintain a constant reference which enables the application of direct torque control, since in the (d, q) reference frame the equation for the torque is:

$$T \propto \varphi_R i_q \tag{12}$$

where φ_R is the amplitude of rotor flux and i_q is the q-axis stator current. A linear relationship between torque and current (i_q) is obtained by maintaining the amplitude of the rotor flux (φ_R) at a fixed value. We can then control the torque by controlling the torque component of stator current vector. Thus by using FOC, torque and flux can be independently controlled.

5 FOC of PMSM with PI Controller

Vector controlled PMSM drives offers better dynamic response and slighter torque pulsations, and needs only a constant switching frequency. The performance of the system is greatly affected by the outer loop in vector control. PI controllers are usually chosen for control applications. When PI controllers are used, systems with open loop transfer functions of type 1 or more have zero error at steady state for a step input [10]. A PI controller can be represented in the s-domain as,

$$G_{PI}(s) = K_P + \frac{K_I}{s} \tag{13}$$

where, K_P is the Proportional Gain term and K_I is the integral coefficient of speed loop. The MATLAB model of field oriented control in PMSM drive using PI controller is shown in the Fig. 2

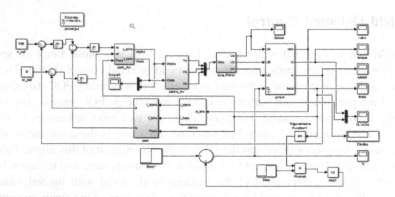

Fig. 2. MATLAB model of FOC in PMSM using PI controller

6 Proposed Scheme

6.1 Resonant Controller

The gain of a proportional resonant current controller $G_{PR(s)}$ is represented by [9]:

$$G_{PR}(s) = K_P + K_I \frac{2\omega_C s}{s^2 + 2\omega_C s + \omega_0^2} \tag{14}$$

where, K_P is the Proportional Gain and K_I is the Integral gain. The dynamics of the system; bandwidth, phase and gain margins are determined by the K_P term and ω_0 is the resonant frequency, ω_C is the bandwidth around the ac frequency ω_0. Now the PR controller have finite gain at the ac frequency ω_0. So that the controller can provide a very lesser steady state error [11]. The PR controller can be easily realizable in digital system using the above equation due to their finite precision. At the resonant frequency

ω, $G_{PR(s)}$ delivers infinite gain in open loop. When implemented in closed loop, it enables perfect tracking of components oscillating at ω. When $G_{PR(s)}$ controllers and $G_{PI(s)}$ are engaged in parallel for $G_{PI-RES(s)}$, only a single gain K_P should be tuned [12]. In $G_{PI-RES(s)}$, K_{ri} represents the resonance coefficient, and ω_C denotes the damping coefficient.

$$G_{PI-RES}(s) = K_P + \frac{K_I}{s} + \frac{2K_{ri}\omega_C s}{s^2 + 2\omega_c s + \omega_0^2} \qquad (15)$$

6.2   FOC of PMSM with PI-RES Controller

Due to the dynamics of the integral component, PI controllers cannot provide a sinusoidal reference without steady state error. A proportional resonant (PR) controller is more suitable to operate with sinusoidal references [13]. Also, it is free from the above mentioned demerits. The conventional outer speed control loop with a PI controller can be shown as in Fig. 3. T_{di} is the inner control loop delay. ω_{ref} is the reference speed, it is usually constant. Here the speed loop with PI controller have only limited bandwidth, and the integrators can achieve better error free control only at zero frequency and it is difficult at other frequencies. So it is difficult to achieve $\Delta T_m \simeq 0$. PI controllers are tuned in such a way as to have high Integral gain [14]. Then the proportional constant is increased to get adequate response.

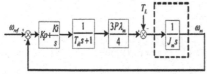

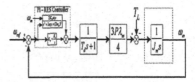

Fig. 3. Block diagram of the outer speed loop with a PI controller

Fig. 4. Block diagram for outer control loop with a PI-RES controller

The block diagram for the outer speed loop by using a proportional resonant controller is shown in Fig. 4. The Proportional resonant controller can provide gain for a specific resonant frequency. At other frequencies also, it cannot provide gain [15]. Since the speed ripples with twice the rotor frequency, a resonant controller resonating at twice the rotor frequency along with the conventional PI controller forms a proportional resonant controller, which can give better results. It can control the harmonics better than that of traditional PI controller [16]. The resonating term is tuned nearer to twice the frequency of the rotor to mitigate the speed ripples. This controller produces a rippled torque current reference which counteracts the torque ripple from the compressor load. Figure 5 shows the MATLAB model of the FOC in permanent magnet-synchronous motor drive using PI-RES controller. The torque current reference is taken from the output of the speed controllers.

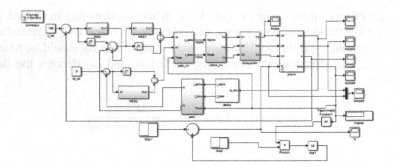

Fig. 5. MATLAB model of FOC in PMSM using PI-RES controller

7 Simulation Results

The proposed method was simulated in MATLAB 2010. A position dependent load torque is applied for representing the compressor. There present lots of ripples in the system without any controller. Figure 6 shows the torque and speed ripples obtained from PMSM without using any controller.

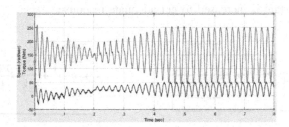

Fig. 6. Output torque and speed response of PMSM

To reduce these torque pulsations, a PI controller based field oriented control is used and the result is as shown in Fig. 7.

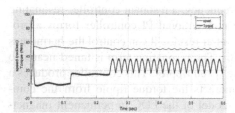

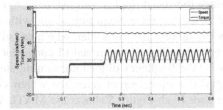

Fig. 7. Output torque and speed ripples in FOC of PMSM using PI controller at 52.36 rad/sec

Fig. 8. Output torque and speed response of FOC of PMSM using PI-RES controller at 52.36 rad/sec

From the results it is clear that the system with PI controller contain 57.89% torque ripples. To reduce this pulsations, a PI-RES controller is implemented and the ripple content is shown as in Fig. 8. It reduce the ripples to 50%. Thus the ripples are reduced by using the new controller. The performance of the system with both PI and PI-RES controllers are analysed at 52.36 rad/sec.

8   Conclusion

A proportional resonant controller based Field oriented control method for mitigating the torque ripple in PMSM drive with a compressor load is simulated in MATLAB and the results are plotted. Torque ripples at low speed is the main disadvantage associated with PMSM which leads to problems such as mechanical vibration, fluctuations in speed and noise. So a parallel combination of a variable frequency resonant controller is combined along with the proportional integral controller as a new PI-RES controller. It enables to reduce the ripples in speed when position dependent load is applied. So that it provides longer lifetime for the system and saves energy to an extent. It is clear from the results that the new technique was more better than the conventional methods using PI controller.

References

1. Shi, J.L., Liu, T.H.: Chang. Y.: Position control of an interior permanent-magnet synchronous motor without using a shaft position sensor. IEEE Trans. Ind. Electron. **54**, 1989–2000 (2007)
2. Holtz, J., Sprngobe, B.: Torque ripple identification and compensation for high-precision permanent magnet motor drives. IEEE Trans. Ind. Electron. **52**(5), 309–320 (1993)
3. Stringa, L., Helmes, J.: Torque-ripple compensation in permanent magnet synchronous machines by high-bandwidth current control. IEEE Trans. Ind. Electron. **47**(6) (1997)
4. Lintor, M., Tredorescu, R., Blaabjero, F.: Control of multiple harmonics for three-phase grid converter systems using a PI-RES current controller. IEEE Trans. **23**(3), 846–851 (2009)
5. Panda, S.K., Joan-Xine, Quan, W.: A review of torque ripple reduction in pm synchronous motor drives. In: Proceedings of IEEE Power and Energy - Conversion and supply of Electrical Energy in the 21st Century, vol. 3, pp. 1–9, July 2007
6. Adhavan, J., Kuppusamy, B., Jayabhaskaran, K., Jagannathan, C.: Fuzzy logic controller based field oriented control of permanent magnet synchronous motor. In: Proceedings RAICS, pp. 585–594 (2012)
7. Lauud, D.N., Helmons, C.G.: Current regulation in Stationary frame for PWM inverters with zero steady-state error. IEEE Trans. **20**(4), pp. 825–833 (2004)
8. Colamartin, F., Machand, C., Razak, A.: Torque ripple minimization in permanent magnet synchronous servodrive. IEEE Trans. Energy Convers. **14**(3), 615–620 (1995)
9. Petrvic, V., Ortga, R., Stankovic, A.S., Tadmor, G.: Design and implementation of an adaptive controller for torque ripple minimization in PMSM. IEEE Trans. Power Electron. **15**(4), 873–882 (2002)
10. Basel, M.C., Mianos, S.J.: Design of PI and PID controllers with transient performances specification. IEEE Educ. **57**, 366–372 (2004)

11. Yapes, A.K., Frejedo, F.D., Lopez, O.: High performance digital resonance controllers implemented with two integrators. IEEE Trans. Power Electron. **28**(4), 562–575 (2012)
12. Castille, M., Martas, J., de Vicunas, L.G., Guerrero, J.M.: Guidelines for designing single-phase grid-connected photovoltaic inverters using damped resonant harmonic compensators. IEEE Trans. Ind. Electron. **59**, 4591–4600 (2006)
13. Teodorescu, R., Blaabjerg, F., Liserre, M., Loh, P.C.: Proportional resonant controllers and filters for grid-connected voltage-source converters. In: IEE Proceedings - Electric Power Applications, vol. 153, no. 5, pp. 750–762 (2006)
14. Qias, W., Panda, S.K.: An iterative learning control based torque ripple minimization in PM synchronous motors. IEEE Trans. **18**(2), (2003)
15. Jahens, T.M., Sunng, W.L.: Pulsating torque minimization techniques for permanent magnet ac drives—a review. IEEE Trans. Ind. Electron. **44**(2), 323–332 (1997)
16. De, D., Ramanarayanan, V.: A proportional multiresonant controller for three-phase four-wire high-frequency link inverter. IEEE Trans. Power Electron. **26**(4), 895–902 (2008)

Comparative Study of Different Materials on Performance of Chevron Shaped Bent-Beam Thermal Actuator

T. Aravind$^{(\boxtimes)}$, R. Ramesh, S. Praveen Kumar, and S. Ramya

ECE, Saveetha Engineering College, Thandalam, Chennai, India
{aravind,ramesh,praveenkumar,ramya}@saveetha.ac.in

Abstract. In this paper, a bent beam thermal microactuator's performance is presented. Displacement of this thermal actuator has been compared for different materials. The proposed bent beam microactuator operates on the code of Joule's heating effect with the able guidance of thermal expansion which brings into play the advantage of the profile (chevron) to augment the cumulative effect of thermal expansion so that it can produce large in-plane displacement with minimum operating voltage. This revision is done by considering the basic benefit in the design of a bent beam microactuator system and thereby simulating this bent beam thermal microactuator using COMSOL Multiphysics5.1 to analyze the response. The behavior of this thermal bent beam microactuator is premeditatedly decided for different materials such as aluminum, gold, argentum, cuprum and nickel and observations are recorded for the displacement response for an applied electric potential 0.2 V between two anchors.

Keywords: MEMS · Bent-beam actuator · In-plane displacement
Materials

1 Introduction

MEMS or the much advanced successor NEMS, both the technologies are garnering additional popularity over the recent past couple of decades [1, 2]. Micro sensors or MEMS Sensors and Microactuators or MEMS actuators are the devices which play a major role in Micro Electro Mechanical Systems (MEMS) [3, 4]. Applications of MEMS take account of widely publicized commercial applications like inkjet-printer, cartridges, in sensitive applications by means of miniature robots, micro engines, in communication applications by means of micro transmissions, optical scanners etc., Microactuators that typically renovates electrical energy input into output mechanical energy in terms of force and motion. Micro-electrothermal actuators are the prominent and most attractive micro-moving device since it can deliver high force and displacement when the outcomes are contended with added category of microactuators namely magnetic, piezoelectric (vibration based) microactuators and electrostatic (electron contention based) microactuators. In particular the thermal actuators have numerous applications, such as 3D optical switching [5], Microengines [6], mechanically tunable photonic crystal lens [7] and various other applications.

© Springer Nature Singapore Pte Ltd. 2018
I. Zelinka et al. (Eds.): ICSCS 2018, CCIS 837, pp. 743–751, 2018.
https://doi.org/10.1007/978-981-13-1936-5_76

Magnetic actuators typically work based on Electro Magnetic effect i.e., a magnetic field generates the force of repulsion or attraction in an electromagnetic material when an electric current is applied. It possesses high current, low voltage but typically needs special materials in fabrication process and has less efficiency. When two plates are separated by a layer of insulation of certain thickness and just by applying voltage across it, one can induce Electrostatic force. Electrostatic materials have the major disadvantage that they are not able to produce large displacement and force [8, 9].

Piezoelectric actuation has its base on the phenomenon of inverse piezoelectric principle in which, the application of electric potential to a crystal lattice of asymmetric nature leads to the deformation in the material that can be observed in a particular direction. The flexibility in this mechanism is the availability of a choice of structures such as a diaphragm, tube, cantilever and stacked multi layer structure can be considered for different applications [10]. But piezoelectric actuators have the inconvenience that it responds for soaring input applied voltage thus they necessitate a peripheral voltage amplifier. The resultant towering voltage is incompatible amid Integrated Chip technology and silicon processes.

Electro-thermal actuators have the advantages of higher displacement and lower operating temperature. Electro thermal actuators are classified into two types such as in-plane (hot cold) actuator and out of plane (bilayer) actuator. The variation in thermal expansion coefficient between two layers of a bilayer structure produces actuation in vertical direction (out of plane) and others that exploit different geometric structure made from unique material to invoke varying expansions and actuation in in-plane direction [11].

The hot-and-cold arm actuator is composed of narrower hot arm and wider cold arm of different lengths and widths. Flexure combines the wider arm with the anchor. Two such above depicted arms are united to form a U-shape at their split ends. When a potential is functional amid the anchors in series, the narrower arm is intensely heated than the wider arm due to the minor resistance which creates the mechanical force and produces deflection towards wider cold arm [12]. Also when a voltage is applied in parallel, it produces deflection in opposite direction. This U-shaped actuator produces force in the order of few micro-Newton each, but can produce large force by linking small tendons.

The advantage of V-shaped bent beam actuator, it produces high force with lower operating voltage. The bent-beam thermal actuator has an arched beam extending between two anchors. This arrangement of arched beam in a symmetric manner with a thin lengthy beam inclined at a diminutive angle α from the center arm so called shuttle. Thus this bent beam is also called a V-shaped beam and as it resembles like chevron shaped structure so called chevron actuator [13].

V-beam thermal actuators are running cascade in comparable to that by using the supplementary beam structures supporting mechanical actuation. V-beam actuators can also embrace a core post called shuttle that connects the multiple arched beams and assists to push that in in-plane direction. In this effort chevron actuator is intended for the design and comparison is done with its performance of using different materials. Materials are chosen based on its electrical and thermal conductivity. V-beam actuators are promising in applications that require in-plane deflection with high force outputs [14, 15].

2 Bent Beam Actuator Principle

A bent beam thermal actuator facilitated on theory of Joule heating effect and the widely accolade thermal expansion principle. The Joule's law states that a material is subjected to an electric current will acquire thermal energy. This leads to the collision of the moving electrons and atoms in the material. This collision of electrons and atoms in turn causes vibration which pump the surrounding atoms. The amplitude of vibration is large which results in increasing dimensions. This phenomenon is called thermal expansion. Chevron actuator with a single beam is shown in Fig. 1.

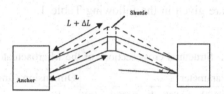

Fig. 1. Chevron beam actuator

The anticipated bent beam (V-beam in general) microactuator which consists of four beams mounted on one over the other supplementary beams [16]. While there are four duos of described arrangement of beams are involved, productivity in terms of force will be amplified in the order of four folds better than that of lone V-beam or bent beam microactuator. On the other hand, raising the amount of beams has to be limited for the reason that it swells up the order of stiffness of the construction and reasonably elongated shuttle. This roots to abate thermal expansion therein results in reduction in displacement. Every single beam is crafted in such an angle which is entitled pre-bending angle. Anchors hold the beams together between the duo of pillars like structure; one is endowed with voltage and the other one is grounded [17].

Beams are coupled at the midpoint or the center pillar mentioned to be as shuttle from now onwards, which stages a key function for receiving displacement in the premeditated in-plane direction which is designed to be parallel to the substrate. Observation can be done with an electric potential is functionally applied to one pillar, the difference in voltage induces the flow of current all the way from end to end of the bent beam assembly and the warmth of the beam mounts as a consequence of ohmic heating and the effect of thermally evoked spreading out of the surface of material causes shuttle to move superficial and ensuring to behave as a voltage controlled actuator competent of bringing in the in-plane displacement [18]. The deflection measurement for the bent beam actuator can be done by means of the subsequent expression (1).

$$\text{Def} = \left[L^2 + 2L(\Delta L) - L\cos^2(\alpha)\right]^{1/2} - L\sin(\alpha) \tag{1}$$

Where L is the length of single beam
α is the pre-bending angle.
ΔL is the elongation of the bent beam due to thermal expansion.

3 Simulation and Analysis of Bent Beam Actuator

The proposed thermal bent beam microactuator design is shown in Fig. 2. It consists of two contact pads, movable shuttle and four twain of thermally expandable beams. Beams are inclined at pre-bending angle 10°. The constituent geometrical cross sections of the structure are given in the following Table 1.

Table 1. Structural cross sections of the Microactuator design

Parameter	Dimension (μm)
Number of the beam pairs (n)	4
Beam length (L)	265
Beam width (W)	10
Gap between the beams	95
Contact pad length	400
Contact pad width	50
Shuttle length	335
Shuttle width	50
Thickness	10

The Chevron electrothermal (thermal) microactuator is designed and the results were obtained and verified successfully by using the simulation platform of COMSOL Multiphysics software's Structural mechanics module. The 2D view of the actuator design as shown in Fig. 2.

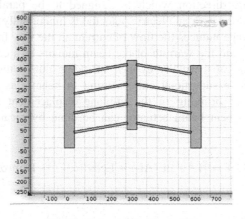

Fig. 2. Actuator design in 2D

The simulations are carried out by applying voltages upto 0.2 V across the anchors at constant temperature of 293.15 K. Material properties of different materials used for simulations in Comsol Multiphysics software [19, 20] are given in the following Table 2.

Table 2. Properties of different materials

Property	Al	Au	Ag	Cu	Ni
Young's Modulus (Gpa)	78	79	83	11	219
Poison's Ratio	0.35	0.44	0.37	0.33	0.31
Electrical Conductivity (s/m)	20	11.8	58.6	53.1	12.8
Thermal conductivity (W/m*K)	235	318	429	401	90.7
Coefficient of thermal expansion (1/K)	23.1	14.2	18.1	16.1	13.4

The material definition once completed then the Boundary Conditions has to be specified. Two anchors are fixed and one anchor is supplied with voltage and other anchor is grounded. The Core part of the structural mechanics is carried out by means of FEM analyses. The Meshing for the structure has been carried out for even distribution of voltage to get accurate results. In this work tetrahedral mesh is chosen.

4   Performance Analysis

The simulation analysis for bent beam actuator is revealed in Fig. 3. The potential difference induced between the two contact pads produces a disturbance in the shuttle which moves forward and hence a uniaxial displacement is produced in the desired direction.

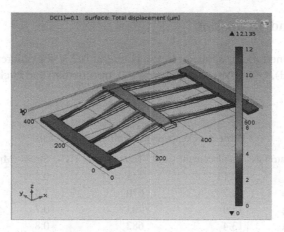

Fig. 3. In-plane displacement for an applied 0.2 V electric potential.

Temperature distributions of a micro-actuator is revealed in the below Fig. 4. It illustrates that temperature is getting gradually increased from the one of the pillars and the peak temperature is obtained in the shuttle which produces displacement.

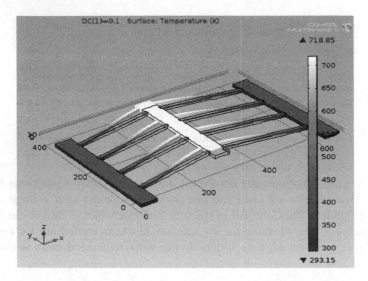

Fig. 4. Temperature distributions in the actuator for 0.2 V applied electric potential

The inference is dealt in such a way that the displacement is increased when we add up to the number of beams. In this manuscript, the number of expandable arms has been limited as four, since the number of beams is more, there is a sharp increase stiffness of the structure resulting in degradation in thermal expansion.

5   Results and Discussion

The results obtained for an applied electric potential 0.2 V for different materials such as Aluminum (Al), Gold (Au), Argentum (Ag), Cuprum (Cu) and Nickel (Ni) as shown in Table 3.

Table 3. Comparison of different materials

Materials	Displacement (μm)	Temperature (°C)	Stress (Mpa)
Al	12.13	445	0.5
Au	10.7	670	1.4
Ag	15.9	703	0.7
Cu	13.4	682	0.8
Ni	11.6	726	1.5

Displacement versus voltage for different materials is shown in Fig. 5. This being the evidence for the finding that - out of the entire materials under consideration here, argentum provides the premier displacement and gold responds with the least displacement but aluminium has better displacement than the silver.

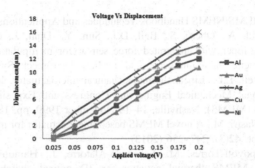

Fig. 5. Voltage Versus Displacement of bent beam actuator for different materials

The comparison of temperature distributions of bent beam actuator for different materials is depicted as in Fig. 6. It shows that out of all materials aluminum has the lowest temperature.

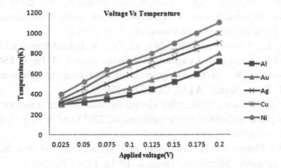

Fig. 6. Voltage Versus Temperature of bent beam actuator for different materials

6 Conclusion

In this work, performance of Chevron shaped bent-beam microactuator is studied for different materials. This study compares the parameters such as displacement, temperature and mises stress, for different materials. The comparative study shows that among different materials such as Aluminum, Gold, Argentum, Cuprum and Nickel, Aluminum has the desired displacement of about 12 μm with the minimum operating temperature about 445 °C which is compatible for biomedical applications. It is also less than that of the melting point of the aluminium (660 °C) for an applied voltage

0.2 V and Mises stress obtained is 0.5 Mpa which is very much lower than the fracture strength of Aluminum (90 Mpa).

References

1. Leondes, C.T.: MEMS/NEMS Handbook Techniques and Applications. Springer, US (2006)
2. Beyeler, F., Neild, A., Oberti, S., Bell, D.J., Sun, Y., Dual, J., et al.: Monolithically fabricated microgripper with integrated force sensor for manipulating micro objects and biological cells aligned in an ultrasonic field. J. Microelectromech. Syst. 16(1), 7–15 (2007)
3. Cragun, R., Howell, L.L.: Linear thermomechanical microactuator. In: Proceedings of the ASME International Mechanical Engineering Congress and Exposition, Microelectromechanical Systems (MEMS), Nashville, 14–19 November 1999, pp. 181–188 (1999)
4. Atashzaban, E., Nasiri, M.: A novel MEMS based linear actuator for mirror shape correction application. J. Opt. 42(3), 247–256 (2013)
5. Varona, J., Tecpoyotl-Torres, M., Escobedo-Alatorre, J., Hamoui, A.A.: Design and fabrication of a MEMS thermal actuator for 3D optical switching applications. In: Proccedings of the Digest of the IEEE/LEOS Summer Topical Meetings, Acapulco, Mexico, 21–23 July 2008, pp. 31–32 (2008)
6. Jae-Sung, P., Chu, L.L., Oliver, A.D., Gianchandani, Y.B.: Bent-beam electrothermal actuators—Part II: Linear and rotary microengines. J. Microelectromechan. Syst. 10, 255–262 (2001)
7. Cui, Y., Tamma, V.A., Lee, J.-B., Park, W.: Mechanically tunable negative-index photonic crystal lens. IEEE Photonics J. 2(6), 1003–1012 (2010)
8. Kalaiarasi, A.R., Thilagar, S.H.: Design and modeling of electrostatically actuated microgripper. In: Proceedings of IEEE/ASME International Conference on Mechatronics and Embedded Systems and Applications, pp. 7–11 (2012)
9. Chen, T., Sun, L., Chen, L., Rong, W., Li, X.: A hybird-type electrostatically driven Microgripper with an integrated vacuum tool. Sens. Actuat. A-Phys. 158, 320–327 (2010)
10. Nah, S.K., Zhong, Z.W.: A microgripper using piezoelectric actuation for micro-object manipulation. J. Sens. Actuat. A133, 218–224 (2009)
11. Wei, J., Duc, T.C., Sarro, P.M.: An electro-thermal silicon-polymer micro-gripper for simultaneous in plane and out-of-plane motions. In: 22nd Euro Sensors Conference, Dresden, Germany, pp. 7–10 (2008)
12. Kalaiarasi, A.R., Thilagar, S.H.: Modeling and characterization of a SOIMUMP's hybrid electro thermal actuator. Microsyst. Technol. 19(1), 113–120 (2013)
13. Que, L., Park, J.S., Gianchandani, Y.B.: Bent-beam electro-thermal actuators for high force applications. In: Proceedings of the Twelfth IEEE International Conference on Micro Electro Mechanical Systems, Orlando, 17–21 January 1999, pp. 31–36 (1999)
14. Kwan, A.M.H., et al.: Improved design for an electrothermal in-plane microactuator. J. Microelectromech. Syst. 21(3), 586–593 (2012)
15. Varona Jorge, V., Tecpoyotl-Torres, M., Hamoui, A.A.: Design of MEMS vertical–horizontal chevron thermal actuators. Sens. Actuat. A Phys. 153(1), 127–130 (2009)
16. Sinclair, M.J.: A high force low area MEMS Thermal Actuaor. Sev. Intersoc. Conf. Thermal Thermomech. Phenom. Electron. Syst. ITHERM 1, 127–132 (2000)
17. Que, L., Park, J.S., Gianchandani, Y.B.: Bent-beam electrothermal actuators-Part I: single beam and cascaded devices. J. Microelectromech. Syst. 10(2), 247–254 (2001)

18. Baracu, A., et al.: Design and fabrication of a MEMS chevron-type thermal actuator. In: International Conferences and Exhibition on Nanotechnologies and Organic Electronics, NANOTEXNOLOGY (2014)
19. Aziz, A.A., Rehman, M.S., Aris, H.: Finite element analysis of a lateral micro-electrothermal actuator. In: RSM Proceedings, pp. 354–358 (2011)
20. Jaina, N., Arora, D., Kumar, S.: Effect of residual stress of different materials on performance of chevron beam actuator. Int. J. Curr. Eng. Technol. 3(1), March 2013

Investigation on Four Quadrant Operation of BLDC MOTOR Using Spartan-6 FPGA

C. Gnanavel[1](✉), T. Baldwin Immanuel[1], P. Muthukumar[3],
and Padma Suresh Lekshmi Kanthan[2]

[1] Department of EEE, AMET Deemed to be University,
Chennai 603112, Tamil Nadu, India
gnana2007@gmail.com, bimmanuelt@gmail.com
[2] Baselios Mathew II College of Engineering,
Sasthamkotta 690521, Kerala, India
suresh_lps@yahoo.co.in
[3] Prasad V. Potluri Siddhartha Institute of Technology,
Vijayawada 520007, Andhra Pradesh, India
muthukumarvlsi@gmail.com

Abstract. This paper proposes an FPGA based four quadrants operation of Brushless DC (BLDC) motor control using FPGA-SPARTAN-6 device. This control practice for four-quadrant operation identifies the rotor rotating constraint and fluctuations the quadrant of operation accordingly. The motor controlling methodology designed to work in all the four quadrants without any deprivation of power. A low-cost, easy to use improvement and assessment platform for Spartan-6 FPGA designs. This paper presents modern BLDC motor drives with an importance on FPGA Spartan-6 control of these motors. The effectiveness of the proposed technique established complete experimental results.

Keywords: Field programmable gate array · BLDC · Three phase
Four quadrant · AC to DC · DC to AC

1 Introduction

In industrial control, there are certain processes are required of adjustment from normal operation for accuracy position/speed control with respect to load or supply voltage. Such fine-tunings are usually accomplished with variable speed drive and it consists of (1) Electric motor, (2) Power Converter (3) Controller. Electric Motor: It is coupled indirectly/directly to the load to furnish particular application. Power Converter: It supervises the power flow from an AC source to the motor by suitable control algorithm implemented to trigger the power semiconductor switches. Controller: The controller generates PWM signal to the converter & hence forms the heart of the Variable speed system (VSDs). Brushless DC (BLDC) motors are referred by many aliases: Brushless Permanent Magnet Motors, Permanent Magnet AC motors, Permanent Magnet Synchronous Motors. The confusion comes up because a BLDC motor does not unswervingly run by a DC voltage source [1].

© Springer Nature Singapore Pte Ltd. 2018
I. Zelinka et al. (Eds.): ICSCS 2018, CCIS 837, pp. 752–763, 2018.
https://doi.org/10.1007/978-981-13-1936-5_77

BLDC motors have one of the motor types which becomes most popularity. The stator of a BLDC motor involves of stacked steel laminations with windings located in the slots that are axially cut along the inner boundary or about stator salient poles. The rotor has constructed of permanent magnets and can differ from two to eight pole pairs with alternate north (N) and south (S) poles. In sequence to rotate a BLDC motor, the stator windings ought to be bracing, in a categorization. It is vital to recognize the rotor location in the directive to appreciate as to which winding must be energized [2].

The basic principle difference between BLDC with dc motor is, A BLDC motor has a rotor made up of permanent magnets with windings. The brushes and mechanical commutator have been eliminated and the windings are connected to the regulator electronics. The control electronics trade the occupation of the commutator and energize the suitable winding. The energized stator winding leads the rotor crowd-puller and switches just as the rotor aligns with the stator. The superior characteristics is the BLDC motor does not exhibit any sparks.

The brushes of a dc Motor have several disadvantages; brush life, brush deposit, greatest speed, and low electrical noise. BLDC Motors are potentially cleaner, more rapidly, more efficient, less noisy and more dependable. The Spartan-6 family is made on a 45-nanometer [nm], 9-metal layer, and dual-oxide process technology. Convolution method Digital Control of Four Quadrant Operation of BLDC Motor using dsPIC30F4011 was not obtained regenerative braking due to heat losses [3]. This Proposes four quadrant operation of using advanced FPGA Spartan-6 is to reduced heat losses at braking time. In fact, vigor is preserved during the regenerative period. The Spartan-6 series objectives applications with a low-power track dangerous cost sensitivity and high-volume.

2 Block Diagram of Four Quadrant Operation

Some of the modern applications request visit speed inversion at least cost. In such cases, a drive framework working in four quadrants is exceptionally fundamental for effective and smooth speed inversion. Numerous control plans have been proposed by various creators, however, the major disadvantage is their expanded unpredictability. In this segment, a basic and effective control plot for four-quadrant activity is displayed. The four quadrants of the task are forward motoring, forward braking, reverse motoring and reverse braking respectively. In motoring mode, the magnitude of the supply voltage is more than the back EMF and the energy flows from supply to motor whereas in generating mode, the greatness of the supply voltage is not as much as the back EMF and the vitality streams from the engine to supply. Another essential certainty to be noted is, in forwarding motoring and invert braking modes (Quad-I and IV), the heading of current inside the motor terminal is same though in sending braking and turn around motoring (Quad-II and III), the current is in the inverse direction [4].

BLDC engines are by and large powered by a three-stage voltage source inverter (VSI) which is controlled utilizing rotor position. The rotor position can be detected utilizing Hall sensors, resolvers, or optical encoders. As of late some extra uses of BLDC engines have been accounted for in electric vehicles (EVs) and cross breed electric vehicles (HEVs) because of natural worries of vehicular emanations. BLDC motor has been discovered more reasonable for EVs/HEVs and other low power applications, because of high power thickness, diminished volume, high torque, high effectiveness, simple to control, straightforward equipment low maintenance. The controllers can additionally be isolated based on strong state switches and control methodologies. The BLDCM needs rotor-position detecting just at the substitution focuses, Every 60° electrical in the three-stages; in this way, a relatively straightforward controller is required for recompense and current control Fig. 1 details explain about block diagram.

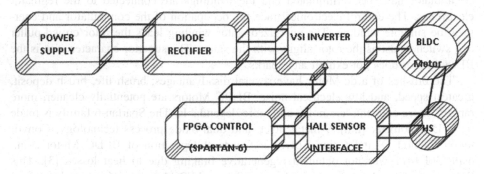

Fig. 1. General block diagram

This paper presents cutting-edge four quadrant task of BLDC Motor. In view of the blend of these three Hall sensor flags, the correct movement of replacement can be tenacious. These signs are decoded by combinational rationale to give the terminating signs to 120° conduction on every one of the three stages. The rotor area decoder has six yields which control the upper and lower stage leg IGBTs the rotor shaft position is detected by a Hall Effect sensor motions as spoke to in Table 1. Method of the task, at a particular moment, one IGBTs in the upper leg and on IGBTs in the lower leg will be exchanged ON.

3   Closed-Loop Control of BLDC

Closed loop control is important in this application; there is a scope of open-loop control standards accessible, their ease of use is however restricted rather to static load and non-slowed down the task. The usable control strategies normally contrast in fundamental parameters including torque control quality, usable precise speed extends, execution unpredictability and others. For the most part, every technique can be found in its sensor and senseless flavors; sensor-less implies that there are no pole position sensors, for the most part, supplanted by twisting back-EMF estimation [5] Fig. 2 detailed mode of operation in BLDC motor.

The BLDC motor having four possible modes or quadrants of operation. In Quadrant-I is forward motoring operation which implies forward torque and speed. Torque is proportional to the speed. Conversely, Quadrant-III is reverse motoring operation at the time motor work in reverse speed and torque. Now the motor is "motoring" in the reverse direction, revolving backwards with the reverse torque. Quadrant-II is forward breaking at that time motor is stopped, but torque is being practical in reverse. Torque is being used to "break" the motor, and the motor is now generating power as a result. Finally, Quadrant-IV is accurately the conflicting. The motor works in reverse rotation to produce the reverse torque. Again, torque is being practical to effort to leisurely the motor and change its direction to forward again. Afresh, power is organism generated by the motor. The BLDC motor is rotating in clockwise direction at the starting mode, but when the speed reversals way is obtained, the manage goes into the clockwise regeneration mode, which brings the rotor to the idle position. Instead of coming up for the fixed fester position, uninterrupted energization of the main phase is attempted. This swiftly slows down the rotor to a fester position. Therefore, there is the inevitability for determining the instant when the rotor of the machine is ideally positioned for turnaround Hall-effect sensors are worn to establish the rotor position and from the Hall sensor outputs, it is unwavering whether the machine has overturned its direction. This is the idyllic flash for energizing the stator phase so that the machine can start motoring in the counter clockwise direction [5].

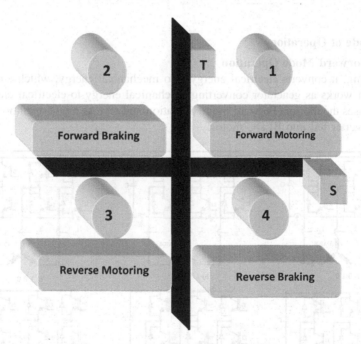

Fig. 2. Four quadrant operation of BLDC Motor

Table 1. Switching table of BLDC motor inverter for the mode of 120°

S. no	Hall sensor interface			Forwarding PWM output signals						Reversing PWM output signals						Breaking	Rotation angle (θ)
	H_a	H_b	H_c	1	2	3	4	5	6	1	2	3	4	5	6		
1	0	1	0	1	0	0	0	0	1	0	1	0	0	1	0	0	0°–30°
2	1	1	0	0	0	1	0	0	1	0	0	0	1	1	0	0	30°–60°
3	1	0	0	0	1	1	0	0	0	1	0	0	1	0	0	0	60°–90°
4	1	0	1	0	1	0	0	1	0	1	0	0	0	0	1	0	90°–120°
5	0	0	1	0	0	0	1	1	0	0	0	1	0	0	1	0	120°–150°
6	0	1	1	1	0	0	1	0	0	0	1	1	0	0	0	0	150°–180°
7	0	1	0	1	0	0	0	0	1	0	1	0	0	1	0	0	180°–210°
8	1	1	0	0	0	0	1	0	0	1	0	0	0	1	1	0	210°–240°
9	1	0	0	0	1	1	0	0	0	1	0	0	1	0	0	0	240°–270°
10	1	0	1	0	1	0	0	1	0	1	0	0	0	0	1	0	270°–300°
11	0	0	1	0	0	0	1	1	0	0	0	1	0	0	1	0	300°–330°
12	0	1	1	1	0	0	1	0	0	0	1	1	0	0	0	0	330°–360°
13	x	x	x	1	0	1	0	1	0	1	0	1	0	1	0	1	-

3.1 Mode of Operations

3.1.1 Forward Mode Operation

In motoring, it converts electrical energy into mechanical energy, which supports its motion. It works as generator converting mechanical energy to electrical energy, and thus opposes the motion. Forward mode operation and reverse mode operation, voltage source inverter fed into BLDC motor.

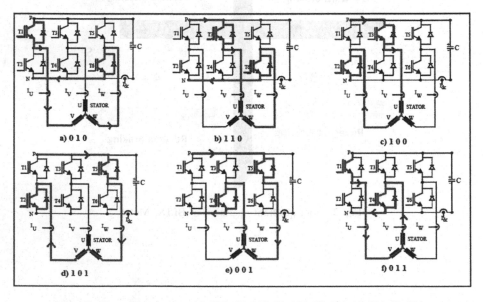

Fig. 3. Forward motoring (a) 010 (b) 110 (c) 100 (d) 101 (e) 001 (f) 001

Various PWM methods are available to control the VSI [6]. Forward and Reverse PWM Signal have been controlled BLDC motor and sense hall sensor output signal at mode operation of 120° forwarding PWM output signal is based on various switching signal (T_1, T_6), (T_3, T_6), (T_2, T_3), (T_2, T_5), (T_4, T_5), (T_1, T_4) Fig. 3 Forward motoring operations [7].

3.1.2 Reverse Mode Operation

Reversing PWM Output (T_2, T_5), (T_4, T_5), (T_1, T_4), (T_1, T_6), (T_3, T_6), (T_2, T_3) practically to see hall sensor output in forward and reverse operation hall sensor output (010), (110), (100), (101), (001), (011), (010). Shown in Fig. 4 Reverse motoring. The torque and speed coordinates for both forward (positive) and reverse (negative) motions. In quadrant-I, developed power is positive at the time torque also rotating in clockwise direction. Hence machine works as a motor provide mechanical energy. Operation in quadrant-I is called forward motoring mode. In quadrant-II, power is negative at reverse direction. Hence, machine moving parts under braking divergent the motion (Fig. 5).

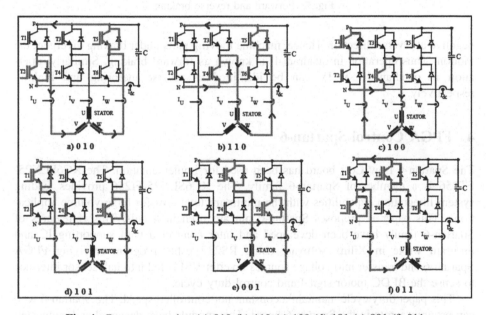

Fig. 4. Reverse motoring (a) 010 (b) 110 (c) 100 (d) 101 (e) 001 (f) 011

Thus operation in quadrant-II is known as forward braking. Similarly, operation in quadrant-III & IV can be identified as reverse motoring and braking respectively. The torque and speed coordinates for both forward (positive) and reverse (negative) motions. In quadrant I, developed power is positive at the time torque also rotating in clockwise direction. Hence machine works as a motor supplying mechanical energy. Operation in quadrant-I is called forward motoring mode. In quadrant-II, power is

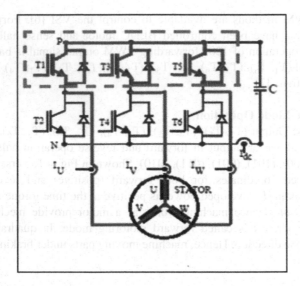

Fig. 5. Forward and reverse braking

negative at reverse direction. Hence, machine moving parts under braking divergent the motion. Thus operation in quadrant-II is known as forward braking. Similarly, operation in quadrant-III & IV can be identified as reverse motoring and braking respectively.

4 FPGA Control Spartan-6

The Spartan-6 Low Cost board has been used for implementation. The XC6SLX25-FT256 is a member of Spartan-6 family. The XC6SLX25-FI256 provides leading system integration capabilities with the minimum total cost for huge volume applications. This paper is proposes Spartan-6 pin are used in details. This is features of Spartan-6 family have been developed and inter connected with supporting IC and program wrote in Xilinx software using RS232 cable program receiving FPGA Spartan-6 and transfer into voltage sources inverter.VSI is fed into hall sensor interface to sense the BLDC motor signal and control duty cycle.

This paper duty cycle maintain a constant not controlled speed. The Spartan-6 who can program wrote in the Spartan-6 board. This module have designed and connected with voltage sources inverter. Spartan-6 is produced PWM pulse data have been transmitted (TXD) VSI using RS 232. Receiving data (RXD) to VSI Fig. 6 illustrated PWM pulse (Fig. 7).

The Spartan-6 development board provides an RS232 port that can be driven by the Spartan-6 FPGA [6–9]. Here we having one 9-pin D MALE serial port connector are used for external connection. Subsets of the RS232 signals are used on the Spartan-6 development board to implement this interface (RxD and TxD signals) for done a serial communication operation we have a separate IC named as ICL3232CBN.RXD -Data

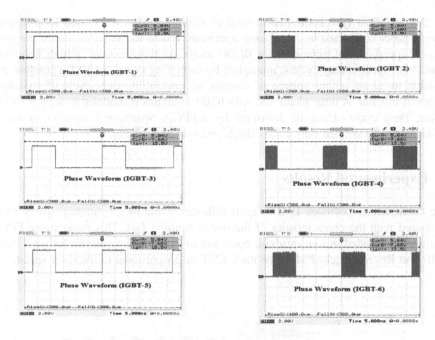

Fig. 6. Experimental pulse waveform

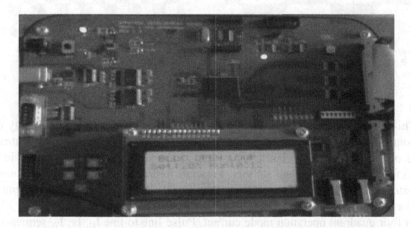

Fig. 7. Spartan-6 FPGA board

received by FPGA, TXD - Data transmitted by FPGA. Voltage sources inverter connected RS232 with Spartan-6. They are Spartan-6 PWM pulse generated totally six IGBT switches used VSI. That PWM pulse practically measure sin DSO three positive and negative PWM is generated. VSI fed into BLDC. This have controlled PWM pulse that means to vary duty cycle to control of speed of the BLDC Motor. This is paper minimum of duty cycle 20% have tested at low speed operation.

Two different type controlled method is available open and closed loop system. This is paper contracted in closed loop operation. PWM technique is one of the most popular speed control techniques for BLDC motor. In this technique a high frequency signal with specific duty cycle is multiplied by switching signals of VSI. Therefore it is possible to adjust output voltage of inverter by controlling duty cycle of switching pulses of inverter. A three phase VSI with IGBT switches is modeled to supply BLDC motor. Duty cycle of can be determine by a FPGA Spartan-6 Controller or toggle between two predefined duty cycles (high and low duty cycle).

5   Experimental Results

The characteristics of back EMF is quite different while motor running slowly when compared with the characteristics of the motor running fast. This paper using FPGA spartan-6 controlled in four quadrant operation to Fig. 8 see waveform forward Back-EMF and Reverse back- EMF. The back EMF is proportional to the rotor speed.

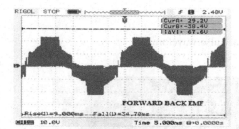 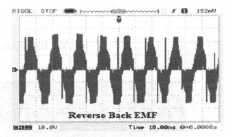

Fig. 8. Experimental of forward and reverse BACK EMF

This research results are obtained while maintaining minimum of 20% duty cycle for controlling four quadrant operations. The research is not focusing to control the speed of the BLDC motor, but it focused to measure the speed of the motor in both directions by using three hall sensors. Forward and reverse braking time motor has been stopped. Back-EMF also is Zero at the time BLDC motor hall sensor output may be 0 or 1.

In four quadrant operation mode current. Pulse line to line I_U, I_V, I_W sequence can be varied it generated current Fig. 9 illustrated phase current. Three phase BLDC motor current pulse is measured in between each two phase.

When T_1 and T_5 switch is ON at time current flow through T_1-U-V-T_6 current is positive I_U. When T_3 and T_6 switch is ON at time current flow through T_3-V-W-T_6 current is positive I_V. When T6 andT_1 switch is ON at time current flow through T_6-W-U-T_1 current is positive I_W.

When (T2, T5), (T4, T5),(T4, T1) this switch is ON at the time negative phase current is flowed like that forward mode operation only change switch sequence same current is flow in reverse mode of operation I_U, I_V, I_W. Forward and reverse braking

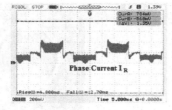

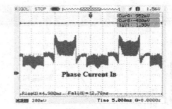

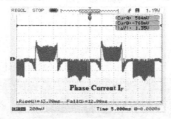

Fig. 9. Phase current I_U, I_V, I_W

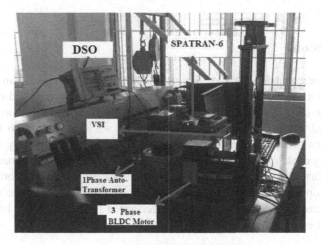

Fig. 10. Hardware photo

operation current flow is only positive partially switches only ON. This paper is done by four quadrant operation in BLDC motor by using Spartan-6 controller interconnected VSI with hall sensor interface (Fig. 10).

These sensor signals have sense the forward direction. Controller using RS232 connected with PC. Xilinx software using wrote program and deployment controller. Control the BLDC motor perform has been show typical output waveform. A permanent Magnet AC motor, which has a trapezoidal BACK EMF, is referred to as brushless DC motor (BLDC). The graph is drawn between actual speed V_s Set speed (Fig. 11).

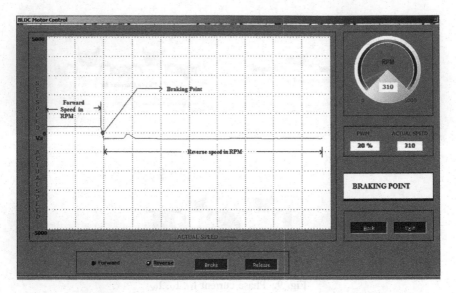

Fig. 11. Four quadrant operation of BLDC motor waveform at duty cycle 20%, rated speed 310 rpm

6 Conclusion

This paper presented the state of the art BLDC motor drives with an emphasis on implementation of FPGA Spartan-6 controller. In this system for speed control, PWM pulses are not used instead the four quadrant operations at constant speed and torque has performed. Duty cycle can be determined by a FPGA Spartan-6 Controller by toggle between two predefined duty cycles (such as high and low duty cycle). In this work, the duty cycle of the closed loop operation is fixed as 20% at a constant speed of 310 rpm in forward motoring, reverse motoring, forward and reverse braking running at low speed operation. At the time of braking point, a few seconds, the motor have to be the generated heat. In future work, the generated heat energy has been converted into electrical energy saved in the battery.

Appendix-Motor Rating

Type	Trapezoidal motor
Voltage	310 V DC
Poles	4
Current	4.52 A
PeakTorque	6.60 NM
Speed	4600 rpm

References

1. Kumar, V.V., Muruganandham, J.: Speed control of BLDC motor for four quadrant operation without loss of power. In: 2015 International Conference on Computation of Power, Energy, Information and Communication. IEEE (2015)
2. Joice, C.S., Paranjothi, S.R., Kumar, J.S.: Practical implementation of four quadrant operation of three phase brushless DC motor using dsPIC. In: 2011 International Conference on Recent Advancements in Electrical, Electronics and Control Engineering. IEEE (2011)
3. Joice, C.S., Paranjothi, S.R., Kumar, V.J.S.: Digital control strategy for four quadrant operation of three phase BLDC motor with load variations. IEEE Trans. Ind. Inf. 9(2), 974–982 (2013)
4. Krishnakumar, V., Jeevanandhan, S.: Four switch three phase inverter control of BLDC motor. In: 1st International Conference on Electrical Energy Systems, pp. 139–144 (2011)
5. Murphy, J.M.D., Turnbull, F.G.: Power Electronic Control of AC Motors. Pergamon Press, Oxford (1988)
6. Paramasivan, M., Paulraj, M.M., Balasubramanian, S.: Assorted carrier-variable frequency-random PWM scheme for voltage source inverter. IET Power Electron. 10(14), 1993–2001 (2017)
7. Muthukumar, P., Mary, P.M., Jeevananthan, S.: An improved hybrid space vector PWM technique for IM drives. Circ. Syst. 7(09), 2120–2131 (2016)
8. Gnanavel, C., Alexander, S.A.: Experimental validation of an eleven level symmetrical inverter using genetic algorithm and queen bee assisted genetic algorithm for solar photovoltaic applications. J. Circ. Syst. Comput. 27(13), 185–0212 (2018)
9. Jacob, T., Krishna, A., Suresh, L.P., Muthukumar, P.: A choice of FPGA design for three phase sinusoidal pulse width modulation. In: International Conference on Emerging Technological Trends (2016). https://doi.org/10.1109/icett.2016

Modeling of Brushless DC Motor Using Adaptive Control

N. Veeramuthulingam$^{(\boxtimes)}$, A. Ezhilarasi, M. Ramaswamy, and P. Muthukumar

Department of Electrical Engineering, Annamalai University, Annamalai Nagar, Chidambaram, Tamil Nadu, India
sethukumark@gmail.com, jee.ezhiljodhi@yahoo.co.in, aupowerstaff@gmail.com

Abstract. The paper proposes the introduction of an adaptive controller for enhancing the performance of a brushless DC motor (BLDC). It involves the use of the state space model for the motor in an effort to articulate the theory of model reference adaptive control and follow the principles of sensorless feedback mechanism. While the measure of the back EMF reflects the repository speed and enables the computation of the speed error, the reference frame fosters the estimate of the ripple in the torque of the motor. The two stage converter interfaces attach support to assuage the most intriguing of the corrective action through changes in the duty cycle and the modulation index, respectively. The methodology forges its adaptability to various speed ranges and remains immune to source and load distubances without affecting the accuracy of the results. The MATLAB based response illustrates the merits of the algorithm in terms of its speed regulating capability and minimizing the torque ripple to allow the motor find a place in the utility industry.

Keywords: Modelling · BLDC motor · Model reference adaptive control

1 Introduction

The modern world continues to promote the resurgence of solid state drives and invite a thrust towards enhancing its performance. The applications among the many include the brushless dc engines, electrical vehicles, HVAC industry, military application and medical equipments. The advantages rely on the higher power capacity, better effectiveness and lower maintenance.

The brushless dc motor (BLDC) motor driven by direct current, operates on electronic commutation rather the mechanical commutation system based on brushes. The motor torque, current; voltage and speed experience a linear relationship with each other and the significance owes to the removal of the brushes, which swing sparking and brush maintenance. It evokes considerable interest in the industry and finds a wide scope for its use. The modeling and simulation form part of the design of the BLDC motor and the tuning of the parameters of the controller evince a paradigm shift to manifest fresh utility areas.

I. Zelinka et al. (Eds.): ICSCS 2018, CCIS 837, pp. 764–775, 2018.
https://doi.org/10.1007/978-981-13-1936-5_78

The control of the BLDC motor emphasizes the location of the magnitude of the back EMF through its rotor speed operations. At low speed operating range it becomes difficult to measure the speed because of inverter and nonlinearities in the parameter. Though the normal PID based speed control solutions show to be simple, stable, and highly reliable, still several modern control results continue to materialize [1]. However currently many degrees of nonlinear approaches occupy prominence in controlling the motor [2].

A number of closed loop control algorithms have been recommended to design an optimal converter [3] and a torque ripple minimization strategy discussed in [4] to address the trapezoidal shape of back EMF and in addition enliven an easy active method to design an angular speed controller and build a torque controller. The model reference adaptive controller for the motor has been handled by [5] and seen to depend on the correct parameters and the sensor position [6, 8]. The experimental methods have been to shown to involve a PLL [7] and arrive at driving nearly constant speed. However the trend augurs exploring further new control methodologies to enunciate improved performance for the drive motor.

2 Modeling of Brushless DC Motor

The transfer function offers powerful and simple design techniques although it suffers from certain problems in the sense the model appears to be defined under zero initial conditions. Besides it behoves its applicability to linear time-invariant systems and remains limited to single input single output systems. The other restriction relates to the fact that it provides the system output only for a given input and does not reveal any data regarding the internal state of the system.

The approach involves the following assumptions in the modeling of the BLDC motor

- Star wound type stator.
- Symmetric in nature of inductance, resistance and mutual inductance.
- No change in rotor reluctance angle.
- Stator winding aligned with proper orientation.
- Free from saturation effects.
- Identical back- EMF shape in all three phase.
- Inverter remains ideal.
- Negligible hysteresis, eddy current effect and iron losses.

It comprises of both mechanical and electrical equations

The electrical equations:

$$V_a = R * i_a + L * \frac{di_a}{dt} + emf_a \tag{1}$$

$$V_b = R * i_b + L * \frac{di_b}{dt} + emf_b \tag{2}$$

$$V_c = R * i_c + L * \frac{di_c}{dt} + emf_c \tag{3}$$

$$emf_a = K_e * w_m * F(\theta_e)$$
$$emf_b = K_e * w_m * F(\theta_e - 2\pi/3)$$
$$emf_c = K_e * w_m * F(\theta_e + 2\pi/3)$$
$$T_e = [emf_a * i_a + emf_b * i_b + emf_c * i_c]/w_m$$

The mechanical equation relates as is Eq. (4)

$$T_e = B * w_m + J * \frac{dw_m}{dt} + T_l \qquad (4)$$

V_k – Phase voltage, I_k – Phase Current, emf_k – Phase back emf
R, L – Phase resistance and inductance of the stator winding
T_e – Total Electromagnetic torque and T_l – load at motor
θ_e – Electrical angle of the rotor
θ_m – Mechanical angle of the rotor
W_m – angular speed of rotor
P – number of poles on rotor
K_e – motor back emf current (V/rad/Sec)
J – moment of inertia (kg-m^2)
B – Dynamic Frictional Torque constant (NM/rad/Sec)

Mechanical and Electrical angle of the rotor can be related as in Eq. (5)

$$\theta_e = \frac{P}{2}\theta_m \qquad (5)$$

F (θ_e) follows the back emf reference which bears a trapezoidal shape and magnitude in $\pm$ in Eq. (6)

$$F(\theta_e) = \begin{cases} 1 & 0 \le \theta_e \le 2\pi/3 \\ 1 - \frac{6}{\pi}(\theta_e - 2\pi/3) & 2\pi/3 \le \theta_e \le \pi \\ -1 & \pi \le \theta_e \le 5\pi/3 \\ -1 + \frac{6}{\pi}(\theta_e - 5\pi/3) & 5\pi/3 \le \theta_e \le 2\pi \end{cases} \qquad (6)$$

On solving the equation, the complete state-space model can expressed as in Eq. (7)

$$\bar{x} = Ax + Bu$$

$$\begin{bmatrix} \frac{di_a}{dt} \\ \frac{di_b}{dt} \\ \frac{di_c}{dt} \\ \frac{dw_m}{dt} \\ \frac{d\theta_m}{dt} \end{bmatrix} = \begin{bmatrix} -\frac{R}{L_1} & 0 & 0 & -\frac{\lambda_p}{L_1}f_a(\theta_r) & 0 \\ 0 & -\frac{R}{L_1} & 0 & -\frac{\lambda_p}{L_1}f_b(\theta_r) & 0 \\ 0 & 0 & -\frac{R}{L_1} & -\frac{\lambda_p}{L_1}f_c(\theta_r) & 0 \\ \frac{\lambda_p}{J}f_c(\theta_r) & \frac{\lambda_p}{J}f_c(\theta_r) & \frac{\lambda_p}{J}f_c(\theta_r) & -\frac{B}{J} & 0 \\ 0 & 0 & 0 & \frac{P}{2} & 0 \end{bmatrix} \begin{bmatrix} i_a \\ i_b \\ i_c \\ \omega_m \\ \theta_r \end{bmatrix} + \begin{bmatrix} \frac{1}{L_1} & 0 & 0 & 0 \\ 0 & \frac{1}{L_1} & 0 & 0 \\ 0 & 0 & \frac{1}{L_1} & 0 \\ 0 & 0 & 0 & 0 \\ 0 & 0 & 0 & 0 \end{bmatrix} \begin{bmatrix} V_a \\ V_b \\ V_c \\ T_1 \end{bmatrix}$$

$$(7)$$

The Eq. (7) is a state space representation of complete linear modeling in which, linear analysis and control methods can be applied to it.

3 Proposed Methodology

The primary focus owes to elicit the theory of adaptive control through the modeling equations and assuage measures for regulating the speed of the BLDC motor and minimizing the ripple in the torque. The principle of an adaptive control system generates an error from two models using which it allows computing the unknown parameter. The measured quantity from the output of the adaptive system serves as the feedback and enables the closed loop stability through the Popov's Hyper stability criterion [15].

Depending on the quantity (i.e. the functional candidate), it formulates the error signal and develops the model reference adaptive system (MRAS) with d-q components of flux [11]. It relies on the stator resistance variation and experiences from the integrator problems like saturation and drift. The MRAS with on-line stator resistance evaluation reported in [12], reactive power-based MRAS [13, 14] and neural network (NN) based MRAS outline the efforts to attenuate the solutions.

The instability of drive parameters can be fulfilled using MRAC and gainschedling operation [9], Self-turning [14]. The MRAC with parameter transformation [9] involves proportional and integral parts of the algorithm and requires iteration for the optimization of tuning of the adaptation algorithm. The most important merit of the MRAS with signal adaptation is that it does not limit integral parts and essentially turning of controller parameters used for changing plant parameters [10].

The error between the estimated quantities obtained from the models leaves way to drive a fit adaptive mechanism for creating the estimated rotor speed [1] and finds the error and adaptation controller parameters by MIT Rules as seen from Figs. 1 and 2.

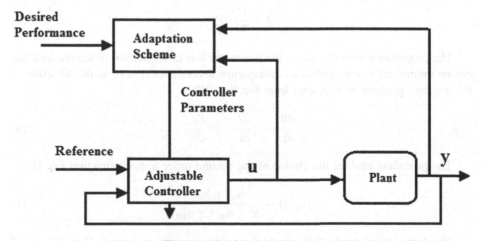

Fig. 1. Adaptive control system

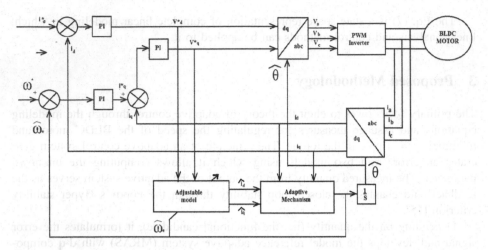

Fig. 2. Model reference adaptive control simulation block diagram of BLDC

The MIT Rule

The MIT rule methods aims to reduce the squared model cost function and owing to the error being minimum it forms an accurate tracking between the actual output and the reference output.

Designing Steps

The adaptation error computed as the difference between the parameter output and the model output as in Eq. (8)

$$\varepsilon = y_p(t) - y_m(t) \tag{8}$$

Where, Y_m, Y_p is the output of the model, plant
The Eq. (9) shows the cost function J

$$J = \frac{1}{2}\varepsilon^2(t) \tag{9}$$

The procedure evolves to adjust the parameter θ in order that the objective function can be minimized to zero and as a consequence necessitates θ to be in the direction of the negative gradient of J as seen from the Eq. (10)

$$\frac{d\theta}{dt} = -\frac{dJ}{dt} = \gamma\varepsilon\frac{\partial\varepsilon}{\partial\theta} \tag{10}$$

The procedure enables the choice of the second order transfer function Eq. (11)

$$G_m(s) = \frac{b_{m1} + b_{m0}}{S^2 + a_{m1}s + a_{m0}} \tag{11}$$

The formulations tracks the error in a manner as specified through Eqs. 12 and 13

$$e = r - y_p \tag{12}$$

$$\frac{de}{dt} = -\frac{dy_p}{dt} \tag{13}$$

The Eq. (14) represents the control law of system for PI controller

$$u(t) = k_p e(t) + k_i \int e(t) dt \tag{14}$$

The Eq. (15) gives the Laplace Transform of Eq. (14)

$$U(s) = K_p E(s) + \frac{K_i}{s} E(s) \tag{15}$$

The closed loop transfer function of control law deduces as in Eq. (16)

$$y_p = \frac{G_p K_p r + \frac{G_p K_i r}{s}}{1 + G_p K_p \frac{G_p K_i}{s}} \tag{16}$$

By solving for Y_p in terms of r and substitute Y_p in Eq. (8), the adaptation error can be obtained from Eq. (17)

$$\varepsilon = \frac{(G_p K_p s + G_p K_i) r}{S(1 + G_p K_p) + G_p K_i} - y_m \tag{17}$$

The Eq. (18) shows the adaption error view to MIT rules for K_p, K_i

$$\frac{d\varepsilon}{dk_p k_i} = \frac{dy_p}{dk_p k_i} \tag{18}$$

The Eq. (20) can be obtained by rewriting the Eq. (17)

$$\varepsilon = \frac{G_p K_p r + \frac{G_p K_i r}{S}}{1 + G_p K_p + \frac{G_p K_i}{S}} \tag{19}$$

$$\varepsilon = \left(G_p K_p r + \frac{G_p K_i r}{S} \right) \left(1 + G_p K_p r + G_p K_p r + \frac{G_p K_i r}{S} \right)^{-1} \tag{20}$$

Applying MIT rules for obtaining K_p, the gradient obtains a form as in Eq. (21)

$$\frac{d\varepsilon}{dK_p} = \frac{G_p r}{1 + G_p K_p + \frac{G_p K_i}{S}} - \frac{G_p \left(G_p K_p r + \frac{G_p K_i r}{S} \right)}{\left(1 + G_p K_p + \frac{G_p K_i}{S} \right)^2} \tag{21}$$

Substituting Eqs. (16) in (21) to get the Eq. (22)

$$\frac{d\varepsilon}{dK_p} = \frac{G_p r}{1 + G_p K_p + \frac{G_p K_i}{S}} - \frac{G_p Y_p}{\left(1 + G_p K_p + \frac{G_p K_i}{S}\right)} \tag{22}$$

The Eq. (23) can be derived by rearranging Eq. (22)

$$\frac{d\varepsilon}{dK_p} = \frac{G_p E}{1 + G_p K_p + \frac{G_p K_i}{S}} \tag{23}$$

Applying MIT rules for obtaining K_i, it becomes as in Eq. (24)

$$\frac{d\varepsilon}{dK_i} = \frac{\frac{G_p}{S} r}{1 + G_p K_p + \frac{G_p K_i}{S}} - \frac{\frac{G_p}{S}\left(G_p K_p r + \frac{G_p K_i r}{S}\right)}{\left(1 + G_p K_p + \frac{G_p K_i}{S}\right)^2} \tag{24}$$

Substituting Eqs. (16) in (24) it gets the form as in Eq. (25)

$$\frac{d\varepsilon}{dK_i} = \frac{\frac{G_p}{S} r}{1 + G_p K_p + \frac{G_p K_i}{S}} - \frac{\frac{G_p}{S} Y_p}{\left(1 + G_p K_p + \frac{G_p K_i}{S}\right)} \tag{25}$$

The Eq. (26) can be obtained by rearranging the above equation

$$\frac{d\varepsilon}{dK_i} = \frac{\frac{G_p}{S} E}{1 + G_p K_p + \frac{G_p K_i}{S}} \tag{26}$$

Under the usual approximations it follows that the parameters fedback relate to the ideal value and the plant becomes the model reference.

$$den\left(\frac{G_p}{1 + G_p K_p + \frac{G_p K_i}{s}}\right) = s^2 + a_{m1} + a_{m0} \tag{27}$$

Applying MIT rules for adjusting the parameters θ_1, θ_2, and equating it to K_P and K_i. in order that the Eq. (10), gives the adjustment parameters

$$\frac{d\theta_1}{dt} = \frac{dk_p}{dt} = -\lambda\varepsilon\frac{d\varepsilon}{dk_p} = -\left(\frac{\lambda_p}{s}\right)\varepsilon\left(\frac{s}{a_0 s^2 + a_{m1} s + a_{m2}}\right)e \tag{28}$$

$$\frac{d\theta_2}{dt} = \frac{dk_i}{dt} = -\lambda\varepsilon\frac{d\varepsilon}{dk_i} = -\left(\frac{\lambda_p}{s}\right)\varepsilon\left(\frac{1}{a_0 s^2 + a_{m1} s + a_{m2}}\right)e \tag{29}$$

Also the second order transfer function of the model reference is given by the Eq. (30)

$$H_m(s) = \frac{16}{S^2 + 4S + 16} \tag{30}$$

4 Simulation

The exercise compares the performance of BLDC motor speed control using MRAC with the normal PI controller through MATLAB based simulation as reflected in Fig. 3. The Table 1 shows the parameters of BLDC motor.

Table 1. Simulation parameter

S. no	Name of parameter	Rating of parameter
1.	Stator resistance (Rs)	2.8750 ohms
2.	Stator inductance (Ls)	8.5e−3 H
4.	DC voltage (Vdc)	146.6077 volts
5.	Rotor flux (λ)	0.175
6.	Moment of inertia (J)	0.0008 kgm^2
7.	Friction (B)	0.001 Nm/rad
8.	Poles (P)	4
9.	Load Torque (TL)	1.4 NM/RPM
10.	Speed (N)	1500 RPM

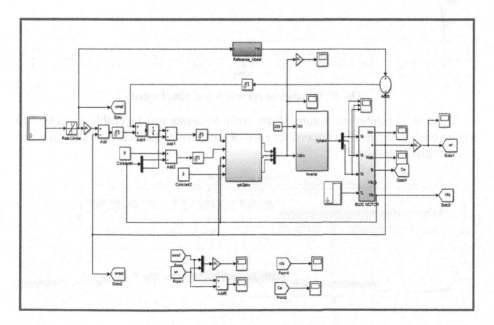

Fig. 3. Simulation model of BLDC-MRAC

The Fig. 4 explain the steady state characteristics of the BLDC motor and shows the ability of the control algorithm to reach the reference speed smoothly.

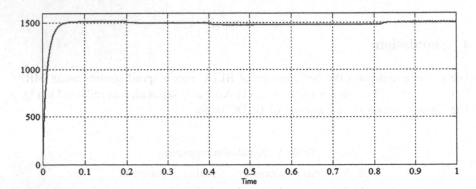

Fig. 4. Speed characteristics of BLDC motor

The Fig. 5 depicts the variation of torque in the motor being subjected to sudden changes in load intervals of 100 percent at 0.2 sec, 200 percent at 0.4 sec, 100 percent at 0.82 respectively and in each case brings out the feature to settle at the new steady state operating values.

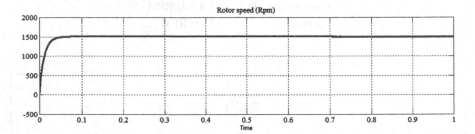

Fig. 5. Comparison ref speed and actual speed

The Fig. 6 exhibit the nature of the control mechanism to enable the speed to restore back to the reference speed for each of the three sudden changes in load at the chosen time intervals.

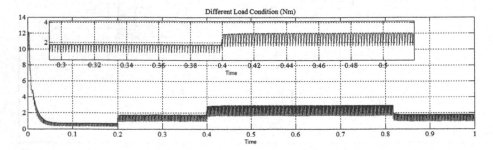

Fig. 6. Torque characteristics

The Fig. 7 displays the Speed characteristics of BLDC motor and brings out the ability to different load condition. Besides the response relates to rejecting the sudden change in torque of 200 percent and establishes the ability of control algorithm is being able to settle at the chosen reference speed even at the new operating point.

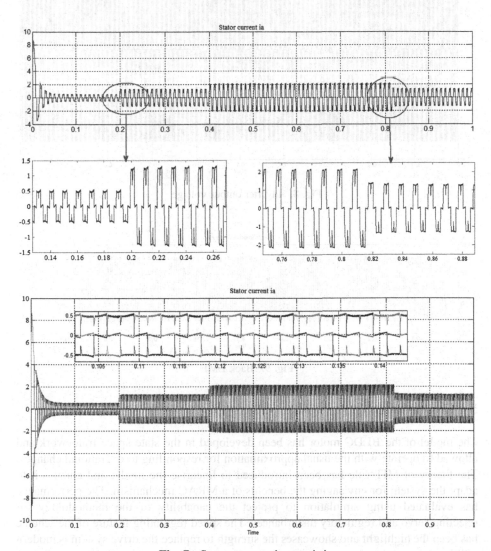

Fig. 7. Stator current characteristics

Figures 8 and 9 displays the inverter output voltage characteristics and the EMF characteristics of BLDC motor has respectively at the chosen operating load.

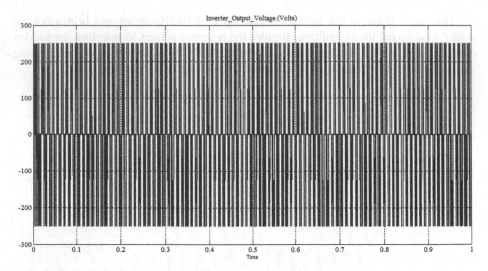

Fig. 8. Inverter output voltage

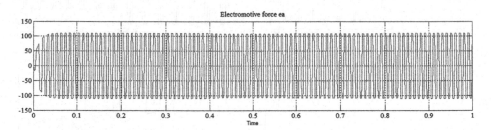

Fig. 9. Back EMF

5 Conclusion

The model of the BLDC motor has been developed in the state space framework and allowed to operate with the usual approximation for responding to operational changes. The theory of MIT rule has been introduced to linearise the model and enables it to adapt the system for envisaging the benefits of a MRAC mechanism. The performance has evaluated using simulation to project the capability of the methodology for rejecting servo and regulatory disturbances. The speed regulating feature of the scheme has been the highlight and showcases the strength to replace the drive system is modern utilities.

Acknowledgement. This publication is an outcome of the R&D work undertaken project under the Visvesvaraya Ph.D Scheme of Ministry of Electronics & Information Technology, Government of India, being implemented by Digital India Corporation.

References

1. Ansari, U., Alam, S., un Nabi Jafri, S.M.: Modeling and control of three phase BLDC motor using PID with genetic algorithm. In: UK Sim 13th International Conference on Modeling and Simulation, pp. 189–194 (2011)
2. Al-Mashakbeh, A.S.O.: Proportional integral and derivative control of brushless DC motor. Eur. J. Sci. Res. **35**, 198–218 (2009)
3. Krishnan, R.: A novel single-switch-per-phase converter topology for four-quadrant PM brushless DC motor drive. IEEE Trans. Ind. Appl. **33**(5), 1154–1162 (1997)
4. Kim, I., Nakazwa, N., Kim, S., Park, C., Yu, C.: Compensation of torque ripple in high performance BLDC motor drives. Control Eng. Pract. **18**, 1166–1172 (2010)
5. Aghili, F.: Adaptive reshaping of excitation currents for accurate torque control of brushless motors IEEE Trans. Control Syst. Technol. **16**(2), 356–365 (2008)
6. Feng, Z., Acarnley, P.P.: Extrapolation technique for improving the effective resolution of position encoders in permanent-magnet motor drives. IEEE/ASME Trans. Mech. **13**, 410–415 (2008)
7. Emura, T., Wang, L.: A high resolution interpolator for incremental encoders based on the quadrature PLL method. IEEE Trans. Ind. Electron. **47**, 84–90 (2000)
8. Nasr, C., Brimaud, R., Glaize, C.: Amelioration of the resolution of a shaft position sensor. In: Proceedings of Power Electronics Applied Conference, Antwerp, Belgium, vol. 1, pp. 2.237–2.241 (1985)
9. Aström, K.J., Wittenmark, B.: Adaptive Control. Addison-Wesley Publishing Company, Reading (1989)
10. Crnosija, P., Ban, Z., Bortsov, Y.A.: Implementation of modified MRAC to drive control. In: 9th European Conference on Power Electronics and Applications, Graz (2001)
11. Watanable, H., et al.: A sensorless detecting strategy of rotor position and speed on permanent magnet synchronous motor. IEEJ Trans. **113D**(11), 1193–1200 (1990)
12. Ohtani, T., et al.: Approach of vector controlled induction motor drive without speed sensor. IEEJ Trans. **10**(2), 199–219 (1987)
13. Kulkarni, A.B., Ehsani, M.: A novel position sensor elimination technique for the interior permanent magnet synchronous motor drive. IEEE Trans. Ind. Appl. IA-28 **1**, 144–150 (1992)
14. Uezato, K., et al.: Vector control of synchronous reluctance motors without position sensor. In: Proceeding of Japan Industry Applications Society Conference, pp. 59–64 (1994)
15. Li, W., Venktesan, R.: A new adaptive control scheme for the indirect vector control system. In: IEEE Conference Proceeding of IAS (1992)

Power Converter Interfaces for Wind Energy Systems - A Review

R. Boopathi[✉] and R. Jayanthi[✉]

Electrical Engineering, Annamalai University, Annamalai Nagar, Tamil Nadu, India
rboopathiyadav@gmail.com, rrjay_pavi@yahoo.co.in

Abstract. The paper attempts to bring out a wide-ranging analysis the performance of electronic converter interfaces that assuage to meritoriously control a
Wind Energy Conversion System (WECS). The system seems to gather significance in an effort to exploit the natural resources and offer support to associate
the generation demand gap. The effective practice of the wind power augurs to
carry down the cost of generation and encourage to satisfy the needs of the utilities. The WECS exist in the form of the variable speed generators and the types
include the Permanent Magnet Synchronous Generator (PMSG), Doubly-fed
Induction Generator (DFIG), Synchronous Generator (SG) and Wound Rotor
Induction Generator (WRIG). The DFIG invites a greater attention owing to its
smaller requirement of the partially rated converters and enables to regulate the
system through a compliance for the transfer the real and reactive power.

Keywords: Power electronics converters · Wind turbines · Power quality
Grid connection

1 Introduction

The wind energy appears to be a fast developing sector surrounded by other renewable
sources and be obligated its merits to being clean, green, in naturally available, low cost
and particularly beneficial to the rural areas. However, it is intermittent and requirement
of initial asset, broadcast aspect, and property area all finds a large amount of cost.

The WECS shown in Fig. 1 relates to a physical system with three main components
where the first one shows the rotor connected in blades. As wind goes through blades,
it allows the rotor to rotate and therefore generates mechanical energy. The next component describes the transfer energy from the rotor to generator and the electric generator
establishes the third to convert the mechanical energy to electric power.

The WECS technique involves, wind tower and a rotor linked with blades then the
hub. The furthermost designer horizontal-axis (HAWTs) comprises of 3-blades, located
windward of the stronghold and contains exactly the mechanical decelerating system,
generators, controlling systems and yaw system [1].

The most popular horizontal system converter with two or more blades consist of a
yaw machinery that turns the rotor blades to wind sides. The next vertical axis converter
consists of blades positioned straight up serves to capture wind freely from wind directions as shown in Fig. 2. However due to lower effectiveness, complex to repairs and

© Springer Nature Singapore Pte Ltd. 2018
I. Zelinka et al. (Eds.): ICSCS 2018, CCIS 837, pp. 776–788, 2018.
https://doi.org/10.1007/978-981-13-1936-5_79

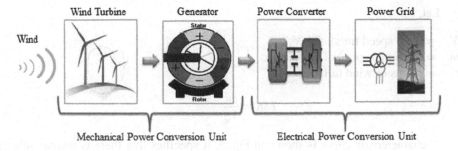

Fig. 1. Main elements of WECS

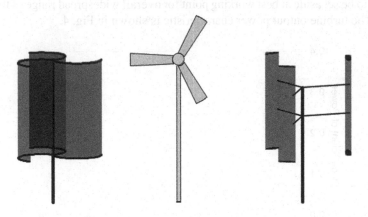

Fig. 2. Various vertical axis winds turbines

huge property occupation, and the application of this converter is deteriorating through the past few spans. The collected power from a wind turbine is particular to every system is given by equation 1 [2]:

Let,

V_w = Wind Speed (m/s),
A = Area of Swept Turbine (m^2),
P_t = Turbine Power (W),
ρ = Air Density (kg/m^3) and
C_P = Performance Coefficient.

$$P_t = \frac{1}{2}\rho A C_p V_w^3 \tag{1}$$

Consequently, if constant value of air density, swept area and velocity of wind, the power C_P of turbine is used toward determine the output energy value of wind. The aerodynamic physical appearance of blades is more supportive to healthier power coefficient. The turbine coefficient remains predisposed by TSR. The equation is shown below (2),

Let,

V_w = wind speed (m/s),
ω = rotational speed of turbine and
r = radius of wind turbine.

$$TSR = \frac{\omega r}{V_w} \tag{2}$$

A characteristic curve is shown in Fig. 3, it specifies that there is unique ratio at which the wind turbine performance is greatest. To attain maximum energy, the TSR is required to be set aside at best working point for overall widespread range of the turbine speeds. The turbine output power characteristic is shown in Fig. 4.

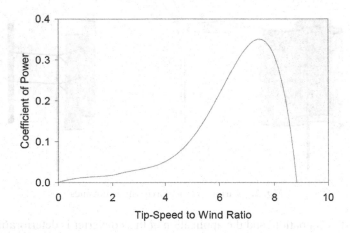

Fig. 3. Typical co-efficient of power curve

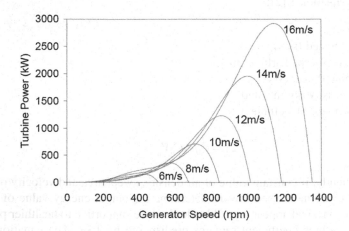

Fig. 4. Turbine output power characteristic

2   Power Converter Interfaces

The power electronic device remains to undergo progress with evolution of self-commutated Insulated Gate Bipolar Transistors (IGBTs), Metal-Oxide-Semiconductor Field-Effect Transistor (MOSFETs) and some more advanced semi-conductor devices are implemented in modern power converter techniques [2]. When interruption of line voltage and the line current carry ability of device is continuing to increase and the significant study strives to explore to alter some other material (silicon carbide), with an emphasis on increasing the value of power density in the power converters.

Besides there exists the variety of power converter topologies to agree with application needs and the directions of power flow. The thyristor converters form the grid commutated power converters through maximum power capability of 6, 12 or more switching pulses. It ingests the inductive power and finds use in maximum voltage and high power systems [3]. The self-commutated converter methods typically operate using Pulse Width Modulation (PWM) method and enable the transfer of mutually active and reactive power. In PWM converters, if switching frequency is maximum, then they will produce harmonics and inter-harmonics up to few kHz and reduce this harmonic spectrum minimum range filters may be used at the inverter output.

In WECS different promising power electronics techniques are available to develop the vibrant and quality output [4]. The major constraint in energy distribution zone transmits to the accurate governor in real and reactive power flow for maintaining system voltage stability. It engages modern power converter and satisfies its ability of converting the electrical energy from ac to dc and reverse process also. The different types of three-phase converter configurations ensemble the conversion process and facilitate to achieve the requirements.

2.1   Voltage Source Inverter (VSI)

The PWM - VSI find an extensive scope in the operation of electrical drives owing to their voltage-buck conversion capability characteristics. Wang, Wu, Dai, Zargari, and Xu brought in a "Low Cost Current Source Converter Solutions for Variable Speed Wind Energy Conversion Systems" in the year 2011 [5] and illustrated that the diode converter with CSI-PWM, it can attain full-range real power regulator but the minimum range of reactive power control over the PWM VSI based topologies, though it slightly increases the rectifier cost the scheme is shown in Fig. 5.

The efforts of Arunkumaran, Raghavendra Rajan, Ajinsekhar, Tejesvi and Sasikumar, offer a "Comparison of Space Vector PWM (SVPWM) and SVM Controlled Smart VSI fed PMSG" containing a 3ϕ diode bridge converter and 3ϕ IGBT inverter for Wind Generation System. The Sine PWM system serves to reduce the harmonics with an increase in the switching losses and the SVM method enables to minimize Harmonic distortion to be at 43.23% and to control the switching losses [6].

Davidson, Gitau, Adam and Hamatwi have introduced "Modeling and Control of Voltage Source Converters for Grid Integration of a Wind Turbine System", in 2016 a pitch angle of blade regulator function on wind model, a rotor speed of field-oriented regulator useful in rectifier for extraction of maximum power in wind system, and vector

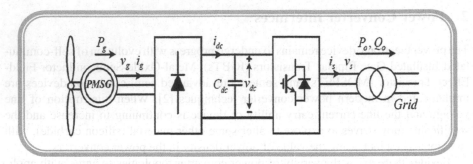

Fig. 5. Diode rectifier and PWM VSI for WECSc

based dc link voltage control implemented on the grid-side inverter to maintain the dc-link voltage in constant and to make sure unity power factor.

2.2 Back to Back Converter

"A New Five-Phase to Three-Phase Back-to-Back Current Source Converter Based Wind Energy Conversion System" was introduced in the year 2013 by Massoud, Ahmed, Abdel-Khalik and Elgenedy. The design in WECS experiences challenges on account of amplified perception of wind into the grid due to voltage and frequency variabilities, ripples of generator torque and grid faults to power converter.

While the predictable SVPWM is implemented to regulate the real and injected reactive power in 3ϕ grid side, then 5ϕ generator side converter is well-ordered using SPWM for keeping 1000A of dc-link current. The "Modified back-to-back current source converter" was suggested in 2015 by Holliday, Adam, Abdelsalam, and Williams came up with novel CSC-BTB converter to operate the wind energy conversion system. Though the BTB CSCs suffer from sag voltages are regularly experience via switches in commutation, there is zero switching loss in the inverter side converter. The main advantages accrue to being simple, offer to control is less complex, switching frequency is minimized and controllable grid having maximum power point tracking to collect real power [7] block diagram of modified back to back converter is shown in Fig. 6.

Sumina, Sacic, Mrcela, and Barisa have thought of "A wind turbine two level back-to-back converter power loss study" in 2016, where the WECS is coupled in grid with the support of converter, *LC* filter with transformer (step-up). In his proposed method, the converter is used to produce the maximum fundamental output voltage with lower harmonics (THD) and to improve the overall system performance [8]. Then the *LC* filter and step-up transformer are used to minimize the losses in converter switches and as well as the grid voltage/current THD.

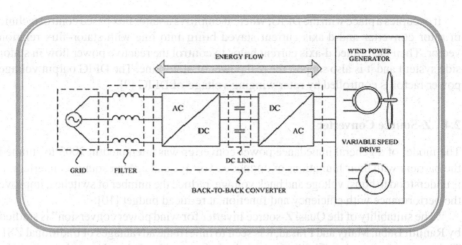

Fig. 6. Block diagram of modified back to back converter

2.3 Matrix Converter

The matrix converter is used for exchanging the variable ac from a fixed ac as seen from the Fig. 7. The bi-directional switches are arranged to any input point possibly will be connected to any output point at any time. The double SVPWM is employed to control the input current and output voltage of the system. The distinct advantages augur due to absence of massive power storing devices and dc-link. However, a larger number of switches is portrayed to be a serious disadvantage.

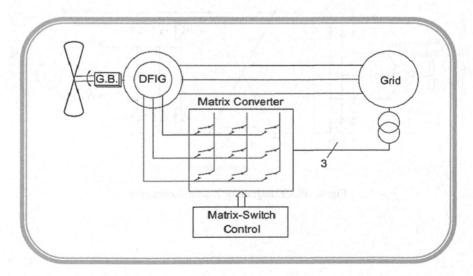

Fig. 7. Block diagram of matrix converter

It occupies a place with the DFIG, where it employs a stator flux based control technic in rotor converter and d-axis current stayed bring into line with stator–flux relation vector. Then the refined d-axis current helps to control the reactive power flow in stator side system and it is also support the real power of stator line. The DFIG output voltage power factor is controlled by rotor winging voltage of DFIG [9].

2.4 Z-Source Converter

The model of a general impedance power converter was attempted in 2003 to stunned the barriers with their limitations of outmoded voltage and current source converters. It is understood to boost voltage and buck voltage, reduce the number of switches, improve the performance with efficiency and function at reduced budget [10].

"The suitability of the Quasi Z-source Inverter for wind power conversion" is studied by Ranjith Babu, Maity and Prasad, it is seen to inherit the advantages of traditional ZSI. The two-port network in Fig. 8 involves the inductors (L_1, L_2) and capacitors (C_1, C_2) are allied in the form of X-shape to arrange for produce quality an impedance coupling. The system is seen to operate with a controller for dynamic and static control of converter through a change in the switching cycle and enhancement of load voltage terminal at the mandatory range and its achieves stability. Though it's low cost and high efficiency claims its suitability for the wind power generation system [11], its unidirectional nature makes it unpopular in DFIG system.

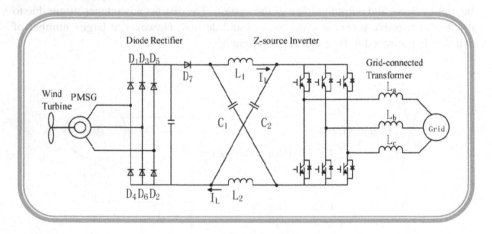

Fig. 8. Block diagram of Z-source converter

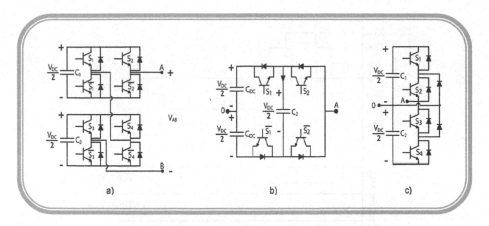

Fig. 9. Block diagram of multilevel converters

2.5 Multilevel Converters

The theory of multilevel converter technic was suggested in past few years; more than a few multilevel converter technics are patented. It is seen to involve the switching devices in series with low level of dc voltage sources to succeed maximum voltage by spending minimum amount of voltage to devices (switches).

The increasing interest in multi-level power converters goes to evolve a mechanism for feeding the wind turbine generator system directly from the prevailing 3-ϕ power grid. There are few different varieties of the multi-level converters frequently executed in maximum energy source and ideal voltage uses. The converters are,

(a) Series Connected H-Bridge (SCHB) converter method,
(b) Flying Capacitor (FC) converter method and
(c) Neutral Point Clamped (NPC) converter method.

as shown in Figs. 9(a), (b) and (c) respectively.

Accumulation to $(N - 1)$ dc link capacitors, the N-level flying capacitor converter have need of secondary capacitors in each power line $[(N^2 - 3N + 2)/2]$, it rises capacity, load, difficulty, and rate of the converter and reduce the performance and lifespan (duration). The NPC method practices, capacitors are connected in series because to split the DC bus voltage keen on static voltage stages. It engages the prospect to bond with neutral line to intermediate power of dc-link point, to form the bi-direction connection in the power converter to reduce the earth drip currents [12].

The bi-direction connection of the NPC converter established WECS generation structure illustrated in Fig. 9(c), it is found to expand the static and transitory action of bi-directional NPC method [13]. On other hand, the amount of holding diodes essential in quadratic ally correlated with number of levels. Where, N is adequately great, more diodes are essential to produce the circuit impossible to execute and more diodes required in each line is $(2N - 4)$ the output voltage waveform of multilevel inverter is shown in Fig. 10.

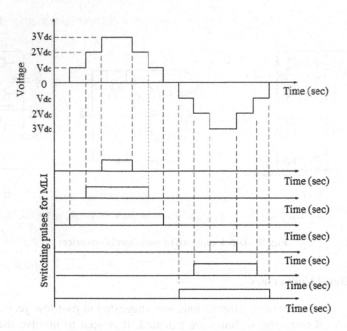

Fig. 10. Output voltage waveform for MLI

2.6 Tandem Converter

The tandem converter contains, the Current Source Converter (CSC) with principal converter and bi-directional PWM-VSI secondary converter. Meanwhile, tandem converter brings about four well-regulated inverters as exposed in Fig. 11 more than a few units of choice is envisaged which qualify sinusoidal input/output Signals.

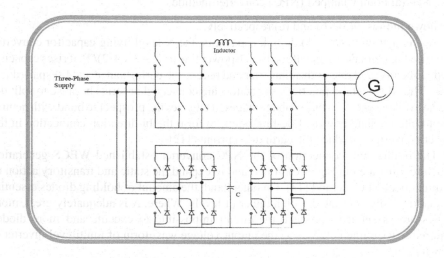

Fig. 11. Block diagram of tandem converter

Dissimilar to the principal converter, the minor converter consumes toward control at a high switching frequency. Benefits of this power converter are, the primary converter requires minimum range of switching frequency, and the secondary converter consume less switching current. It's specified that, the tandem inverter minimize up-to 70% of switching losses. The tandem converter has higher conduction losses in comparison of all other equivalent VSI method but performance efficiency of this converter possibly will increased.

In addition, the main function that relates to reimburse the distortion current both in the source side and the load side converter possibly shall perform similar to an active resistor for providing a restraining action to the main inverter in commercial load circumstances. The increase in the cost and complexity boils down to the maximum volume of apparatuses and measuring device inherent and presented as an obstacle in using the tandem converter is required. The use of the CSI as source side converter shows that, solitary 0.866% of utility voltage is employed and hence the production currents for the tandem converter essentially high to reach the similar energy is observed from the curve as shown in Fig. 12.

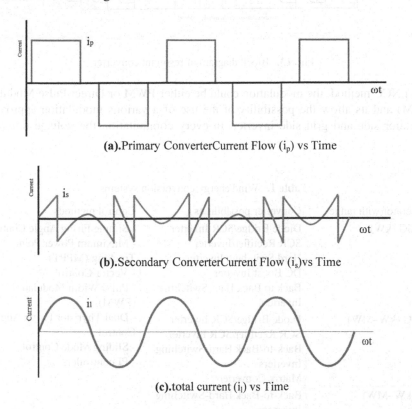

(a).Primary ConverterCurrent Flow (i_p) vs Time

(b).Secondary ConverterCurrent Flow (i_s)vs Time

(c).total current (i_l) vs Time

Fig. 12. (a). Primary converter current flow (i_p) vs Time, (b). Secondary converter current flow (i_s) vs Time, (c). Total current (i_l) vs Time

2.7 Resonant Converter

The resonant converter well-looked in this segment, this type of converter model is designated as Natural Clamped Converter (NCC), its shown in Fig. 13. In this converter, the inductance is coupled in middle of the system to retain the resonance and controls the current from the supply grid by make better dc link voltage.

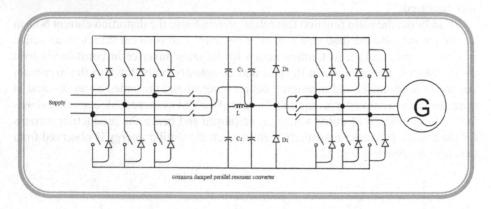

common damped parallel resonant converter

Fig. 13. Block diagram of resonant converter

In NCC method, the modulation could be either PWM or Direct Pulse Modulation (DPM) and its allow the possibility of the use of a various modulation approach in generator side and grid side inverter. In every commutation the voltage difference/

Table 1. Wind energy conversion systems

Generator with range	Converter possibilities	Control methods
PMSG (kW)	Diode Bridge/SCR Inverter SCR Rectifier/Inverter Hard Switching Inverter DC Boost Inverter Back to Back Hard Switching Inverters	- Simple Firing Angle Control - Maximum Power Point Tracking (MPPT) - Vector Control - Pulse Width Modulation (PWM)
DFIG (kW–MW)	Diode Bridge/SCR Inverter SCR Rectifier/SCR Inverter Back-to-Back Hard-Switching Inverters Matrix Converter	- Dual Thyristor Firing Angle Control - Sliding Mode Control - PI controllers
IG (kW–MW)	Back-to-Back Hard-Switching Inverters	
SG (kW–MW)	Diode Bridge/DC Boost/Hard- Switching Inverter Back-to-Back Hard-Switching Inverters	

inequality is established in the main capacitors (C_1 and C_2), then any one of this capacitor will discharge their stored energy. In the situation, the irregular commutations will affect the switching characteristics in very tiny variations.

The various converter topologies are offered for WECS is presented in the Table 1. To increase the efficiency of the converters, the advanced modern controllers are required. Then the overall cost of the system is also will increase and the complication of converter are also increases. If to minimize the control complexity of the inverter, the boost inverter is coupled to the system but the cost is some more high. Now, to expand the benefits of WECS, a finding the middle ground between efficiency and cost is essential.

3 Conclusion

A brief evaluation of the several WECS are articulated with possible converter topologies that can be implement in different combination of PMSG, DFIG, IG and SG. An insight has been thrown on the suitability of the different control schemes and their benefits detailed. The control methods have been focussed with try to achieve extreme output power pass from the wind turbine to the utility grid. The all those efforts are focused on to produce efficient control schemes for the converter and cost minimization in the perspective to address economically feasible solution and overcome aggregate eco-friendly problems. The wind generation has been seen to grow at startling level in earlier era and motivation is carry on to experience leaps and bounds with the advances in the power electronic technology. The drive has been ordained to nurture an enviable progress and ensure the wind energy contributes to arrive at an energy balance economy for the nation.

Acknowledgement. This publication is an outcome of the R&D work undertaken project under the Visvesvaraya Ph.D Scheme of Ministry of Electronics & Information Technology, Government of India, being implemented by Digital India Corporation.

References

1. Chen, Z., Guerrero, J.M., Blaabjerg, F.: A review of the state of the art of power electronics for wind turbines. IEEE Trans. Power Electr. **24**(8), 1859–1875 (2009)
2. Song, S.-H., Kang, S., Hahm, N.-K.: Implementation and control of grid connected AC-DC-AC power converter for variable speed wind energy conversion system. In: Proceedings of the IEEE, pp. 154–158 (2003)
3. Mohammad, S.N., Das, N.K., Roy, S.: Power converters and control of wind energy conversion systems. In: Proceedings of International Conference on Electrical Information and Communication Technology (EICT) (2013)
4. Ackermann, T., Soder, L.: Wind energy Technology and current status: a review. Renew. Sustain. Energy Rev. **4**, 315–374 (2000)
5. Wang, J., Wu, B., Dai, J., Zargari, N.R., Xu, D.: Low cost current source converter solutions for variable speed wind energy conversion systems. In: Proceedings of the IEEE, pp. 825–830 (2011)

6. Arunkumaran, B., Raghavendra Rajan, V., Ajinsekhar, C.S., Tejesvi, N., Sasikumar, M.: Comparison of SVPWM and SVM controlled smart VSI fed PMSG for wind power generation system. In: Proceedings of the IEEE, pp. 221–226 (2014)

7. Abdelsalam, I., Adam, G.P., Holliday, D., Williams, B.W.: Modified back-to-back current source converter and its application to wind energy conversion systems. IET Power Electr. 8(1), 103–111 (2015). ISSN 1755 - 4535

8. Mrcela, I., Sumina, D., Sacic, F., Barisa, T.: A wind turbine two level back-to-back converter power loss study. In: Proceedings of the IEEE, pp. 308–314 (2016)

9. Huang, K., He, Y.: Investigation of a matrix converter-excited brushless doubly-fed machine wind-power generation system. In: Proceedings of the IEEE, pp. 743–748 (2003)

10. Peng, F.Z.: Z-Source inverter. IEEE Trans. Ind. Appl. 39(02), 504–510 (2003)

11. Ranjith Babu, V., Maity, T., Prasad, H.: Study of the suitability of recently proposed Quasi Z-source Inverter for wind power conversion. In: Proceedings of the IEEE - International Conference on Renewable Energy Research and Applications, pp. 837–841 (2014)

12. Alepuz, S., Calle, A., Busquets-Monge, S., Kouro, S., Wu, B.: Use of stored energy in PMSG rotor inertia for low voltage ride through in back to back NPC converter based wind power systems. In: Proceedings of the IEEE, pp. 01–10 (2011)

13. Islam, M.R., Guo, Y.G., Zhu, J.G.: Performance and cost comparison of NPC, FC and SCHB multilevel converter topologies for high-voltage applications. In: Proceedings of the IEEE, pp. 1–6 (2011)

Salp Swarm Optimized Multistage PDF Plus (1+PI) Controller in AGC of Multi Source Based Nonlinear Power System

Prakash Chandra Sahu[✉], Ramesh Chandra Prusty, and Sidhartha Panda

Department of EE, VSSUT, Burla, Odisha, India
Prakashsahu.iter@gmail.com

Abstract. This research article focuses on Automatic Generation Control (AGC) with application of a multistage controller constructed by adding (1+Proportional Integral) with Proportional Derivative with Filter (PDF) in multi source type two area non-linear electrical network. To make system non-linearity and realistic a Generation Rate Constraint (GRC) of 3% is considered in each area. In this multi area power system each area comprises a Thermal unit, Hydro unit and a Gas power generation unit. For obtaining globally optimal solution a Salp Swarm Optimization (SSO) algorithm is applied to tune the gain parameters of above proposed multistage controller. For comparative analysis some controllers like PID and PI controllers are implemented individually apart from this proposed controller. Superiority analysis of above nature inspired SSO algorithm is carried out by comparing it with Cuckoo search algorithm (CSA), Genetic Algorithm (GA) and Particle Swarm Optimization (PSO) algorithm. While optimizing different parameters of system (if any) and above controller, Integral of Time multiplied Absolute Error (ITAE) has been implementing as objective function.

Keywords: Automatic Generation Control (AGC)
Generation Rate Constraint (GRC) · Proportional Derivative with Filter (PDF)
ITAE · Salp Swarm Optimization (SSO)

1 Introduction

The reliable and economic operation of an interlinked electrical network depends on its network frequency and scheduled power exchange. Accordingly when system frequency and tie-line power exchange are kept with their nominal values, there will be a reliable and scheduled operation of an interlinked power network. To achieve better values of area frequency with power exchange research in the area of AGC [1, 2] is carried out. So AGC keeps area frequency and tie-line power close to their nominal values. So research in AGC of multi-area [3, 4] with both isolated and interconnected system is carried out successfully. Based upon the literature in multi-area system by Elgerd and Fosha research in two area, three area and five area is carried out tremendously [5]. Comparative analysis is done among different classical controllers [6] like PID and PI controllers. In this research article a multistage type filter based controller is implemented to keep deviation in net exchange and network frequency within their actual

© Springer Nature Singapore Pte Ltd. 2018
I. Zelinka et al. (Eds.): ICSCS 2018, CCIS 837, pp. 789–800, 2018.
https://doi.org/10.1007/978-981-13-1936-5_80

values. In this proposed controller filter component is introduced to improve the stability of different transient based responses of the network. Though classical methods are simple but it consumes more time and also gives suboptimal results. To avoid these drawbacks now a day's newly developed deterministic algorithms have been implemented for optimizing controller gain parameters. The genetic algorithm (GA) [7] and Particle Swarm Optimization (PSO) [8] have been implemented for optimizing the controller gains. Due to advanced in research some of the deficiencies are found in GA and also local optimum trapped problem in PSO is identified. Due to some drawbacks in above two optimization techniques, to get better optimum values bacterial foraging optimization (BFO) technique have been implemented by researcher in present optimization scenario. BFO [9] having better performances like convergence and low operating time in comparing with GA and PSO. LC Saikia et al. developed another met heuristic algorithm called Firefly algorithm (FA) [10] and applied in the area of AGC for better stabilization. Besides this in this article a proposed newly developed algorithm based on nature called Salp Swarm Optimization (SSO) [11] algorithm is implemented in the area of multi-area AGC for improving dynamic response of system. This method is most robustness and accurate for solving multilevel optimization problem. For superiority analysis the SSO technique is compared with GA, PSO and cuckoo search algorithm. The system investigated for this research is a multi-source type two area power system in which each area comprises thermal, hydro and gas unit [12]. In this proposed model each unit has different stages and are expressed with single order transfer function and is shown Fig. 1.

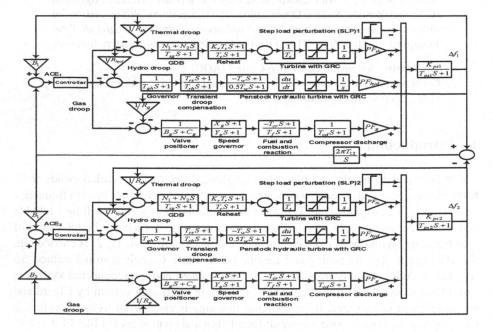

Fig. 1. Two area multi-source based power network

2   Proposed Controller Structure

2.1   (1+PI) with PDF Controller

Controller has most important role in the area of stabilization. The most simple and conventional controllers like P, PI and PID have been favoured among all researchers in the area of stabilization due to their simpler structure and easy of implementation. Transfer function of simple PID controller is expressed by

$$TF_{PID} = K_P + \frac{K_I}{s} + K_D s \tag{1}$$

In PID controller the integral gain parameter tries to reduce the steady state error of responses. Though Integral part decreases steady state error but it makes system more sluggish and takes more time to settle to transient type responses. So to provide both benefits like minimization of steady state error and increase speed of response a multistage (two stage) configuration type controller is proposed which holds first block as PD section and second block as PI section [13]. Here to satisfy high frequency oriented disturbance single order derivative filter is used in this controller. Figure 2 gives detailed structure of multistage controller.

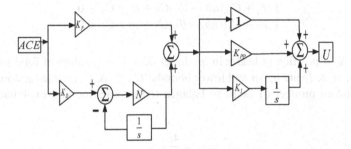

Fig. 2. Block diagram of multistage PDF plus (1+PI) controller

The above controller is expressed with equivalent transfer function

$$TF multi\ stage\_PID = \left(K_P + K_D \left(\frac{N_s}{N + s} \right) \right) \left(1 + K_{PP} + \frac{K_I}{s} \right) \tag{2}$$

3   Objective Function and Algorithm

The fitness or objective function is required in optimization research scenario which helps to develop improved stability of network. The work selects ITAE as fitness function due to its superiority over some other fitness function. In the outline of improvement based controller, the objective function [14] is characterized in view of different execution conditions and constraints which are Integral of Time multiplied Absolute Error

(ITAE), Integral of Time multiplied Squared Error (ITSE), Integral of Absolute Error (IAE) and Integral of Squared Error (ISE)

$$J = ITAE = \int_0^T \left(|\Delta F_i| + |\Delta P_{tie\ i-j}| \right) t\ dt \tag{3}$$

Where ΔF_i is deviation in frequency of ith area, $\Delta P_{tie\ i-j}$ is the net power exchange deviation between i[th] and j[th] area and T = simulation time.

3.1 Salp Swarm Optimization (SSO) Algorithm

The population of salp is splitted in two categories i.e. leader and follower. In salp chain the leader remains at front of the chain to which other salps follow (follower). Similar to other type of swarm based optimization technique [15], the salp occupies a place and is derived in n-dimensional search space and n stands for number of variables. There is a matrix 'X' which stores all the position of salps. In search space the main target of salps is of food location and is noted by 'F'. The below equation helps to update the position of leader

$$X_j^1 = \begin{cases} F_j + C_1((ub_j - lb_j)C_2 + lb_j) & C_3 \geq 0 \\ F_J - C_1((ub_j - lb_j)C_2 + lb_j) & C_3 < 0 \end{cases} \tag{4}$$

Where X_j^1 = position of leader in *jth* dimension; F_j = position of food source in jth dimension, ub & lb are upper and lower bounds; C_1, C_2 & C_3 are the random numbers.

The random number c1 helps to balance the exploration and exploitation and is equated as

$$C_1 = 2e^{-(\frac{4l}{L})^2} \tag{5}$$

Where l = current iteration; L = max number of iteration.

Below equation helps to revise the position of followers and is formed according to Newton's Law's of motion.

$$X_j^i = \frac{1}{2}at^2 + V_0 t \tag{6}$$

Where $i \geq 2$; X_j^i = Position of ith follower salp in j^{th} dimension at time t sec; V_0 = initial speed

$$a = \frac{V_{finital}}{V_0}; v = \frac{x - x_0}{t}; \tag{7}$$

Considering V_0 and discrepancy between iteration the modification of above equation will be

$$X_j^i = \frac{1}{2}(X_j^i + X_j^i - 1), i \geq 2 \tag{8}$$

Finally simulation of salp chain is occurred with the above equation.

Pseudo code of SSO Algorithm

Initialization of salp population   Xi (I = 1,2,3.........n), ub & lb
While (condition not satisfied)
Fitness of each salp is calculated
F = Best salp
Update equation (5)
for each salp (Xi)
if (i == 1
update leading salp by eq(4)
else
update follower by eq(8)
end
end
Amend salps according to upper bound and lower bound
end
return F

4 Simulation Results and Analysis

For this research work the model was run in simulink of MATLAB 2014b environment and the coding was written in .m file. The optimized values of SSO based proposed PDF (1+PI) controller and other controller gains obtained are provided in Table 1.

Table 1. Optimal value of gain parameters of different controllers with SSO Algorithm

Area genco.		Controller gains		
		PDF (1+PI)	PID	PI
		K_P, K_{PP}, K_I, K_D, N	K_P, K_I, K_D	K_P, K_I
Area-1	Thermal	1.9202;1.0155;0.884; 0.5422;62.540	−1.664;0.5676; 1.0880	1.8988; 1.2202
	Hydro	1.0550; −0.9876;1.0000; 0.8656;71.0980	0.9822;1.980; −1.0050	1.6060; 1.0000
	Gas	1.6566; 1.0920; −0.9898;1.0200; 80.9082	1.7660; 1.0020; 0.2044	1.0262; −1,6740
Area-2	Thermal	−1.8650; −0.2288;0.1222;1.0540; 74.0880	−1.9840;1.0882; −0.865	−1.004; 1.5420
	Hydro	1.0064;1.8650; 1.0000; 1.4544; 38.8034	1.0068; −1.000; 1.7620	1.9090; 1.0440
	Gas	1.9880; 1.3454; 1.6588; 1.3800; 42.2020	1.8088; −1.080; 1.5480	1.8200; 1.6720

Figure 3 Deviation in dynamic responses due to slp of 2% at area 1 only with different controllers (a) frequency change of area.1 (b) frequency change of area.3 (c) Net power exchange variation between area.1 and area.2. Figure 4 Different responses are compared and are produced by different optimization techniques due to SLP of 2% at area1 only, (a) frequency change of area.1 (b) frequency change of area.3 (c) Net power exchange variation between area.1 and area.3, Fig. 5(a) Net power exchange variation between area.1 and area.3 due to different magnitude of SLP at different positions for sensitive analysis, Fig. 5(b) frequency change of area.1 due to ±50% and ±25% of normal loading for sensitive analysis. Figure 6(a) Random Load Pattern (RLP), Fig. 6(b) dynamic response of change in frequency of area.1 due to RLP at area.1 only. Figure 7 Convergence curve to show superiority of SSO algorithm.

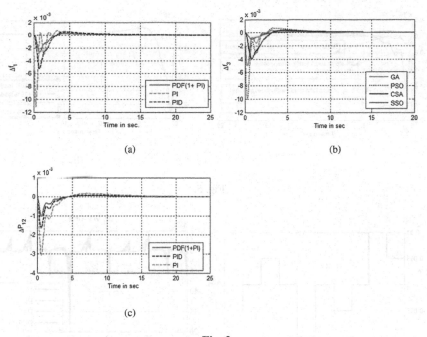

(a)

(b)

(c)

Fig. 3.

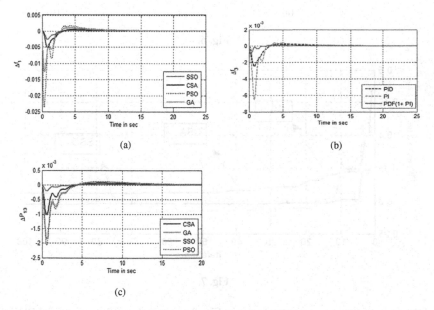

(a)

(b)

(c)

Fig. 4.

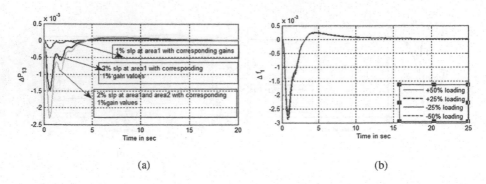

(a)   (b)

Fig. 5.

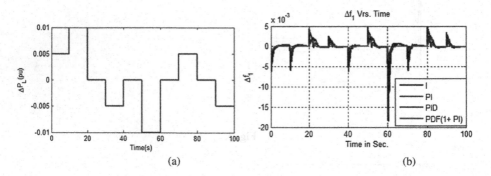

(a)   (b)

Fig. 6.

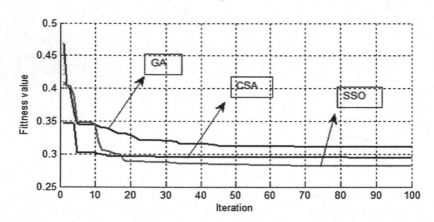

Fig. 7.

Different dynamic responses shown in Fig. 3 suggest that SSO optimized PDF plus (1+PI) controller exhibits better performance in comparing with other implemented controllers. Figure 4 suggests superior performance of SSO algorithm. Figure 5(a) shows dynamic responses obtained due to different magnitude and location SLP. Figure 5(b) exploits robust nature of proposed multistage controller and Fig. 6 suggests better performance of PDF (1+PI) controller with application of Random Load Pattern (RLP).

Comparison of Performances of PID controller, PI controller and proposed multistage PDF plus (1+PI) controller along with different optimization techniques like SSO, CSA, PSO and GA are discussed briefly in this section. Besides this simplicity and robustness of the proposed multistage controller is discussed and presented through responses. The proposed SSO algorithm is used for optimizing the gains of controllers individually with different conditions. Superiority performance of proposed optimization technique is reflected through different dynamic responses The optimized values of SSO based proposed PDF (1+PI) controller and other controller gains obtained are provided in Table 1. Table 2 depicts settling time, peak overshoot and undershoot of different responses and are developed with different controllers. So it is more evident from Table 2 that proposed SSO optimized PDF (1+PI) controller exhibits better performances and their gain parameters need not to change again for awide variation in system parameter.

Table 2. Settling time, peak overshoot, peak undershoot of different responses

Technique	PDF(I+PI with SSO)			PID with SSO			PI with SSO		
	Settling time in Sec. * 10^{-3}	OverShoot in Pu. * 10^{-3}	Under Shoot in Pu * 10^{-3}	Settling Time in Sec. * 10^{-3}	Over Shoot in Pu. * 10^{-3}	Undersho ot in Pu * 10^{-3}	Settling Time in Sec. * 10^{-3}	Over Shoot in Pu. * 10^{-3}	Under shoot in Pu * 10^{-3}
ΔF1	7.5024	0.185	-2.0888	8.8500	0.3665	-4.8484	11.220	0.5542	-10.8464
ΔF2	6.8078	0.126	-1.6256	8.8788	0.2244	-3.8200	10.205	0.4166	-11.8644
ΔF3	4.5060	0.013	-0.8256	7.5608	0.2518	-2.1826	10.508	0.3456	-6.2568
ΔP12	5.8878	0.012	-1.0178	11.208	0.1544	-1.5464	13.509	0.2144	-3.0262
ΔP23	7.2022	0.021	-1.4004	10.654	0.1876	-2.2200	12.987	0.3256	-5.8654
ITAE	38.22×10^{-2}			44.84×10^{-2}			92.56×10^{-2}		

For sensitive analysis of above proposed controller it is tested with varying magnitude and location of SLP. Figure 5(a) suggests response of deviation of frequency in area exhibits least settling time and undershoot when SLP is 1% (least) and applied in area.1 (single area) only. Also the quality of the controllers are checked by regulating some network constant variables, like governor time constant, turbine time constants etc. Here only the turbine time constant (T_t) is taken in to consideration. Here large change in turbine time constant is taken, the time constant is changed to $0.5 \times T_t$, $0.75 \times T_t$ and $1.25 \times T_t$. The dynamic responses are given in Fig. 8. Here it is observed that all the four changes overlap with each other which confers us that the variation of turbine time constant has no impact in frequency deviation, tie line power deviation and generator output power change (Fig. 8). It imposes that the proposed controller is most robust in nature.

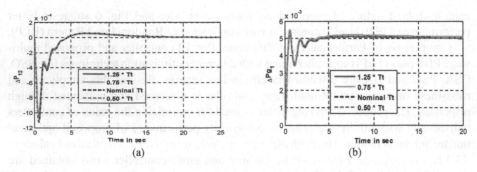

Fig. 8. (a) Deviation of tie-line power due to different Tt values (b) Deviation of power of area1 due to different Tt

Both Figs. 9 and 10 exploits superior performance of SSO optimized PDF plus (1+PI) controller when noise pattern is applied at area.1 only.

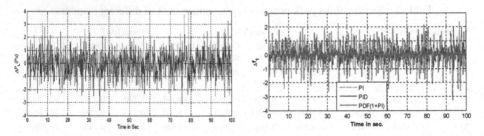

Fig. 9. Noise pattern

Fig. 10. Dynamic response of ar1 frequency due to noise

5   Conclusion

In this paper, a SSO technique is proposed to tune the multistage PDF plus (1+PI) controller parameters for AGC of two area multi source nonlinear electrical network. The nonlinearities like GRC is added to make network realistic. It is observed that proposed SSO optimized PDF plus (1+PI) controller develops improved performance in related to PID and PI controllers. Proposed optimization technique has been put in different testing environment for reflecting its superiority (exploration and exploitation) in comparing with other optimization techniques. It is revealed from the analysis that the proposed SSO optimized multistage plus (1+PI) controller exhibits better perform-ance in terms of average value, best value and standard deviations obtained in 30 runs. From convergence curve it is observed that the proposed SSO algorithm has better convergence in comparing with other optimization techniques.

Appendix

P_{ri} = 2000 MW(Each area rating); P_L^0 = 1740 MW (Avg. load on each area); f = 60 Hz, H = 5MWsec/MVA (Inertia), D = 0.0145 pu MW/Hz; K_{PS} = 68.9655 Hz/pu MW, T_{PS} = 11.49 s; T_{SG} = 0.06 s, T_T = 0.3 s, T_{12} = 0.0433; R_i= 2.4 Hz/pu MW (Regulation); B = 0.4312, a_{12} = −1, K_g = 0.3, T_g = 10.2 s; T_W = 1.1 s, T_{RS} = 4.9 s, T_{RH} = 28.749 s; b_g = 0.049 s; T_{GH} = 0.20 s, X_G = 0.60 s, Y_G = 1.10 s; C_g = 1, T_F = 0.239 s; T_{CR} = 0.010 s, T_{CD} = 0.20 s.

References

1. Kundur, P.: Power System Stability and Control. Tata McGraw Hill, New Delhi (2009)
2. Hota, P.K., Mohanty, B.: Automatic generation control of multi source power generation under deregulated environment. Int. J. Electr. Power Energy Syst. **75**, 205–214 (2016)
3. Sahu, R.K., Panda, S., Padhan, S.: A hybrid firefly algorithm and pattern search technique for automatic generation control of multi area power systems. Int. J. Electr. Power Energy Syst. **64**, 9–23 (2015)
4. Arya, Y., Kumar, N.: AGC of a multi-area multi-source hydrothermal power system interconnected via AC/DC parallel links under deregulated environment. Int. J. Electr. Power Energy Syst. **75**, 127–138 (2016)
5. Elgerd, O.I., Fosha, C.E.: Optimum megawatt frequency control of multi-area electric energy system. IEEE Trans. Power Appar. Syst. **89**, 556–563 (1970)
6. Dash, P., Saikia, L.C., Sinha, N.: Flower pollination algorithm optimized PI-PD cascade controller in automatic generation control of a multi-area power system. Int. J. Electr. Power Energy Syst. **82**, 19–28 (2016)
7. Jain, V., Saini, D., Dinesh Babu, K.N., Saini, J.S.: An intelligent GA-optimized fuzzy controller for automatic generation control for a two-area interconnected system. In: Singh, R., Choudhury, S. (eds.) Proceeding of International Conference on Intelligent Communication, Control and Devices. AISC, vol. 479. Springer, Singapore (2017). https://doi.org/10.1007/978-981-10-1708-7_82
8. Ghoshal, S.P.: Optimizations of PID gains by particle swarm optimizations in fuzzy based automatic generation control. Electr. Power Syst. Res. **72**(3), 203–212 (2004)
9. Nasiruddin, I., Bhatti, T.S., Hakimuddin, N.: Automatic generation control in an interconnected power system incorporating diverse source power plants using bacteria foraging optimization technique. Electr. Power Compon. Syst. **43**(2), 189–199 (2015)
10. Jagatheesan, K., Anand, B., Samanta, S., Dey, N., Ashour, A.S., Balas, V.E.: Design of a proportional-integral-derivative controller for an automatic generation control of multi-area power thermal systems using firefly algorithm. IEEE/CAA J. Automatica Sinica 2017
11. Mirjalili, S., Gandomi, A.H., Mirjalili, S.Z., Saremi, S., Faris, H., Mirjalili, S.M.: Salp swarm algorithm: a bio-inspired optimizer for engineering design problems. Adv. Eng. Softw. **114**, 163–191 (2017)
12. Morsali, J., Zare, K., Hagh, M.T.: Applying fractional order PID to design TCSC-based damping controller in coordination with automatic generation control of interconnected multi-source power system. Eng. Sci. Technol. Int. J. **20**(1), 1–17 (2017)
13. Sivalingam, R., Chinnamuthu, S., Dash, S.: A hybrid stochastic fractal search and local unimodal sampling based multistage PDF plus (1+PI) controller for automatic generation control of power systems. J. Franklin Inst. **354**(12), 4762–4783 (2017)

14. Gül, O., Tan, N.: Analysis of output voltage harmonics of voltage source inverter used pi and pid controllers optimized with ITAE performance criteria. In: ITM Web of Conferences, vol. 13, 01033 (2018)
15. Mahi, M., Baykan, Ö.K., Kodaz, H.: A new hybrid method based on particle swarm optimization, ant colony optimization and 3-opt algorithms for traveling salesman problem. Appl. Soft Comput. **30**, 484–490 (2015)
16. Eason, G., Noble, B., Sneddon, I.N.: On certain integrals of Lipschitz-Hankel type involving products of Bessel functions. Phil. Trans. R. Soc. London **A247**, 529–551 (1955)
17. Maxwell, J.C.: A Treatise on Electricity and Magnetism, 3rd edn., vol. 2, pp. 68–73. Clarendon, Oxford (1892)
18. Jacobs, I.S., Bean, C.P.: Fine particles, thin films and exchange anisotropy. In: Rado, G.T., Suhl, H. (eds.) Magnetism, vol. III, pp. 271–350. Academic, New York (1963)
19. Elissa, K.: Title of paper if known (Unpublished)
20. Nicole, R.: Title of paper with only first word capitalized. J. Name Stand. Abbrev. (in press)
21. Yorozu, Y., Hirano, M., Oka, K., Tagawa, Y.: Electron spectroscopy studies on magneto-optical media and plastic substrate interface. IEEE Transl. J. Magn. Jpn. **2**, 740–741 (1987). [Digests 9th Annual Conf. Magnetics Japan, p. 301, 1982]
22. Young, M.: The Technical Writer's Handbook. University Science, Mill Valley (1989)

Bridgeless Canonical Switching Cell (CSC) Converter Fed Switched Reluctance Motor Drive for Enhancing the PQ Correction

Najma Habeeb[(✉)] and Juna John Daniel

Electrical and Electronics, Sree Buddha College of Engineering, Pattoor, India
najmahabeeb27@gmail.com, junadaniel@gmail.com

Abstract. A Switched Reluctance Motor (SRM) drive has wide benefits such as high ratio of torque to volume, less industrial cost, high dynamic response and range of wide speed is required. In a conventional switched reluctance motor (SRM) drive, the mains power factor is low and the overall Total Harmonic Distortion (THD) rises as a result of harmonics. In instruction to diminish the THD and progress the power factor of the supply of SRM, a Bridgeless canonical switching cell (BL-CSC) converter configuration is employed in Discontinuous inductor current mode and it's recycled to governor the yield voltage of DC and for enhancing the power factor at the supply mains. Furthermore, it reins the voltage of split-phase capacitor and for finding the speed controller of motor sideways. Here, SRM with diode bridge rectifier (DBR) and also, with and without converter is analysed. The THD and power factor of the above systems was analysed. The power quality directories are obtained in the array of World-wide power quality principles like IEC 61000-3-2. The drive performance is evaluated by using MATLAB/SIMULINK software.

Keywords: Switched reluctance motor (SRM) drive · Power quality (PQ)
Discontinuous inductor current mode (DICM) · Total Harmonic Distortion (THD)
Bridgeless canonical switching cell (BL-CSC) converter · Mid-point converter

1 Introduction

Efficiency as well as cost are the greatest dynamic features which express a crucial part in the plan as well as progress of small control motor drives directing near households uses, for example: water pumps, fan, washing machine, vacuum cleaner etc. SRM have significant pole assembly so that stator is delivered with concentrated windings, so it is motivated with the benefit of split-phase converter [1]. Power quality improvement is individual of the essential requirement in today's world. The specified typical average of IEC 6100-3-2 on power quality [4], mentions the power factor (PF) is high and little total harmonics distortion (THD) on the mains. The system of predictable Switched Reluctance Motor nourished by a diode bridge rectifier (DBR) shows a not a sine wave-form current from the AC supply, it shows the grades in large THD current of the instruction of 75% to 85% in AC side. This displays the control plus plan of SRM with front-end of Power Factor Correction (PFC) converter [5]. In order to consuming the

© Springer Nature Singapore Pte Ltd. 2018
I. Zelinka et al. (Eds.): ICSCS 2018, CCIS 837, pp. 801–808, 2018.
https://doi.org/10.1007/978-981-13-1936-5_81

converter, a noble motor dynamic presentation is achieved. Figure 1 displays the conservative Power Factor Correction created Canonical Switching Cell (CSC) converter. In this, a mixture of switch (Sw), and diode (D) & capacitor ($C1$) is known as a 'canonical switching cell (CSC),' and this lockup, mutual with an inductor (Li) and a dc link capacitor (Cd), it is identified as CSC converter. With correct selection and strategy of factors, this grouping is recycled to accomplish PFC operation once nourished by a 1 phase ac mains over Diode Bridge Rectifier and a DC filter. This effort goals at the growth of a bridgeless configuration of CSC converter, which shows the inadequate removal of DBR at front-end converter for falling the transmission losses related with it.

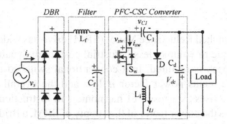

Fig. 1. Conventional CSC converter fed PFC

2   BL-CSC Converter Fed SRM Drive for PFC

Figure 2 displays the planned BL-CSC-converter-based SRM. As revealed in the representation, the Diode Bridge Rectifier is eradicated in this BL-CSC converter, so it plummeting the switching losses related with it. This BL-CSC converter nourished SRM drive has intended to run in a discontinuous inductor current mode (DICM) so the current is curving through the inductors L_{i1} and L_{i2} are irregular, where the voltage transversely the midway capacitors C_1 and C_2 ruins unremitting in a switching period. A method of a moveable dc link voltage for regulatory the speed of SRM is used. The procedure, plan, and controller of BL-CSC converter fed SRM drive are described in the subsequent chapters. Presentation of future drive is confirmed without comes gained on an advanced model with better power quality at the source side for a wide range of speed and voltage.

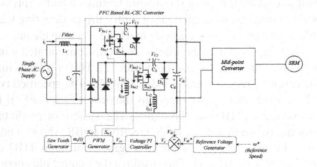

Fig. 2. Anticipated Bridgeless CSC (BL-CSC) converter fed SRM drive

3   Working Principle of BL-CSC PFC Converter

The anticipated BL-CSC converter is plotted to function in a DICM so that the current in inductors L_{i1} and L_{i2} develops irregular for a period of switching. Figure 3(a)–(f) displays dissimilar modes of procedure through the switching phase completely for positive and negative mid-cycles of the voltage of supply, accordingly. Figure 3(b) displays the related diagrams of three kinds of actions.

- *Mode 1-A:* By the way Fig. 3(a) is shown, when control S_{w1} is curved ON, the response side of inductor L_{i1} charging through diode D_p, and current i_{Li} enhances, where the intermediate capacitor C_1 starts settling through the switch S_{w1} to control the dc link capacitor C_d. So that the voltage of intermediate capacitor V_{C1} declines, where the dc link voltage V_{dc} increases.
- *Mode 1-B:* When control S_{w1} is OFF, the energy stowed in the inductor L_{i1} releases to split-phase capacitor C_d through the diode D_1, as revealed in Fig. 3(b). The current i_{Li} diminishes, where the split-phase voltage remains to rise in this process. Midway capacitor C_1 twitches charging, and voltage V_{C1} goes high, as exposed in Fig. 3(b).
- *Mode 1-C:* This method is used in Discontinuous Induction Current Mode (DICM) of process as the current of input inductor L_{i1} is nil, as exposed in Fig. 3(c). The midway capacitor C_1 remains to hold energy and maintains the charge of capacitor, where the split-phase capacitor C_d materials the energy to the capacity. The related behaviour of the converter is obtained for the additional negative mid-cycle of the mains. An inductor L_{i2}, a midway capacitor C_2, and diodes D_n and D_2 conducts in a same way, as presented in Fig. 3(d)–(f).

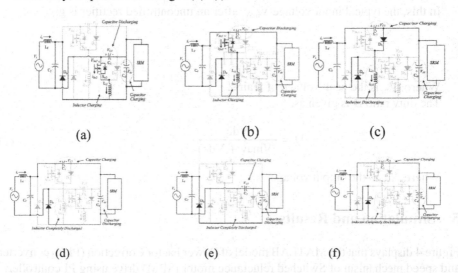

(a)   (b)   (c)

(d)   (e)   (f)

Fig. 3. Different approaches of modes of operation of the proposed BL-CSC converter. (a) Mode 1 - A. (b) Mode 1 - B. (c) Mode 1 - C. (d) Mode 2 - A. (e) Mode 2 - B. (f) Mode 2 - C.

4   Design of PFC BL-CSC Converter

A SR motor of 750 W (Range of the motor) is nourished with front - end PFC converter of 850 W. Wide-ranging of speed regulator is realized by fluctuating the Split- phase link voltage from small value (v_{dcmin} = 100 V) to regarded rate of the DC link voltage (v_{dcmax} = 320 V). The mains voltage (vs) is measured as 160 V to 220 V AC. Table 1 indicates the calculated the standards of Bridgeless CSC converter [9].

Table 1. Calculated values of Bridgeless CSC converter

Parameters	Expressions	Design data	Value
Inductor, L	$\dfrac{V_{in}(t)D(t)}{2I_{in}(t)f_s}$	$\dfrac{1}{2*20000}(\dfrac{170^2}{850})\left(\dfrac{320}{320+170\sqrt{2}}\right)$	485.35 μH
Capacitor, C	$\dfrac{V_{dc}D}{\Delta V_{c1}f_sR_L}$	$\dfrac{850*0.507}{0.5*20000*631.12*120.4}$	566 nF
DC-Link Capacitor, C_d	$\dfrac{P_{min}}{2\omega\delta V^2_{dcmin}}$	$\dfrac{144.6}{2*314*.05*100^2}$	460.73 μF
LC Filter, C_{max}	$\dfrac{I_m}{\omega_L V_m}(\tan\theta)$	$\dfrac{\left(\dfrac{850\sqrt{2}}{220}\right)}{314*220\sqrt{2}}\tan(0.5)$	488.09 nF
LC Filter, L_{req}	$\dfrac{1}{4\Pi^2f_c^2C_f}$	$\dfrac{1}{4*4222.1^2*10^{-9}}9*10^{-3}$	3.91 mH

In this, the typical input voltage V_{inav}, after an uncontrolled rectifier is given as

$$V_{inavg} = \frac{2Vm}{\Pi} \tag{1}$$

Where, V_m is the peak voltage of mains.
The duty ratio D is given as,

$$D = \frac{Vdc}{(Vinav + Vdc)} \tag{2}$$

Where, V_{dc} is the output voltage.

5   Simulation and Results

Figure 4 displays that the MATLAB model of Power Factor Correction (PFC) converter and speed mechanism of Switched reluctance motor (SRM) drive using PI controller.

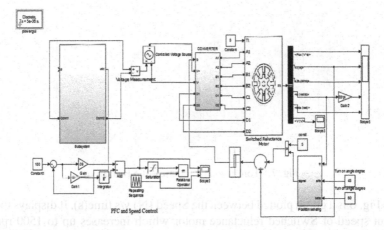

Fig. 4. MATLAB/Simulink model of power factor correction converter and speed control of SRM using PI controller

The model results of this projected system such as input current in the AC mains, speed and torque graph of SRM are as illustrated in figures from Figs. 5, 6 and 7 correspondingly. Figure 5 displays the waveform of input current in AC supply which increases up to 2.6 A.

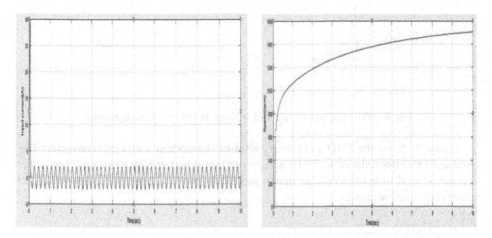

Fig. 5. Simulated input current curve in AC mains

Fig. 6. Speed response of SRM drive

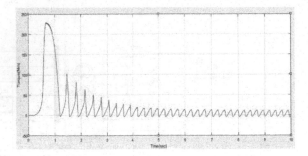

Fig. 7. Torque characteristics of SRM drive

The Fig. 6 is a graph plotted between the Speed (N) v/s time(s). It displays the result of output speed of Switched reluctance motor which increases up to 1500 rpm. The Fig. 7 is a graph plotted between the torque (N-m) v/s time(s). It shows the result of output torque of Switched reluctance motor which increases up to 4.8 N-m.

Figure 8 shows the THD and power factor of without CSC converter. The obtained THD is 19.58% and power factor is 0.8601. This shows that the system harmonics has increased.

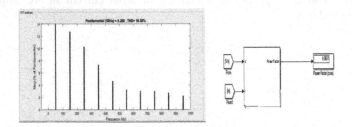

Fig. 8. THD and power factor of without Bridgeless CSC converter

Figure 9 shows the THD and power factor of the system with CSC converter. The obtained THD has reduced to 7.21% and power factor is 0.9639 when a CSC converter was introduced in between the ac mains and Midpoint converter. This shows that the system has low harmonics.

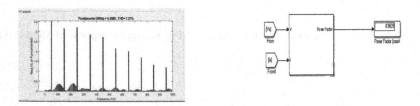

Fig. 9. THD and power factor of with Bridgeless CSC converter

The model of the Bridgeless Canonical switching cell (CSC) Converter developed in MATLAB is given in the Fig. 10.

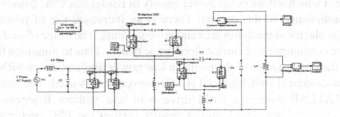

Fig. 10. MATLAB/Simulink model of Bridgeless CSC converter

Once simulated in MATLAB the output waveforms were plotted and below graphs are obtained. The output current and output voltage of Bridgeless CSC Converter is shown in Figs. 11 and 12. The output voltage of this converter is about 320 V. The output current of this converter is about 2.9A.

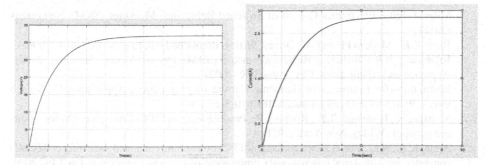

Fig. 11. Voltage versus time graph of Bridgeless CSC converter

Fig. 12. Current versus time graph of Bridgeless CSC converter

Table 2 shows the SRM drive show over wide range of speed control.

Table 2. SRM motor drive performance under larger speed control range

V_{dc}(V)	Speed (rpm)	Is(A) (Peak)	THD of Is (%)	PF
160	1350	1.03	19.58	0.8665
180	1400	1.22	13.67	0.9127
200	1475	1.94	9.70	0.9446
225	1523	2.31	7.21	0.9639
260	1579	2.67	4.07	0.9834
280	1662	3.18	2.99	0.9883
300	1736	3.82	2.44	0.9917
320	1784	4.27	1.81	0.9932

6 Conclusion

A new system which advances the power quality in Bridgeless CSC converter for SRM drive was anticipated in this project. There is an increased use of power electronic devices in the electrical machines area and drive systems. The usage of such devices has caused in the consumption of current from the ac mains. Thus to minimize the THD and to enhance the power factor of the mains in low power applications, a SRM drive with a Bridgeless canonical switching cell (CSC) converter was used. The simulations are done on MATLAB using the SRM drive with and without Bridgeless canonical switching cell converter. The Simulink results verifies the PFC performance of the Bridgeless CSC converter and THD about 7.21% is obtained with 0.963 PF. The presentation of the future system is estimated under fluctuating source AC voltages and initiate agreeable. The indices power quality in speed regulator as well as fluctuating AC mains voltage is attained inside the limits by IEC 61000-3-2.

References

1. Miler, J. (ed.): Electronics Control of Switched Reluctance Machines (SRM). Newmaans (2011)
2. Boss, K.K.: Modern PE and AC Drives. Willey, Prentices-Hall, Englewood Cliffs (2008)
3. Krishnaan, R.R.: Switched Reluctance motor (SRM) Drives: Model, Simulink, Study, Approach and Applications. CPC Press, UK (2002)
4. Restrictions for Emissions of Harmonic Current (supply current Apparatus ≤ 17 A each stage), World-wide Quality Standard IEC610-3-2 (2004)
5. Mohaan, N., Undeland, T.K., Robbins, P.: Power Electronic (PE) Converters, Presentations and Strategy. Willey, New York (2011)
6. Ando, K.A., et al.: Power factor correction (PFC) consuming canonical switching cell (CSC) converter. In: 27th Anual International Communications Energy Conference, INCLEC 2004, 18–23 August 2005, pp. 117–124 (2005)
7. Singh, S., Bisth, V., Bhuvaneswari, G.: Power factor correction (PFC) in switched mode power supply for computers using canonical switching cell (CSC) converter. In: IETT PE, vol. 8, no. 2, pp. 234–244 (2015)
8. Bist, V., Singh, B.: A PFC-based BLDC motor drive using a canonical switching cell converter. IEEE J. Ind. Inform. **10**(2), 1207–1215 (2015)
9. Singh, B., Bist, V.: A BL-CSC converter-fed BLDC motor drive with power factor correction. IET Trans. Ind. Electr. **43**(1), 162–193 (2015)
10. Bist, V., Singh, B.: PFC CUK converter-fed BLDC motor drive. IEEE Trans. PE **28**(2), 861–887 (2015)
11. Fard, H.F., Sree, M.M.: Bridgeless (BL) CUK power factor correction (PFC) converter with decreasing the transmission losses. IEEE Power Electron. **6**(8), 1640 (2012)
12. Sabsali, A.K., Ismail, E.H., Al-Safar, M.A., Fibroin, A.A.: New BL Discontinuous Induction Conduction Mode(DICM) Sepic and Cuk PFC converters with low transmission and changing losses. IEEE Trans. Ind. Appl. **47**(2), 873–881 (2011)
13. Mahdavi, M., Fard, H.F.: Bridgeless SEPIC power factor correction (PFC) rectifier with reduced components and conduction losses in BLDC. IEEE Trans. Ind. Electron. **70**(10), 4153–4160 (2012)

Modeling and Simulation of Cantilever Based RF MEMS Switch

Raji George[1(✉)], C. R. Suthikshn Kumar[2],
and Shashikala A. Gangal[1,2]

[1] Electronics Engineering Department,
Defence Institute of Advanced Technology, Pune, India
raji_saml@yahoo.co.in
[2] Electronic Science Department, Savitribai Phule Pune University, Pune, India

Abstract. In this paper, Cantilever based RF MEMS switches are designed and simulated for applications in reconfigurable antennas. Due to the low cost and ease of fabrication Cantilever based RF MEMS switch is preferred. The CoventorWare software (FEM) is used for the simulation of the proposed RF MEMS switch. The pull in voltage is calculated using the formulas and analyzed using MATLAB. Based on the analysis of the dependence of these parameters on the geometry of the cantilever, it is designed and simulated using CoventorWare software. Using this software the optimization of the design of the cantilever switch is performed for minimum pull in voltage. The pull in voltage for the proposed design is obtained as 34 V.

Keywords: Cantilever · MEMS · MATLAB · CoventorWare
Pull-in voltage

1 Introduction

Micro Electro Mechanical Systems are the integration of mechanical components, sensors, actuators and electronics on the same substrate. MEMS switches have excellent performance in microwave and mm wave frequencies. Compared to the semiconductor switches MEMS switches have lower insertion loss, high isolation, zero power consumption, small size and low inter-modulation distortion. The disadvantage of MEMS switches are low switching speed and high actuation voltage. Due to its low loss and high isolation MEMS switches are used in reconfigurable antennas, filters, tuners and phase shifters.

RFMEMS switches can be classified according to their actuation method as electro-static, magnetic, piezoelectric or thermal, based on the movement as vertical or lateral, based on contact types as metal-metal or metal-insulator-metal and based on circuit configuration as series or shunt. In this paper a series circuit, metal to metal contact, vertical, electro-statically actuated RF MEMS switch is discussed. In this paper a detailed analysis of an RF MEMS switch is done, to obtain pull-in voltage with CoventorWare FEM model and MATLAB software.

© Springer Nature Singapore Pte Ltd. 2018
I. Zelinka et al. (Eds.): ICSCS 2018, CCIS 837, pp. 809–816, 2018.
https://doi.org/10.1007/978-981-13-1936-5_82

2　Literature Review

2.1　Reconfigurable Antenna

Re-configurability can be achieved in antenna by using an RF switch with the antenna. Figure 1 shows the structure of a dual-band frequency reconfigurable antenna containing two micro-strip patches connected by an RF switch. When Patch 1 is fed with the switch OFF, it resonates at a frequency of f1. When the Patch 1 and Patch 2 are connected when RF switch is ON, it resonates at a new second frequency f2. Thus the length of the antenna is decided by the (On/Off) status of the RF switch [2]. The RF switches used with reconfigurable antennas are semiconductor switches like PIN diode, FETs, varactor diodes and RF MEMS switches. Compared to the semiconductor switches, RF MEMS switches are preferred due to their lower insertion loss, better isolation, linearity, zero power consumption and high Q factor.

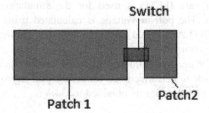

Fig. 1.　Dual patch reconfigurable antenna

2.2　RF MEMS Switch

Figure 2 shows the structure of the proposed RFMEMS switch. The substrate is silicon with a dielectric layer of silicon dioxide on it. The signal lines and the electrodes are of Gold. The tip of the cantilever is made of Copper. The Cantilever Beam is of Silicon Nitride.

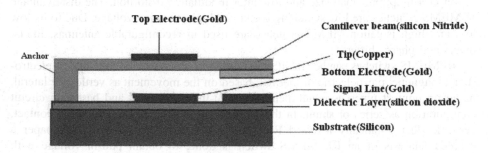

Fig. 2.　Structure of the proposed RF MEMS switch

There are different optimizing methods for MEMS switches. One of the important parameters for optimizing the RF MEMS switch is the Pull in voltage. In the design of electro-statically actuated MEMS switch, the analysis of pull in voltage is very important. For analysis, the MEMS switch is modeled using lumped parameters. It is modeled as a parallel plate capacitor constrained by a spring. [3] Following are the steps followed for cantilever design:

1. Understanding the design of the surface micro machined cantilever through calculation.
2. Studying the effect of cantilever dimensions on various design parameters (Analysis using MATLAB)
3. Carrying out FEM based simulation for further optimization and extraction of final cantilever dimensions from simulation results. (Simulation in CoventorWare Software) [4] (Fig. 3).

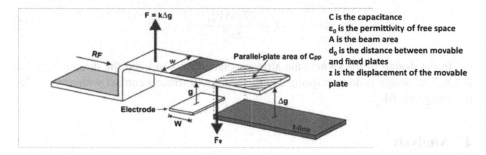

Fig. 3. Cantilever structure

3   Design

The system with one degree of freedom is given as:

$$m\frac{d^2z}{dt^2} + kz = f_E \tag{1}$$

Where m is the mass of the spring and k is the spring constant.

The electrostatic force f_E is found by differentiating the stored energy of the capacitor w.r.t. the position of the movable plate.

$$f_E = -\frac{d\left(\frac{CV^2}{2}\right)}{dz} = \frac{\varepsilon_0 A V^2}{2(d_0 - z)^2} \tag{2}$$

$$C = \frac{\varepsilon_0 A}{d_0 - z} \tag{3}$$

$$f_M = kz \tag{4}$$

f_M is the mechanical elastic restoring force at equilibrium.

At static equilibrium,

$$f_M = f_E \tag{5}$$

To determine the pull-in voltage $V_{PI} = V\left(\frac{2\Delta g}{3}\right)$,

$$V_{PI} = \sqrt{\frac{8kd_0^3}{27\varepsilon_0 A}} \tag{6}$$

$$k = \frac{27\varepsilon_0 A V_{PI}^2}{8d_0^3} \tag{7}$$

In the design of the switch, the air gap is taken as 3 μm and the dimple height as 2 μm. The design is done ensuring that the beam will make a contact with a stable Pull in voltage [5, 6].

4 Analysis

4.1 MATLAB Analysis

The Pull in voltage given by the Eq. (8) is analyzed using Matlab

$$V_{PI} = \sqrt{\frac{8k(\lambda_r)d_0^3}{27\varepsilon_0 A_{eff}}} \tag{8}$$

It is observed from the plots Fig. 4(a)-(d) that the Pull in voltage decreases exponentially with the cantilever length. The pull in voltage increases with increase in cantilever thickness and the gap height. The cantilever width has no effect on the pull in voltage.

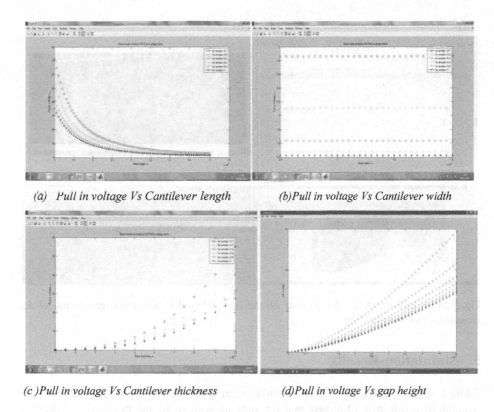

(a) Pull in voltage Vs Cantilever length   (b)Pull in voltage Vs Cantilever width

(c)Pull in voltage Vs Cantilever thickness   (d)Pull in voltage Vs gap height

Fig. 4. Variation of pull; in voltage wrt to switch dimensions

4.2 COVENTORWARE Simulation

The optimization of the cantilever is performed for minimum Pull in Voltage. Figure 5 (a)-(c) shows the model of the RF MEMS switch for simulation in CoventorWare. Simulations are performed for different dimensions and the Pull in voltage is observed. Cantilever length is varied from 150 μm to 450 μm, cantilever width is varied from 50 μm to 150 μm and cantilever thickness is varied from 0.5 to 3 μm. Simulation of RF MEMS design using CoSolve EM in CoventorWare software will determine the voltage at which the electrostatic force and mechanical restoring diverges and will determine the voltage bias range in which the consistency between forcing and restoring becomes unstable.

814   R. George et al.

(a)   (b)

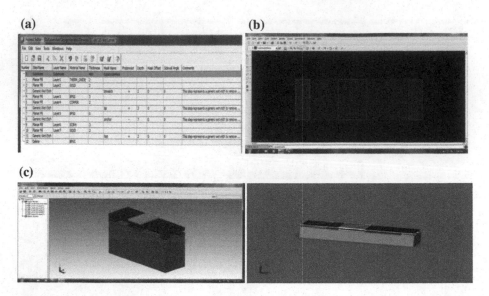

(c)

Fig. 5. (a): Process flow, (b): Layout of the switch model, (c): Preprocessor model for simulation in CoventorWare

5   Simulation Results

Table 1 shows the summary of the simulation in CoSolve in CoventorWare. From the simulaterd result it is observed that the pull in voltage of the design is 34.0625 V (Fig. 6).

Fig. 6. Simulated result in MemMech and CoSolve

Thus, the optimized dimensions of the switch for the obtained Pull in voltage of 34.06 V are as follows (Table 2, Figs. 7 and 8):

Table 1. Summary of the simulation

Summary	Voltage	Iterations	Status	Contact	Disp	Disp change
Step_1	0	2	Converged	No	0	0
Step_2	10	2	Converged	No	0.053793	0.0003432
Step_3	20	3	Converged	No	0.235317	0.000249
Step_4	30	6	Converged	No	0.669551	0.000686
Step_5	40	3	Diverged	No	5.189447	2.800891
Step_6	35	6	Diverged	No	1.471796	0.079897
Step_7	32.5	8	Converged	No	0.906489	0.000909
Step_8	33.75	15	Converged	No	1.138873	0.000858
Step_9	34.375	7	Diverged	No	1.379778	0.025576
Step_10	34.0625	37	Converged	No	1.311321	0.0009908

Table 2. Optimized dimensions of the simulated switch

Parameter	Dimension
Length of the beam	450 μm
Width of the beam	150 μm
Thickness of the beam	3 μm
Length of the actuating electrodes	50 μm
Area of the tip	50 μm × 150 μm
Air gap	3 μm
Distance between bottom electrode an anchor	50 μm

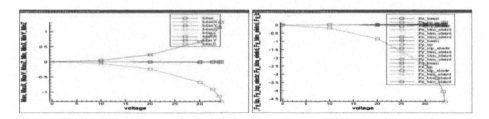

Fig. 7. Variation of displacement and electrostatic force wrt actuation voltage

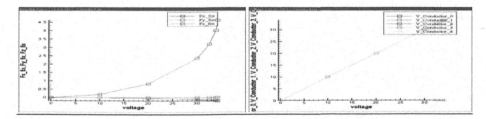

Fig. 8. Variation of reactive force and conductor voltage wrt actuation voltage

6   Conclusion

The results provide analysis of the Pull in voltage of the cantilever based RF MEMS switch which is a key factor in the performance of RF MEMS switches. The performance of the RFMEMS switch could further be improved to get a lower Pull in voltage. The actual fabrication and testing of the proposed switches can be done to experimentally justify the simulation results achieved in this paper.

Acknowledgement. The authors would like to gratefully acknowledge the Electronic Science Department, Savitribai Phule Pune University, Pune for providing the software tool CoventorWare software required for simulation and evaluation for the switch design.

References

1. Guo, F.M., et al.: The experimental model of micro-cantilever switch and its optimization model. In: 2010 5th IEEE International Conference on Nano/Micro Engineered and Molecular Systems (NEMS), pp. 1–4 (2010)
2. Vakilian, M., et al.: Optimization of cantilever-based MEMS switch used in reconfigurable antennas. In: IEEE-ICSE 2012 Proceedings, Kuala Lumpur, Malaysia (2012)
3. Rebeiz, G.M., Muldavin, J.B.: RF MEMS switches and switch circuits. IEEE Microw. Mag. **2** (4), 59–71 (2001)
4. Kshirsagar, A.V., Duttagupta, S.P., Gangal, S.A.: Design of MEMS cantilever - hand calculation. Sens. Transducers J. **91**(4), 55–69 (2008)
5. Varadan, V.K., Vinoy, K.J., Jose, K.A.: RF MEMS and Their Applications. Wiley, New York (2003)
6. Peroulis, D., Pacheco, S.P.: Electromechanical considerations in developing low voltage RF MEMS switches. IEEE Trans. Microw. Theory Tech. **51**, 259–270 (2003)

Stability Study of Integrated Microgrid System

B. V. Suryakiran[✉], Vinit kumar Singh, Ashu Verma, and T. S. Bhatti

Centre for Energy Studies, IIT Delhi, New Delhi, India
{skiran.jes16,esz158498,averma,tsb}@ces.iitd.ac.in

Abstract. India has a high potential of renewable energy and the electrical power produced from these sources can be utilized for various applications. One of the application is that these sources can be clustered to form a micro grid for improving the availability of power supply especially in rural areas. In this paper, an attempt has been made to study and model renewable sources based micro grid having limited grid support. Renewable sources considered while modeling are Photo Voltaic system (PVS), Wind Energy Conversion System (WECS) and Biogas-genset. Grid provides constant power to the system (Limited grid) which is having power from the renewable sources to ensure continuity of the power supply. A STATCOM has been introduced for the reactive power support of the micro grid. Stability study has been performed to validate the results.

Keywords: Renewable sources · PVS · WECS · Biogas-genset · Limited grid
STATCOM

1 Introduction

Electrification in India is not completely realized in number of states like Nagaland, Telangana, Odisha, Jharkhand, Bihar, Arunachal Pradesh, Assam and Tripura [1]. However, these states possess a good potential of renewable energy. This paper takes into account a cluster of 4 medium sized villages while modeling the micro grid under the scheme of Integrated Rural Electrification System (IRES). The sources that are considered while modeling this micro grid are Photo Voltaic System (PVS), Wind Energy Conversion System (WECS), Biogas-genset and Limited grid. As PVS and WECS are intermittent in nature and limited grid supplies constant power, therefore Biogas-genset is used for supplying the power to maintain balance between generation and load. STATCOM has been used for the reactive power support of the overall system.

The percentage of installed capacity of WECS has witnessed a tremendous growth over the years [2]. Various types, their configurations and comparison have been presented by Li and Chen [3]. The analysis of WECS coupled with PMSG and Lead acid battery system for residential purposes has been presented by L. Barote and C. Marinescu [4]. PVS is one of the most rapidly increasing renewable source for distributed generation in the world. For damping out the intermittency of the generation and the source, Battery Energy storage System (BESS) is being used in practice. One of the currently trending topics include the integration of PVS with the grid [5–7]. India being agriculture- dominated country, has a wide scope of biomass. Research has been going

© Springer Nature Singapore Pte Ltd. 2018
I. Zelinka et al. (Eds.): ICSCS 2018, CCIS 837, pp. 817–825, 2018.
https://doi.org/10.1007/978-981-13-1936-5_83

on harvesting of this energy using IC engines with electrical power conversion by PMSG/ Induction Generators (IGs) [8, 9].The concept of the Limited grid presented in the paper uses PI controllers for controlling the voltages and voltage angles between the Grid and the common bus in order to maintain constant power flow from the grid to the common bus [10].

2 Integrated Microgrid System

The Micro-grid system has been designed with the various renewable sources, viz., WECS, PVS and Biogas-genset. It has also been designed with the Limited grid support and a STATCOM for reactive power support.

The WECS considered uses a Permanent Magnet Synchronous Generator (PMSG) for power generation. The output power obtained from the PMSG is then passed through AC/DC/AC power electronic interface to obtain a power at constant frequency and voltage which is then passed through transformer for obtaining power at appropriate voltage. The PVS generates dc power which varies with the variation of solar irradiance and it is further passed through DC/DC/AC power electronic interface to obtain the power at constant frequency and constant voltage. By using transformer it is brought to appropriate voltage level for coupling it with the common bus. The Biogas-genset uses an IC engine which is coupled with Synchronous Generator (SG) for the power generation. AVR/exciter is used for the control of excitation to provide power at desired voltage level.

Limited grid has been connected to the micro-grid by bringing the power to an appropriate voltage level using transformers. Grid has been made Limited for ensuring

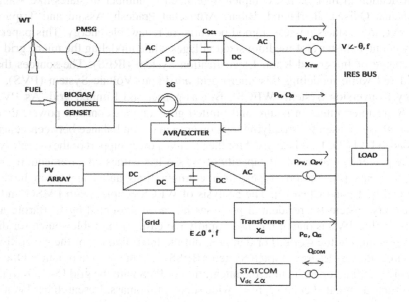

Fig. 1. Single line diagram of the microgrid.

the continuity of the power flow to the micro-grid. This concept is much appreciable for the areas facing power cuts. While modeling, Limited grid is taken as the reference for the micro-grid. This has been done by using PI controllers to control the voltage and the angle of the common bus.

Inverters connected with the PVS and WECS are fixed at 0.9 pf (lagging). Since the inverters are connected to intermittent sources, supply of reactive power is unreliable which led to the usage of STATCOM. The single line diagram of the micro-grid is shown in Fig. 1.

3 Mathematical Modelling

3.1 Incremental Power Balance

At any instant the real and reactive power generations must be equal to the real and reactive power demand respectively. Mathematical expression of the above statement is given by Eqs. (1) and (2).

$$P_W + P_{grid} + P_{PV} + P_{bg} = P_D \tag{1}$$

$$Q_W + Q_{grid} + Q_{PV} + Q_{bg} = Q_D \tag{2}$$

Following a small disturbance, the Eqs. (1) and (2) can be written as:

$$\Delta P_W + \Delta P_{grid} + \Delta P_{PV} + \Delta P_{bg} = \Delta P_D \tag{3}$$

$$\Delta Q_W + \Delta Q_{grid} + \Delta Q_{PV} + \Delta Q_{bg} = \Delta Q_D \tag{4}$$

Change in demand in terms of real power is directly proportional to change in the frequency. Hence, change in frequency can be given by Eq. (5).

$$\Delta F(s) = \frac{K_{FS}}{1 + s * T_{FS}}(\Delta P_W + \Delta P_{grid} + \Delta P_{PV} + \Delta P_{bg} - \Delta P_D) \tag{5}$$

Similarly, the change in the reactive power is proportion to the change in system voltage. The mathematical expression is given by Eq. (6).

$$\Delta V(s) = \frac{K_{VS}}{1 + s * T_{VS}}(\Delta Q_W + \Delta Q_{grid} + \Delta Q_{PV} + \Delta Q_{bg} - \Delta Q_D) \tag{6}$$

3.2 Modelling of WECS with PMSG/PVS

The transfer function block diagram of the WECS has been shown in the Fig. 2. The power flow equations through the transformer connecting inverter output terminal and microgrid are given by Eqs. (7) and (8)

$$P'_W = \frac{V_{inW} * V * sin(\delta_{inW} + \theta)}{X_{TW}} \tag{7}$$

$$Q_W = \frac{V_{inW} * V * cos(\delta_{inW} + \theta) - V^2}{X_{TW}} \tag{8}$$

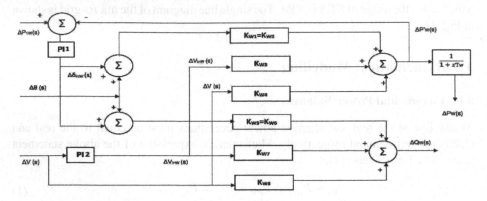

Fig. 2. Transfer function block diagram representation of WECS.

Following a small disturbance, Eqs. (7) and (8) are given by

$$\Delta P'_W(s) = K_{W1}\Delta V_{inW}(s) + K_{W2}\Delta V(s) + K_{W3}\Delta\delta_{inW}(s) + K_{W4}\Delta\theta(s) \tag{9}$$

$$\Delta Q_W(s) = K_{W5}\Delta V_{inW}(s) + K_{W6}\Delta V(s) + K_{W7}\Delta\delta_{inW}(s) + K_{W8}\Delta\theta(s) \tag{10}$$

Where,

$$K_{W1} = K_{W2} = \frac{V_{inW} * Vcos(\delta_{inW} + \theta)}{X_{TW}} \quad K_{W3} = \frac{Vsin(\delta_{inW} + \theta)}{X_{TW}} \quad K_{W4} = \frac{V_{inW}sin(\delta_{inW} + \theta)}{X_{TW}}$$

$$K_{W5} = K_{W6} = -\frac{V_{inW}sin(\delta_{inW} + \theta)}{X_{TW}} \quad K_{W7} = \frac{Vcos(\delta_{inW} + \theta)}{X_{TW}} \quad K_{W8} = \frac{V_{inW}cos(\delta_{inW} + \theta) - 2V}{X_{TW}}$$

Considering the delay caused due to the presence of mechanical system

$$\Delta P_W(s) = \frac{1}{1 + sT_w}\Delta P'_W(s) \tag{11}$$

The Modelling for the PVS is carried out in the same way as WECS.

3.3 Modelling of Biogas-Genset

The transfer function block diagram representation for real power and reactive power generation has been shown in Figs. 3 and 4 respectively. IEEE type-1 excitation system has been considered while modeling.

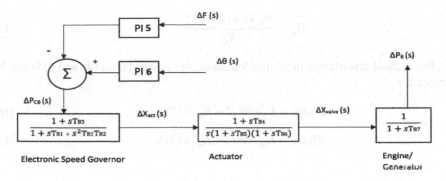

Fig. 3. Transfer function block diagram representation of real power generation of Biogas-genset.

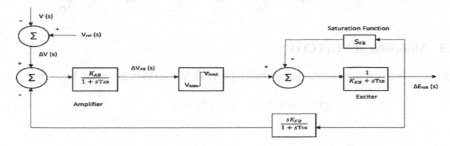

Fig. 4. Transfer function block diagram representation of reactive power generation of Biogas-genset.

Dynamic state equations after neglecting saturation function can be represented by Eqs. (12) and (13).

$$\Delta E'_{qB}(s) = \frac{1}{1 + sT_{GB}}[K_{1B}\Delta E_{fdB}(s) + K_{2B}\Delta V(s)] \tag{12}$$

$$\Delta Q_B(s) = K_{3B}\Delta E'_{qB}(s) + K_{4B}\Delta V(s) \tag{13}$$

Where,

$$K_{1B} = \frac{X'_d}{X_d} \quad K_{2B} = \frac{(X_d - X'_d)\cos(\delta + \theta)}{X_d} \quad K_{3B} = \frac{V\cos(\delta + \theta)}{X'_d} \quad K_{4B} = \frac{E'_{qB}\cos(\delta + \theta) - 2V}{X'_d}$$

3.4 Modelling of Limited Grid

The power flow equations through the transformer are given by Eqs. (14) and (15).

$$P_G = \frac{E_G V \sin(\theta)}{X_G} \tag{14}$$

$$Q_G = \frac{E_G V cos(\theta) - V^2}{X_G} \tag{15}$$

For a small disturbance in control variables, the power Eqs. (14) and (15) can be written as:

$$\Delta P_G(s) = K_{1G}\Delta\theta(s) + K_{2G}\Delta V(s) \tag{16}$$

$$\Delta Q_G(s) = K_{3G}\Delta\theta(s) + K_{4G}\Delta V(s) \tag{17}$$

Where,

$$K_{1G} = \frac{-E_G V cos\theta}{X_G} \quad K_{2G} = \frac{E_G sin\theta}{X_G} \quad K_{3G} = \frac{E_G V sin\theta}{X_G} \quad K_{4G} = \frac{E_G V cos\theta - 2V}{X_G}$$

3.5 Modelling of STATCOM

The reactive power generated/absorbed by the STATCOM is given by Eq. (18).

$$Q_{COM} = k^2 V_d^2 B - kV_d VB cos(\alpha) \tag{18}$$

Where, $k = \frac{p\sqrt{6}}{6\pi}$, p = 6 pulse inverter has been considered for the modelling. A disturbance in Q_{COM} is given by the Eq. (19) and the dynamic model used for modelling purpose has been shown in Fig. 5.

$$\Delta Q_{COM}(s) = K_{C0}\Delta\alpha(s) + K_{C1}\Delta V(s) \tag{19}$$

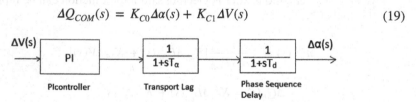

Fig. 5. Transfer function block diagram representation of STATCOM in dynamic state.

Where, $K_{C0} = kV_d VB sin\alpha$, $K_{C1} = 2k^2V_d B - kVd B cos\alpha$.

4 Results

The dynamic variation of system parameters are shown in Fig. 6. The integrated micro-grid, IRES has been simulated with a 1% change in the active power demand and it is observed that Voltage of the system settles in about 4 s, whereas the frequency of the system settles in about 2.5 s. The increase in the active power demand is supplied by the biogas-genset. The real and reactive power change of the grid settles with zero steady state error in about 10 s. Since, the power factor of the synchronous generator of BG is constant, it also generates excessive reactive power. A very small part of this reactive power is consumed by the inverters connected with the PV and WECS and remaining part is consumed by the STATCOM connected to the microgrid.

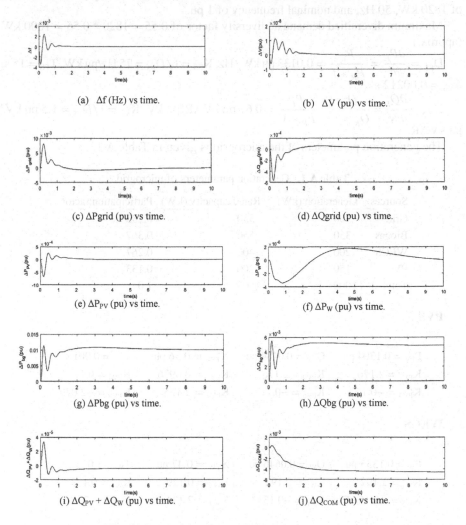

(a) Δf (Hz) vs time.

(b) ΔV (pu) vs time.

(c) ΔPgrid (pu) vs time.

(d) ΔQgrid (pu) vs time.

(e) ΔP_{PV} (pu) vs time.

(f) ΔP_W (pu) vs time.

(g) ΔPbg (pu) vs time.

(h) ΔQbg (pu) vs time.

(i) $\Delta Q_{PV} + \Delta Q_W$ (pu) vs time.

(j) ΔQ_{COM} (pu) vs time.

Fig. 6. Dynamic response of the system parameters.

5 Conclusion

The microgrid with various renewable sources along with the support of the limited grid has been modelled and simulated. The model has been validated by performing stability study. It has been observed that for a small disturbance, the system frequency and voltage settle with zero steady state error within 4 s. The participation of the STATCOM is much higher than the inverters connected in the system.

Appendix

The microgrid has been modelled for 4 medium sized villages having contract demand of 1820 kW, 50 Hz, and nominal frequency of 1 pu.

Maximum diversified demand at diversity factor of $0.55 = 1820 * 0.55 = 1000$ kW (approx.)

$$D_{FS} = \frac{\partial P_L}{\partial f} = \frac{P_L}{P_R * f} = 0.0133 \text{ pu kW/Hz}, K_{FS} = 1/D_{FS} = 75 \text{ Hz/pu kW}, T_{FS} = 15 \text{ s},$$

$T_{VS} = 0.00212$ s.

$$D_{VS} = \frac{\partial Q_L}{\partial V} = \frac{Q_L}{Q_R * V} = \frac{P_L}{P_R * V} = 0.67 \text{ pu kVAR/pu kV}, K_{VS} = 1/D_{VS} = 1.5 \text{ pu kV/}$$
pu kVAR.

The generation parameters of the microgrid is given in Table A.1.

Table A.1. Generation parameters of microgrid.

Sources	Generation (kW)	Rated capacity (kW)	Participation factor
Grid	300	350	0.233
Biogas	350	550	0.367
WECS	200	400	0.267
PV	150	200	0.133
Total	1000	1500	1.0

PVS

$P_{PV} = 0.1304$ pu	$Q_{PV} = 0.0484$ pu	$X_{TPV} = 0.36$ pu	$T_{PV} = 0.001$ s
$K_{PV1} = 2.176$	$K_{PV2} = 2.176$	$K_{PV3} = 0.0976$	$K_{PV4} = 0.1$
$K_{PV5} = -0.1$	$K_{PV6} = -0.1$	$K_{PV7} = 2.125$	$K_{PV8} = -2.079$

WECS

$P_W = 0.1333$ pu	$Q_W = 0.0646$ pu	$X_{TW} = 0.47$ pu	$T_W = 1.0$ s
$K_{W1} = 2.192$	$K_{W2} = 2.192$	$K_{W3} = 0.129$	$K_{W4} = 0.1333$
$K_{W5} = -0.1333$	$K_{W6} = -0.1333$	$K_{W7} = 2.124$	$K_{W8} = -2.063$

BIOGAS

$P_B = 0.2062$ pu		$Q_B = 0.0998$ pu	
$X_d = 1$ pu	$X'_d = 0.15$ pu	$T'_{d0} = 5.0$ s	$T_B = 0.75$ s
$T_{B1} = 0.01$ s	$T_{B2} = 0.02$ s	$T_{B3} = 0.15$ s	$T_{B4} = 0.2$ s
$T_{B5} = 0.014$ s	$T_{B6} = 0.04$ s	$T_{B7} = 0.036$ s	$K_{AB} = 200$
$T_{AB} = 0.05$ s	$K_{EB} = 1$	$T_{EB} = 2.0$ s	$K_{FB} = 0.5$
$T_{FB} = 1.0$ s	$K_{1B} = 0.15$	$K_{2B} = 0.8472$	$K_{3B} = 2.437$
$K_{4B} = -2.336$			

GRID

$P_G = 0.2$ pu		$Q_G = 0.09686$ pu	
$K_{1G} = -2.8746$	$K_{2G} = 0.2$	$K_{3G} = 0.2$	$K_{4G} = -2.6809$

STATCOM

$Q_G = 0.1771$pu	$K_{C0} = 1.9039$	$\alpha = 23.8°$	$K_{C1} = 3.4318$

References

1. State-wise village electrification status as on date, National Data Sharing and Accessibility Policy (NDSAP), Government of India, August 2017. https://data.gov.in/resources/state-wise-village-electrification-status-date
2. Indian Wind Energy, A brief outlook, Global Wind Energy Council (GWEC) (2016)
3. Yadav, D.K., Bhatti, T.S., Verma, A.: Study of integrated rural electrification system using wind-biogas based hybrid system and limited grid supply system. Int. J. Renew. Energy Res. (IJRER) **7**(1), 1–11 (2017)
4. Li, H., Chen, Z.: Overview of different wind generator systems and their comparisons. IET Renew. Power Gener. **2**(2), 123–128 (2008)
5. Barote, L., Marinescu, C.: PMSG wind turbine system for residential applications. In: 2010 International Symposium on Power Electronics Electrical Drives Automation and Motion (SPEEDAM), pp. 772–777. IEEE (2010)
6. Ding, F., Li, P., Huang, B., Gao, F., Ding, C., Wang, C.: Modeling and simulation of grid-connected hybrid photovoltaic/battery distributed generation system. In: 2010 China International Conference on Electricity Distribution (CICED), pp. 1–10. IEEE (2010)
7. Makhlouf, M., Messai, F., Benalla, H.: Modeling and simulation of grid-connected hybrid photovoltaic/battery distributed generation system. Can. J. Electr. Electron. Eng. **3**(1), 1–10 (2012)
8. Shah, R., Mithulananthan, N., Bansal, R., Lee, K.Y., Lomi, A.: Influence of largescale pv on voltage stability of sub-transmission system. Int. J. Electr. Eng. Inform. **4**(1), 148–161 (2012)
9. Kundu, P., Tandon, A.: Capacitor self-excited double-armature synchronous generator for enhanced power output. In: 1998 IEEE Region 10 International Conference on Global Connectivity in Energy, Computer, Communication and Control, TENCON 1998, vol. 2, pp. 391–397. IEEE (1998)
10. Wang, L., Lin, P.-Y.: Analysis of a commercial biogas generation system using a gas engine–induction generator set. IEEE Trans. Energy Convers. **24**(1), 230–239 (2009)

A High Speed Two Step Flash ADC

K. Lokesh Krishna[1]([⊠]) [iD], Yahya Mohammed Ali Al-Naamani[2], and K. Anuradha[3]

[1] S.V.College of Engineering, Tirupati, Andhra Pradesh, India
[2] S.V.C.E.T., Chittoor, Andhra Pradesh, India
[3] L.B.C.E.W., Visakhapatnam, Andhra Pradesh, India
kayamlokesh78@gmail.com

Abstract. As wireless communication equipment's are demanding higher speed of operation, low power and into the digital domain, it becomes essential to design a high speed and low power ADC. This paper presents a novel power efficient, high-speed two step analog to digital converter (ADC) architecture combining two whole Flash ADCs with feed forward circuitry. The proposed circuit has been designed to overcome the drawbacks of the conventional flash ADC which draws more power due to the high speed comparator bank. Also the proposed two-step ADC employs a modified double tail comparator circuit which operates at high speed and consumes less power. The individual block of two step flash ADC is designed, simulated and implemented in CMOS 130 nm N-well technology operated at 1.8 V power supply voltage. The ADC consumes 2.32mW with a resolution of 6-bits for input signal frequencies upto 1 GHz and occupies a silicon area of 0.226 mm^2. The operating speed of the design is 10 GHz and the simulated static INL and DNL is found to remain within 0.15LSB and 0.42LSB respectively.

Keywords: Threshold inverter quantizer · Comparator · Register · Low power
Time interleaving · Residue amplifier and CMOS

1 Introduction

Existing electronic systems such as Software Defined Radios, Aircraft communications, High performance Data Acquisition Systems, Broad band wireless communication systems, Medical Imaging and Diagnostic equipment's, High frequency oscilloscopes, Spectrum analyzers, 4G Long Term Evolution (LTE) systems, Signal generators, Communication test equipment's and High frequency oscilloscopes demand high data rates with low power consumption. The signals that we experience in our daily lives are analog. These signals are non-quantized, low amplitude or high amplitude and continuously vary with time. Nowadays due to the advances in digital processing techniques, utmost processing is carried out in digital domain only. The advantages of digital signal processing techniques are easier to design since exact values of voltages and currents are not required, accurate and highly precise, data storage is easier, logic can be reprogrammed, and less prone to noise. Hence it becomes essential that the received signal must be transformed to digital at some instance of processing.

© Springer Nature Singapore Pte Ltd. 2018
I. Zelinka et al. (Eds.): ICSCS 2018, CCIS 837, pp. 826–836, 2018.
https://doi.org/10.1007/978-981-13-1936-5_84

A data converter is a microelectronic circuit that is used to translate the given analog input voltage signal to output digital voltage signal or a digital input signal to an analog output signal. Hence data converter circuits are extensively used as interface between analog and digital circuits and vice-versa. Broadly there are two various types of data converter circuits such as (i) ADC and (ii) Digital to Analog Converter (DAC). ADC transforms an analog signal (which is continuous time and continuous magnitude) into corresponding digital signals (discrete time and discrete magnitude). Generally there are different ADC architectures available to be used for different emerging applications. Identification of an ADC for a given applications is determined by its design specifications. However no single ADC architecture is found to be appropriate for all these applications. Selection of the correct ADC necessitates tradeoffs between conversion time, silicon area, resolution, static performance, channel count, cost and dynamic performance. Irrespective of the applications, the three key parameters to be considered in the design of ADC circuit are dynamic linearity over large operating bandwidth, die area and power consumption.

Various ADC architectures such as Flash type, Two-step flash type, Dual Slope type, Sigma-Delta, Pipeline and Interleaving are available for various applications. The performance of an ADC is frequently affected by the type of the input signal they process. Because the given input analog signal is continuous in nature, ADCs suffer numerous problems such as clock jitter, nonlinear input impedance, number of bits, signal and clock skew, number of components, chip size, power dissipation etc. These problems limit the use of ADC architectures for various applications. In all the available ADC architectures, Flash analog to digital converter circuits are very fast converters and are well appropriate for large bandwidth applications. Flash ADCs are realized by connecting high speed comparator circuits in series. The number of high speed comparators required to realize a flash ADC is (2^N-1), where N denotes resolution of the converter. But the disadvantage is that they draw a lot of power as the resolution of the converter circuit increases which in turn reflects in the increase in the silicon area. Due to this flash ADCs are restricted to a resolution of (4–6) bits. Also as the chip size increases, more problems associated with signal and clock routing becomes noticeable. Pipelined analog to digital converter architecture are mostly used for sampling signals from hundreds of kilo samples per second (KS/s) upto 800 MS/s with resolutions varying from (6-16) bits. Sigma-Delta ($\sum$-Δ) ADC architectures are used principally in applications requiring lower speed while requiring a trade-off of sampling speed for resolution by oversampling, followed by filtering techniques to decrease quantization noise.

Sigma-Delta ($\sum$-Δ) ADC architecture are used in low speed applications with resolution ranging from (12–24) bits. Dual slope type ADCs deliver high resolution and offer good in-line frequency and noise rejection. Time interleaved ADCs uses few identical analog to digital converters to process regular sampled data series at a faster rate than the operating rate of each individual ADC. Due to the Time inter leaving technique, the operating bandwidth of the converter is increased which allows for easier frequency planning and reduction in circuit complexity and price of the anti-aliasing filter that is typically used at the front stage of analog to digital converter. Successive Approximation Register (SAR) analog to digital converter architectures are used in applications requiring medium to very high resolution and with sample rate of the order of 10MS/s to 100MS/s. Recent improvements in architectural design and scaling of transistors allowed for decrease in power consumption and high speed operation.

Two step analog to digital converter architecture also known as subranging converters offers the best tradeoff between sampling speed, power consumption and latency. The two-step flash ADC generates the digital output in multiple stages using multiple clock cycles. The principal component in the design of two-step flash architecture is the high speed comparator. The number of required high speed comparators reduces which in turn reflects in the reduction of power consumption and silicon area.

Nasri et al. presents the implementation of a low-power 4-bit high sample rate folding-flash ADC which uses a novel unbalanced double-tail dynamic comparator and results in major reduction of the kick-back noise [1]. Wang et al. propose an column-level two-step ADC [2]. A four way time interleaved ADC driving four sets of track and hold switches that suits the requirements of NRZ 10G Ethernet (10GE) standard is proposed [3]. Ferragina et al. describes the design of low-power and high resolution flash architecture which utilizes interpolation technique and V-I converters operated as preamplifier stage [4]. Ritter et al. present a flash ADC that doesn't use a track and hold circuit or time interleaving technique which reflects in reduction of circuit complexity and operates at high speed [5]. A wide input bandwidth and low power ADC operating at 1 GHz appropriate for ultra-wide band system is proposed [6]. Tretter et al. presents the design and characterization of single-core flash ADC, which is capable of reaching high sampling rate without using time interleaving technique [7]. In this proposed work a high speed two step flash architecture is designed, simulated and implemented using CMOS 130 nm technology.

The summary of this work is as follows. Section 2 explains the background of two step flash ADC architecture and various parameters associated with the proposed work. Section 3 presents the design and implementation of the proposed two-step architecture. Section 4 shows the obtained simulation results. Lastly the conclusions are drawn in Sect. 5.

2   Two Step Flash ADC Architecture

Two-step flash architecture also called as the parallel, feed forward architecture is separated into two whole flash ADCs with feed-forward circuitry [9]. Figure 1, shows the diagram of two step flash architecture. The first part of the converter performs a rough evaluation of the value of the input voltage signal, and the second converter performs a fine conversion. The main benefit of this architecture is that the number of high speed comparators are greatly reduced from (2^N-1) to $2*(2^{N/2}-1)$, where N denotes the number of bits of the converter. In the proposed two-step flash ADC 6-bits are considered. So in a conventional flash ADC 63 comparators are essential, whereas the two-step flash ADC requires only 14 comparators. The given input signal voltage is first sampled by the Sample and Hold (S/H) circuit, and then the most significant bits (MSBs) are converted by the first 3-bit flash ADC. This output is then transformed back to an analog signal voltage with the 3-bit (DAC). The end result of this calculation, called as residue is now multiplied by an amplifier by $2^{N/2}$ times and given as input to the second 3-bit flash ADC. This second 3-bit ADC yields the least significant bits (LSB) through another flash conversion.

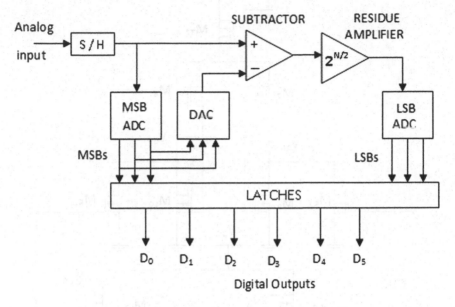

Fig. 1. Proposed diagram of two step Flash ADC

The inner blocks of two step flash ADC are explained below:

2.1 Comparator Circuit

Comparators implemented using CMOS technologies are the most vital basic building blocks in analog and mixed-mode circuit designs. The diagram of comparator circuit is presented in Fig. 2.

A CMOS comparator circuit is used to equate the given analog input signal with a reference signal (V_{ref}) and generates a binary signal output. In this proposed work the design of comparator plays a major role in high speed conversion. There are different kinds of comparator architectures such as clock based dynamic comparator, clock based double tail comparator, continuous time comparator, threshold inverted quantization (TIQ) based comparator and standard cell based comparator.

Clock built dynamic comparators are used in ultra-high speed applications due to the strong regenerative feedback used in the regenerative latch. The main advantage of these circuits are very less static power consumption, high input impedance and complete rail-to-rail output voltage swing is available. But the disadvantages are more delay due to the stacking of several transistors. The double tail comparator circuit offers less delay and operates at lower VDD. Finally standard cell comparators also work at lower supply voltages and gives better stability. In the proposed two step flash ADC, a modified double tail comparator circuit is used.

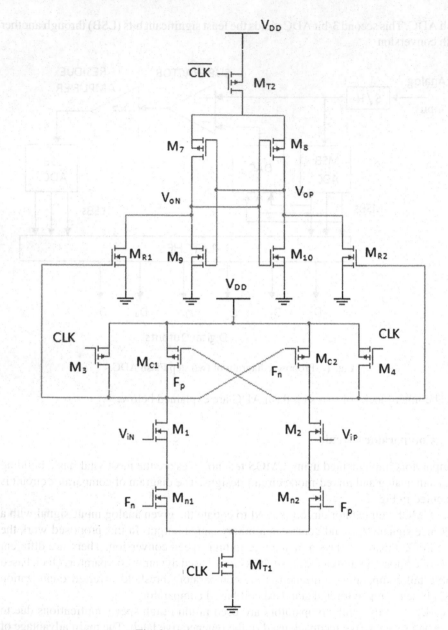

Fig. 2. Schematic of comparator circuit

During the reset stage of operation i.e. when CLK = 0, both M_{T1} and M_{T2} are off. So M_4 and M_3 pulls both F_p and F_n nodes to V_{DD}. Hence M_{C2} and M_{C1} are switched off. The intermediate stage consisting of transistors M_{R1} and M_{R2} resets both latch outputs to ground. Similarly during decision making phase i.e. when CLK = V_{DD}, both M_{T1} and M_{T2} are switched on. At the commencement of this decision making phase, both the transistors are still cut off. Thus F_n and F_p jump to fall with changed rates according to the input signal voltages.

2.2 3-Bit Flash ADC

It contains a resistive divider network with 8 resistors, which offers the reference voltage to each input terminal of comparator. The reference voltage for each comparator is one LSB greater than the reference voltage for the comparator proximately below it. The output of a comparator circuit produces logic high output when its analog input signal voltage exceeds the voltage at another input terminal applied to it. Else, the output voltage of the comparator becomes logic low. A thermometric code is then translated to get the suitable 3-bit digital code [8]. Figure 3, shows the diagram of 3-bit flash ADC.

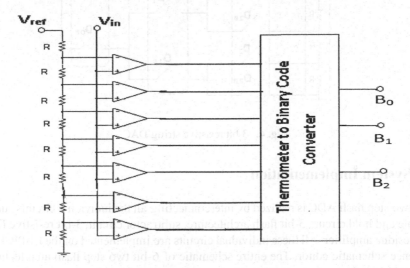

Fig. 3. Block diagram of 3-bit flash architecture

2.3 3-Bit Resistive DAC Circuit

Figure 4, shows a resistor string 3-bit DAC implemented with switches. In this implementation, transmission gates are used rather than n-channel switches. A transmission gate approach has extra source and drain capacitance to ground, but this extra capacitance is offset by the reduced series switch resistance, which is due to the inherent circuit configuration i.e. due to the parallel combination of n-channel and p-channel. Resistive type DACs does not suffer from the charge injection problems as noticed in the capacitive type DACs. Also a resistive type DAC has very small systematic error voltage and is inherently monotonic and compatible with purely digital technologies. Moreover the resistive type DAC occupies less silicon area than the capacitive type DAC.

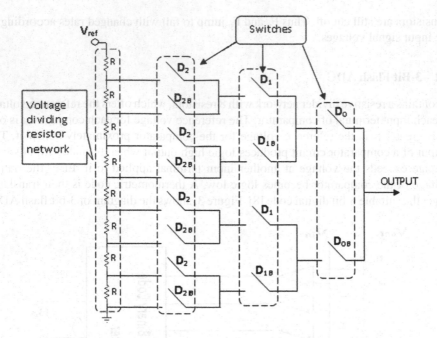

Fig. 4. 3-bit resistive string DAC

3 System Implementation

The two step flash ADC is realized by interconnecting all the individual circuits such as sample and hold circuit, 3-bit flash architecture, subtractor circuit, 3-bit resistive DAC, and residue amplifier. All these individual circuits are implemented on the HSPICE and Cadence schematic editor. The entire schematic of 6-bit two step flash architecture is shown in Fig. 5.

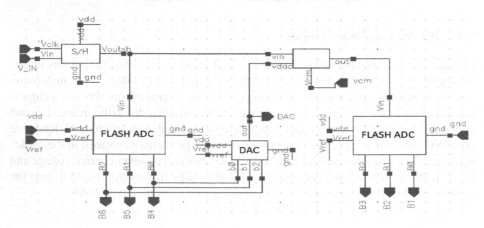

Fig. 5. Schematic diagram of 6-bit two-step flash ADC

4   Simulation

The summary of the proposed 6-bit two-step flash ADC was simulated using Cadence spectre in 130 nm CMOS technology. The simulated results of the double-tail comparator circuit are shown in Fig. 6.

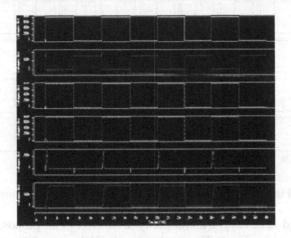

Fig. 6. Simulation results of double tail comparator

Figures 7 and 8 presents the simulated results of 3 bit flash ADC and 6 bit two-step flash architecture

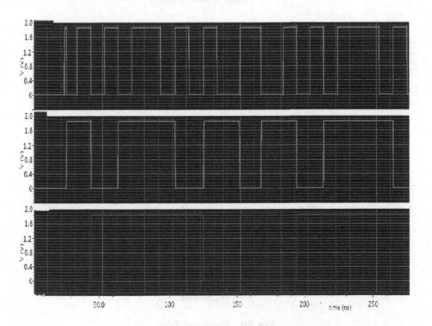

Fig. 7. Simulation results of 3-bit flash architecture

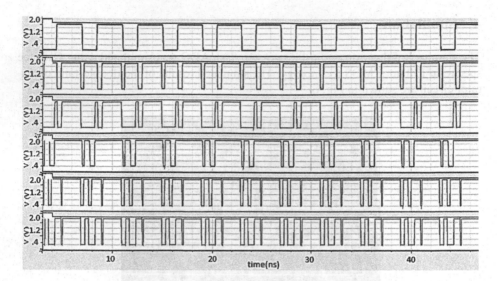

Fig. 8. Simulated results of 6-bit two-step flash architecture

The simulated static integral non-linearity (INL) error of the proposed two-step flash architecture is shown in Fig. 9. The simulated static DNL error of the proposed two-step flash architecture is presented in Fig. 10. Because INL and DNL are less than 0.4LSB, this proposed two step flash ADC exhibits higher linearity.

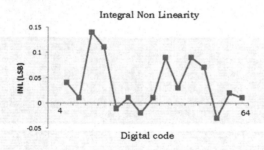

Fig. 9. Simulated INL

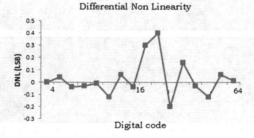

Fig. 10. Simulated DNL

The performance summary of the proposed 6 bit two step flash architecture is given in Table 1.

Table 1. Summary of two–step flash ADC

Technology	CMOS 130 nm
Resolution	6 bits
Input voltage range	1.0 V_{P-P}
Power Supply	1.8 V
Sampling frequency	10 GHz
Input Signal Frequency	upto 1 GHz
Area	0.226 mm^2
Power dissipation	2.32 mW
DNL	-0.18LSB $\leq$ DNL $\geq$ 0.42LSB
INL	-0.02LSB $\leq$ INL $\geq$ 0.15LSB
SNDR	68 dB

5 Conclusion

This work highlights the design, simulation and implementation of a high speed and low powered 6 bit two step flash ADC operating at a clock frequency of 10 GHz. As the sampling frequency of two-step flash ADCs increases, the design of comparators play a key role as it contribute to both low power and high speed. A modified double tail comparator circuit is used in the proposed system to attain the high operating speed and consumes low power. The proposed two-step flash architecture is operated at a power supply of 1.8 V and the obtained power consumption is only 2.32 mW. Operating speed of the proposed ADC is 10 GHz. The simulated static DNL and INL are measured to be between 0.15/$-$0.02LSB and 0.42/$-$0.18LSB respectively. The ADC occupies a silicon die area of 0.226 mm^2 shows a high signal to noise distortion ratio (SNDR) of 68 dB. These attractive specifications of the proposed two step flash architecture make it favorable for evolving applications such as in wireless USB system, where low power and high speed is of utmost importance.

References

1. Nasri, B., Sebastian, S.P., You,K.D., RanjithKumar, R., Shahrjerdi, D.: A 700 µW 1GS/s 4-bit folding-flash ADC in 65 nm CMOS for wideband wireless communications. In: 2017 International Symposium on Circuits and Systems (ISCAS), Baltimore, MD, pp. 1–4. IEEE (2017)
2. Wang, G., Lu, W., Zhang, L., Zhang, Y., Chen, Z.: 14-bit 20 µW column level two step ADC for 640 × 512 IRFPA. Electron. Lett. **51**(14), 1054–1056 (2013)
3. Varzaghani, A., et al.: 10.3GS/s, 6-Bit Flash ADC for 10G ethernet applications. IEEE J. Solid-State Circ. **48**(12), 3038–3048 (2013)

4. Ferragina, V., Ghittori, N., Maloberti, F.: Low-power 6-bit flash ADC for high-speed data converters architectures. In: 2006 International Symposium on Circuits and Systems, Island of Kos. IEEE (2006)
5. Ritter, P., Le Tual, S., Allard, B., Möller, M.: Design considerations for a 6 Bit 20 GS/s SiGe BiCMOS Flash ADC without track-and-hold. IEEE J. Solid-State Circ. **49**(9), 1886–1894 (2014)
6. Lien, Y.-C., Lin, Y.-Z., Chang, S.-J.: A 6-bit 1GS/s low-power flash ADC. In: 2009 International Symposium on VLSI Design, Automation and Test, Hsinchu, pp. 211–214 (2009)
7. Tretter, G., Khafaji, M.M., Fritsche, D., Carta, C., Ellinger, F.: Design and characterization of a 3-bit 24-GS/s flash ADC in 28-nm low-power digital CMOS. IEEE Trans. Microw. Theory Technol. **64**(4), 1143–1152 (2016)
8. Krishna, K.L., Srihari, D., Reena, D., Ramashri, T.: A 4b 40 Gbps 140 mW 2.2 mm^2 0.13 μm pipelined ADC for I-UWB receiver. In: 2013 International Conference on Computing, Communications and Networking Technologies (ICCCNT), pp. 1–6. IEEE (2013)
9. Baker, R.J.: CMOS Circuit Design, Layout and Simulation. IEEE Press Series on Microelectronic systems, vol. 2. Wiley-Interscience, Hoboken (2013)

Design and Implementation of Whale Optimization Algorithm Based PIDF Controller for AGC Problem in Unified System

Priyambada Satapathy, Sakti Prasad Mishra, Binod Kumar Sahu, Manoj Kumar Debnath$^{(\boxtimes)}$, and Pradeep Kumar Mohanty

Siksha 'O' Anusandhan (Deemed to be University), Bhubaneswar, Odisha, India
lirasatapathy@gmail.com,
saktiprasadmishra9@gmail.com, binoditer@gmail.com,
mkd.odisha@gmail.com, pkmohanty68@rediffmail.com

Abstract. Design aspect of controller mechanism to regulate the load frequency in an interconnected generating system is usually regarded as an optimization issue with various constraints. As per the need of the designer, the controller must be designed to fulfill the desired requirements to yield a suitable solutions. This paper presents whale optimization algorithm (WOA) based conventional proportional integral derivative type controller where a filter is employed in the derivative part of the controller for a two area combined thermal power system. The derivative filter is included here to reduce the unwanted signal in the input side of the controller. In the optimization process time based error function is taken as fitness function (ITAE) for designing controller parameters. The performance ability of the controller is modeled with MATLAB Simulink package. The efficiency and dominace of recommended WOA tuned PIDF controller is compared with Jaya based PIDF controller using time domain simulations which validates the effectiveness of WOA based proposed controller employed in the aforesaid dual area combined power system.

Keywords: Whale optimization algorithm
PID with derivative filter controller · Automatic generation control

1 Introduction

For load frequency control (LFC) the main aim is to match the overall power generation as compared to load demand. Any deviation between demand and generation yields the deviation of system frequency from the nominal value [1]. The tie-lines exchange the contract power with in areas and afford inter area provision during aberrant condition. Any type of anomalous circumstances and variations in load in area cause misleading in swapping of power between the generating stations. The severe frequency abnormality results moderately collapse to system. So as to maintain the active power output of generators within a predefined limit, load frequency control (LFC) system is implemented [2, 3] which take care about the mismatching in scheduled power exchanges and frequency of system. A mechanism is required which balance the desired tie-line power flow and system frequency as well as helps the

© Springer Nature Singapore Pte Ltd. 2018
I. Zelinka et al. (Eds.): ICSCS 2018, CCIS 837, pp. 837–846, 2018.
https://doi.org/10.1007/978-981-13-1936-5_85

system to regain its stability. The popular employed controller for AGC is the PI (proportional integral) type controller. The PID (proportional integral derivative) and PI controllers are extremely modest for application having no cost, easy realization, robust in nature and yields better dynamics response. However, when complexity in the system hikes their performance goes down. It is quite oblivious that by the application of fixed gain conventional PI and PID controllers, although the system stability is reduced to zero but its performance becomes worst when complexity of power system rises because of introduction of nonlinearities in system and increase in correcting areas. Considering the wide range of operations the fixed gain controllers are architected for the failures in the normal operating environment for providing better results. The articles [4–7] discussed about various classical as well as adaptive controllers. The algorithms of these controllers are very complex with regular identifications of some online models and there by getting desired outputs. Topics including Artificial Neural Networks (ANNs) and Fuzzy Neural Networks (FNNs) as well as Fuzzy Logic Systems (FLSs) are discussed in [8–10]. In fuzzy techniques, the salient features are that it furnishes a description that is free of models to control the system. The controllers optimized by the ANN technique are very promising in increasing the performance of the system once it is utilized with proper quantities of neural network along with plant identification. FNN takes the advantages of both ANN in learning the process and FLS, in handing unpredictable information's [11–17].

2 Proposed System

The layout of a dual area combined system consisting of two non-reheat type thermal power plant [15] is shown in Fig. 1. The tie-line in the power system connects various control areas that are interconnected with each other. Different blocks in the power system are used to represent various complex as well as components of the system in the model. minimum phase along with time dependant non-linear systems are included in various structures. All the generators in each area are supposed to form coherent groups. Each area requires its frequency of system & tie line power to be monitored. In Fig. 1, ACE1 & ACE2 represent area control errors; B1 & B2 frequency bias factors; R1 & R2 governor speed regulation constants; Tg1 & Tg2 governor time constants in sec; u1 & u2 are the control input and output to speed governing system; ΔPg1 & ΔPg2 are change in valve positions of governor; ΔPt1 & ΔPt2 are changes in valve positions of output & represent turbine time coefficient in sec; TP1 & TP2 are power systems time coefficient in sec; KP1 & KP2 are gains of power systems; & are changes in load demands; T12 is synchronization constants; ΔPtie, 1–2 denotes change in tie-line power in pu and Δf1 & Δf2 are the system frequency deviation in Hertz. The inputs for the speed governing system are control outputs u1 & u2.

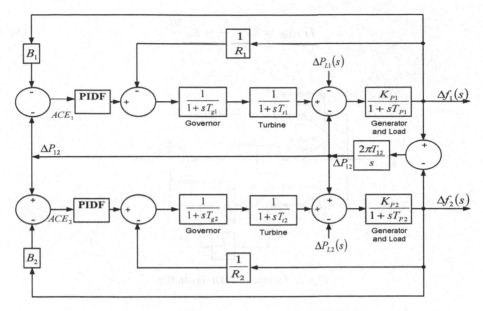

Fig. 1. Two area interconnected power system with PIDF controller

3 Proposed Controller

PID-controller is generally used for getting both stability as well as true system output. It is combination of three control actions namely P-mode, I-mode and D-mode. The input to PID controller is the discrepancy between reference and actual output. The reduction of oscillations, ensure of zero steady error and making system faster are taken care by P-mode, I-mode and D-mode control action respectively. Sensors usually produce noise which is typically of higher frequency showing higher values of derivative of that produced noise. Noise may be produced by tie line telemetering system and tie-line power present in area control error [ACE]. Placement of a filter along with derivative part will result in reduction of unwanted sgnal of high frequency. Since noise has effect on the derivative part, a tuned filter is employed in the derivative part of the regulator instead of the output part of PID controllers. For two equal areas, two similar PID controllers with derivative filters (PIDF) are employed. Figure 2 shows structure of PIDF controller.

$$ACE_1 = B_1 \Delta f_1 + \Delta P_{tie1-2} \tag{1}$$

$$ACE_2 = B_2 \Delta f_2 + \Delta P_{tie2-1} \tag{2}$$

u_1 & u_2 are the output of proposed controllers & input to the system model. Transfer function of said controller is expressed by:

$$TF_{PIDN} = K_p + \frac{K_i}{s} + K_d \frac{Ns}{s} \tag{3}$$

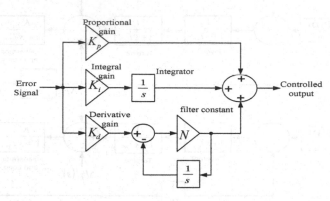

Fig. 2. Outline of PIDF controller

4 Woa Optimization Technique

WOA algorithm is developed by Mirjalili and Lewis [18] in 2016. This algorithm is mimics the bubble-net hunting strategy of humpback whales. Various processes included in WOA algorithm are:

i. Initialization: - This phase is common for all most all optimization algorithms. In this phase a first population having dimension [NP × D] is produced, where 'NP' signifies the number of population and 'D' indicates the dimension. After evaluating the initial population, the best performing solution is taken as global best or optimum location of the search agent (humpback whales) to target prey.

ii. Encircling prey: - In this phase other search agents (humpback whales) tries to improve their corresponding locations in the direction of the best agent. Mathematically the updated positions are conveyed as:

$$X_{new} = X_{best} - A \times D \tag{4}$$

$$D = |C \times X_{best} - X| \tag{5}$$

Where, 'C' and 'A' are coefficient vectors generated as:

$$C = 2r \tag{6}$$

$$A = 2ar - a \tag{7}$$

Here during the course of iteration the value of 'a' decreases linearly from 2 to and the arbitrary number 'r' is assigned in the range [0, 1].

iii. Attacking method: - In the exploitation phase the bubble-net natue of humpback whales, the following two methodologies are developed:

    a. Shrinking enclosing method: - This activity is realized by updating the initial population using Eqs. (4–7).

    b. Spiral updating position: - In this stage a spiral equation is used between the individual solution and the best performing solution to mimic the helix-shaped motion of humpback whales which is given by

$$X_{new} = D'e^{bl}\cos(2\pi l) + X_{best} \tag{8}$$

Where $D' = |X_{best} - X|$ constant 'b' is used to define the form of logarithmic spiral and 'l' is an arbitrary number defined between [0, 1].

Humpback whales spin nearby the target following both circular and spiral paths. This behavior is modeled mathematically by assuming 50% probability for both the circular and spiral paths as follows:

$$X_{new} = \begin{cases} X_{best} - A.D & \text{if } p < 0.5 \\ D'.e^{bl}.\cos(2\pi l) + X_{best} & \text{if } p \geq 0.5 \end{cases} \tag{9}$$

Here 'p' stands for an arbitrary value between 0 to 1.

iv. Search for prey phase (exploration phase): - In this phase the humpback whale position is updated by arbitrarily preferred search individual rather than finest search individual. Mathematically this phase is conveyed as

$$X_{new} = X_{rand} - A.D \tag{10}$$

Where,

$$D = |C.X_{rand} - X| \tag{11}$$

And $C = 2r$ same as Eq. (7).

5 Simulation Result and Discussion

For this analysis, a non-reheat based dual area thermal power plants is simulated under MATLAB simuink surroundings. Nominal constraints of the power system under study are taken from reference [15]. The controller parameters for PIDF are considered in the range of [0–5] for Kp, Ki and Kd and [0–500] for filter coefficient N. Size of the population and total number of iterations are both considered as 100. Two different cases are considered for performance study of the proposed controller.

    Case-1: In area-1 an abrupt disturbance of 0.05 pu and in area-2 nominal loading
    Case-2: In area-and area-2 an abrupt disturbance of 0.05 pu.

    The simulation results of WOA optimized PIDF controller with above two cases are listed in Tables 1, 2, 3 and 4. The Tables 1 and 2 show the values of objective function (ITAE) and controller parameters i.e Kp, K_i, K_d and N for both the cases with WOA based- and Jaya based PIDF controllers. Overviews of deviation in frequencies and tie line power with regard to least undershoot and settling times for both the case studies are presented in Tables 3 and 4.

From Table 1, it is clearly visible that smaller value of ITAE of 0.0252 is obtained with WOA optimized PIDF controller as compared to Jaya based PIDF controller. Hence ITAE of proposed WOA optimized PIDF controller improves by 73% in comparison with Jaya optimization based controller in case-1. Also Table 3 reveals that the settling time in sec of frequency deviation (Δf_1 and Δf_2) and tie-line power deviation (ΔP_{tie}) of both the areas are minimum for WOA optimized controller.

From Table 3, it is noticed that % improvement of settling time of Δf_1, Δf_2 and ΔP_{tie} with WOA based PIDF controller as compared to Jaya based controller are 48.5%, −15.7% and 6% respectively. Similarly from Table 3 it is clear that % improvement of peak undershoot of Δf_1, Δf_2 and ΔP_{tie} with WOA optimization based PIDF controller as compared to Jaya based controller are 50.9%, 76.4% and 72.3% respectively.

Tables 2 and 4 depicts the simulation results for case-2. The recorded results in Table 2 in terms of objective function shows that better objective function (ITAE = 0.1407) is obtained with proposed algorithm as compared to Jaya optimized controller (ITAE = 0.4020) i.e. an improvement of 65%. Similar good results in terms of frequency deviations and tie-line power deviations in terms of settling time and peak undershoot are noted in Table 4 with proposed WOA based PIDF controller as compared to Jaya based PIDF controller.

For further deep analysis, the time response based simulation of frequency deviation in both areas under two dissimilar circumstances are exposed in figures [3–8]. In unified system the application of any load perturbation in any one of the areas results in oscillations of frequencies in both the area. Under case-1, Figs. 3 and 4 demonstrate the deviations in frequencies of all the associated areas and Fig. 5 represents the tie-line power deviation. Figures 3, 4 and 5 conclude that frequency and tie-line power deviations settle down to zero very quickly with proposed algorithm. Figures 6, 7 and 8 show the oscillations of frequency and interline power with case-2 where heavy load changes are considered in both areas. In case-2 also the figures reveal that proposed WOA based controller show better result.

Table 1. Optimum value of ITAE and gain parameters for case-1.

Controller	K_P	K_I	K_D	N	ITAE
WOA based PIDF	Area-1				
	4.8305	5.0000	1.4306	244.9165	0.0252
	Area-2				
	4.2456	3.0681	4.2080	125.6532	
Jaya based PIDF [15]	Area-1				0.0935
	0.8375	2.0216	0.3694	368.92	
	Area-2				
	0.8375	2.0216	0.3694	368.92	

Table 2. Finest values of ITAE and gain constraints for case-2.

Controller		K_P	K_I	K_D	N	ITAE
WOA based PIDF	Area-1					
		5.0000	3.8245	3.5024	414.6503	0.1407
	Area-2					
		5.0000	4.3302	3.8304	380.9821	
Jaya based PIDF [15]	Area-1					0.4020
		1.9505	1.9211	0.8168	439.05	
	Area-2					
		1.9505	1.9211	0.8168	139.05	

Table 3. Simulation results for case-1.

Controller	Undershoots			Setting time in sec		
	$\Delta f_1 \times 10^{-3}$ in Hz	$\Delta f_2 \times 10^{-3}$ in Hz	$\Delta P_{tie} \times 10^{-3}$ in pu	Δf_1	Δf_2	ΔP_{tie}
WOA based PIDF	0.0294	0.0085	0.0034	1.47	3.13	2.52
Jaya based PIDF [15]	0.0599	0.0361	0.0123	2.85	2.70	2.68

Table 4. Simulation results for case-2.

Controller	Undershoots			Setting time in sec		
	$\Delta f_1 \times 10^{-3}$ in Hz	$\Delta f_2 \times 10^{-3}$ in Hz	$\Delta P_{tie} \times 10^{-3}$ in pu	Δf_1	Δf_2	ΔP_{tie}
WOA based PIDF	0.0451	0.0063	0.0134	3.52	4.46	4.19
Jaya based PIDF [15]	0.0946	0.0946	0.0345	4.28	4.27	5.18

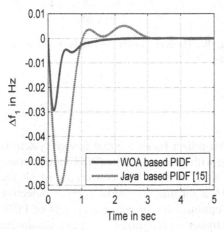

Fig. 3. Frequency oscillation in area-1 (case-1).   **Fig. 4.** Frequency oscillation in area-2 (case-1).

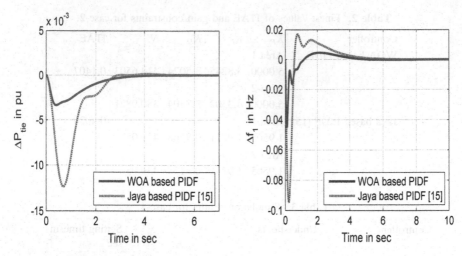

Fig. 5. Tie-line power oscillation (case-1).   **Fig. 6.** Frequency oscillation in area-1 (case-2).

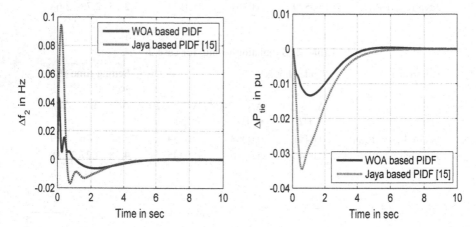

Fig. 7. Frequency oscillation in area-2 (case-2).   **Fig. 8.** Tie-line power oscillation in area-2 (case-2).

6   Conclusion

In this study, to face challenges of AGC a non-reheat based thermal power system is addressed. Conventional PID controllers along with derivative filter (PIDF) are adopted in both the areas to enrich the activities of AGC system. The constraints of all the controllers are optimally designed using WOA algorithm. The most popular objective function in the form of ITAE is selected for optimization of gain parameters of PIDF controller. The advantage of the projected controller is recognized under two dissimilar circumstances. Time domain analysis for the response of the system frequencies and interline power of all associated areas demonstrates the preeminence of the projected controller for the inspected system. The effectiveness of WOA algorithm is also successfully compared with Jaya algorithm.

References

1. Behera, N.: Load frequency control of power system. Diss. (2013)
2. Christie, R.D., Bose, A.: Load frequency control issues in power system operations after deregulation. IEEE Trans. Power Syst. **11**(3), 1191–1200 (1996)
3. Pan, C.-T., Liaw, C.-M.: An adaptive controller for power system load-frequency control. IEEE Trans. Power Syst. **4**(1), 122–128 (1989)
4. Sahu, R.K., Panda, S., Rout, U.K.: DE optimized parallel 2-DOF PID controller for load frequency control of power system with governor dead-band nonlinearity. Int. J. Electr. Power Energy Syst. **49**, 19–33 (2013)
5. Kocaarslan, I., Çam, E.: Fuzzy logic controller in interconnected electrical power systems for load-frequency control. Int. J. Electr. Power Energy Syst. **27**(8), 542–549 (2005)
6. Ali, E.S., Abd-Elazim, S.M.: BFOA based design of PID controller for two area load frequency control with nonlinearities. Int. J. Electr. Power Energy Syst. **51**, 224–231 (2013)
7. Behera, A., et al.: Hybrid ITLBO-DE optimized fuzzy pi controller for multi-area automatic generation control with generation rate constraint. In: Satapathy, S., Bhateja, V., Das, S. (eds.) Smart Computing and Informatics. Smart Innovation, Systems and Technologies, vol. 77, pp. 713–722. Springer, Singapore (2018)
8. Çam, E.: Application of fuzzy logic for load frequency control of hydroelectrical power plants. Energy Convers. Manag. **48**(4), 1281–1288 (2007)
9. Talaq, J., Al-Basri, F.: Adaptive fuzzy gain scheduling for load frequency control. IEEE Trans. Power Syst. **14**(1), 145–150 (1999)
10. Yeşil, E., Güzelkaya, M., Eksin, I.: Self tuning fuzzy PID type load and frequency controller. Energy Convers. Manag. **45**(3), 377–390 (2004)
11. Demiroren, A., Zeynelgil, H.L., Sengor, N.S.: The application of ANN technique to load-frequency control for three-area power system. In: Power Tech Proceedings, 2001 IEEE Porto, vol. 2. IEEE (2001)
12. Shayeghi, H., Shayanfar, H.A.: Application of ANN technique based on μ-synthesis to load frequency control of interconnected power system. Int. J. Electr. Power Energy Syst. **28**(7), 503–511 (2006)
13. Bahgaat, N.K., et al.: Load frequency control in power system via improving PID controller based on particle swarm optimization and ANFIS techniques. In: Research Methods: Concepts, Methodologies, Tools, and Applications: Concepts, Methodologies, Tools, and Applications, vol. 462 (2015)
14. Sahu, B.K., Pati, S., Mohanty, P.K., Panda, S.: Teaching–learning based optimization algorithm based fuzzy-PID controller for automatic generation control of multi-area power system. Appl. Soft Comput. **27**, 240–249 (2015)
15. Singh, S.P., Prakash, T., Singh, V.P.: M G Babu Analytic hierarchy process based automatic generation control of multi area interconnected power system using jaya algorithm. Eng. Appl. Artif. Intell. **60**, 35–44 (2017)
16. Lal, D.K., Barisal, A.K., Tripathy, M.: Grey wolf optimizer algorithm based Fuzzy PID controller for AGC of multi-area power system with TCPS. Procedia Comput. Sci. **92**, 99–105 (2016)

17. Demiroren, A., Yesil, E.: Automatic generation control with fuzzy logic controllers in the power system including SMES units. Int. J. Electr. Power Energy Syst. **26**(4), 291–305 (2004)
18. Mirjalili, S., Lewis, A.: The whale optimization algorithm. Adv. Eng. Softw. **95**, 51–67 (2016)

PMSM Control by Deadbeat Predictive Current Control

R. Reshma[✉] and J. Vishnu

Electrical and Electronics, Sree Buddha College of Engineering, Pattoor, India
reshmarekhanaluthengil@gmail.com, vishnu052@gmail.com

Abstract. Considering classical framework imbalance and one-leap lag, an revised current controller is introduced in this work. Synchronous Motor using permanent magnet (PMSM) current control of is boosted up using this controller. By improving the current control, losses in the machine are minimized and machine performance can be improved. For doing this, traditional prognostic current control performance is figured out. Depending on sliding mode aggressive reaching law, a new disturbance observer which takes stator current as input, is being proposed. Stator Current Disturbance Observer (SCDO) can concurrently forecast approaching stator current's value and record system disturbance due to framework imbalance. For compensating the value of voltage reference determined by deadbeat predictive controller, a feed ahead value is considered. This feed ahead value is the prediction currents based on SCDO. These are used for restoring the fragmented current in DPCC. By connecting the control part using deadbeat predictive technique, current forecasting and feed forward rectification part based on SCDO, a compound control approach is developed in the work. This paper focuses an robust SCDO based on new modifying law which can increase the control performance of the existing arrangement. Electric Vehicle control can be used as an application.

Keywords: Classical framework · Deadbeat · Disturbance observer
Permanent magnet synchronous motor (PMSM) · Predictive control
Sliding mode

1 Introduction

For smooth torque control in PMSM drives, large performance of current control is necessary. For this, a novel predictive current controller which is robust and discrete is proposed for drive systems using PMSM. Deadbeat structure is mainly used to design controller and current prediction schemes. This deadbeat control has good varying response but it contains parametric variations and unmodelled dynamics. For providing good condition, a discrete time integral term is added to the deadbeat current forecasting. The controller is easy to apply and suitable for achieving high performance in PMSM applications. Digital control systems of PMSM are now frequently used in industrial applications. For a realistic PMSM system, current control performance affects the performance of the system. Many current control methods have been studied for getting high constant state precision and fast progressive torque response. Hysteresis control [2],

© Springer Nature Singapore Pte Ltd. 2018
I. Zelinka et al. (Eds.): ICSCS 2018, CCIS 837, pp. 847–854, 2018.
https://doi.org/10.1007/978-981-13-1936-5_86

proportional – integral control [3] and predictive control [4] are the most common current control methods. For enhancing the control performance and satisfying the effects of system disturbances, the blending of control using prediction technique and disturbance observer has been studied.

2   Modelling of PMSM

The following equations have been applied for modeling of PMSM [9]. The direct and quadrature-axis voltages are given as,

$$V_q = R_q i_q + P\lambda_q + \omega_r \lambda_d \tag{1}$$

$$V_d = R_d i_d + P\lambda_d - \omega_r \lambda_q \tag{2}$$

The direct and quadrature axis fluxes are,

$$\lambda_q = Li_q \tag{3}$$

$$\lambda_d = L_d i_d + \lambda_f \tag{4}$$

The d- and q-axis currents are given as

$$i_d = \int \frac{1}{L_d}(V_d - i_d R_{d+} \omega_r L_q i_q) \tag{5}$$

$$i_q = \int \frac{1}{L_q}(V_q - i_q R_q - \omega_r (L_d i_{d+} L_m i_{dr}) \tag{6}$$

The torque produced is given as,

$$T_e = \frac{3P}{4}\left(\lambda_d i_q - \lambda_q i_d\right) \tag{7}$$

Mechanical torque developed is given by,

$$T_e = T_L + B\omega_m + J\frac{d\omega_m}{dt} \tag{8}$$

The rotor mechanical speed of the motor is given by,

$$\omega_m = \int \frac{1}{J}(T_e - T_L - B\omega_m) \tag{9}$$

and

$$\omega_e = \frac{P}{2}\omega_m \tag{10}$$

The modelling of Permanent Magnet Synchronous Motor is executed using the above Eq. [10]. Figure 1 shows the modelling of PMSM. Table 1 depicts the parameters of the machine.

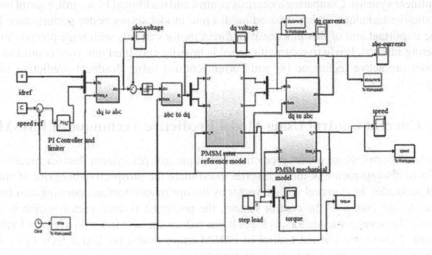

Fig. 1. MATLAB model of PMSM

Table 1. Machine parameters

Power_rated	4 kW
Voltage_rated	400 V
Speed_rated	1500 rpm
Current_rated	10A
Flux linkage	0.175 Wb
Number of poles	4
d-axis & q-axis inductance	9.5 mH
Stator resistance	2.875 Ω

The output of electrical model is the direct axis and quadrature axis currents and the electrical torque [11]. Speed of the machine is the output of mechanical model.

3   Predictive Current Control

Hysteresis control has been used for the current control of PMSM [5]. This has much influence such as rapid current responses, good heftiness, and easy breakthrough application. The main problems of this method are large current swell and fluctuating switching constancy. The advantages of Proportional Integral(PI) based current control are high constant-state control precision and invariable switching frequency [6]. This advantages have made possible for improving the deadlock remnant control system performance for various applications. Control system of PMSM consists of inescapable disturbances as well as data changes. This has made it impossible for above control algorithms to obtain a useful varying performance over an entire operating range for nonlinear systems. Comparing hysteresis control and traditional PI control, control using predictive technique which is based on discrete model shows better performance [7]. The important aim of this practice is to direct motor currents with huge precision in a fleeting interval. Predictive current control is broadly classified into two: control using model predictive technique [8] and current control using deadbeat predictive technique [9].

4 Current Control Using Model Predictive Technique of PMSM

Current control using model predictive technique utilizes system distinct model and built-in discrete nature of motor inverter to estimate the prospective behavior of states and persuades the eventual voltage vector by the optimization of an operating cost function. As the output of the control system, the preferred voltage vector, which is one among the seven basic vectors, is used. It can reduce the cost function to a small value. Figure 2 shows the current control of PMSM using model predictive technique. The voltage signals for the motor are generated from an inverter. The inverter consists of six switches. The switching signals for the switches are generated by Pulse Width Modulation using Space Vector technique (SVPWM). SVPWM switching signals are created from Model Predictive Current Control. The currents from the motor are controlled by MPCC and the switching sequences are generated accordingly.

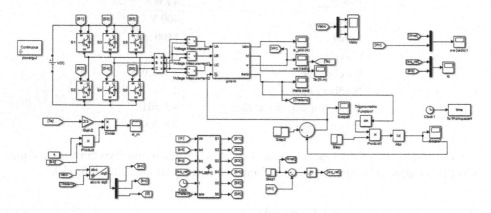

Fig. 2. MPCC of PMSM

5 Deadbeat Predictive Current Control

The block diagram of the proposed system is shown in Fig. 3 [1]. First, we have to measure stator current, speed of motor and angular displacement. Then we have to predict the future values of direct axis and quadrature axis currents which is performed by adaptive stator current and disturbance observer (ASCDO). Predicted values of d and q-axis currents, the reference values of currents and speed of motor are the inputs to DPCC. DPCC calculates the voltage vectors. ASCDO also estimates the predicted parameter disturbances. Both these become the inputs to the sum block. The voltage output of the controller becomes the input of Space Vector Pulse Width Modulation (SVPWM) block and these modulation schemes makes the switching sequences to inverter and inverter is connected to PMSM.

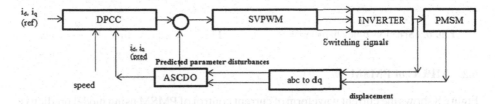

Fig. 3. Deadbeat predictive current control of PMSM

6 Simulation Results and Discussions

6.1 Modelling of PMSM

The following graphs show the simulation results of PMSM modeling in MATLAB. Figure 4 shows the voltage waveforms of PMSM. The peak value of PMSM is 325 V. Figure 5 depicts the current waveforms of PMSM. The peak value of current is 10 A.

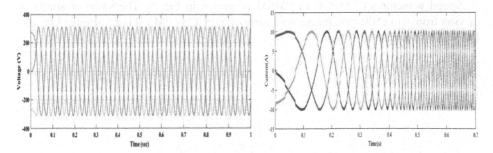

Fig. 4. Voltage vs time characteristics   **Fig. 5.** Current vs time characteristics

Figure 6 shows the change of speed with time. Speed increases from 0 to 1500 rpm and remains steady after 0.2 s. The steady state speed is the same as that of the commanded reference speed which validates the simulation result. Figure 7 depicts the developed torque of the motor. Developed torque follows the load torque after 0.3 s

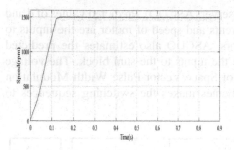

Fig. 6. Speed vs time characteristics   **Fig. 7.** Torque vs time characteristics

6.2 MPCC of PMSM

Figure 8 shows the current waveform of current control of PMSM using model predictive technique. The rms value of current is 14.14 A.

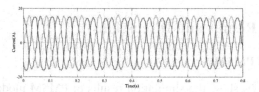

Fig. 8. Current vs time characteristics

Speed waveform of MPCC of PMSM is shown in Fig. 9. The value of speed is varying from 0 to 1500 rpm. torque waveform of MPCC of PMSM is depicted in Fig. 10. The value of torque is 10 Nm.

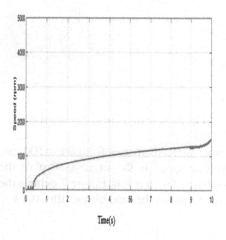
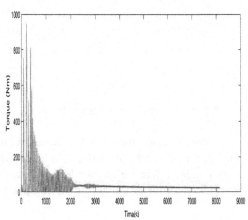

Fig. 9. Speed vs time characteristics   **Fig. 10.** Torque vs time characteristics

7   Conclusion

The modelling of machine is completed and waveforms are obtained successfully. Current control using Model Predictive technique of PMSM is done. The predictive current control algorithm of permanent magnet synchronous motor (PMSM) drives have clear-cut current tracking and consistent switching frequency. The proposed algorithm is very easy and can be effectively implemented. This current control based on deadbeat predictive with adaptive SCDO (ASCDO) can be applied for Electric Vehicle (EV) control. To improve the control performance of PMSM system, a complex algorithm combining Deadbeat Predictive Current Controller and Stator Current and Disturbance Observer is used. A novel sliding mode exponential reaching law is introduced to suppress the sliding-mode chattering of SCDO. A DPCC + ASCDO method is developed to further improve the performance of DPCC + SCDO method. The novelty of this paper is that deadbeat predictive current control with adaptive SCDO (ASCDO) can be applied for Electric Vehicle (EV) control. The main advantages of motor drives are high responsivity, ease of control, and low noise. By the above said advantages, we can give vehicles improved stable flexibility, security, and bankability.

References

1. Zang, X., Hhou, B., Meaii, Y.: Deadbeat predictive current control of PMSM with stator current and disturbance observer. IEEE Trans. Power Electron. **34**(6), 3819–3834 (2017)
2. Sheuaul, J.A., Lijøeekelsøy, K., Midteosund, T., Undeiaaland, T.: Synchronous reference frame hysteresis current-control for grid converter application. IEEE Trans. Ind. Electron. **48**(5), 2185–2195 (2011)
3. Kazsmirkowseki, M.P., Malleasani, L.: Current control techniques for three phase voltage source PWM converters: a survey. IEEE Trans. Ind. Electron. **46**(5), 681–793 (1998)
4. Xaeiae, W., Whang, X., Wahneg, F., Xuei, W., Kenneeal, R.M., Lhorenz, R.D.: Finite control set MPTC with a deadbeat solution for PMSM drives. IEEE Trans. Ind. Electron. **72**(9), 5402–5410 (2015)
5. Khaeng, B.J., Lheiaw, C.M.: A robust hysteresis current controlled PWM inverter for linear PMSM driven magnetic suspended positioning system. IEEE Trans. Power Electron. **48**(5), 957–967 (2001)
6. Shong, W., Mai, J., Zheou, L., Feing, X.: Deadbeat predictive power control of single-phase three-level neutral-point-clamped converters using space-vector modulation for electric railway traction. IEEE Trans. Power Electron. **31**(1), 722–732 (2016)
7. Lien, C.K., Lieu, T.H., Yui, J.T., Fui, L.C., Hsiaoe, C.F.: Model free predictive current control for interior permanent-magnet synchronous motor drives based on current difference detection technique. IEEE Trans. Ind. Electron. **61**(2), 668–682 (2014)
8. Prendli, M., Schaletz, E.: Sensorless model predictive direct current control using novel second-order PLL observer for PMSM drive systems. IEEE Trans. Ind. Electron. **58**(9), 4088–4096 (2011)

9. Sebbastian, T., Salemon, G., Rahhman, M.: Modelling of permanent magnet synchronous motors. IEEE Trans. Magn. **22**, 1070–1072 (1986)

10. Pillayi, P., Krishna, R.: Modelling of permanent magnet motor drives. IEEE Trans. Ind. Electron. **35**, 538–542 (1988)

11. Ceui, B., Zheou, J., Rhen, Z.: Modelling and simulation of permanent magnet synchronous motor drives (2001)

Green Energy

Improving the Performance of Sigmoid Kernels in Multiclass SVM Using Optimization Techniques for Agricultural Fertilizer Recommendation System

M. S. Suchithra[✉] and Maya L. Pai

Department of Computer Science and IT, Amrita School of Arts and Sciences,
Kochi Amrita Vishwa Vidyapeetham, Edappally, India
suchithrams194@gmail.com, mayalpai@gmail.com

Abstract. Support Vector Machines (SVM) are advancing rapidly in the field of machine learning due to their enhancing performance in categorization and prediction. But it is also known that the performance of SVM can be affected by different kernel tricks and regularization parameters like Cost and Gamma. The polynomial kernel seems to be more suitable for performing multiclass SVM classification for the dataset used here. In this study, we propose an improved sigmoid kernel SVM classifier by adjusting the cost and gamma parameters with which a better performance can be achieved. The study is conducted for a multiclass soil fertilizer recommendation system for paddy fields. Furthermore, different optimization methods like Genetic Algorithm and Particle Swarm Optimization are used to tune the SVM parameters. Finally, a comparative study on the performance is also done for the different choices of the parameters, pointing out their accuracies.

Keywords: SVM · Sigmoid kernel tricks · Tuning parameters
Genetic algorithm · Particle swarm optimization · Soil fertilizers

1 Introduction

Earlier studies have shown that Support Vector Machines (SVMs) perform well in solving multiclass classification problems [1, 2]. Mainly two types of approaches are used to solve the same. One is by using the optimization problem, and the other by decomposing it into several binary SVM training problems. Structural SVMs (SSVMs) solve the multiclass problem as one optimization problem however one versus one (OVO) and one versus all (OVA) decomposition solve using several binary SVM training problems. With several approaches to multi-class SVM training and classification, choosing a proper approach to solve a particular problem is difficult, since different training methods commonly produce SVMs with significantly different effectiveness. Existing studies have shown that no single method outperforms others in solving every problem [2, 3]. Improving the efficiency of these classifiers have been a wide study space in machine learning above the previous two decades. SVM is a strong classification tool, which efficiently overcomes numerous traditional classification

© Springer Nature Singapore Pte Ltd. 2018
I. Zelinka et al. (Eds.): ICSCS 2018, CCIS 837, pp. 857–868, 2018.
https://doi.org/10.1007/978-981-13-1936-5_87

difficulties like the curse of dimensionality and local optimum. In SVM one usually surfaces with hindrances in Quadratic Optimization, Kernel Mapping and Maximum Margin Classifiers [4]. The focus of this paper is the first issue, Kernel Mapping. Multiclass SVM such as OAO-SVM and OAA-SVM decomposes multiclass labels into several two class labels and it trains an SVM classifier to reconstruct the obtained solutions from each two class labels of the problem, by combining them [5].

Soil problems resulting from either excess or shortage of certain elements, as a consequence of specific soil properties, are global and significant constraints to agriculture production. Though soil research and soil test based advisory services stated in the Kerala State decades back, their effectiveness was not appreciated in the light of maintenance of soil health and hence enhancing crop productivity. The current status of soil testing services by the laboratories under Department of Agriculture in Kerala is effective enough to meet the requirements of the farmers. These labs analyze different soils and give recommendations regarding the nutrients based on the results [6]. According to the reports of 2012, the contribution of agriculture to the Gross State Domestic Product (GSDP) of the Kerala state has been steadily declining from 36.99% in 1980–81 to 9.14% in 2011–12, in spite of the fact that the share of irrigated area in Gross Cropped Area (GCA) which was 13.20% in 1980–81 increased to 18.43% in 2011–12. The above statistics gives us the conclusion of decreasing productivity of food crops [7].

The cropping pattern has also undergone significant changes over time. The share of food crops in GCA was 37.46% in 1980–81 but dropped to 11.50% by 2011–12. The share of major cash crops in GCA however, increased from 44.59% to 62.51% during the same period. The shift from food crops to cash crops is clearly discernable. The percentage of agricultural laborers which was 28.23% in 1981 declined to 16.10% in 2011. The total area under cultivation increased by 24.87% from 1960–61(23.48 lakh hectare-GCA) to 1970–71 (29.33 lakh hectare) in the State and reached 30.21 lakh hectare in 2000–01. However, the last decade observed a decrease and the trend still continues, and the GCA declined to 26.62 lakh hectare in 2011–12. One of the key motives for the drop in GCA was due to decline in area under paddy cultivation over the years and was replaced by more remunerative and less labor absorbing crops like coconut, banana, vegetables etc. Concerted efforts are required to augment rice production to meet at least 20% of our requirement through appropriate support and incentive mechanisms.

Intensive cultivation often with incorrect soil and unproductive crop management practices have given rise to a heavy loss in soil quality. Organic manure application has been greatly neglected in the State in recent years due to lack of availability and high cost. This may have an adverse effect on soil quality parameters [8]. Improper or excessive use of chemical fertilizers has created imbalances in plant available nutrients in the soil. All these have affected the productivity of Kerala soils. The Machine Learning (ML), offer new possibilities in the field of agriculture and may help in data evaluation and decision making. It helps to predict the application rate of fertilizers for Kerala soils. In this study, the main fertilizers selected for prediction are Urea, MOP (Muriate of Potash), Phosphate and Organic Fertilizers.

In this paper, we do not concentrate on deriving a general method for multiclass classification but intend to optimize the sigmoid kernel SVM classifier, in order to make it a better option over the others. Our key idea is to train the data using the sigmoid kernel and also to tune the regularization parameters of SVM to find their optimized values. Besides, we also check the possibility of the latter using computational intelligence techniques to obtain the most accurate prediction model.

2 Support Vector Machines

2.1 Review

Support Vector Machine is the main algorithm for regression and classification. In relating to enormous data categorization, old-fashioned optimization algorithms such as Quasi-Newton Method or Newton Method are outdated and inefficient due to enormous space necessity. It is hypothetically sharp and shows a worthy generalization outcome for numerous practical problems. The success lies in its suitable design of kernel functions and the selection of appropriate parameters. SVMs belong to a class of linear classification with generalization. A unique feature of SVM is that it can simultaneously maximize the geometric margin and minimizes the observed classification errors [9]. Thus SVM is also known as Maximum Margin Classifiers. In SVM, the input data (vector) can be mapped to a complex information space. There it constructs a maximal separating hyperplane and two other parallel hyperplanes are made on each side of it. The separating hyperplane is a hyperplane that maximizes the margin between the two parallel hyperplanes. If the margin between these parallel hyperplanes were increased, then the generalization error of the classifier will be decreased [10].

Consider the data points $<x_i, y_i>$ where i = 1 to n and y_n is a constant representing the class to which x_n belongs. Each x_n denotes the n-dimensional real vector. The training data can be viewed as dividing the hyperplane i.e.;

$$W.X + b = 0 \tag{1}$$

Where W is the n-dimensional vector and b is a scalar value. The real vector W represented as vertical to the separating hyperplane. Inserting the balance factor b permits to add the boundary [11]. Suppose, the training data are linearly separable, it can choose these hyperplanes and thus attempt to maximize the distance between hyperplanes. The distance between the hyperplane is $2/|w|$. Thus it needs to minimize $|w|$. Training samples along the hyperplane are knowns as support vectors [12]. The Fig. 1 explains these concepts.

2.2 Multiclass Support Vector Machine

To permit multiclass classification, one against one technique from Libsvm is used by fixing all binary sub-classifiers and identifying the accurate class by a voting method. Libsvm implements the algorithm called Sequential Minimal Optimization (SMO) for the working of kernelized SVMs which supports regression and classification [14]. Since several real-life datasets contain multiclass information, the SVM classification can be expanded from binary classification to multiclass classification. To solve the Multiclass SVM, the binary classification decision functions are combined with each

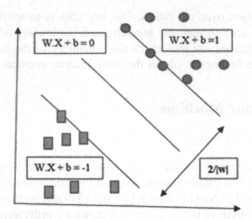

Fig. 1. SVM Maximum margin hyperplanes for two class problem.

other and this can be done by two methods, OVA decomposition and OVO decomposition [15]. The OVA decomposition and OVO decomposition converts the multiclass task into a sequence of binary subtasks. These subtasks can be trained by the simple binary SVM [13, 16].

2.3   Multiclass SVM Kernels

Kernel functions create the features of SVM simulation and level of nonlinearity. A basic and necessary criterion for a common kernel to be effective is that it essentially satisfies Mercer's theorem [17]. The sigmoid kernel was well known for SVMs due to its beginning from the neural network. Kernel functions help the user for converting nonlinear spaces into linear spaces. Most packages which deal SVM includes numerous nonlinear kernels ranging from straightforward polynomial basis functions to sigmoid functions [18, 19]. The user has to choose the suitable kernel function from this list. The R software will take care of converting the data, categorizing it and retransforming the outcomes back into the original space [31]. Unfortunately, with a huge amount of attributes in a dataset, it is very hard to know which kernel would work best. The most usually used ones are the radial basis and polynomial functions [20, 21].

2.4   Optimization in Multiclass Support Vector

The data obtained from the analysis of soil testing is classified with SVM for predicting fertilizer application rate. Classification performance is investigated for polynomial and sigmoid kernel functions used in SVM. The performance of sigmoid kernel can be increased with the help of regularization parameter optimization [22]. In SVM, this can be achieved by direct tuning of the parameters like Cost and Gamma by Grid Search [21]. The advanced optimization of parameters with computational intelligence approaches like Genetic Algorithm (GA) and Particle Swarm Optimization (PSO) will aid to get the optimum values for these parameters [23, 24].

3 Data

The geographical study area of this paper is the region of Thrissur, in the state of Kerala, one of the most prominent agricultural states in India, lying between north latitudes 10^0 31' and 10^0 52' and east longitudes 76^0 13' and 76^0 21'. Major soil types in the district include laterite soil, hydromorphic saline soils, brown hydromorphic soils, riverine alluvium, coastal alluvium and forest loamy soil. The Kole land agro-ecological unit is spread over the coastal parts of Thrissur district. The major crop cultivated in the Kole land is rice. The main challenge is to increase crop yield for solving food security problem. However, soil quality and crop yield are negatively affected by changing trends of temperature and rainfall, insufficient water and light, agriculture practices and absence of nutrients. It is important to develop an effective nutrient management system by means of adequate soil analysis and proper application of fertilizers [25].

Fertilizer has an important role in future development of Indian agriculture sector as the cropping pattern has undergone significant changes over time. To guarantee the accessibility of required quantity and the right quality of fertilizers is a major concern to farmers. The price of urea is much lesser than that of other fertilizers. This has resulted in the excessive use of urea for Rice crop, thereby inverting the balanced principles of fertilizer application. To sustain and improve soil health and its productivity, soil test based judicious application of fertilizers in accordance with nutrients is of prime concern. Balanced fertilization would assure the availability of necessary nutrients in the soil to satisfy the requirements of plants at different stages of growth. ML methods provide a more efficient way of predicting fertilizers under different crop varieties. This study describes the development of fertilizer application rate prediction model for Rice Crop by making use of SVM and its optimization techniques. This study is on the basis of variables including soil pH, EC, Soil nutrients (OC; P; K; S; Zn; B; CU; Mn and Fe) and Fertilizers Viz. Urea, MOP, Phosphate and Organic Fertilizer. The collected data from Thrissur region of the State Government of Kerala (India), (which is depicted in Fig. 2), during the years 2015 to 2017 is used as the dataset for this study. Details about standardization of the amount of each input are publically available in soil health card website [26]. The soil pH is denoted as the decimal logarithm of the hydrogen concentration. The EC is denoted in milli-Siemens per centimeter (mS/cm), while OC is represented as mass percentages (denoted as a percent). The values of K and P are expressed in kilograms per hector (kg/ha), while Fe, Cu, Zn, Mn, B, and S are expressed as parts per million (ppm).

Fig. 2. The geographical area of study. (Source: http://www.onefivenine.com)

4 Methodology

This work is mainly focused on the prediction accuracy of fertilizer application rate in Paddy fields. The SVM algorithm with the sigmoid kernel function is treated as the basic methodology for prediction, and optimization strategies are used in this model for

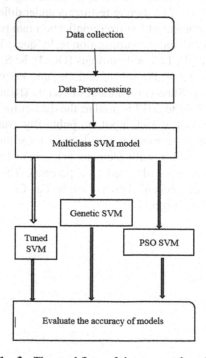

Fig. 3. The workflow of the proposed work.

assessing the performance of sigmoid kernels. The workflow of the proposed study is represented in Fig. 3.

The collected data is preprocessed for analyzing the effectiveness of the parameters and to identify the significant parameters. The improved dataset is divided into three percentages; 70% of data is for training, 15% is for validation and next 15% is for testing. To analyze the effect of kernels in SVM modeling, the data is trained and tested with polynomial kernel and sigmoid kernel. The analysis shows that the performance of polynomial kernel is better when we compare it with the sigmoid kernel. The effect of using the polynomial kernel in this study for fertilizer application rate prediction is described in Table 1.

Table 1. Training and testing accuracy for polynomial kernel fertilizer rate prediction.

Fertilizer	No of support vectors	Parameters	Polynomial kernel accuracy		
		Cost and gamma	Training	Validation	Testing
Urea	1217	1 and 0.09	89.27	89.98	88.91
MOP	787	1 and 0.09	97.03	97.52	96.62
Phosphate	846	1 and 0.09	97.78	98.76	97.29
Organic	434	1 and 0.09	95.83	97.75	96.7

The performance of sigmoid kernel SVM is closely related to modeling parameters Cost and Gamma. The result of SVM model with the sigmoid kernel is described in Table 2.

Table 2. Training and testing accuracy for sigmoid kernel fertilizer rate prediction.

Fertilizer	No of support vectors	Parameters	Sigmoid kernel accuracy		
		Cost and gamma	Training	Validation	Testing
Urea	1441	1 and 0.09	68.6	69.26	66.55
MOP	1200	1 and 0.09	75.03	77.68	73.7
Phosphate	1207	1 and 0.09	76.6	73.92	71.97
Organic	747	1 and 0.09	82.16	81.63	83.82

Thus here we propose different optimization techniques that can be used with sigmoid multiclass SVM to increase its performance. In SVM, the tuning of Cost and Gamma parameters of sigmoid kernel function will increase the accuracy of the prediction. The commonly used parameter tuning method with SVM is Grid search. In this method, a particular range of values is assigned for Cost and Gamma with SVM training. The Grid search tuning will produce an optimal value for these parameters from this range. This value is used with SVM model validation and testing and it increases the accuracy rate and Kappa value of each fertilizer prediction [27]. This result is shown in the tables from Tables 3, 4, 5 and 6. To perform the parameter selection with GAs requires the mapping of the problem to the genetic framework of natural selection and random variation [28].

Table 3. Training and Test Accuracy for Urea fertilizer rate prediction.

Methods	Parameters	Sigmoid kernel accuracy		
	Cost and gamma	Training	Validation	Testing
Multiclass SVM	1 and 0.09	68.60	69.26	66.55
Tuncd Multiclass SVM	5 and 0.01	89.71	89.98	91.68
Genetic SVM	76.50 and 0.0098	95.78	96.55	96.88
PSO SVM	159.58 and 0.0047	96.59	97.41	97.57

The GA parameter selection procedure on the training dataset will create a subclass of the training data with the selected parameters and train the SVM procedure on this subclass. Then we evaluate the working of these selected parameters by validating and testing the data. The GA control procedure with the 10-fold cross validation is selected as the GA modeling procedure because of which the entire genetic algorithm is run 10 times. The performance accuracy for all fertilizers is shown in Tables 3, 4, 5 and 6.

Table 4. Training and test error rate for MOP fertilizer rate prediction.

Methods	Parameters	Sigmoid kernel accuracy		
	Cost and gamma	Training	Validation	Testing
Multiclass SVM	1 and 0.09	75.04	77.68	73.7
Tuned multiclass SVM	4 and 0.01	92.75	91	92.91
Genetic SVM	76.50 and 0.0098	98.59	97.23	97.92
PSO SVM	159.58 and 0.0047	98.71	98.1	98.44

PSO is a population-based optimization technique in which the particles update themselves with the internal velocity and it will help to get the optimum values for Cost and Gamma parameters.

Table 5. Training and test error rate for phosphate fertilizer rate prediction.

Methods	Parameters	Sigmoid kernel accuracy		
	Cost and gamma	Training	Validation	Testing
Multiclass SVM	1 and 0.09	76.69	73.92	71.97
Tuned multiclass SVM	5 and 0.01	88.90	87.91	87.02
Genetic SVM	76.50 and 0.0098	96.06	98.7	97.92
PSO SVM	159.58 and 0.0047	99.33	99.65	99.13

During this optimization process, the key element is the selection of the fitness function. In this method, k-fold cross-validation error is applied to estimate the generalization capability of SVM simulation. Especially, the data elements should be divided into K parts and one of them is removed as a test data. The remaining K-1 parts are taken as training data. The small error value indicates stronger generalization capacity and a number of appropriate parameters [29].

Table 6. Training and test error rate for organic fertilizer rate prediction.

Methods	Parameters	Sigmoid kernel accuracy		
	Cost and gamma	Training	Validation	Testing
Multiclass SVM	1 and 0.09	82.16	81.63	83.82
Tuned multiclass SVM	5 and 0.01	93.79	94.8	95.65
Genetic SVM	76.50 and 0.0098	96.2	95.6	96.4
PSO SVM	159.58 and 0.0047	97.84	98.8	97.92

It is clear that the SVM parameter tuned with PSO provides better accuracy for prediction and it reduces the number of support vectors used for classification which is shown in Fig. 4. Kappa statistic helps to quantify the percent agreement above chance. If it is greater than 0.75 then it is an excellent agreement [30]. The Kappa rating for fertilizer dataset in validation and testing is depicted in Figs. 5 and 6.

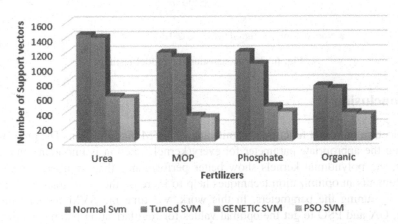

Fig. 4. The number of support vectors generated with models.

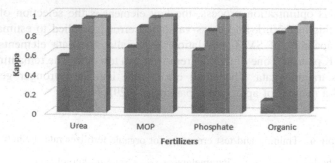

Fig. 5. Kappa statistics for model validation.

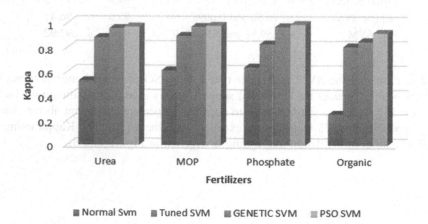

Fig. 6. Kappa statistics for model testing.

5 Conclusion

The selection of kernel function is significant for SVM classification. The procedure of selecting the appropriate parameter for every kernel function is important. In our soil dataset, the polynomial kernels show better performance than sigmoid kernels. The advancements in optimization techniques help to increase the performance of sigmoid kernel by tuning the parameters. In this work, we tune the SVM model with Grid Search, GA and PSO to get the optimal values for regularization parameters like Cost and Gamma. The most optimal values are obtained when we combine the SVM with PSO. The proposed method can work as a superior processing technique for optimizing the feature selection method since it increases the categorization and prediction accuracy. It can also be useful to handle difficulties in other lands and crops in the future.

References

1. Schlkopf, B., et al.: Support vector methods in learning and feature extraction (1998)
2. Burges, C.J.C.: A tutorial on support vector machines for pattern recognition. Data Min. Knowl. Discov. **2**(2), 121–167 (1998)
3. Hsu, C.-W., Lin, C.-J.: A comparison of methods for multiclass support vector machines. IEEE Trans. Neural Netw. **13**(2), 415–425 (2002)
4. Chapelle, O., et al.: Choosing multiple parameters for support vector machines. Mach. Learn. **46**(1-3), 131–159 (2002)
5. Crammer, K., Singer, Y.: On the learnability and design of output codes for multiclass problems. Mach. Learn. **47**(2-3), 201–233 (2002)
6. Rani, P.S., Latha, A.: Effect of calcium, magnesium, and boron on nutrient uptake and yield of rice in Kole lands of Kerala. Indian J. Agric. Res. **51**(4), 388–391 (2017)
7. Sehgal, J.L., Mandal, D.K., Mandal, C., Vadivelu, S.: Agro-ecological regions of India, vol. 24. NBSS Publication (1990)
8. Kumar, R.A., Aslam, M.M., Raj, V.J., Radhakrishnan, T., Kumar, K.S., Manojkumar, T.K.: A statistical analysis of soil fertility of Thrissur district, Kerala. In: 2016 International Conference on Data Science and Engineering (ICDSE), pp. 1–5. IEEE (2016)
9. Vapnik, V.: An overview of statistical learning theory. IEEE Trans. Neural Netw. **10**(5), 988–999 (1999)
10. Cristianini, N., Shawe-Taylor, J.: Introduction to Support Vector Machines. Cambridge University Press, Cambridge (2000)
11. Wang, L. (ed.): Support Vector Machines: Theory and Applications, vol. 177. Springer, Heidelberg (2005). https://doi.org/10.1007/b95439
12. Schlkopf, B., Smola, A.: Learning with Kernels. MIT Press, Cambridge (2001)
13. Sangeetha, R., Kalpana, B.: A comparative study and choice of an appropriate kernel for support vector machines. In: Das, V.V., Vijaykumar, R. (eds.) ICT 2010. CCIS, vol. 101, pp. 549–553. Springer, Heidelberg (2010). https://doi.org/10.1007/978-3-642-15766-0_93
14. Chang, C.-C., Lin, C.-J.: LIBSVM: a library for support vector machines (2001). http://www.csie.ntu.edu.tw/cjlin/libsvm
15. Weston, J., Watkins, C.: Multi-class support vector machines, Technical report (1998)
16. Burges, C.J.C.: A tutorial on support vector machines for pattern recognition. Data Mining Knowl. Discov. **2**(2), 121–167 (1998)
17. Herbrich, R.: Learning Kernel Classifiers: Theory and Algorithms. MIT Press, Cambridge (2001). ISBN 026208306X
18. Sangeetha, R., Kalpana, B.: Optimizing the kernel selection for support vector machines using performance measures. In: Proceedings of the 1st Amrita ACM-W Celebration on Women in Computing in India. ACM (2010)
19. Sangeetha, R., Kalpana, B.: Performance evaluation of kernels in multiclass support vector machines. Training **2**, 2 (2011)
20. Xia, G., Shao, P.: Factor analysis algorithm with mercer kernel. In: IEEE Second International Symposium on Intelligent Information Technology and Security Informatics (2009)
21. Hsu, C.-W., Chang, C.-C., Lin, C.-J.: A practical guide to support vector classification, pp. 1–16 (2003)
22. Duan, K.-B., Keerthi, S.S.: Which is the best multiclass SVM method? An empirical study. In: Oza, N.C., Polikar, R., Kittler, J., Roli, F. (eds.) MCS 2005. LNCS, vol. 3541, pp. 278–285. Springer, Heidelberg (2005). https://doi.org/10.1007/11494683_28

23. Wu, C.-H., et al.: A real-valued genetic algorithm to optimize the parameters of support vector machine for predicting bankruptcy. Expert Syst. Appl. **32**(2), 397–408 (2007)
24. Lin, S.-W., et al.: Particle swarm optimization for parameter determination and feature selection of support vector machines. Expert Syst. Appl. **35**(4), 1817–1824 (2008)
25. Gajbhiye, K.S., Mandal, C.: Agro-ecological zones, their soil resource and cropping systems. In: Status of Farm Mechanization in India, Cropping Systems, Status of Farm Mechanization in India, pp. 1–32 (2000)
26. Soil Health Card. http://soilhealth.dac.gov.in/
27. Kuhn, M., Johnson, K.: Applied predictive modeling, vol. 26. Springer, New York (2013)
28. Tu, C.-J., Chuang, L.-Y., Chang, J.-Y., Yang, C.-H.: Feature selection using PSO-SVM. Int. J. Comput. Sci. **33**(1), 111–116 (2007)
29. Zhang, X., Guo, Y.: Optimization of SVM parameters based on PSO algorithm. In: Fifth International Conference on Natural Computation, ICNC 2009, vol. 1. IEEE (2009)
30. Sim, J., Wright, C.C.: The kappa statistic in reliability studies: use, interpretation, and sample size requirements. Phys. Ther. **85**(3), 257–268 (2005)
31. R-Team: R: A Language and Environment for Statistical Computing. R Foundation for Statistical Computing, version 3.1.2, Vienna, Austria (2016)

Author Index

Printed in the United States
By Bookmasters